THE
TRANSPORTER
FactsBook

THE TRANSPORTER
FactsBook

Jeffrey Griffith
Department of Biochemistry and Molecular Biology
University of New Mexico School of Medicine

Clare Sansom
Department of Crystallography
Birkbeck College, University of London

Academic Press
Harcourt Brace & Company, Publishers
SAN DIEGO   LONDON   BOSTON   NEW YORK
SYDNEY   TOKYO   TORONTO

Academic Press
525 B Street, Suite 1900, San Diego, California 92101-4495, USA
http://www.apnet.com

Academic Press Limited
24–28 Oval Road, London NW1 7DX, UK
http://www.hbuk.co.uk/ap/

ISBN 0-12-303965-7

Library of Congress Cataloging-in-Publication Data

Griffith, Jeffrey.
    The transporter factsbook / by Jeffrey Griffith, Clare Sansom.
        p.   cm.
    Includes index.
    ISBN 0-12-303965-7 (alk. paper)
    1. Carrier proteins.   I. Sansom, Clare.   II. Title.
OP552.C34075   1997
572′69–dc21   97-44438
                                                                              CIP

A catalogue record for this book is available from the British Library

Typeset in Great Britain by Alden Group, Oxford.

Printed and bound in the United Kingdom
Transfered to Digital Printing, 2011

Contents

Preface

The Transporter FactsBook had its inception early in 1996 when Tessa Picknett at Academic Press approached the authors with the idea of preparing a volume on transport proteins. Recognizing that the book would contain several different types of transporters, and that additional transporter species were being described almost daily, it was decided that the only way to make the volume comprehensive would be to base the chapters on families of related transporters, rather than individual proteins. Using this method, we have been able to include nearly 800 transport proteins in this volume. More important, this comparative approach, which stresses the structural, mechanistic and biological properties that are common to closely related proteins, provides an objective basis for identifying potential evolutionary relationships between distantly related groups of proteins and establishes a system for classifying and characterizing newly described transporters. The authors hope that this basis for identification and classification will continue to make the volume a valuable resource even after the compilation of transporters it contains is no longer comprehensive.

An undertaking of this scope and complexity would not have been possible without the help and advice of many people. In particular, the authors would like to thank Jennifer Bryant and Peggy Moran at the University of New Mexico School of Medicine for cheery and able assistance in establishing the relationships between the nearly 800 transporters described in *The Transporter FactsBook*, Dr Mark Platt, now at the Louisiana State University School of Medicine, for wizardry in editing and modifying the phylogenetic trees, and Tessa Picknett and her staff at Academic Press for encouragement, support and patience in getting the manuscript into press.

There will undoubtedly be omissions and errors in this volume although we hope that they will be infrequent. We would greatly appreciate being informed of any inaccuracies by writing to the Editor, The Transporter FactsBook, Academic Press, 24–28 Oval Road, London NW1 7DX, UK, so that these can be rectified in future editions.

Jeff Griffith   Clare Sansom

Abbreviations

Å	angstrom unit
ABC	ATP binding cassette
ADP	adenosine diphosphate
Asn	asparagine
Asp	aspartic acid
ATP	adenosine triphosphate
C-	carboxyl-
C-4	4-carbon
CFTR	cystic fibrosis transmembrane conductance regulator
CNS	central nervous system
3-D	three-dimensional
DNA	deoxyribonucleic acid
EB	ethidium bromide
FAD	flavin adenine dinucleotide
GABA	4-amino butyric acid
Gln	glutamine
Glu	glutamic acid
HMA	heavy metal binding sequence
kDa	kilodalton
M	molar
mol. wt	molecular weight
MDR	multidrug resistance
MFS	major facilitator superfamily
mV	millivolts
N-	amino
NADH	nicotinamide adenine dinucleotide (reduced form)
Δp	proton-motive force
ΔpH	transmembrane pH gradient
$\Delta\psi$	transmembrane charge gradient
NMR	nuclear magnetic resonance
PEP	phosphoenolpyruvate
PIR	protein identification resource
PTS	Phosphoenolpyruvate-dependent sugar phosphotransferase system
QUAC	quaternary ammonium compound
SD	standard deviation
USA	uniporter-symporter-antiporter

Section I

THE INTRODUCTORY CHAPTERS

1 Function and Structure of Membrane Transport Proteins

Peter J. F. Henderson (Department of Biochemistry and Molecular Biology,
University of Leeds, Leeds LS2 9JT, UK)

INTRODUCTION

The hydrophobic bilayer membrane that bounds cells is inherently impermeable to the great majority of hydrophilic solutes required for cell nutrition and to many of the waste products and/or toxins that must be excreted. Accordingly, the membrane contains proteins, the sole function of which is to catalyze the translocation of substrates through the membrane. As the substrates for many membrane processes can be obtained in radioisotope-labeled form, it has been technically feasible to characterize the functions of many of these transport proteins. The structures of the proteins themselves, however, have proved to be difficult to elucidate: they are of low natural abundance in the membrane; they are very hydrophobic and refractory to isolation methods in aqueous solutions; and, even when purified, usually in non-denaturing detergents, they are very difficult to crystallize. Where the proteins happen to be abundant – bacteriorhodopsin from *Halobacterium halobium*, K^+/Na^+ ATPases in nerve and Ca^{2+} ATPase from muscle, cytochrome oxidases in bacteria and mitochondria, glucose transporter from human erythrocytes, for example – progress has been made in elucidating the structure–function relationship. Yet, of these proteins the three-dimensional structure has only been determined for bacterio-rhodopsin and the oxidases [1-3], and this is just the beginning of determining their molecular mechanisms of operation.

Free-living microorganisms (bacteria, algae, yeasts, parasitic protozoa) often inhabit environments where nutrients are in short supply, and different species must compete with each other for the available metabolites. Accordingly, they couple expenditure of metabolic energy to inward transport of essential nutrients (K^+, NH_4^+, P_i, SO_4^{2-}, sugars, vitamins, etc.) to achieve intracellular concentrations sufficient for optimal growth rates. This expenditure can amount to 20–30% of the organism's available energy when a carbohydrate is fermented under anaerobic conditions to yield only 2–3 moles ATP per mole sugar [4,5]. Since the efficiencies of the transport steps may therefore influence cell yield and growth rate [4,6,7] an understanding of the transport processes is important to both the academic researcher seeking to understand bacterial cell physiology, and the industrial manager trying to maintain the profitability of a fermentation process. Furthermore, the process of eliminating metabolic wastes and/or toxins such as antibiotics is often coupled to the expenditure of metabolic energy, an indication of its importance for survival. Motility appears to be driven by transport processes also, although this may not consume so much energy [8].

In higher organisms, where survival functions are distributed between different organs, the energization of nutrient capture and waste efflux may be confined to specific tissues, e.g. the gut and the kidney. As a result of their activities, cells in other tissues enjoy an unchallenging environment in which their energy reserves can be channeled into other functions. Thus, their transport processes more often occur by facilitated diffusion.

As approximately 5–15% of all proteins, revealed by the current efforts in genome sequencing, are membrane transport proteins[9], we anticipate the need for a huge effort in the new millennium to determine the structures of these proteins that are vital for the capture of nutrients and hence the first stage in cell growth. Their additional roles in antibiotic resistance, toxin secretion, ATP synthesis, ion balance, generation of action potentials, synaptic neurotransmission, kidney function, intestinal absorption, tumor growth and other diverse cell functions in organisms from microbe to man presage a major investigative effort to elucidate their molecular mechanisms of action. This effort to elucidate vectorial processes can be compared to the continuing efforts to understand enzyme-mediated catalysis, though there is the possibility of an underlying uniformity of translocation mechanism despite the huge numbers of independent transport proteins that exist.

The advent of recombinant DNA technology has enabled the study of membrane transport proteins to be furthered in at least four major directions. The first is the burgeoning appearance of an enormous number of amino acid sequences of the proteins predicted from the DNA sequences of their genes in the genome mapping projects. This sequence information has enabled a second advance: the unambivalent exposure of the evolutionary relationships between proteins not thought hitherto to be related. The third is the manipulation of the genes to expedite amplified expression and purification of the proteins. Finally the ability to mutagenize individual amino acids and to make chimeric proteins is being used to elucidate the relationship of function to structure.

A number of transport proteins play a role in human health and disease. The study of "ABC" transport systems (see later) in mammalian cells was intensified with the discovery that cystic fibrosis, the commonest inherited disease in the western world, was caused by a defect in the Cl^- transport protein[10]. The significance of a multidrug resistance protein, "Mdr" that catalyzes secretion of cytotoxins and the failure of anti-tumor chemotherapy similarly focused attention on a different ABC system. In both cases their similarity as ABC-type systems would have been completely obscured without the amino acid sequence information derived from the cloning and sequencing of their genes. Other transport proteins are involved in glucose/galactose malabsorption, albinism, adrenoleukodystrophy.

This FactsBook is intended to catalyze this new age of exploration of membrane transport protein structure. It is our major goal to arrive at a sensible classification of transport systems based upon both evolutionary and mechanistic considerations. The numbers of protein sequences now known is too large to include them all, and the expected appearance of legions more from the genome sequencing programs makes it timely to formulate a systematic approach to their classification. First it is important to describe current concepts of their functions. The treatment below is necessarily brief, and the reader is referred to the appropriate chapters in standard biochemistry textbooks[11,12] for a fuller introduction.

A watershed in the field occurred when Peter Mitchell[13–15] showed that transport processes were intimately associated with the mechanism of oxidative and photo-synthetic ATP synthesis, a process which is central to energy metabolism in almost all organisms. However, because of the difficulties in studying the hydrophobic membrane proteins involved we know very little about the molecular mechanism of such vectorial events; this contrasts with the wealth of information on the molecular mechanisms of chemical events catalyzed by water-soluble enzymes. It is quite possible that there is an underlying unity in the molecular mechanism of the translocation process, even when the direction of solute movement and any energization steps are completely different. This question is likely to be illuminated only when we elucidate the 3D structures and determine the structure–activity relationships of the transport proteins. By far the most central question in the transport field is precisely this – what are the 3D structures of the proteins involved?

Before reaching this question it is useful to define some terms often used in the characterization of transport processes.

USEFUL CONCEPTS

Passive diffusion

Passive diffusion is the translocation of a solute across a membrane down its electrochemical gradient without the participation of a transport protein. The process follows Fick's law, and so obeys the relationship below in which the velocity has a linear relationship to the [solute]:

$$v = PAc$$

where v is velocity, P is the permeability coefficient for the particular solute, A is the area, and c is the difference in solute concentration across the cell membrane.

Diffusion has a low temperature coefficient ($v \propto$ absolute temperature) and is non-specific. Typical biologically important compounds that follow this mechanism are O_2, CO_2, NH_3, HCO_2H, CH_3CO_2H, $CH_2OH.CHOH.CH_2OH$ – small, neutral molecules that are soluble in lipid membranes.

Facilitated diffusion

Facilitated diffusion is the translocation of a solute across a membrane down its electrochemical gradient catalyzed by a transport protein. The Michaelis–Menten relationship [11,12] often adequately relates the initial rate of transport (v) to initial substrate concentration ($[S] = c$ at zero time):

$$v = V_{max}.[S]/(K_m + [S])$$

(V_{max} is maximum velocity, $K_m = [S]$ where v is $V_{max}/2$). As with enzyme reactions, there is a high temperature coefficient and, usually, strong substrate specificity. Biological substrates that follow this mechanism are typically charged and/or larger than about the size of glycerol, with a very low inherent solubility in biological membranes. Mitchell classified such transport of a single substrate as "uniport", and glycerol transport is an example of such facilitated diffusion in E. coli [16,17]. However, in free-living single-cell organisms, e.g. bacteria, yeasts, algae, the rate of capture of nutrient from the environment by this mechanism is probably too slow at the dilute concentrations that prevail in their normal environments to support competitive growth. Therefore, we usually find that transport of their vital nutrients is coupled to consumption of metabolic energy by active transport (see below) rather than facilitated diffusion. Presumably, during the course of evolution of such organisms the expenditure of precious energy reserves on transport has been a very significant survival factor.

In contrast, transport of solutes between intracellular organelles in eukaryotes, or into tissue cells from the blood, often occurs by facilitated diffusion since high concentrations of solute are already established, for example by the Na^+-glucose symport system (below) so that facilitated diffusion by the tissue glucose uniporters is sufficient to support cell metabolism. The seminal example of such a facilitated transport was the GLUT1 glucose transport protein in human erythrocytes [18,19].

Active transport

The term active transport is used to describe the net transport of a solute across a biological membrane from a low to a high electrochemical potential. Active transport shows the following characteristics.

- Accumulation of solute occurs against a concentration gradient.
- The solute is not chemically modified during translocation.
- Saturable steady-state kinetics are observed, often following the Michaelis–Menten relationship (above).
- There is a high temperature coefficient typical of enzyme-catalyzed reactions.
- Substrate specificity is restricted.
- An input of metabolic energy is required.

Active transport processes embrace a variety of molecular mechanisms, in which energy may be derived from light, oxidoreduction, ATP hydrolysis, or pre-existing solute gradients. It is conceptually helpful to classify them further into "primary" and "secondary" mechanisms [20] (Fig. 1). Secondary transport can be subdivided into "symport" or "antiport", terms introduced by Mitchell [13,14] (Fig. 1).

Primary active transport

Primary transport involves the direct conversion of chemical or photosynthetic energy into an electrochemical potential of solute across the membrane barrier. Thus, translocation of protons driven by oxidation of respiratory substrates [15,21,22], by hydrolysis of ATP [15,22–24] or by light energy absorbed by bacteriorhodopsin [1] all fall into this category. Many nutrient transport systems involving binding proteins in bacteria are of the primary type, directly energized by ATP (see below). All these examples transport one substrate in one direction and so are described as "uniport" [14].

Secondary active transport

Secondary transport involves the conversion of a pre-existing electrochemical gradient, usually of H^+ or Na^+ ions, into a new electrochemical gradient of the transported species. Thus the ultimate energy source for secondary transport systems is a primary chemical or photochemical conversion. In *E. coli* primary proton ejection by respiration or ATPase powers secondary sugar–H^+ *symport* (obligatory coupling of H^+ and solute movement in the *same* direction [14]; see Fig. 1) or secondary Na^+/H^+ *antiport* (the obligatory coupling of H^+ and solute movement in the opposite direction [14]; Fig. 2). For example, the resulting Na^+ gradient can be further coupled to melibiose transport by a melibiose–Na^+ symport, so that net melibiose accumulation is driven by respiration (or ATPase) via H^+ and Na^+ gradients (Fig. 2).

In *E. coli* the transmembrane H^+ gradient would appear to be the "common currency" of many energized transport reactions, and the Na^+ gradient of relatively few. However, in other organisms living in salt environments the Na^+ gradient is the dominant factor maintained by a primary Na^+ pump [25–27], as it is in multicellular eukaryotes.

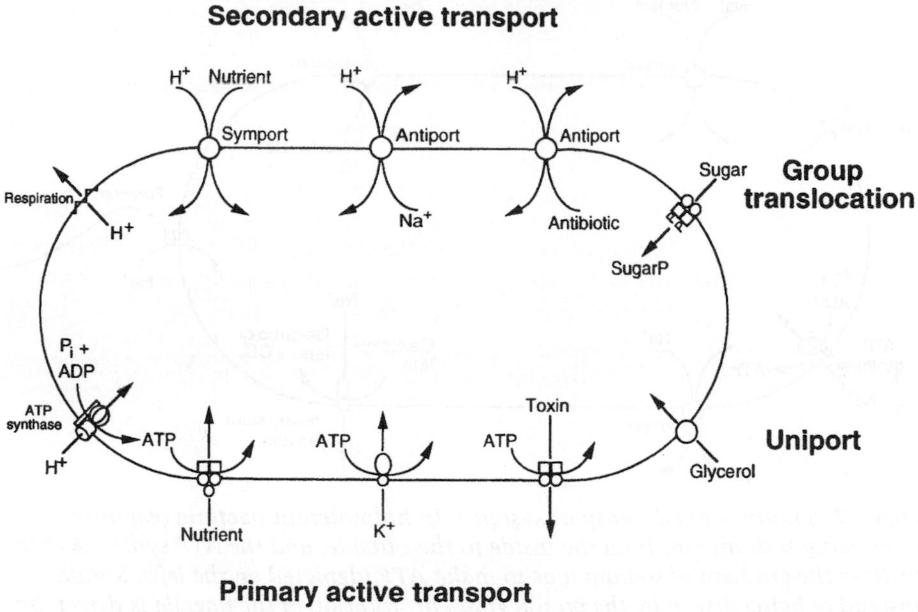

Secondary active transport

Figure 1 *Energization of sugar transport in* E. coli. *The large oval represents the cytoplasmic membrane of the microorganism. A transmembrane electrochemical gradient of protons is generated by respiration or ATP hydrolysis depicted on the left. This can be utilized by the proton–nutrient symport or proton–substrate antiport systems shown along the top. Some sugars can be accumulated by an alternative mechanism involving ATP, a binding protein, and two or three other proteins shown along the bottom, with other ATP-dependent primary transport systems for uptake of K^+ or efflux of toxin. A phosphotransferase mechanism involving PEP and two or three proteins for sugar accumulation is shown on the upper right, and facilitated transport of glycerol on the lower right.*

Group translocation

All the above mechanisms operate without chemical modification of the solute. Group translocation systems catalyze both the translocation and concomitant chemical modification of the solute. For a range of carbohydrates in many species of bacteria, phosphoenol pyruvate (PEP) is the donor to produce internal sugar-phosphate from external free sugar [28] (Fig. 1); the glucose phosphotransferase system is particularly widespread amongst anerobic organisms.

CLASSIFICATION OF MEMBRANE TRANSPORT SYSTEMS ACCORDING TO THEIR ENERGETICS

Although thousands of transport processes, each catalyzed by its own protein, have been identified, the strategies found coupling metabolic energy to the translocation process are relatively few in number. These are now described to provide a formal

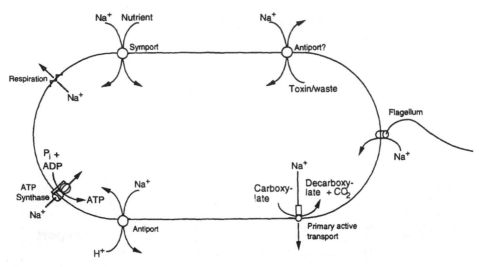

Figure 2 *Sodium-linked transport systems. In halotolerant bacteria respiration may pump sodium ions from the inside to the outside, and the ATP synthase then utilizes the gradient of sodium ions to make ATP (depicted on the left). Similarly, instead of being driven by the proton gradient, rotation of the flagella is driven by the sodium gradient (right). Nutrients may be accumulated by a Na$^+$ substrate symport system (top left) and toxins excreted by a Na$^+$ substrate antiport (top right). In many bacteria, sodium ions are excreted by a Na$^+$/H$^+$ antiport (bottom left). In a few species there are sodium-secreting active transport systems driven by decarboxylation reactions (bottom right).*

basis for a preliminary classification of all the processes. While recent work indicates that a single transport system might employ more than one energization mechanism[29,30], or even that at least one novel mechanism may exist (TonB), the vast majority of biological transport systems so far fall conveniently into one of these classes. Their operation is illustrated in Fig. 1.

While previous investigators made many fundamental contributions to understanding transport processes (see review[31]), it was Peter Mitchell who showed how vital are vectorial processes to the totality of energy metabolism in living organisms. Accordingly, we will now sketch in the chemiosmotic approach before focusing on individual mechanisms of solute translocation.

The chemiosmotic theory of oxidative and photosynthetic phosphorylation

In 1961, Peter Mitchell proposed his Chemiosmotic Theory of Oxidative and Photosynthetic Phosphorylation[32]. This sought to explain how ATP synthesis is coupled to oxidative or photosynthetic electron transfer by the use of an electrochemical gradient of protons across the membrane as a high-energy intermediate between the processes. This brilliant concept generated a wealth of productive experimental investigations that, not without some controversy, arrived at an acceptance that proton transport is a fundamental feature of ATP synthesis in virtually all organisms. The molecular mechanism of these processes is just beginning to be understood, with the very recent elucidations of the structures of proton-translocating proteins and electron transfer proteins[2,3]. There has also been the realization that rotation of the proteins in the membrane is a key feature of energy transmission for flagella[8,33], and ATP synthase[23,34].

The four basic parts of the chemiosmotic system, corresponding to the four postulates of the Chemiosmotic Hypothesis, can be paraphrased as follows[31,32]:

1. The proton-translocating reversible ATPase system.
2. The proton-translocating oxido-reduction or light-driven electron transfer chain.
3. The exchange diffusion systems, coupling proton translocation to that of anions and cations.
4. The ion-impermeable coupling membrane, in which systems 1, 2 and 3 reside.

The chemiosmotic view of substrate transport mechanism

It is postulate 3, which predicts the involvement of transport systems in the process of balancing charge and osmolarity across the membrane, that led Peter Mitchell to consider the energetics of solute uptake into bacteria. In 1963 he suggested that the uptake of sugars into microbial cells might be energized by a transmembrane proton gradient[13]. The idea required that an individual transport system catalyze the simultaneous translocation of protons with a substrate molecule, "symport", or the experimentally indistinguishable "antiport" of hydroxyl ions[13,31]. In this hypothesis energy released by respiration or ATP hydrolysis and "stored" as the electrochemical gradient of protons, could drive accumulation of the nutrient[13-15,31]. The principle is illustrated in Fig. 1. However, this brilliant prediction remained untested until 1970, when Ian West devised experimental conditions in which the movement of lactose or substrate analogs into cells of Escherichia coli containing the lactose transport protein (LacY) evoked an alkaline pH change showing proton movement in the same direction[35-37]. Since then the structure–activity relationship of the LacY protein has been explored by every practicable method of modern molecular biology[38-40]. Several other sugar–H^+ systems have been characterized, but, most importantly, the principles enunciated by Mitchell[13,14] have been shown to apply to diverse bacterial transport systems responsible, not just for the capture of nutrients like sugars, amino acids, vitamins and ions, but also for the extrusion of wastes and toxins including lactate, Na^+ or antibiotics[31].

Many transport systems are not ion-linked

Although the Chemiosmotic Theory formed a framework to unify ideas on mechanisms of transport, it became evident that not all transport sytems were linked to ion translocation. In bacteria, the seminal experiments of Berger and Heppel[42,43] showed that transport systems associated with periplasmic binding proteins were energized "directly", probably by ATP. These early ideas have been reinforced by the subsequent discovery of numerous ATP binding cassette, "ABC", transport systems in all types of organism that function to transport substrates into, or out of, whole cells or subcellular compartments. They are reviewed most recently by Higgins[10] and Boos and Lucht[44].

Furthermore, the uptake of some carbohydrates into bacteria, including most importantly, glucose, was accompanied by simultaneous phosphorylation[28]. This chemical conversion occurred at the expense of phosphoenol pyruvate (PEP) via a cascade of phosphate transfer reactions[28]. The operation of such vectorial "group translocation" reactions was considered in detail by Mitchell[14,45], and the subsequent elucidation of these interesting systems has been reviewed most recently by Postma et al.[28].

CLASSIFICATION OF TRANSPORT SYSTEMS ACCORDING TO THE AMINO ACID SEQUENCES OF THEIR PROTEINS

Proteins catalyzing a single type of transport function and/or energization mechanism do not *necessarily* exhibit homology at the primary sequence level. Note that they might nevertheless have similar secondary and tertiary structures. Thus the sequences of the rhamose–H^+ and fucose–H^+ symport proteins of E. coli are not homologous to that of the arabinose–H^+, xylose–H^+ or galactose–H^+ symport protein of the same organism[46,47] and none of the sugar–H^+ symporters are homologous to the sugar–Na^+ symporters[48]. In addition, some phosphotransferase enzymes II are homologous while others are not[28].

More important, perhaps, is that some proteins catalyzing *different* types of transport according to the above classifications exhibit a high degree of primary sequence homology. One example is the similarity of E. coli sugar–H^+ symport proteins for arabinose, xylose, or galactose to the mammalian non-energized glucose uniporter, GLUT1 (Fig. 3[47]). Another example is the similarity of bacterial K^+ ATPase uniport to mammalian Na^+/K^+ ATPase antiport and Ca^{2+} ATPase uniport proteins. The mitochondrial H^+–P_i symport, ADP/ATP antiport, and oxoglutarate/malate antiport proteins[49] also show homology to one another.

It seems likely that our understanding of the molecular mechanisms of transport processes will be much enhanced by this rapid proliferation of information about the amino acid sequences of membrane transport proteins. In this book the transport proteins are arranged according to such evolutionary families. At least 28 families can be identified already (Table 1), and there are likely to be many more as the sequence databases grow.

TRANSPORT ACROSS PROKARYOTIC CELL MEMBRANES

Penetration of the cell wall by solutes

The cell walls of gram-negative bacteria have a complex multilayered structure that includes lipopolysaccharide, an outer lipid membrane, peptidoglycan, the periplasm and an inner phospholipid bilayer membrane (Fig. 4[50]). This wall can be regarded as having at least two global functions, that are to an extent antagonistic. In the first instance the wall has to protect the cell against external toxins and environmental changes inimical to life; secondly, it has to permit the uptake of vital nutrients. The wall must also confer mechanical strength to maintain the integrity of the cell, for example when there are changes in osmotic pressure.

In E. coli and a number of other species the evidence suggests that compounds of molecular weight less than about 900 penetrate to the inner membrane at rates that

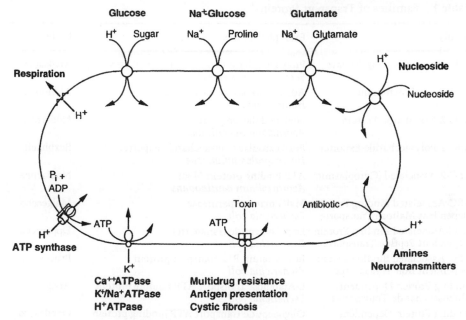

Figure 3 *Mammalian homologues of bacterial transport proteins. The bacterial transporters are depicted as in Figs 1 and 2 with their mammalian homologues indicated in bold type around.*

do not limit cell growth[50,51]. This is achieved by at least three factors. First, the lipopolysaccharide layer is permeable to hydrophilic solutes, though it may be impermeable to more hydrophobic molecules including antibiotics[50,52]. Secondly, the outer membrane contains channel-forming trimeric proteins ("porins"[52]), acting as molecular sieves that permit simple diffusion of solutes of M_r up to 900, including di- and trisaccharides[50-52]. Thirdly, the outer membrane also contains other porin-like proteins which exhibit some specificity for the permeant molecule, and pass the substrate (we presume) to high-affinity binding proteins in the periplasm[50,52].

In general the porins can be regarded as forming a "pore" or "channel" that enables passive diffusion of solute into the periplasm at a rate sufficient for growth. However, not all porins are non-specific. A clear example of this is the maltoporin, LamB, that aids the entry into the cell of oligosaccharides containing up to six glucose units. The molecular basis of this specificity has recently been elucidated with the characterization of a "greasy slide" in the pore that interacts with the hydrophobic face of the sugar molecules[53]. Similarly, the preference of one porin protein for anions, of which a most important nutrient is inorganic phosphate ions, is explained by a positively charged region in the molecule. Thus, the porins may reflect an evolutionary bridge between passive and facilitated modes of diffusion of nutrients into the cell.

Importantly, the inner and outer membranes also have to function as conduits for secretion. Included amongst their substrates are: protein, carbohydrate and lipid components of outer layers of the cell wall; proteins and toxins secreted by

Table 1 Families of Transport Proteins[1]

Family	Example: Species	Code
Calcium-transporting ATPase	Probable calcium-transporting ATPase 4 *Saccharomyces cerevisiae*	Atc4sacce
Peroxisomal Membrane	Adrenoleukodystrophy protein *Homo sapiens*	Aldhomsa
ABC-2 Nodulation Protein	Nodj nodulation protein *Azorhizobium caulinodans*	Nodjazoca
ABC-2 Polysaccharide Exporter	BexB capsular polysaccharide exporter *Haemophilus influenzae*	Bexbhaein
ABC-2 Associated (Cytoplasmic)	ATP-binding protein NodI *Azorhizobium caulinodans*	Nodiazoca
ABC-Associated Binding Protein Dependent Maltose Transporter	MalG maltose permease *Escherichia coli*	Malgescco
ABC-Associated Binding Protein Dependent Peptide Transporter	DppC dipeptide transporter *Escherichia coli*	Dppcescco
ABC-Associated Binding Protein Dependent Iron Transporter	Btuc vitamin B12 transport protein *Escherichia coli*	Btucescco
Binding Protein Dependent Monosaccharide Transporter	L-Arabinose transport ATP binding protein *Escherichia coli*	Aragescco
Binding Protein Dependent Peptide Transporter	Oligopeptide transport ATP binding protein *Escherichia coli*	Oppdescco
Heme Exporter	Heme exporter CycV *Bradorhizobium japonicum*	Cycvbraja
Plasma Membrane Cation-Transporting ATPase	Calcium-transporting ATPase *Homo sapiens*	Atchomsa
Macrolide-Streptogramin-Tylosin Resistance	Erythromycin resistance protein MsrA *Staphylococcus epidermalis*	Msrastaep
H^+–Sugar Symporter or Sugar Uniporter	Glut1 facilitative glucose transporter *Homo sapiens*	Glut1homsa
H^+–Rhamnose Symporter	RhaT rhamnose–H^+ symporter *Escherichia coli*	Rhatescco
H^+–Amino Acid Symporter	PheP phenylalanine transporter *Escherichia coli*	Phepescco
H^+–Lactose–Sucrose–Nucleoside Symporter	LacY lactose–H^+ symporter *Escherichia coli*	Lacyescco
H^+–Galactoside–Pentose–Hexuronide Symporter	MelB melibiose–H^+ symporter *Escherichia coli*	Melbescco
H^+–Oligopeptide Symporter	Pet1 oligopeptide–H^+ symporter *Homo sapiens*	Pet1homsa
H^+–Fucose Symporter	FucP fucose–H^+ symporter *Escherichia coli*	Fucpescco
H^+–Carboxylate Symporter	KgtP α-ketoglutarate–H^+ symporter *Escherichia coli*	Kgtpescco
H^+–Nucleoside Symporter	NupC pyrimidine nucleoside–H^+ symporter *Escherichia coli*	Nupcescco
Heavy Metal-Transporting ATPase	Copper-transporting ATPase 1 *Homo sapiens*	At7ahomsa

[1] Data kindly provided by J.K. Griffith and C.E. Sansom.

Table 1 Continued

Family	Example: Species	Code
Sugar Phosphate Transporter	UhpT hexose phosphate transporter *Escherichia coli*	Uhptescco
H$^+$/Amine Vesicular Antiporter	Vesicular amine transporter 2 (VAT2) *Homo sapiens*	Vat2homsa
14-Helix H$^+$/Multidrug Antiporter	QacA multidrug resistance protein *Staphylococcus aureus*	Qacastaau
4-Helix H$^+$/Multidrug Antiporter	QacC multidrug resistance protein *Staphylococcus aureus*	Ebrstaau
12-Helix H$^+$/Multidrug Antiporter	TetA(C) tetracycline antiporter *Escherichia coli*	Tcr2escco
Acriflavin-Cation Resistance	AcrB acriflavin resistance protein *Escherichia coli*	Acrbescco
Yeast Multidrug Resistance	Bmr benomyl-methotrexate resistance *Candida albicans*	Bmrpcanal
Na$^+$/Ca$^+$ Exchanger	Cardiac sodium/calcium exchanger *Homo sapiens*	Nac1homsa
Na$^+$–Proline Symporter	PutP proline–Na$^+$ symporter *Escherichia coli*	Putpescco
Na$^+$–Glucose Symporter	Sglt1 glucose–Na$^+$ symporter *Homo sapiens*	Nagchomsa
Vacuolar ATPase	Vacuolar ATPase subunit *Homo sapiens*	Vph1homsa
Na$^+$–Dicarboxylate Symporter	DctA dicarboxylate–Na$^+$ symporter *Escherichia coli*	Dctaescco
Na$^+$–PO$_4$ Symporter	Npt1 phosphate–Na$^+$ cotransporter *Homo sapiens*	Npt1homsa
Na$^+$–Branched Amino Acid Symporter	Brnq branched chain amino acid transporter *Salmonella typhimurium*	Brnqsalty
Na$^+$–Citrate Symporter	CitN citrate transporter *Klebsiella pneumoniae*	Citnklepn
Na$^+$–Alanine–Glycine Symporter	ACP alanine transporter Thermophilic bacterium PS-3	Alcpthep3
Na$^+$–Neurotransmitter Symporter	Net1 noradrenalin–Na$^+$ symporter *Homo sapiens*	Ntnohomsa
Na$^+$/H$^+$ Antiporter	Nhe1 Na$^+$/H$^+$ antiporter *Homo sapiens*	Nhe1homsa
Phosphenolpyruvate-Dependent Sugar Phosphotransferase System (PTS)	PtaA N–acetyl glucosamine permease II *Escherichia coli*	Ptaaescco
Anion Exchanger	AE1 anion exchange protein 1 *Homo sapiens*	B3athomsa
Mitochondrial Adenine Nucleotide Translocator	Ant1 ADP/ATP carrier protein *Homo sapiens*	Ant1homsa
White	White protein *Drosophila melanogaster*	Whitdrome
Mitochondrial Phosphate Carrier	PHC phosphate carrier protein *Homo sapiens*	Mpcphomsa

Table 1 Continued

Family	Example: Species	Code
Nitrate Transporter I	NarK nitrate–nitrite facilitator protein *Escherichia coli*	Narkescco
Nitrate Transporter II	CmA nitrate transporter *Emericella nidulans*	Crnaemeni
Spore Germination	Spore germination protein GraII *Bacillus subtilis*	Gra2bacsu
Vacuolar Membrane Pyrophosphatase	Pyrophospate-energized vacuolar proton pump *Arabidopsis thaliana*	Avp3arath
Gluconate Transporter	GntP gluconate transporter *Bacillus subtilis*	Gntpbacsu
ABC 1 & 2	ATP binding protein ABC1 *Mus musculus*	Abc1musmu
Yeast Multidrug Resistance	Multidrug resistance protein Cdr1 *Candida albicans*	Cdr1canal
Cystic Fibrosis Transmembrane Conductance Regulator	Cystic fibrosis transmembrane conductance regulator *Homo sapiens*	Cftrhomsa
P-Glycoprotein	Multidrug resistance protein Mdr1 *Homo sapiens*	Mdr1homsa

pathogenic organisms that aid their infection of host cells; enzymes required for the digestion of extracellular macromolecules such as cellulose, proteins, nucleic acids and lipids present as the result of the death of other organisms; and the active secretion of "assault" agents such as antibiotics.

Penetration of the inner cell membrane by solutes

The inner cell membrane, a protein-containing phospholipid bilayer (Fig. 4[54]) is the barrier preventing the entry of most ambient solutes into the bacterial cell. Nutrient uptake is therefore effected by integral membrane transport proteins, either singly or in complexes, the majority of which are synthesized only in the presence of their substrate (see below). Energization of transport is effected at this inner membrane. Amongst its many other functions are the processes of respiration, ATP synthesis, maintenance of the K^+ gradient, motility and osmoregulation, which are themselves transport processes [15,55,56]. The membrane is therefore a dynamic entity of transport proteins, some of which are dependent on others (Figs 1 and 2). For example, only one *inducible* protein is required for lactose transport[40], but the energization of its accumulation requires the respiratory chain or ATPase activity (see Figs 1 and 2), which are more permanent features of the membrane [56,57].

The importance of proton transport across the inner membranes

The Chemiosmotic Theory of Mitchell proposed that the respiratory enzymes pump protons across the inner bacterial membrane so that energy released by substrate

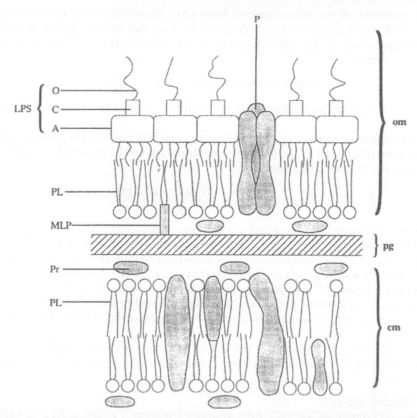

Figure 4 *Schematic drawing of the gram-negative bacterial cell envelope. The outer membrane (om) consists of lipopolysaccharide, phospholipid and proteins, most of which are porins. Inside the outer membrane is a peptidoglycan layer (pg), which is noncovalently bonded to the outer membrane via murein lipoproteins, themselves covalently attached to the peptidoglycan. The cell membrane (cm) is composed of phospholipid and protein, and is the location of the integral membrane proteins involved in transport. The region between the outer membrane and the cell membrane is called the periplasm. The wavy lines are fatty acid residues that anchor the phospholipids and lipid A into the membrane. LPS, lipopolysaccharide; O, oligosaccharide; C, core; A, lipid A; P, porin; PL, phospholipid; MLP, murein lipoprotein; Pr, protein; om, outer membrane; pg, peptidoglycan; cm, cell membrane. [Copied, with permission, from White, D. (1995) The Physiology and Biochemistry of Prokaryotes. Oxford University Press, New York.]*

oxidation is conserved as an electrochemical proton gradient [58-60]. This "proton-motive force" could then be used as an energy "currency" for expenditure on ATP synthesis, nutrient transport, chemotaxis, osmoregulation, etc. (Fig. 1). In organisms without respiratory enzymes an H^+ ATPase could maintain the proton-motive force utilizing ATP generated by fermentative metabolism.

The existence of the proton-motive force (Δp) across the inner membrane has been conclusively established in a diversity of bacterial species. Its magnitude is usually equivalent to 200–300 mV, made up of both electrical ($\Delta\Psi$) and osmotic (ΔpH) components.

$$\text{Proton-motive force } \Delta p = \Delta\Psi - Z\Delta\text{pH}$$

where Z is RT/zF, the factor that converts pH units to millivolts, usually calculated at 25°C.

When the proton-motive force is used to energize solute transport by proton-coupled mechanisms (Figs 1 and 2), the gradients of solute that can be achieved are related to the Δp by the following equation

$$\log[S_i]/[S_0] = \frac{(n+m)\,\Delta\Psi - nZ\,\Delta\text{pH}}{Z}$$

where m is the substrate charge and n is the proton/substrate ratio.

As already described, the Chemiosmotic Theory has been an invaluable guide for the elucidation of transport mechanisms. It is important to note that in some organisms living in alkaline and/or high salt environments, the Na^+ ion has replaced the H^+ as the coupling cation [61,62]. While in most examples the "conventional" oxidases and ATP synthase components seem simply to have adapted to pump Na^+ instead of H^+, in some organisms Na^+-pumping decarboxylase enzymes generate an electrochemical gradient of Na^+ [62] (Fig. 2).

The diagram in Fig. 5 illustrates the following mechanisms by which bacteria are known to effect the transport of some nutrients into their cells, and some solutes out.

1. Facilitated diffusion.
2. The "ATP-Binding-Cassette" ABC systems ("uniport") utilizing ATP to capture nutrient or drive efflux (Figs 1 and 5).
3. The group translocation mechanism utilizing PEP as energy source (Figs 1 and 5).

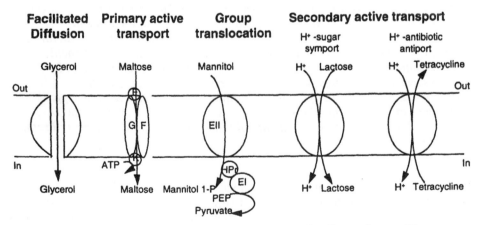

Figure 5 *Mechanisms of transport across the bacterial cell membrane. The different types of transport activity are described in the text.*

4. The H^+ nutrient coupled ("symport") systems utilizing the transmembrane electrochemical gradient of protons generated by respiration or ATPase (Figs 1 and 5).
5. Coupled transport of similarly charged compounds – anions or cations – in opposite directions ("antiport", Figs 1, 2 and 5), which may effect either accumulation of desired substrate or efflux of enzyme, waste or toxin.

The best-understood membrane transport processes have been studied in the gram-negative organisms *Escherichia coli* and *Salmonella typhimurium*, which are convenient because of their unicellular nature. Furthermore, most of their transport mechanisms appear to occur in many other microorganisms and even man himself.

TRANSPORT ACROSS EUKARYOTIC CELL MEMBRANES

The considerations that apply to understanding transport in prokaryotes extend to eukaryotes with important exceptions. It is more difficult with multicellular organisms where cells occur in tissues. Also, eukaryote cells have subcellular compartments bounded by membranes. Obviously, the transport reactions involved in ATP synthesis are localized in mitochondria and chloroplasts, which use an H^+ electrochemical gradient for energy coupling. In order to accommodate solute–H^+ symporters or antiporters in the cell membrane, therefore, organisms like yeast have an H^+ ATPase located there[63]. Mammalian cells, however, utilize a transmembrane Na^+ gradient generated by an Na^+/K^+ ATPase to accommodate solute–Na^+ symporters or antiporters[64]. Quite often the maintenance of high concentrations of nutrient in the extracellular fluid (from the blood in mammals or vascular system in plants) obviates the need for energized transport into the cell, so higher organisms can utilize facilitated diffusion systems in their cell membranes rather than active transport.

Translocation of substrates between intracellular and extracellular compartments can have sophisticated functional implications. Examples are the release and recapture of neurotransmitter substances in nerve[65]; sucrose mobilization in plants[66]; antigen peptide presentation in lymphocytes[10]; protein targeting in plants and animal cells[67]. Since little is known about each of the individual proteins that contributes to these processes our understanding remains superficial at the present time.

THE NUMBER OF MEMBRANE PROTEIN COMPONENTS AND/OR DOMAINS INVOLVED IN A TRANSPORT SYSTEM

Facilitated diffusion transport systems usually contain a single protein. Similarly, secondary active transport systems usually contain one protein, if we discount those that generate the driving ion gradient. Primary active transport systems may occasionally contain one protein, for example bacteriorhodopsin. However, most appear to comprise a protein complex, involving from as few as two polypeptides (X^+ ATPase[68]) through six (histidine transport system) to 20 ($F_1 F_0$ ATPase) and more in, for example, NADH dehydrogenase[69].

Both the ABC and phosphotransferase systems illustrate how transport systems that contained several separate polypeptides in primitive organisms may become fused together during the course of evolution so that one polypeptide with functionally distinguishable domains effects translocation. This has been particularly well illustrated by Higgins [70] and by Postma *et al.* [28] (see Fig. 6).

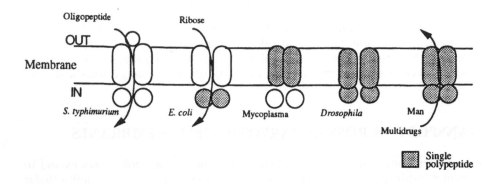

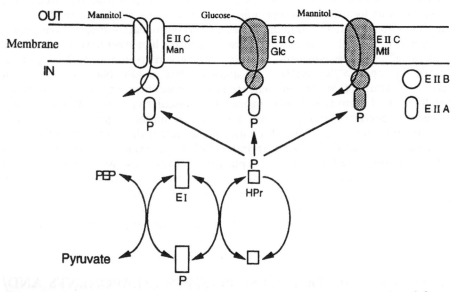

Figure 6 *Proteins of multicomponent transport systems may become fused during evolution. The transport systems illustrated are discussed in the text. The upper part of the figure shows schematically various ABC primary active transport systems and the organisms in which they are found; the different polypeptides are unfused in the example on the left, and the shading indicates different types of fusion between functionally discrete domains that has occurred in other examples. The lower part of the figure shows different group translocation transport systems, all of which are phosphotransferases found in* Escherichia coli; *the different polypeptides, E IIA and E IIB, associated with the membrane component, E IIC, are unfused in the example on the left and the shading indicates different types of fusion in other examples. The figures are derived from information in refs.* [28,44,70]

MANY MEMBRANE TRANSPORT PROTEINS ARE PREDICTED TO CONTAIN 12 TRANSMEMBRANE DOMAINS

Hydropathy plots are widely used to predict if regions of a protein might span the membrane as an α helix [71]. This method is particularly applicable when a protein is predicted to contain a high proportion of hydrophobic amino acids. Some examples are shown in Fig. 7. The only authentication of their validity is the reasonable correspondence of predicted α helices with those actually observed in bacteriorhodopsin and membrane proteins of the photosynthetic reaction centre and light-harvesting complex [72,73], and more recently cytochrome oxidases [2,3]. There is discussion over which algorithm, if any, is satisfactory [74–77]. Despite these uncertainties, the majority of the transport protein sequences in Table 1 are predicted to contain 12 hydrophobic regions of sufficient length (19+ amino acids) to span the membrane as α helices [78,79]. The possible exceptions are the transporters for methylenomycin and quaternary ammonium compounds, which may have 14 [80,81], and the rhamnose-H^+ transporter, predicted to have 10 [82]. Many of the sugar transport proteins have an extensive central (i.e. between transmembrane domains 6 and 7) hydrophilic region of about 65 amino acids which is predicted to contain a substantial proportion of α helix. Most of the other transport proteins also have a central hydrophilic region, although it is usually shorter than that of the sugar transport proteins. Taken with the evidence of some sequence duplication in the two halves of many of the proteins [48], it seems reasonable to propose the existence of internal dimerization, originally resulting, perhaps, from gene duplication. This also accords with the same proposal by Lancaster [83], based on kinetic and inhibitor studies of the LacY porter.

Despite the differences in individual sequences, an underlying similarity between transport proteins from otherwise dissimilar groups seems to exist [79], even though some catalyze mechanistically rather different types of transport reaction – uniport, antiport, or symport (influx or efflux). One example of a 12-helix arrangement is shown in Fig. 8.

In this context, it is interesting that many other groups of membrane transport proteins are predicted to have 12 membrane-spanning α helices. One is the series of phosphate antiporters in prokaryotes [79]. Another is the "ABC" group typified by the Mdr, multiple drug resistance factor [10,44,70]; some individual members of this group catalyze influx of substrate and others catalyze efflux. Yet another group is the family of mitochondrial transporters, which are thought to function as a dimer, each subunit having six α helices [84,85] (see discussion below); here again transporters of similar sequence catalyze different types of transport reaction – uniport, antiport, or symport, influx or efflux. A fourth group contains the homologous transporters for noradrenaline and gamma-aminobutyric acid [86].

Within the family of mitochondrial transport proteins each is predicted to contain six hydrophobic regions and transmembrane helices [84,85]. However, in several examples there is evidence for dimerization to form a functional unit with 12 predicted helices. Interestingly, this family has strong evidence of internal triplication in each polypeptide [84], implying that there are six equivalent domains in the functional dimer.

It is important to consider the possible arrangements of 12 helices in the membrane, and several groups have obtained evidence for the nearest-neighbor relationships of predicted transmembrane helices in individual membrane transport proteins, using fluorescence energy transfer, second site revertants of mutants, cysteine

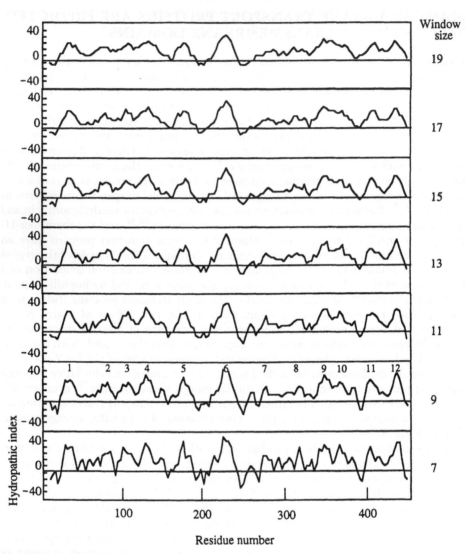

Figure 7a *Hydropathy plots of the L-fucose–H⁺ symport protein, FucP, of* E. coli. *The algorithm of Kyte and Doolittle (1982, see text) was used with window sizes of 7–19 amino acid residues to generate a series of hydropathy plots of FucP; the putative positions of 12 helices are indicated in the plot with a window of nine residues. [Copied, with permission, from Gunn et al. (1995) Molec. Microbiol. 15, 771–783.]*

mutagenesis and other techniques. In addition, many reviewers have hypothesized as to how the arrangement might be. However, until we determine the actual 3D structures of some of these proteins such models should perhaps be regarded with caution.

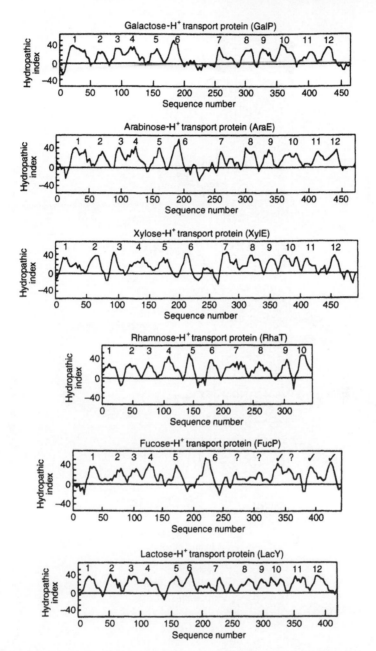

Figure 7b *Hydropathic profiles of membrane transport proteins. The amino acid sequence of each of the indicated transport proteins was analyzed for hydropathy using the algorithm of Kyte and Doolittle with a window of 11 residues. The majority can be interpreted in terms of 12 putative transmembrane helices, but the L-rhamnose–H⁺ symport protein appears to have 10. [Copied, with permission, from Henderson (1991) Bioscience Reports 11, 477–538, ref.[31]].*

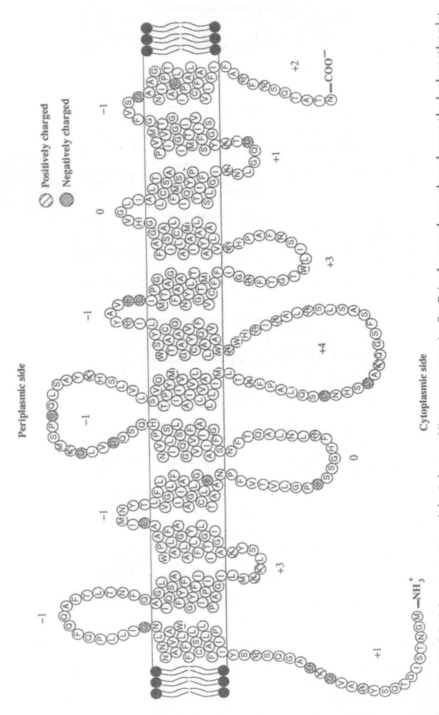

Figure 8a *Model of the orientation of the L-fucose–H⁺ symport protein, FucP, in the membrane, based upon the hydropathy plot (Fig. 7a). Note the predominance of positive residues inside the membrane, which follows the rule of von Heijne (see text). [Copied, with permission, from Gunn et al. (1995) Molec. Microbiol. 15, 771–783.]*

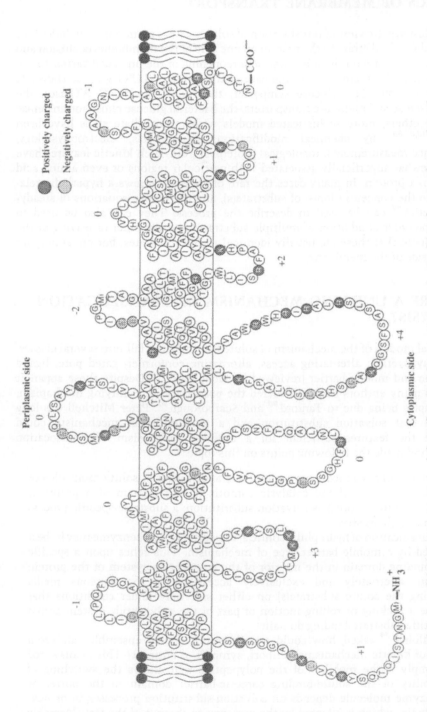

Figure 8b *Modified model of the orientation of the L-fucose–H⁺ symport protein, FucP, in the membrane, based upon β-lactamase fusion data. [Copied, with permission, from Gunn et al. (1995) Molec. Microbiol. 15, 771–783.]*

KINETICS OF MEMBRANE TRANSPORT

The simplest kinetic view of translocation of solutes across membranes includes four steps. Binding of substrate to the protein on one side of the membrane; occlusion and translocation; release on the other side; and reopening of the unloaded carrier to the original side of the membrane. In the case of the human GLUT1 glucose transport protein, the rabbit Na^+–glucose symporter, the Na^+/K^+ and Ca^{2+} ATPases, the bacterial lactose and melibiose transporters, the bacterial glucose phosphotransferase and some others, more sophisticated models with intermediate steps have been advanced [18,87,88]. By chemical modification, mutagenesis, electrophysiology, fluorescence measurements, topological proteolysis, etc. such kinetic features have been somewhat superficially associated with particular regions or even amino acid residues in a protein. In many cases the rate of transport shows a hyperbolic relationship to the concentration(s) of substrate(s), and the classic equations of steady-state kinetics [89] can be used to describe the process. They can also be used to analyze the order of addition of multiple substrates and the order of leaving of the products (note that these are usually identical to the substrates, but are simply on the other side of the membrane).

IS THERE A UNIFYING MECHANISM OF TRANSLOCATION CATALYSIS?

Theoretical models of the mechanism of solute translocation fall into several classes, which may overlap: alternating access; alternating conformer; gated pore; ligand conduction and mobile barrier (reviewed by Henderson [31]). Despite these apparent variations many authors have considered the possibility of a unifying mechanism, two examples being due to Tanford [90] and Scarborough [91]. Peter Mitchell has long advocated that solvation substitution and a mobile barrier mechanism could constitute the features sufficient for a unifying mechanism of translocation catalysis, and made the following points on this topic [60].

1) The dominant process governing the translocation of solute molecules or ions from one side of the catalytic osmotic barrier domain of a porter or osmoenzyme to the other is solvation substitution: a substrate-specific process of secondary chemistry.

2) Translocation of hydrophilic solute(s) in a porter or osmoenzyme may be best explained by a mobile barrier type of mechanism. This relies upon a specific solute-binding domain in the interior of the polypeptide system of the protein becoming alternately and exclusively accessible to the aqueous media containing the solute substrate(s) on either side only under conditions that facilitate a rocking or rolling motion of part of the polypeptide system across the specific substrate-binding domain.

3) Maloney [79] asked, how could a uniform (12 α-helix) ensemble catalyze a variety of kinetic mechanisms: uniport, symport or antiport? This is answered very simply if the mobility of the polypeptide that allows the switching of accessibility of the solute-binding osmotic-barrier domain in the porter or osmoenzyme molecule depends on solvation-substitution processes in or near that domain, which is affected by the presence or absence of the translocatable

solute(s). In the case of osmoenzymes, this may also be effected by the binding of other ligands. Thus, barrier mobility would be activated: in a uniport whether solute was bound or not; in a symporter, only when both or neither of the solutes were bound; in an antiporter, only when either one or the other solute, but not both or neither, were bound; in an osmoenzyme, under appropriate conditions of binding of the translocatable solute(s) and also of other chemical group-donating and group-accepting ligands.

...The alternating access model of transport proteins...attributed to Tanford, resembles, in some respects, my mobile barrier model. But the model discussed by Tanford[90], like the gated pore type of model considered by Brooker[92], which seems to be consistent with the concept of a proton relay, discussed by Roepe et al.[93] misses the fundamental importance of solvation substitution in the proposed motion of the osmotic barrier over the solute-binding domain.

One of the most attractive properties of the mobile barrier type of mechanism of solute translocation arises from its presumed dependence on the subtle secondary chemical processes of solvation substitution, both with respect to the binding of its solute substrate(s) and with respect to the kinetic activation of the mobility of the barrier across the catalytic substrate-binding domain. Thus, it would be expected to show the close interrelationships between changes of organic substrate specificity, changes of cation specificity, and changes in translocational kinetics induced by certain amino acid substitutions, already described for several transporters.

The tendency for the active species of solute-translocating proteins to contain 12 α-helical components...may possibly be relevant to the mobile barrier type of mechanism. Invoking the concept of close packing in hexagonal arrays of the cylindrical α-helices, and assuming the requirement for a cleft opening alternately above and below the catalytically active molecule, imagined with the plane of the membrane lying flat on the page, one is tempted to suggest a binary hexagonal arrangement with two hexagonal lobes sharing a pair of α-helices (the cylindrical α-helices appearing as circles from above). The catalytic solute-binding domain would lie in the region between the two shared α-helices, and extend to the neighbouring helices on either side. One of the shared helices might act as a hinge, allowing slight relative movement of the two lobes, while the other shared helix would cant outwards from its partner alternately at top and bottom, allowing accessibility of a centrally positioned solute-binding domain alternately and exclusively from above and below. Or perhaps both of the shared helices might cant outward from its partner alternately at top and bottom to give a relatively symmetrical cleft opening alternately and exclusively from above and below.

From this eclectic viewpoint transport systems can be envisaged as modular in construction, with a basic porter unit capable of carrying out solvation substitution and the molecular events of translocation (by mobilization of an internal barrier[59,60], or by conformational changes effecting alternate access to each side of the membrane[90,94]). There can be additional proteins/domains to bring the initial solvation substitution under independent control, e.g. with a binding protein type of system. And/or there may be different proteins/domains to bring the translocation events under control of an ATP-hydrolyzing protein as in the ABC transport systems or the P-type ATPases [10,44,70,95], or under control of a decarboxylating reaction as in the bacterial Na^+ transporters [26,62].

CAN THE THREE-DIMENSIONAL STRUCTURES OF TRANSPORT PROTEINS BE DETERMINED?

Our understanding of the molecular mechanisms of membrane transport proteins is still severely handicapped by our ignorance of their three-dimensional structures [94]. The problems of determining such information for membrane proteins have been admirably reviewed by Pattus [96]. Nevertheless, there are recent advances that raise hopes of determining the complete three-dimensional structure of a membrane transport protein by physical methods.

The first is the elucidation of the structure of bacterial and mitochondrial cytochrome oxidases at atomic level resolution by X-ray crystallography [2,3]. The bacterial enzyme is a four-subunit protein, but the most intriguing component is subunit I, which contains the heme groups and comprises 12 membrane-spanning α helices. These are arranged in three groups of four helices, each group of which can be hypothesized to form a "pore" suitable for transmembrane conduction of H^+. This is the structure of a primary active transport system for protons. It is a useful exercise to model the unsophisticated 12-helix representation of other transport proteins (Fig. 8), around such a structure as an aid to hypothesizing how larger substrates might have their passage through the membrane catalyzed.

Continued refinement has occurred of the application of electron diffraction techniques and data analysis, so that high-resolution structures, e.g. of light-harvesting protein [73] and of visual rhodopsin [97], can be achieved. These techniques should be capable of further refinement to higher levels of resolution [60].

There has also been improvement in crystals of the bacterial porin proteins, enabling X-ray crystallography to be improved .

Finally, there is the application of biophysical techniques to determine the structure of small membrane-spanning peptides [98,99]. If individual, or a small number of combined, transmembrane domains of the lactose–H^+ (or any other) transporter can be expressed, purified and reconstituted in the native form, as already partly achieved, it may be possible to determine the structure of parts of the protein separately by NMR and build up an overlapping picture of the whole.

CONCLUSIONS

We are entering an era when the amino acid sequences of a huge number of transport proteins will be available from the DNA sequence databases. For reasons of scientific curiosity and/or biomedical utility a select number of these will be chosen for detailed investigation. It is very important that we learn how to determine the three-dimensional structures of these proteins. We will then be in a position to define the structure–activity relationship of the protein to the point where it can be manipulated for the good of humanity – to design a new generation of antimicrobials, perhaps, to devise molecular-size electronic components of nanocomputers, to cure cystic fibrosis by gene therapy, and unconceived applications. This book collates the information that is currently available to us and arranges it in a manner to expedite future developments.

References

[1] Henderson, R. et al. (1990) J. Mol. Biol. 213, 899–929.
[2] Iwata, S. et al. (1995) Nature 376, 660–669.
[3] Tsukihawa, T. et al. (1996) Science 272, 1136–1144.
[4] Muir, M. et al. (1985) J. Bacteriol. 163, 1237–1242.
[5] White, D. (1995) The Physiology and Biochemistry of Prokaryotes. Oxford University Press, Oxford.
[6] Koch, A. (1971) Adv. Microb. Physiol. 6, 147–217.
[7] Button, D.K. (1985) Microbiol. Rev. 49, 270–297
[8] McNab, R.M. (1996) In Escherichia coli and Salmonella (Neidhardt, N.C., ed.). ASM Press, Washington DC, pp. 123–145.
[9] Goffeau, A. et al. (1997) Yeast 13, 43–54.
[10] Higgins, C.F. (1992) Annu. Rev. Cell. Biol. 8, 67–113.
[11] Mathews, C.K. and van Holde, K.E. (1996) Biochemistry. Benjamin/Cummings, Redwood City, CA.
[12] Voet, D. and Voet, J.G. (1996) Biochemistry. John Wiley, Chichester.
[13] Mitchell, P. (1963) Biochem. Soc. Symp. 22, 142–169.
[14] Mitchell, P. (1973) Bioenergetics 4, 63–91.
[15] Mitchell, P. (1966) Chemiosmotic Coupling in Oxidative and Photosynthetic Phosphorylation. Glynn Research, Bodmin.
[16] Heller, K.B. et al. (1980) J. Bacteriol. 144, 274–278.
[17] Maloney, P.C. and Wilson, T.H. (1996) In Escherichia coli and Salmonella (Neidhardt, N.C., ed.). ASM Press, Washington DC, pp. 1130–1148.
[18] Stein, W.D. (1986) Transport and Diffusion across Cell Membranes. Academic Press, Orlando, FL.
[19] Baldwin, S.A. (1993) Biochim. Biophys. Acta 1154, 17–50.
[20] Harold, F.M. and Maloney, P.C. (1996) In Escherichia coli and Salmonella (Neidhardt, N.C., ed.). ASM Press, Washington DC, pp. 283–306.
[21] Wikstrom, M. (1989) Nature 338, 776–778.
[22] Kagawa, Y. (1984). Bioenergetics (Ernster, L., ed.). Elsevier, Amsterdam, pp. 149–186.
[23] Abrahams, J.P. et al. (1994) Nature 370, 621–628.
[24] Fillingame, R.H. (1996) Curr. Opin. Struct. Biol. 6, 491–498.
[25] Tokuda, H. (1986) Methods Enzymol. 125, 520–530.
[26] Dimroth, P. (1986) Methods Enzymol. 125, 530–540.
[27] Dimroth, P. (1990) Philos. Trans. R. Soc. Lond. B 326, 465–477.
[28] Postma, P. et al. (1996) In Escherichia coli and Salmonella (Neidhardt, N.C., ed.). ASM Press, Washington DC, pp. 1149–1174.
[29] Forward, J.A. et al. (1997) J. Bacteriol. 179 (in press).
[30] Lewis, K. (1994) Trends Biochem. Sci. 19, 119–123.
[31] Henderson, P.J.F. (1991) Biosci. Reports 11, 477–538.
[32] Mitchell, P. (1961) Nature 191, 144–148.
[33] Meister, M. et al. (1987) Cell 49, 643–650.
[34] Noji, H. et al. (1997) Nature 386, 299–302.
[35] West, I.C. (1970) Biochem. Biophys. Res. Commun. 41, 655–661.
[36] West, I.C. and Mitchell, P. (1972) Bioenergetics 3, 445–462.

[37] West, I.C. and Mitchell, P. (1973) Biochem. J. 132, 587–592.

[38] Kaback, H.R. (1986) Methods Enzymol. 125, 214–230.

[39] Kaback, H.R. et al. (1990) Trends Biochem. Sci. 15, 309–314.

[40] Kaback, H.R. (1997) Proc. Natl. Acad. Sci. USA 94, 5539–5543.

[41] Henderson, P.J.F. (1990) J. Bioenerg. Biomembr. 22, 525–569.

[42] Berger, E.A. (1973) Proc. Natl Acad. Sci. USA 70, 1514–1518.

[43] Berger, E.A. and Heppel, L.A. (1974) J. Biol. Chem. 249, 7747–7750.

[44] Boos, W. and Lucht, J.M. (1996) In *Escherichia coli* and *Salmonella* (Neidhart, F.C., ed.). ASM Press, Washington DC, pp. 1175–1209.

[45] Mitchell, P. (1977) In Microbial Energetics (Haddock, B.A. and Hamilton, W.A., eds) pp. 383–423. Cambridge U.P., Cambridge, UK.

[46] Henderson, P.J.F. and Maiden, M.C.J. (1990) Phil. Trans. R. Soc. Lond. B 326, 391–410.

[47] Maiden, M.C.J. et al. (1987) Nature 325, 641–643.

[48] Griffith, J.K. et al. (1992) Curr. Topics Cell Biol. 4, 684–695.

[49] Runswick, M.J. et al. (1987) EMBO J. 6, 1367–1373.

[50] Nikaido, H. (1996) In *Escherichia coli* and *Salmonella* (Neidhart, F.C., ed.). ASM Press, Washington DC, pp. 29–47.

[51] Engel, A. et al. (1985) Nature 317, 643–645.

[52] Cowan, S.W. et al. (1994) In Bacterial Cell Wall, New Comprehensive Biochemistry, vol. 27. Elsevier, Amsterdam, pp. 353–362.

[53] Schirmer, T. et al. (1995) Science 267, 512–514.

[54] Kadner, R.J. (1996) In *Escherichia coli* and *Salmonella* (Neidhart, F.C., ed.). ASM Press, Washington DC, pp. 58–87.

[55] West, I.C. and Mitchell, P. (1974). Biochem. J. 144, 87–90.

[56] Harold, F.M. and Maloney, P.C. (1996) In *Escherichia coli* and *Salmonella* (Neidhart, F.C., ed.). ASM Press, Washington DC, pp. 283–306.

[57] Gennis, R.B. and Stewart, V. (1996) In *Escherichia coli* and *Salmonella* (Neidhart, F.C., ed.). ASM Press, Washington DC, pp. 217–261.

[58] Mitchell, P. (1970) Symp. Soc. Gen. Microbiol. 20, 121–166.

[59] Mitchell, P. (1990) Res. Microbiol. 141, 286–289.

[60] Mitchell, P. (1990) Res. Microbiol. 141, 384–385.

[61] Skulachev, V.P. (1985) Eur. J. Biochem. 151, 199–208.

[62] Dimroth, P. (1990) Philos. Trans. R. Soc. Lond. B 326, 465–477.

[63] Kruckeberg, A.L. (1996) Arch. Microbiol.166, 283–292.

[64] Hirayama, B. et al. (1996) Am. J. Physiol – Gastrointestinal and Liver Physiology 33, G919–G926.

[65] Schuldiner, S. (1997) Physiol. Rev. 75, 369–392.

[66] Subbaiah, C.C. et al. (1994) Plant Cell, 6, 1747–1762.

[67] High, S. et al. (1997) In Membrane Protein Assembly (von Heijne, G., ed.). Springer-Verlag, Heidelberg, pp. 119–134.

[68] Mcintosh, I. and Cutting, G.R. (1992) FASEB J. 6, 2775–2782.

[69] Weiss, H. et al. (1991) Eur. J. Biochem. 197, 563–576.

[70] Higgins, C.F. (1995) Cell 82, 693–696.

[71] von Heijne, G. (1994) Annu. Rev. Biophys. Biomol. Struct. 23, 167–192.

[72] Deisenhofer, J. et al. (1984) J. Mol. Biol. 180, 385–398.

[73] Kuhlbrandt, W. and Wang, D.N. (1991) Nature 350, 130–134.

[74] Lodish, H.F. (1988) Trends Biochem. Sci. 13, 332–334.

[75] von Heijne, G. (1988) Biochim. Biophys. Acta. 947, 307–333.

[76] White, S.H. and Jacobs, R.E. (1990) J. Membr. Biol. 115, 145–158.
[77] Crimi, M. and Esposti, M.D. (1991) Trends Biochem. Sci. 16, 119.
[78] Baldwin, S.A. (1990) Biotech. Appl. Biochem. 12, 512–516.
[79] Maloney, P.C. (1990) Res. Microbiol. 141, 374–383.
[80] Neal, R.J. and Chater, K.F. (1987) Gene 58, 229–241.
[81] Paulsen, I.T. et al. (1996) Microbiol. Rev. 60, 575–608.
[82] Tate, C.G. and Henderson, P.J.F. (1992) J. Biol. Chem. 268, 26850–26857.
[83] Lancaster, J.R. (1982) FEBS Lett. 150, 9–18.
[84] Palmieri, F. et al. (1990) Biochim. Biophys. Acta 1018, 147–150.
[85] Runswick, M.J. et al. (1994) DNA Sequence 4, 281–291.
[86] Pacholczyk, T. et al. (1991) Nature 350, 350–354.
[87] Cloherty, E.K. (1995) Biochemistry 34, 15395–15406.
[88] Pourcher, T. et al. (1990) Philos. Trans. R. Soc. Lond. B 326, 411–423.
[89] Henderson, P.J.F. (1992) In Enzyme Assays: A practical approach. (Eisenthal, R. and Danson, M.J., ed.). Oxford U.P., Oxford, pp. 277–316.
[90] Tanford, C. (1983) Annu. Rev. Biochem. 52, 379–409.
[91] Scarborough, G.A. (1985) Microbiol. Rev. 49, 214–231.
[92] Brooker, R.J. (1990) Res. Microbiol. 141, 309–315.
[93] Roepe, P.D. et al. (1990) Res. Microbiol. 141, 290–308.
[94] Karlin, A. (1997) Proc. Natl. Acad. Sci. USA 94, 5508–5509.
[95] Jorgensen, P.L. and Anderson, J.P. (1988) J. Membr. Biol. 103, 95–120.
[96] Pattus, F. (1990) Curr. Opin. Cell Biol. 2, 681–685.
[97] Schertler, G.F. (1997) Molec. Biol. of the Cell 7, 970.
[98] Barsukov, I.G. et al. (1990) Eur. J. Biochem. 192, 321–327.
[99] Lemmon, M.A. et al. (1994) Nature Struct. Biol. 1, 157–163.

2 Amino Acid Sequence Comparisons

For many years there was little information about the amino acid sequences of membrane transport proteins, owing to the difficulty of obtaining sufficient purified quantities for conventional protein sequencing. This changed during the past decade with the cloning and sequencing of ever increasing numbers of genes and, more recently, entire genomes, from which the amino acid sequences of many integral membrane proteins have been deduced. These transport proteins have been grouped by a number of functional criteria, including mechanism (e.g. sodium-solute symporters[1]), topology (e.g. 12-transmembrane helix transporters[2]), intracellular location (e.g. mitochondrial transporters[3]), and possession of amino acid sequence domains (e.g. ATP binding cassette (ABC) transporters[4]). In some instances, the amino acid sequences of proteins grouped by functional criteria are related, for example the mitochondrial phosphate carrier and adenine nucleotide translocator families[3]. In other instances, the amino acid sequences of proteins grouped by functional criteria have no apparent relationship to one another, for example most families of sodium-solute symporters[1].

It is potentially instructive to group transport proteins by the relationships between their amino acid sequences because overall similarity between amino acid sequences can indicate similar three-dimensional structures, implying similar mechanisms of action. Algorithms such as FASTA[5] and BLASTP[6] search amino acid sequence databases, for example SwissProt, PIR and GenPept, and list in order of local relatedness proteins whose amino acid sequences are similar to that of the query sequence. The relationships between the amino acid sequences of the proteins identified in this fashion then can be quantitated by pairwise comparison. The statistical significance of the alignment score for each pairwise comparison is evaluated by comparing it to the mean score obtained from comparison of each sequence to random permutations of the other, and is expressed as the number of standard deviations (SD) by which the maximum score for the real comparison exceeds the mean of the scores for the comparisons to randomized sequences[7]. If an alignment score is greater than 9 SD above the mean of randomly permuted sequences the proteins are very likely homologous, scores of 6–9 SD are taken to indicate likely relatedness, and 3–6 SD possible relatedness.

The probability of obtaining an alignment score of 9 SD by chance is approximately 10^{-18}. Therefore, it is likely that homologous members of a family share a common evolutionary origin, implying similar three-dimensional structures and functional properties. None the less, there are often unexpected differences between the functional attributes of homologous transporters within a family. For example, passive glucose transporters of eukaryotes and proton-dependent sugar transporters of prokaryotes are members of the same family of homologous sugar transporters, implying that the passive transporters evolved relatively recently without extensive sequence modification[2,8]. There are also several families in which there is neither a structural nor chemical relationship between many of the substrates recognized by homologous transporters[8,9]. Thus, a perceived difference in function need not be a consequence of a profound difference in structure. In these instances, the structure–activity relationships between functionally dissimilar members of a family can be investigated using algorithms which cluster sequences by similarity to produce a dendrogram representing the clustering relationships. The PILEUP algorithm[10], used herein, first aligns the two most similar sequences to produce a cluster of two sequences, then aligns this cluster with another cluster of the next two most similar sequences and so on until all sequences have been included in the dendrogram.

Amino acid sequence comparisons also reveal unexpected relationships amongst seemingly dissimilar *families* of proteins. For example, amino acid sequence elements that are highly conserved in the family that contains facilitative sugar transport proteins of mammals also occur in the family that contains proton-dependent tetracycline antiporters of bacteria[8,9]. Although there is not significant similarity between all members of all of these families, there is significant similarity (>3 SD) between many members of different families. When the amino acid sequences of multiple families are significantly similar, the families are presumed to be derived from a common ancestor and are considered subgroups of a superfamily of related transporters [8,9,11].

One of the most functionally diverse superfamilies, the uniporter-symporter-antiporter (USA) or major facilitator (MFS) superfamily, contains uniporters, symporters, and antiporters of structurally dissimilar sugars, sugar phosphate esters, antibiotics, antiseptics, disinfectants carboxylated compounds, catecholamines and indolamines [8,9,11]. The significance of about 40% of the pairwise comparisons between families of the superfamily exceed 3 SD and the ALIGN scores for certain pairwise comparisons between families are as high 8.7 SD, reflecting their presumed common ancestry [8]. This predicts that they also have similar three-dimensional structures, suggests fundamentally similar molecular mechanisms, and implies that relatively subtle structural differences account for the differences in the functional properties of the proteins, such as the recognition of structurally dissimilar substrates, or the vectorial mechanism. As pointed out previously, a perceived profound difference in function need not be a consequence of a profound difference in structure.

Multiple sequence alignments generated in this manner often reveal highly conserved "signature motifs"[8,9]. These may be unique to either the family or a subgroup of the family, or common to a group of families which share a functional attribute. Signature motifs of the first category can have great utility in assessing the potential relatedness of transporters which are not homologous by the criterion of the alignment score. Signature motifs of the second and third categories, i.e. those that are highly conserved in proteins with a common functional attribute, for example substrate specificity, mode of energization or vectorial mechanism, are predicted to be necessary for that attribute. These predictions can then be tested with site-directed mutagenesis and other molecular–genetic approaches. In only a few instances has it been possible to crystallize integral membrane proteins for molecular structural analysis. Therefore, most investigations of the structure–activity relationships of transport proteins have been founded on amino acid sequence comparisons of this sort.

Signature motifs that are conserved in all transporters of a superfamily may dictate structural or functional attributes that are common to all members of the superfamily. For example, alignment of the consensus sequences of the several families comprising the USA/MFS superfamily identifies several amino acid sequence motifs which are highly conserved in all or some of these diverse transporters. A "G-X-X-X-D-R/K-X-G-R-R/K" motif, which is strongly predicted to form a β-turn in most cases, is highly conserved between the second and third predicted helices of transporters in all families of the USA/MFS superfamily[2,8,9]. The "G-X-X-X-D-R/K-X-G-R-R/K" motif has been proposed to act as a cytoplasmic gate that limits the flow of substrate into and out of the cytoplasm. Site-directed and insertional mutagenesis of the TETA(B) tetracycline/H^+ antiporter and LACY lactose/H^+ symporter have demonstrated that several of these conserved residues of the motif are necessary for function. Similarly, a "R-X-X-X-G-X-X-X-G/A" motif is conserved in the fourth predicted helix and the preceding predicted extracellular hydrophilic loop of transporters in all families of the USA/MFS superfamily [2,8,9]. The "R-X-X-X-G-X-X-X-G/A" motif has been proposed to function in energy coupling. In the ATP binding cassette (ABC) superfamily, the "G-H-S-G-A-G-K-S-T" and "I-L-L-D-E" motifs, the so-called Walker A and B motifs, define the superfamily. These motifs, the first of which is known to be involved in phosphoryl transfer, are shared by many nucleotide binding proteins [4].

Although overall amino acid sequence relatedness and the conservation of highly conserved signature motifs provides strong presumptive evidence that two proteins have related functions, this is not always the case. For example, signature motifs corresponding the ATP binding domains define the ATP binding cassette (ABC) transporter superfamily [4]. However, these domains are also found in at least two families of the ABC superfamily that are neither associated with the membrane nor implicated in transport. These are the UVRA family of DNA excision repair proteins and the EF3 family of translational elongation factors. Thus, the conservation of an extended functional domain in two proteins, in this case the ATP binding cassette, does not by itself indicate that the two proteins have related functions, although in most instances this is true.

The second category of signature motif is conserved in, and thereby can define, subgroups of a superfamily. These motifs may dictate the shared structural or functional properties of the subset, such as substrate specificity or vectorial mechanism, predictions that also can be tested by site-specific mutagenesis. For example, a "G-X-X-X-G-P-X-X-G" motif is highly conserved in the fifth predicted membrane-spanning region of transporters of all families of the USA/MFS superfamily which direct substrate export, but not in any of the transporter families which direct substrate uptake [8,9,12]. Molecular modeling of the so-called "antiporter motif" predicts that a "kink" at approximately the position of the GP dipeptide, resulting in a change in helix axis direction of approximately 20 degrees, would be more stable than a regular helical conformation. The repeating pattern of glycine residues in the antiporter motif also forms a pocket, devoid of side-chains, on the surfaces of the fifth predicted helices. Site-directed mutagenesis experiments indicate that even very slight alterations in the structure of this motif, for example replacement of the hydrogen of glycine with either the small methyl side-chain of alanine or the methylol side-chain of serine, has profound and specific effects on resistance to tetracycline [12].

Intramolecular amino acid sequence comparisons are also useful in investigating structure–activity relationships. For example, there are significant similarities between the amino acid sequences of the N- and C-terminal halves of transporters in many families and superfamilies, including the acriflavin-cation resistance family and the USA/MFS superfamily [8]. This implies that these proteins arose by the duplication of a half-sized ancestor, suggesting that the N- and C-terminal halves of the transporters might have evolved to contain independent functional domains. This prediction was confirmed for the USA/MFS superfamily by demonstrating that paired in-frame deletion constructs of the *E. coli* LACY lactose/H^+ symporter complement each other functionally[13]. Using similar methods, two functional complementation groups also have been defined in the TETA(B) tetracycline/H^+ antiporter, which belongs to a different family from LACY[14,15].

Intramolecular amino acid sequence comparisons have also shown that the N-terminal halves of distantly related transporters of the USA/MFS superfamily are generally much more similar than the C-terminal halves, provided the proteins being compared have structurally dissimilar substrates[8]. Thus, the greater conservation of the N-terminal halves of transporters that recognize structurally dissimilar substrates has been interpreted to reflect the conservation of structures which confer the substrate binding-induced conformational change that is proposed to be common to these transporters' mechanism of action.

The C-terminal halves of transporters that recognize structurally dissimilar substrates are much less conserved than their N-terminal halves, a situation frequently reversed when transporters that recognize structurally similar substrates are considered. These observations support the interpretation that substrate specificity is determined by sequence motifs contained in the C-terminal halves of these transporters. Consistent with this possibility, inhibitor, photo-affinity labeling and domain exchange studies suggest that the substrate binding sites for the USA/MFS superfamily's sugar transporters are located in their C-terminal halves[16]. Likewise, mutations resulting in altered substrate specificities in various antibiotic antiporters have been found primarily in the C-terminal halves of the proteins[9].

References
1   Reizer, J. et al. (1994) Biochim. Biophys. Acta 1197, 133–166.
2   Henderson, P.J.F. (1993) Curr. Opin. Cell Biol. 5, 708–721.
3   Kuan, J. and Saier, M. (1993) CRC Crit. Rev. Biochem. Mol. Biol. 28, 209–233.
4   Higgins, C.F. (1992) Annu. Rev. Cell Biol. 8, 67–113.
5   Lipman, D. and Pearson, W (1985) Science 227, 1435–1441.
6   Altschul, S. et al. (1990) J. Mol. Biol. 215, 403–410.
7   Dayhoff, M. et al. (1983) Methods Enzymol. 91, 524–545.
8   Griffith, J. et al. (1992) Curr. Opin. Cell Biol. 4, 684–695.
9   Paulsen, I. et al. (1996) Microbiol. Rev. 60, 575–608.
10   Devereaux, J. et al. (1984) Nucleic Acids Res. 12, 387–395.
11   Marger, M.D. and Saier, M. (1993) Trends Biochem. Sci. 18, 13–20.
12   Varela, M. et al. (1995) Mol. Memb. Biol. 12, 313–319.
13   Bibi, E. and Kaback, H.R. (1990) Proc. Natl Acad. Sci. USA 87, 4325–4329.
14   Rubin, R.A. and Levy, S.B. (1991) J. Bacteriol. 173, 4503–4509.
15   Yamaguchi, A. et al. (1993) FEBS Lett. 324, 131–135.
16   Carruthers, A. (1990) Physiol. Rev. 70, 1135–1176.

3 Organization of the Data

INTRODUCTION

Two kinds of information are provided in *The Transporter FactsBook*. The first is a compilation of the physical and biological properties of nearly 800 transport proteins. Although every attempt was made to make this compilation comprehensive, some sequences were not included, either by design (see below) or by unintentional omission. Moreover, new transporter sequences are being added to the databases on a near daily basis. Thus, this information is best viewed as a representative, rather than an exhaustive, overview of the characteristics of membrane transport proteins.

The second kind of information is a comparison of the physical and biological properties of more than 50 families of transport proteins defined by the relatedness of their amino acid sequences. These data provide rationale bases for grouping proteins and identifying relationships between their structures and functions. A key feature of these data is the consensus amino acid sequence that has been provided for each transporter family or group of families. These are displayed in the multiple amino acid sequence alignments and also in the plots of the predicted topologies. The former indicates what kinds of substitutions are permitted at a conserved residue while the latter presents the conserved residues in the context of predicted structure. The consensus sequences provide means to classify newly identified transporters, particularly when they are not closely related to known proteins. They also define sequence elements that are conserved in multiple families with a common functional characteristic, and therefore may be necessary for the expression of that characteristic. This data is useful in predicting the locations of individual structural or functional domains, and designing experiments to test these predictions with site-directed mutagenesis or other techniques. Because the predictive value of the correlation between a signature sequence and a specific functional characteristic increases with the addition of each new sequence to the family, this information, rather than becoming outdated, will in fact become even more valuable as it is refined by the addition of new transporter sequences.

DEFINITION OF FAMILY

The FASTA and BLASTP algorithms [1,2] were used with default parameters to search the SwissProt, Protein Identification Resource (PIR) and Genbank/EMBL Genpept protein sequence databases for transport proteins that share local similarity with any of several query sequences representative of known classes of transport proteins.

The overall (versus local) similarities of the proteins identified in each search were then quantified by pairwise comparisons using the ALIGN [3] algorithm. ALIGN calculates a score for the best alignment between any pair of sequences using an empirically derived scoring matrix and two types of penalties for breaking a sequence. The first, the gap penalty, is applied every time a gap is inserted, regardless of the length of the gap. The second, the bias, is applied according to the length of the gap. The ALIGN program utilized the normalized Dayhoff 250 PAM mutational matrix, a gap penalty of 6.0 and a bias of 6.0.

The statistical significance of each alignment score was evaluated by comparing it to the mean score obtained from comparison of each sequence to 100 random permutations of the other sequence, and is expressed as the number of standard deviations (SD) by which the maximum score for the real comparison exceeds the mean of the scores for the randomized sequences. Pairs of proteins with ALIGN scores in excess of 9 SD were considered homologous, i.e. having a common evolutionary origin[4], and together constituted a "family"[5]. Hypothetical proteins, the open reading frames of unidentified genes, and partial sequences are not included.

Proteins identified in each FASTA or BLASTP search that had ALIGN scores less than 9 SD with the query sequence were used as query sequences for succeeding FASTA and BLASTP searches. Additional families of homologous sequences were again identified by pairwise comparisons using ALIGN. This process was repeated until all transport proteins identified by the successive FASTA and BLASTP searches were assigned to families. "Orphan transporters", proteins which are not homologous to any other transporter in the database, were not included.

GROUPING OF FAMILIES

Families with seemingly similar activities, e.g. "H$^+$-dependent symporters" or "P-type ATPases" were grouped together in a section. However, the reader should bear in mind that transporters with similar functions do not necessarily have related amino acid sequences and vice versa.

ORGANIZATION OF THE DATA

Summary

The summary provides an overview of the physical and biological properties of the family, its distribution in nature, its relationship to other families, and known disease associations.

Nomenclature, biological sources and substrates

Each sequence in a family was assigned an eight- or nine-character alphanumeric code. This code was derived from three or four characters taken from the protein name, the first three characters from the genus name and the first two characters from the species name. For example, the code for the XYLE transporter of *Escherichia coli* is Xyleescco. In a few cases, where the species is unknown, the last two characters are "sp". In many sequences found in the SwissProt database – the main exceptions being sequences from very common higher eukaryotes (e.g. human, rat, cattle) – the sequence code is equivalent to the SwissProt code without the underscore separating the parts describing the protein and its source. Tabulated information for sequences only currently present in the EMBL/GENBANK databases refers to the GenPept translations of the gene sequences.

The "Description" of each protein, taken directly from the sequence database, is listed in the second column. All known synonyms, including gene names, are included within square brackets below the description in the second column. "Organism", listed in the third column, refers to the Latin name of the species; the common name of the species, or (for most unicellular organisms) a classification such as "gram-negative bacterium" or "yeast" is included within square brackets in the third column. Substances listed in the "Substrate" column are known to be transported across the membrane. Where a protein is only known to confer resistance to a toxic compound, the compound's name is given in this column in square brackets. Where the mechanism of transporter action is known to be symport or antiport, the coupled ions are also listed here.

Phylogenetic trees

Phylogenetic trees were constructed for all families containing more than two members using the PILEUP algorithm [7] with default parameters. Proteins more than 90% identical to at least one other member of the family are indicated in the text by italics and are not included in the phylogenetic trees.

Topology plots

Each topology plot is derived from a single, typical member of a transporter family. In most cases, the predicted membrane-spanning regions, indicated in the figures by the shaded rectangles, and the interhelical loops, indicated in the figures by thin solid lines, are identified from hydropathy plots and analysis of α helix-forming propensity; in a few cases, these predictions are supported by experimental evidence derived from reporter fusions, susceptibility to proteolytic cleavage, reactivity with peptide-specific antibodies or scanning glycosylation mutagenesis. The number of the first and last residue of each predicted membrane helix is boxed. In families with more than two members, and unless there is a very high percentage identity between all family members (more than 50% of the sequence is identical in at least 75% of the proteins), the locations and identities of residues conserved in more than 75% of family members are indicated on the topology plots. All residues that are conserved in a family are not necessarily conserved in the representative transporter shown in the topology plot. In these instances, the residue is indicated with an asterisk.

In the ABC transporter superfamily, the active transporters consist of four domains: two ATP binding domains and two transmembrane domains. These four domains may be expressed as separate chains or fused to form multidomain proteins: almost every conceivable type of domain fusion has been found [6]. The sequence motifs characteristic of this superfamily are found in the ATP binding domains. In families in which the ATP binding domains are expressed separately from the transmembrane domains, the tables and alignments describe the cytoplasmic ATP binding domains associated with the transmembrane domains. Since the former chains do not cross the membrane, no topology plots are included for these families. There is great variability in the relatedness of the separately expressed transmembrane domains. Some of the chains containing these transmembrane domains constitute discrete families of homologous proteins, for example the ABC-associated binding protein-dependent maltose, peptide and iron transporter families. Other chains are no more similar to one another than would be expected for non-related transmembrane proteins which contain many highly hydrophobic regions. These are not included.

Physical and genetic characteristics

Molecular weights and sequence length (in amino acids) are listed for all proteins. When available, the proteins' principal expression sites (tissue or organ specificity), Michaelis constants (K_m) and chromosomal loci are listed. Where a bacterial sequence is known to be plasmid-encoded, this is also indicated. The chromosomal loci for humans, *Escherichia coli*, *Haemophilus influenzae*, *Saccharomyces cerevisiae*, and *Bacillus subtilis* are taken from the Online Mendelian Inheritance in Man, Encyclopedia of *E. coli* Genes and Metabolism, Encyclopedia of *Haemophilus influenzae* Genes and Metabolism, Saccharomyces Genomic Information Resource and the *Bacillus subtilis* Genomic Databases, respectively.

Multiple amino acid sequence alignments

Multiple amino acid sequence alignments were calculated using the PILEUP algorithm [7] with default parameters. The consensus sequences list residues present in at least 75% of the aligned sequences. Conservative substitutions were not taken into account. To ensure that the consensus sequences are not biased by the contribution of very closely related sequences, proteins more than 90% identical to at least one other member of the family (indicated in the text in italics) were not included in the alignments. Residues within the consensus sequence that are also conserved in at least one other family are indicated in bold type.

Database accession numbers

Information for each transporter was abstracted from the files in the SwissProt, PIR and EMBL/GENBANK databases identified by the accession numbers. No more than two accession numbers for each database are included. SwissProt was used as the primary data source as it is an extremely well annotated database.

References

Supplemental references cited in the summary and recent reviews, when available, are listed at the end of each chapter. Reviews are shown in bold type.

References
[1] Lipman, D. and Pearson, W. (1985) Science 227, 1435–1441.
[2] Altschul, S. et al. (1990) J. Mol. Biol. 215, 403–410.
[3] Dayhoff, M. et al. (1983) Methods Enzymol. 91, 524–545.
[4] Reeck, G. et al. (1987) Cell 40, 667.
[5] **Griffith, J. et al. (1992) Curr. Opin. Cell Biol. 4, 684–695.**
[6] **Higgins, C.F. (1992) Annu. Rev. Cell Biol. 8, 67–113.**
[7] Devereaux, J. et al. (1984) Nucleic Acids Res. 12, 387–395.

THE
MEMBRANE
TRANSPORT
PROTEINS

Part 1

P-Type ATPases

Summary

Transporters of the calcium-transporting ATPase family, the example of which is the probable calcium-transporting ATPase 4 from *Saccharomyces cerevisiae* (Atc4sacce), mediate active transport of calcium ions, driven by ATPase activity (EC 3.6.1.38). ATPase 4 may be involved in ribosome assembly[1]. Members of this family are only found in yeasts.

Statistical analysis of multiple amino acid sequence comparisons places the calcium-transporting ATPase family in the P-type ATPase superfamily (also known as E1-E2 ATPases[2,3]). Proteins in this superfamily use the energy of ATP hydrolysis to pump ions across cell membranes. P-Type ATPases are all predicted to contain at least six transmembrane helices by the hydropathy of their amino acid sequences. They have two large cytoplasmic loops separating three pairs of transmembrane helices; the larger of these loops contains the ATP binding domain. The sequences are usually extended by one or more pairs of helices. The calcium-transporting ATPase from *Schizosaccharomyces pombe*[4] is predicted to contain a total of 12 transmembrane helices.

Many residues and some short sequence motifs are completely conserved within the calcium-transporting ATPase family, including motifs unique to the family and signature motifs of the P-type ATPase superfamily.

Nomenclature, biological sources and substrates

CODE	DESCRIPTION [SYNONYMS]	ORGANISM [COMMON NAMES]	SUBSTRATE(S)
Atc4sacce	Probable calcium-transporting ATPase 4 [DRS2, YAL026C, FUN38]	*Saccharomyces cerevisiae* [yeast]	Ca^{2+}
Atc5sacce	Probable calcium-transporting ATPase 5 [YER166W, SYGP-ORF7]	*Saccharomyces cerevisiae* [yeast]	Ca^{2+}
Atcxschpo	Probable calcium-transporting ATPase	*Schizosaccharomyces pombe* [yeast]	Ca^{2+}

Phylogenetic tree

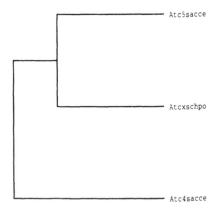

Proposed orientation of ATCX[4] in the membrane

The model is based on predictions of membrane-spanning regions and α-helical content. The N-terminus of the protein is illustrated on the inside and is folded ten times through the membrane. The predicted membrane-spanning helices are portrayed as rectangles. The numbers corresponding to the first and last residue of each membrane-spanning helix are boxed. Residues that are conserved in more than 75% of the aligned transporters (see below) are shown.

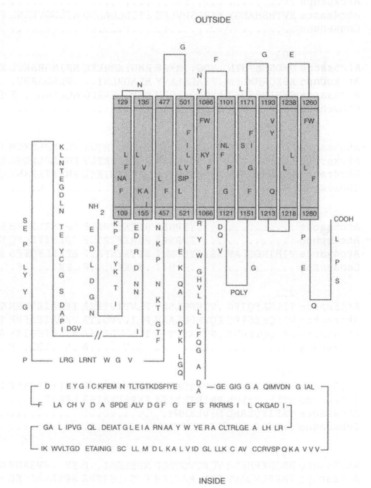

Physical and genetic characteristics

	AMINO ACIDS	MOL. WT	CHROMOSOMAL LOCUS
Atc4sacce	1355	153 844	Chromosome 1
Atc5sacce	1571	177 797	
Atcxschpo	1402	159 355	

Multiple amino acid sequence alignments

```
          1                                                      50
Atc5sacce MSGTFHGDGH APMSPFEDTF QFEDNSSNED THIAPTHFDD GATSNKYSRP
Atcxschpo .......... .......... .......... .......... ..........
Atc4sacce .......... ....MNDDRE TPPKRKPGED DTLFDIDFLD DTTSHSGSRS
Consensus .......... .......... .......... .......... ..........

          51                                                     100
Atc5sacce QVSFNDETPK NKREDAEEFT FNDDTEYDNH SFQPTPKLNN GSGTFDDVEL
Atcxschpo .......... .......... .......... .......... ....MESVEE
Atc4sacce KVTNSHANGY YIPPSHVLPE ETIDLDADDD NIENDVHENL FMSNNHDDQT
Consensus .......... .......... .......... .......... ..........

          101                                                    150
Atc5sacce DNDSGEPHTN .YDG..MKRF RMGTKRNKKG NPIMGRSKTL KWARKNIPNP
Atcxschpo KSKQRRWLPN .FKALRLKVY RLADRLNI.. .PLADAARV. .........E
Atc4sacce SWNANRFDSD AYQPQSLRAV KPPGLFARFG NGLKNA...F TFKRKKGPES
Consensus .......... .......... .......... .......... ..........

          151                                                    200
Atc5sacce FE.....DFT KDDIDPGAIN RAQEL.RTVY YN.MPLPKDM IDEEGNPIMQ
Atcxschpo LE.....EY. .DGSDPQSLR GLQKLPRTLY FG.LPLPDSE LDDTGEAKRW
Atc4sacce FEMNHYNAVT NNELDDNYLD SRNKFNIKIL FNRYILRKNV GDAEGNGEPR
Consensus .E........ .......D... .......... ....L..... .D..G.....

          201                                                    250
Atc5sacce .......... ...YPRNKI RTTKYTPLTF LPKNILFQFH NFANVYFLVL
Atcxschpo .......... ...FPRNKI RTAKYTPIDF IPKNIFLQFQ NVANLFFLFL
Atc4sacce VIHINDSLAN SSFGYSDNHI STTKYNFATF LPKFLFQEFS KYANLFFLCT
Consensus .......... ......N.I .T.KY....F .PK.....F. ..AN..FL..

          251                                                    300
Atc5sacce IILGAFQIFG .VTNPGLSAV PLVVIVIITA IKDAIEDSRR TVLDLEVNNT
Atcxschpo VILQSISIFG EQVNPGLAAV PLIVVVGITA VKDAIEDFRR TMLDIHLNNT
Atc4sacce SAIQQVPHVS .PTNRYTTIG TLLVVLIVSA MKECIEDIKR ANSDKELNNS
Consensus .......... ...N...... .L.V.....A .K..IED..R ...D...NN.

          301                                                    350
Atc5sacce KTHILEGVEN ENVSTDNISL WRRFKKANSR LLFKFIQYCK EHLTEEGKKK
Atcxschpo PTLRLSHYQN PNIRTEYISY FRRFKKRISA LFRVF...... ..LAKQEEKK
Atc4sacce TAEIFSEAHD DFVEKRWI.. .......... .......... ..........
Consensus .......... ......I... .......... .......... ..........

          351                                                    400
Atc5sacce RMQRKRHELR VQKTVGTSGP RSSLDSI..D SY..RVSADY GRPSLDYDNL
Atcxschpo RAKRLNDAVP LED.MAGSES RPSYDSIFRE SFEAKRSFED SKGKVPLSAL
Atc4sacce .......... .......... .......... .......... ..........
Consensus .......... .......... .......... .......... ..........

          401                                                    450
Atc5sacce EQGAG..... EANIVDRSLP PRTDCKFAKN YWKGVKVGDI VRIHNNDEIP
Atcxschpo DGTATILQSR PMDIIDYEAE ATGECHFKKT YWKDVRVGDF VKVMDNDEIP
Atc4sacce .......... .......... .......... ...DIRVGDI IRVKSEEPIP
Consensus .......... .......... .......... ......VGD. .........IP
```

```
          451                                              500
Atc5sacce ADIILLSTSD TDGACYVETK NLDGETNLKV RQSLKCTNTI RTSKDIARTK
Atcxschpo ADIVIINSSD PEGICYIETK NLDGETNLKM RHALTCGKNV VDEASCERCR
Atc4sacce ADTIILSSSE PEGLCYIETA NLDGETNLKI KQSRVETAKF IDVKTLKNMN
Consensus AD......S. ..G.CY.ET. NLDGETNLK. ................ ....

          501                                              550
Atc5sacce FWIESEGPHS NLYTYQGNMK W..RNLADG. ...EIRNEPI TINNVLLRGC
Atcxschpo FWIESEPPHA NLYEYNGACK SFVHSEAGGS DTSQTVSEPI SLDSMLLRGC
Atc4sacce GKVVSEQPNS SLYTYEGTM. ........... .TLNDRQIPL SPDQMILRGA
Consensus ....SE.P... .LY.Y.G... ........... ....P. ......LRG.

          551                                              600
Atc5sacce TLRNTKWAMG VVMFTGGDTK IMLNSGITPT KKSRISRELN FSVVINFVLL
Atcxschpo VLRNTKWVIG VVVFTGDDTK IMLNSGAPPL KRSRITRNLN WNVYLNFIIL
Atc4sacce TLRNTAWIFG LVIFTGHETK LLRNATATPI KRTAVEKIIN RQIIRLFTVL
Consensus .LRNT.W..G .V.FTG..TK ...N....P. K........N ......F..L

          601                                              650
Atc5sacce FILCFVSGIA NGVYYDKKGR S.RFSYEFGT IAGSAATNGF VSFWVAVILY
Atcxschpo FSMCFVCAVV EGIAWRGHSR S.SYYFEFGS IGGSPAKDGV VTFFFTGVILF
Atc4sacce IVLILISSIG NVIMSTADAK HLSYLYLEGT NKAGLFFKDF LTFW...ILF
Consensus ................. ........G. ..........F....IL.

          651                                              700
Atc5sacce QSLVPISLYI SVEIIKTAQA AFIYGDVLLY NAKLDYPCTP KSWNISDDLG
Atcxschpo QNLVPISLYI SIEIVKTIQA IFIYFDKDMY YKKLKYACTP KSWNISDDLG
Atc4sacce SNLVPISLFV TVELIKYYQA FMIGSDLDLY YEKTDTPTVV RTSSLVEELG
Consensus ..LVPISL... .E..K..QA ..I..D...Y ..K...........LG

          701                                              750
Atc5sacce QVEYIFSDKT GTLTQNVMEF KKCTINGVSY GRAYTEALAG LRKRQGIDVE
Atcxschpo QVEYIFSDKT GTLTQNVMEF KKCTINGVAY GEAFTEAMAG MAKREGKDTE
Atc4sacce QIEYIFSDKT GTLTRNIMEF KSCSIAGHCY IDKIPE.... ..... .....
Consensus Q.EYIFSDKT GTLT.N.MEF K.C.I.G..Y .....E.... .....

          751                                              800
Atc5sacce TEGRREKAEI AKDRDTMIDE LRALSGNSQF YPEEVTFVSK EFVRDLKGAS
Atcxschpo ELTLQKQSFI ERDRMQMISQ MRNMHDNKYL VDDNLTFISS QFVHDLAGKA
Atc4sacce ........... ..DKTATVED ........... .GIEVGYRKF DDLKKKLNDP
Consensus ........... ..D...... ........... ........ .....

          801                                              850
Atc5sacce GEVQQRCCEH FMLALALCHS VLVEANPDNP KKLDLKAQSP DEAALVATAR
Atcxschpo GEEQSLACYE FFLALALCHS VVADRVGD.. .RIVYKAQSP DEAALVGTAR
Atc4sacce SDEDSPIIND FLTLLATCHT VIPEFQSDGS ..IKYQAASP DEGALVQGGA
Consensus .......... F...LA.CH. V......D.. ......A.SP DE.ALV....

          851                                              900
Atc5sacce DVGFSFVGKT KK..GLIIEM QGIQKEFEIL NILEFNSSRK RMSCIVKIPG
Atcxschpo DVGFVFLDQR RD..IMVTRA LGETQRFKLM DTIEFSSARK RMSVIVK...
Atc4sacce DLGYKFIIRK GNSVTVLLEE TGEEKEYQLL NICEFNSTRK RMSAIFRF..
Consensus D.G..F.... ........... .G......... ....EF.S.RK RMS.I.....
```

```
          901                                              950
Atc5sacce LNPGDEPRAL LICKGADSII YSRLSRQSGS NSEAILEK.T ALHLEQYATE
Atcxschpo ...GPDNRYV LICKGADSII FERL....EP NEQVELRKTT SEHLRIFALE
Atc4sacce ....PDGSIK LFCKGADTVI LERLDDEANQ YVEATMR... ..HLEDYASE
Consensus .......... L.CKGAD..I ..RL..... ......... ..HL...A.E

          951                                             1000
Atc5sacce GLRTLCIAQR ELSWSEYEKW NEKYDIAAAS LANREDELEV VADSIERELI
Atcxschpo GLRTLCIAKR ELTEEEYYEW KEKYDIAASA IENREEQIEE VADLIESHLT
Atc4sacce GLRTLCLAMR DISEGEYEEW NSIYNEAATT LDNRAEKLDE AANLIEKNLI
Consensus GLRTLC.A.R .....EY..W ...Y..AA.. ..NR...... .A..IE....

         1001                                             1050
Atc5sacce LLGGTAIEDR LQDGVPDCIE LLAEAGIKLW VLTGDKVETA INIGFSCNLL
Atcxschpo LLGGTAIEDR LQEGVPDSIA LLAQAGIKLW VLTGDKMETA INIGFSCNLL
Atc4sacce LIGATAIEDK LQDGVPETIH TLQEAGIKIW VLTGDRQETA INIGMSCRLL
Consensus L.G.TAIED. LQ.GVP..I. .L..AGIK.W VLTGD..ETA INIG.SC.LL

         1051                                             1100
Atc5sacce NNEMELLVIK TTGDDVKEFG SEPSEIVDAL LSKYLKEYFN LTGSEEEIFE
Atcxschpo DAGMDMIKF. ....DVDQEV STPE..LEVI LADYLYRYFG LSGSVEELEA
Atc4sacce SEDMNLLIIN EETRDDTE.. .......... .RNLLEKINA LNEHQLSTHD
Consensus ...M...... ....D..... ......... ....L..... ..........

         1101                                             1150
Atc5sacce AKKDHEFPKG NYAIVIDGDA LKLALYGEDI RRKFLLLCKN CRAVLCCRVS
Atcxschpo AKKDHDTPSG SHALVIDGSV LKRVL.DGPM RTKFLLLCKR CKAVLCCRVS
Atc4sacce MK........ SLALVIDGKS LGFAL.EPEL EDYLLTVAKL CKAVICCRVS
Consensus .K........ ..ALVIDG.. L...L..... ....L...K. C.AV.CCRVS

         1151                                             1200
Atc5sacce PSQKAAVVKL VKDSLDVMTL AIGDGSNDVA MIQSADVGIG IAGEEGRQAV
Atcxschpo PAQKADVVQL VRESLEVMTL AIGDGANDVA MIQKADIGVG IVGEEGRAAA
Atc4sacce PLQKALVVKM VKRKSSSLLL AIASGANDVS MIQAAHVGVG ISGMEGMQAA
Consensus P.QKA.VV.. V.......L AI..G.NDV. MIQ.A..G.G I.G.EG..A.

         1201                                             1250
Atc5sacce MCSDYAIGQF RYLARLVLVH GRWSYKRLAE MIPEFFYKNM IFALALFWYG
Atcxschpo MSADYAIGQF RFLSKLVLVH GRWDYNRVAE MVNNFFYKSV VWTFTLFWYQ
Atc4sacce RSADIALGQF KFLKKLLLVH GSWSYQRISV AILYSFYKNT ALYMTQFWYV
Consensus ...D.A.GQF ..L..L.LVH G.W.Y.R... .....FYK.. ......FWY.

         1251                                             1300
Atc5sacce IYNDFDGSYL YEYTYMMFYN LAFTSLPVIF LGILDQDVND TISLVVPQLY
Atcxschpo IYNNFDANYL FDYTYVMLFN LIFSSLPVIV MGVYDQDVNA DLSLRIPQLY
Atc4sacce FANAFSGQSI MESWTMSFYN LFFTVWPPFV IGVFDQFVSS RLLERYPQLY
Consensus ..N.F..... ......N L.F...P... .G..DQ.V.. .......PQLY

         1301                                             1350
Atc5sacce RVGILRKEWN QRKFLWYMLD GLYQSIICFF FPYLVYHKNM IVTSNGLGLD
Atcxschpo KRGILQLNSA RKIFIGYMLD GFYQSVICFF FSFLVINNVT TAAQNGRDTM
Atc4sacce KLGQKGQFFS VYIFWGWIIN GFFHSAIVFI GTILIYRYGF ALNMHGELAD
Consensus ..G....... ...F...... G...S.I.F. ...L...... .....G....
```

```
             1351                                            1400
Atc5sacce HR.YFVGVYV TTIAVISCNT YVLLHQY.RW DWFSGLFIAL SCLVVF.AWT
Atcxschpo AV.QDLGVYV AAPTIMVVDT YVILNQS.NW DVFSIGLWAL SCLTFW.FWT
Atc4sacce HWSWGVTVYT TSVIIVLGKA ALVTNQWTKF TLIAIPGSLL FWLIFFPIYA
Consensus .......VY. .......... .....Q.... .........L ..L.......


             1401                                            1450
Atc5sacce GIWSSAIASR EFFKAAARIY GAPSFWAVFF VAVLFCLLPR FTYDSFQKFF
Atcxschpo GVYSQSLYTY EFYKSASRIF RTPNFWAVLC GTIVSCLFPK FLFMTTQKLF
Atc4sacce SIFPHANISR EYYGVVKHTY GSGVFWLTLI VLPIFALVRD FLWKYYKRMY
Consensus .......... E......... ....FW.... ....L... F.........


             1451                                            1500
Atc5sacce YPTDVEIVRE MWQHGHFDHY PPGYDPTDPN RPKVTKAGQH GEKIIEGIAL
Atcxschpo WPYDVDIIRE SYRTKRLHEL DEEEE..... ...IENAEQS PDWASSTLQV
Atc4sacce EPETYHVIQE MQKYNISDSR PHVQQF.... ........QN AIRKVRQVQR
Consensus .P.......E ....A..... .......... ......Q. ...........


             1501                                            1550
Atc5sacce SDNLGGSNYS RDSVVTEEIP MTFMHGEDGS PSGYQKQETW MTSPKETQDL
Atcxschpo P..FNASSSS LATPKKEPLR L......DTN SLTLTSSMPR SFTPSYTPSF
Atc4sacce MKKQRGFAFS QAEEGGQE.K IVRMYDTTQK RGKYGELQDA SANPFNDNNG
Consensus .........S .......... .......... .......... ...P......


             1551                                            1600
Atc5sacce LQSPQFQQAQ TFGRGPSTNV RSSLDRTREQ MIATNQLDNR YSVERARTSL
Atcxschpo LEGSPVFSDE ILNRGEYMPH RGSISSSEQP LRP....... ..........
Atc4sacce LGSNDFESAE PF........ ......IENP FADGNQNSNR FSSSRDDISF
Consensus L......... .......... .......... .......... ..........


             1601      1618
Atc5sacce DLPGVTNAAS LIGTQQNN
Atcxschpo .......... ........
Atc4sacce DI........ ........
Consensus .......... ........
```

Residues listed in the consensus sequence are present in at least 75% of the aligned transporter sequences. Residues indicated in boldface type are also conserved in at least one other family of the P-type ATPase family.

Database accession numbers

	SWISSPROT	PIR	EMBL/GENBANK
Atc4sacce	P39524	S30768	L01795; G171114
Atc5sacce	P32660	S30822	U18922; G603407
Atcxschpo	Q09891		Z67757; E208899

References
[1] Ripmaster, T.L. et al. (1993) Mol. Cell. Biol. 13, 7901–7912.
[2] **Green, N.M. and MacLennan, D.H. (1989) Biochem. Soc. Trans. 17, 819–822; Green, N.M. (1989) Biochem. Soc. Trans. 17, 970–972.**
[3] **Fagan, M.J. and Saier, M.H. Jr. (1994) J. Mol. Evol. 38, 57–99.**
[4] Barrell, B.G. et al. unpublished; EMBL/GenBank/DDBJ databases.

Summary

Transporters of the plasma membrane cation-transporting ATPase family, examples of which are the human calcium-transporting ATPase [1] (Atcdhomsa) and the plasma membrane proton pump from *Arabidopsis thaliana* [2] (Pma2arath), mediate active transport of cations – sodium, potassium, or calcium ions – or protons, driven by ATPase activity (EC 3.6.1.-). Members of this family may mediate influx, efflux or exchange of cations: for example, the human gastric potassium-transporting ATPase [3] mediates the exchange of protons and potassium ions across the plasma membrane and is responsible for acid production in the stomach. In plants and fungi, plasma membrane proton pumps [4,5] drive the active transport of nutrients by proton symport. Plasma membrane cation-transporting ATPases are widely distributed throughout both eukaryotic and prokaryotic taxa.

Statistical analysis of multiple amino acid sequence comparisons places the plasma membrane cation-transporting ATPase family in the P-type ATPase superfamily (also known as E1-E2 ATPases [6,7]). Proteins in this superfamily use the energy of ATP hydrolysis to pump ions across cell membranes. P-Type ATPases are all predicted to contain at least six transmembrane helices by the hydropathy of their amino acid sequences. They have two large cytoplasmic loops separating three pairs of transmembrane helices; the larger of these loops contains the ATP binding domain. The sequences are usually extended by one or more pairs of helices. Members of the plasma membrane cation-transporting ATPase family are predicted to contain a total of eight or ten transmembrane helices: all plasma membrane proton pumps contain eight helices, but some calcium transporters have ten. Some members of this family contain other sequence motifs – for example, the human plasma membrane calcium-transporting ATPase [1] contains two calmodulin binding domains towards the C-terminus. Some proteins may be glycosylated.

Only a few amino acid residues and short sequence motifs are conserved within the plasma membrane cation-transporting ATPase family, including motifs unique to the family and signature motifs of the P-type ATPase superfamily.

Nomenclature, biological sources and substrates

CODE	DESCRIPTION [SYNONYMS]	ORGANISM [COMMON NAMES]	SUBSTRATE(S)
Atal synsp	Cation-transporting ATPase [PMA1]	*Synechocystis* sp. [cyanobacterium]	Metal ions
Atcartsf	Calcium-transporting ATPase; sarcoplasmic/endoplasmic reticulum type [Calcium pump]	*Artemia sanfranciscana* [brine shrimp]	Ca^{2+}
Atcplafa	Calcium-transporting ATPase [Calcium pump]	*Plasmodium falciparum* [protozoan]	Ca^{2+}
Atctrybr	Probable calcium-transporting ATPase, [Calcium pump, TBA1]	*Trypanosoma brucei* [trypanosome]	Ca^{2+}
Atcl sacce	Calcium-transporting ATPase 1, [PMR1, SCC1, BSD1 YGL167C, G1666]	*Saccharomyces cerevisiae* [yeast]	Ca^{2+}

CODE	DESCRIPTION [SYNONYMS]	ORGANISM [COMMON NAMES]	SUBSTRATE(S)
Atc3sacce	Calcium-transporting ATPase 3 [PMC1, YGL006W]	Saccharomyces cerevisiae [yeast]	Ca^{2+}
Atc3schpo	Calcium-transporting ATPase 3 [CTA3]	Schizosaccharomyces pombe [yeast]	Ca^{2+}
Atcaorycu	Calcium-transporting ATPase, sarcoplasmic reticulum type, calcium pump, neonatal isoform [ATP2A1]	Oryctolagus cuniculus [rabbit]	Ca^{2+}
Atcbdrome	Calcium-transporting ATPase, sarcoplasmic/endoplasmic reticulum type [Calcium pump, CA-P60A]	Drosophila melanogaster [fruit fly]	Ca^{2+}
Atcbgalga	Calcium-transporting ATPase sarcoplasmic/endoplasmic reticulum type [Calcium pump, SERCA1]	Gallus gallus [chicken]	Ca^{2+}
Atcborycu	Calcium-transporting ATPase, sarcoplasmic reticulum type, adult isoform [Calcium pump, ATP2A1]	Oryctolagus cuniculus [rabbit]	Ca^{2+}
Atcdfelca	Calcium-transporting ATPase, sarcoplasmic reticulum type [Calcium pump, SERCA2]	Felix cattus [cat]	Ca^{2+}
Atcdhomsa	Calcium-transporting ATPase, sarcoplasmic reticulum type [ATP2A2, ATP2B]	Homo sapiens [human]	Ca^{2+}
Atcdorycu	Calcium-transporting ATPase, sarcoplasmic reticulum type [ATP2A2]	Oryctolagus cuniculus [rabbit]	Ca^{2+}
Atcdratno	Calcium-transporting ATPase, sarcoplasmic reticulum type [ATP2A2]	Rattus norvegicus [rat]	Ca^{2+}
Atcdsussc	Calcium-transporting ATPase	Sus scrofa [pig]	Ca^{2+}
Atcehomsa	Calcium-transporting ATPase, endoplasmic reticulum type [ATP2A2]	Homo sapiens [human]	Ca^{2+}
Atcesussc	Calcium-transporting ATPase, endoplasmic reticulum type [ATP2A2]	Sus scrofa [pig]	Ca^{2+}
Atceorycu	Calcium-transporting ATPase, endoplasmic reticulum type [ATP2A2]	Oryctolagus cuniculus [rabbit]	Ca^{2+}
Atceratno	Calcium-transporting ATPase, endoplasmic reticulum type [ATP2A2]	Rattus norvegicus [rat]	Ca^{2+}
Atcfratno	Calcium-transporting ATPase 3 [Calcium pump, ATP2A3]	Rattus norvegicus [rat]	Ca^{2+}
Atclmycge	Probable cation-transporting P-type ATPase [PACL, MG071]	Mycoplasma genitalium [gram-negative bacterium]	Metal ions $(Ca^{2+}?)$
Atclsynsp	Cation-transporting ATPase [PACL]	Synechococcus sp. [cyanobacterium]	Metal ions $(Ca^{2+}?)$
Atcphomsa	Calcium-transporting ATPase plasma membrane, isoform 1B [Calcium pump, ATP2B1]	Homo sapiens [human]	Ca^{2+}

CODE	DESCRIPTION [SYNONYMS]	ORGANISM [COMMON NAMES]	SUBSTRATE(S)
Atcporycu	Calcium-transporting ATPase plasma membrane, isoform 1B [Calcium pump, ATP2B1, PMCA1B]	*Oryctolagus cuniculus* [rabbit]	Ca^{2+}
Atcpratno	Calcium-transporting ATPase plasma membrane, isoform 1B [Calcium pump, ATP2B1]	*Rattus norvegicus* [rat]	Ca^{2+}
Atcpsussc	Calcium-transporting ATPase plasma membrane, isoform 1B [Calcium pump, PMCA1B]	*Sus scrofa* [pig]	Ca^{2+}
Atcqhomsa	Calcium-transporting ATPase plasma membrane, brain isoform 2 [Calcium pump, ATP2B2]	*Homo sapiens* [human]	Ca^{2+}
Atcqratno	Calcium-transporting ATPase plasma membrane, brain isoform 2 [Calcium pump, ATP2B2]	*Rattus norvegicus* [rat]	Ca^{2+}
Atcrhomsa	Calcium-transporting ATPase plasma membrane, isoform 4 [Calcium pump, ATP2B4]	*Homo sapiens* [human]	Ca^{2+}
Athahomsa	Potassium-transporting ATPase α chain [ATP4A]	*Homo sapiens* [human]	K^+/H^+
Athaorycu	Potassium-transporting ATPase α chain [Gastric H^+/K^+ ATPase α subunit, ATP4A]	*Oryctolagus cuniculus* [rabbit]	K^+
Atharatno	Potassium-transporting ATPase α chain [Gastric H^+/K^+ ATPase α subunit, ATP4A]	*Rattus norvegicus* [rat]	K^+
Athasussc	Potassium-transporting ATPase α chain [Gastric H^+/K^+ ATPase α subunit, ATP4A]	*Sus scrofa* [pig]	K^+
Atmaescco	Mg^{2+} transport ATPase, P-type 1 [MGTA, MGT, CORB]	*Escherichia coli* [gram-negative bacterium]	Mg^{2+}
Atmasalty	Mg^{2+} transport ATPase, P-type 1 [MGTA]	*Salmonella typhimurium* [gram-negative bacterium]	Mg^{2+}
Atmbsalty	Mg^{2+} transport ATPase, P-type 2 [MGTB]	*Salmonella typhimurium* [gram-negative bacterium]	Mg^{2+}
Atn1bufma	Sodium/potassium-transporting ATPase α1 chain [Sodium pump, Na^+/K^+ ATPase]	*Bufo marinus* [toad]	Na^+, K^+
Atn1equca	Sodium/potassium-transporting ATPase α1 chain [Sodium pump, Na^+/K^+ ATPase]	*Equus caballus* [horse]	Na^+, K^+
Atn1galga	Sodium/potassium-transporting ATPase α1 chain [Sodium pump, Na^+/K^+ ATPase]	*Gallus gallus* [chicken]	Na^+, K^+
Atn1oviar	Sodium/potassium-transporting ATPase α1 chain [Sodium pump, Na^+/K^+ ATPase, ATP1A1]	*Ovis aries* [sheep]	Na^+, K^+

CODE	DESCRIPTION [SYNONYMS]	ORGANISM [COMMON NAMES]	SUBSTRATE(S)
Atn1ratno	Sodium/potassium-transporting ATPase α1 chain [Sodium pump, Na$^+$/K$^+$ ATPase, ATP1A1]	*Rattus norvegicus* [rat]	Na$^+$, K$^+$
Atn1sussc	Sodium/potassium-transporting ATPase α1 chain [Sodium pump, Na$^+$/K$^+$ ATPase, ATP1A1]	*Sus scrofa* [pig]	Na$^+$, K$^+$
Atn1homsa	Sodium/potassium-transporting ATPase α1 chain [Sodium pump, Na$^+$/K$^+$ ATPase, ATP1A1]	*Homo sapiens* [human]	Na$^+$, K$^+$
Atn1sacce	Sodium transport ATPase 1 [ENA1, PMR2, HOR6, YDR040C, YD6888.02C]	*Saccharomyces cerevisiae* [yeast]	Na$^+$
Atn2galga	Sodium/potassium-transporting ATPase α1 chain [Sodium pump]	*Gallus gallus* [chicken]	Na$^+$, K$^+$
Atn2homsa	Sodium/potassium-transporting ATPase α1 chain [Sodium pump, ATP1A2]	*Homo sapiens* [human]	Na$^+$, K$^+$
Atn2ratno	Sodium/potassium-transporting ATPase α1 chain [Sodium pump, ATP1A2]	*Rattus norvegicus* [rat]	Na$^+$, K$^+$
Atn2sacce	Sodium transport ATPase 2 [ENA2, PMR2B, YDR039C]	*Saccharomyces cerevisiae* [yeast]	Na$^+$
Atn3galga	Sodium/potassium-transporting ATPase α3 chain [Sodium pump]	*Gallus gallus* [chicken]	Na$^+$, K$^+$
Atn3homsa	Sodium/potassium-transporting ATPase α3 chain [Sodium pump, ATP1A3]	*Homo sapiens* [human]	Na$^+$, K$^+$
Atn3sussc	Sodium/potassium-transporting ATPase α3 chain [Sodium pump, ATP1A3]	*Sus scrofa* [pig]	Na$^+$, K$^+$
Atn3ratno	Sodium/potassium-transporting ATPase α3 chain [Sodium pump, ATP1A3]	*Rattus norvegicus* [rat]	Na$^+$, K$^+$
Atnaartsa	Sodium/potassium-transporting ATPase α chain [Sodium pump]	*Artemia salina* [brine shrimp]	Na$^+$, K$^+$
Atnaartsf	Sodium/potassium-transporting ATPase α chain [Sodium pump]	*Artemia sanfranciscana* [brine shrimp]	Na$^+$, K$^+$
Atnacatco	Sodium/potassium-transporting ATPase α chain [Sodium pump]	*Catostomus commersoni* [sucker]	Na$^+$, K$^+$
Atnadrome	Sodium/potassium-transporting ATPase α chain [Sodium pump, NA-P]	*Drosophila melanogaster* [fruit fly]	Na$^+$, K$^+$
Atnahydat	Sodium/potassium-transporting ATPase α chain] [Sodium pump, NA-P]	*Hydra attenuata* [hydra]	Na$^+$, K$^+$
Atnatorca	Sodium/potassium-transporting ATPase α chain [Sodium pump]	*Torpedo californica* [ray]	Na$^+$, K$^+$

CODE	DESCRIPTION [SYNONYMS]	ORGANISM [COMMON NAMES]	SUBSTRATE(S)
Atxaleido	Probable E1-E2 type cation ATPase 1A	*Leishmania donovani* [trypanosome]	Metal ions
Atxbleido	Probable E1-E2 type cation ATPase 1A	*Leishmania donovani* [trypanosome]	Metal ions
Pma1ajeca	Plasma membrane ATPase [Proton pump, PMA1]	*Ajellomyces capsulata* [gram-negative bacterium]	H^+
Pma1arath	Plasma membrane ATPase [Proton pump, AHA1]	*Arabidopsis thaliana* [mouse-ear cress]	H^+
Pma1canal	Plasma membrane ATPase [Proton pump, PMA1]	*Candida albicans* [yeast]	H^+
Pma1klula	Plasma membrane ATPase [Proton pump, PMA1]	*Kluyveromyces lactis* [yeast]	H^+
Pma1lyces	Plasma membrane ATPase [Proton pump, LHA1]	*Lycopersicon esculentum* [tomato]	H^+
Pma1neucr	Plasma membrane ATPase [Proton pump, PMA1]	*Neurospora crassa* [mold]	H^+
Pma1nicpl	Plasma membrane ATPase 1 [Proton pump, PMA1]	*Nicotiana plumbaginifolia* [tobacco]	H^+
Pma1schpo	Plasma membrane ATPase 1 [Proton pump, PMA1]	*Schizosaccharomyces pombe* [yeast]	H^+
Pma1sacce	Plasma membrane ATPase 1 [Proton pump, PMA1, YGL008C]	*Saccharomyces cerevisiae* [yeast]	H^+
Pma1zygro	Plasma membrane ATPase [Proton pump]	*Zygosaccharomyces rouxii* [yeast]	H^+
Pma2arath	Plasma membrane ATPase 2 [Proton pump, AHA1]	*Arabidopsis thaliana* [mouse-ear cress]	H^+
Pma2sacce	Plasma membrane ATPase 2 [Proton pump, PMA2, YPL036W]	*Saccharomyces cerevisiae* [yeast]	H^+
Pma2schpo	Plasma membrane ATPase 2 [Proton pump, PMA2]	*Schizosaccharomyces pombe* [yeast]	H^+
Pma3arath	Plasma membrane ATPase 3 [Proton pump, AHA3]	*Arabidopsis thaliana* [mouse-ear cress]	H^+
Pma3nicpl	Plasma membrane ATPase 1 [Proton pump, PMA3]	*Nicotiana plumbaginifolia* [tobacco]	H^+
Pma4nicpl	Plasma membrane ATPase 4 [Proton pump, PMA4]	*Nicotiana plumbaginifolia* [tobacco]	H^+

Phylogenetic tree

Proteins listed subsequently in italics are at least 90% identical to the paired transporters listed in parenthesis and are therefore not included in the phylogenetic tree: *Atcbgalga, Atcaorycu* (Atcborycu); *Atcdfelca, Atcdsussc, Atcdorycu, Atceorycu, Atcdratno, Atceratno, Atcehomsa, Atcesussc* (Atcdhomsa); *Atcporycu, Atcpratno, Atcpsussc* (Atcphomsa); *Atcqratno* (Atcqhomsa); *Athasussc, Athaorycu, Atharatno* (Athahomsa); *Atmasalty* (Atmaescco); *Atn1bufma, Atn1galga, Atn1oviar, Atn1equca, Atn1sussc, Atn1ratno* (Atn1homsa); *Atn2sacce* (Atn1sacce); *Atn2galga, Atn2ratno* (Atn2homsa); *Pma2arath* (Pma1arath); *Pma3nicpl* (Pma1nicpl).

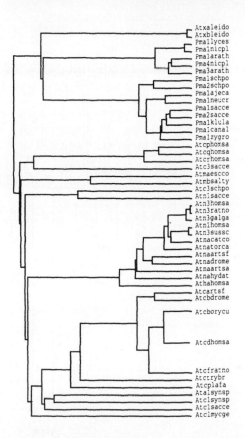

Atxaleido
Atxbleido
Pmallyces
Pmalnicpl
Pmalarath
Pma4nicpl
Pma3arath
Pmalschpo
Pma2schpo
Pmalajeca
Pmalneucr
Pmalsacce
Pma2sacce
Pmalklula
Pmalcanal
Pmalzygro
Atcphomsa
Atcqhomsa
Atcrhomsa
Atc3sacce
Atmaescco
Atmbsalty
Atc3schpo
Atn1sacce
Atn3homsa
Atn3ratno
Atn3galga
Atn1homsa
Atn3sussc
Atnacatco
Atnatorca
Atnaartsf
Atnadrome
Atnaartsa
Atnahydat
Athahomsa
Atcartsf
Atcbdrome
Atcborycu
Atcdhomsa
Atcfratno
Atctrybr
Atcplafa
Atalsynsp
Atclsynsp
Atclsacce
Atclmycge

Proposed orientation of ATC1 [8] in the membrane

The model is based on predictions of membrane-spanning regions and α-helical content. The N-terminus of the protein is illustrated on the inside and is folded ten times through the membrane. The predicted membrane-spanning helices are portrayed as rectangles. The numbers corresponding to the first and last residue of each membrane-spanning helix are boxed. Residues that are conserved in more than 75% of the aligned transporters (see below) are shown.

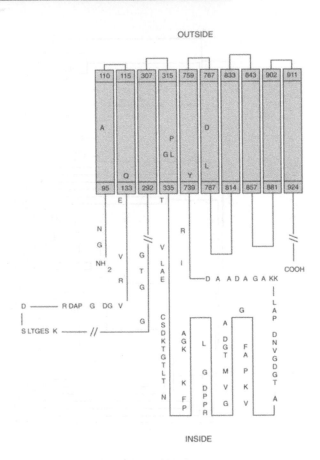

Physical and genetic characteristics

	AMINO ACIDS	MOL. WT	EXPRESSION SITES	CHROMOSOMAL LOCUS
Ata1synsp	915	98 902		
Atcartsf	1003	110 343		
Atcplafa	1228	139 414		
Atctrybr	1011	110 313		
Atc1sacce	950	104 570		
Atc3sacce	1173	130 860		Chromosome 7
Atc3schpo	1037	115 327		
Atcaorycu	1001	110 458	skeletal muscle	
Atcbdrome	1002	109 597		
Atcbgalga	994	109 023	skeletal muscle	
Atcborycu	994	109 489	skeletal muscle	
Atcdfelca	997	109 712	heart	
Atcdhomsa	997	109 690	kidney	12q23–q24.1
Atcdorycu	997	109 644	muscle	
Atcdratno	997	109 680	heart, stomach	
Atcdsussc	997	109 726	stomach smooth muscle	
Atcehomsa	1042	114 756	kidney	12q23–q24.1
Atcesussc	1042	114 791	stomach smooth muscle	
Atceorycu	1042	114 705	smooth muscle	

	AMINO ACIDS	MOL. WT	EXPRESSION SITES	CHROMOSOMAL LOCUS
Atceratno	1043	114 767	brain	
Atcfratno	999	109 359	kidney (& others)	
Atclmycge	874	96 317		
Atclsynsp	926	99 696		
Atcphomsa	1220	134 684	erythrocyte	12q21–q23
Atcporycu	1220	134 650	stomach, smooth muscle & others	
Atcpratno	1176	129 510	brain	
Atcpsussc	1220	134 709	stomach, smooth muscle	
Atcqhomsa	1198	132 722	brain	3p26-p25
Atcqratno	1198	132 615	brain	
Atcrhomsa	1205	133 930	erythrocyte	1q25–q32
Athahomsa	1035	114 090	stomach	19q13.1
Athaorycu	1035	114 201	stomach	
Atharatno	1033	114 037	stomach	
Athasussc	1034	114 286	stomach	
Atmaescco	898	99 466		92.8–100 minutes
Atmasalty	902	99 782		
Atmbsalty	908	100 428		mgtB locus
Atn1bufma	1023	112 599	kidney, bladder & others	
Atn1equca	1021	112 696	kidney	
Atn1galga	1021	112 231	kidney	
Atn1homsa	1023	112 895		1p13-p11
Atn1ratno	1023	113 054	brain	
Atn1sussc	1021	112 680	kidney	
Atn1oviar	1021	112 657	kidney	
Atn1sacce	1091	120 357		Chromosome 4
Atn2galga	1017	112 050		
Atn2homsa	1020	112 265		1q21–q23
Atn2ratno	1020	112 217	brain	
Atn2sacce	1091	120 317		Chromosome 4
Atn3galga	1010	111 284		
Atn3homsa	1013	111 692	brain	19q12–q13.2
Atn3sussc	1021	112 653	kidney	
Atn3ratno	1013	111 735	brain	
Atnaartsa	996	111 022		
Atnaartsf	1004	110 699		
Atnacatco	1027	113 313	hypothalamus	
Atnadrome	1038	115 342	tubules, muscles, nervous system	
Atnahydat	1031	114 161		
Atnatorca	1022	112 429		
Atxaleido	974	107 449		
Atxbleido	974	107 305		
Pma1ajeca	916	98 884		
Pma1arath	948	104 182		
Pma1canal	895	97 459		Chromosome 3
Pma1klula	899	98 259		
Pma1lyces	956	105 103		
Pma1neucr	920	99 886		
Pma1nicpl	957	105 155	leaf, stem, flower, root	
Pma1schpo	919	99 883		
Pma1sacce	918	99 619		Chromosome 7
Pma1zygro	920	100 061		
Pma2arath	947	104 270	root	
Pma2sacce	947	102 171		Chromosome 16
Pma2schpo	1010	110 127		
Pma3arath	948	104 318		
Pma3nicpl	956	105 112	leaf, stem, flower, root	
Pma4nicpl	952	105 188	leaf, stem, flower, root	

Multiple amino acid sequence alignments

```
          1                                                   50
Pma2schpo MQRNNGEGRP EGMHRISRFL HGNPFKNNAS PQDDSTTRTE VYEEGGVEDS

          51                                                  100
Pma2schpo A VDYDNASGNA APRLTAAPNT HAQQANLQ SGNTSITHET QSTSRGQEAT
Pma2sacce .......... .......... ........MS STEAKQYKEK PSKEYLHASD
Atnadrome .......... .......... .......... .......... ........MA

          101                                                 150
Atxaleido .......... ...... .....MSSKK YELDAAAFED KPESHSDAEM
Atxbleido .......... ...... .....MSSKK YELDAAAFED KPESHSDAEM
Pma1lyces .......... ...... ....MAEKP EVLDAVLKET VDLENIPIEE
Pma1nicpl .......... ...... ....MGEEKP EVLDAVLKEA VDLENIPIEE
Pma1arath .......... ...... ... SGLEDIKNET VDLEKIPIEE
Pma4nicpl .......... ...... ...MAKA ISLEEIKNET VDLEKIPIEE
Pma3arath .......... ...... ....A SGLEDIVNEN VDLEKIPIEE
Pma1schpo ......MADN AGEYHDAEKH APEQQAPPPQ QPAHAAAPAQ DDEPDDDIDA
Pma2schpo .........TS PSLSASHEKP ARPQTGEGSD NEDEDEDIDA
Pma1ajeca .MAHSAASGA AS...... AA HFEKKTPEVA HEEKKPPLPE EEDEDEDMDA
Pma1neucr .MADHSASGA PALSTNIESG KFDEKAAEAA AYQPKPKV.. EDDEDEDIDA
Pma1sacce ....MTDTSS SSSSSSASSV SAHQPTQEKP AKTYDDAASE SSD.DDDIDA
Pma2sacce GDDPANNSAA SSSSSSSTST SASSSAAAVP RKAAAASAAD DSDSDEDIDQ
Pma1klula .......... ...MSAATEP TKEKPVNNQD SDDEDEDIDQ
Pma1canal .......... M SATEPTNEKV DKIV...... SDDEDEDIDQ
Pma1zygro ..MSDERITE KPPHQQPESE GEPVPEEEVE EETEEEVPDE QSSEDDDIDG
Atcphomsa .......... MGDM ANNSVAYSGV KNSLKE..AN HDGDFGITLA
Atcqhomsa .......... MGDM T.NSDFYS.. KNQRNE..SS HGGEFGCTME
Atcrhomsa .......... M TNPSD..RVL PANSMA..ES REGDFGCTVM
Atc3sacce .......... MSRQ DENSALLANN ENNKPSYTGN ENGVYDNFKL
Atmaescco .......... MFKEIF TRLIRHLPSR LVHRDPLPGA QQTVNTVVPP
Atmbsalty .......... .MTDMNIENR KLNR.PASEN DKQHKKVFPI
Atc3schpo .......... .......... .......... ....MVTINI
Atn1sacce .......... .......... .......... ..MGEGTTKE
Atn3homsa .......... ....MGDK KDDKDSPKKN KGKERRDLDD
Atn3ratno .......... ....MGDK KDDKSSPKKS KAKERRDLDD
Atn3galga .......... ...MGD. KGEKESPKKG KGK..RDLDD
Atn1homsa .......... ...MGKGVGR DKYEPAAVSE QGDK...KGK KGKKDRDMDE
Atn2homsa .......... ...MGRGAGR EYSPAATTAE NGGG.....K KKQKEKELDE
Atn3sussc .......... ...MGKGVGP DKYEPAAVSE HGDK.....K KAKKERDMDE
Atnacatco .......... ...MGVGDGR DQYELAAMSE QSGKKKSKNK KEKKEKDMDE
Atnatorca .......... ...MGKGAAS EKYQPAATSE NAKNSKKSKS KTT...DLDE
Atnaartsf .......... .....MAKG KQKKGKDLNE
Atnadrome LRSDYEHGRA DSYRVATVIA TDDDNRTADG QYKSRRKMPA KVNKKENLDD
Atnaartsa .......... .......M GKKQGKQLSD
Atnahydat .......... MADPGDLESR GKADSYSVAE KKSAPKKISK KNANKAKLED
Athahomsa .....MGKAE NYELYSVELG PGPGGDMAAK MSKKKKAGGG GGKRKEKLEN
Atctrybr  .......... .......... .......... .........M
Atc1synsp .......... .......... .......... .....MKGA
Atc1sacce .......... .......... MSDNPFNASL LDEDSNRERE
Consensus .......... .......... .......... .........
```

```
           151                                                   200
Atxaleido  TPQKPQRRQS VLSKAVSEHD ERATGPATDP VPPSK.......  ..GLTTEEA
Atxbleido  TPQKPQRRQS VLSKAVSEHD ERATGPATDL LPPSK.......  ..GLTTEEA
Pma1lyces  VFENLRC... ..........  .......... .......      ..TREGLTATAA
Pma1nicpl  VFENLRC... ..........  .......... .......      ..TKEGLTATAA
Pma1arath  VFQQLKC... ..........  .......... .......      ..TREGLTTQEG
Pma4nicpl  VFEQLKC... ..........  .......... .......      ..TREGLSADEG
Pma3arath  VFQQLKC... ..........  .......... .......      ..SREGLSGAEG
Pma1schpo  LIEELFSEDV QEEQEDNDDA PA.AGEA..K AVPEELLQTD    MNTGLTMSEV
Pma2schpo  LIEDLYSQDQ EEEQVEEEES PGPAGAA..K VVPEELLETD    PKYGLTESEV
Pma1ajeca  LIEELESQDG HIDIEDDEDG EPGGA....R PVPDELLTTD    TRHGLTDAEV
Pma1neucr  LIEDLESHDG HDAEEEEEA  TPGGG....R VVPEDMLQTD    TRVGLTSEEV
Pma1sacce  LIEELQSNHG VDDEDSDNDG PVAAGEA..R PVPEEYLQTD    PSYGLTSDEV
Pma2sacce  LIDELQSNYG EGDESGEEEV RTDGVHAGQR VVPEKDLSTD    PAYGLTSDEV
Pma1klula  LIEDLQSHHG LDDE.SEDDE HVAAGSA..R PVPEELLQTD    PSYGLTSDEV
Pma1canal  LVADLQSNPG AGDEEEEEEN .....DSSFK AVPEELLQTD    PRVGLTDDEV
Pma1zygro  LIDELQSQ.E AHEEAEEDDG PAAAGEA..R KIPEELLQTD    PSVGLSSDEV
Atcphomsa  ELRALMELRS TDALRKIQES YG.DVYGICT KLKTSPNEGL    S.GNPADLER
Atcqhomsa  ELRSLMELRG TEAVVKIKET YG.DTEAICR RLKTSPVEGL    P.GTAPDLEK
Atcrhomsa  ELRKLMELRS RDALTQINVH YG.GVQNLCS RLKTSPVEGL    S.GNPADLEK
Atc3sacce  SKSQLSDLHN PKSIRSFVRL FGYESNSLFK YLKTDKNAGI    SLPEISNYRK
Atmaescco  SLSAHCLKMA VMPEEELWKT FDTHPEG... ..........    ..LNQAEVS
Atmbsalty  EAE......A FHSPEETLAR LNSHRQG... ..........    ..LTIEEASE
Atc3schpo  SNPVYFSDIK DVESE..... FLTSIPNG.. ..........    ..LTHEEAQN
Atn1sacce  NNNAEFNAYH TLTAEEAAEF IGTSLTEG.. ..........    ..LTQDEFVH
Atn3homsa  LKKEVAMTEH KMSVEEVCRK YNTDCVQG.. ..........    ..LTHSKAQE
Atn3ratno  LKKEVAMTEH KMSVEEVCRK YNTDCVQG.. ..........    ..LTHSKAQE
Atn3galga  LKKEVAMTEH KMSIEEVCRK YNTDCVQG.. ..........    ..LTHSKAQE
Atn1homsa  LKKEVSMDDH KLSLDELHRK YGTDLSRG.. ..........    ..LTSARAAE
Atn2homsa  LKKEVAMDDH KLSLDELGRK YQVDLSKG.. ..........    ..LTNQRAQD
Atn3sussc  LKKEVSMDDH KLSLDELHRK YGTDLSRG.. ..........    ..LTPARAAE
Atnacatco  LKKEVDLDDH KLSLEELHHK YGTDLSKG.. ..........    ..LSNSRAEE
Atnatorca  LKKEVSLDDH KLNLDELHQK YGTDLTQG.. ..........    ..LTPARAKE
Atnaartsf  LKKELDIDFH KIPIEECYQR LGSNPETG.. ..........    ..LTNAQARS
Atnadrome  LKQELDIDFH KISPEEMYQR FQTHPENG.. ..........    ..LSHARAKE
Atnaartsa  LKKELELDQH KIPLEELCRR LGTNTETG.. ..........    ..LTSSQAKS
Atnahydat  LKKELEMTEH SMKLESLLSM YETSLEKG.. ..........    ..LSENIVAR
Athahomsa  MKKEMEINDH QLSVAELEQK YQTSATKG.. ..........    ..LSASLAAE
Atcartsf   .....MEDAH AKKWEEVVDY FGVDPERG.. ..........    ..LALEQVKK
Atcbdrome  .....MEDGH SKTVEQSLNF FGTDPERG.. ..........    ..LTLDQIKA
Atcborycu  .....MEAAH SKSTEECLAY FGVSETTG.. ..........    ..LTPDQVKR
Atcdhomsa  .....MENAH TKTVEEVLGH FGVNESTG.. ..........    ..LSLEQVKK
Atcfratno  .....MEEAH LLSAADVLRR FSVTAEGG.. ..........    ..LTLEQVTD
Atctrybr   LPENLPTDPA AMTPAAVAAA LRVDTKVG.. ..........    ..LSSNEVEE
Atcplafa   .MEEVIKNAH TYDVEDVLKF LDVNKDNG.. ..........    ..LKNEELDD
Ata1synsp  MGAFPLPPNQ YGFPHLKF.. LPPSPSTRGR HSCRFAHRSR    FRSDSGAVAQ
Atc1synsp  IVSASLTDVR QPIAHWHS.. LTVEECHQQL D....AHRNG    LTAEVA..AD
Atc1sacce  ILDATAEALS KPSPSLEYCT LSVDEALEKL D....TDKNG    GLRSSNEANN
Atc1mycge  .......... ..........  .......... ...MNSW     TGLSEQAAIK
Consensus  .......... ..........  .......... .......      ..........
```

```
           201                                                   250
Atxaleido  EELLKKYGRN EL.PEKKTPS WLIYVRGLWG PMPAALWI..  .........A
Atxbleido  EELLKKYGRN EL.PEKKTPS WLIYVRGLWG PMPAALWI..  .........A
```

```
Pmallyces QERLSIFGYN KL.EEKKESK FLKFLGFMWN PLSWVMEA.. ........A
Pmalnicpl QERLAIFGYN KL.EEKKDSK LLKFLGFMWN PLSWVMEA.. ........A
Pmalarath EDRIVIFGPN KL.EEKKESK ILKFLGFMWN PLSWVMEA.. ........A
Pma4nicpl ASRLQIFGPN KL.EEKNESK ILKFLGFMWN PLSWVMEA.. ........A
Pma3arath ENRLQIFGPN KL.EEKKESK LLKFLGFMWN PLSWVMEA.. ........A
Pmalschpo EERRKKYGLN QM.KEELENP FLKFIMFFVG PIQFVMEM.. ........A
Pma2schpo EERKKKYGLN QM.KEEKTNN IKKFLSFFVG PIQFVMEL.. ........A
Pmalajeca VARRKKYGLN QM.KEEKENL VLKFLSYFVG PIQFVMEA.. ........A
Pmalneucr VQRRRKYGLN QM.KEEKENH FLKFLGFFVG PIQFVMEG.. ........A
Pmalsacce LKRRKKYGLN QM.ADEKESL VVKFVMFFVG PIQFVMEA.. ........A
Pma2sacce ARRRKKYGLN QM.AEENESL IVKFLMFFVG PIQFVMEA.. ........A
Pmalklula TKRRKKYGLN QM.SEETENL FVKFLMFFIG PIQFVMEA.. ........A
Pmalcanal TKRRKRYGLN QM.AEEQENL VLKFVMFFVG PIQFVMEA.. ........A
Pmalzygro VNRRKKYGLN QM.REESENL LVKFLMFFIG PIQFVMEA.. ........A
Atcphomsa ..REAVFGKN FIPPKKPKTF LQLVWEALQD VTLIILEI.. ........A
Atcqhomsa ..RKQIFGQN FIPPKKAKPF LQLVWEALQD VTLIILEI.. ........A
Atcrhomsa ..RRQVFGHN VIPPKKPKTF LELVWEALQD VTLIILEI.. ........A
Atc3sacce TNRYKNYGDN SLPERIPKSF LQLVWAAFND KTMQLLTV.. ........A
Atmaescco ..AREQHGEN KLPAQQPSPW WVHLWVCYRN PFNILLTI.. ........L
Atmbsalty ..RLKVYGRN EVAHEQVPPA LIQLLQAFNN PFIYVLMA.. ........L
Atc3schpo ..RLSEYGEN RLEADSGVSA WKVLLRQVLN AMCVVLIL.. ........A
Atnlsacce ..RLKTVGEN TLGDDTKIDY KAMVLHQVCN AMIMVLLI.. ........S
Atn3homsa ..ILARDGPN ALTPPPTTPE WVKFCRQLFG GFSILLWI.. ........G
Atn3ratno ..ILARDGPN ALTPPPTTPE WVKFCRQLFG GFSILLWI.. ........G
Atn3galga ..ILARDGPN ALTPPPTTPE WVKFCRQLFG GFSILLWI.. ........G
Atnlhomsa ..ILARDGPN ALTPPPTTPE WIKFCRQLFG GFSMLLWI.. ........G
Atn2homsa ..VLARDGPN ALTPPPTTPE WVKFCRQLFG GFSILLWI.. ........G
Atn3sussc ..ILARDGPN ALTPPPTTPE WVKFCRQLFG GFSMLLWI.. ........G
Atnacatco ..ILARDGPN ALTPPPTTPE WVKFCKQMFG GFSMLLWT.. ........G
Atnatorca ..ILARDGPN ALTPPPTTPE WIKFCRQLFG GFSILLWT.. ........G
Atnaartsf ..NIERDGPN CLTPPKTTPE WIKFCKNLFG GFALLLWT.. ........G
Atnadrome ..NLERDGPN .LTPPKQTPE WVKFCEDLF. GVAMLLWI.. ........G
Atnaartsa ..HLEKYGPN ALTPPRTTPE WIKFCKQLFG GFQMLLWI.. ........G
Atnahydat ..NLERDGLN ALTPPKQTPE WVKFCKQMFG GFSMLLWI.. ........G
Athahomsa ..LLLRDGPN ALRPPRGTPE YVKFARQLAG GLQCLMWV.. ........A
Atcartsf  ..NQEKYGPN ELPAEEGKSL LTLILEQFDD LLVKILLL.. ........A
Atcbdrome ..NQKKYGPN ELPTEEGKSI WQLVLEQFDD LLVKILLL.. ........A
Atcborycu ..HLEKYGHN ELPAEEGKSL WELVIEQFED LLVRILLL.. ........A
Atcdhomsa ..LKERWGSN ELPAEEGKTL LELVIEQFED LLVRILLL.. ........A
Atcfratno ..ARERYGPN ELPTEEGKSL WELVVEQFED LLVRILLL.. ........A
Atctrybr  ..RRQAFGIN ELPSEPPTPF WKLVLAQFED TLVRILLL.. ........A
Atcplafa  ..RRLKYGLN ELEVEKKKSI FELILNQFDD LLVKILLL.. ........A
Atalsynsp ..RYEQYGRN ELKFKPGKPA WLRFLLQFHQ PLLYILLI.. ........A
Atclsynsp ..RLALYGPN ELVEQAGRSP LQILWDQFAN IMLLMLLA.. ........V
Atclsacce ..RRSLYGPN EITVEDDESL FKKFLSNFIE DRMILLLI.. ........G
Atclmycge ..SRQEHGAN FLPEKKATPF WLLFLQQFKS LVVILLLL.. ........A
Consensus ......G.N .......... .......... .......... .........
```

```
          251                                          300
Atxaleido IIIEFAL... ....E..... .......... ....NWPDG AILFAIQIAN
Atxbleido IIIEFAL... ....E..... .......... ....NWPDG AILFAIQIAN
Pmallyces AIMAIALANG GGKPP..... .......... ....DWQDF VGIITLLIIN
Pmalnicpl AIMAIALANG GGKPP..... .......... ....DWQDF VGIITLLIIN
Pmalarath ALMAIALANG DNRPP..... .......... ....DWQDF VGIICLLVIN
```

```
Pma4nicpl AVMAIALANG DGKPP............... .....DWQDF IGIICLLVIN
Pma3arath AIMAIALANG GGKPP............... .....DWQDF VGIVCLLVIN
Pma1schpo AALAAGL... ....R............. .....DWVDF GVICALLMLN
Pma2schpo AALAAGL... ....R.,,,,,,,,,,,, .....DWVDF GVICALLLLN
Pma1ajeca AILAAGL... ....E............. .....DWVDF GVICALLLLN
Pma1neucr AVLAAGL... ....E............. .....DWVDF GVICGLLLLN
Pma1sacce AILAAGL... ....S............. .....DWVDF GVICGLLMLN
Pma2sacce AILAAGL... ....S............. .....DWVDV GVICALLLLN
Pma1klula AILAAGL... ....E............. .....DWVDF GVICGLLFLN
Pma1canal AVLAAGL... ....E............. .....DWVDF GVICALLLLN
Pma1zygro AVLAAGL... ....E............. .....DWVDF GVICGLLFLN
Atcphomsa AIVSLGLSFY QPPEGDNALC GEVSVG.EEE GEGETGWIEG AAILLSVVCV
Atcqhomsa AIISLGLSFY HPPGEGNEGC ATAQGGAEDE GEAEAGWIEG AAILLSVICV
Atcrhomsa AIISLVLSFY RPAGEENELC GQVATTPEDE NEAQAGWIEG AAILFSVIIV
Atc3sacce AVVSFVLGLY ......ELWM QPPQYDPEGN KIKQVDWIEG VAIMIAVFVV
Atmaescco GAISYATE.......................DLFAA GVIALMVAIS
Atmbsalty AGVSFITDYW LPLRRGE... .......E.. ....TDLTGV LIILTMVSLS
Atc3schpo AALSFGT... ................... .....TDWIEG GVISAIIVLN
Atn1sacce MIISFAM... ................... .....HDWITG GVISFVIAVN
Atn3homsa AILCFLAYGI QAGTE.D... .......DPS GD...NLYLG IVLAAVVIIT
Atn3ratno AILCFLAYGI QAGTE.D... .......DPS GD...NLYLG IVLAAVVIIT
Atn3galga AILCFLAYGI QAGTE.D... .......EPS ND...NLYLG IVLAAVVIIT
Atn1homsa AILCFLAYSI QAATE.E... .......EPQ ND...NLYLG VVLSAVVIIT
Atn2homsa AILCFLAYGI QAAME.D... .......EPS ND...NLYLG VVLAAVVIVT
Atn3sussc AILCFLAYGI QAATE.E... .......EPQ ND...NLYLG VVLSAVVIIT
Atnacatco AVLCFLAYGI LAAME.D... .......EPA ND...NLYLG VVLSAVVIIT
Atnatorca AILCFLAYGI QVATV.D... .......NPA ND...NLYLG VVLSTVVIIT
Atnaartsf AILCFLAYGI EASSGNE... .......DML KD...NLYLG IVLATVVIVT
Atnadrome AILCFVAYSI QASTS.E... .......EPA DD...NLYLG IVLSAVVIVT
Atnaartsa SILCFIAYTM EKYKNPD... ......VL GD...NLYLG LALLFVVIMT
Atnahydat AILCFFAFGI RAVRD.T... .......NPN MD...ELYLG IVLSVVVIIT
Athahomsa AAICLIAFAI QAS.EGD... .......LTT DD...NLYLA IALIAVVVVT
Atcartsf  AIISLVLA.. ...LFEEH......... ...DDE AEQLTAYVEP FVILLILIAN
Atcbdrome AIISFVLA.. ...LFEEH......... ...E... ETFTAFVEP LVILLILIAN
Atcborycu ACISFVLA.. ...WFEEG......... ...E... ETITAFVEP FVILLILIAN
Atcdhomsa ACISFVLA.. ...WFEEG......... ...E... ETITAFVEP FVILLILVAN
Atcfratno ALVSFVLA.. ...WFEEG......... ...E... ETTTAFVEP LVIMLILVAN
Atctrybr  ATVSFAMA.. ...VVENN......... ...A... ....ADFVEP FIILLILILN
Atcplafa  AFISFVLT.. ...LLDMK......... ...HKK IE.ICDFIEP LVIVLILILN
Atalsynsp GTVKAFLG.. ...SWTNA......... ...W.... .VIWGVTLVN
Atclsynsp AVVSGALD.. ...LRDGQ......... ...FPKDA IAILVIVVLN
Atclsacce SAVVSLFM.. ...GNIDD......... ...AVSIT LAIFIVVTV.
Atclmycge SLLSFVVAIV SGLRSNW... .......NFN HDLIIEWVQP FIILLTVFAN
Consensus A............................ ..............
```

```
                301                                                 350
Atxaleido ATIGWYETIK AGDAVAALKN SLKPTATVYR ......DSKW QQIDAAVLVP
Atxbleido ATIGWYETIK AGDAVAALKN SLKPTATVYR ......DSKW QQIDAAVLVP
Pma1lyces STISIFIEENN AGNAAAALMA RLAPKAKVLR ......DGKW DEEDASVLVP
Pma1nicpl STISFIEENN AGNAAAALMA RLAPKAKVLR ......DGRW KEEDAAVLVP
Pma1arath STISFIEENN AGNAAAALMA GLAPKTKVLR ......DGKW SEQEAAILVP
Pma4nicpl STISFIEENN AGNAAAALMA GLAPKTKVLR ......DGRW SEQEAAILVP
Pma3arath STISFVEENN AGNAAAALMA GLAPKTKVLR ......DGKW SEQEASILVP
Pma1schpo AVVGFVQEYQ AGSIVDELKK SLALKAVVIR ......EGQV HELEANEVVP
```

```
Pma2schpo ATVGFVQEYQ AGSIVDELKK TMALKASVLR ......DGRV KEIEASEIVP
Pma1ajeca ACVGFVQEFQ AGSIVDELKK TLALKAVVLR ......NGRL TEVEAPEVVP
Pma1neucr AVVGFVQEFQ AGSIVDELKK TLALKAVVLR ......DGTL KEIEAPEVVP
Pma1sacce AGVGFVQEFQ AGSIVDELKK TLANTAVVIR ......DGQL VEIPANEVVP
Pma2sacce ASVGFIQEFQ AGSIVDELKK TLANTATVIR ......DGQL IEIPANEVVP
Pma1klula AAVGFIQEYQ AGSIVDELKK TLANSAVVIR ......DGNL VEVPSNEVVP
Pma1canal AFVGFIQEYQ AGSIVDELKK TLANSALVVR ......NGQL VEIPANEVVP
Pma1zygro AGVGFIQEFQ AGSIVEELKK TLANTATVIR ......DGSV QEAPANEIVP
Atcphomsa VLVTAFNDWS KEKQFRGLQS RIEQEQKFTV IRGGQ....V IQIPVADITV
Atcqhomsa VLVTAFNDWS KEKQFRGLQS RIEQEQKFTV VRAGQ....V VQIPVAEIVV
Atcrhomsa VLVTAFNDWS KEKQFRGLQC RIEQEQKFSI IRNGQ....L IQLPVAEIVV
Atc3sacce VLVSAANDYQ KELQFAKLNK KKE.NRKIIV IRNDQ....E ILISIHHVLV
Atmaescco TLLNFIQEAR STKAADALKA MVSNTATVLR VINDKGENGW LEIPIDQLVP
Atmbsalty GLLRFWQEFR TNRAAQALKK MVRTTATVLR RGPGNIGAVQ EEIPIEELVP
Atc3schpo ITVGFIQEYK AEKTMDSLRT LASPMAHVTR ....SSKTD. .AIDSHLLVP
Atn1sacce VLIGLVQEYK ATKTMNSLKN LSSPNAHVIR ....NGKSE. .TINSKDVVP
Atn3homsa GCFSYYQEAK SSKIMESFKN MVPQQALVIR ....EGEK.. MQVNAEEVVV
Atn3ratno GCFSYYQEAK SSKIMESFKN MVPQQALVIR ....EGEK.. MQVNAEEVVV
Atn3galga GCFSYYQEAK SSKIMESFKN MVPQQALVIR ....EGEK.. MQLNAEEVVV
Atn1homsa GCFSYYQEAK SSKIMESFKN MVPQQALVIR ....NGEK.. MSINAEEVVV
Atn2homsa GCFSYYQEAK SSKIMDSFKN MVPQQALVIR ....EGEK.. MQINAEEVVV
Atn3sussc GCFSYYQEAK SSKIMESFKN MVPQQALVIR ....NGEK.. MSINAEEVVV
Atnacatco GCFSYYQDAK SSKIMDSFKN LVPQQALVVR ....DGEK.. KQINAEEVVI
Atnatorca GCFSYYQEAK SSKIMDSFKN MVPQQALVIR ....DGEK.. SSINAEQVVV
Atnaartsf GIFSYYQENK SSRIMDSFKN LVPQYALALR ....EGQR.. VTLKAEELTM
Atnadrome GVFSYYQESK SSKIMESFKN MVPQFATVIR ....EGEK.. PSLRAEDLVL
Atnaartsa GCFAYYQDHN ASKIMDSFKN LMPQFAFVIR ....DGKK.. IQLKAEEVTV
Atnahydat GCFSYYQESK SSKIMESFKK MIPQEALVLR ....DGKK.. ITINAEQCVV
Athahomsa GCFGYYQEFK STNIIASFKN LVPQQATVIR ....DGDK.. FQINADQLVV
Atcartsf  AVVGVWQEKN AESAIEALKE YEPEMGKVIR ....ADKTGI QKIKARDLVP
Atcbdrome AVVGVWQERN AESAIEALKE YEPEMGKVVR ....QDKSGI QKVRAKEIVP
Atcborycu AIVGVWQERN AENAIEALKE YEPEMGKVYR ....ADRKSV QRIKARDIVP
Atcdhomsa AIVGVWQERN AENAIEALKE YEPEMGKVYR ....QDRKSV QRIKAKDIVP
Atcfratno AIVGVWQERN AESAIEALKE YEPEMGKVIR ....SDRKGV QRIRARDIVP
Atctrybr  ATVGVWQENR AEGAIEALKS FVPKTAVVLR ....DG..DI KTVNAEELVP
Atcplafa  AAVGVWQECN AEKSLEALKE LQPTKAKVLR ....DGKWEI ..IDSKYLYV
Atalsynsp AIIGYIQEAK AEGAIASLAK AVTTEATVLR ....DGQN.. LRIPSQDLVI
Atclsynsp AVLGYLQESR AEKALAALKG MAAPLVRVRR ....DNRD.. QEIPVAGLVP
Atclsacce ...GFVQEYR SEKSLEALNK LVPAECHLMR ....CGQE.. SHVLASTLVP
Atclmycge SLIGSIQEFK AQKSASALKS LTKSFTRVFR ....NG..EL ISINVSEVVV
Consensus ......QE.. .......... ......V.R .......... ........V.
```

```
                351                                        400
Atxaleido GDLVKLASGS AVPADCSI.. .NEGVIDVDE AALTGESLPV TM........
Atxbleido GDLVKLASGS AVPADCSI.. .NEGVIDVDE AALTGESLPV TM........
Pma1lyces GDIISIKLGD IIPADARLLE .GDP.LKIDQ SALTGESLPV TK........
Pma1nicpl GDIISIKLGD IIPADARLLE .GDP.LKIDQ SALTGESLPV TK........
Pma1arath GDIVSIKLGD IIPADARLLE .GDP.LKVDQ SALTGESLPV TK........
Pma4nicpl GDIISVKLGD IIPADARLLE .GDP.LKIDQ SALTGESLPV TK........
Pma3arath GDIVSIKLGD IIPADARLLE .GDP.LKVDQ SALTGESLPA TK........
Pma1schpo GDILKLDEGT IICADGRVVT .PDVHLQVDQ SAITGESLAV DK........
Pma2schpo GDILHLDEGT ICPADGRLIT .KDCFLQVDQ SAITGESLAV DK........
Pma1ajeca GDILQVEEGT IIPADGRIVT .EEAFLQVDQ SAITGESLAV DK........
Pma1neucr GDILQVEEGT IIPADGRIVT .DDAFLQVDQ SALTGESLAV DK........
```

```
Pma1sacce GDILQLEDGT VIPTDGRIVT .EDCFLQIDQ SAITGESLAV DK.......
Pma2sacce GEILQLESGT IAPADGRIVT .EDCFLQIDQ SAITGESLAA EK.......
Pma1klula GDILQLEDGV VIPADGRLVT .EDCFIQIDQ SAITGESLAV DK.......
Pma1canal GDILQLEDGT VIPTDGRIVS .EDCLLQVDQ SAITGESLAV DK.......
Pma1zygro GDILKLEDGT VIPADGRLVT .EECFLQVDQ SSITGESLAV DK.......
Atcphomsa GDIAQVKYGD LLPADGILIQ GND..LKIDE SSLTGESDHV KK.......
Atcqhomsa GDIAQVKYGD LLPADGLFIQ GND..LKIDE SSLTGESDQV RK.......
Atcrhomsa GDIAQVKYGD LLPADGILIQ GND..LKIDE SSLTGESDHV KK.......
Atc3sacce GDVISLQTGD VVPADCVMIS GKC...EADE SSITGESNTI QKFPVDNSLR
Atmaescco GDIIKLAAGD MIPADLRILQ ARD..LFVAQ ASLTGESLPV EK.......
Atmbsalty GDVVFLAAGD LVPADVRLLA SRD..LFISQ SILSGESLPV EK.......
Atc3schpo GDVVVLKTGD VVPADLRLVE TVN..FETDE ALLTGESLPV IK.......
Atn1sacce GDICLVKVGD TIPADLRLIE TKN..FDTDE SLLTGESLPV SK.......
Atn3homsa GDLVEIKGGD RVPADLRIIS AHGC..KVDN SSLTGESEPQ TR.......
Atn3ratno GDLVEIKGGD RVPADLRIIS AHGC..KVDN SSLTGESEPQ TR.......
Atn3galga GDLVEVKGGD RVPADLRIIS AHGC..KVDN SSLTGESEPQ TR.......
Atn1homsa GDLVEVKGGD RIPADLRIIS ANGC..KVDN SSLTGESEPQ TR.......
Atn2homsa GDLVEVKGGD RVPADLRIIS SHGC..KVDN SSLTGESEPQ TR.......
Atn3sussc GDLVEVKGGD RIPADLRIIS ANGC..KVDN SSLTGESEPQ TR.......
Atnacatco GDLVEVKGGD RIPADLRIIS SHGC..KVDN SSLTGESEPQ TR.......
Atnatorca GDLVEVKGGD RIPADLRIIS ACSC..KVDN SSLTGESEPQ SR.......
Atnaartsf GDIVEVKFGD RVPADLRVLE ARSF..KVDN SSLTGESEPQ AR.......
Atnadrome GVLVELEFGD LIPLVYRIIE ARDF..KVDN SSLTGESEPQ SR.......
Atnaartsa GDLVEVKFGD RIPADIRITS CQSM..KVDN SSLTGESEPQ SR.......
Atnahydat GDVVFVKFGD RIPADIRIVE CKGL..KVDN SSLTGESEPQ SR.......
Athahomsa GDLVEMKGGD RVPADIRILA AQGC..KVDN SSLTGESEPQ TR.......
Atcartsf  GDIVEISVGD KIPADLRLIS ILSTTLRIDQ SILTGESVSV IK.......
Atcbdrome GDLVEVSVGD KIPADIRITH IYSTTLRIDQ SILTGESVSV IK.......
Atcborycu GDIVEVAVGD KVPADIRILS IKSTTLRVDQ SILTGESVSV IK.......
Atcdhomsa GDIVEIAVGD KVPADIRLTS IKSTTLRVDQ SILTGESVSV IK.......
Atcfratno GDIVEVAVGD KVPADLRLIE IKSTTLRVDQ SILTGESVSV TK.......
Atctrybr  GDVVEVAVGN RVPADMRVVE LHSTTLRADQ SILNGESVEA MK.......
Atcplafa  GDIIELSVGN KTPADARIIK IYSTSLKVEQ SMLTGESCSV DK.......
Ata1synsp GDIVSLASGD KVPADLRL.. LKVRNLQVDE SALTGEAVPV EK.......
Atc1synsp GDLILLEAGD QVPADARL.. VESANLQVKE SALTGEAEAV QK.......
Atc1sacce GDLVHFRIGD RIPADIRI.. IEAIDLSIDE SNLTGENEPV HK.......
Atc1mycge GDIIFVDAGD IIPADGKLLQ VN..NLRCLE SFLTGESTPV DK.......
Consensus GD......G. ..PAD.R... ........D. S.LTGES... .K.......
```

```
               401                                         450
Atxaleido  ...................................GP EHMPKMGSNV VRGEVEGTVQ
Atxbleido  ...................................GP EHMPKMGSNV VRGEVEGTVQ
Pma1lyces  ...................................GP GDGVYSGSTC KQGEIEAVVI
Pma1nicpl  ...................................GP GDGVYSGSTC KQGEIEAIVI
Pma1arath  ...................................HP GQEVFSGSTC KQGEIEAVVI
Pma4nicpl  ...................................NP GDEVFSGSTC KQGELEAVVI
Pma3arath  ...................................GP GEEVFSGSTC KQGEIEAVVI
Pma1schpo  ...................................HY GDPTFASSGV KRGEGLMVVT
Pma2schpo  ...................................HQ NDTMYSSSTV KRGEAFMVVT
Pma1ajeca  ...................................HK GDTCYASSAV KRGEAFMVIT
Pma1neucr  ...................................HK GDQVFASSAV KRGEAFVVIT
Pma1sacce  ...................................HY GDQTFSSSTV KRGEGFMVVT
Pma2sacce  ...................................HY GDEVFSSSTV KTGEAFMVVT
Pma1klula  ....................... .......RF GDSTFSSSTV KRGEAFMIVT
```

```
Pma1canal .......................... .......... RS GDSCYSSSTV KTGEAFMIVT
Pma1zygro .......................... .......... HY GDEVFSSSTV KRGEGFMIVT
Atcphomsa ......SLDK .................... DPLLLSGTHV MEGSGRMVVT
Atcqhomsa ......SVDK .................... DPMLLSGTHV MEGSGRMLVT
Atcrhomsa ......SLDK .................... DPMLLSGTHV MEGSGRMVVT
Atc3sacce DFKKFNSIDS HNHSKPLDIG DVNEDGNKIA DCMLISGSRI LSGLGRGVIT
Atmaescco .......... AATTRQPEH. ....SNPLEC DTLCFMGTTV VSGTAQAMVI
Atmbsalty .....YDVMA DVAGKDSEQL PDKDKSLLDL GNICLMGTNV TSGRAQAVVV
Atc3schpo .....DAHAT F......Q.. MNEDVPIGDR INLAYSSSIV TKGRAKGICY
Atn1sacce .....DANLV F......G.. KEEETSVGDR LNLAFSSSAV VKGRAKGIVI
Atn3homsa .....SPD.. .......... .CTHDNPLET RNITFFSTNC VEGTARGVVV
Atn3ratno .....SPD.. .......... .CTHDNPLET RNITFFSTNC VEGTARGVVV
Atn3galga ....SPD.. .......... .CTHDNPLET RNITFFSTNC VEGTARGVVI
Atn1homsa .....SPD.. .......... .FTNENPLET RNIAFFSTNC VEGTARGIVV
Atn2homsa .....SPE.. .......... .FTHENPLET RNICFFSTNC VEGTARGIVI
Atn3sussc .....SPD.. .......... .FTNENPLET RNIAFFSTNC VEGTARGIVV
Atnacatco .....SPD.. .......... .FSNDNPLET KNIAFFSTNC VEGTARGIVI
Atnatorca .....SPE.. .......... .YSSENPLET KNIAFFSTNC VEGTARGIVI
Atnaartsf .....SPE.. .......... .FTNDNPLET KNLAFFSTNA VEGTMRGIVI
Atnadrome .....GAE.. .......... .FTHENPLET KNLAFFSTNA VEALPKGVVI
Atnaartsa .....STE.. .......... .CTNDNPLET KNLAFFFTNT LEGTGRGIVI
Atnahydat .....AVD.. .......... .FTHENPLET KNLAFFSTNA VEGTATGIVV
Athahomsa .....SPE.. .......... .CTHESPLET RNIAFFSTMC LEGTAQGLVV
Atcartsf .....HTDPV ......PD.. P..RAVNQDK KNMLFSGTNV SAGKARGVVM
Atcbdrome .....HTDAI ......PD.. P..RAVNQDK KNILFSGTNV AAGKARGVVI
Atcborycu .....HTEPV ......PD.. P..RAVNQDK KNMLFSGTNI AAGKALGIVA
Atcdhomsa ....HTDPV ......PD.. P..RAVNQDK KNMLFSGTNI AAGKAMGVVV
Atcfratno .....HTDAI ......PD.. P..RAVNQDK KNMLFSGTNI ASGKALGVAV
Atctrybr .....QIEAV ......KG.. R..QERFPA. .CMVYSGTAI VYGKALCVVV
Atcplafa .....YAEKM ......ED.. SYKNCEIQLK KNILFSSTAI VCGRCIAVVI
Atalsynsp ......AV.. ......EL.. LPEETPLAER LNMAYAGSFV TFGQGTGVVV
Atclsynsp .....LAD.. ......QQ.. LPTDVVIGDR TNCLFQGTEV LQGRGQALVY
Atc1sacce .....TSQTI EKSSFNDQ.. PNSIVPISER SCIAYMGTLV KEGHGKGIVV
Atclmycge .....TI... .......D.. SNEKATILEQ TNLVFSGAQV VYGSGVFQVE
Consensus ........... ........... ........... ...........G.......


                 451                                              500
Atxaleido YTGSLTFFGK TAALLQSVES DLGN...... ...... ....
Atxbleido YTGSLTFFGK TAALLQSVES DLGN...... ...... ....
Pma1lyces ATGVHTFFGK AAHLVDSTNQ V.GH...... ...... ....
Pma1nicpl ATGVHTFFGK AAHLVDSTNQ V.GH...... ...... ....
Pma1arath ATGVHTFFGK AAHLVDSTNQ V.GH...... ...... ....
Pma4nicpl ATGVHTFFGK AAHLVDSTNN V.GH...... ...... ....
Pma3arath ATGVHTFFGK AAHLVDSTNQ V.GH...... ...... ....
Pma1schpo ATGDSTFVGR AASLVNAAAG GTGH...... ...... ....
Pma2schpo ATADSTFVGR AASLVGAAGQ SQGH...... ...... ....
Pma1ajeca ATGDNTFVGR GPALVNAASA GTGH...... ...... ....
Pma1neucr ATGDNTFVGR AAALVNAASG GSGH...... ...... ....
Pma1sacce ATGDNTFVGR AAALVNKAAG GQGH...... ...... ....
Pma2sacce ATGDNTFVGR AAALVGQASG VEGH...... ...... ....
Pma1klula ATGDSTFVGR AAALVNAASG GSGH...... ...... ....
Pma1canal ATGDSTFVGR AAALVNKASA GTGH...... ...... ....
Pma1zygro ATGDNTFVGR AASLVNAAAG GQGH...... ...... ....
Atcphomsa AVGVNSQTGI IFTLLGAGGE EEEKKDEKKK EKKNKKQDGA IENRNKAKAQ
```

```
Atcqhomsa AVGVNSQTGI IFTLLGAGGE EEEKKD.....KKAKQQDGA..........
Atcrhomsa AVGVNSQTGI ILTLLGVNED DEGEK......KKKGKKQGV PENRNKAKTQ
Atc3sacce SVGINSVYGQ TMTSLNAEP..................................
Atmaescco ATGANTWFGQ LAGRVSEQES EPN.........................
Atmbsalty ATGSRTWFGS LAKSIVGTRT QT.........................
Atc3schpo ATGMQTQIGA IAAGLRQKGK LFQRPEKDEP NYRRKL....
Atn1sacce KTALNSEIGK IAKSLQGDSG LISR....DP SKSWLQ....
Atn3homsa ATGDRTVMGR IAT..LASGL EVGK......
Atn3ratno ATGDRTVMGR IAT..LASGL EVGK......
Atn3galga ATGDRTVMGR IAT..LASGL EVGK......
Atn1homsa YTGDRTVMGR IAT..LASGL EGGQ......
Atn2homsa ATGDRTVMGR IAT..LASGL EVGR......
Atn3sussc YTGDRTVMGR IAT..LASGL EGGQ......
Atnacatco STGDRTVMGR IAT..LASGL EVGR......
Atnatorca NIGDHTVMGR IAT..LASGL EVGQ......
Atnaartsf GIGDNTVMGR IAG..LASGL DTGE......
Atnadrome SCGDHTVMGR IAA..LASGL DTG.......
Atnaartsa NVGDDSVMGR IAC..LASSL DSGK......
Atnahydat RIGDNTVMGR IAN..LASGL GSGK......
Athahomsa NTGDRTIIGR IAS..LASGV ENEK......
Atcartsf  GTGLNTAIGS IRTQMFE..T EEMK......
Atcbdrome GTGLSTAIGK IRTEMSE..T EEIK......
Atcborycu TTGVSTEIGK IRDQMAA..T EQDK......
Atcdhomsa ATGVNTEIGK IRDEMVA..T EQER......
Atcfratno ATGLHTELGK IRSQMAA..V EPER......
Atctrybr  RTGASTEIGT IERDVRE..Q EEVK......
Atcplafa  NIGMKTEIGH IQHAVIESNS EDTQ......
Atalsynsp ATANATEMGQ ISQSMEK..Q VSLM......
Atclsynsp ATGMNTELGR IATLLQS..V ESEK......
Atc1sacce GTGTNTSFGA VFEMMNN..I EKPK......
Atclmycge AVGIKTQVGK IAKTVDDSVT KL.........
Consensus ..G..T..G. ...........................
```

```
                501                                           550
Atxaleido ..............................................IHVILRRV
Atxbleido ..............................................IHVILRRV
Pma1lyces .................................................FQKV
Pma1nicpl .................................................FQKV
Pma1arath .................................................FQKV
Pma4nicpl .................................................FQKV
Pma3arath .................................................FQKV
Pma1schpo .................................................FTEV
Pma2schpo .................................................FTEV
Pma1ajeca .................................................FTEV
Pma1neucr .................................................FTEV
Pma1sacce .................................................FTEV
Pma2sacce .................................................FTEV
Pma1klula .................................................FTEV
Pma1canal .................................................FTEV
Pma1zygro .................................................FTEV
Atcphomsa DGAAMEMQPL KSEEGGDGDE KDKKKANLPK KEKSVLQGKL TKLAVQIGKA
Atcqhomsa ..AAMEMQPL KSAEGGDAD. .DRKKASMHK KEKSVLQGKL TKLAVQIGKA
Atcrhomsa DGVALEIQPL NSQEGIDNEE KDKKAVKVPK KEKSVLQGKL TRLAVQIGKA
Atc3sacce ...................................ESTPLQLHL SQLADNISVY
```

63

```
Atmaescco  ..........................  .................  .AFQQGISRV
Atmbsalty  ..........................  .................  .AFDRGVNSV
Atc3schpo  ...............NKY Y..LKVTSYY VQRVLGLNVG TPLQRKLTVL
Atn1sacce  ...............NTW ISTKKVTGAF ....LGTNVG TPLHRKLSKL
Atn3homsa  ..........................  ............. TPIAIEIEHF
Atn3ratno  ..........................  ............. TPIAIEIEHF
Atn3galga  ..........................  ............. TPIAVEIEHF
Atn1homsa  ..........................  ............. TPIAAEIEHF
Atn2homsa  ..........................  ............. TPIAMEIEHF
Atn3sussc  ..........................  ............. TPIAAEIEHF
Atnacatco  ..........................  ............. TPISIEIEHF
Atnatorca  ..........................  ............. TPIAAEIEHF
Atnaartsf  ..........................  ............. TPIAKEIAHF
Atnadrome  ..........................  ............. TPIAKEIHHF
Atnaartsa  ..........................  ............. TPIAREIEHF
Atnahydat  ..........................  ............. TPIALEIEHF
Athahomsa  ..........................  ............. TPIAIEIEHF
Atcartsf   ..........................  ............. TPLQQKLDEF
Atcbdrome  ..........................  ............. TPLQQKLDEF
Atcborycu  ..........................  ............. TPLQQKLDEF
Atcdhomsa  ..........................  ............. TPLQQKLDEF
Atcfratno  ..........................  ............. TPLQRKLDEF
Atctrybr   ..........................  ............. TPLQVKLDEF
Atcplafa   ..........................  ............. TPLQIKIDLF
Atalsynsp  ..........................  ............. TPLTRKFAKF
Atclsynsp  ..........................  ............. TPLQQRLDKL
Atc1sacce  ..........................  ............. TPLQLTMDKL
Atclmycge  ..........................  ............. SPLQQKLEKI
Consensus  ..........................  .............  ..........
```

```
               551                                          600
Atxaleido MFSLCAISFM LCMCCFIYLL A.......... ...RFY.... ET.....FRH
Atxbleido MLALCAISFI LCMCCFIYLL A.......... ...RFY.... ET.....FRH
Pma1lyces LTAIGNFCIC SIAVGMIIEI I.......... ...VMYPIQH RK.....YRP
Pma1nicpl LTAIGNFCIC SIAVGMIIEI I.......... ...VMYPIQH RA.....YRP
Pma1arath LTSIGNFCIC SIAIGIAIEI V.......... ...VMYPIQH RK.....YRD
Pma4nicpl LTAIGNFCIC SIAIGMLVEI I.......... ...VMYPIQH RK.....YRD
Pma3arath LTAIGNFCIC SIAVGIAIEI V.......... ...VMYPIQR RH.....YRD
Pma1schpo LNGIGTILLV LVLLTLFCIY T.......... ...AAF.YRS VR.....LAR
Pma2schpo LNGIGTILLV LVILTLLCIY T.......... ...AAF.YRS VR.....LAA
Pma1ajeca LNGIGTVLLI LVILTLLVVW V.......... ...SSF.YRS NS.....IVT
Pma1neucr LNGIGTILLI LVIFTLLIVW V.......... ...SSF.YRS NP.....IVQ
Pma1sacce LNGIGIILLV LVIATLLLVW T.......... ...ACF.YRT NG.....IVR
Pma2sacce LNGIGIILLV LVIATLLLVW T.......... ...ACF.YRT VG.....IVS
Pma1klula LNGIGTILLI LVIVTLLLVW V.......... ...ASF.YRT NK.....IVR
Pma1canal LNGIGTTLLV FVIVTLLVVW V.......... ...ACF.YRT VR.....IVP
Pma1zygro LNGIGVILLV LVVITLLLIW T.......... ...ACF.YRT VR.....IVP
Atcphomsa GLLMSAITVI ILVLYFVIDT FWVQKRPWLA ECTPIYIQY. FV.....KFF
Atcqhomsa GLVMSAITVI ILVLYFTVDT FVVNKKPWLP ECTPVYVQY. FV.....KFF
Atcrhomsa GLLMSALTVF ILILYFVIDN FVINRRPWLP ECTPIYIQY. FV.....KFF
Atc3sacce GCV.SAI.IL FLVLFTRYLF YIIPEDGRFH DLDPAQKGSK FM.....NIF
Atmaescco SMLLIRFMLV MAPVVLLING YTKGD..... ....WW EA.....ALF
Atmbsalty SWLLIRFMLI MVPVVLLING FSKGD..... ....WV EA.....SLF
Atc3schpo AYILFCIAII LAIIVMAAHS FHVTN..... ....  EV.....SIY
```

```
Atn1sacce AVLLFWIAVL FAIIVMASQK FDVDK.............  ........ RV.....AIY
Atn3homsa IQLITGVAVF LGVSFFILSL I..........  ....LGYTWL EA.....VIF
Atn3ratno IQLITGVAVF LGVSFFILSL I..........  ....LGYTWL EA.....VIF
Atn3galga IQLITGVAVF LGISFFVLSL I..........  ....LGYTWL EA.....VIF
Atn1homsa IHIITGVAVF LGVSFFILSL I..........  ....LEYTWL EA.....VIF
Atn2homsa IQLITGVAVF LGVSFFVLSL I..........  ....LGYSWL EA.....VIF
Atn3sussc IHIITGVAVF LGVSFFILSL I..........  ....LEYTWL EA.....VIF
Atnacatco IHIITGVAVF LGVSFLLLSL V..........  ....LGYSWL EA.....VIF
Atnatorca IHIITGVAVF LGVSFFILSL I..........  ....LGYTWL EA.....VIF
Atnaartsf IHIITGVAVF LGVTFFIIAF V..........  ....LGYHWL DA.....VVF
Atnadrome IHLITGVAVF LGVTFFVIAF I..........  ....LGYHWL DA.....VIF
Atnaartsa IHIITAMAVS LAAVFAVISF L..........  ....YGYTWL EA.....AIF
Atnahydat IHIVTGVAVF LGVSFLIISL A..........  ....MGYHWL EA.....IIF
Athahomsa VDIIAGLAIL FGATFFIVAM C..........  ....IGYTFL RA.....MVF
Atcartsf  GEQLSKVISV ICVAVWAINI GHFNDPAHGG S.......WI KG.....AIY
Atcbdrome GEQLSKVISV ICVAVWAINI GHFNDPAHGG S.......WI KG.....AIY
Atcborycu GEQLSKVISL ICVAVWLINI GHFNDPVHGG S.......WI RG.....AIY
Atcdhomsa GEQLSKVISL ICIAVWIINI GHFNDPVHGG S.......WI RG.....AIY
Atcfratno GRQLSHAISV ICVAVWVINI GHFADPAHGG S.......WL RG.....AVY
Atctrybr  GVLLSKVIGY ICLVVFAVNL VRWYATHKPT KNETFFTRYI QP.....SVH
Atcplafa  GQQLSKIIFV ICVTVWIINF KHFSDPIHGS ........FL YG.....CLY
Atalsynsp SHTLLYVIVT LAAFTFAVGW ......GRGG SPLEM.............
Atclsynsp GNVLVSGALI LVAIVVGLGV ......LNGQ SWEDL.............
Atclsacce GKDLSLVSFI VIGMICLVGI ......IQGR SWLEM.............
Atclmycge GKWFSWFGLG LFAVVFLVQT ALLG......  ......FDNFT NN.....WSI
Consensus ...............................................
```

```
              601                                          650
Atxaleido ALQFAVVVLV VSIPIALEIV VTTTLAVGSK HLSKHKIIVT KLSAIEMMSG
Atxbleido ALQFAVVVLV VSIPIALEIV VTTTLAVGSK HLSKHKIIVT KLSAIEMMSG
Pma11yces GIDNLLVLLI GGIPIAMPTV LSVTMAIGSH RLAQQGAITK RMTAIEEMAG
Pma1nicpl GIDNLLVLLI GGIPIAMPTV LSVTMAIGSH RLAQQGAITK RMTAIEEMAG
Pma1arath GIDNLLVLLI GGIPIAMPTV LSVTMAIGSH RLSQQGAITK RMTAIEEMAG
Pma4nicpl GIDNLLVLLI GGIPIAMPTV LSVTMAIGSH RLSQQGAITK RMTAIEEMAG
Pma3arath GIDNLLVLLI GGIPIAMPTV LSVTMAIGSH KLSQQGAITK RMTAIEEMAG
Pma1schpo LLEYTLAITI IGVPVGLPAV VTTTMAVGAA YLAEKQAIVQ KLSAIESLAG
Pma2schpo LLEYTLAITI IGVPVGLPAV VTTTMAVGAA YLAKKKAIVQ KLSAIESLAG
Pma1ajeca ILEFTLAITI IGVPVGLPAV VTTTMAVGAA YLAKKKAIVQ KLSAIESLAG
Pma1neucr ILEFTLAITI IGVPVGLPAV VTTTMAVGAA YLAKKKAIVQ KLSAIESLAG
Pma1sacce ILRYTLGITI IGVPVGLPAV VTTTMAVGAA YLAKKQAIVQ KLSAIESLAG
Pma2sacce ILRYTLGITI IGVPVGLPAV VTTTMAVGAA YLAKKQAIVQ KLSAIESLAG
Pma1klula ILRYTLAITI VGVPVGLPAV VTTTMAVGAA YLAKKQAIVQ KLSAIESLAG
Pma1canal ILRYTLAITI IGVPVGLPAV VTTTMAVGAA YLAKKQAIVQ KLSAIESLAG
Pma1zygro ILRYTLGITI VGVPVGLPAV VTTTMAGGAA YLAKKQAIVQ KLSAIESLAG
Atcphomsa I..IGVTVLV VAVPEGLPLA VTISLAYSVK KMMKDNNLVR HLDACETMGN
Atcqhomsa I..IGVTVLV VAVPEGLPLA VTISLAYSVK KMMKDNNLVR HLDACETMGN
Atcrhomsa I..IGITVLV VAVPEGLPLA VTISLAYSVK KMMKDNNLVR HLDACETMGN
Atc3sacce I..TSITVIV VAVPEGLPLA VTLALAFATT RMTKDGNLVR VLRSCETMGS
Atmaescco ....ALSVAV GLTPEMLPMI VTSTLARGAV KLSKQKVIVK HLDAIQNFGA
Atmbsalty ....ALAVAV GLTPEMLPMI VSSNLAKGAI AMSRRKVIVK RLNAIQNFGA
Atc3schpo ....AISLGI SIIPESLIAV LSITMAMGQK NMSKRRVIVR KLEALEALGG
Atn1sacce ....AICVAL SMIPSSLVVV LTITMSVGAA VMVSRNVIVR KLDSLEALGA
Atn3homsa L....IGIIV ANVPEGLLAT VTVCLTVTAK RMARKNCLVK NLEAVETLGS
Atn3ratno L....IGIIV ANVPEGLLAT VTVCLTLTAK RMARKNCLVK NLEAVETLGS
```

```
Atn3galga L....IGIIV ANVPEGLLAT VTVCLTLTAK RMARKNCLVK NLEAVETLGS
Atn1homsa L....IGIIV ANVPEGLLAT VTVCLTLTAK RMARKNCLVK NLEAVETLGS
Atn2homsa L....IGIIV ANVPEGLLAT VTVCLTLTAK RMARKNCLVK NLEAVETLGS
Atn3sussc L....IGIIV ANVPEGLLAT VTVCLTLTAK RMARKNCLVK NLEAVETLGS
Atnacatco L....IGIIV ANVPEGLLAT VTVCLTLTAK RMAKKNCLVK NLEAVETLGS
Atnatorca L....IGIIV ANVPEGLLAT VTVCLTLTAK RMARKNCLVK NLEAVETLGS
Atnaartsf L....IGIIV ANVPEGLLAT VTVCLTLTAK RMASKNCLVK NLEAVETLGS
Atnadrome L....IGIIV ANVPEGLLAT VTVCLTLTAK RMASKNCLVK NLEAVETLGS
Atnaartsa M....IGIIV AKVPEGLLAT VTVCLTLTAK RMAKKNCLVR NLEAVETLGS
Atnahydat L....IGIIV ANVPEGLLAT VTVCLTLTAK KMAKKNCLVK HLEAVETLGS
Athahomsa F....MAIVV AYVPEGLLAT VTVCLSLTAK RLASKNCVVK NLEAVETLGS
Atcartsf   YFKIAVALAV AAIPEGLPAV ITTCLALGTR RMAKKNAIVR SLPSVETLGC
Atcbdrome YFKIAVAVAV AAIPEGLPAV ITTCLALGTR RMAKKNAIVR SLPSVETLGC
Atcborycu YFKIAVALAV AAIPEGLPAV ITTCLALGTR RMAKKNAIVR SLPSVETLGC
Atcdhomsa YFKIAVALAV AAIPEGLPAV ITTCLALGTR RMAKKNAIVR SLPSVETLGC
Atcfratno YFKIAVALAV AAIPEGLPAV ITTCLALGTR RMARKNAIVR SLPSVETLGC
Atctrybr   CLKVAVALAV AAIPEGLPAV VTTCLALGTR RMAQHNALVR DLPSVETLGR
Atcplafa   YFKISVALAV AAIPEGLPAV ITTCLALGTR RMVKKNAIVR KLQSVETLGC
Atalsynsp .FEAAVALAV SGIPEGLPAV VTVTLAIGVN RMAKRNAIIR KLPAVEALGS
Atclsynsp .LSVGLSMAV AIVPEGLPAV ITVALAIGTQ RMVQRESLIR RLPAVETLGS
Atclsacce .FQISVSLAV AAIPEGLPII VTVTLALGVL RMAKRKAIVR RLPSVETLGS
Atclmycge ALIGAIALVV AIIPEGLVTF INVIFALSVQ KLTKQKAIIK YLSVIETLGS
Consensus .......... ...P.GL... .T......... ........V. .L.A.E....
```

```
           651                                              700
Atxaleido VNMLCSDKTG TLTLNKMEIQ .......... .......... ..........
Atxbleido VNMLCSDKTG TLTLNKMEIQ .......... .......... ..........
Pma1lyces MDVLCSDKTG TLTLNKLTVD .......... .......... ..........
Pma1nicpl MDVLCSDKTG TLTLNKLTVD .......... .......... ..........
Pma1arath MDVLCSDKTG TLTLNKLSVD .......... .......... ..........
Pma4nicpl MDVLCSDKTG TLTLNKLSVD .......... .......... ..........
Pma3arath MDVLCSDKTG TLTLNKLSVD .......... .......... ..........
Pma1schpo VEVLCSDKTG TLTKNKLSLG .......... .......... ..........
Pma2schpo VEILCSDKTG TLTKNRLSLG .......... .......... ..........
Pma1ajeca VEILCSDKTG TLTKNKLSLA .......... .......... ..........
Pma1neucr VEILCSDKTG TLTKNKLSLH .......... .......... ..........
Pma1sacce VEILCSDKTG TLTKNKLSLH .......... .......... ..........
Pma2sacce VEILCSDKTG TLTKNKLSLH .......... .......... ..........
Pma1klula VEILCSDKTG TLTKNKLSLH .......... .......... ..........
Pma1canal VEILCSDKTG TLTKNKLSLH .......... .......... ..........
Pma1zygro VEILCSDKTG TLTKNKLSLH .......... .......... ..........
Atcphomsa ATAICSDKTG TLTMNRM..T VVQAYINEKH YKK.....VP EPEAIPPNIL
Atcqhomsa ATAICSDKTG TLTTNRM..T VVQAYVGDVH YKE.....IP DPSSINTKTM
Atcrhomsa ATAICSDKTG TLTMNRM..T VVQAYIGGIH YRQ.....IP SPDVFLPKVL
Atc3sacce ATAVCSDKTG TLTENVM..T VVRGFPGNSK FDDSKSLPVS EQRKLNSKKV
Atmaescco MDILCTDKTG TLTQDKIVLE .......... .......... ..........
Atmbsalty MDVLCTDKTG TLTQDNIFLE .......... .......... ..........
Atc3schpo VTDICSDKTG TITQGKMITR RVWI...... .......... ..........
Atn1sacce VNDICSDKTG TLTQGKMLAR QIWI...... .......... ..........
Atn3homsa TSTICSDKTG TLTQNRM..T VAHMW..... .......... .......FDN
Atn3ratno TSTICSDKTG TLTQNRM..T VAHMW..... .......... .......FDN
Atn3galga TSTICSDKTG TLTQNRM..T VAHMW..... .......... .......FDN
Atn1homsa TSTICSDKTG TLTQNRM..T VAHMW..... .......... .......FDN
Atn2homsa TSTICSDKTG TLTQNRM..T VAHMW..... .......... .......FDN
```

```
Atn3sussc TSTICSDKTG TLTQNRM..T VAHMW.......  ............FDN
Atnacatco TSTICSDKTG TLTQNRM..T VAHMW.......  ............FDN
Atnatorca TSTICSDKTG TLTQNRM..T VAHMW.......  ............FDN
Atnaartsf TSTICSDKTG TLTQNRM..T VAHMW.......  ............FDG
Atnadrome TSTICSDKTG TLTQNRM..T VAHMW.......  ............FDN
Atnaartsa TSTICSDKTG TLTQNRM..T VAHMW.......  ............FDQ
Atnahydat TSVICSDKTG TLTQNRM..T VAHMW.......  ............FDK
Athahomsa TSVICSDKTG TLTQNRM..T VSHLW.......  ............FDN
Atcartsf  TSVICSDKTG TLTTNQM..S VSRMF.......  ...............
Atcbdrome TSVICSDKTG TLTTNQM..S VSRMF.......  ...............
Atcborycu TSVICSDKTG TLTTNQM..S VCKMF.......  ...............
Atcdhomsa TSVICSDKTG TLTTNQM..S VCRMF.......  ...............
Atcfratno TSVICSDKTG TLTTNQM..S VCRMF.......  ...............
Atctrybr  CTVICSDKTG TLTTNMM..S VLHAF.......  ...............
Atcplafa  TTVICSDKTG TLTTNQMTTT VFHLFRESDS LTEYQLCQKG DTYYFYESSN
Atalsynsp ATVVCSDKTG TLTENQM..T VQAVYAGGKH YEVSGGGYSP KGEFWQVMGE
Atclsynsp VTTICSDKTG TLTQNKM..V VQQIHTLDHD FTVTGEGYVP AGHF..LIGG
Atc1sacce VNVICSDKTG TLTSNHM..T VSKLWCLDSM SNKLNVLSLD KNKKTKNSNG
Atclmycge VQIICTDKTG TLTQNQMKV. .......... .......... ..........
Consensus ....CSDKTG TLT.N..... .......... .......... ..........
```

```
          701                                            750
Atxaleido .......... .......... .......... .......... ..........
Atxbleido .......... .......... .......... .......... ..........
Pma1lyces .......... .......... .......... .......... ..........
Pma1nicpl .......... .......... .......... .......... ..........
Pma1arath .......... .......... .......... .......... ..........
Pma4nicpl .......... .......... .......... .......... ..........
Pma3arath .......... .......... .......... .......... ..........
Pma1schpo .......... .......... .......... .......... ..........
Pma2schpo .......... .......... .......... .......... ..........
Pma1ajeca .......... .......... .......... .......... ..........
Pma1neucr .......... .......... .......... .......... ..........
Pma1sacce .......... .......... .......... .......... ..........
Pma2sacce .......... .......... .......... .......... ..........
Pma1klula .......... .......... .......... .......... ..........
Pma1canal .......... .......... .......... .......... ..........
Pma1zygro .......... .......... .......... .......... ..........
Atcphomsa .......... SYLVTGISVN C......... .......... ..........
Atcqhomsa .......... ELLINAIAIN S......... .......... ..........
Atcrhomsa .......... DLIVNGISIN S......... .......... ..........
Atc3sacce FEENCSSSLR NDLLANIVLN STAFENRDYK KNDKNTNGSK NMSKNLSFLD
Atmaescco .......... ...NHTDISG KTSE...... .......... ..........
Atmbsalty .......... ...HHLDVSG VKSS...... .......... ..........
Atc3schpo .PSYGYLSVD TSDAN.NPTI GTVSGL.... ........EAA MQDVLKEKKQ
Atn1sacce .PRFGTITIS NSDDPFNPNE GNVSLIPRFS PYEYSHNEDG DVGILQNFKD
Atn3homsa QIHEADTTED QSGTSFDKSS HTWV...... .......... ..........
Atn3ratno QIHEADTTED QSGTSFDKSS HTWV...... .......... ..........
Atn3galga QIHEADTTED QSGTSFDKSS ATWV...... .......... ..........
Atn1homsa QIHEADTTEN QSGVSFDKTS ATWL...... .......... ..........
Atn2homsa QIHEADTTED QSGATFDKRS PTWT...... .......... ..........
Atn3sussc QIHEADTTEN QSGVSFDKTS ATWL...... .......... ..........
Atnacatco QIHEADTTEN QSGTSFDRSS DTWA...... .......... ..........
Atnatorca QIHEADTTEN QSGISFDKTS LSWN...... .......... ..........
```

```
Atnaartsf TITEADTTED QSGAQFDKSS AGWK...... ..........
Atnadrome QIIEADTTED QSGVQYDRTS PGFK..... ...........
Atnaartsa KIVTADTTEN QSGNQLYRGS KGFP..... ...........
Atnahydat MIVEADTTED QSGIAHDKGS LTWK..... ...........
Athahomsa HIHTADTTED QSGQTFDQSS ETWR..... ...........
Atcartsf  VFKDIPDDAA PELYQFELTG STYE...........PI GE.TFMQGQK
Atcbdrome IFDKV.EGND SSFLEFEMTG STYE...........PI GE.VFLNGQR
Atcborycu IIDKV.DGDF CSLNEFSITG STYA...........PE GE.VLKNDKP
Atcdhomsa ILDRV.EGDT CSLNEFTITG STYA...........PI GE.VHKDDKP
Atcfratno VVAEA.EAGA CRLHEFTISG TTYT...........PE GE.VRQGEQL
Atctrybr  TLK.....GD GSIKEYELKD SRFN...........IV SNSVTCEGRQ
Atcplafa  LTNDIYAGES SFFNKLKDEG NVEALTDDGE EGSIDEADPY SDYFSSDSKK
Atalsynsp EVDNVLLDGL PPVLEECLLT G.........  ..........
Atclsynsp EI..IVPNDY RDLMLL.LAA G.........  ..........
Atclsacce NLKNYLTEDV R....ETLTI G.........  ..........
Atclmycge .......VDH FCFNSTTQTD LARA.....  ..........
Consensus ..........  ..........  .........  ..........

                 751                                    800
Atxaleido .........  ..........  .........  ..........
Atxbleido .........  ..........  .........  ..........
Pma1lyces .........  ..........  .........  ..........
Pma1nicpl .........  ..........  .........  ..........
Pma1arath .........  ..........  .........  ..........
Pma4nicpl .........  ..........  .........  ..........
Pma3arath .........  ..........  .........  ..........
Pma1schpo .........  ..........  .........  ..........
Pma2schpo .........  ..........  .........  ..........
Pma1ajeca .........  ..........  .........  ..........
Pma1neucr .........  ..........  .........  ..........
Pma1sacce .........  ..........  .........  ..........
Pma2sacce .........  ..........  .........  ..........
Pma1klula .........  ..........  .........  ..........
Pma1canal .........  ..........  .........  ..........
Pma1zygro .........  ..........  .........  ..........
Atcphomsa ......AYTS K.........  .........  ..........
Atcqhomsa ......AYTT K.........  .........  ..........
Atcrhomsa ......AYTS K.........  .........  ..........
Atc3sacce KCKSRLSFFK K.........  .........  ..........
Atmaescco .........  ..........  .........  ..........
Atmbsalty .........  ..........  .........  ..........
Atc3schpo EMKNID.PSN QP........  .........  ..........
Atn1sacce RLYEKDLPED ID........  .........  ..........
Atn3homsa .........  ..........  .........  ..........
Atn3ratno .........  ..........  .........  ..........
Atn3galga .........  ..........  .........  ..........
Atn1homsa .........  ..........  .........  ..........
Atn2homsa .........  ..........  .........  ..........
Atn3sussc .........  ..........  .........  ..........
Atnacatco .........  ..........  .........  ..........
Atnatorca .........  ..........  .........  ..........
Atnaartsf .........  ..........  .........  ..........
Atnadrome .........  ..........  .........  ..........
Atnaartsa .........  ..........  .........  ..........
```

```
Atnahydat .......................................................
Athahomsa .......................................................
Atcartsf  INA..ADYDA .............VKEI. ...TTIC...........
Atcbdrome IKA..ADYDT .............LQEL. ...STIC...........
Atcborycu IRS..GQFDG .............LVEL. ...ATIC...........
Atcdhomsa VNC..HQYDG .............LVEL. ...ATIC...........
Atcfratno VRC..GQFDG .............LVEL. ...ATIC...........
Atctrybr  VSSPLEQDGA .............LTKL. ...ANIA...........
Atcplafa  MKNDLNNNNN NNNNSSRSGA KRNIPLKEMK SNENTIISRG SKILEDKINK
Atalsynsp .......................................................
Atclsynsp .......................................................
Atclsacce .......................................................
Atclmycge .......................................................
Consensus .......................................................


            801                                              850
Atxaleido .......EQC F.TFEEGNDL KSTLVLAALA AKWREPPRDA LDTMVLG...
Atxbleido .......EQC F.TFEEGNDL KSTLVLAALA AKWREPPRDA LDTMVLG...
Pma1lyces .......KAL IEVFAKGIDA DTVVLMAARA S..RIENQDA IDTAIVGML.
Pma1nicpl .......KNL IEVFAKGVDA DMVVLMAARA S..RTENQDA IDAAIVGML.
Pma1arath .......KNL VEVFCKGVEK DQVLLFAAMA S..RVENQDA IDAAMVGML.
Pma4nicpl .......RNL VEVFAKGVDK EYVLLLAARA S..RVENQDA IDACMVGML.
Pma3arath .......KNL IEVYCKGVEK DEVLLFAARA S..RVENQDA IDAAMVGML.
Pma1schpo .......EPF T...VSGVSG DDLVLTACLA ASRKRKGLDA IDKAFLKALK
Pma2schpo .......EPY C...VEGVSP DDLMLTACLA SSRKKKGLDA IDKAFLKALR
Pma1ajeca .......EPY C...VSGVDP EDLMLTACLA ASRKKKGIDA IDKAFLKSLR
Pma1neucr .......DPY T...VAGVDP EDLMLTACLA ASRKKKGIDA IDKAFLKSLK
Pma1sacce .......EPY T...VEGVSP DDLMLTACLA ASRKKKGLDA IDKAFLKSLK
Pma2sacce .......EPY T...VEGVSP DDLMLTACLA ASRKKKGLDA IDKAFLKSLI
Pma1klula .......EPY T...VEGVDP DDLMLTACLA ASRKKKGLDA IDKAFLKSLI
Pma1canal .......EPY T...VEGVEP DDLMLTACLA ASRKKKGLDA IDKAFLKSLI
Pma1zygro .......EPY T...VEGVSS DDLMLTACLA ASRKKKGLDA IDKAFLKSLA
Atcphomsa ............ILPPEK EGGLPRHV.. .......GNK TECALLGLL.
Atcqhomsa ............ILPPEK EGALPRQV.. .......GNK TECGLLGFV.
Atcrhomsa ............ILPPEK EGGLPRQV.. .......GNK TECALLGFV.
Atc3sacce ............ ....GNREDD EDQLFKNVNK GRQEPFIGSK TETALLSLAR
Atmaescco ........... .RVLHSAWLN SHYQTGLKNL LDTAVLEGTD EESA......
Atmbsalty ........... .RVLMLAWLN SSSQSGARNV MDRAILRFGE GRIA......
Atc3schpo ...SDQFIPL LKTCALCNLS TV.NQTETGE ...WVVKGEP TEIALHVFSK
Atn1sacce ...MDLFQKW LETATLANIA TVFKDDATDC ...WKAHGDP TEIAIQVFAT
Atn3homsa ........AL SHIAGLCNRA VFKGGQDNIP VLKRDVAGDA SESALLKCIE
Atn3ratno ........AL SHIAGLCNRA VFKGGQDNIP VLKRDVAGDA SESALLKCIE
Atn3galga ........AL SHIAGLCNRA VFKGGQENVP ILKRDVAGDA SESALLKCIE
Atn1homsa ........AL SRIAGLCNRA VFQANQENLP ILKRAVAGDA SESALLKCIE
Atn2homsa ........AL SRIAGLCNRA VFKAGQENIS VSKRDTAGDA SESALLKCIE
Atn3sussc ........AL SRIAGLCNRA VFQANQENLP ILKRAVAGDA SESALLKCIE
Atnacatco ........SL ARIAGLCNRA VFLAEQIDVP ILKRDVAGDA SESALLKCIE
Atnatorca ........AL SRIAGLCNRA VFQAGQDSVP ILKRSVAGDA SESALLKCIE
Atnaartsf ........AL VKIAALCSRA EFKPNQSTTP ILKREVTGDA SEAAILKCVE
Atnadrome ........AL SRIATLCNRA EFKGGQDGVP ILKKEVSGDA SEAALLKCME
Atnaartsa ........EL IRVASLCSRA EFKTEHAHLP VLKRDVNGDA SEAAILKFAE
Atnahydat ........SL AKVAALCSRA EFKPNQNDVA VLRKECTGDA SETAILKFVE
Athahomsa ........AL CRVLTLCNRA AFKSGQDAVP VPKRIVIGDA SETALLKFSE
Atcartsf  ............ ....MMCNDS AIDFNEYKQA FEK...VGEA TETALIVLGE
```

```
Atcbdrome ........... ....IMCNDS AIDYNEFKQA FEK...VGEA TETALIVLAE
Atcborycu ........... ....ALCNDS SLDFNETKGV YEK...VGEA TETALTTLVE
Atcdhomsa ........... ....ALCNDS ALDYNEAKGV YEK...VGEA TETALTCLVE
Atcfratno ........... ....ALCNDS ALDYNEAKGV YEK...VGEA TETALTCLVE
Atctrybr  ........... ....VLCNDA SLHHNAATVQ VEK...IGEA TEAALLVMSE
Atcplafa  YCYSEYDYNF YMCLVNCNEA NIFCNDNSQI VKK...FGDS TELALLHFVH
Atalsynsp ............ ...MLCNDS QLEHRGDDWA V.....VGDP TEGALLASAA
Atclsynsp ............ ...AVCNDA ALVASGEHWS I.....VGDP TEGSLLTVAA
Atclsacce ............ ...NLCNNA SF..SQEHAI F.....LGNP TDVALL..EQ
Atclmycge ............ ..LCLCNNA SISKDA.... .NK...TGDP TEIALLEWKD
Consensus ........... .......... .......... ......... ...A......

          851                                               900
Atxaleido .......... .......... .......... .......... ..........
Atxbleido .......... .......... .......... .......... ..........
Pma1lyces .......... .......... .......... .......... ..........
Pma1nicpl .......... .......... .......... .......... ..........
Pma1arath .......... .......... .......... .......... ..........
Pma4nicpl .......... .......... .......... .......... ..........
Pma3arath .......... .......... .......... .......... ..........
Pma1schpo NY........ .......... .......... .......... ..........
Pma2schpo NY........ .......... .......... .......... ..........
Pma1ajeca YY........ .......... .......... .......... ..........
Pma1neucr YY........ .......... .......... .......... ..........
Pma1sacce QY........ .......... .......... .......... ..........
Pma2sacce EY........ .......... .......... .......... ..........
Pma1klula SY........ .......... .......... .......... ..........
Pma1canal NY........ .......... .......... .......... ..........
Pma1zygro QY........ .......... .......... .......... ..........
Atcphomsa ..LDLKR.DY QDVR...... .......... .......... ..........
Atcqhomsa ..LDLKQ.DY EPVR...... .......... .......... ..........
Atcrhomsa ..TDLKQ.DY QAVR...... .......... .......... ..........
Atc3sacce LSLGLQPGEL QYLR...... .......... .......... ..........
Atmaescco .......... .......... .......... .......... ..........
Atmbsalty .......... .......... .......... .......... ..........
Atc3schpo RFNY...... .......... .......... .......... ..........
Atn1sacce KMDL...... .......... .......... .......... ..........
Atn3homsa .......... .......... .......... .......... ..........
Atn3ratno .......... .......... .......... .......... ..........
Atn3galga .......... .......... .......... .......... ..........
Atn1homsa .......... .......... .......... .......... ..........
Atn2homsa .......... .......... .......... .......... ..........
Atn3sussc .......... .......... .......... .......... ..........
Atnacatco .......... .......... .......... .......... ..........
Atnatorca .......... .......... .......... .......... ..........
Atnaartsf .......... .......... .......... .......... ..........
Atnadrome .......... .......... .......... .......... ..........
Atnaartsa .......... .......... .......... .......... ..........
Atnahydat .......... .......... .......... .......... ..........
Athahomsa .......... .......... .......... .......... ..........
Atcartsf  KLNPYNL... .......... ...SKAGKDR RSAALVVRED
Atcbdrome KLNSFSV... .......... ...NKSGLDR RSAAIACRGE
Atcborycu KMNVFNT... .......... ...EVRNLSK VERANACNSV
Atcdhomsa KMNVFDT... .......... ...ELKGLSK IERANACNSV
```

```
Atcfratno KMNVFDT... ................. ...DLKGLSR VERAGACNSV
Atctrybr  KFANIKG... ................. ...D...... .SAVNAFRTL
Atcplafa  NFDILPTFSK NNKMPAEYEK NTTPVQSSNK KDKSPRGINK FFSSKNDNSH
Atalsynsp KAGF....... ................. ....SQAGLAS QKP.......
Atclsynsp KAGI....... ................. ....DPEGLQR VLP.......
Atclsacce LANF....... ................. ....EMPDIRN TVQ.......
Atclmycge RSQL....... ................. ....DLKTYYR V.........
Consensus ........... ................. ................. ...
```

```
          901                                              950
Atxaleido ........... ................. ................. ...
Atxbleido ........... ................. ................. ...
Pmallyces ........... ................. ................. ...
Pmalnicpl ........... ................. ................. ...
Pmalarath ........... ................. ................. ...
Pma4nicpl ........... ................. ................. ...
Pma3arath ........... ................. ................. ...
Pmalschpo ........... ................. ................. ...
Pma2schpo ........... ................. ................. ...
Pmalajeca ........... ................. ................. ...
Pmalneucr ........... ................. ................. ...
Pmalsacce ........... ................. ................. ...
Pma2sacce ........... ................. ................. ...
Pmalklula ........... ................. ................. ...
Pmalcanal ........... ................. ................. ...
Pmalzygro ........... ................. ................. ...
Atcphomsa ........... ................. ................. ...
Atcqhomsa ........... ................. ................. ...
Atcrhomsa ........... ................. ................. ...
Atc3sacce ........... ................. ................. ...
Atmaescco ........... ................. ................. ...
Atmbsalty ........... ................. ................. ...
Atc3schpo ........... ................. ................. ...
Atnlsacce ........... ................. ................. ...
Atn3homsa ........... ................. ................. ...
Atn3ratno ........... ................. ................. ...
Atn3galga ........... ................. ................. ...
Atnlhomsa ........... ................. ................. ...
Atn2homsa ........... ................. ................. ...
Atn3sussc ........... ................. ................. ...
Atnacatco ........... ................. ................. ...
Atnatorca ........... ................. ................. ...
Atnaartsf ........... ................. ................. ...
Atnadrome ........... ................. ................. ...
Atnaartsa ........... ................. ................. ...
Atnahydat ........... ................. ................. ...
Athahomsa ........... ................. ................. ...
Atcartsf  MDTRW...... ................. ................. ...
Atcbdrome IETKW...... ................. ................. ...
Atcborycu IRQLM...... ................. ................. ...
Atcdhomsa IKQLM...... ................. ................. ...
Atcfratno IKQLM...... ................. ................. ...
Atctrybr  CEGKW...... ................. ................. ...
Atcplafa  ITSTLNENDK NLKNANHSNY TTAQATTNGY EAIGENTFEH GTSFENCFHS
```

```
Ata1synsp .......... .......... .......... .......... ..........
Atc1synsp .......... .......... .......... .......... ..........
Atc1sacce .......... .......... .......... .......... ..........
Atc1mycge .......... .......... .......... .......... ..........
Consensus .......... .......... .......... .......... ..........

          951                                                1000
Atxaleido .......... ......... AAD LDECDNYQQL NFVPFDPTTK
Atxbleido .......... ......... AAD LDECDNYQQL NFVPFDPTTK
Pma1lyces .......... ......... ADP KEARAGIREI HFLPFNPTDK
Pma1nicpl .......... ......... ADP KEARAGIREI HFLPFNPTDK
Pma1arath .......... ......... ADP KEARAGIREV HFLPFNPVDK
Pma4nicpl .......... ......... ADP KEARAGIREV HFLPFNPVDK
Pma3arath .......... ......... ADP KEARAGIREI HFLPFNPVDK
Pma1schpo .......... ......... PGP RSMLTKYKVI EFQPFDPVSK
Pma2schpo .......... ......... PKA KDQLSKYKVL DFHPFDPVSK
Pma1ajeca .......... ......... PRA KSVLTQYKVL EFHPFDPVSK
Pma1neucr .......... ......... PRA KSVLSKYKVL QFHPFDPVSK
Pma1sacce .......... ......... PKA KDALTKYKVL EFHPFDPVSK
Pma2sacce .......... ......... PKA KDALTKYKVL EFHPFDPVSK
Pma1klula .......... ......... PRA KAALTKYKLL EFHPFDPVSK
Pma1canal .......... ......... PRA KAALPKYKVI EFQPFDPVSK
Pma1zygro .......... ......... PKA KGALTKYKVL EFHPFDPVSK
Atcphomsa .......... ......... NEI PE....EALY KVYTFNSVRK
Atcqhomsa .......... ......... SQM PE....EKLY KVYTFNSVRK
Atcrhomsa .......... ......... NEV PE....EKLY KVYTFNSVRK
Atc3sacce .......... ......... DQP MEKFNIEKVV QTIPFESSRK
Atmaescco .......... ........ RSLASRWQKI DEIPFDFERR
Atmbsalty .......... ........ PSTKARFIKR DELPFDFVRR
Atc3schpo ...GKEDLLK TNT....... .......FV REYPFDSEIK
Atn1sacce ...PHNALTG EKSTNQSNEN DQSSLSQHNE KPGSAQFEHI AEFPFDSTVK
Atn3homsa .......... .....LSSGSV KLMRERNKKV AEIPFNSTNK
Atn3ratno .......... .....LSSGSV KLMRERNKKV AEIPFNSTNK
Atn3galga .......... .....LSSGSV KVMRERNKKV AEIPFNSTNK
Atn1homsa .......... .....LCCGSV KEMRERYAKI VEIPFNSTNK
Atn2homsa .......... .....LSCGSV RKMRDRNPKV AEIPFNSTNK
Atn3sussc .......... .....LCCGSV KEMRERYTKI VEIPFNSTNK
Atnacatco .......... .....LCCGSV KEMREKFTKV AEIPFNSTNK
Atnatorca .......... .....LCCGSV SQMRDRNPKI VEIPFNSTNK
Atnaartsf .......... .....LTTGET EAIRKRNKKI CEIPFNSANK
Atnadrome .......... .....LALGDV MNIRKRNKKI AEVPFNSTNK
Atnaartsa .......... ....MSTGSV MNIRSKQKKV SEIPFNSANK
Atnahydat .......... .....LSVGNV MDIRAKNKKV TEIPFNSTNK
Athahomsa .......... .....LTLGNA MGYRDRFPKV CEIPFNSTNK
Atcartsf  .......... .......... KKE FTLEFSRDRK
Atcbdrome .......... .......... KKE FTLEFSRDRK
Atcborycu .......... .......... KKE FTLEFSRDRK
Atcdhomsa .......... .......... KKE FTLEFSRDRK
Atcfratno .......... .......... QKE FTLEFSRDRK
Atctrybr  .......... .......... KKN ATLEFTRKRK
Atcplafa  KLGNKINTTS THNNNNNNNN NSNSVPSECI SSWRNECKQI KIIEFTRERK
Ata1synsp .......... .......... .........RL DSIPFESDYQ
Atc1synsp .......... .......... .........RQ DEIPFTSERK
Atc1sacce .......... .......... .........KV QELPFNSKRK
```

```
Atclmycge ........................ .............. ...... YEKAFDSIRK
Consensus ........................ ............... ........PF....K


          1001                                               1050
Atxaleido RTAATLVDRR SGEK....... ..FDVTKGAP HVILQMV.......YNQDE
Atxbleido RTAATLVDRR SGEK....... ..FDVTKGAP HVILQMV.......YNQDE
Pma1lyces RTALTYLD.G EGKM....... ..HRVSKGAP EQILNLA... ...HNKSD
Pma1nicpl RTALTYLD.G EGKM....... ..HRVSKGAP EQILNLA... ...HNKSD
Pma1arath RTALTYID.S DGNW....... ..HRVSKGAP EQILDLA... ...NARPD
Pma4nicpl RTALTYID.N NNNW....... ..HRASKGAP EQILDLC... ...NAKED
Pma3arath RTALTFID.S NGNW....... ..HRVSKGAP EQILDLC... ...NARAD
Pma1schpo KVTAYVQA.P DGTR....... ..ITCVKGAP LWVLKTV... ...EEDHP
Pma2schpo KITAYVEA.P DGQR....... ..ITCVKGAP LWVFKTV... ...QDDHE
Pma1ajeca KVSAVVLS.P QGER....... ..ITCVKGAP LSVLKTV... ...EEDHP
Pma1neucr KVVAVVES.P QGER....... ..ITCVKGAP LFVLKTV... ...EEDHP
Pma1sacce KVTAVVES.P EGER....... ..IVCVKGAP LFVLKTV... ...EEDHP
Pma2sacce KVTAVVES.P EGER....... ..IVCVKGAP LFVLKTV... ...EEDHP
Pma1klula KVTAIVES.P EGER....... ..IICVKGAP LFVLKTV... ...EEEHP
Pma1canal KVTAIVES.P EGER....... ..IICVKGAP LFVLKTV... ...EDDHP
Pma1zygro KVTAVVES.P EGER....... ..IICVKGAP LFVLKTV... ...EEDHP
Atcphomsa SMSTVLKNSD GS......... .YRIFSKGAS EIILKKCFKI LSANGEAKVF
Atcqhomsa SMSTVIKLPD ES......... .FRMYSKGAS EIVLKKCCKI LNGAGEPRVF
Atcrhomsa SMSTVIRNPN GG......... .FRMYSKGAS EIILRKCNRI LDRKGEAVPF
Atc3sacce WAGLVVKYKE GKN....KKP FYRFFIKGAA EIVSKNCSYK RNSDDTLEEI
Atmaescco RMSVVVAE.N TEHHQ...... ...LVCKGAL QEILNVCSQV RHN..GEIVP
Atmbsalty RVSVLVEDAQ HGDRC...... ...LICKGAV EEMMMVATHL REG..DRVVA
Atc3schpo RM.AVIYEDQ QG........ QYTVYAKGAV ERILERCSTS NG......ST
Atn1sacce RMSSVYYNNH NE........ TYNIYGKGAF ESIISCCSSW YGKDGVKITP
Atn3homsa YQLSIHE.TE DPN.....DN RYLLVMKGAP ERILDRCSTI LLQ..GKEQP
Atn3ratno YQLSIHE.TE DPN.....DN RYLLVMKGAP ERILDRCATI LLQ..GKEQP
Atn3galga YQLSIHE.TE DPN.....DN RYLLVMKGAP ERILDRCSTI LLQ..GKEQP
Atn1homsa YQLSIHKNPN TS.....EP QHLLVMKGAP ERILDRCSSI LLH..GKEQP
Atn2homsa YQLSIHE.RE DS.....PQ SHVLVMKGAP ERILDRCSTI LVQ..GKEIP
Atn3sussc YQLSIHKNPN TA.....EP RHLLVMKGAP ERILDRCTSI LIH..GKEQP
Atnacatco YQLSVHKIPS GGK.....ES QHLLVMKGAP ERILDRCATI MIQ..GKEQL
Atnatorca YQLSIHE..N DKA.....DS RYLLVMKGAP ERILDRCSTI LLN..GEDKP
Atnaartsf FQVSIHE.NE DKS.....DG RYLLVMKGAP ERILERCSTI FMN..GKEID
Atnadrome YQVSIHE.TE DTN.....DP RYLLVMKGAP ERILERCSTI FIN..GKEKV
Atnaartsa YQVSVHERED KSG....... .YFLVMKGAP ERILERCSTI LID..GTEIP
Atnahydat YQVSVHEQEN SSG....... .YLLVMKGAP EKVLERCSTI LIN..GEEQP
Athahomsa FQLSIHTL.E DPR.....DP RHLLVMKGAP ERVLERCSSI LIK..GQELP
Atcartsf  SMSSYCVPL. KAG...LLSN GPKMFVKGAP EGVLDRCTHV RVG.TKKV.P
Atcbdrome SMSSYCTPL. KAS...RLGT GPKLFVKGAP EGVLERCTHA RVG.TTKV.P
Atcborycu SMSVYCSPA. KSS...RAAV GNKMFVKGAP EGVIDRCNYV RVG.TTRV.P
Atcdhomsa SMSVYCTPN. KPS...RTSM S.KMFVKGAP EGVIDRCTHI RVG.STKV.P
Atcfratno SMSVYCTPT. RAD...PKAQ GSKMFVKGAP ESVIERCSSV RVG.SRTV.P
Atctrybr  SMSVHVTSTV TGS...PASS TNNLFVKGAP EEVLRRSTHV MQDNGAVV.Q
Atcplafa  LMSVIVENKK K......... EIILYCKGAP ENIIKNCKYY .LTKNDIR.P
Atalsynsp YMA.........TLHDGD GRTIYVKGSV ESLLQRCESM LLDDG.QMVS
Atclsynsp RMSVVVADLG ETTLTIREGQ PYVLFVKGSA ELILERCQHC .FGNA.QLES
Atc1sacce LMATKILN.. ......PVDN KCTVYVKGAF ERILEYSTSY LKSKGKKTEK
Atclmycge LMTVVVQKDN R.........FIVIVKGAP DVLL.............P
Consensus .................. ........KGAP ...L...............
```

```
          1051                                         1100
Atxaleido INDEVVDI.. ..IDSL...A ARGVRCLSVA KTD....... .QQGRWHMA.
Atxbleido INDEVVDI.. ..IDSL...A ARGVRCLSVA KTD....... .QQGRWHMA.
Pma1lyces IERRVHTV.. ..IDKF...A ERGLRSLGVA YQEVPEGRKE SAGGPWQFI.
Pma1nicpl IERRVHAV.. ..IDKF...A ERGLRSLGVA YQEVPEGRKE SAGGPWQFI.
Pma1arath LRKKVLSC.. ..IDKY...A ERGLRSLAVA RQVVPEKTKE SPGGPWEFV.
Pma4nicpl VRRKVHSM.. ..MDKY...A ERGLRSLAVA RRTVPEKSKE SPGGRWEFV.
Pma3arath LRKRVHST.. ..IDKY...A ERGLRSLAVS RQTVPEKTKE SSGSPWEFV.
Pma1schpo IPEDVLSAYK DKVGDL...A SRGYRSLGVA RK........ IEGQHWEIM.
Pma2schpo VPEAITDAYR EQVNDM...A SRGFRSLGVA RK........ ADGKQWEIL.
Pma1ajeca IPDEVDSAYK NKVAEF...A TRGFRSLGVA RK........ RGEGSWEIL.
Pma1neucr IPEEVDQAYK NKVAEF...A TRGFRSLGVA RK........ RGEGSWEIL.
Pma1sacce IPEDVHENYE NKVAEL...A SRGFRALGVA RK........ RGEGHWEIL.
Pma2sacce IPEDVHENYE NKVAEL...A SRGFRALGVA RK........ RGEGHWEIL.
Pma1klula IPEDVRENYE NKVAEL...A SRGFRALGVA RK........ RGEGHWEIL.
Pma1canal IPEDVHENYQ NTVAEF...A SRGFRSLGVA RK........ RGEGHWEIL.
Pma1zygro IPEDVHENYE NKVAEL...A SRGFRALGVA RK........ RGEGHWEIL.
Atcphomsa RPRDRDDIVK TVIEPM...A SEGLRTICLA FRDFPAGE.. .PEPEWDNEN
Atcqhomsa RPRDRDEMVK KVIEPM...A CEWLRTICVA YRDFPSS... .PEPDWDNEN
Atcrhomsa KNKDRDDMVR TVIEPM...A CDGLRTICIA YRDF..DD.. .TEPSWDNEN
Atc3sacce NEDNKKE.TD DEIKNL...A SDALRAISVA HKDFCECDSW PPEQLRDKDS
Atmaescco LDDIMLRKIK RVTDTLNRQG ...LRVVAVA TKYLPAREGD ..YQRAD...
Atmbsalty LTETRRELLL AKTEDYNAQG ...FRVLLIA TRKLDGSGNN PTLSVED...
Atc3schpo LEEPDRELII AQMETLAAEG LRVL.ALATK VIDKADNWE. .....TLPRD
Atn1sacce LTDCDVETIR KNVYSLSNEG LRVL.GFASK SFTKDQVNDD QLKNITSNRA
Atn3homsa LDEEMKEAFQ NAYLELGGLG ERVL.GFCHY YLPEEQYPQG FAFDC.DDVN
Atn3ratno LDEEMKEAFQ NAYLELGGLG ERVL.GFCHY YLPEEQFPKG FAFDC.DDVN
Atn3galga LDEEMKEAFQ NAYLELGGLG ERVL.GFCHF YLPEEQYPKG FAFDC.DDVN
Atn1homsa LDEELKDAFQ NAYLELGGLG ERVL.GFCHL FLPDEQFPEG FQFDT.DDVN
Atn2homsa LDKEMQDAFQ NAYMELGGLG ERVL.GFCQL NLPSGKFPRG FKFDT.DELN
Atn3sussc LDEELKDAFQ NAYLELGGLG ERVL.GFCHL FLPDEQFPEG FQFDT.DDVN
Atnacatco LDDEIKESFQ NAYLELGGLG ERVL.GFCHF YLPDEQFPEG FQFDA.DDVN
Atnatorca LNEEMKEAFQ NAYLELGGLG ERVL.GFCHL KLSTSKFPEG YPFDV.EEPN
Atnaartsf MTEELKEAFN NAYMELGGLG ERVL.GFCDY LLPLDKYPHG FAFNA.DDAN
Atnadrome LDEEMKEAFN NAYMELGGLG ERVL.GFCDF MLPSDKYPNG FKFNT.DDIN
Atnaartsa LDNHMKECFN NAYMELGGMG ERVL.GFCDF ELPSDQYPRG YVFDA.DEPN
Atnahydat LKDDVIEIYN KAYDELGGLG ERVL.GFCHY YLPVDQYPKG FLFKTEEEQN
Athahomsa LDEQWREAFQ TAYLSLGGLG ERVL.GFCQL YLNEKDYPPG YAFDV.EAMN
Atcartsf  MTPAIMDKIL EVTRAYG.TG RDTLRCLALA TIDDPMDPKD MDIIDSTKFV
Atcbdrome LTSALKAKIL ALTGQYG.TG RDTLRCLALA VADSPMKPDE MDLGDSTKFY
Atcborycu MTGPVKEKIL SVIKEWG.TG RDTLRCLALA TRDTPPKREE MVLDDSSRFM
Atcdhomsa MTSGVKQKIM SVIREWG.SG SDTLRCLALA THDNPLRREE MHLEDSANFI
Atcfratno LSATSREHIL AKIRDWG.SG SHTLRCLALA TRDTPPRKED MQLDDCSQFV
Atctrybr  LSATHRKRII EQLDKIS.GG ANALRCIGFA FKPTKA.VQH VRLNDPATFE
Atcplafa  LNETLKNEIH NKIQNM...G KRALRTLSFA YKK..LSSKD LNIKNTDDYY
Ata1synsp I...DRGEIE ENVE..D.MA QQGLRVLAFA KKTVEPHHHA IDHGD.....
Atc1synsp LTAATRQQIL AAGE..A.MA SAGMRVLGFA YR...PSAIA DVDED.....
Atc1sacce LTEAQKATIN ECAN..S.MA SEGLRVFGFA KLTLSDSSTP LT.ED.....
Atc1mycge LCNNVQNEVK NIENLLDQSA GQGLRTLAVA LKVL....YK FDQNDQKQID
Consensus ........ .. ......... .......... .......... ..........


          1101                                         1150
Atxaleido .......... ........G ILTFLDPPRP DTKDTIRRSK EYGVDVKMIT
Atxbleido .......... ........G ILTFLDPPRP DTKDTIRRSK EYGVDVKMIT
```

```
Pma1lyces .................A LLPLFDPPRH DSAETIRRAL NLGVNVKMIT
Pma1nicpl .................G LLPLFDPPRH DSAETIRRAL NLGVNVKMVT
Pma1arath .................G LLPLFDPPRH DSAETIRRAL NLGVNVKMIT
Pma4nicpl .................G LLPLFDPPRH DSAETIRRAL NLGVNVKMIT
Pma3arath .................G VLPLFDPPRH DSAETIRRAL DLGVNVKMIT
Pma1schpo .................G IMPCSDPPRH DTARTISEAK RLGLRVKMLT
Pma2schpo .................G IMPCSDPPRH DTARTIHEAI GLGLRIKMLT
Pma1ajeca .................G IMPCSDPPRH DTAKTINEAK TLGLSIKMLT
Pma1neucr .................G IMPCMDPPRH DTYKTVCEAK TLGLSIKMLT
Pma1sacce .................G VMPCMDPPRD DTAQTVSEAR HLGLRVKMLT
Pma2sacce .................G VMPCMDPPRD DTAQTINEAR NLGLRIKMLT
Pma1klula .................G VMPCMDPPRD DTAQTVNEAR HLGLRVKMLT
Pma1canal .................G IMPCMDPPRD DTAATVNEAR RLGLRVKMLT
Pma1zygro .................G VMPCMDPPRD DTAATVNEAK RLGLSVKMLT
Atcphomsa DIVTGLTCI. ........A VVGIEDPVRP EVPDAIKKCQ RAGITVRMVT
Atcqhomsa DILNELTCI. ........C VVGIEDPVRP EVPEAIRKCQ RAGITVRMVT
Atcrhomsa EILTELTCI. ........A VVGIEDPVRP EVPDAIAKCK QAGITVRMVT
Atc3sacce PNIAALDLLF NSQKGLILDG LLGIQDPLRA GVRESVQQCQ RAGVTVRMVT
Atmaescco ..ESDLILE. ........G YIAFLDPPKE TTAPALKALK ASGITVKILT
Atmbsalty ..ETELTIE. ........G MLTFLDPPKE SAGKAIAALR DNGVAVKVLT
Atc3schpo VAESSLEFV. ........S LVGIYDPPRT ESKGAVELCH RAGIRVHMLT
Atn1sacce TAESDLVFL. ........G LIGIYDPPRN ETAGAVKKFH QAGINVHMLT
Atn3homsa FTTDNLCFV. ........G LMSMIGPPRA AVPDAVGKCR SAGIKVIMVT
Atn3ratno FTTDNLCFV. ........G LMSMIDPPRA AVPDAVGKCR SAGIKVIMVT
Atn3galga FATDNLCFV. ........G LMSMIDPPRA AVPDAVGKCR SAGIKVIMVT
Atn1homsa FPIDNLCFV. ........G LISMIDPPRA AVPDAVGKCR SAGIKVIMVT
Atn2homsa FPTEKLCFV. ........G LMSMIDPPRH AVPDAVGKCR SAGIKVIMVT
Atn3sussc FPLDNLCFV. ........G LISMIDPPRA AVPDAVGKCR SAGIKVIMVT
Atnacatco FPTENLCFV. ........G LMSMIDPPRA AVPDAVGKCR SAGIKVIMVT
Atnatorca FPITDLCFV. ........G LMSMIDPPRA AVPDAVGKCR SAGIKVIMVT
Atnaartsf FPLTGLRFA. ........G LMSMIDPPRA AVPDAVAKCR SAGIKVIMVT
Atnadrome FPIDNLRFV. ........G LMSMIDPPRA AVPDAVAKCR SAGIKVIMVT
Atnaartsa FPISGLRFV. ........G LMSMIDPPRA AVPDAVSKCR SAGIKVIMVT
Atnahydat FPLEGLCFL. ........G LLSMIDPPRA AVPDAVSKCR SAGIKVIMVT
Athahomsa FPSSGLCFA. ........G LVSMIDPPRA TVPDAVLKCR TAGIRVIMVT
Atcartsf  KYEQNCTFV. ........G VVGMLDPPRK EVLDAIERCR AAGIRVIVIT
Atcbdrome QYEVNLTFV. ........G VVGMLDPPRK EVFDSIVRCR AAGIRVIVIT
Atcborycu EYETDLTFV. ........G VVGMLDPPRK EVMGSIQLCR DAGIRVIMIT
Atcdhomsa KYETNLTFV. ........G CVGMLDPPRI EVASSVKLCR QAGIRVIMIT
Atcfratno QYETGLTFV. ........G CVGMLDPPRP EVAACITRCS RAGIRVVMIT
Atctrybr  DVESDLTFV. ........G ACGMLDPPRE EVRDAIVKCR TAGIRVVVIT
Atcplafa  KLEQDLIYL. ........G GLGIIDPPRK YVGRAIRLCH MAGIRVFMIT
Ata1synsp .IETGLIFL. ........G LQGMIDPPRP EAIAAVHACH DAGIEVKMIT
Atc1synsp .AETDLTWL. ........G LMGQIDAPRP EVREAVQRCR QAGIRTLMIT
Atc1sacce .LIKDLTFT. ........G LIGMNDPPRP NVKFAIEQLL QGGVHIIMIT
Atc1mycge ELENNLEFL. ........G FVSLQDPPRK ESKEAILACK KANITPIMIT
Consensus ......... .........G .....DPPR. .......... ..G..V.M.T
```

```
              1151                                        1200
Atxaleido GDHLLIAKEM CRMLDLDPN. ..........................IL
Atxbleido GDHLLIAKEM CRMLDLDPN. ..........................IL
Pma1lyces GDQLAIGKET GRRLGMGTN. ..........................MY
Pma1nicpl GDQLAIGKET GRRLGMGTN. ..........................MY
Pma1arath GDQLAIGKET GRRLGMGTN. ..........................MY
```

```
Pma4nicpl GDQLAIAKET GRRLGMGTN. ..................MY
Pma3arath GDQLAIAKET GRRLGMGSN. ..................MY
Pma1schpo GDAVDIAKET ARQLGMGTN. ..................IY
Pma2schpo GDAVGIAKET ARQLGMGTN. ..................VY
Pma1ajeca GDAVGIARET SRQLGLGTN. ..................VY
Pma1neucr GDAVGIARET SRQLGLGTN. ..................IY
Pma1sacce GDAVGIAKET CRQLGLGTN. ..................IY
Pma2sacce GDAVGIAKET CRQLGLGTN. ..................IY
Pma1klula GDAVGIAKET CRQLGLGTN. ..................IY
Pma1canal GDAVGIAKET CRQLGLGTN. ..................IY
Pma1zygro GDAVGIAKET CRQLGLGTN. ..................IY
Atcphomsa GDNINTARAI ATKCGILH.. .PGED.............FLC
Atcqhomsa GDNINTARAI AIKCGIIH.. .PGED.............FLC
Atcrhomsa GDNINTARAI ATKCGILT.. .PGDD.............FLC
Atc3sacce GDNILTAKAI ARNCAILSTD ISSEA............YSA
Atmaescco GDSELVAAKV CHEVGLDAGE ...................V
Atmbsalty GDNPVVTARI CLEVGIDTHD ...................I
Atc3schpo GDHPETAKAI AREVGIIPP. ...........FIS DRDPNMSWMV
Atn1sacce GDFVGTAKAI AQEVGILPTN ...........LYH YSQEIVDSMV
Atn3homsa GDHPITAKAI AKGVGIISEG NETVEDIAAR LNIP...VSQ VNPRDAKACV
Atn3ratno GDHPITAKAI AKGVGIISEG NETVEDIAAR LNIP...VSQ VNPRDAKACV
Atn3galga GDHPITAKAI AKGVGIISEG NETVEDIAAR LNIP...VSQ VNPRDAKACV
Atn1homsa GDHPITAKAI AKGVGIISEG NETVEDIAAR LNIP...VSQ VNPRDAKACV
Atn2homsa GDHPITAKAI AKGVGIISEG NETVEDIAAR LNIP...MSQ VNPREAKACV
Atn3sussc GDHPITAKAI AKGVGIISEG NETVEDIAAR LNIP...VSQ VNPRDAKACV
Atnacatco GDHPITAKAI AKGVGIISEG NETVEDIAAR LNIP...VNE VNPRDAKACV
Atnatorca GDHPITAKAI AKGVGIISEG NETVEDIAAR LNIP...VNQ VNPRDAKACV
Atnaartsf GDHPITAKAI AKSVGIISEG NETVEDIAAR LNIP...VSE VNPRDAKAAV
Atnadrome GDHPITAKAI AKSVGIISEG NETVEDIAQR LNIP...VSE VNPREAKAAV
Atnaartsa GDHPITAKAI ARQVGIISEG HETVDDIAAR LNIP...VSE VNPRSAQAAV
Atnahydat GDHPITAKAI AKGVGIISEG NECEEDIALR LNIPLEDLSE DQKKSAKACV
Athahomsa GDHPITAKAI AASVGIISEG SETVEDIAAR LRVP...VDQ VNRKDARACV
Atcartsf  GDNKATAEAI CRRIGVFGED ENTEGM.... ..............A
Atcbdrome GDNKATAEAI CRRIGVFAED EDTTGK.... ..............S
Atcborycu GDNKGTAIAI CRRIGIFGEN EEVADR.... ..............A
Atcdhomsa GDNKGTAVAI CRRIGIFGQD EDVTSK.... ..............A
Atcfratno GDNKGTAVAI CRRLGIFGDT EDVLGK.... ..............A
Atctrybr  GDRKETAEAI CCKLGLLSST ADTTGL.... ..............S
Atcplafa  GDNINTARAI AKEINILNKN EGDDEKD... .........NY TNNKNTQICC
Ata1synsp GDHISTAQAI AKRMGIAAEG DGIA...... ...............
Atc1synsp GDHPLTAQAI ARDLGITEVG HPV....... ...............
Atc1sacce GDSENTAVNI AKQIGIPVID PKLS...... ...............V
Atc1mycge GDHLKTATVI AKELGILTLD NQ........ ..............A
Consensus GD....A... ....G..... ...............
```

```
              1201                                    1250
Atxaleido TADKLPQIKD ANDLPEDLGE KYGDMMLSVG GFAQVFPEHK FMIVETL...
Atxbleido TADKLPQIKD ANDLPEDLGE KYGDMMLSVG GFAQVFPEHK FMIVETL...
Pma1lyces PSSALLGQTK DESIA...AL PIDELIEKAD GFAGVFPEHK YEIVKRL...
Pma1nicpl PSSALLGQTK DESIS...AL PIDELIEKAD GFAGVFPEHK YEIVKRL...
Pma1arath PSAALLGTDK DSNIA...SI PVEELIEKAD GFAGVFPEHK YEIVKKL...
Pma4nicpl PSASLLGQDK DSAIA...SL PIEELIEKAD GFAGVFPEHK YEIVKKL...
Pma3arath PSSSLLGKHK DEAMA...HI PVEDLIEKAD GFAGVFPEHK YEIVKKL...
Pma1schpo .NAERLGLTG GGNMP...GS EVYDFVEAAD GFGEVFPQHK YAVVDIL...
```

```
Pma2schpo .NAERLGLSG GGDMP...GS EVNDFVEAAD GFAEVFPQHK YAVVDIL...
Pma1ajeca .NAERLGLGG GGTMP...GS EVYDFVEAAD GFAEVFPQHK YNVVEIL...
Pma1neucr .NAERLGLGG GGDMP...GS EVYDFVEAAD GFAEVFPQHK YNVVEIL...
Pma1sacce .NAERLGLGG GGDMP...GS ELADFVENAD GFAEVFPQHK YRVVEIL...
Pma2sacce .NAERLGLGG GGDMP...GS ELADFVENAD GFAEVFPQHK YRVVEIL...
Pma1klula .NAERLGLGG GGDMP...GS ELADFVENAD GFAEVFPQHK YNVVEIL...
Pma1canal .DADRLGLSG GGDMA...GS EIADFVENAD GFAEGFPTNK YNAVEIL...
Pma1zygro .DAERLGLGG GGSMP...GS EMYDFVENAD GFAEVFPQHK FAVVDIL...
Atcphomsa LEGKDFNRRI RNEKGEIEQE RIDKIWPKLR VLARSSPTDK HTLVKGIID.
Atcqhomsa LEGKEFNRRI RNEKGEIEQE RIDKIWPKLR VLARSSPTDK HTLVKGIID.
Atcrhomsa LEGKEFNRLI RNEKGEVEQE KLDKIWPKLR VLARSSPTDK HTLVKGIID.
Atc3sacce MEGTEFRKLT KNER.......IRILPNLR VLARSSPEDK RLLVE.....
Atmaescco VIGSDIETLS DDELANLAQR ......TT..LFARLTPMHK ERIVTLL...
Atmbsalty LTGTQVEAMS DAELASEVEK ......RA..VFARLTPLQK TRILQAL...
Atc3schpo MTGSQFDALS DEEVDSL..K ......ALCL VIARCAPQTK VKMIEAL...
Atn1sacce MTGSQFDGLS EEEVDDL..P ......VLPL VIARCSPQTK VRMIEAL...
Atn3homsa IHGTDLKDFT SEQIDEILQN .....HTEI VFARTSPQQK LIIVEGC...
Atn3ratno IHGTDLKDFT SEQIDEILQN .....HTEI VFARTSPQQK LIIVEGC...
Atn3galga IHGTDLKDMS SEQIDEILQN .....HTEI VFARTSPQQK LIIVEGC...
Atn1homsa VHGSDLKDMT SEQLDDILKY .....HTEI VFARTSPQQK LIIVEGC...
Atn2homsa VHGSDLKDMT SEQLDEILKN .....HTEI VFARTSPQQK LIIVEGC...
Atn3sussc VHGSDLKDMT SEQLDDILKY .....HTEI VFARTSPQQK LIIVEGC...
Atnacatco VHGGDLKDLS CEQLDDILKY .....HTEI VFARTSPQQK LIIVEGC...
Atnatorca VHGTDLKDLS HENLDDILHY .....HTEI VFARTSPQQK LIIVEGC...
Atnaartsf VHGGELRDIT PDALDEILRH .....HPEI VFARTSPQQK LIIVEGC...
Atnadrome VHGAELRDVS SDQLDEILRY .....HTEI VFARTSPQQK LIIVEGC...
Atnaartsa IHGNDLKDMN SDQLDDILRH .....YREI VFARTSPQQK LIIVEGV...
Atnahydat IHGAKLKDIK NEELDKILCD .....HTEI VFARTSPQQK LIIVEGC...
Athahomsa INGMQLKDMD PSELVEALRT .....HPEM VFARTSPQQK LVIVESC...
Atcartsf  YTGREFDDLS VEGQRDAVAR ......SR..LFARVEPFHK SKIVEYL...
Atcbdrome YSGREFDDLS PTEQKAAVAR ......SR..LFSRVEPQHK SKIVEFL...
Atcborycu YTGREFDDLP LAEQREACRR ......AC..CFARVEPSHK SKIVEYL...
Atcdhomsa FTGREFDELN PSAQRDACLN ......AR..CFARVEPSHK SKIVEFL...
Atcfratno YTGREFDDLS PEQQRQACRT ......AR..CFARVEPAHK SRIVENL...
Atctrybr  YTGQELDAMT PAQKREAVLT ......AV..LFSRTDPSHK MQLVQLL...
Atcplafa  YNGREFEDFS LEKQKHILKN .....TPRI VFCRTEPKHK KQIVKVL...
Ata1synsp FEGRQLATMG PAELAQAAED ......S..C VFARVAPAQK LQLVEAL...
Atc1synsp LTGQQLSAMN GAELDAAVRS ......V..E VYARVAPEHK LRIVESL...
Atc1sacce LSGDKLDEMS DDQLANVIDH ......V..N IFARATPEHK LNIVRAL...
Atc1mycge VLGSELDEKK ILDYR....... VFARVTPQQK LAIVSAW...
Consensus ............................FA...P..K ...V......
```

```
           1251                                              1300
Atxaleido .......RQR GYTCAMTGDG VNDAPALKRA DVGIAVH.GA TDAARAAADM
Atxbleido .......RQR GYTCAMTGDG VNDAPALKRA DVGIAVH.GA TDAARAAADM
Pma1lyces .......QAR KHICGMTGDG VNDAPALKKA DIGIAVD.DA TDAARSASDI
Pma1nicpl .......QAR KHICGMTGDG VNDAPALKKA DIGIAVD.DA TDAARSASDI
Pma1arath .......QER KHIVGMTGDG VNDAPALKKA DIGIAVA.DA TDAARGASDI
Pma4nicpl .......QER KHIVGMTGDG VNDAPALKKA DIGIAVA.DA TDAARGASDI
Pma3arath .......QER KHICGMTGDG VNDAPALKKA DIGIAVA.DA TDAARGASDI
Pma1schpo .......QQR GYLVAMTGDG VNDAPSLKKA DTGIAVE.GA TDAARSAADI
Pma2schpo .......QQR GYLVAMTGDG VNDAPSLKKA DAGIAVE.GA SDAARSAADI
Pma1ajeca .......QQR GYLVAMTGDG VNDAPSLKKA DTGIAVE.GA SDAARSAADI
Pma1neucr .......QQR GYLVAMTGDG VNDAPSLKKA DTGIAVE.GS SDAARSAADI
```

```
Pma1sacce .......QNR GYLVAMTGDG VNDAPSLKKA DTGIAVE.GA TDAARSAADI
Pma2sacce .......QNR GYLVAMTGDG VNDAPSLKKA DTGIAVE.GA TDAARSAADI
Pma1klula .......QQR GYLVAMTGDG VNDAPSLKKA DTGIAVE.GA TDAARSAADI
Pma1canal .......QSR GYLVAMTGDG VNDAPSLKKA DTGIAVE.GA TDAARSAADI
Pma1zygro .......QQR GYLVAMTGDG VNDAPSLKKA DTGIAVE.GA TDAARSAADI
Atcphomsa ....STVSDQ RQVVAVTGDG TNDGPALKKA DVGFAMGIAG TDVAKEASDI
Atcqhomsa ....STHTEQ RQVVAVTGDG TNDGPALKKA DVGFAMGIAG TDVAKEASDI
Atcrhomsa ....STVGEH RQVVAVTGDG TNDGPALKKA DVGFAMGIAG TDVAKEASDI
Atc3sacce .....TLKGM GDVVAVTGDG TNDAPALKLA DVGFSMGISG TEVAREASDI
Atmaescco .......KRE GHVVGFMGDG INDAPALRAA DIGISVD.GA VDIAREAADI
Atmbsalty .......QKN GHTVGFLGDG INDAPALRDA DVGISVD.SA ADIAKESSDI
Atc3schpo .......HRR KAFVAMTGDG VNDSPSLKQA NVGIAMGQNG SDVAKDASDI
Atn1sacce .......HRR KKFCTMTGDG VNDSPSLKMA NVGIAMGING SDVSKEASDI
Atn3homsa .......QRQ GAIVAVTGDG VNDSPALKKA DIGVAMGIAG SDVSKQAADM
Atn3ratno .......QRQ GAIVAVTGDG VNDSPALKKA DIGVAMGIAG SDVSKQAADM
Atn3galga .......QRQ GAIVAVTGDG VNDSPALKKA DIGVAMGIRG SDVSKQAADM
Atn1homsa .......QRQ GAIVAVTGDG VNDSPALKKA DTGVAMGIAG SDVSKQAADM
Atn2homsa .......QRQ GAIVAVTGDG VNDSPALKKA DIGIAMGISG SDVSKQAADM
Atn3sussc .......QRQ GAIVAVTGDG VNDSPALKKA DIGVAMGIAG SDVSKQAADM
Atnacatco .......QRT GAIVAVTGDG VNDSPALKKA DIGVAMGIAG SDVSKQAADM
Atnatorca .......QRQ GAIVAVTGDG VNDSPALKKA DIGVAMGIAG SDVSKQAADM
Atnaartsf .......QRQ GAIVAVTGDG VNDSPALKKA DIGVAMGIAG SDVSKQAADM
Atnadrome .......QRM GAIVAVTGDG VNDSPALKKA DIGVAMGIAG SDVSKQAADM
Atnaartsa .......QRQ GEFVAVTGDG VNDSPALKKA DIGVAMGIAG SDVSKQAADM
Atnahydat .......QRQ GAIVAVTGDG VNDSPALKKA DIGVAMGIAG SDVSKQAADM
Athahomsa .......QRL GAIVAVTGDG VNDSPALKKA DIGVAMGIAG SDAAKNAADM
Atcartsf  .......QGM GEISAMTGDG VNDAPALKKA EIGIAMG.SG TAVAKSAAEM
Atcbdrome .......QSM NEISAMTGDG VNDAPALKKA EIGIAMG.SG TAVAKSAAEM
Atcborycu .......QSY DEITAMTGDG VNDAPALKKA EIGIAMG.SG TAVAKTASEM
Atcdhomsa .......QSF DEITAMTGDG VNDAPALKKA EIGIAMG.SG TAVAKTASEM
Atcfratno .......QSF NEITAMTGDG VNDAPALKKA EIGIAMG.SG TAVAKSAAEM
Atctrybr  .......KDE RLICAMTGDG VNDAPALKKA DIGIAMG.SG TEVAKSASKM
Atcplafa  .......KDL GETVAMTGDG VNDAPALKSA DIGIAMGING TEVAKEASDI
Ata1synsp .......QEK GHIVAMTGDG VNDAPALKRA DIGIAMGKGG TEVARESSDM
Atc1synsp .......QRQ GEFVAMTGDG VNDAPALKQA NIGVAMGITG TDVSKEASDM
Atc1sacce .......RKR GDVVAMTGDG VNDAPALKLS DIGVSMGRIG TDVAKEASDM
Atc1mycge .......KEA GFTVSVTGDG VNDAPALIKS DVGCCMGITG VDIAKDASDL
Consensus ............ ...A.TGDG VND.PALKKA ..G.A..... .D.A..A.D.
```

```
                  1301                                          1350
Atxaleido VLT......E PGLSVVVEAM LVSREVFQRM LSFLTYRISA TL.QLVCFFF
Atxbleido VLT......E PGLSVVVEAM LVSREVFQRM LSFLTYRISA TL.QLVCFFF
Pma1lyces VLT......E PGLSVIISAV LTSRAIFQRM KNYTIY..AV SI.TIRIVLG
Pma1nicpl VLT......E PGLSVIISAV LTSRAIFQRM KNYTIY..AV SI.TIRIVLG
Pma1arath VLT......E PGLSVIISAV LTSRAIFQRM KNYTIY..AV SI.TIRIVFG
Pma4nicpl VLT......E PGLSVIISAV LTSRAIFQRM KNYTIY..AV SI.TIRIVFG
Pma3arath VLT......E PGLSVIISAV LTSRAIFQRM KNYTIY..AV SI.TIRIVFG
Pma1schpo VFL......A PGLSAIIDAL KTSRQIFHRM YSYVVYRIAL SL.HLEIFLG
Pma2schpo VFL......A PGLSAIIDAL KTSRQIFHRM YAYVVYRIAL SL.HLEIFLG
Pma1ajeca VFL......A PGLSAIIDAL KTSRQIFHRM YAYVVYRIAL SL.HLEIFLG
Pma1neucr VFL......A PGLGAIIDAL KTSRQIFHRM YAYVVYRIAL SI.HLEIFLG
Pma1sacce VFL......A PGLSAIIDAL KTSRQIFHRM YSYVVYRIAL SL.HLEIFLG
Pma2sacce VFL......A PGLSAIIDAL KTSRQIFHRM YSYVVYRIAL SL.HLEIFLG
Pma1klula VFL......A PGLSAIIDAL KTSRQIFHRM YSYVVYRIAL SL.HLEIFLG
```

```
Pma1canal VFL......A PGLSAIIDAL KTSRQIFHRM YSYVVYRIAL SL.HLELFLG
Pma1zygro VFL......A PGLSAIIDAL KTSRQIFHRM YAYVVYRIAL SL.HLEIFLG
Atcphomsa ILT......D DNFTSIVKAV MWGRNVYDSI SKFLQFQLTV NVVAVIVAFT
Atcqhomsa ILT......D DNFSSIVKAV MWGRNVYDSI SKFLQFQLTV NVVAVIVAFT
Atcrhomsa ILT......D DNFTSIVKAV MWGRNVYDSI SKFLQFQLTV NVVAVIVAFT
Atc3sacce ILM......T DDFSAIVNAI KWGRCVSVSI KKFIQFQLIV NITAVILTFV
Atmaescco ILL......E KSLMVLEEGV IEGRRTFANM LKYIKMTASS NFGNVFSVLV
Atmbsalty ILL......E KDLMVLEEGV IKGRETFGNI IKYLNMTASS NFVNVFSVLV
Atc3schpo VLT......D DNFSSIVNAI EEGRRMFDNI MRFVLHLLVS NVGEVILLVV
Atn1sacce VLS......D DNFASILNAV EEGRRMTDNI QKFVLQLLAE NVAQALYLII
Atn3homsa ILL......D DNFASIVTGV EEGRLIFDNL KKSIAYTLTS NIPEITPFLL
Atn3ratno ILL......D DNFASIVTGV EEGRLIFDNL KKSIAYTLTS NIPEITPFLL
Atn3galga ILL......D DNFASIVTGV EEGRLIFDNL KKSIAYTLTS NIPEITPFLL
Atn1homsa ILL......D DNFASIVTGV EEGRLIFDNL KKSIAYTLTS NIPEITPFLI
Atn2homsa ILL......D DNFASIVTGV EEGRLIFDNL KKSIAYTLTS NIPEITPFLL
Atn3sussc ILL......D DNFASIVTGV EEGRLIFDNL KKSIAYTLTS NIPEITPFLI
Atnacatco ILL......D DNFASIVTGV EEGRLIFDNL KKSIAYTLTS NIPEITPFLF
Atnatorca ILL......D DNFASIVTGV EEGRLIFDNL KKSIAYTLTS NIPEITPFLV
Atnaartsf ILL......D DNFASIVTGV EEGRLIFDNL KKSIVYTLTS NIPEISPFLL
Atnadrome ILL......D DNFASIVTGV EEGRLIFDNL KKSIAYTLTS NIPEISPFLA
Atnaartsa ILL......D DNFASIVTGV EEGRLIFDNI KKSIAYTLTS KIPELSPFLM
Atnahydat ILL......D DNFASIVTGV EEGRLIFDNL KKSIVYTLTS NIPEISPFLM
Athahomsa ILL......D DNFASIVTGV EQGRLIFDNL KKSIAYTLTK NIPELTPYLI
Atcartsf  VLA......D DNFSTIVAAV EEGRAIYNNM KQFIRYLISS NIGEVVSIFL
Atcbdrome VLA......D DNFSSIVSAV EEGRAIYNNM KQFIRYLISS NIGEVVSIFL
Atcborycu VLA......D DNFSTIVAAV EEGRAIYNNM KQFIRYLISS NVGEVVCIFL
Atcdhomsa VLA......D DNFSTIVAAV EEGRAIYNNM KQFIRYLISS NVGEVVCIFL
Atcfratno VLS......D DNFASIVAAV EEGRAIYNNM KQFIRYLISS NVGEVVCIFL
Atctrybr  VLA......D DNFATVVKAV QEGRAIYNNT KQFIRYLISS NIGEVVCILV
Atcplafa  VLA......D DNFNTIVEAI KEGRCIYNNM KAFIRYLISS NIGEVASIFI
Atalsynsp LLT......D DNFASIEAAV EEGRTVYQNL RKAIAFLLPV NGGESMTILI
Atclsynsp VLL......D DNFATIVAAV EEGRIVYGNI RKFIKYILGS NIGELLTIAS
Atc1sacce VLT......D DDFSTILTAI EEGKGIFNNI QNFLTFQLST SVAALSLVAL
Atclmycge IIS......D DNFATIVNGI EEGRKTFLTC KRVLLNLFLT SIAGTVVVLL
Consensus ......... .....I.... ...R...... .....Y.... .........
```

```
              1351                                           1400
Atxaleido IACFSLTPKA YGSVDPHFQF FHLPVLMFML ITLLNDG... CLMTIGYDHV
Atxbleido IACFSLTPKA YGSVDPNFQF FHLPVLMFML ITLLNDG... CLMTIGYDHV
Pma1lyces FMLLALIWK. .......... FDFPPFMVLI IAILNDG... TIMTISKDRV
Pma1nicpl FMLLALIWK. .......... FDFPPFMVLI IAILNDG... TIMTISKDRV
Pma1arath FMLIALIWE. .......... FDFSAFMVLI IAILNDG... TIMTISKDRV
Pma4nicpl FMFIALIWK. .......... YDFSAFMVLI IAILNDG... TIMTISKDRV
Pma3arath FMLIALIWK. .......... FDFSPFMVLI IAILNDG... TIMTISKDRV
Pma1schpo LWLIIRNQL. .......... LNLE..LVVF IAIFADV... ATLAIAYDNA
Pma2schpo LWLIIRNQL. .......... LNLE..LIVF IAIFADV... ATLAIAYDNA
Pma1ajeca LWIAILNTS. .......... LNLQ..LVVF IAIFADI... ATLAIAYDNA
Pma1neucr LWIAILNRS. .......... LNIE..LVVF IAIFADV... ATLAIAYDNA
Pma1sacce LWIAILDNS. .......... LDID..LIVF IAIFADV... ATLAIAYDNA
Pma2sacce LWIAILNNS. .......... LDIN..LIVF IAIFADV... ATLTIAYDNA
Pma1klula LWIAILNRS. .......... LNID..LVVF IAIFADV... ATLAIAYDNA
Pma1canal LWIAILNRS. .......... LDIN..LIVF IAIFADV... ATLAIAYDNA
Pma1zygro LWIAILNHS. .......... LDID..LIVF IAIFADV... ATLAIAYDNA
Atcphomsa G..ACIT... .........QD SPLKAVQMLW VNLIMDTLAS LALATEPPTE
```

```
Atcqhomsa G..ACIT... ........QD SPLKAVQMLW VNLIMDTFAS LALATEPPTE
Atcrhomsa G..ACIT... ........QD SPLKAVQMLW VNLIMDTFAS LALATEPPTE
Atc3sacce SSVASSD... ........ET SVLTAVQLLW INLIMDTLAA LALATDKPDP
Atmaescco ASAF...... .....LPF LPMLPLHLLI QNLLYD.VSQ VAIPFDNVDD
Atmbsalty ASAF...... .....IPF LPMLAIHLLI QNLMYD.ISQ LSLPWDKMDK
Atc3schpo GLAFR..... ...DEVHLSV FPMSPVEILW CNMITSSFPS MGLGMELAQP
Atn1sacce GLVFR..... ...DENGKSV FPLSPVEVLW IIVVTSCFPA MGLGLEKAAP
Atn3homsa FIMANI.... ........P LPLGTITILC IDLGTDMVPA ISLAYEAAES
Atn3ratno FIMANI.... ........P LPLGTITILC IDLGTDMVPA ISLAYEAAES
Atn3galga FIMANI.... ........P LPLGTITILC IDLGTDMVPA ISLAYEAAES
Atn1homsa FIIANI.... ........P LPLGTVTILC IDLGTDMVPA ISLAYEQAES
Atn2homsa FIIANI.... ........P LPLGTVTILC IDLGTDMVPA ISLAYEAAES
Atn3sussc FIIANI.... ........P LPLGTVTILC IDLGTDMVPA ISLAYEQAES
Atnacatco FIIANI.... ........P LPLGTVTILC IDLGTDMLPA ISLAYEAAES
Atnatorca FIIANV.... ........P LPLGTVTILC IDLGTDMVPA ISLAYERAES
Atnaartsf FILFDI.... ........P LPLGTVTILC IDLGTDMVPA ISLAYEEAES
Atnadrome SILCDI.... ........P LPLGTVTILC IDLGTDMVPA ISLAYDHAEA
Atnaartsa YILFDL.... ........P LAIGTVTILC IDLGTDVVPA ISMAYEGPEA
Atnahydat FILFGI.... ........P LPLGTITILC IDLGTDMVPA ISLAYEKAES
Athahomsa YITVSV.... ........P LPLGCITILF IELCTDIFPS VSLAYEKAES
Atcartsf  TAAL..... .......GLPE .ALIPVQLLW VNLVTDGLPA TALGFNPPDL
Atcbdrome TAAL..... .......GLPE .ALIPVQLLW VNLVTDGLPA TALGFNPPDL
Atcborycu TAAL..... .......GLPE .ALIPVQLLW VNLVTDGLPA TALGFNPPDL
Atcdhomsa TAAL..... .......GFPE .ALIPVQLLW VNLVTDGLPA TALGFNPPDL
Atcfratno TAIL..... .......GLPE .ALIPVQLLW VNLVTDGLPA TALGFNPPDL
Atctrybr  TGLF..... .......GLPE .ALSPVQLLW VNLVTDGLPA TALGFNAPDR
Atcplafa  TALL..... .......GIPD .SLAPVQLLW VNLVTDGLPA TALGFNPPEH
Ata1synsp SVLL..... ......ALN. LPILSLQVLW LNMINSITMT VPLAFEAKSP
Atc1synsp APLL..... ......GLGA VPLTPLQILW MNLVTDGIPA LALAVEPGDP
Atc1sacce STAF..... ......KLPN .PLNAMQILW INILMDGPPA QSLGVEPVDH
Atc1mycge GLFILGQVFK TNLLQQGHDF QVFSPTQLLI INLFVHGFPA VALAVQPVKE
Consensus ......... ......... ......... .....D.... ..L.......
```

```
          1401                                              1450
Atxaleido IPSERPQKWN LPVVFVSASI LAAVACGSSL MLLWIGLEGY SSQYYENSWF
Atxbleido IPSERPQKWN LPVVFVSASI LAAVACGSSL MLLWIGLEGY SSQYYENSWF
Pma1lyces KPSPLPDSWK LAEIFTTGVV LGGYLAMMTV IFFWAAYKTN FFPRIFGVST
Pma1nicpl KPSPLPDSWK LAEIFTTGIV LGGYLAMMTV IFFWAAYKTN FFPHVFGVST
Pma1arath KPSPTPDSWK LKEIFATGIV LGGYQAIMSV IFFWAAHKTD FFSDKFGVRS
Pma4nicpl KPSPMPDSWK LKEIFATGVV LGGYQALMTV VFFWAMHDTD FFSDKFGVKS
Pma3arath KPSPTPDSWK LKEIFATGVV LGGYMAIMTV VFFWAAYKTD FFPRTFHVRD
Pma1schpo PYSMKPVKWN LPRLWGLSTV IGIVLAIGTW ITNTTMI... .......AQG
Pma2schpo PYAMKPVKWN LPRLWGLATI VGILLAIGTW IVNTTMI... .......AQG
Pma1ajeca PFSKTPVKWN LPKLWGMSVL LGIVLAVGTW ITLTTML... .......VGS
Pma1neucr PYSQTPVKWN LPKLWGMSVL LGVVLAVGTW ITVTTMY... .......AQG
Pma1sacce PYSPKPVKWN LPRLWGMSII LGIVLAIGSW ITLTTMF... .......LP.
Pma2sacce PYAPEPVKWN LPRLWGMSII LGIVLAIGSW ITLTTMF... .......LP.
Pma1klula PYSPKPVKWN LRRLWGMSVI LGIILAIGTW ITLTTMF... .......VP.
Pma1canal PYDPKPVKWN LPRLWGMSLV IGIILAIGTW ITLTTML... .......LP.
Pma1zygro PFSPSPVKWN LPRLWGMSIM MGIILAAGTW ITLTTMF... .......LP.
Atcphomsa SLLLRKP.YG RNKPLISRTM MKNILGHAFY QLVVV..... ..FTLLFAGE
Atcqhomsa TLLLRKP.YG RNKPLISRTM MKNILGHAVY QLALI..... ..FTLLFVGE
Atcrhomsa SLLKRRP.YG RNKPLISRTM MKNILGHAFY QLIVI..... ..FILVFAGE
Atc3sacce NIMDRKP.RG RSTSLISVST WKMILSQATL QLIVT..... ..FILHFYGP
```

```
Atmaescco EQIQKPQRWN PADL...GRF MIFFGPISSI FDILTFCLMW WVFHANTPET
Atmbsalty EFLRKPRKWD AKNI...GRF MLWIGPTSSI FDITTFALMW YVFAANNVEA
Atc3schpo DVMERLPHDN KVGIFQKSLI VDMM..... .........V YGFFLGVVSL
Atn1sacce DLMDRPPHDS EVGIFTWEVI IDTF..... .........A YGIIMTGSCM
Atn3homsa DIMKRQPRNP RTDKLVNERL ISMAYGQ... IGMIQALGGF FSYFVILAEN
Atn3ratno DIMKRQPRNP RTDKLVNERL ISMAYGQ... IGMIQALGGF FSYFVILAEN
Atn3galga DIMKRQPRNP RSDKLVNERL ISMAYGQ... IGMIQALGGF FSYFVILAEN
Atn1homsa DIMKRQPRNP KTDKLVNERL ISMAYGQ... IGMIQALGGF FTYFVILAEN
Atn2homsa DIMKRQPRNS QTDKLVNERL ISMAYGQ... IGMIQALGGF FTYFVILAEN
Atn3sussc DIMKRQPQNP KTDKLVNEQL ISMAYGQ... IGMIQALGGF FTYFVILAEN
Atnacatco DIMKRQPRNP KTDKLVNERL ISIAYGQ... IGMIQALAGF FTYFVILAEN
Atnatorca DIMKRQPRNP KTDKLVNERL ISMAYGQ... IGMIQALGGF FSYFVILAEN
Atnaartsf DIMKRRPRNP VTDKLVNERL ISLAYGQ... IGMIQASAGF FVYFVIMAEC
Atnadrome DIMKRPPRDP FNDKLVNSRL ISMAYGQ... IGMIQAAAGF FVYFVIMAEN
Atnaartsa D..PRKPRDP VKEKLVNERL ISMAYGQ... IGVMQAFGGF FTYFVIMGEC
Atnahydat DIMKRHPRNP IRDKLVNERL ISLAYGQ... IGMMQATAGF FTYFIILAEN
Athahomsa DIMHLRPRNP KRDRLVNEPL AAYSYFQ... IGAIQSFAGF TDYFTAMAQE
Atcartsf  DIMNKPPRRA D.EGLITGWL FFRYMAIGTY VGAATVGAAA HWFMMSPTGP
Atcbdrome DIMEKPPRKA D.EGLISGWL FFRYMAIGFY VGAATVGAAA WWFVFSDEGP
Atcborycu DIMDRPPRSP K.EPLISGWL FFRYMAIGGY VGAATVGAAA WWFMYAEDGP
Atcdhomsa DIMNKPPRNP K.EPLISGWL FFRYLAIGCY VGAATVGAAA WWFIAADGGP
Atcfratno DIMEKLPRNP R.EALISGWL FFRYLAIGVY VGLATVAAAT WWFLYDAEGP
Atctrybr  DIMEQRPRRM E.EPIVNGWL FMRYMVIGVY VGLATVGGFL WWFLRHG...
Atcplafa  DVMKCKPRHK N.DNLINGLT LLRYIIIGTY VGIATVSIFV YWFLFYPDSD
Ata1synsp GIMQQAPRNP N.EPLITKKL ....LHRILL VSLFNW.... ..........
Atc1synsp TIMQRRPHNP Q.ESIFARGL GTYMLRVGVV FSAFTI.... ..........
Atc1sacce EVMKKPPRKR T.DKILTHDV MKRLLTTAAC IIVGTV.... ..........
Atc1mycge KLM..VGSFS T.KNLFYNRQ GFDLIWQSLF LSFLTL.... ..........
Consensus ......... .......... .......... .......... ..........
```

```
                1451                                          1500
Atxaleido HRLGLAQLPQ GKLVTMMYLK ISISDFLTLF SSRTGGHFFF YMPPSPILFC
Atxbleido HRLGLAQLPQ GKLVTMMYLK ISISDFLTLF SSRTGGHFFF YVPPSPILFC
Pma1lyces LEKTATD.DF RKLASAIYLQ VSTISQALIF VTRSRSWSFV ERPGL..LLV
Pma1nicpl LEKTATD.DF RKLASAIYLQ VSIISQALIF VTRSRSWSFV ERPGF..LLV
Pma1arath IRDNNDE... .LMGAVYLQ VSIISQALIF LTRSRSWYFV ERPGA..LLM
Pma4nicpl LRNSDEE... .MMSALYLQ VSIISQALIF VTRSRSWSFL ERPGM..LLV
Pma3arath LRGSEHE... .MMSALYLQ VSIVSQALIF VTRSRSWSFT ERPGY..FLL
Pma1schpo QNRGIVQ.NF GVQDEVLFLE ISLTENWLIF VTRCNGPFWS SIPSW..QLS
Pma2schpo QNRGIVQ.NF GVQDEVLFLQ ISLTENWLIF ITRCSGPFWS SFPSW..QLS
Pma1ajeca ENGGIVQ.NF GRTHPVLFLE ISLTENWLIF ITRANGPFWS SIPSW..QLS
Pma1neucr ENGGIVQ.NF GNMDEVLFLQ ISLTENWLIF ITRANGPFWS SIPSW..QLS
Pma1sacce .KGGIIQ.NF GAMNGIMFLQ ISLTENWLIF ITRAAGPFWS SIPSW..QLA
Pma2sacce .NGGIIQ.NF GAMNGVMFLQ ISLTENWLIF VTRAAGPFWS SIPSW..QLA
Pma1klula .KGGIIQ.NF GSIDGVLFLQ ISLTENWLIF ITRAAGPFWS SIPSW..QLS
Pma1canal .KGGIIQ.NF GGLDGILFLQ ISLTENWLIF VTRAQGPFWS SIPSW..QLS
Pma1zygro .KGGIIQ.NF GSIDGILFLE ISLTENWLIF ITRAVGPFWS SIPSW..QLA
Atcphomsa KFF....... ........DI DSGRNAPLHA PPSEHYTIVF
Atcqhomsa KMF....... ......QI DSGRNAPLHS PPSEHYTIIF
Atcrhomsa KMF....... ........DI DSGRKAPLHS PPSQHYTIVF
Atc3sacce ELF....... ...FKKHEDEI TSHQQQQLNA .......MTF
Atmaescco QTLFQSGWFV VGLLSQTLIV HM......I RTRRVPFIQS CASWPLMIMT
Atmbsalty QALFQSGWFI EGLLSQTLVV HM......L RTQKIPFIQS RATLPVLLTT
Atc3schpo MTWVVIMYGF GTGNLSYDCN AHYHAGCNDV FKARSAVFAV VTFCILIMAV
```

```
Atn1sacce ASFTGSLYGI NSGRLGHDCD GTYNSSCRDV YRSRSAAFAT MTWCALILAW
Atn3homsa GFLPGNLVGI RLNWDDRTVN DLEDSYGQQW TYEQRKVVEF TCH...TAFF
Atn3ratno GFLPGNLVGI RLNWDDRTVN DLEDSYGQQW TYEQRKVVEF TFH...TAFF
Atn3galga GFLPSCLVGI RLSWDDRTIN DLEDSYGQQW TYEQRKVVEF TCH...TAFF
Atn1homsa GFLPIHLLGL RVDWDDRWIN DVEDSYGQQW TYEQRKIVEF TCH...TAFF
Atn2homsa GFLPSRLLGI RLDWDDRTMN DLEDSYGQEW TYEQRKVVEF TCH...TAFF
Atn3sussc GFLPIHLLGL RVNWDDRWIN DVEDSYGQQW TYEQRKIVEF TCH...TAFF
Atnacatco GFLPPRLLGI RMNWDDKYIN DLEDSYGQQW TYEQRKIVEF TCH...TAFF
Atnatorca GFLPIDLIGI REKWDELWTQ DLEDSYGQQW TYEQRKIVEY TCH...TSFF
Atnaartsf GFLPWDLFGL RKHWDSRAVN DLTDSYGQEW TYDARKQLES SCH...TAYF
Atnadrome GFLPKKLFGI RKMWDSKAVN DLTDSYGQEW TYRDRKTLEY TCH...TAFF
Atnaartsa GFLPNRLFGL RKWWESKAYN DLTDSYGQEW TWDARKQLEY TCH...TAFF
Atnahydat GFLPSYLFGL RSQWDDMSNN NLLDSFGSEW TYFQRKEIEL TCQ...TAFF
Athahomsa GWFPLLCVGL RAQWEDHHLQ DLQDSYGQEW TFGQRLYQQY TCY...TVFF
Atcartsf  G...LNFYQL SHHLQCTPEN E.......YF EGIDCEIFSD P.H.PMTMAL
Atcbdrome K...LSYWQL THHLSCLGGG D.......EF KGVDCKIFSD P.H.AMTMAL
Atcborycu G...VTYHQL THFMQCTEDH P.......HF EGLDCEIFEA P.E.PMTMAL
Atcdhomsa R...VSFYQL SHFLQCKEDN P.......DF EGVDCAIFES P.Y.PMTMAL
Atcfratno Q...VTFHQL RNFLKCSEDN P.......LF AGIDCEVFES R.F.PTTMAL
Atctrybr  ....FSWHDL TTYTAC... S.......DM TNGTCLLLAN PQT.ARAIAL
Atcplafa  MHTLINFYQL SHYNQCKAWN NFRVNKVYDM SEDHCSYFSA GKIKASTLSL
Atalsynsp .................. ...ILIFGMF EWVNRTYDDL ALAR..TMAI
Atclsynsp .................. ...VLMVIAY QYTQVPLPGL DPKRWQTMVF
Atclsacce .................. ...YIFV... ..KEMAEDGK VTARDTTMTF
Atclmycge .....LFYSL GIIYAINNRD LQTSGDLINR AGSTCGFF.. ..........
Consensus .................. ........ .......... ..........
```

```
                1501                                               1550
Atxaleido GAIISLLVST MAASFWHKSR PDNVLTEGLA WGQTNAEKLL PLWVWIYCIV
Atxbleido GAIISLLVST MAASFWHKSR PDNVLTEGLA WGQTNAEKLL PLWVWIYCIV
Pma1lyces FAFFVAQLVA TLIAVYANWS FAAI...... ...EGIGWGW AGVIWLYNIV
Pma1nicpl IAFVIAQLVA TLIAVYANWS FAAI...... ...EGIGWGW AGVIWIYNLV
Pma1arath IAFVIAQLVA TLIAVYADWT FAKV...... ...KGIGWGW AGVIWIYSIV
Pma4nicpl IAFMIAQLVA TLIAVYANWA FARV...... ...KGCGWGW AGVIWLYSII
Pma3arath IAFWVAQLIA TAIAVYGNWE FARI...... ...KGIGWGW AGVIWLYSIV
Pma1schpo GAVLAVDILA TMFCIFGWFK GGHQ...... ...TSI..VA VLRIWMYSFG
Pma2schpo GAVLVVDILA TLFCIFGWFK GGHQ...... ...TSI..VA VIRIWMYSFG
Pma1ajeca GAILLVDIIA TLFTIFGWFV GGQ....... ...TSI..VA VVRIWVFSFG
Pma1neucr GAIFLVDILA TCFTIWGWFE HSD....... ...TSI..VA VVRIWIFSFG
Pma1sacce GAVFAVDIIA TMFTLFGWWS ENW....... ...TDI..VT VVRVWIWSIG
Pma2sacce GAVFAVDIIA TMFTLFGWWS ENW....... ...TDI..VS VVRVWIWSIG
Pma1klula GAVLIVDIIA TMFCLFGWWS QNW....... ...NDI..VT VVRVWIFSFG
Pma1canal GAVLIVDIIA TCFTLFGWWS QNW....... ...TDI..VT VVRTWIWSFG
Pma1zygro GAVFVVDVVA TMFTLFGWWS QNW....... ...TDI..VT VVRIYIWSIG
Atcphomsa NTFVLMQLFN EINARKIHGE R.......... .....NVFEG IFNNAIFCTI
Atcqhomsa NTFVMMQLFN EINARKIHGE R.......... .....NVFDG IFRNPIFCTI
Atcrhomsa NTFVLMQLFN EINSRKIHGE K.......... .....NVFSG IYRNIIFCSV
Atc3sacce NTFVWLQFFT MLVSRKLDEG DGISNWRGRI SAANLNFFQD LGRNYYFLTI
Atmaescco VIVMIVGIAL PFSPLASYLQ .........L QALP....LS YFPWLVAILA
Atmbsalty GLIMAIGIYI PFSPLGAMVG .........L EPLP....LS YFPWLVATLL
Atc3schpo EVKNFDNSLF NLHGIPWGEW .........N FR...YFLHT LVENKFLAWA
Atn1sacce EVVDMRRSFF RMH..PDTDS .........P VK...EFFRS IWGNQFLFWS
Atn3homsa VSIVVVQWAD LIICKTRRNS .........V FQ.....QG .MKNKILIFG
Atn3ratno VSIVVVQWAD LIICKTRRNS .........V FQ.....QG .MKNKILIFG
```

```
Atn3galga VSIVVVQWAD LIICKTRRNS .........V FQ......QG .MKNKILIFG
Atn1homsa VSIVVVQWAD LVICKTRRNS .........V FQ.....QG .MKNKILIFG
Atn2homsa ASIVVVQWAD LIICKTRRNS .........V FQ.....QG .MKNKILIFG
Atn3sussc VSIVVVQWAD LVICKTRRNS .........V FQ.....QG .MKNKILIFG
Atnacatco TSIVIVQWAD LIICKTRRNS .........V FQ.....QG .MKNKILIFG
Atnatorca VSIVIVQWAD LIICKTRRNS .........I FQ.....QG .MKNKILIFG
Atnaartsf VSIVIVQWAD LIISKTRRNS ..........V FQ.....QG .MRNNILNFA
Atnadrome ISIVVVQWAD LIICKTRRNS .........I FQ.....QG .MRNWALNFG
Atnaartsa ISIVIVQWTD LIICKTRRLS .........L FQ.....QG .MKNGTLNFA
Atnahydat TTIVVVQWAD LIISKTRRLS .........L FQ.....QG .MTNWFLNFG
Athahomsa ISIEVCQIAD VLIRKTRRLS .........A FQ.....QG FFRNKILVIA
Atcartsf  SVLVTIEMLN AINSLSENQS .........L LVMPP..... .WSNIWLISA
Atcbdrome SVLVTIEMLN AMNSLSENQS .........L ITMPP..... .WCNLWLIGS
Atcborycu SVLVTIEMCN ALNSLSENQS .........L MRMPP..... .WVNIWLLGS
Atcdhomsa SVLVTIEMCN ALNSLSENQS .........L LRMPP..... .WENIWLVGS
Atcfratno SVLVTIEMCN ALNSVSENQS .........L LRMPP..... .WLNPWLLGA
Atctrybr  SILVVVEMLN ALNALSENAS .........L IVSRP..... .SSNVWLLFA
Atcplafa  SVLVLIEMFN ALNALSEYNS .........L FEIPP..... .WRNMYLVLA
Atalsynsp QALVAARVIY LLSISQLGRS .........F LGYVTGKRQT ITKASILLLG
Atclsynsp TTLCLAQMGH AIAVR...SD .........L LTIQTPMR...TNPWLWLS
Atc1sacce TCFVFFDMFN ALACRHNTKS .........I FEI......G FFTNKMFNYA
Atclmycge .ILGASAALN SLNLMVDKPL .........L MTNP...... WFFKLVWIGS
Consensus ........................................................
```

```
          1551                                              1600
Atxaleido WWFVQDVVKV LAHICMDAVD LFGCVSDASG SGPIKPYSDD MKVNGFEPVK
Atxbleido WWFVQDVVKV LAHICMDAVD LFGCVSDASG SGPIKPYSDD MKVNGFEPVK
Pma1lyces TYIPLDLIKF LIRYALSGKA WDLVLEQRIA FTRKKDFGKE L..RELQWAH
Pma1nicpl FYIPLDIIKF FIRYALSGRA WDLVFERRIA FTRKKDFGKE Q..RELQWAH
Pma1arath TYFPQDILKF AIRYILSGKA WASLFDNRTA FTTKKDYGIG E..REAQWAQ
Pma4nicpl FYLPLDIMKF AIRYILSGKA WNNLLDNKTA FTTKKDYGKE E..REAQWAL
Pma3arath FYFPLDIMKF AIRYILAGTA WKNIIDNRTA FTTKQNYGIE E..REAQWAH
Pma1schpo IFCIMAGTYY ILS...ESAG FDRMMNGK.P KESRNQRSIE DLVVALQRTS
Pma2schpo IFCLIAGVYY ILS...ESSS FDRWMHGK.H KERGTTRKLE DFVMQLQRTS
Pma1ajeca CFCVLGGLYY LLQ...GSAG FDNMMHGKSP KKNQKQRSLE DFVVSLQRVS
Pma1neucr IFCIMGGVYY ILQ...DSVG FDNLMHGKSP KGNQKQRSLE DFVVSLQRVS
Pma1sacce IFCVLGGFYY EMS...TSEA FDRLMNGKPM KEKKSTRSVE DFMAAMQRVS
Pma2sacce IFCVLGGFYY IMS...TSQA FDRLMNGKSL KEKKSTRSVE DFMAAMQRVS
Pma1klula VFCVMGGAYY MMS...ESEA FDRFMNGKSR RDKPSGRSVE DFLMAMQRVS
Pma1canal VFCVMGGAYY LMS...TSEA FDNFCNGRKP QQHTDKRSLE DFLVSMQRVS
Pma1zygro IFCCLGGAYY LMS...ESET FDRLMNGKPL KENKSTRSVE DFLASMRRVS
Atcphomsa VLGTFVVQII IVQFGGKPFS CSELSIEQWL WSIFLGMGTL LWGQLISTIP
Atcqhomsa VLGTFAIQIV IVQFGGKPFS CSPLQLDQWM WCIFIGLGEL VWGQVIATIP
Atcrhomsa VLGTFICQIF IVEFGGKPFS CTSLSLSQWL WCLFIGIGEL LWGQFISAIP
Atc3sacce MAIIGSCQVL IMFFGGAPFS IARQTKSMWI TAVLCGMLSL IMGVLVRICP
Atmaescco GYMTLTQLVK GFYSRRYGWQ .........................
Atmbsalty SYCLVAQGMK RFYIKRFGQW F........................
Atc3schpo IALAAVSVFP TIYIPVINRD VFKHTYIGWE WGVVA..... ...........
Atn1sacce IIFGFVSAFP VVYIPVINDK VFLHKPIGAE WGLAI..... ...........
Atn3homsa LFEETALAAF LSYCPGMDVA LRMYPL.... ...... ....KPSWW
Atn3ratno LFEETALAAF LSYCPGMDVA LRMYPL.... ...... ....KPSWW
Atn3galga LFEETALAAF LSYCPGMDVA LRMYPL.... ...... ....KPSWW
Atn1homsa LFEETALAAF LSYCPGMGVA LRMYPL.... ...... ....KPTWW
Atn2homsa LLEETALAAF LSYCPGMGVA LRMYPL.... ...... ....KVTWW
```

```
Atn3sussc LFEETALAAF LSYCPGMGVA LRMYPL.... ............KPTWW
Atnacatco LFEETALAAF LSYCPGMDVA LRMYPL.... ............KPNWW
Atnatorca LFEETALAAF LSYTPGTDIA LRMYPL.... ............KPSWW
Atnaartsf LVFETCLAAF LSYTPGMDKG LRMYPL.... ............KINWW
Atnadrome LVFETVLAAF LSYCPGMEKG LRMYPL.... ............KLVWW
Atnaartsa LVFETCVAAF LSYTPGMDKG LRMYPL.... ............KIWWW
Atnahydat LFFETALAAF LQYTPGVNTG LRLRPM.... ............NFTWW
Athahomsa IVFQVCIGCF LCYCPGMPNI FNFMPI.... ............RFQWW
Atcartsf  ICLSMTLHFV ILYVEILSTV FQICPL.... ............TLTEW
Atcbdrome MALSFTLHFV ILYVDVLSTV FQVTPL.... ............SAEEW
Atcborycu ICLSMSLHFL ILYVDPLPMI FKLKAL.... ............DLTQW
Atcdhomsa ICLSMSLHFL ILYVEPLPLI FQITPL.... ............NVTQW
Atcfratno VVMSMALHFL ILLVPPLPLI FQVTPL.... ............SGRQW
Atctrybr  IFSSLSLHLI IMYVPFFAKL FNIVPLGVDP HVVQQAQPWS ILTPTNFDDW
Atcplafa  TIGSLLLHVL ILYIPPLARI FGVVPL.... ............SAYDW
Atalsynsp IAVAIALQIG FSQLPFMNVL FKTAPM.... ............DWQQW
Atclsynsp VIVTALLQLA LVYVVSPLQKF FGTHSL.... ............SQLDL
Atc1sacce VGLSLLGQMC AIYIPFFQSI FKTEKL.... ............GISDI
Atclmycge LA.SILVFLL IIFINPLGLV FNVLQ..... ............DLTNH
Consensus .......... .......... .......... .......... ..........


               1601                                            1650
Atxaleido KPAEKSTEKA LNSSVSSASH KALEGLREDT HSPIEEASPV NVYVSRDQK.
Atxbleido KPAEKSTEKA LNLSVSSGPH KALEGLREDT HVLNESTSPV NAFSPKVKK.
Pma1lyces AQRTLHGLQV PD.PKIFSET TNFNELNQLA EEAKRRAEIA RLRELHTLKG
Pma1nicpl AQRTLHGLQV PD.TKLFSEA TNFNELNQLA EEAKRRAEIA RLRELHTLKG
Pma1arath AQRTLHGLQP KEDVNIFPEK GSYRELSEIA EQAKRRAEIA RLRELHTLKG
Pma4nicpl AQRTLHGLQP PEATNLFNEK NSYRELSEIA EQAKRRAEMA RLRELHTLKG
Pma3arath AQRTLHGLQN TETANVVPER GGYRELSEIA NQAKRRAEIA RLRELHTLKG
Pma1schpo TRHEKGDA.. .......... .......... .......... ..........
Pma2schpo THHEAEGKVT S......... .......... .......... ..........
Pma1ajeca TQHEKSS... .......... .......... .......... ..........
Pma1neucr TQHEKSQ... .......... .......... .......... ..........
Pma1sacce TQHEKET... .......... .......... .......... ..........
Pma2sacce TQHEKSS... .......... .......... .......... ..........
Pma1klula TQHEKEN... .......... .......... .......... ..........
Pma1canal TQHEKST... .......... .......... .......... ..........
Pma1zygro TQHEKGN... .......... .......... .......... ..........
Atcphomsa TSRLKFLKEA GHGTQKEEIP EEELAEDVEE IDHAERELRR GQILWFRGLN
Atcqhomsa TSRLKFLKEA GRLTQKEEIP EEELNEDVEE IDHAERELRR GQILWFRGLN
Atcrhomsa TRSLKFLKEA GHGTTKEEIT KD..AEGLDE IDHAEMELRR GQILWFRGLN
Atc3sacce .......... .....DEVA VKVFPAAFVQ .......... .RFKYVFGLE
Atc3schpo VAVMFYFFYV EIWKSIRRSL TNPQKKGKFR RTL....... .......SNT
Atn1sacce AFTIAFWIGA ELYKCGKRRY FKTQRAHNPE NDLESNNKRD PFEAYSTSTT
Atn3homsa FCAFPYSFLI FVYDEIRKLI LRRNPGGWVE KETYY..... ..........
Atn3ratno FCAFPYSFLI FVYDEIRKLI LRRNPGGWVE KETYY..... ..........
Atn3galga FCAFPYSFLI FVYDEIRKLI LRRNPGGWVE KETYY..... ..........
Atn1homsa FCAFPYSLLI FVYDEVRKLI IRRRPGGWVE KETYY..... ..........
Atn2homsa FCAFPYSLLI FIYDEVRKLI LRRYPGGWVE KETYY..... ..........
Atn3sussc FCAFPYSLLI FVYDEVRKLI IRRRPGGWVE KETYY..... ..........
Atnacatco FCAFPYSLLI FIYDEIRKLI LRRNPGGWME RETYY..... ..........
Atnatorca FCAFPYSLII FLYDEARRFI LRRNPGGWVE QETYY..... ..........
Atnaartsf FPALPFSFLI FVYDEARKFI LRRNPGGWVE QETYY..... ..........
Atnadrome FPAIPFALAI FIYDETRRFY LRRNPGGWLE QETYY..... ..........
```

```
Atnaartsa FPPMPFSLLI LVYDECRKFL MRRNPGGFLE RETYY............
Atnahydat LPGLPFSLLI FVYDEIRRYL LRKNPGGWVE KETYY............
Athahomsa LVPLPYGILI FVYDEIRKLG VRCCPGSWWD QELYY............
Atcartsf  IVVLKISFPV LLL....DEV LKFVARKYTD EFSFIK........
Atcbdrome ITVMKFSIPV VLL....DET LKFVARKIAD VPDVVVDRM.....
Atcborycu LMVLKISLPV IGL....DEI LKFIARNYLE G.............
Atcdhomsa LMVLKISLPV ILM....DET LKFVARNYLE PAILE.........
Atcfratno GVVLQMSLPV ILL....DEA LKYLSRHHVD EKKDLK........
Atctrybr  KAVIVFSVPV IFL....DEL LKFITRRMEK AQEKKKD......
Atcplafa  FLVFLWSFPV IIL....DEI IKFYAKRKLK EEQRTKKIKI D.....
Ata1synsp AICLLPMIPM VPV....RIL ANRLDP............
Atc1synsp AIC.LGFSLL LFV....YLE AEKWVRHGRY ............
Atc1sacce LLLLLISSSV FIV....DEL RKLWTRKKNE EDSTYFSNV. .....
Atc1mycge PVLISYSFGG VILYMGMNEV VKLIRLGYGN I..........
Consensus ....................
```

```
                 1651                                    1700
Pma1lyces HVESVVKLKG LDIETIQQSY TV.................
Pma1nicpl HVESVVKLKG LDIETIQQAY TV.................
Pma1arath HVESVAKLKG LDIDTAGHHY TV.................
Pma4nicpl HVESVVKLKG LDIETIQQHY TV.................
Pma3arath HVESVVKLKG LDIETAG.HY TV.................
Atcphomsa RIQTQIRVVN AFRSSLYEGL EKPESRSSIH NFMTHPEFRI EDSEPHIPLI
Atcqhomsa RIQTQIRVVK AFRSSLYEGL EKPESRTSIH NFMAHPEFRI EDSQPHIPLI
Atcrhomsa RIQTQIKVVK AFHSSLHESI QKPYNQKSIH SFMTHPEFAI EEELPRTPLL
Atc3sacce FLRKNHTGKH DDEEALLEES DSPESTAFY. .............
Atc3schpo ITTESKLSEK DLEHRLFLQS RRA........
Atn1sacce IHTEVNIGIK Q.............
Atceorycu EGVSWPFVLL IVPLVMWVYS TDTNFSDLLW S...........
Atceratno DGISWPFVLL IMPLVVWVYS TDTNFSDMFW S...........
Atcehomsa DGISWPFVLL IMPLVIWVYS TDTNFSDMFW S...........
Atcesussc DGISWPFVLL IMPLVIWVYS TDTNFSDMFW S...........
Consensus ....................
```

```
                 1701                                    1750
Atcphomsa DDTDAEDDAP TKR........ .NSSPPPSPN KNNNAVDSGI HLTIEMNKSA
Atcqhomsa DDTDLEEDAA LKQ........ .NSSPPSSLN KNNSSIDSGI NLTTDTSKSA
Atcrhomsa DEEEEENPDK ASKFGTRVLL LDGEVTPYAN TNNNAVDCN. ..QVQLPQS.
Consensus ....................
```

```
                 1751        1766
Atcphomsa TSSSPGSPLH SLETSL
Atcqhomsa TSSSPGSPIH SLETSL
Atcrhomsa .....DSSLQ SLETSV
Consensus ..............
```

Proteins listed subsequently in italics are at least 90% identical to the paired transporters listed in parenthesis and are therefore not included in the alignment: *Atcbgalga, Atcaorycu* (Atcborycu); *Atcdfelca, Atcdsussc, Atcdorycu, Atceorycu, Atcdratno, Atceratno, Atcehomsa, Atcesussc* (Atcdhomsa); *Atcporycu, Atcpratno, Atcpsussc* (Atcphomsa); *Atcqratno* (Atcqhomsa); *Athasussc, Athaorycu, Atharatno* (Athahomsa); *Atmasalty* (Atmaescco); *Atn1bufma, Atn1galga, Atn1oviar, Atn1equca, Atn1sussc, Atn1ratno*

(Atn1homsa); *Atn2sacce* (Atn1sacce); *Atn2galga, Atn2ratno* (Atn2homsa); *Pma2arath* (Pma1arath); *Pma3nicpl* (Pma1nicpl). Residues listed in the consensus sequence are present in at least 75% of the aligned transporter sequences. Residues indicated in boldface type are also conserved in at least one other family of the P-type ATPase superfamily.

Database accession numbers

	SWISSPROT	PIR	EMBL/GENBANK
Ata1synsp	P37367	S40440; S33207	X71022; G296568
Atcartsf	P35316	S07526	X51674; G665604
Atcplafa	Q08853		X71765; G402222
Atctrybr	P35315	A45598	M73769; G162201
Atc1sacce	P13586	S05787; PWBYR1	M25488; G172199
Atc3schpo	P22189	A36096	J05634; G173355
Atc3sacce	P38929		U03060; G454003
Atcaorycu	P04191	A01075; PWRBFC	M12898; G164779
Atcbdrome	P22700	A36691; S07050	M62892; G158416
Atcbgalga	P13585	A32792	M26064; G211224
Atcborycu	P11719		M12898
Atcdfelca	Q00779	S23444	Z11500; G1081
Atcdhomsa	P16614	B31981	M23115; G306851
Atcdorycu	P04192	A01076; PWRBSC	X02814; G1469
Atcdratno	P11508	B31982; S04269	J04023; G203059
Atcdsussc	P11606	S04651	X15073; G1921
Atcehomsa	P16615	A31981	M23114; G306850
Atcesussc	P11607	S04652	X15074; G1923
Atceorycu	P20647	S10335; PWRBMC	J04703; G164739
Atceratno	P11507	A31982	J04022; G203057
Atcfratno	P18596	A34307	M30581; G206899
Atclmycge	P47317		U39687; G1045747
Atclsynsp	P37278	S36742	D16436; G435123
Atcphomsa	P20020	A30802	J04027; G190133
Atcporycu	Q00804	S17179	X59069; G1675
Atcpratno	P11505	A28065	J03753; G203047
Atcpsussc	P23220	S13057	X53456; G2061
Atcqhomsa	Q01814	A38871	L20977; G404702
Atcqratno	P11506	B28065	J03754; G203049
Atcrhomsa	P23634	A35547	M25874; G179163
Athahomsa	P20648	A35292; A36558	J05451; G561634
Athaorycu	P27112	S23406	X64694; G1471
Atharatno	P09626	A25344	J02649; G203037
Athasussc	P19156	A31671; A24228	M22724; G164384
Atmaescco	P39168		U14003; G537084
Atmasalty	P36640		U07843; G468207
Atmbsalty	P22036	B39083	M57715; G397973
Atn1bufma	P30714	S24650; A43451	Z11798; G62492
Atn1equca	P18907	S04630	X16773; G871026
Atn1galga	P09572	A28199	J03230; G211220
Atn1ratno	P06685	A24639; S00460	D10359; G220824
Atn1sussc	P05024	B24862	X03938; G1898
Atn1oviar	P04074	A01074; PWSHNA	X02813; G1206
Atn1homsa	P05023	A24414	D00099; G219942
Atn1sacce	P13587	S05788; PWBYR2	U24069; G790261
Atn2galga	P24797		M59959; G212406
Atn2homsa	P50993		J05096; G179165
Atn2ratno	P06686	B24639	M14512; G203029
Atn2sacce	Q01896	S25007	X67136; G5513
Atn3galga	P24798	B37227	M59960; G212408

	SWISSPROT	PIR	EMBL/GENBANK
Atn3homsa	P13637	S00801	M37457; G497763
Atn3sussc	P18874		M38445; G164382
Atn3ratno	P06687	C24639	M14513; G203031
Atnaartsa	P17326	S06635	Y07513; G5670
Atnaartsf	P28774	JH0470	X56650; G10934
Atnacatco	P25489	S14740; PWCCNM	X58629; G62642
Atnadrome	P13607	S03632	X14476; G732656
Atnahydat	P35317		M75140; G159258
Atnatorca	P05025	S00503	X02810; G64400
Atxaleido	P11718	A27124; PXLNPD	M17889; G159294
Atxbleido	P12522		J04004; G159295
Pma1ajeca	Q07421		L07305; G409249
Pma1arath	P20649	A32326; PXMUP1	M24107; G166746
Pma1canal	P28877	A41336; PXCKP	M74075; G170818
Pma1klula	P49380		L37875; G598435
Pma1lyces	P22180	A45506	M60166; G170464
Pma1neucr	P07038	A26497; PXNCP	M14085; G168761
Pma1nicpl	Q08435	A41779	M80489; G170289
Pma1schpo	P09627	A28454; PXZP1P	J03498; G173429
Pma1sacce	P05030	A25823; PXBY1P	X03534; G4187
Pma1zygro	P24545	JX0181; PXKZP	D10764; G218531
Pma2arath	P19456	A37116; PXMUP2	J05570; G166629
Pma2sacce	P19657	A32023; PXBY2P	J04421; G295644
Pma2schpo	P28876	A40945; PXZP2P	M60471; G173431
Pma3arath	P20431	A33698; PXMUP3	J04737; G166625
Pma3nicpl	Q08436		M80490; G170295
Pma4nicpl	Q03194	S24959; S33548	X66737; G19704

References

1. Lytton, J. and MacLennan, D.H. (1988) J. Biol. Chem. 263, 15024–15031.
2. Harper, J.F. et al. (1989) Proc. Natl Acad. Sci. USA 86, 1234–1238.
3. Maeda, M. et al. (1990) J. Biol. Chem. 265, 9027–9032.
4. **Sussman, M.R. (1994) Annu. Rev. Plant Physiol. Plant Mol. Biol. 45, 211–234.**
5. **Assmann, S.M. and Haubrick, L.L. (1996) Curr. Opin. Cell Biol. 8, 458–467.**
6. **Green, N.M. and MacLennan, D.H. (1989) Biochem. Soc. Trans. 17, 819–822; Green, N.M. (1989) Biochem. Soc. Trans. 17, 970–972.**
7. **Fagan, M.J. and Saier, M.H. Jr. (1994) J. Mol. Evol. 38 57–99.**
8. Rudolph, H.K. et al. (1989) Cell 58, 133–145.

Summary

Transporters of the heavy metal-transporting ATPase family, examples of which are heavy-metal transporting P-type ATPases from bacteria such as *Enterococcus* (Atkaentfa)[1] and human copper-transporting ATPase 1[2] (At7ahomsa), mediate active transport of heavy metal ions driven by ATPase activity. Where the natural substrate is known it is usually divalent copper or cadmium. The nitrogen fixation protein FIXI from *Rhizobium meliloti*[3] is also a member of this family. In humans, mutations in copper-transporting ATPases cause hereditary Menkes' disease (Cu-transporting ATPase 1[2]) and Wilson's disease (Cu-transporting ATPase 2[4]). Members of the heavy metal-transporting ATPase family have a broad biological distribution that includes gram-positive and gram-negative bacteria, yeast and humans. Heavy metal-transporting ATPases from bacteria may be chromosomal or plasmid-encoded.

Statistical analysis of multiple amino acid sequence comparisons places the heavy metal-transporting ATPase family in the P-type ATPase superfamily (also known as E1-E2 ATPases[5,6]). Proteins in this superfamily use the energy of ATP hydrolysis to pump ions across cell membranes. P-Type ATPases are all predicted to contain at least six transmembrane helices by the hydropathy of their amino acid sequences. They have two large cytoplasmic loops separating three pairs of transmembrane helices; the larger of these loops contains the ATP binding domain. The sequences are usually extended by one or two more pairs of helices[5]. Members of the heavy metal-transporting ATPase family are predicted to contain eight transmembrane helices[7]. They also have an N-terminal cytoplasmic domain which contains one or more repeats of a sequence associated with heavy metal binding, the HMA sequence[7]. In the human copper-transporting proteins[2,4] this domain contains six tandem HMA sequences. Eukaryotic proteins may be glycosylated.

A few short sequence motifs are very highly conserved within the heavy metal-transporting ATPase family of transporters, including motifs unique to the family and signature motifs of the P-type ATPase superfamily.

Nomenclature, biological sources and substrates

CODE	DESCRIPTION [SYNONYMS]	ORGANISM [COMMON NAMES]	SUBSTRATE(S)
At7ahomsa	Copper-transporting ATPase 1 [Copper pump 1, Menkes' disease-associated protein, ATP7A, MNK, MC1]	*Homo sapiens* [human]	Cu^{2+}
At7acrigr	Copper-transporting ATPase 1	*Cricetulus griseus* [hamster]	Cu^{2+}
At7bhomsa	Copper-transporting ATPase 2 [Copper pump 2, Wilson's disease-associated protein, ATP7B, WND, PWD, WC1]	*Homo sapiens* [human]	Cu^{2+}
Atc2sacce	Probable calcium-transporting ATPase [PCA1, YBR295W, YBR2112]	*Saccharomyces cerevisiae* [yeast]	Ca^{2+}

CODE	DESCRIPTION [SYNONYMS]	ORGANISM [COMMON NAMES]	SUBSTRATE(S)
Atcssynsp	Cation-transporting ATPase [PACS]	*Synechococcus* sp. [cyanobacterium]	Metal ions
Atkaentfa	Potassium/copper-transporting ATPase A [ATKA]	*Enterococcus faecalis* [gram-positive bacterium]	Cu^{2+}, K^+
Atkbentfa	Potassium/copper-transporting ATPase A [ATKB]	*Enterococcus faecalis* [gram-positive bacterium]	Cu^{2+}, K^+
Atsyescco	Probable copper-transporting ATPase	*Escherichia coli* [gram-negative bacterium]	Cu^{2+}
Atsysynsp	Probable copper-transporting ATPase [SYNA]	*Synechococcus* sp. [cyanobacterium]	Cu^{2+}
Atu1sacce	Probable copper-transporting ATPase [Cu^{2+}-ATPase, CCC2]	*Saccharomyces cerevisiae* [yeast]	Cu^{2+}
Cadabacfi	Probable cadmium-transporting ATPase [Cadmium efflux ATPase, CADA]	*Bacillus firmus* [gram-positive bacterium]	Cd^{2+}
Cadastaau	Probable cadmium-transporting ATPase [Cadmium efflux ATPase, CADA]	*Staphylococcus aureus* [gram-positive bacterium]	Cd^{2+}
Caddstaau	Probable cadmium-transporting ATPase [Cadmium efflux ATPase, CADA]	*Staphylococcus aureus* [gram-positive bacterium]	Cd^{2+}
Ctpbraja	P-Type ATPase	*Bradyrhizobium japonicum* [gram-negative bacterium]	Metal ions
Ctppromi	Heavy metal-transporting P-type ATPase	*Proteus mirabilis* [gram-negative bacterium]	Metal ions
Ctpamycle	Cation-transporting P-type ATPase A [CTPB]	*Mycobacterium leprae* [gram-negative bacterium]	Mg^{2+}
Ctpbmycle	Cation-transporting P-type ATPase A [CTPB]	*Mycobacterium leprae* [gram-negative bacterium]	Mg^{2+}
Fixirhime	Nitrogen fixation protein [FIXI]	*Rhizobium meliloti* [gram-negative bacterium]	Metal ions

Phylogenetic tree

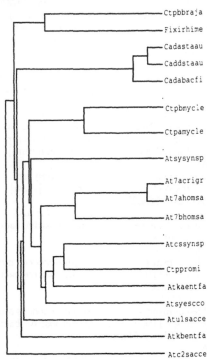

Proposed orientation of AT7A ^{2} in the membrane

The model is based on predictions of membrane-spanning regions and α-helical content. The N-terminus of the protein is illustrated on the inside and is folded eight times through the membrane. The predicted membrane-spanning helices are portrayed as rectangles. The numbers corresponding to the first and last residue of each membrane-spanning helix are boxed. Residues that are conserved in more than 75% of the aligned transporters (see below) are shown.

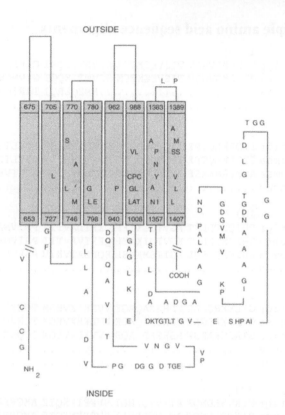

Physical and genetic characteristics

	AMINO ACIDS	MOL. WT	EXPRESSION SITES	CHROMOSOMAL LOCUS
At7ahomsa	1500	163 334	endothelial cells	Xq13.3
At7acrigr	1476	160 335		
At7bhomsa	1443	154 776	liver, kidneys	13q14.3
Atc2sacce	1216	131 838		
Atcssynsp	747	79 732		
Atkaentfa	727	78 388		copAB operon
Atkbentfa	745	81 522		copAB operon
Atsyescco	834	87 782		
Atsysynsp	790	83 694		
Atu1sacce	1004	109 828		Chromosome 4
Cadabacfi	723	78 207		
Cadastaau	727	78 811		
Caddstaau	804	86 882		
Ctpbraja	730	77 337		
Ctppromi	829	87 859		
Ctpamycle	780	82 384		
Ctpbmycle	750	78 224		
Fixirhime	757	79 559		

Multiple amino acid sequence alignments

```
            1                                                      50
At7acrigr   MEPSMDVNSV TISVEGMTCI SCVRTIEQKI GKENGIHHIK VSLEEKSATI
At7ahomsa   MDPSMGVNSV TISVEGMTCN SCVWTIEQQI GKVNGVHHIK VSLEEKNATI
At7bhomsa   .......... .......... ..MPEQERQI TAREGASRKI LS.KLSLPTR
Consensus   .......... .......... .......... .......... ..........

            51                                                    100
At7acrigr   IYDPKLQTPK TLQEAIDDMG FDALLHNANP LPVLTDTLFL TVTASLTLPW
At7ahomsa   IYDPKLQTPK TLQEAIDDMG FDAVIHNPDP LPVLTDTLFL TVTASLTLPW
At7bhomsa   AWEPAMKKSF AFDNVGYEGG LDGLGPSSQV ATSTVRILGM TCQSCV....
Consensus   .......... .......... .......... .......... ..........

            101                                                   150
At7acrigr   DHIQSTLLKT KGVTDIKIFP QKRTLAVTII PSIVNANQIK ELVPELSLET
At7ahomsa   DHIQSTLLKT KGVTDIKIYP QKRTVAVTII PSIVNANQIK ELVPELSLDT
At7bhomsa   KSIEDRISNL KGIISMKVSL EQDSATVKYV PSVVCLQQVC HQIGDMGFEA
Consensus   .......... .......... .......... .......... ..........

            151                                                   200
At7acrigr   GTLEKRSGAC EDHSMAQAGE VVLKIKVEGM TCHSCTSTTE GKIGKLQGVQ
At7ahomsa   GTLEKKSGAC EDHSMAQAGE VVLKMKVEGM TCHSCTSTIE GKIGKLQGVQ
At7bhomsa   SIAEGKAASW PSRSLP.AQE AVVKLRVEGM TCQSCVSSIE GKVRKLQGVV
Atc2sacce   .......... .......... .......... .......... MKPEKLFSGL
Consensus   .......... .......... .......... .......... ..........

            201                                                   250
At7acrigr   RIKVSLDNQE ATIVYQPHLI SVEEIKKQIE AMGFPAFVKK QPKYLKLGAI
At7ahomsa   RIKVSLDNQE ATIVYQPHLI SVEEMKKQIE AMGFPAFVKK QPKYLKLGAI
At7bhomsa   RVKVSLSNQE AVITYQPYLI QPEDLRDHVN DMGFEAAIKS KVAPLSLGPI
Atc2sacce   GTSDGEYGVV NSENISIDAM QDNRGECHRR SIEMHANDNL GLVSQRDCTN
Consensus   .......... .......... .......... .......... ..........

            251                                                   300
At7acrigr   DVERLKNT.. ..PVKSLEGS QQR.PSYPSD S....TATFI IEGMHCKSCV
At7ahomsa   DVERLKNT.. ..PVKSSEGS QQRSPSYTND S....TATFI IDGMHCKSCV
At7bhomsa   DIERLQSTNP KRPLSSANQN FNNSETLGHQ GSHVVTLQLR IDGMHCKSCV
Atc2sacce   RPKITPQECL SETEQICHHG ENRTKAGLDV DDAETGGDHT NESRVDECCA
Consensus   .......... .......... .......... .......... ..........

            301                                                   350
At7acrigr   SNIESALPTL QYVSSIAVSL ENRSAIVKYN ASSVTPEMLI KAIEAVSPGQ
At7ahomsa   SNIESTLSAL QYVSSIVVSL ENRSAIVKYN ASSVTPESLR KAIEAVSPGL
At7bhomsa   LNIEENIGQL LGVQSIQVSL ENKTAQVKYD PSCTSPVALQ RAIEALPPGN
Atc2sacce   EKVNDTETGL DVDSCCGDAQ TGGDHTNESC VDGCCVRDSS VMVEEVTGSC
Consensus   .......... .......... .......... .......... ..........

            351                                                   400
At7acrigr   YRVSIANEVE STSS...SPS SSSLQKMPLN VVSQPLTQET VINISGMTCN
At7ahomsa   YRVSITSEVE STSN...SPS SSSLQKIPLN VVSQPLTQET VINIDGMTCN
At7bhomsa   FKVSLPDGAE GSGTDHRSSS SHSPGSPPRN QV.QGTCSTT LIAIAGMTCA
Atc2sacce   EAVSSKEQLL TSFEVVPSKS EGLQSIHDIR ETTRCNTNSN QHTGKGRLCI
Consensus   .......... .......... .......... .......... ..........
```

```
          401                                                    450
At7acrigr SCVQSIEGVV SKKPGVKSIH VSLANSFGTV EYDPLLTAPE TLREVIVDMG
At7ahomsa SCVQSIEGVI SKKPGVKSIR VSLANSNGTV EYDPLLTSPE TLRGAIEDMG
At7bhomsa SCVHSIEGMI SQLEGVQQIS VSLAEGTATV LYNPSVISPE ELRAAIEDMG
Atc2sacce ESSDSTLKKR SCKVSRQKIE VSSKPECCNI SCVERIASRS CEKRTFKGST
Consensus .......... .......... .......... .......... ..........


          451                                                    500
At7acrigr FDAVLPDMSE PLVVIAQPSL ETPLLPSTND .......... ..........
At7ahomsa FDATLSDTNE PLVVIAQPSS EMPLLTSTNE FYTKG..... ..........
At7bhomsa FEASVVSESC STNPLGNHSA GNSMVQTTDG TPTSVQEVAP HTGRLPANHA
Atsyescco .......... .......... .......... .....MSQTI DLTLDGLSCG
Atc2sacce NVGISGSSST DSLSEKFFSE QYSRMYNRYS SILKNLGCIC NYLRTLGKES
Consensus .......... .......... .......... .......... ..........


          501                                                    550
Caddstaau ......MDSS TKTLTEDKQV YRVEGFSCAN CAGKFEKNVK ELSGVHDAKV
At7acrigr .......QDN MMTAVHSKCY IQVSGMTCAS CVANIERNLR REEGIYSVLV
At7ahomsa ...MTPVQDK EEGKNSSKCY IQVTGMTCAS CVANIERNLR REEGIYSILV
At7bhomsa PDILAKSPQS TRAVAPQKCF LQIKGMTCAS CVSNIERNLQ KEAGVLSVLV
Ctppromi  ......MNTP TTLSSANRLS LPVEGMTCAS CVGRVERALK AVPEIKDAVV
Atsyescco HCVKRVKESL EQRPDVEQAD VSITEAHVTG TASAEQLIET IKQAGYDASV
Atulsacce .......... .....MREVI LAVHGMTCSA CTNTINTQLR ALKGVTKCDI
Atc2sacce CCLPKVRFCS GEGASKKTKY SYRNSSGCLT KKKTHGDKER LSNDNGHADF
Consensus .......... .......... .......... .......... ..........


          551                                                    600
Ctpbraja  .......... .......... ....MHVT RDFSHY.... .VRTAGEGIK
Fixirhime ........MS CCASSAAIMV AEGGQASPAS EELWLA.... .SRDLGGGLR
Cadastaau .......... .......... .......MS EQKVK..... ....LMEEE
Caddstaau NFGASKIDVF GSATVEDLEK AGAFENLKVA PEKARR.... .RVEPVVTED
Cadabacfi .......... .......... ....MS DQKA...... ...ITSEQE
Ctpbmycle .......... .......... ....MT ASLVED.... .TNNNHESVR
Ctpamycle .......... .......... .......... .......... ......MQ
Atsysynsp .......... .......... ...MPAAI.. ..VHSADPSST
At7acrigr ALMAGKAEVR YNPAVIQ..P PVIAEFIREL GFGATV.... .MENADEGDG
At7ahomsa ALMAGKAEVR YNPAVIQ..P PMIAEFIREL GFGATV.... .IENADEGDG
At7bhomsa ALMAGKAEIK YDPEVIQ..P LEIAQFIQDL GFEAAV.... .MEDYAGSDG
Atcssynsp .......... .......... .......... .......... ....MVNQQ
Ctppromi  NLATERADIT FSSTPNP..V ........L AVSAIE.... .SSGYKVPEE
Atkaentfa .......... .......... .......... .......... ...MATNTK
Atsyescco SHPKAKPLAE SSIPSEA.L ..........T AVSEAL.... .PAATADDDD
Atulsacce SLVTNECQVT YDNEVTADSI KEIIEDCGFD CEILRD.... .SEITAISTK
Atkbentfa .......... .......... .......M.... .NNGIDPENE
Atc2sacce VCSKSCCTKM KDCAVTSTIS GTSSSEISRI VSMEPIENHL NLEAGSTGTE
Consensus .......... .......... .......... .......... ..........


          601                                                    650
Ctpbraja  HIDLAVEGVH CAGCMAKIER GLSAIPDVTL ARVNLTDRRV ALEWKAGT..
Fixirhime QTELSVPNAY CGTCIATIEG ALRAKPEVER ARVNLSSRRV SIVWKEEVGG
Cadastaau MNVYRVQGFT CANCAGKFEK NVKKIPGVQD AKVNFGASKI DVYGNASVEE
Caddstaau KNVYRVEGFS CANCAGKFEK NVKQLAGVQD AKVNFGASKI DVYGNASVEE
Cadabacfi MKAYRVQGFT CANCAGKFEK NVKQLSGVED AKVNFGASKI AVYGNATIEE
Ctpbmycle RIQLDVAGML CAACASRVET KL.NKIPGVR ASVNFATRVA TI...DAVDV
```

```
Ctpamycle RIQLNITGMS CSCCAPNGWN NLPNKLSDFS TLVNSATRVA RL...TSAR.
Atsysynsp SILVEVEGMK CAGCVAAVER RLQQTAGVEA VSVNLITRLA KVDYDAALIE
At7acrigr ILKLVVRGMT CASCVHKIES TLTKHKGIFY CSVALATNKA HIKYDPEIIG
At7ahomsa VLELVVRGMT CASCVHKIES SLTKHRGILY CSVALATNKA HIKYDPEIIG
At7bhomsa NIELTITGMT CASCVHNIES KLTRTNGITY ASVALATSKA LVKFDPEIIG
Atcssynsp ..TLTLRGMG CAACAGRIEA LIQALPGVQE CSVNFGAEQA QVCYDPALTQ
Ctppromi  ITELAIEEMT CASCVGRVEK ALAQIPGVLE ATVNLATERA RVRHLSGVVS
Atkaentfa METFVITGMT CANCSARIEK ELNEQPGVMS ATVNLATEKA SVKYTDTTTE
Atsyescco SQQLLLSGMS CASCVTRVQN ALQSVPGVTQ ARVNLAERTA LVM...GSAS
Atu1sacce EGLLSVQGMT CGSCVSTVTK QVEGIEGVES VVVSLVTEEC HVIYEPSKT.
Atkbentfa TNKKGAIGKN PEEKITVEQT NTKNNLQEHG KMENMDQHHT HGHMERHQQM
Atc2sacce HIVLSVSGMS CTGCESKLKK SFGALKCVHG LKTSLILSQA EFNLDLAQGS
Consensus .......G.. C..C..... ..... ......V..... ..... ....


            651                                               700
Ctpbraja  ..LDPGRFID RLEELGYKAY PFETESAEVA EVAES..... ......RF
Fixirhime RRTNPCDFLH AIAERGYQTH LFSPGEEEGD DLLKQ..... ......
Cadastaau ...LEK..... ...AGAFENL KVSPEKLANQ TIQRVKDDTK AHKEEKTPFY
Caddstaau ...LEK..... ...AGAFENL KVIPEKLANP SIQAVKEDTK APKEEKIPFY
Cadabacfi ...LEK..... ...AGAFENL KVTPEKSARQ ASQEVKEDT. ..KEDKVPFY
Ctpbmycle ...AVDELRQ VIEQAGYRAT ......... .AHAESAVEE IDPDADYARN
Ctpamycle ...SPRPLRY VKAVRRAALC ......... .TDGGEALQR RQADADNARY
Atsysynsp ...DPTVLTT EITGLGFRAQ LRQDDNPLTL PIAEIPPLQQ QR........
At7acrigr ...PRDIIHT .IGSLGFEAS LVKKDRSASH LDHKREIKQW RSS.......
At7ahomsa ...PRDIIHT .IESLGFEAS LVKKDRSASH LDHKREIRQW RRS.......
At7bhomsa ...PRDIIKI .IEEIGFHAS LAQRNPNAHH LDHKMEIKQW KKS.......
Atcssynsp ...VAAIQAA .IEAAGYHAF PLQDPWDN.. EVEAQERHRR ARSQRQLAQR
Ctppromi  ...ITDLEVA .VVHAGYKPR RLSDNPANTR DLSEERREKE ARS.......
Atkaentfa .....RLIKS .VENIGYGAI LYDEAHKQKI AEEKQTYLRK MKFD......
Atsyescco ...PQDLVQA .VEKAGYGAK RLKMTLNAAS ASKKPPSLAM KR........
Atu1sacce ...TLETARE MIEDCGFDSN IIMDGNGNAD MTEKTVILKV TKAFEDESPL
Atkbentfa ...DHGHMSG .MDHSHMDHE DMSGMNHSHM GHENMSGMDH SMHMGNFKQK
Atc2sacce VKDVIKHLSK TTEFKYEQIS NHGSTIDVVV PYAAKDFINE EWPQGVTELK
Consensus .......... ..... ..... ..... ..... ..... ..... .....


            701                                               750
Ctpbraja  LLRCLGVAAF ATMNVMMLSI PVWSGNVSDM LPEQRDFF.. ..........
Fixirhime LILAVAVSGF AATNIMLLSV SVWSGAD.... .AATRDLF.. ..........
Cadastaau KKHSTLLFAT LLIAFGYLSH FVNGE..... ..... ..... .....
Caddstaau KKHSTLLFAT LLIAFGYLSH FVNGE..... ..... ..... .....
Cadabacfi KKHSTLLYAS LLITFGYLSS YVNGE..... ..... ..... .....
Ctpbmycle LLRRLIVAAL LFVPLADLST ..... ..... ..... .....
Ctpamycle LLIRLAVAAA LFVPLAHLSV ..... ..... ..... .....
Atsysynsp .......... LQLAIAAFLL ..... ..... ..... .....
At7acrigr FLVSLFFCTP VMGLMMYMMA ..... ..... ..... .....
At7ahomsa FLVSLFFCIP VMGLMTYMMV ..... ..... ..... .....
At7bhomsa FLCSLVFGIP VMALMIYMLI ..... ..... ..... .....
Atcssynsp VWVSGLIASL LVIGSLPMML ..... ..... ..... .....
Ctppromi  LRRALLIATI FTLPVFVIEM ..... ..... ..... .....
Atkaentfa LIFSAILTLP LMLAMIAMML ..... ..... ..... .....
Atsyescco FRWQAIVALA VGIPVMVWGM ..... ..... ..... .....
Atu1sacce ILSSVSERFQ FLLDLGVKSI EISDDMHTLT IKYCCNELGI RDLLRHLERT
Atkbentfa FWLSLILAIP IILFSPMMGM SF........ ..... .....
```

```
Atc2sacce IVERNIIRIY FDPKVIGARD LVNEGWSVPV SIAPFSCHPT IEVGRKHLVR
Consensus ......... ......... ......... ......... .........


          751                                                800
Ctpbraja  ......... ......... ......... ......... .........
Fixirhime ......... ......... ......... ......... .........
Cadastaau ......... ......... ......... ......... .........
Caddstaau ......... ......... ......... ......... .........
Cadabacfi ......... ......... ......... ......... .........
Ctpbmycle .....MFAIV PTNR..... ......... ......... FPGWGYLL..
Ctpamycle .....MFAVL PSTH..... ......... ......... FPGWEWML..
Atsysynsp .....IVSSW GHLGHWLDHP LPGTDQL... ......... ...WFH..
At7acrigr .....MEHHF ATIHHNQSMS NEEMIKNHSS MFLERQILPG LSIMNLLS..
At7ahomsa .....MDHHF ATLHHNQNMS KEEMINLHSS MFLERQILPG LSVMNLLS..
At7bhomsa ......... .......PS NEP....HQS MVLDHNIIPG LSILNLIF..
Atcssynsp .....GIS.I PGIPMWLHHP G......... ......... ..LQ..
Ctppromi  .....GSHFI PGVHHWVTQT LGQQ..... ......... ..LNWYIQ..
Atkaentfa .....GSH.. GPIVSFFHLS L......... ......... ....VQ..
Atsyescco ......... ..IGDNMMVT ADNR..... ......... ..SLWLVI..
Atu1sacce GYKFTVFSNL DNTTQLRLLS KEDEIRFWKK NSIKSTLLAI ICMLLYMIVP
Atkbentfa ......... ......... ......... ....PFQVT FPGSNWVV..
Atc2sacce VGCTTALSII LTIPILVMAW APQLREKIST IS........ .........
Consensus ......... ......... ......... ......... .........


          801                                                850
Ctpbraja  ......... ......HWLS ALIALPAAAY AGQPFFRSAW
Fixirhime ......... ......HWIS ALIAGPALIY AGRFFYKSAW
Cadastaau ......... ......DNLVT SMLFVGSIVI GGYSLFKVGF
Caddstaau ......... ......DNLVT SMLFVSSIVI GGYSLFKVGF
Cadabacfi ......... ......ENIVT TLLFLASMFI GGLSLFKVGL
Ctpbmycle ......... .... TALAAPIVTW AAWPFHRVAL
Ctpamycle ......... .... TALAIPVVTW AAWPFHRVAI
Atsysynsp ......... .... ALLATWALLG PGRSILQAGW
At7acrigr ......... .... LLLCLPVQFF GGWYFYIQAY
At7ahomsa ......... .... FLLCVPVQFF GGWYFYIQAY
At7bhomsa ......... .... FILCTFVQLL GGWYFYVQAY
Atcssynsp ......... .... LGLTLPVLWA .GRSFFINAW
Ctppromi  ......... .... FVLATIVMFG PGLRFFKKGI
Atkaentfa ......... .... LLFALPVQFY VGWRFYKGAY
Atsyescco ......... .... GLITLAVMVF AGGHFYRSAW
Atu1sacce MMWPTIVQDR IFPYKETSFV RGLFYRDILG VILASYIQFS VGFYFYKAAW
Atkbentfa ......... .... LVLATILFIY GGQPFLSGAK
Atc2sacce ......... .......AS MVLATIIQFV IAGPFYLNAL
Consensus ......... ......... ..L...... .G..F.....


          851                                                900
Ctpbraja  RALS.AKTTN MDVPISIGVI LALGMSVVET I......... ....HHAE
Fixirhime NAIR.HGRTN MDVPIALAVS LSYGMSLHET I......... ...GHGE
Cadastaau QNLI.RFDFD MKTLMTVAVI GATIIGK... ......... .....
Caddstaau QNLI.RFDFD MKTLMTVAVI GAAIIGE... ......... .....
Cadabacfi QNLL.RFEFD MKTLMTVAVI GGAIIGE... ......... .....
Ctpbmycle RNAR.YRAAS METLISAGIL AATGWSLSTI FVDKEPRQTH GIWQAILHSD
Ctpamycle HNAR.YHGAS METLISTGIT AATIWSLYTV FGHHQSTEHR GVWRALLGSD
Atsysynsp QGLR.CGAPN MNSLVLLGTG SAYLASLVAL LW.......P QL......GW
```

```
At7acrigr KALK.HKTAN MDVLIVLATT IAFAYSLII. LL.......V AMYERAKVNP
At7ahomsa KALK.HKTAN MDVLIVLATT IAFAYSLII. LL.......V AMYERAKVNP
At7bhomsa KSLR.HRSAN MDVLIVLATS IAYVYSLVI. LV.......V AVAEKAERSP
Atcssynsp KAFR.QNTAT MDTLVAVGTG AAFLYSLAVT LF.......P QWLTRQGLPP
Ctppromi  PALL.RGAPD MNSLVSVGTV AAYGYSVVST FI.......P QVL..PAGTA
Atkaentfa HALK.TKAPN MDVLVAIGTS AAFALSIYNG FF.......P ......SHSH
Atsyescco KSLL.NGAAT MDTLVALGTG VAWLYSMSVN LW.......P QWFPMEA..R
Atulsacce ASLK.HGSGT MDTLVCVSTT CAYTFSVFSL VHNMFHPSST GKLPR.....
Atkbentfa MELK.QKSPA MMTLIAMGIT VAYVYSVYSF I........A NLINPHTHVM
Atc2sacce KSLIFSRLIE MDLLIVLSTS AAYIFSIVSF GY........ .FVVGRPLST
Consensus .......... M.L....... .A...S.... ..........

          901                                        950
Ctpbraja  HAYFDAAIML LTFLLVGRFL DQNMRRRTRA VAGNLAALKA ETAAKFVGPD
Fixirhime HAWFDASVTL LFFLLIGRTL DHMMRGRART AISGLARLSP RGATVVHPDG
Cadastaau ...WAEASIV VILFAISEAL ERFSMDRSRQ SIRSLMDIAP KEALVRRNG.
Caddstaau ...WAEASIV VILFAISEAL ERFSMDRARQ SIRSLMDIAP KEALVRRNG.
Cadabacfi ...WAEVAIV VILFAISEAL ERFSMDRARQ SIRSLMDIAP KEALVKRNG.
Ctpbmycle SIYFEVAAGV TVFVLAGRFF EARAKSKAGS ALRALAARGA KNVEVLLPNG
Ctpamycle AIYFEVAAGI TVFVLAGKYY TARAKSHASI ALLALAALSA KDAAVLQPDG
Atsysynsp VCFFDEPVML LGFILLGRTL EEQARFRSQA ALQNLLALQP ETTQLLTAPS
At7acrigr ITSFDTPPML FVFIALGRWL EHIAKGKTSE ALAKLISLQA TEATIVT...
At7ahomsa ITFFDTPPML FVFIALGRWL EHIAKGKTSE ALAKLISLQA TEATIVT...
At7bhomsa VTFFDTPPML FVFIALGRWL EHLAKSKTSE ALAKLMSLQA TEATVVT...
Atcssynsp DVYYEAIAVI IALLLLGRSL EERAKGQTSA AIRQLIGLQA KTARVLR...
Ctppromi  NIYFEAAVVI VTLILLGRNL EAKAKGNTSQ AIKRLVGLQA KTARVSR...
Atkaentfa DLYFESSSMI ITLILLGKYL EHTAKSKTGD AIKQMMSLQT KTAQVLR...
Atsyescco HLYYEASAMI IGLINLGHML EARARQRSSK ALEKLLDLTP PTARLVT...
Atulsacce .IVFDTSIMI ISYISIGKYL ETLAKSQTST ALSKLIQLTP SVCSII....
Atkbentfa DFFWELATLI .VIMLLGHWI EMNAVSNASD ALQKLAELLP ESVKRLKKDG
Atc2sacce EQFFETSSLL VTLIMVGRFV SELARHRAVK SI.SVRSLQA SSAILVDKTG
Consensus .......... ......G..L E.......... ....L..L.. ..A.......

          951                                       1000
Ctpbraja  .......... EISQVPVAAI SPGDIVLLRP GERCAVDGTV IEGRSEIDQS
Fixirhime .......... SREYRAVDEI NPGDRLIVAA GERVPVDGRV LSGTSDLDRS
Cadastaau .......... QEIIIHVDDI AVGDIMIVKP GEKIAMDGII VNGLSAVNQA
Caddstaau .......... QEIMIHVDDI AVGDIMIVKP GEKIAMDGII INGVSAVNQA
Cadabacfi .......... QEIMIHVDDI AVGDIMIVKP GQKIAMDGVV VSGYSAVNQT
Ctpbmycle .......... AELTIPAGEL KKQQHFLVRP GETITADGVV IDGTATIDMS
Ctpamycle .......... SEMVIPANEL NEQQRFVVRP GQTIAADGLV IDGSATVSMS
Atsysynsp SIAPQDLLEA PAQIWPVAQL RAGDYVQVLP GDRIPVDGCI VAGQSTLDTA
At7acrigr ..LDSDNILL SEEQVDVELV QRGDIIKVVP GGKFPVDGRV IEGHSMVDES
At7ahomsa ..LDSDNILL SEEQVDVELV QRGDIIKVVP GGKFPVDGRV IEGHSMVDES
At7bhomsa ..LGEDNLII REEQVPMELV QRGDIVKVVP GGKFPVDGKV LEGNTMADES
Atcssynsp ..QGQ..... .ELTLPITEV QVEDWVRVRP GEKVPVDGEV IDGRSTVDES
Ctppromi  ..HGE..... .ILEIPLDQV MMGDIVVVRP GEKIPVDGEV VEGHSYVDES
Atkaentfa ..DGK..... .EETIAIDEV MIDDILVIRP GEQVPTDGRI IAGTSALDES
Atsyescco ..DEG..... .EKSVPLAEV QPGMLLRLTT GDRVPVDGEI TQGEAWLDEA
Atulsacce ....SDVERN ETKEIPIELL QVNDIVEIKP GMKIPADGII TRGESEIDES
Atkbentfa .......... TEETVSLKEV HEGDRLIVRA GDKMPTDGTI DKGHTIVDES
Atc2sacce .......... KETEINIRLL QYGDIFKVLP DSRIPTDGTV ISGSSEVDEA
Consensus .......... .......... ...D...V.P G.....DG.. ..G....D..
```

```
            1001                                                    1050
Ctpbraja    LITGETLYVT AEQGTPVYAG SMNISGTLRV RVSAASEATL LAEIARLLDN
Fixirhime   VVNGESSPTV VTTGDTVQAG TLNLTGPLTL EATAAARDSF IAEIIGLMEA
Cadastaau   AITGESVPVS KAVDDEVFAG TLNEEGLIEV KITKYVEDTT ITKIIHLVEE
Caddstaau   AITGESVPVA KTVDDEVFAG TLNEEGLLEV KITKYVEDTT ISKIIHLVEE
Cadabacfi   AITGESVPVE KTVDNEVFAG TLNEEGLLEV EITKLVEDTT ISKIIHLVEE
Ctpbmycle   AITGEARPVH ASPASTVVGG TTVLDGRLVI EATAVGGDTQ FAAMVRLVED
Ctpamycle   PITGEAKPVR VNPGAQVIGG TVVLNGRLIV EAAAVGDETQ LAGMVRLVEQ
Atsysynsp   MLTGEPLPQP CQVGDRVCAG TLNLSHRLVI RAEQTGSQTR LAAIVRCVAE
At7acrigr   LITGEAMPVA KKPGSTVIAG SINQNGSLLI CATHVGADTT LSQIVKLVEE
At7ahomsa   LITGEAMPVA KKPGSTVIAG SINQNGSLLI CATHVGADTT LSQIVKLVEE
At7bhomsa   LITGEAMPVT KKPGSTVIAR SINAHGSVLI KATHVGNDTT LAQIVKLVEE
Atcssynsp   MVTGESLPVQ KQVGDEVIGA TLNKTGSLTI RATRVGRETF LAQIVQLVQQ
Ctppromi    MITGEPVPVA KEIGAEVVGG TINKTGTFSF KVTKVGANTI LAQIIRLVEE
Atkaentfa   MLTGESVPVE KKEKDMVFGG TINTNGLIQI QVSQIGKDTV LAQIIQMVED
Atsyescco   MLTGEPIPQQ KGEGDSVHAG TVVQDGSVLF RASAVGSHTT LSRIIRMVRQ
Atulsacce   LMTGESILVP KKTGFPVIAG SVNGPGHFYF RTTTVGEETK LANIIKVMKE
Atkbentfa   AVTGESKGVK KQVGDSVIGG SINGDGTIEI TVTGTGENGY LAKVMEMVRK
Atc2sacce   LITGESMPVP KKCQSIVVAG SVNGTGTLFV KLSKLPGNNT ISTIATMVDE
Consensus   ..TGE..PV. ......V..G ..N..G.... ......T. ...I...V..
```

```
            1051                                                    1100
Ctpbraja    ALQARSRYMR LADRASRLYA PVVHATALIT ILGWVIA... .........
Fixirhime   AEGGRARYRR IADRAARYYS PAVHLLALLT FVGWMLV... .........
Cadastaau   AQGERAPAQA FVDKFAKYYT PIIMVIAALV AVVPPLFFGG SWDTW.....
Caddstaau   AQGERAPAQA FVDKFAKYYT PIIMVIAALV AVVPPLFFGG SWDTW.....
Cadabacfi   AQGERAPSQA FVDKFAKYYT PIIMIIATLV AIVPPLFFDG SWETW.....
Ctpbmycle   AQVQKARVQH LADRIAAVFV PMVFVIAGLA GASWLLAG.. .........
Ctpamycle   AQQQNANAQR LADRIASVFV PCVFAVAALD ...RCWMA.. .........
Atsysynsp   AQQRKAPVQR FADAIAGRFV YGVCAIAALT FGFWATLGSR WWPQVLQQPL
At7acrigr   AQTSKAPIQQ FADKLGGYFV PFIVLVSIAT LLVWIIIGFQ NFT.......
At7ahomsa   AQTSKAPIQQ FADKLSGYFV PFIVFVSIAT LLVWIVIGFL NFE.......
At7bhomsa   AQMSKAPIQQ LADRFSGYFV PFIIIMSTLT LVVWIVIGFI DFG.......
Atcssynsp   AQASKAPIQR LADQVTGWFV PAVIAIAILT FLLWFNWI.. .........
Ctppromi    AQGSKLPIQA LVDKVTMWFV PAVMIGATIT FFIWLAFG.. .........
Atkaentfa   AQGSKAPIQQ IADKISGIFV PIVLFLALVT LLVTGWLT.. .........
Atsyescco   AQSSKPEIGQ LADKISAVFV PVVVVIALVS AAIWYFFG.. .........
Atulsacce   AQLSKAPIQG YADYLASIFV PGILILAVLT FFIWCFI... .........
Atkbentfa   AQGEKSKLEF LSDKVAKWLF YVALVVGIIA FIAWLFLA.. .........
Atc2sacce   AKLTKPKIQN IADKIASYFV PTIIGITVVT FCVWIAVG.. .........
Consensus   AQ......Q. ..D...... P........ ......... .........
```

```
            1101                                                    1150
Ctpbraja    .................G ASWHDAIVTG VAVLIITCPC ALGLAIPTVQ
Fixirhime   .................E GDVRHAMLVA VAVLIITCPC ALGLAVPVVQ
Cadastaau   ...................VYQG LAVLVVGCPC ALVISTPISI
Caddstaau   ...................VYQG LAVLVVGCPC ALVITTPISI
Cadabacfi   ...................IYQG LAVLVVGCPC ALVISTPISI
Ctpbmycle   ...............ASP DRAFSVVLG. ..VLVIACPC TLGLATPTAM
Ctpamycle   ...............DRR ERTRPSVLGA IAVLVIACPC ALGLATPTAM
Atsysynsp   PGLLIHAPHH GMEMAHPHSH SPLLLALTLA ISVLVVACPC ALGLATPTAI
At7acrigr   ...IVETYFP GYSRSISRTE TIIRFAFQAS ITVLCIACPC SLGLATPTAV
At7ahomsa   ...IVETYFP GYNRSISRTE TIIRFAFQAS ITVLCIACPC SLGLATPTAV
At7bhomsa   ...VVQKYFP NPNKHISQTE VIIRFAFQTS ITVLCIACPC SLGLATPTAV
```

```
Atcssynsp ........... ........GN ..VTLALITA VGVMIIACPC ALGLATPTSI
Ctppromi  ............ ........PE PALTFALINA VAVLIIACPC AMGLATPTSI
Atkaentfa ............ ....KD WQ..LALLHS VSVLVIACPC ALGLATPTAI
Atsyescco ............ ......PA PQIVYTLVIA TTVLIIACPC ALGLATPMSI
Atulsacce ...LNISANP PVAFTANTKA DNFFICLQTA TSVVIVACPC ALGLATPTAI
Atkbentfa ............ ...... .NLPDALERM VTVFIIACPH ALGLAIPLVV
Atc2sacce .......... .IRVEKQSRS DAVIQAIIYA ITVLIVSCPC VIGLAVPIVF
Consensus ............ ........... ..VL...CPC .LGLATP...


                1151                                          1200
Ctpbraja  TVASGAMFKS GVLLNSGDAI ERLAEADHVI FDKTGTLTLP DLEVMNAADI
Fixirhime VVAAGRLFQG GVMVKDGSAM ERLAEIDTVL LDKTGTLTIG KPRLVNAHEI
Cadastaau VSAIGNAAKK GVLVKGGVYL EKLGAIKTVA FDKTGTLTKG VPVVTDFEVL
Caddstaau VSAIGNAAKK GVLIKGGVYL EELGAIKAIA FDKTGTLTKG VPVVTDFKVL
Cadabacfi VSAIGNAAKK GVLVKGGVYL EEMGALKAIA FDKTGTLTKG VPAVTDYNVL
Ctpbmycle MVASGRGAQL GIFIKGYRAL ETINAIDTVV FDKTGTLTLG QLSVSTVTST
Ctpamycle MVASGRGAQL GILLKGHESF EATRAVDTVV FDKTGTLTTG QLKVSAVTAA
Atsysynsp LVATGLAAEQ GILVRGGDVL EQLARIKHFV FDKTGTLTQG QFELIEIQPL
At7acrigr MVGTGVGAQN GILIKGGEPL EMAHKVKVVV FDKTGTITHG TPVVNQVKVL
At7ahomsa MVGTGVGAQN GILIKGGEPL EMAHKVKVVV FDKTGTITHG TPVVNQVKVL
At7bhomsa MVGTGVAAQN GILIKGGKPL EMAHKIKTVM FDKTGTITHG VPRVMRVLLL
Atcssynsp MVGTGKGAEY GILIKSAESL ELAQTIQTVI LDKTGTLTQG QPSVTDFLAI
Ctppromi  MVGTGRAAEL GILFRKGEAL QALRDVSVVA LDKTGTLTKG RPELTDLIP.
Atkaentfa MVGTGVGAHN GILIKGGEAL EGAAHLNSII LDKTGTITQG RPEVTDVIGP
Atsyescco ISGVGRAAEF GVLVRDRDAL QRASTLDTVV FDKTGTLTEG KPQVVAVKTF
Atulsacce MVGTGVGAQN GVLIKGGEVL EKFNSITTFV FDKTGTLTTG FMVVKKFLKD
Atkbentfa ARSTSIAAKN GLLLKNRNAM EQANDLDVIM LDKTGTLTQG KFTVTGIEIL
Atc2sacce VIASGVAAKR GVIFKSAESI EVAHNTSHVV FDKTGTLTEG KLTVVHETVR
Consensus ....G..A.. G.L.K..... E......... .DKTGTLT.G ...V......


                1201                                          1250
Ctpbraja  PA........ DIFELAGRLA LSSHHPVAAA VAQAAGARSP IV........
Fixirhime SP........ GRLATAAAIA VHSRHPIAVA IQNSAGAASP IA........
Cadastaau ND...QVEEK ELFSIITALE YRSQHPLASA IMKKAEQDNI PYSNVQV...
Caddstaau ND...QVEEK ELFSIITALE YRSQHPLASA IMKKAEQDNI TYSDVRV...
Cadabacfi NK...QINEK ELLSIITALE YRSQHPLASA IMKKAEEENI TYSDVQV...
Ctpbmycle GGW.CSGE.. .VLALASAVE AASEHSVATA IV......AA YADPRPV...
Ctpamycle PGW.QANE.. .VLQMAATVE SASEHAVALA IA......AS TTHREPV...
Atsysynsp AD....VDPD RLLQWAAALE ADSRHPLATA LQT..AAQAA NLAPIAA...
At7acrigr VES.NKIPRS KILAIVGTAE SNSEHPLGAA VTKYCKQELD TETLGTC...
At7ahomsa TES.NRISHH KILAIVGTAE SNSEHPLGTA ITKYCKQELD TETLGTC...
At7bhomsa GDV.ATLPLR KVLAVVGTAE ASSEHPLGVA VTKYCKEELG TETLGYC...
Atcssynsp GD...RDQQQ TLLGWAASLE NYSEHPLAEA IVRY..GEAQ GITLSTV...
Ctppromi  AE...KFEYN EILSLVASIE TYSEHPIAQS IVNA..ANEA KLTLASV...
Atkaentfa KE......... .IISLFYSLE HASEHPLGKA IVAY..GAKV GAKTQPI...
Atsyescco AD...VDEAQ A.LRLAAALE QGSSHPLARA IL....DKAG DMRLPQV...
Atulsacce SNWVGNVDED EVLACIKATE SISDHPVSKA IIRYCDGLNC NKALNAVVLE
Atkbentfa DE...AYQEE EILKYIGALE AHANHPLAIG IMNYLKEKKI TPYQAQ....
Atc2sacce GDRHNSQ... ...SLLLGLT EGIKHPVSMA IASYLKEKGV SAQNVSNTKA
Consensus .............. .......E .S.HP..A I.......... .........


                1251                                          1300
Ctpbraja  GAVEE.AGQG VRADVDGAE. ...........................
Fixirhime GDIREIPGAG IEVKTEDGV. ...........................
```

```
Cadastaau EEFTSITGRG IKGIVNGTT...................................
Caddstaau KDFTSITGRG IQGNIDGTT...................................
Cadabacfi EDFSSITGKG IKGIVNGTT...................................
Ctpbmycle ADFVAFAGCG VSGVVAEHH...................................
Ctpamycle ANFRAVPGHG VSGTVAERA...................................
Atsysynsp SDRQQVPGLG VSGTCDGR...................................
At7acrigr TDFQVVPGCG ISCKVTNIEG LLHKSNLKIE ENNTKNASLV QIDAINEQSS
At7ahomsa IDFQVVPGCG ISCKVTNIEG LLHKNNWNIE DNNIKNASLV QIDASNEQSS
At7bhomsa TDFQAVPGCG IGCKVSNVEG ILAHSERPL..................SA
Atcssynsp TDFEAIPGSG VQGQVEGI.........................
Ctppromi  DNFEAIPGFG VSATVDGR.........................
Atkaentfa TDFVAHPGAG ISGTINGV.........................
Atsyescco NGFRTLRGLG VSGEAEGH.........................
Atu1sacce SEYVLGKGIV SKCQVNG....................
Atkbentfa .EQKNLAGVG LEATVEDKD.........................
Atc2sacce VTGKRVEGTS YSG.......................
Consensus .......G.G.................................
```

```
          1301                                            1350
Ctpbraja  ...................IRLGRPS FCGAEALVGD GTRLDP....
Fixirhime ...................YRLGSRD F....AVGGS GPDGRQ....
Cadastaau ...................YYIGSPK LFKELNVSDF SLGFENNVKI
Caddstaau ...................YYIGSPR LFKELNVSDF SLEFENKVKV
Cadabacfi ...................YYIGSPK LFKELLTNDF DKDLEQNVTT
Ctpbmycle ...................VKIGKPS WVTRNA..PC DVVLESARRR
Ctpamycle ...................VRVGKPS WIASRC..NS TTLV.TARRN
Atsysynsp ..................SLRLGNPT WV.........QVATAKLP
At7acrigr TSSSMIIDAP LSNAVDT..Q QYKVLIGNRE WMIRNGL.VI SNDVDDSMID
At7ahomsa TSSSMIIDAQ ISNALNA..Q QHKVLIGNRE WMIRNGL.VI NNDVNDFMTE
At7bhomsa PASHLNEAGS LPAEKDAVPQ TFSVLIGNRE WLRRNGL.TI SSDVSDAMTD
Atcssynsp ..................WLQIGTQR WLGELGI.ET S.ALQNQWED
Ctppromi  ..................SVSVGADR FMKQLGL.DV S.QFASSAQK
Atkaentfa ..................HYFAGTRK RLAEMNL.SF D.EFQEQALE
Atsyescco ..................ALLLGNQA LLNEQQV.GT K.AIEAEITA
Atu1sacce ...................N TYDICIGNEA LILEDAL.KK SGFINSNVDQ
Atkbentfa ..................VKIINEK EAKRLGL.KI D...PERLKN
Atc2sacce ..................LKLQGGNCR WLGHNNDPDV RKALE.....
Consensus ...........................G..............
```

```
          1351                                            1400
Ctpbraja  .....EASIV AFSKGAEKFI LWVRQGLRPD AQAVIAALKA RNI.GIEILS
Fixirhime .....SEAIL SL.DFRELAC FRFEDQPRPA SRESIEALGR LGI.ATGILS
Cadastaau LQNQGKTAMI IGTEKTILGV IAVADEVRET SKNVIQKLHQ LGIKQTIMLT
Caddstaau LQNQGKTAMI IGTDQTILGV IAVADEVRET SKNVILKLHQ LGIKQTIMLT
Cadabacfi LQNQGKTAMI IGTEKEILAV IAVADEVRES SKEILQKLHQ LGIKKTIMLT
Ctpbmycle RRITGETVVF VSVDGVACGA VAIADTVKDS AADAISALCS RGL.HTILLT
Ctpamycle AELRGETAVF VEIDGEQCGV IAVADAVKAS AADAVAALHD RGF.RTALLT
Atsysynsp TGSAAATSIW LADDQQLLAC FWLQDQPRPE AAEVVQALRS RGA.TVQILS
At7acrigr HGRKGRPAVL VTIDDELCGL IAIADTVKPE AELAVHILKS MGL.EVVLMT
At7ahomsa HERKGRTAVL VAVDDELCGL IAIADTVKPE AELAIHILKS MGL.EVVLMT
At7bhomsa HEMKGQTAIL VAIDGVLCGM IAIADAVKQE AALAVHTLQS MGV.DVVLIT
Atcssynsp WEAAGKTVVG VAADGHLQAI LSIADQLKPS SVAVVRSLQR LGL.QVVMLT
Ctppromi  LGEQGKTPLY TAIDGRLAAI IAVADPIKET TPEAIKALHA LGL.KVAMIT
Atkaentfa LEQAGKTVMF LANEEQVLGM IAVADQIKED AKQAIEQLQQ KGV.DVFMVT
```

```
Atsyescco QASQGATPVL LAVDGKAVAL LAVRDPLRSD SVAALQRLHK AGY.RLVMLT
Atu1sacce ....GNTVSY VSVNGHVFGL FEINDEVKHD SYATVQYLQR NGY.ETYMIT
Atkbentfa YEAQGNTVSF LVVSDKLVAV IALGDVIKPE AKEFIQAIKE KNI.IPVMLT
Atc2sacce ...QGYSVFC FSVNGSVTAV YALEDSLRAD AVSTINLLRQ RGI.SLHILS
Consensus ....G.T... ........... ....D..... .......L.. .G......T
```

```
          1401                                               1450
Ctpbraja  GDREPAVKAA AHALAI..PE WRAGVTPADK IARIEEL... .....KRRG.
Fixirhime GDRAPVVAAL ASSLGI..SN WYAELSPREK VQVCAAA... .....AEAG.
Cadastaau GDNQGTANAI GTHVGV..SD IQSELMPQDK LDYIKKM... .....QSE..
Caddstaau GDNQGTAEAI GAHVGV..SD IQSELLPQDK LDYIKKM... .....KAE..
Cadabacfi GDNKGTANAI GGQVGV..SD IEAELMPQDK LDFIKQL... .....RSE..
Ctpbmycle GDNQAAARAV AAQVGI..DT VIADMLPEAK VDVIQRL... .....RDQG.
Ctpamycle GDNPASAAAV ASRIGI..DE VIADILPEDK VDVIEQL... .....RDRG.
Atsysynsp GDRQTTAVAL AQQLGLESET VVAEVLPEDK AAAIAAL... .....QSQG.
At7acrigr GDNSKTARSI ASQVGI..TK VFAEVLPSHK VAKVKQL... .....QEEG.
At7ahomsa GDNSKTARSI ASQVGI..TK VFAEVLPSHK VAKVKQL... .....QEEG.
At7bhomsa GDNRKTARAI ATQVGI..NK VFAEVLPSHK VAKVQEL... .....QNKG.
Atcssynsp GDNRRTADAI AQAVGI..TQ VLAEVRPDQK AAQVAQL... .....QSRG.
Ctppromi  GDNKATAKAI AKQLGI..DE IVAEVLPDGK VAALKQL... .....SQKG.
Atkaentfa GDNQRAAQAI GKQVGIDSDH IFAEVLPEEK ANYVEKL... .....QKAG.
Atsyescco GDNPTTANAI AKEAGI..DE VIAGVLPDGK AEAIKHL... .....QSEG.
Atu1sacce GDNNSAAKRV AREVGISFEN VYSDVSPTGK CDLVKKI... .....QDKEG
Atkbentfa GDNPKAAQAV AEYLGI..NE YYGGLLPDDK EAIVQRY... .....LDQG.
Atc2sacce GDDDGAVRSM AARLGIESSN IRSHATPAEK SEYIKDIVEG RNCDSSSQSK
Consensus GDN...A.A. A...GI.... .......P..K ..........
```

```
          1451                                               1500
Ctpbraja  .ARVLMVGDG MNDAPSLAAA HVSMS.PISA AHLSQATADL VFL......G
Fixirhime .HKALVVGDG INDAPVLRAA HVSMA.PATA ADVGRQAADF VFM......H
Cadastaau YDNVAMIGDG VNDAPALAAS TVGIAMGGAG TDTAIETADI ALM......G
Caddstaau HGNVAMIGDG VNDAPALAAS TVGIAMGGAG TDTAIETADI ALM......G
Cadabacfi YGNVAMVGDG VNDAPALAAS TVGIAMGGAG TDTALETADV ALM......G
Ctpbmycle .HTVAMVGDG INDGPALACA DLGLAM.GRG TDVAIGAADL ILV......R
Ctpamycle .HVVAMVGDG INDGPALARA DLGMAI.GRG TDVAIGAADI ILV......R
Atsysynsp .DAVAMIGDG INDAPALATA AVGISL.AAG SDIAQDSAGL LLS......R
At7acrigr .KRVAMVGDG INDSPALAMA NVGIAI.GTG TDVTIEAADV VFI......R
At7ahomsa .KRVAMVGDG INDSPALAMA NVGIAI.GTG TDVAIEAADV VLI......R
At7bhomsa .KKVAMVGDG VNDSPALAQA DMGVAI.GTG TDVAIEAADV VLI......R
Atcssynsp .QVVAMVGDG INDAPALAQA DVGIAI.GTG TDVAIAASDI TLI......S
Ctppromi  .DKVAFVGDG INDAPALAQA DVGLAI.GTG TDVAIEAADV VLM......S
Atkaentfa .KKVGMVGDG INDAPALRLA DVGIAM.GSG TDIAMETADV TLM......N
Atsyescco .RQVAMVGDG INDAPALAQA DVGIAM.GGG SDVAIETAAI TLM......R
Atu1sacce NNKVAVVGDG INDAPALALS DLGIAI.STG TEIAIEAADI VILCGNDLNT
Atkbentfa .KKVIMVGDG INDAPSLARA TIGMAI.GAG TDIAIDSADV VLT......N
Atc2sacce RPVVVFCGDG TNDAIGLTQA TIGVHI.NEG SEVAKLAADV VML......K
Consensus ...V.MVGDG .ND.PALA.A ..G.A....G .D.A...AD. ..........
```

```
          1501                                               1550
Ctpbraja  RPLAPVAAAI DSARKALHLM RQNLWLAIGY NVLAVPVAIS GV.......V
Fixirhime ERLSAVPFAI ETSRHAGQLI RQNFALAIGY NVIAVPIAIL GY.......A
Cadastaau DDLSKLPFAV RLSRKTLNII KANITFAIGI KIIALLLVIP GWLTLWIAIL
Caddstaau DDLSKLPFAV RLSRKTLNII KANITFAIGI KIIALLLVIP GWLTLWIAIL
Cadabacfi DDLRKLPSTV KLSRKTLNII KANITFAIAI KFIASLLVIP GWLTLWIAIL
```

```
Ctpbmycle DSLGVVPVAL DLARATMRTI RINMIWAFGY NVAAIPIASS GL.......L
Ctpamycle DNLDVVPITL DLAAATMRTI KFNMVWAFGY NIAAIPIAAA GL.......L
Atsysynsp DRLDSVLVAW NLSQMGLRTI RQNLTWALGY NVVMLPLAAG AFLPAYGLAL
At7acrigr NDLLDVVASI DLSRKTVKRI RINFLFPLIY NLVGIPIAAG VFLPI.GLVF
At7ahomsa NDLLDVVASI DLSRKTVKRI RINFVFALIY NLVGIPIAAG VFMPI.GLVL
At7bhomsa NDLLDVVASI HLSKRTVRRI RINLVLALIY NLVGIPIAAG VFMPI.GIVL
Atcssynsp GDLQGIVTAI QLSRATMTNI RQNLFFAFIY NVAGIPIAAG ILYPLLGWLL
Ctppromi  GDLRGVVDAI ALSQATIRNI KQNLFWTFAY NALLIPVAAG MLYPINGMLL
Atkaentfa SHLTSINQMI SLSAATLKKI KQNLFWAFIY NTIGIPFAA. .....FG.FL
Atsyescco HSLMGVADAL AISRATLHNM KQNLLGAFIY NSIGIPVAAG ILWPFTGTLL
Atu1sacce NSLRGLANAI DISLKTFKRI KLNLFWALCY NIFMIPIAMG VLIP.WGITL
Atkbentfa SDPKDILHFL ELAKETRRKM IQNLWWGAGY NIIAIPLAAG ILAPI.GLIL
Atc2sacce PKLNNILTMI TVSQKAMFRV KLNFLWSFTY NLFAILLAAG AFV...DFHI
Consensus ..L........ ..S..T...I ..N...A..Y N....P.A.. .........L
```

```
          1551                                            1600
Ctpbraja  TPLIAAAAMS GSSILVMLNS LR.......A RSDSREIV.. ..........
Fixirhime TPLVAAVAMS SSSLVVVFNA LRLKRSLAAG RGATPGTLIH SGAVTS....
Cadastaau SDMGA..... ..TILVALNS LRLMRVKDK. .......... ..........
Caddstaau SDMGA..... ..TILVALNS LRLMRVKDK. .......... ..........
Cadabacfi SDMGA..... ..TLLVALNG LRLMRVKE.. .......... ..........
Ctpbmycle NPLIAGAAMA FSSFFVVSNS LRLSNFGLSQ TSD....... ..........
Ctpamycle NPLVAGAAMA FSSFFVVSNS LRLRNFGAIL SCGTSRHRTV KRWRCPPPTR
Atsysynsp TPAIAGACMA VSSLAVVSNS LLLRYWFRRS LNHSVSV... ..........
At7acrigr QPWMGSAAMA ASSVSVVLSS LFLKLYRKPT YDNYELRTRS HTGQRSPSEI
At7ahomsa QPWMGSAAMA ASSVSVVLSS LFLKLYRKPT YESYELPARS QIGQKSPSEI
At7bhomsa QPWMGSAAMA ASSVSVVLSS LQLKCYKKPD LERYEAQAHG HMKPLTASQV
Atcssynsp SPMLAGAAMA FSSVSVVTNA LRLRQFQPR. .......... ..........
Ctppromi  SPIFAAAAMA LSSVFVLGNA LRLKRFQAPM KTH....... ..........
Atkaentfa NPIIAGGAMA FSSISVLLNS LSLNRKTIK. .......... ..........
Atsyescco NPVVAGAAMA LSSITVVSNA NRLLRFKPKE .......... ..........
Atu1sacce PPMLAGLAMA FSSVSVVLSS LMLKKWTPPD IESHGISDFK SKFSIGNFWS
Atkbentfa SPAVGAVLMS LSTVVVALNA LTLK...... .......... ..........
Atc2sacce PPEYAGLGEL VSILPVIFVA ILLRYAKI.. .......... ..........
Consensus .P..A...M. .S...V.... L.L....... .......... ..........
```

```
          1601                                            1650
Ctpamycle LRSTACSPVD ASPLRPVAHR TGVKPPTHR. .......... ..........
At7acrigr SVHVGIDDAS RNSPRLGLLD RIVNYSRASI NSLLSDKRSL NS.VVNSEPD
At7ahomsa SVHVGIDDTS RNSPKLGLLD RIVNYSRASI NSLLSDKRSL NS.VVTSEPD
At7bhomsa SVHIGMDDRW RDSPRATPWD QVSYVSQVSL SSLTSDKPSR HSAAADDDGD
Atu1sacce RLFSTRAIAG EQDIESQAGL MSNEEVL... .......... ..........
Consensus .......... .......... .......... .......... ..........
```

```
          1651       1667
At7acrigr KHS....... ........
At7ahomsa KHSLLVGDFR EDDDTAL
At7bhomsa KWSLLLNGRD EEQYI..
Consensus .......... ........
```

Residues listed in the consensus sequence are present in at least 75% of the aligned transporter sequences. Residues indicated in boldface type are also conserved in at least one other family of the P-type ATPases.

Database accession numbers

	SWISSPROT	PIR	EMBL/GENBANK
At7ahomsa	Q04656	S37287	L06133; G179253
At7bhomsa	P35670		U11700; G551502
Atc2sacce	P38360	S46177	Z29332; G547580
Atcssynsp	P37279	S36741	D16437; G435125
Atkaentfa	P32113	A45995	L13292; G290642
Atkbentfa	P05425	A29576; B45995	L13292; G290643
Atsyescco			U58330
Atsysynsp	P37385		U04356; G436954
Atu1sacce	P38995	S48298	L36317; G538515
Cadabacfi	P30336	D42707	M90750; G143753
Cadastaau	P20021	A32561	J04551; G150719
Caddstaau	P37386		L10909; G152978
Ctpbraja			X95634
Ctppromi			U42410
Ctpamycle	P46839		Z46257; G559907
Ctpbmycle	P46840		Z46257; G559912
Fixirhime	P18398	C32052; S39994	M24144; Z21854

References

[1] Odermatt, A. et al. (1993) J. Biol. Chem. 268, 12775–12779.

[2] Vulpe, C. et al. (1993) Nature Genet. 3, 7–13.

[3] Kahn, D. et al. (1989) J. Bacteriol. 171, 929–939.

[4] Petrukhin, K. et al. (1993) Nature Genet. 5, 338–343.

[5] **Green, N.M. and MacLennan, D.H. (1989) Biochem. Soc. Trans. 17, 819–822; Green, N.M. (1989) Biochem. Soc. Trans. 17, 970–972.**

[6] **Fagan, M.J. and Saier, M.H. Jr. (1994) J. Mol. Evol. 38, 57–99.**

[7] Bull, P.C. and Cox, D.W. (1994) Trends Genet. 10, 246–252.

Part 2

Vacuolar ATPases

Summary

Transporters of the vacuolar ATPase family, examples of which are vacuolar ATPase and vacuolar proton pump subunits from humans [1] (Vph1homsa), rodents [2] (Vpp1ratno) and yeast [3] (Stv1sacce), mediate proton transport by ATPase (H+-transporting ATP synthase; EC 3.6.1.34) activity. Other members of the vacuolar ATPase family include the mouse immune suppression factor TJ6 [4] (Tj6musmu). This ATPase subunit is required for assembly as well as for ATPase activity [3]. Members of the vacuolar ATPase family have only been found in eukaryotes.

Statistical analysis of multiple amino acid sequence comparisons reveals no apparent relationship between these transporters and any other ATPase or transporter family. Members of the vacuolar ATPase family contain two domains: a hydrophilic N-terminal domain containing many charged residues, and a hydrophobic C-terminal domain. The hydrophobic domain is predicted to contain six transmembrane helices by the hydropathy of amino acid sequences. Unusual for any transporter family, both the N-terminal domain and the C-terminus are predicted to be extracellular. They are also known to be glycosylated.

Many amino acids, and several long sequence motifs, are conserved throughout this family. These conserved sequence motifs are more prevalent in the hydrophobic, membrane-spanning C-terminal domain of the proteins.

Nomenclature, biological sources and substrates

CODE	DESCRIPTION [SYNONYMS]	ORGANISM [COMMON NAMES]	SUBSTRATE(S)
Stv1sacce	Vacuolar ATP synthase 101 kDa subunit [V-ATPase subunit AC115, STV1, YMR054W, YM9796.07]	Saccharomyces cerevisiae [yeast]	H+
Tj6musmu	Immune suppressor factor j6b7	Mus musculus [mouse]	H+
Vph1homsa	Vacuolar proton pump subunit [OC-116 kDa, VPP1]	Homo sapiens [human]	H+
Vph1neucr	Vacuolar ATPase 98 kDa subunit [VPH1]	Neurospora crassa [mold]	H+
Vph1sacce	Vacuolar ATP synthase 95.5 kDa subunit [VPH1, YOR270C]	Saccharomyces cerevisiae [yeast]	H+
Vpp1caeel	Putative clathrin-coated vesicle/synaptic vesicle proton pump subunit [ZK637.8]	Caenorhabditis elegans [nematode]	H+
Vpp1ratno	Clathrin-coated vesicle/synaptic vesicle proton pump 116 kDa subunit	Rattus norvegicus [rat]	H+

Phylogenetic tree

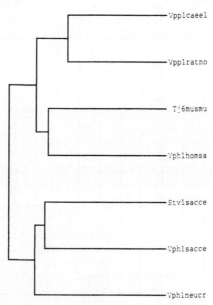

Proposed orientation of VPH1 in the membrane

The model is based on predictions of membrane-spanning regions and α-helical content. The N-terminus of the protein is illustrated on the outside and is folded six times through the membrane. The predicted membrane-spanning helices are portrayed as rectangles. The numbers corresponding to the first and last residue of each membrane-spanning helix are boxed. Residues that are conserved in more than 75% of the aligned transporters (see below) are shown.

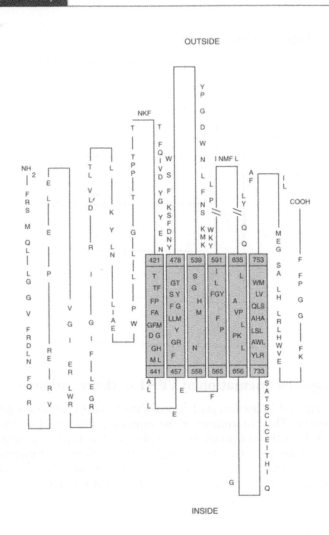

Physical and genetic characteristics

	AMINO ACIDS	MOL. WT	EXPRESSION SITES	CHROMOSOMAL LOCUS
Stv1sacce	890	101 660		ADH3 to centromere
Tj6musmu	855	98 048	thymus	
Vph1homsa	829	93 011		17q21
Vph1neucr	856	97 992		
Vph1sacce	840	95 528		Chromosome 15
Vpp1caeel	1030	117 544		ZK637.8
Vpp1ratno	838	96 327	brain	

Multiple amino acid sequence alignments

```
          1                                                  50
Vpp1caeel MGDYVTPGEE PPQPGIYRSE QMCLAQLYLQ SDASYQCVAE LGELGLVQFR
Vpp1ratno .......... ..MGELFRSE EMTLAQLFLQ SEAAYCCVSE LEELGKVQFR
Tj6musmu  .......... ..MGSLFRSE SMCLAQLFLQ SGTAYECLSA LGEKGLVQFR
Vph1homsa .......... ..MGSMFRSE EVALVQLFLP TAAAYTCVSR LGELGLVEFR
Stv1sacce .........M NQEEAIFRSA DMTYVQLYIP LEVIREVTFL LGKMSVFMVM
Vph1sacce .......MA EKEEAIFRSA EMALVQFYIP QEISRDSAYT LGQLGLVQFR
Vph1neucr .......MA PKQDTPFRSA DMSMVQLYIS NEIGREVCNA LGELGLVHFR
Consensus .......... .......FRS. .M...QL... .......... LG..G.V.FR

          51                                                 100
Vpp1caeel DLNPDVSSFQ RKYVNEVRRC DEMERKLRYL EREIKKDQIP M.........
Vpp1ratno DLNPDVNVFQ RKFVNEVRRC EEMDRKLRFV EKEIRKANIP I.........
Tj6musmu  DLNQNVSSFQ RKFVGEVKRC EELERILVYL VQEITRADIP L.........
Vph1homsa DLNASVSAFQ RRFVVDVWRC EELEKTFTFL QEEVRRAGLV L.........
Stv1sacce DLNKDLTAFQ RGYVNQLRRF DEVERMVGFL NEVVEKHAAE TWKYILHIDD
Vph1sacce DLNSKVRAFQ RTFVNEIRRL DNVERQYRYF YSLLKKHDIK LY.......E
Vph1neucr DLNSELSAFQ RAFTQDIRRL DNVERQLRYF HSQMEKAGIP LRKF.....D
Consensus DLN.....FQ R..V....R. ...ER..... .......... ..........

          101                                                150
Vpp1caeel .......... LDTGENPDAP LPREMIDLEA TFEKLENELR EVNKNEETLK
Vpp1ratno .......... MDTGENPEVP FPRDMIDLEA NFEKIENELK EINTNQEALK
Tj6musmu  .......... PEGEASPPAP PLKHVLEMQE QLQKLEVELR EVTKNKEKLR
Vph1homsa .......... PPPKGRLPAP PPRDLLRIQE ETERLAQELR DVRGNQQALR
Stv1sacce EGNDIAQPDM ADLINTMEPL SLENVNDMVK EITDCESRAR QLDESLDSLR
Vph1sacce GDTDKYLDGS GELY...VPP SGSVIDDYVR NASYLEERLI QMEDATDQIE
Vph1neucr PDVDI..... ......LTPP TTTEIDELAE RAQTLEQRVS SLNESYETLK
Consensus .......... .........P .......... ....E..... ..........

          151                                                200
Vpp1caeel KNFSELTELK HILRKTQTFF EEVDHDRWRI LEGGSGRRGR STEREETRPL
Vpp1ratno RNFLELTELK FILRKTQQFF DEMADP..DL LEESSS.... .........L
Tj6musmu  KNLLELVEYT HMLRVTKTFL KRNVEFEPTY EEFPAL.... ....ENDSL
Vph1homsa AQLHQLQLHA AVLRQG.... .HEPQLAAAH TD.GAS.... ....ERTPL
Stv1sacce SKLNDLLEQR QVIFECSKFI EVNPGIAGRA TNPEIEQEER DVDEFRMTPD
Vph1sacce VQKND.LEQY RFILQSG... .......... .......... ..DEFFLKGD
Vph1neucr KREVELTEWR WVLREAGGFF DRAHG..... .......... NVEEIRASTD
Consensus .....L.E.. .......... .......... .......... ..........

          201                                                250
Vpp1caeel IDIGDMDDDS AARMSAQAAM LRLGYVVLGK MDRPESATIA KRDLVYVVLF
Vpp1ratno LEPNEM.... .....GRGAP LRLG...... .......... ..........
Tj6musmu  LDYSCMQ... .......... .......... .......... ..........
Vph1homsa LQAPGGP... .......... .......... .......... ..........
Stv1sacce DISETLSDAF SFDDETPQDR GALG...... .......... ..........
Vph1sacce NTDST..... SYMDEDMIDA NGEN...... .......... ..........
Vph1neucr NDDAPL.... ....LQDV EQHN...... .......... ..........
Consensus .......... .......... .......... .......... ..........

          251                                                300
Vpp1caeel VSFSFCIPLV FFPDSFLHED MIASSAESSG IGEVLSADEE ELSGRFSDAM
Vpp1ratno .......... .......... .......... .......... ..........
```

```
Tj6musmu    ........... ........... ........... ........... ...........
Vph1homsa   ........... ........... ........... ........... ...........
Stv1sacce   ........... ........... ........... .....NDLTR NQSVEDLSFL
Vph1sacce   ........... ........... ........... .....IAAAI GASVN.....
Vph1neucr   ........... ........... ........... .....TAADV ERSFSGMNIG
Consensus   ........... ........... ........... ........... ...........

             301                                              350
Vpp1caeel   SPLKLQLRFV AGVIQRERLP AFERLLWRAC RGNVFLRTSE IDDVLNDTVT
Vpp1ratno   ........FV AGVINRERIP TFERMLWRVC RGNVFLRQAE IENPLEDPVT
Tj6musmu    .RLGAKLGFV SGLIQQGRVE AFERMLWRAC KGYTIVTYAE LDECLEDPET
Vph1homsa   .HQDLRVNFV AGAVEPHKAP ALERLLWRAC RGFLIASFRE LEQPLEHPVT
Stv1sacce   EQGYQHRYMI TGSIRRTKVD ILNRILWRLL RGNLIFQNFP IEEPLLEGKE
Vph1sacce   .......YV TGVIARDKVA TLEQILWRVL RGNLFFKTVE IEQPVYDVKT
Vph1neucr   ........FV AGVIGRDRVD AFERILWRTL RGNLYMNQAE IPEPLIDPTI
Consensus   .........V .G.I...... ..ER.LWR.. RG.......E ...L.....

             351                                              400
Vpp1caeel   GDPVNKCVFI IFFQGDHLKT KVKKICEGFR ATLYPCPDTP QERREMSIGV
Vpp1ratno   GDYVHKSVFI IFFQGDQLKN RVKKICEGFR ASLYPCPETP QERKEMASGV
Tj6musmu    GEVIKWYVFL ISFWGEQIGH KVKKICDCYH CHIYPYPNTA EERREIQEGL
Vph1homsa   GEPATWMTFL ISYWGEQIGQ KIRKITDCFH CHVFPFLQQE EARLGALQQL
Stv1sacce   K..VEKDCFI IFTHGETLLK KVKRVIDSLN GK...IVSLN TRSSELVDTL
Vph1sacce   REYKHKNAFI VFSHGDLIIK RIRKIAESLD ANLYDVDSSN EGRSQQLAKV
Vph1neucr   NEPVLKNVFV IFAHGKEILA KIRRISESMG AEVYNVDEHS DLRRDQVHEV
Consensus   ........F. I...G..... ....I..... .......... ..R......

             401                                              450
Vpp1caeel   MTRIEDLKTV LGQTQDHRHR VLVAASKNVR MWLTKVRKIK SIYHTLNLFN
Vpp1ratno   NTRIDDLQMV LNQTEDHRQR VLQAAAKNIR VWFIKVRKMK AIYHTLNLCN
Tj6musmu    NTRIQDLYTV LHKTEDYLRQ VLCKAAESVC SRVVQVRKMK AIYHMLNMCS
Vph1homsa   QQQSQELQEV LGETERFLSQ VLGRVLQLLP PGQVQVHKMK AVYLALNQCS
Stv1sacce   NRQIDDLQRI LDTTEQTLHT ELLVIHDQLP VWSAMTKREK YVYTTLNK..
Vph1sacce   NKNLSDLYTV LKTTSTTLES ELYAIAKELD SWFQDVTREK AIFEILNKSN
Vph1neucr   NARLEDVQNV LRNTQQTLEA ELAQISQSLS AWMITISKEK AVYNTLNLFS
Consensus   .....DL..V L..T...... .L........ .........K ..Y..LN...

             451                                              500
Vpp1caeel   IDVTQKCLIA EVWCPIAELD RIKMALKRGT DESGSQVPSI LNRMETNEAP
Vpp1ratno   IDVTQKCLIA EVWCPVTDLD SIQFALRRGT EHSGSTVPSI LNRMQTNQTP
Tj6musmu    FDVTNKCLIA EVWCPEVDLP GLRRALEEGS RESGATIPSF MNTIPTKETP
Vph1homsa   VSTTHKCLIA EAWCSVRDLP ALQEALRDSS MEEG..VSAV AHRIPCRDMP
Stv1sacce   FQQESQGLIA EGWVPSTELI HLQDSLKDYI ETLGSEYSTV FNVILTNKLP
Vph1sacce   YDTNRKILIA EGWIPRDELA TLQARLGEMI ARLGIDVPSI IQVLDTNHTP
Vph1neucr   YDRARRTLIA EGWCPTNDLP LIRSTLQDVN NRAGLSVPSI INEIRTNKTP
Consensus   .......LIA E.W.P...L. .....L.... ..G....... ....T...P

             501                                              550
Vpp1caeel   PTYNKTNKFT KGFQNIVDAY GIATYREINP APYTMISFPF LFAVMFGDMG
Vpp1ratno   PTYNKTNKFT HGFQNIVDAY GIGTYREINP APYTVITFPF LFAVMFGDFG
Tj6musmu    PTLIRTNKFT EGFQNIVDAY GVGSYREVNP ALFTIITFPF LFAVMFGDFG
Vph1homsa   PTLIRTNRFT ASFQGIVDRY GVGRYQEVNP APYTIITFPF LFAVMFGDVG
Stv1sacce   PTYHRTNKFT QAFQSIVDAY GIATYKEINA GLATVVTFPF MFAIMFGDMG
Vph1sacce   PTFHRTNKFT AGFQSICDCY GIAQYREINA GLPTIVTFPF MFAIMFGDMG
```

```
Vph1neucr PTYLKTNKFT EAFQTIVNAY GTATYQEVNP AIPVIVTFPF LFAVMFGDFG
Consensus PT...TNKFT ..FQ.IVD.Y G...Y.E.N. ...T..TFPF .FA.MFGD.G

          551                                              600
Vpp1caeel HGAIMLLAAL FFILKEKQLE AARIKDEIFQ TFFGGRYVIF LMGAFSIYTG
Vpp1ratno HGILMTLFAV WMVLRESRIL SQKNENEMFS MVFSGRYIIL LMGLFSIYTG
Tj6musmu  HGFVMFLFAL LLVLNENHPR LSQSQ.EILR MFFDGRYILL LMGLFSVYTG
Vph1homsa HGLLMFLFAL AMVLAENRPA VKAAQNEIWQ TFFRGRYLLL LMGLFSIYTG
Stv1sacce HGFILFLMAL FLVLNERKFG .AMHRDEIFD MAFTGRYVLL LMGAFSVYTG
Vph1sacce HGFLMTLAAL SLVLNEKKIN .KMKRGEIFD MAFTGRYIIL LMGVFSMYTG
Vph1neucr HALIMLCAAL AMIYWEKPLK .KVTF.ELFA MVFYGRYIVL VMAVFSVYTG
Consensus HG..M.L.AL ...L.E.... ......E... .F.GRY..L LMG.FS.YTG

          601                                              650
Vpp1caeel FMYNDVFSKS INTFGSSW.. .......QNT IPESVIDYYL DDEKRSESQL
Vpp1ratno LIYNDCFSKS LNIFGSSW.. ........SV RPMFTIGNWT EETLLGSSVL
Tj6musmu  LIYNDCFSKS VNLFGSGWNV CAMYSSSHSP EEQRKMVLWN DSTIRHSRTL
Vph1homsa FIYNECFSRA TSIFPSGWSV AAMANQSG.. ........WS DAFLAQHTML
Stv1sacce LLYNDIFSKS MTIFKSGWQW ..PSTFRKG. ............E
Vph1sacce FLYNDIFSKT MTIFKSGWKW ..PDHWKKG. ............E
Vph1neucr LIYNDVFSKS MTLFDSQWKW VVPENFKEG. ............M
Consensus ..YND.FSK. ...F.S.W.. ................

          651                                              700
Vpp1caeel IL.PPETAFD GNPYPIGVDP VWNLAEGNKL SFLNSMKMKM SVLFGIAQMT
Vpp1ratno QLNPAIPGVF GGPYPFGIDP IWNIA.TNKL TFLNSFKMKM SVILGIIHML
Tj6musmu  QLDPNIPGVF RGPYPFGIDP IWNLA.TNRL TFLNSFKMKM SVILGIFHMT
Vph1homsa TLDPNVTGVF LGPYPFGIDP IWSLA.ANHL SFLNSFKMKM SVILGVVHMA
Stv1sacce SIEAKKTGV. ...YPFGLDF AWH.GTDNGL LFSNSYKMKL SILMGYAHMT
Vph1sacce SITATSVGT. ...YPIGLDW AWH.GTENAL LFSNSYKMKL SILMGFIHMT
Vph1neucr TVKAVLREPN GYRYPFGLDW RWH.GTENEL LFINSYKMKM AIILGWAHMT
Consensus ..........YP.G.D. .W.....N.L .F.NS.KMK. S...G..HM.

          701                                              750
Vpp1caeel FGVLLSYQNF IYFKSDLDIK YMFIPQMIFL SSIFIYLCIQ ILSKWLFFGA
Vpp1ratno FGVSLSLFNH IYFKKPLNIY FGFIPEIIFM SSLFGYLVIL IFYKWTAYDA
Tj6musmu  FGVVLGIFNH LHFRKKFNVY LVSVPEILFM LCIFGYLIFM IIYKWLAYSA
Vph1homsa FGVVLGVFNH VHFGQRHRLL LETLPELTFL LGLFGYLVFL VIYKWLCVWA
Stv1sacce YSFMFSYINY RAKNSKVDII GNFIPGLVFM QSIFGYLSWA IVYKW.....
Vph1sacce YSYFFSLANH LYFNSMIDII GNFIPGLLFM QGIFGYLSVC IVYKW.....
Vph1neucr YSLCFSYINA RHFKRPIDIW GNFVPGMIFF QSIFGYLVLC IIYKW.....
Consensus ........N. ..F...... ...P...F. ...FGYL... I.YKW.....

          751                                              800
Vpp1caeel VGGTVLGYKY PGSNCAPSLL IGLINMFMMK SRNAGFVDDS GETYPQCYLS
Vpp1ratno .......... HSSRNAPSLL IHFINMFLF. ...........SYPESGNA
Tj6musmu  .......... ETSREAPSIL IEFINMFLFP ...........TSKTHG
Vph1homsa .......... ARA.ASPSIL IHFINMFLFS ...........HSPSNR
Stv1sacce .....SKDWI KDDKPAPGLL NMLINMFLAP GTIDD..Q...........
Vph1sacce .....AVDWV KDGKPAPGLL NMLINMFLSP GTIDD..E...........
Vph1neucr .....SVDWF GTGRQPPGLL NMLIYMFLQP GTLDGGVE...........
Consensus ..............P..L ...INMFL..............

          801                                              850
Vpp1caeel TWYPGQSFFE TIFVLVAIAC VPVMLFGKPY FLWKEEKERR EGGHRQLATI
Vpp1ratno MLYSGQKGIQ CFLIVVAMLC VPWMLLFKPL ILRHQYLRKK HLGTLNFGGI
```

```
Tj6musmu  .LYPGQAHVQ RVLVALTVLA VPVLFLGKPL FLLWLHNGRN CFGMSRSG..
Vph1homsa LLYPRQEVVQ ATLVVLALAM VPILLLGTPL HLLHRHRRR. ...LRRRP..
Stv1sacce .LYSGQAKLQ VVLLLAALVC VPWLLLYKPL TLRRLNKNGG GGRPHGYQSV
Vph1sacce .LYPHQAKVQ VFLLLMALVC IPWLLLVKPL HFKFTHKK.. ....KSHEPL
Vph1neucr .LYPGQATVQ VILLLLAVIQ VPILLFLKPF YLRWENNRAR AKGYRGIGER
Consensus .LY..Q...Q ..L...A... VP..L..KP. .L........ ..........
```

```
              851                                            900
Vpp1caeel EIILVVLALV QVPIMLFAKP YFLYRRDKQQ SRYSTLTAES NQHQSVRADI
Vpp1ratno ......... .......... .......... .......... ...RVGNGP
Tj6musmu  ......... .......... .......... .......... ...YTLVRKD
Vph1homsa ......... .......... .......... .......... ......ADRQ
Stv1sacce GNI....... .......... .......... ......EH .EEQIAQQRH
Vph1sacce PST....... .......... .......... ......EA .DA.......
Vph1neucr SRV....... .......... .......... ......SA LDEDDEEDPS
Consensus ......... .......... .......... .......... ..........
```

```
              901                                            950
Vpp1caeel NQDDAEVVHA PEQTPKPSGH GHGHGDGP.. .....LEMGD VMVYQAIHTI
Vpp1ratno TEEDAEIIQH DQLSTHSEDA EEPTEDEV.. .....FDFGD TMVHQAIHTI
Tj6musmu  SEEEVSLLGN QDIE.EGNSR MEEGCREVTC E...EFNFGE ILMTQAIHSI
Vph1homsa EENKAGLLDL PDASVNGWSS DEEKAGGLDD EEEAELVPSE VLMHQAIHTI
Stv1sacce SAEGFQGMII SDVASVADSI NESVGGG.... .EQGPFNFGD VMIHQVIHTI
Vph1sacce SSEDLEAQQL ISAMDADDAE EEEVGSG.... .SHGE.DFGD IMIHQVIHTI
Vph1neucr NGDDYEGAAM LT........ HDEHGDG.... .EHEEFEFGE VMIHQVIHTI
Consensus ......... .......... .......... ....G. ....Q.IHTI
```

```
              951                                           1000
Vpp1caeel EFVLGCVSHT ASYLRLWALS LAHAQLSDVL WTMVFRNAFV LDGYTGAIAT
Vpp1ratno EYCLGCISNT ASYLRLWALS LAHAQLSEVL WTMVIHIGLH VRSLAGGLGL
Tj6musmu  EYCLGCISNT ASYLRLWALS LAHAQLSDVL WAMLMRVGLR VDTTYG...V
Vph1homsa EFCLGCVSNT ASYLRLWALS LAHAQLSEVL WAMVMRIGLG LGREVGVAAV
Stv1sacce EFCLNCISHT ASYLRLWALS LAHAQLSSVL WDMTISNAFS SKNSGSPLAV
Vph1sacce EFCLNCVSHT ASYLRLWALS LAHAQLSSVL WTMTIQIAFG FRGF...VGV
Vph1neucr EFCLNSVSHT ASYLRLWALS LAHQQLSAVL WSMTMAKALE SKGLGG..AI
Consensus E.CL.C.S.T ASYLRLWALS LAHAQLS.VL W.M...... ..........
```

```
             1001                                           1050
Vpp1caeel YI...LFFIF GSLSVFILVL MEGLSAFLHA LRLHWVEFQS KFYGGLGYEF
Vpp1ratno FF...IFAAF ATLTVAILLI MEGLSAFLHA LRLHWVEFQN KFYTGTGFKF
Tj6musmu  LL.LPVMAFF AVLTIFILLV MEGLSAFLHA IRLHWVEFQN KFYVGAGTKF
Vph1homsa VL.VPIFAAF AVMTVAILLV MEGLSAFLHA LRLHWVEFQN KFYSGTGYKL
Stv1sacce MKVVFLFAMW FVLTVCILVF MEGTSAMLHA LRLHWVEAMS KFFEGEGYAY
Vph1sacce FMTVALFAMW FALTCAVLVL MEGTSAMLHS LRLHWVESMS KFFVGEGLPY
Vph1neucr FLVVA.FAMF FVLSVIILII MEGVSAMLHS LRLAWVESFS KFAEFGGWPF
Consensus ......FA.. ......IL.. MEG.SA.LH. LRLHWVE... KF..G.G...
```

```
             1051       1073
Vpp1caeel APFSFEKILA EEREAEENL. ...
Vpp1ratno LPFSFEHIRE GKFDE..... ...
Tj6musmu  VPFSFSLLSS KFSNDDSIA. ...
Vph1homsa SPFTFAATDD ......... ...
Stv1sacce EPFSFR.... ..AIIE.... ...
```

```
Vph1sacce EPFAFEYKDM EVAVASASSS ASS
Vph1neucr TPFSFKQQLE ESEELKEYIG ...
Consensus .PF.F.... .......... ...
```

Residues listed in the consensus sequence are present in at least 75% of the aligned transporter sequences.

Database accession numbers

	SWISSPROT	PIR	EMBL/GENBANK
Stv1sacce	P37296	A54081	U06465; G460160
Tj6musmu	P15920	JH0287	M31226; G293678
Vph1homsa			U45285
Vph1neucr			U36396
Vph1sacce	P32563	A42970	M89778; G173173
Vpp1caeel	P30628	S15795	Z11115; G1067097
Vpp1ratno	P25286	B38656	M58758; G206430

References

[1] Li, Y.P. et al. (1996) Biochem. Biophys. Res.. Commun. 218, 813–821.

[2] Perin, M.S. et al. (1991) J. Biol. Chem. 266, 3877–3881.

[3] Manolson, M.F. et al. (1992) J. Biol. Chem. 267, 14294–14303.

[4] Lee, C.-K. and Ghoshal, K.K.D. (1990) Mol. Immunol. 27, 1137–1144.

Part 3

ABC Multidrug Resistance Proteins

Summary

Typical transporters of the white family, the example of which is the white [1] protein of *Drosophila melanogaster* (Whitdrome), mediate the import of pigment precursors into cells in the compound eye by acting as ATP-dependent efflux pumps. In *Drosophila* the white protein dimerizes with the brown protein (Browdrome) to import guanine [2] and with the scarlet protein (Scrtdrome) to import tryptophan [3]. Members of the white transporter family are also found in mammals and a few other eukaryotes. The human homolog of the white protein [4] is located on chromosome 21 and may be implicated in Down's syndrome (trisomy 21).

Statistical analysis of multiple amino acid sequence comparisons places the white transporter family in the multidrug resistance subdivision of the ATP binding cassette (ABC) superfamily [5]. Proteins in this superfamily use the energy of ATP hydrolysis to pump substrates across cell membranes. Transporters of the white family consist of a single ATP binding domain (containing the sequence patterns characteristic of the ABC transporter superfamily) fused to a transmembrane domain, with the ATP binding domain towards the N-terminus [2]. The functional transporter complex is formed from a dimer – in the case of the *Drosophila* pigment proteins, a heterodimer of white with either brown or scarlet. The transmembrane domains are predicted to contain six membrane-spanning helices by the hydropathy of their amino acid sequences.

Several amino acids are conserved within the white transporter family, including motifs unique to the family, signature motifs of the ABC superfamily, and motifs necessary for function by the criterion of site-directed mutagenesis.

Nomenclature, biological sources and substrates

CODE	DESCRIPTION [SYNONYMS]	ORGANISM [COMMON NAMES]	SUBSTRATE(S)
Adp1sacce	Probable ATP-dependent dermease precursor [ADP1, YCR11C, YCR105]	*Saccharomyces cerevisiae* [yeast]	
Browdrome	Brown protein [BW]	*Drosophila melanogaster* [fruit fly]	Guanine
Scrtdrome	Scarlet protein [ST]	*Drosophila melanogaster* [fruit fly]	Tryptophan
Whitanoal	Eye pigment protein [White]	*Anopheles albimanus* [mosquito]	Pigment precursors?
Whitdrome	White protein [W]	*Drosophila melanogaster* [fruit fly]	Guanine tryptophan
Whithomsa	White protein homolog [WHIT1]	*Homo sapiens* [human]	Pigment precursors?
Whitmusmu	White protein homolog	*Mus musculus* [mouse]	Pigment precursors?

Phylogenetic tree

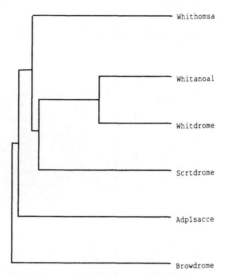

Whithomsa

Whitanoal

Whitdrome

Scrtdrome

Adplsacce

Browdrome

Proteins listed subsequently in italics are at least 90% identical to the paired transporters listed in parenthesis and are therefore not included in the phylogenetic tree: *Whitmusmu* (Whithomsa).

Proposed orientation of white protein [1] in the membrane

The model is based on predictions of membrane-spanning regions and α-helical content. The N-terminus of the protein is illustrated on the inside and is folded six times through the membrane. The predicted membrane-spanning helices are portrayed as rectangles. The numbers corresponding to the first and last residue of each membrane-spanning helix are boxed. Residues that are conserved in more than 75% of the aligned transporters (see below) are shown.

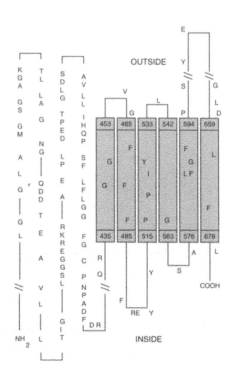

Physical and genetic characteristics

	AMINO ACIDS	*MOL. WT*	*EXPRESSION SITES*	*CHROMOSOMAL LOCUS*
Adp1sacce	1049	117 231		Chromosome 3
Browdrome	675	75 943		
Scrtdrome	666	74 506		
Whitanoal	709	79 052		
Whitdrome	687	75 672	head	
Whithomsa	674	75 169	retina	21q22.3
Whitmusmu	666	74 032		

Multiple amino acid sequence alignments

```
           1                                                50
Adp1sacce MGSHRRYLYY SILSFLLLSC SVVLAKQDET PFFEGTSSKN SRLTAQDKGN

           51                                               100
Adp1sacce DTCPPCFNCM LPIFECKQFS ECNSYTGRCE CIEGFAGDDC SLPLCGGLSP

           101                                              150
Adp1sacce DESGNKDRPI RAQNDTCHCD NGWGGINCDV CQEDFVCDAF MPDPSIKGTC

           151                                              200
Adp1sacce YKNGMIVDKV FSGCNVTNEK ILQILNGKIP QITFACDKPN QECNFQFWID
```

```
            201                                                         250
Whithomsa   .......... .......... .......... .......... ..........MA
Whitanoal   .......... .......... .......... .......... ...MTINTDD
Whitdrome   .......... .......... .......... .......... ...MGQEDQE
Scrtdrome   .......... .......... .......... .......... ..........
Adp1sacce   QLESFYCGLS DCAFEYDLEQ NTSHYKCNDV QCKCVPDTVL CGAKGSIDIS
Browdrome   .......... .......... .......... .......... ..........
Consensus   .......... .......... .......... .......... ..........


            251                                                         300
Whithomsa   AFSVGTAMNA SSYSAEMTEP KS........ ......VCVS VDEVVSSNME
Whitanoal   QYADGESKTT ISSNRRYSTS SF........ ......QDQS MEDDGINATL
Whitdrome   LLIRGGSKHP SAEHLNNGDS GA........ ......ASQS CINQGFGQ..
Scrtdrome   .MSDSDSKRI DVEAPERVEQ HE........ ......LQVM PVGSTIEVPS
Adp1sacce   DFLTETIKGP GDFSCDLETR QCKFSEPSMN DLILTVFGDP YITLKCESGE
Browdrome   .......... .......... .......... .......... ..........
Consensus   .......... .......... .......... .......... ..........


            301                                                         350
Whithomsa   ATETDLLNGH LKKVDNNLTE AQRFSSLPRR AAVNIEFRDL S.YSVPEGPW
Whitanoal   TNDKATL.IQ VWRPKSY... GSVKGQIPAQ DRLTYTWREI DVFGQAAIDG
Whitdrome   AKNYGTL.LP PSPPEDS... GSGSGQLA.. ENLTYAWHNM DIFGAVNQPG
Scrtdrome   LDSTPKL.SK RNSSERSLPL RSYSKWSPTE QGATLVWRDL CVYTNVGGSG
Adp1sacce   CVHYSEIPGY KSPSKDPTVS WQGKLVLALT AVMVLALFTF ATFYISKSPL
Browdrome   .......... ....MQ ESGGSSGQGG PSLCLEWKQL NYYVPDQEQS
Consensus   .......... .......... .......... .......... ..........


            351                                                         400
Whithomsa   WRKKGY.... .......... .......... .......... ..........
Whitanoal   KSREPLCSRL RHCFTRQRLV KDFNPR.... .......... ..........
Whitdrome   SGWRQLVNRT RGLFCNERHI PA..PR.... .......... ..........
Scrtdrome   .......... .......QRM.... .......... ..........
Adp1sacce   FRNGLGSSKS PIRLPDEDAV NNFLQNEDDT LATLSFENIT YSVPSINSDG
Browdrome   .......... .......... .......... .......N YSFWNECRKK
Consensus   .......... .......... .......... .......... ..........


            401                                                         450
Whithomsa   ..KTLLKGIS GKFNSGELVA IMGPSGAGKS TLMNILAGYR ETGMKGAV..
Whitanoal   ..KHLLKNVT GVARSGELLA VMGSSGAGKT TLLNELAFRS PPGVKISPNA
Whitdrome   ..KHLLKNVC GVAYPGELLA VMGSSGAGKT TLLNALAFRS PQGIQVSPSG
Scrtdrome   ..KRIINNST GAIQPGTLMA LMGSSGSGKT TLMSTLAFRQ PAGTVVQGDI
Adp1sacce   VEETVLNEIS GIVKPGQILA IMGGSGAGKT TLLDILAMKR KTG...HVSG
Browdrome   RELRILQDAS GHMKTGDLIA ILGGSGAGKT TLLAAISQRL RGNLTGDV..
Consensus   .....L.... G....G.L.A .MG.SGAGKT TL...LA.... .G.......


            451                                                         500
Whithomsa   ..LINGLPRD LRCFRKVSCY IMQDDMLLPH LTVQEAMMVS AHLKLQ...E
Whitanoal   IRTLNGVPVT AEQMRARCAY VQQDDLFIPS LTTKEHLMFQ AMLRMGRDVP
Whitdrome   MRLLNGQPVD AKEMQARCAY VQQDDLFIGS LTAREHLIFQ AMVRMPRHLT
Scrtdrome   L..INGRRIG PF.MHRNHGY VYQDDLFLGS VSVLEHLNFM AHLRLDRRVS
Adp1sacce   SIKVNGISMD RKSFSKIIGF VDQDDFLLPT LTVFETVLNS ALLRLPKALS
Browdrome   ..VLNGMAME RHQMTRISSF LPQFEINVKT FTAYEHLYFM SHFKMHRRTT
Consensus   ....NG.... .......... ..QDD..... .T..E.... A.........
```

```
             501                                                   550
Whithomsa  KDEGRREMVK EILTALGLLS CANTRTGS.. ....LSGGQR KRLAIALELV
Whitanoal  ATPIKMHRVD EVLQELSLVK CADTIIGVAG RVKGLSGGER KRTAFRSETL
Whitdrome  YRQ.RVARVD QVIQELSLSK CQHTIIGVPG RVKGLSGGER KRLAFASEAL
Scrtdrome  KEERRLI.IK ELLERTGLLS AAQTRIGSGD DKKVLSGGER KRLAFAVELL
Adp1sacce  F.EAKKARVY KVLEELRIID IKDRIIG.NE FDRGISGGEK RRVSIACELV
Browdrome  KAE.KRQRVA DLLLAVGLRD AAHTRI.... ..QQLSGGER KRLSLAEELI
Consensus  ........V. ..L....L.. ...T.IG... ....LSGGER KR...A.E..


             551                                                   600
Whithomsa  NNPPVMFFDE PTSGLDSASC FQVVSLMKGL A......... .........
Whitanoal  TDPHLLLCDE PTSSLDSFMA QSVLQVLKGM A......... .........
Whitdrome  TDPPLLICDE PTSGLDSFTA HSVVQVLKKL S......... .........
Scrtdrome  NNPVILFCDE PTTGLDSYSA QQLVATLYEL A......... .........
Adp1sacce  TSPLVLFLDE PTSGLDASNA NNVIECLVRL S......... .........
Browdrome  TDPIFLFCDE PTTGLDSFSA YSVIKTLRHL CTRRRIAKHS LNQVYGEDSF
Consensus  ..P..L..DE PT.GLDS..A ..V...L..L .......... .........


             601                                                   650
Whithomsa  .......... .......... .......... ..........QG.
Whitanoal  .......... .......... .......... ..........MK.
Whitdrome  .......... .......... .......... ..........QK.
Scrtdrome  .......... .......... .......... ..........QK.
Adp1sacce  .......... .......... .......... ..........SDY
Browdrome  ETPSGESSAS GSGSKSIEME VVAESHESLL QTMRELPALG VLSNSPNGTH
Consensus  .......... .......... .......... ..........


             651                                                   700
Whithomsa  GRSIICTIHQ PSAKLFELFD QLYVLSQGQC VYRGKVCNLV PYLRDLGLNC
Whitanoal  GKTIILTIHQ PSSELYCLFD RILLVAEG.V AFLGSPYQSA DFFSQLGIPC
Whitdrome  GKTVILTIHQ PSSELFELFD KILLMAEGRV AFLGTPSEAV DFFSYVGAQC
Scrtdrome  GTTILCTIHQ PSSQLFDNFN NVMLLADGRV AFTGSPQHAL SFFANHGYYC
Adp1sacce  NRTLVLSIHQ PRSNIFYLFD KLVLLSKGEM VYSGNAKKVS EFLRNEGYIC
Browdrome  KKAAICSIHQ PTSDIFELFT HIILMDGGRI VYQGRTEQAA KFFTDLGYEL
Consensus  .......IHQ P.S..F.LF. ...L...G.. ...G...... .F....G..C


             701                                                   750
Whithomsa  PTYHNPADFV MEV..ASGEY GDQN...SRL VRAVREGMCD SDH.......
Whitanoal  PPNYNPADFY VQMLAIAPNK ETEC...RET IKKICDSFAV SPI.......
Whitdrome  PTNYNPADFY VQVLAVVPGR EIES...RDR IAKICDNFAI SKV.......
Scrtdrome  PEAYNPADFL IGVLATDPGY EQAS...QRS AQHLCDQFAV SSA.......
Adp1sacce  PDNYNIADYL IDITFEAG.. PQGK...RRR IRNISDLEAG TDTNDIDNTI
Browdrome  PLNCNPADFY LKTLADKEGK ENAGAVLRAK YEHETDGLYS GS........
Consensus  P...NPADF. .......... ........D. ..........


             751                                                   800
Whithomsa  .......... .......... .......... KRDLGGDAEV
Whitanoal  .......... .......... .......... ARDI..IETA
Whitdrome  .......... .......... .......... ARDM..EQLL
Scrtdrome  .......... .......... .......... AKQR..DMLV
Adp1sacce  HQTTFTSSDG TTQREWAHLA AHRDEIRSLL RDEEDVEGTD GRRGATEIDL
Browdrome  .......... .......... .......... ...WLL ARSYSGDYLK
Consensus  .......... .......... .......... .R........
```

```
           801                                                    850
Whithomsa  NPFLWHRPSE EVKQTKRLKG LRKDSSSMEG CHSF........ ....SASC.L
Whitanoal  SQVNGDGGIE LTRTKHTTDP YFLQPMEGVD STGY....... ....RA.SWW
Whitdrome  ATKNLEKPLE .......... ......QPEN GYTY....... ....KA.TWF
Scrtdrome  N......LE IHMAQSGNFP F......DTE VESF........ ....RGVAWY
Adp1sacce  NTKLLHDKYK DSVYYAELSQ EIEEVLSEGD EESNVLNGDL PTGQQSAGFL
Browdrome  HVQNFKK... .......... .......... .......... ......IRWI
Consensus  .......... .......... .......... .......... ..........

           851                                                    900
Whithomsa  TQFCILFKRT FLSIMRDSVL THLRITSHIG IGLLIGLLYL GIGNETKK.V
Whitanoal  TQFYCILWRS WLSVLKDPML VKVRLLQTAM VASLIGSIYF G.QVLDQDGV
Whitdrome  MQFRAVLWRS WLSVLKEPLL VKVRLIQTTM VAILIGLIFL G.QQLTQVGV
Scrtdrome  KRFHVVWLRA IVTLLRDPTI QWLRFIQKIA MAFIIGACFA GTTEPSQLGV
Adp1sacce  QQLSILNSRS FKNMYRNPKL LLGNYLLTIL LSLFLGTLYY NVSNDI.SGF
Browdrome  YQVYLLMVRF MTEDLRNIRS GLIAFGFFMI TAVTLSLMYS GIGGLTQRTV
Consensus  .Q......R. .......... .......... ....G.... G.......V

           901                                                    950
Whithomsa  LSNSGFLFFS MLFLMFAALM PTVLTFPLEM GVFLREHLNY WYSLKAYYLA
Whitanoal  MNINGSLFLF LTNMTFQNVF AVINVFSAEL PVFLREKRSR LYRVDTYFLG
Whitdrome  MNINGAIFLF LTNMTFQNVF ATINVFTSEL PVFMREARSR LYRCDTYFLG
Scrtdrome  QAVQGALFIM ISENTYHPMY SVLNLFPQGF PLFMRETRSG LYSTGQYYAA
Adp1sacce  QNRMGLFFFI LTYFGFVT.F TGLSSFALER IIFIKERSNN YYSPLAYYIS
Browdrome  QDVGGSIFML SNEMIFTFSY GVTYIFPAAL PIIRREVGEG TYSLSAYVA
Consensus  ....G..F.. .....F.... .....F..... ..F.RE..... .Y....Y...

           951                                                   1000
Whithomsa  KTMAD.VPFQ IMFPVAYCSI VYWMTSQPSD AVRFVLFAAL GTMTSLVAQS
Whitanoal  KTIAE.LPLF IAVPFVFTSI TYPMIGLKAA ISHYLTTLFI VTLVANVSTS
Whitdrome  KTIAE.LPLF LTVPLVFTAI AYPMIGLRAG VLHFFNCLAL VTLVANVSTS
Scrtdrome  NILAL.LPGM IIEPLIFVII CYWLTGLRST FYAFGVTAMC VVLVMNVATA
Adp1sacce  KIMSEVVPLR VVPPILLSLI VYPMTGLNMK DNAFFKCIGI LILF.NLGIS
Browdrome  LVLS.FVPVA FFKGYVFLSV IYASIYYTRG FLLYLSMGFL MSLSAVAAVG
Consensus  ......P. ...P.....I .Y........ ......... ..L.......

          1001                                                   1050
Whithomsa  LGLL.IGAAS TSLQVATFVG PVTAIPVLLF SGFFVSFDTI P.TYLQWMSY
Whitanoal  FGYL.ISCAS SSISMALSVG PPVVIPFLIF GGFFLNSASV P.AYFKYLSY
Whitdrome  FGYL.ISCAS SSTSMALSVG PPVIIPFLLF GGFFLNSGSV P.VYLKWLSY
Scrtdrome  CGCF.FSTAF NSVPLAMAYL VPLDYIFMIT SGIFIQVNSL P.VAFWWTQF
Adp1sacce  LEILTIGIIF EDLNNSIILS VLVLLGSLLF SGLFINTKNI TNVAFKYLKN
Browdrome  YGVF.LSSLF ESDKMASECA APFDLIFLIF GGTYMNVDTV PG.....LKY
Consensus  .G........ .S...A.... .......L.F .G.F...... P.........

          1051                                                   1100
Whithomsa  ISYVRYGFEG VILSIY.GLD REDLHCDIDE TCHFQKSEAI LRELDVENAK
Whitanoal  LSWFRYANEA LLINQWADHR DGEIGCTRAN VTCPASGEII LETFNFRVED
Whitdrome  LSWFRYANEG LLINQWADVE PGEISCTSSN TTCPSSGKVI LETLNFSAAD
Scrtdrome  LSWMLYANEA MTAAQWSGVQ NITCFQESAD LPCFHTGQDV LDKYTFNESN
Adp1sacce  FSVFYYAYES LLINEVKTLM LKERKYGLNI EV...PGATI LSTFGFVVQN
Browdrome  LSLFFYSNEA LMYKFWIDID NIDCPVN.ED HPCIKTGVEV LQQGSYRNAD
Consensus  .S...Y..E. .......... .......... ......G... L.........
```

```
              1101                        1135
Whithomsa ..LYLDFIVL GIFFISLRLI AYLVLRYKIR AER..
Whitanoal ..FALDIGCL FALIVLFRLG ALFCLWLRSR SKE..
Whitdrome ..LPLDYVGL AILIVSFRVL AYLALRLRAR RKE..
Scrtdrome ..VYRNLLAM VGLYFGFHLL GYYCLWRRAR KL...
Adplsacce ..LVFDIKIL ALFNVVFLIM GYLALKWIVV EQK..
Browdrome YTYWLDCFSL VVVAVIFHIV SFGLVRRYIH RSGYY
Consensus .....D...L ......F... ....L..... .....
```

Residues listed in the consensus sequence are present in at least 75% of the aligned transporter sequences. Proteins listed subsequently in italics are at least 90% identical to the paired transporters listed in parenthesis and are therefore not included in the alignment: *Whitmusmu* (Whithomsa). Residues indicated in boldface type are also conserved in at least one other family of the ABC transporter superfamily.

Database accession numbers

	SWISSPROT	PIR	EMBL/GENBANK
Adplsacce	P25371	S19421; S40914	X59720; G5381
Browdrome	P12428	A31399; FYFFB	M20630; G157014
Scrtdrome	P45843		U39739; G1079665
Whitanoal			L76302
Whitdrome	P10090	S07263; FYFFW	X51749; G8826
Whithomsa	P45844		X91249; E218444
Whitmusmu			U34920

References
[1] Pepling, M. and Mount, S.M. (1990) Nucleic Acids Res. 18, 1633.
[2] Dreesen, T.D. et al. (1988) Mol. Cell Biol. 8, 5206–5215.
[3] Tearle, R.G. et al. (1989) Genetics 122, 595–606.
[4] Chen, H. et al. (1996) Am. J. Hum. Genet. 59, 66–75.
[5] **Higgins, C.F. (1992) Annu. Rev. Cell Biol. 8, 67–113.**

Summary

Transporters of the ABC 1 & 2 family, the example of which is the novel mouse ATP binding protein ABC 1 [1] (Abc1musmu), are believed to act as transporters, although their natural substrate is unknown. The two known members of this family are found only in mammals.

Statistical analysis of multiple amino acid sequence comparisons places the ABC 1 & 2 transporter family in the multidrug resistance subdivision of the ATP binding cassette (ABC) superfamily [2]. Proteins in this superfamily use the energy of ATP hydrolysis to pump substrates across cell membranes. Transporters of the ABC 1 & 2 transporter family consist of a single polypeptide chain made up of four domains. The N- and C-terminal halves of the protein are homologous, and each half is made up of a transmembrane domain followed by an ATP binding domain. Each transmembrane domain is predicted to contain six membrane-spanning helices by the hydropathy of the amino acid sequences and may be glycosylated.

Nomenclature, biological sources and substrates

CODE	DESCRIPTION [SYNONYMS]	ORGANISM [COMMON NAMES]	SUBSTRATE(S)
Abc1musmu	ATP binding cassette transporter 1	Mus musculus [mouse]	Unknown
Abc2musmu	ATP binding cassette transporter 2	Mus musculus [mouse]	Unknown

Proposed orientation of ABC1 [1] in the membrane

The model is based on predictions of membrane-spanning regions and α-helical content. The N-terminus of the protein is illustrated on the inside and is folded twelve times through the membrane. The predicted membrane-spanning helices are portrayed as rectangles. The numbers corresponding to the first and last residue of each membrane-spanning helix are boxed.

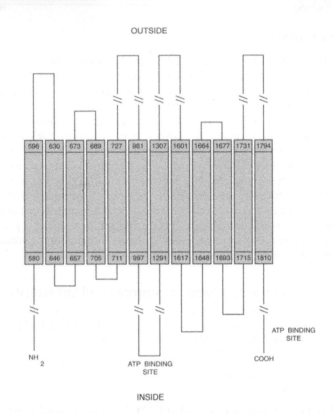

OUTSIDE

| 596 | 630 | 673 | 689 | 727 | 981 | 1307 | 1601 | 1664 | 1677 | 1731 | 1794 |

| 580 | 646 | 657 | 705 | 711 | 997 | 1291 | 1617 | 1648 | 1693 | 1715 | 1810 |

ATP BINDING
SITE

NH₂

ATP BINDING
SITE

COOH

INSIDE

Physical and genetic characteristics

	AMINO ACIDS	MOL. WT	EXPRESSION SITES	CHROMOSOMAL LOCUS
Abc1musmu	2201	246 686	uterus, many others	4A5-B3
Abc2musmu	1472	163 140	uterus, many others	2A2-B

Multiple amino acid sequence alignments

```
          1                                                  50
Abc1musmu -GCEYFALFE EQGIGVQWDN LFESPVEEDG FNLTTAVSMM LFDTFLYGVM
Abc2musmu QAC------- ---------- ----AMESRH F--------- ----------
Consensus ..C....... .......... .......E... F......... ..........

          51                                                 100
Abc1musmu TWYIEAVFPG QYGIPRPWYF PCTKSYWFGE EIDEKSHPGS SQKGVSEICM
Abc2musmu ---------- ---------- ---------E ET-------- --RG-----M
Consensus .......... .......... .........E E.......... ...G......

          101                                                150
Abc1musmu EEEPTHLRLG VSIQNLVKVY RDGMKVAVDG LALNFYEGQI TSFLGHNGAG
Abc2musmu EEEPTHLPLV VCVDKLTKVY KNDKKLALNK LSLNLYENQV VSFLGHNGAG
Consensus EEEPTHL.LL V....L.KVY ....K.A... L.LN.YE.Q. .SFLGHNGAG
```

```
           151                                                      200
Abc1musmu  KTTTMSILTG LFPPTSGTAY ILGKDIRSEM SSIRQNLGVC PQHNVLFDML
Abc2musmu  KTTTMSILTG LFPPTSGSAT IYGHDIRTEM DEIRKNLGMC PQHNVLFDRL
Consensus  KTTTMSILTG LFPPTSG.A. I.G.DIR.EM ..IR.NLG.C PQHNVLFD.L


           201                                                      250
Abc1musmu  TVEEHIWFYA RLKGLSEKHV KAEMEQMALD VGLPPSKLKS KTSQLSGGMQ
Abc2musmu  TVEEHLWFYS RLKSMAQEEI RKETDKMIED LELS-NKRHS LVQTLSGGMK
Consensus  TVEEH.WFY. RLK....... ..E...M..D ..L...K..S ....LSGGM.


           251                                                      300
Abc1musmu  RKLSVALAFV GGSKVVILDE PTAGVDPYSR RGIWELLLKY RQGRTIILST
Abc2musmu  RKLSVAIAFV GGSRAIILDE PTAGVDPYAR RAIWDLILKY KPGRTILLST
Consensus  RKLSVA.AFV GGS...ILDE PTAGVDPY.R R.IW.L.LKY ..GRTILLST


           301                                                      350
Abc1musmu  HHMDEADILG DRIAIISHGK L-CCVGSSLF LKNQLGTGYY LTLVKKDVES
Abc2musmu  HHMDEADLLG DRIAIISHGK LKCC-GSPLF LKGAYXDGYR LTLVKQPAEP
Consensus  HHMDEADLLG DRIAIISHGK L.CC.GS.LF LK.....GY. LTLVK...E.


           351                                                      400
Abc1musmu  SLSSCRN-SS STVSCLKKED SVSQSSSDAG LGSDHESDTL TIDVSAISNL
Abc2musmu  GTSQEPGLAS SPSGCPR--- ---------- LSSCSEPQ-- ------VSQF
Consensus  ..S......S S...C..... ..... ..... L.S..E.... ......S..


           401                                                      450
Abc1musmu  IRKHVSEARL VEDIGHELTY VLPYEAAKEG AFVELFHEID DRLSDLGISS
Abc2musmu  IRKHVASSLL VSDTSTELSY ILPSEAVKKG AFERLFQQLE HSLDALHLSS
Consensus  IRKHV....L V.D...EL.Y .LP.EA.K.G AF..LF.... ..L..L..SS


           451                                                      500
Abc1musmu  YGISETTLEE IFLKVAEES- GVDAETSD-- ----GTLPAR RNRRAFGDKQ
Abc2musmu  FGLMDTTLEE VFLKVSEEDQ SLENSEADVK ESRXDVLPGA EGLTAVGGQA
Consensus  .G...TTLEE .FLKV.EE.. .......D.. ......LP.. ....A.G...


           501                                                      550
Abc1musmu  S----C---- -----LHP-- ---------- --FTED-DAV DPN-DSDIDP
Abc2musmu  GNLARCSELA QSQASLQSAS SVGSARGEEG TGYSDGYGDY RPLFDNLQDP
Consensus  .....C.... .....L.... .......... .......... .P..D...DP


           551                                                      600
Abc1musmu  ES---RETDL LS-GMDGKGS YQLKGWKLTQ QQFVALLWKR LLIARRSRKG
Abc2musmu  DNVSLQEAEM EALAQVGQGS RKLEGWWLKM RQFHGLLVKR FHCARRNSKA
Consensus  ......E... ......G.GS ....GW.L.. .QF..LL.KR ...ARR..K.


           601                                                      650
Abc1musmu  FFAQIVLPAV FVCIALVFSL IVPPFGKYPS LELQPWMYNE QYT-----FV
Abc2musmu  LCSQILLPAF FVCVAMTVAL SVPEIGDLPP LVLSPSQYHN -YTQPRGNFI
Consensus  ...QI.LPA. FVC.A....L .VP..G..P. L.L.P..Y.. .Y......F.


           651                                                      700
Abc1musmu  SNDAPEDMGT QELLNALTKD PGFGTRCMEG NPIPDTPCLA GEEDWTISPV
Abc2musmu  PYANEERQEY RLRLSPDASP QQLVSTFRLP SGVGATCVLK SPANGSLGPM
Consensus  .....E.... .....L.... .......... .......... ...T..L......P.
```

```
          701                                                   750
Abc1musmu PQSIVDLFQN GNWTMKNPSP ACQCSSDKIK KMLPVC---P PGAGGLPPPQ
Abc2musmu ----LNLSSG ESRLLAARFF DSMCL-ESFT QGLPLSNFVP PPPSPAPSDS
Consensus ......L... .......... ...C...... ..LP.....P P.....P...

          751                                                   800
Abc1musmu RKQKTADILQ --NLTGRNIS DYLVKTYVQI IAKSLKNKIW ------VNEF
Abc2musmu PVXPDEDSLQ AWNMSLPPTA GPETWTSAPS LPRLVHEPVR CTCSAQGTGF
Consensus ......D.LQ ..N....... .....T.... .......... .........F

          751                                                   800
Abc1musmu RY----GGFS --LGVSNSQA LP--PSHEVN DAIKQMKKLL KLTK------
Abc2musmu SCPSSVGGHP PQMRVVTGDI LTDITGHNVS EYLLFTSDRF RLHRYGAITF
Consensus ......GG.. ....V..... L.....H.V. .......... .L........

          851                                                   900
Abc1musmu ----DTSADR FLSSLGRFMA GLDTKNNVKV WFNNKGWHAI SSFLNVINNA
Abc2musmu GNVQKSIPAS FGARVPPMVR KIAVRRVAQV LYNNKGYHSM PTYLNSLNNA
Consensus .......... F......... .........V ..NNKG.H.. ...LN..NNA

          901                                                   950
Abc1musmu ILRANLQKGE -NPSQYGITA FNHPLNLTKQ QLSEVALMTT SVDVLVSICV
Abc2musmu ILRANLPKSK GNPAAYXITV TNHPMNKTSA SLS-LDYLLQ GTDVVIAIFI
Consensus ILRANL.K.. .NP....IT. .NHP.N.T.. .LS....... ..DV...I..

          951                                                  1000
Abc1musmu IFAMSFVPAS FVVFLIQERV SKAKHLQFIS GVKPVIYWLSN FVWDMCNYV
Abc2musmu IVAMSFVPAS FVVFLVAEKS TKAKHLQFVS GCNPVIYWLAN YVWDMLNYL
Consensus I.AMSFVPAS FVVFL..E.. .KAKHLQF.S G..PVIYWL.N .VWDM.NY.

          1001                                                 1050
Abc1musmu VPATLVIIIF ICFQQKSYVS STNLPVLALL LLLYGWSITPL MYPASFVFK
Abc2musmu VPATCCVIIL FVFDLPAYTS PTNFPAVLSL FLLYGWSITPI MYPASFWFE
Consensus VPAT...II. ..F....Y.S .TN.P....L .LLYGWSITP. MYPASF.F.

          1051                                                 1100
Abc1musmu IPSTAYVVLT SVNLFIGING SVATFVLELF TNNK-LNDIND ILKSVFLIF
Abc2musmu VPSSAYVFLI VINLFIGITA TVATFLLQLF EHDKDLKVVNS YLKSCFLIF
Consensus .PS.AYV.L. ..NLFIGI.. .VATF.L.LF ...K.L...N. .LKS.FLIF

          1101                                                 1150
Abc1musmu PHFCLGRGLI DMVKNQAMAD ALERFGE-NR FVSPLSWDLVG RNLFAMAVE
Abc2musmu PNYNLGHGLM EMAYNEYINE YYAKIGQFDK MKSPFEWDIVT RGLVAMTVE
Consensus P...LG.GL. .M..N..... ......G... ..SP..WD.V. R...AM.VE

          1151                                                 1200
Abc1musmu GVVFFLITVL IQYRFFIRPR PVKAKLPPLN DEDEDVRRERQ RILDGGGQN
Abc2musmu GFVGFFLTIM CQYNFLRQPQ RLPVSTKPV- EDDVDVASERQ RVLRGDADN
Consensus G.V.F..T.. .QY.F...P. ........P. ..D.DV..ERQ R.L.G...N

          1201                                                 1250
Abc1musmu DILEIKELTK IYR-RK--RK PAVDRICIGI -PPGECFGLLG VNGAGKSTT
Abc2musmu DMVKIENLTK VYKSRKIGRI LAVDRLCLGV CVPGECFGLLG VNGAGKTST
Consensus D......LTK .Y..RK..R. .AVDRLCLG. ..PGECFGLLG VNGAGK..T
```

```
          1251                                              1300
Abc1musmu FKMLTGDTPV TRGDAFLNKN SILSNIHEVH QNMGYCPQFDA ITELLTGRE
Abc2musmu FKMLTGDEST TGGEAFVNGH SVLKDLLQVQ QSLGYCPQFDV PVDELTARE
Consensus FKMLTGD... T.G.AF.N.. S.L.....V. Q..GYCPQFD. ....LT.RE

          1301                                              1350
Abc1musmu HVEFFALLRG VPEKEVGKFG EWAIRKLGLV KYGEKYASNYS GGNKRKLST
Abc2musmu HLQLYTRLRC IPWKDEAQVV KWALEKLELT KYADKPAGTYS GGNKRKLST
Consensus H......LR. .P.K...... .WA..KL.L. KY..K.A..YS GGNKRKLST

          1351                                              1400
Abc1musmu AMALIGGPPV VFLDEPTTGM DPKARRFLWN CALSIVKEGRS VVLTSHSME
Abc2musmu AIALIGYPAF IFLDEPTTGM DPKARRFLWN LILDLIKTGRS VVLTSHSME
Consensus A.ALIG.P.. .FLDEPTTGM DPKARRFLWN ..L...K.GRS VVLTSHSME

          1401                                              1450
Abc1musmu ECEALCTRMA IMVNGRFRCL GSVQHLKNRF GDGYTIVVRIA GSNPDLKPV
Abc2musmu ECEALCTRLA IMVNGRLHCL GSIQHLKNRF GDGYMITVRTK SSQ-NVKDV
Consensus ECEALCTR.A IMVNGR..CL GS.QHLKNRF GDGY.I.VR.. .S....K.V

          1451                                              1500
Abc1musmu QEFFGLAFPG SVLKEKHRNM LQYQLPSSLS SLARIFSILSQ SKKRLHIED
Abc2musmu VRFFNRNFPE AHAQGKTPYK VQYQLKSEHI SLAQVFSKMEQ VVGVLGIED
Consensus ..FF...FP. .....K.... .QYQL.S... SLA..FSS... ....L.IED

          1501                                              1550
Abc1musmu YSVSQTTLDQ VFVNFAKDQS DDDHLKD--- ---------LS LHKNQ----
Abc2musmu YSVSQTTLDN VFVNFAKKQS DNVEQQEAEP SSLPSPLGLLS LLRPRPAPT
Consensus YSVSQTTLD. VFVNFAK.QS D......... ........LS L........

          1551                                              1590
Abc1musmu ---TVV---- -DVAV----L TSFLQDEKVK ESYV-----
Abc2musmu ELRALVADEP EDLDTEDEGL ISF-EEERAQ LSFNTDTLC
Consensus .....V.... .D.......L .SF...E... .S.......
```

Residues listed in the consensus sequence are present in both transporter sequences.

Database accession numbers

	SWISSPROT	PIR	EMBL/GENBANK
Abc1musmu	P41233		X75926; G495257
Abc2musmu	P41234		X75927; G495259

References
[1] Luciani, M.F. et al. (1994) Genomics 21, 150–159.
[2] **Higgins, C.F. (1992) Annu. Rev. Cell Biol. 8, 67–113.**

Summary

Transporters of the yeast multidrug resistance family, examples of which are the multidrug resistance protein CDR1 from *Candida albicans*[1] (Cdr1canal) and the brefeldin A resistance protein BFR1 from *Schizosaccharomyces pombe*[2] (Bfr1schpo), mediate resistance to one or, often, many structurally dissimilar antifungal agents by acting as ATP-dependent efflux pumps. Members of the family are only found in yeasts. They may be encoded chromosomally or by plasmids.

Statistical analysis of multiple amino acid sequence comparisons places the yeast multidrug resistance family in the multidrug resistance subdivision of the ATP binding cassette (ABC) superfamily. Proteins in this superfamily use the energy of ATP hydrolysis to pump substrates across cell membranes. Transporters of the yeast multidrug resistance family consist of a single polypeptide chain made up of four domains. The N- and C-terminal halves of the protein are homologous, and each half is made up of an ATP binding domain followed by a transmembrane domain. Each transmembrane domain is predicted to contain six membrane-spanning helices by the hydropathy of the amino acid sequences, so the functional transporters contain 12 such helices. Proteins may be glycosylated.

Many residues, including several long sequence motifs, are well conserved within the yeast multidrug resistance family, including motifs unique to the family, signature motifs of the ABC superfamily, and motifs necessary for function by the criterion of site-directed mutagenesis.

Nomenclature, biological sources and substrates

CODE	DESCRIPTION [SYNONYMS]	ORGANISM [COMMON NAMES]	[RESISTANCE[a]]
Bfr1schpo	Brefeldin A resistance protein [BFR1, HBA2]	*Schizosaccharomyces pombe* [yeast]	[Brefeldin A-like antibiotics]
Cdr1canal	Multidrug resistance protein [CDR1]	*Candida albicans* [yeast]	[Cycloheximide, chloramphenicol, others]
Pdr5sacce	Suppressor of toxicity of sporidesmin [PDR5, STS1, YDR1, LEM1, YOR153W]	*Saccharomyces cerevisiae* [yeast]	[Cycloheximide, sulfomethuron-methyl]
Pdrbsacce	ATP-dependent permease [PDR11, YIL013C]	*Saccharomyces cerevisiae* [yeast]	[4-Nitroquinoline-n-oxide]
Snq2sacce	SNQ2 protein [SNQ2, YDR011W, YD8119.16]	*Saccharomyces cerevisiae* [yeast]	[4-Nitroquinoline-n-oxide, triaziquone]
Ydr1sacce	YDR1 protein	*Saccharomyces cerevisiae* [yeast]	[Multiple antibiotics]

[a] Presumed substrates; protein confers resistance to specified compounds.

Phylogenetic tree

Proteins listed subsequently in italics are at least 90% identical to the paired transporters listed in parenthesis and are therefore not included in the phylogenetic tree: *Ydr1sacce* (Bfr1schpo).

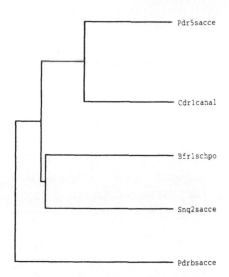

Proposed orientation of CDR1 in the membrane

The model is based on predictions of membrane-spanning regions and α-helical content. The N-terminus of the protein is illustrated on the inside and is folded 12 times through the membrane. The predicted membrane-spanning helices are portrayed as rectangles. The numbers corresponding to the first and last residue of each membrane-spanning helix are boxed. Residues that are conserved in more than 75% of the aligned transporters (see below) are shown.

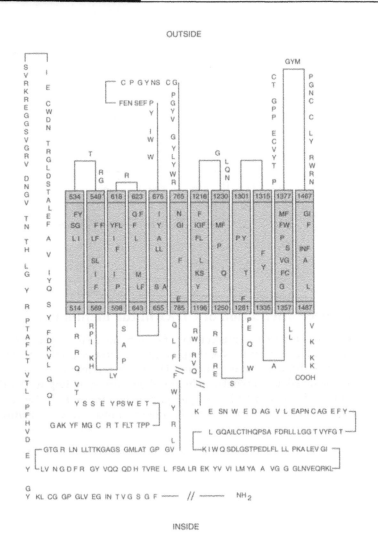

Physical and genetic characteristics

	AMINO ACIDS	MOL. WT	CHROMOSOMAL LOCUS
Bfr1schpo	1530	171750	Plasmid pDB248'
Cdr1canal	1501	169937	
Pdr5sacce	1511	170437	Chromosome 15
Pdrbsacce	1410	160405	Chromosome 9
Snq2sacce	1501	168766	Chromosome 4
Ydr1sacce	1444	163294	

Multiple amino acid sequence alignments

```
          1                                              50
Pdr5sacce ........... ..MPEAKLNN NVNDVTSYSS ASSSTENAAD LHNYNGFDEH
Cdr1canal ........... ..MSDSKMSS QDESKLEKAI SQDSSSENHS INEYHGFDAH
Bfr1schpo MNQNSDTTHG QALGSTLNHT TEVTRISNSS DHFEDSSSNV DESLDSSNPS
Snq2sacce MSNIKSTQDS .......SHN AVARSSSASF AASEESFTGI THDKDEQSDT
Pdrbsacce ........... .......... .......... .......... ..........
Consensus ........... .......... .......... .......... ..........

          51                                             100
Pdr5sacce TEARIQKLAR TLTAQSMQNS TQSAPNKSDA QSIFSSGVEG VNPIFSDPEA
Cdr1canal TSENIQNLAR TFTHDSFKDD SSAGLLKYLT H...MSEVPG VNPY..EHEE
Bfr1schpo SNEKASHTNE EYRSKGNQSY VPSSSNEPSP ESSSNSDSSS SDDSSVDRLA
Snq2sacce PADKLTKMLT G.PARDTASQ ISATVSEMAP DVVSKVE.SF ADALSRHTTR
Pdrbsacce ........... .......... .......... .......... ..........
Consensus ........... .......... .......... .......... ..........

          101                                            150
Pdr5sacce PGYDPKLDPN SENFSSAAWV KNMAHLSAAD PDFYKPYSLG CAWKNLSASG
Cdr1canal INND.QLNPD SENFNAKFWV KNLRKLFESD PEYYKPSKLG IGYRNLRAYG
Bfr1schpo GDPFEL.... GENFNLKHYL RAYKDSLQRD DIITR..SSG VCMRDHSVYG
Snq2sacce SGAFNMDSDS DDGFDAHAIF ESFVRDADEQ GIHIR..KAG VTIEDVSAKG
Pdrbsacce ........... .......... .SLSKYFNPI PDASVTFDGA TVQLEESLGA
Consensus ........... ....F..... .......... .......G.. ....S..G

          151                                            200
Pdr5sacce ASADVAYQST VVNIPYKILK SGLRKFQRSK ETNTFQILKP MDGCLNPGEL
Cdr1canal VANDSDYQPT VTNALWKLAT EGFRHFQKDD DSRYFDILKS MDAIMRPGEL
Bfr1schpo VGSGYEFLKT FPDIF...LQ P.YRAITEKQ VVE.KAILSH CHALANAGEL
Snq2sacce VDASALEGAT FGNILCLPLT I.FKGIKAKR HQKMRQIISN VNALAEAGEM
Pdrbsacce VQNDEESASE FKNVGHLE.. .......... .......ISD ITFRANEGEV
Consensus V........T ..N....... .......... .....I... .......GE.

          201                                            250
Pdr5sacce LVVLGRPGSG CTTLLKSISS NTHGFDLGAD TKISYSGYSG DDIKKHFRGE
Cdr1canal TVVLGRPGAG CSTLLKTIAV NTYGFHIGKE SQITYDGLSP HDIERHYRGD
Bfr1schpo VMVLGQPGSG CSTFLRSVTS DTVHYK.RVE GTTHYDGIDK ADMKKFFPGD
Snq2sacce ILVLGRPGAG CSSFLKVTAG EIDQFAGGVS GEVAYDGIPQ EEMMKRYKAD
Pdrbsacce VLVLGNPTSA ...LFKGLFH GHKHLKYSPE GSIRFKDNEY KQFASKCPHQ
Consensus ..VLG.PG.G C...LK.... .......... ....Y.G... ..........

          251                                            300
Pdr5sacce VVYNAEADVH LPHLTVFETL VTVARLKTPQ NRIKGVDRES YANHLAEVAM
Cdr1canal VIYSAETDVH FPHLSVGDTL EFAARLRTPQ NRGEGIDRET YAKHMASVYM
Bfr1schpo LLYSGENDVH FPSLTTAETL DFAAKCRTPN NRPCNLTRQE YVSRERHLIA
Snq2sacce VIYNGELDVH FPYLTVKQTL DFAIACKTPA LRVNNVSKKE YIASRRDLYA
Pdrbsacce IIYNNEQDIH FPYLTVEQTI DFALSCKFHI PKQERIE... ....MRDELL
Consensus ..Y..E.DVH FP.LTV..TL .FA....TP. .R....... Y.........

          301                                            350
Pdr5sacce ATYGLSHTRN TKVGNDIVRG VSGGERKRVS IAEVSICGSK FQCWDNATRG
Cdr1canal ATYGLSHTRN TNVGNDFVRG VSGGERKRVS IAEASLSGAN IQCWDNATRG
```

```
Bfr1schpo TAFGLTHTFN TKVGNDFVRG VSGGERKRVT ISEGFATRPT IACWDNSTRG
Snq2sacce TIFGLRHTYN TKVGNDFVRG VSGGERKRVS IAEALAAKGS IYCWDNATRG
Pdrbsacce KEFGLSHVKK TYVGNDYVRG VSGGERKRIS IIETFIANGS VYLWDNSTKG
Consensus ...GL.HT.N T.VGND.VRG VSGGERKRVS I.E........CWDN.TRG


          351                                            400
Pdr5sacce LDSATALEFI RALKTQADIS NTSATVAIYQ CSQDAYDLFN KVCVLDDGYQ
Cdr1canal LDSATALEFI RALKTSAVIL DTTPLIAIYQ CSQDAYDLFD KVVVLYEGYQ
Bfr1schpo LDSSTAFEFV NVLRTCANEL KMTSFVTAYQ ASEKIYKLFD RICVLYAGRQ
Snq2sacce LDASTALEYA KAIRIMTNLL KSTAFVTIYQ ASENIYETFD KVTVLYSGKQ
Pdrbsacce LDSATALEFL SITQKMAKAT RSVNFVKISQ ASDKIVSKFD KILMLGDSFQ
Consensus LDS.TALEF. ......A... .....V.IYQ .S...Y..FD K..VL..G.Q


          401                                            450
Pdr5sacce IYYGPADKAK KYFED.MGYV CPSRQTTADF LTSV...... .TSPSERTLN
Cdr1canal IFFGKATKAK EYFEK.MGWK CPQRQTTADF LTSL...... .TNPAEREPL
Bfr1schpo IYYGPADKAK QYFLD.MGFD CHPRETTPDF LTAI...... .SDPKA.RFP
Snq2sacce IYFGLIHEAK PYFAK.MGYL CPPRQATAEF LTAL...... .TDPNGFHLI
Pdrbsacce VFYGTMEECL THFHDTLQIK KNPNDCIIEY LTSILNFKFK ETSNSIVGLD
Consensus I..G....AK .YF...MG.. C..R..T..F LT........ .T.P......


          451                                            500
Pdr5sacce KDML...KKG IHIPQTPKEM NDYWVKSPNY KELMKEV... .. ........
Cdr1canal PGYE...DK. ..VPRTAQEF ETYWKNSPEY AELTKEI... .. ........
Bfr1schpo RKGF...EN. .RVPRTPDEF EQMWRNSSVY ADLMAEMESY DKRWTETTPA
Snq2sacce KPGY...EN. .KVPRTAEEF ETYWLNSPEF AQMKKDIAAY KEK.......
Pdrbsacce TPSVVSEENQ ALNINNETDL HTLWIQSPYY KHWKA..... .. ........
Consensus .......... ...P.T..E. ...W..SP.Y .................


          501                                            550
Pdr5sacce .......DQR LLN.DDEASR EAIKEAHIAK QSKRARPSSP YTVSYMMQVK
Cdr1canal .......DEY FVECERSNTR ETYRESHVAK QSNNTRPASP YTVSFFMQVR
Bfr1schpo SSEAPEKDNF GSDISATTKH ELYRQSAVAE KSKRVKDTSP YTVTFSQQLW
Snq2sacce .......... VNTEKTK EVYDESMAQE KSKYTRKKSY YTVSYWEQVK
Pdrbsacce .......... ...ITSKTVQ ECTR.....K DVNPDDISPI FSIPLKTQLK
Consensus .......... .......... E......... .S......S. YTV....Q..


          551                                            600
Pdr5sacce YLLIRNMWRL RNNIGFTLFM ILGNCSMALI LGSMFFKIMK KGDTSTFYFR
Cdr1canal YGVARNFLRM KGDPSIPIFS VFGQLVMGLI LSSVFYNL.. SQTTGSFYYR
Bfr1schpo YCLARSWERY INDPAYIGSM AFAFLFQSLI IGSIFYDM.. KLNTVDVFSR
Snq2sacce LCTQRGFQRI YGNKSYTVIN VCSAIIQSFI TGSLFYNT.. PSSTSGAFSR
Pdrbsacce TCTVRAFERI IGDRNYLISQ FVSVVVQSLV IGSLFYNI.. PLTTIGSFSR
Consensus ....R...R. .......... ........LI .GS.FY.... ...T.....R


          601                                            650
Pdr5sacce GSAMFFAILF NAFSSLLEIF SLYEARPITE KHRTYSLYHP SADAFASVLS
Cdr1canal GAAMFFAVLF NAFSSLLEIM SLFEARPIVE KHKKYALYRP SADALASIIS
Bfr1schpo GGVLFFSILF CALQSLSEIA NMFSQRPIIA KHRASALYHP AADVISSLIV
Snq2sacce GGVLYFALLY YSLMGLANIS ..FEHRPILQ KHKGYSLYHP SAEAIGSTLA
Pdrbsacce GSLTFFSILF FTFLSLADMP ASFQRQPVVR KHVQLHFYYN WVETLATNFF
Consensus G...FF..LF ....SL..I. ..F..RPI.. KH....LY.P .A....S...
```

```
          651                                                   700
Pdr5sacce EIPSKLIIAV CFNIIFYFLV DFRRNGGVFF FYLLINIVAV FSMSHLFRCV
Cdr1canal ELPVKLAMSM SFNFVFYFMV NFRRNPGRFF FYWLMCIWCT FVMSHLFRSI
Bfr1schpo DLPFRFINIS VFSIVLYFLT NLKRTAGGFW TYFLFLFIGA TCMSAFFRSL
Snq2sacce SFPFRMIGLT CFFIILFFLS GLHRTAGSFF TIYLFLTMCS EAINGLFEMV
Pdrbsacce DCCSKFILVV IFTIILYFLA HLQYNAARFF IFLLFLSVYN FCMVSLFALT
Consensus ..P...I... .F.I..YFL. ...R..G.FF ...L...... ..M..LF...

          701                                                   750
Pdr5sacce GSLTKTLSEA MVPASMLLLA LSMYTGFAIP KKKILRWSKW IWYINPLAYL
Cdr1canal GAVSTSISGA MTPATVLLLA MVIYTGFVIP TPSMLGWSRW INYINPVGYV
Bfr1schpo AGIMPNVESA SALGGIGVLA IAIYTGYAIP NIDVGWWFRW IAYLDPLQFG
Snq2sacce SSVCDTLSQA NSISGILMMS ISMYSTYMIQ LPSMHPWFKW ISYVLPIRYA
Pdrbsacce ALIAPTLSMA NLLAGILLLA IAMYASYVIY MKDMHPWFIW IAYLNPAMFA
Consensus .......S.A ......L.LA ...Y....I. ......W..W I.Y..P....

          751                                                   800
Pdr5sacce FESLLINEFH GIKFPCAE.Y VPRGPAYANI SSTESVCTVV GAVPGQDYVL
Cdr1canal FESLMVNEFH GREFQCAQ.Y VPSGPGYENI SRSNQVCTAV GSVPGNEMVS
Bfr1schpo FESLMINEFK ARQFECSQ.L IPYGSGYDNY PVANKICPVT SAEPGTDYVD
Snq2sacce FESMLNAEFH GRHMDCANTL VPSGGDYDNL SDDYKVCAFV GSKPGQSYVL
Pdrbsacce MEAILSNELF NLKLDCHESI IPRGEYYDNI SFSHKACAWQ GATLGNDYVR
Consensus FES...NEF. .....C.... .P.G.Y.N. S.....C... G..PG..YV.

          801                                                   850
Pdr5sacce GDDFIRGTYQ YYHKDKWRGF GIGMAYVVFF FFVYLFLCEY NEGAKQKGEI
Cdr1canal GTNYLAGAYQ YYNSHKWRNL GITIGFAVFF LAIYIALTEF NKGAMQKGEI
Bfr1schpo GSTYLYISFN YKTRQLWRNL AIIIGYYAFL VFVNIVASET LNFNDLKGEY
Snq2sacce GDDYLKNQFQ YVYKHTWRNF GILWCFLLGY VVLKVIFTEY KRPVKGGGDA
Pdrbsacce GRDYLKSGLK YTYHHVWRNF GIIIGFLCFF LFCSLLAAEY ITPLFTRENL
Consensus G..YL..... Y.....WRN. GI......F. ........E. .......G..

          851                                                   900
Pdr5sacce LVFPR..... ....SIVKRM KKRGVLTEKN ANDPENVGER SDLSSDRKM.
Cdr1canal VLFLK..... ....GSLKKH KRKTAASNKG DIEAGPVAGK LDYQDEAEA.
Bfr1schpo LVFRR..... ....GHAPDA VKAAVNEGGK PLDLETGQD. ...TQGGDV.
Snq2sacce LIFKK..... ....GSKRFI AHADEESPDN VNDIDAKE.. .......QF.
Pdrbsacce LRWNNYLKRY CPFLNSQKKN NKSAITNNDG VCTPKTPIAN FSTSSSSVPS
Consensus L.F....... .......... .......... .......... ..........

          901                                                   950
Pdr5sacce .......... LQESSEEESD TYGEIG.LSK SEAIFHWRNL CYEVQIKAET
Cdr1canal .......... VNNEKFTEKG STGSVD.FPE NREIFFWRDL TYQVKIKKED
Bfr1schpo .......... VKESPDNEEE LNKEYEGIEK GHDIFSWRNL NYDIQIKGEH
Snq2sacce .......... SSESSGANDE VFDDLE.... AKGVFIWKDV CFTIPYEGGK
Pdrbsacce VSHQYDTDYN IKHPDETVNN HTKESVAMET QKHVISWKNI NYTI....GD
Consensus .......... .......... .......... ...F.W... .Y........

          951                                                  1000
Pdr5sacce RRILNNVDGW VKPGTLTALM GASGAGKTTL LDCLAERVTM GVIT.GDILV
Cdr1canal RVILDHVDGW VKPGQITALM GASGAGKTTL LNCLSERVTT GIITDGERLV
Bfr1schpo RRLLNGVQGF VVPGKLTALM GESGAGKTTL LNVLAQRVDT GVVT.GDMLV
Snq2sacce RMLLDNVSGY CIPGTMTALM GESGAGKTTL LNTLAQR.NV GIIT.GDMLV
Pdrbsacce KKLINDASGY ISSG.LTALM GESGAGKTTL LNVLSQRTES GVVT.GELLI
Consensus R..L..V.G. ..PG..TALM G.SGAGKTTL LN.L..R... G..T.G..LV
```

```
          1001                                              1050
Pdr5sacce NGIPRDK..S FPRSIGYCQQ QDLHLKTATV RESLRFSAYL RQPAEVSIEE
Cdr1canal NGHALDS..S FQRSIGYVQQ QDVHLPTSTV REALQFSAYL RQSNKISKKE
Bfr1schpo NGRGLDS..T FQRRTGYVQQ QDVHIGESTV REALRFSAAL RQPASVPLSE
Snq2sacce NGRPIDA..S FERRTGYVQQ QDIHIAELTV RESLQFSARM RRPQHLPDSE
Pdrbsacce DGQPLTNIDA FRRSIGFVQQ QDVHLELLTV RESLEISCVL RG......DG
Consensus NG...D.... F.R..GYVQQ QD.H....TV RE.L.FSA.L R........E

          1051                                              1100
Pdr5sacce KNRYVEEVIK ILEMEKYADA VVGVAGEGLN VEQRKRLTIG VELTAKPKLL
Cdr1canal KDDYVDYVID LLEMTDYADA LVGVAGEGLN VEQRKRLTIG VELVAKPKLL
Bfr1schpo KYEYVESVIK LLEMESYAEA IIGTPGSGLN VEQRKRATIG VELAAKPALL
Snq2sacce KMDYVEKIIR VLGMEEYAEA LVGEVGCGLN VEQRKKLSIG VELVAKPDLL
Pdrbsacce DRDYLGVVSN LLRLPS..EK LVA....DLS PTQRKLLSIG VELVTKPSLL
Consensus K..YV..VI. .L.M..YA.A .VG..G.GLN VEQRK.L.IG VEL.AKP.LL

          1101                                              1150
Pdr5sacce VFLDEPTSGL DSQTAWSICQ LMKKLANHGQ AILCTIHQPS AILMQEFDRL
Cdr1canal LFLDEPTSGL DSQTAWSICK LMRKLADHGQ AILCTIHQPS ALIMAEFDRL
Bfr1schpo LFLDEPTSGL DSQSAWSIVC FLRKLADAGQ AILCTIHQPS AVLFDQFDRL
Snq2sacce LFLDEPTSGL DSQSSWAIIQ LLRKLSKAGQ SILCTIHQPS ATLFEEFDRL
Pdrbsacce LFLDEPTSGL DAEAALTIVQ FLKKLSMQGQ AILCTIHQPS KSVISYFDNI
Consensus LFLDEPTSGL DSQ.AW.I.. ...KL...GQ AILCTIHQPS A.....FDRL

          1151                                              1200
Pdr5sacce LFMQRGGKTV YFGDLGEGCK TMIDYFESHG AHKC..PADA NPAEWMLEVV
Cdr1canal LFLQKGGRTA YFGELGENCQ TMINYFEKYG ADPC..PKEA NPAEWMLQVV
Bfr1schpo LLLQKGGKTV YFGDIGEHSK TLLNYFESHG AVHC..PDDG NPAEYILDVI
Snq2sacce LLLRKGGQTV YFGDIGKNSA TILNYFERNG ARKC..DSSE NPAEYILEAI
Pdrbsacce YLLKRGGECV YFGSLPNAC. ...DYFVAHD RRLTFDREMD NPADFVIDVV
Consensus L.L..GG.TV YFG..G.... T...YFE..G A..C...... NPAE..L.V.

          1201                                              1250
Pdr5sacce GAAPGSHANQ D.................... ....YYEVWR NSEEYRAVQS
Cdr1canal GAAPGSHAKQ D.................... ....YFEVWR NSSEYQAVRE
Bfr1schpo GAGATATTNR D.................... ....WHEVWN NSEERKAISA
Snq2sacce GAGATASVKE D.................... ....WHEKWL NSVEFEQTKE
Pdrbsacce GSGSTNIPMD DAEKPTSSKI DEPVSYHKQS DSINWAELWQ SSPEKVRVAD
Consensus GA........ D................... ......E.W. NS.E......

          1251                                              1300
Pdr5sacce ELDWMERELP KKG..SITAA EDKHEFSQSI IYQTKLVSIR LFQQYWRSPD
Cdr1canal EINRMEAELS KLP..RDNDP EALLKYAAPL WKQYLLVSWR TIVQDWRSPG
Bfr1schpo ELDKINASFS NSEDKKTLSK EDRSTYAMPL WFQVKMVMTR NFQSYWREPS
Snq2sacce KVQDLINDLS KQETKSEVG. DKPSKYATSY AYQFRYVLIR TSTSFWRSLN
Pdrbsacce DLLLLEEEAR KSGVDFTTSV WSPPSYME.. ..QIKLITKR QYICTKRDMT
Consensus .......... K......... ...........Q...V..R .....WR...

          1301                                              1350
Pdr5sacce YLWSKFILTI FNQLFIGFTF FKAGTSLQGL QNQMLAVFMF TVIFNPILQQ
Cdr1canal YIYSKIFLVV SAALFNGFSF FKAKNNMQGL QNQMFSVFMF FIPFNTLVQQ
Bfr1schpo ILMSKLALDI FAGLFIGFTF YNQGLGVQNI QNKLFAVFMA TVLAVPLING
Snq2sacce YIMSKMMLML VGGLYIGFTF FNVGKSYVGL QNAMFAAFIS IILSAPAMNQ
Pdrbsacce YVFAKYALNA GAGLFIGFSF WRTKHNINGL QDAIFLCFMM LCVSSPLINQ
Consensus Y..SK..L.. ...LFIGF.F ........GL QN.....FM. .....P...Q
```

132

```
           1351                                              1400
Pdr5sacce  YLPSFVQQRD LYEARERPSR TFSWISFIFA QIFVEVPWNI LAGTIAYFIY
Cdr1canal  MLPYFVKQRD VYEVREAPSR TFSWFAFIAG QITSEIPYQV AVGTIAFFCW
Bfr1schpo  LQPKFIELRN VFEVREKPSN IYSWVAFVFS AIIVEIPFNL VFGTLFFLCW
Snq2sacce  IQGRAIASRE LFEVRESQSN MFHWSLVLIT QYLSELPYHL FFSTIFFVSS
Pdrbsacce  VQDKALQSKE VYIAREARSN TYHWTVLLIA QTIVELPLAI SSSTLFFLCC
Consensus  ........R. ..E.RE..S. ...W...... Q...E.P... ...T..F...

           1401                                              1450
Pdr5sacce  YYPIGFYSNA SAAGQLHERG ALFWLFSCAF YVYVGSMGLL VISFNQVAES
Cdr1canal  YYPLGLYNNA TPTDSVNPRG VLMWMLVTAF YVYTATMGQL CMSFSELADN
Bfr1schpo  FYPIKFYKHI HHPGD...KT GYAWLLYMFF QMYFSTFGQA VASACPNAQT
Snq2sacce  YFPLRIF... FEASR...SA VYFLNYCIMF QLYYVGLGLM ILYMSPNLPS
Pdrbsacce  YFCCGFETSA RVAG...... .VFYLNYILF SMYYLSFGLW LLYSAPDLQT
Consensus  Y.P...... .......... ..........F ..Y....G.. ..........

           1451                                              1500
Pdr5sacce  AANLASLLFT MSLSFCGVMT TPSAMPRFWI FMYRVSPLTY FIQALLAVGV
Cdr1canal  AANLATLLFT MCLNFCGVLA GPDVLPGFWI FMYRCNPFTY LVQAMLSTGL
Bfr1schpo  ASVVNSLLFT FVITFNGVLQ PNSNLVGFWH WMHSLTPFTY LIEGLLSDLV
Snq2sacce  ANVILGLCLS FMLSFCGVTQ PVSLMPGFWT FMWKASPYTY FVQNLVGIML
Pdrbsacce  AAVFVAFLYS FTASFCGVMQ PYSLFPRFWT FMYRVSPYTY FIETFVSLLL
Consensus  A.....LL.. ....FCGV.. ..S..P.FW. FM....P.TY ..........

           1501                                              1550
Pdr5sacce  ANVDVKCADY ELLEFTPPSG MTCGQYMEPY LQLAK.TGYL TDENATDTCS
Cdr1canal  ANTFVKCAER EYVSVKPPNG ESCSTYLDPY IKFA..GGYF ETRND.GSCA
Bfr1schpo  HGLPVECKSH EMLTINPPSG QTCGEYMSAF LTNNTAAGNL LNPNATTSCS
Snq2sacce  HKKPVVCKKK ELNYFNPPNG STCGEYMKPF L..EKATGYI ENPDATSDCA
Pdrbsacce  HDREVNCSTS EMVPSQPVMG QTCGQFMKPF I..DEFGGKL HINNTYTVCA
Consensus  ....V.C... E.....PP.G .TCG.YM.P. .......G.. ...N....C.

           1551                                              1600
Pdr5sacce  FCQISTTNDY LANVNSFYSE RWRNYGIFIC YIAFNYIAGV FFYWLARVPK
Cdr1canal  FCQMSSTNTF LKSVNSLYSE RWRNFGIFIA FIAINIILTV IFYWLARVPK
Bfr1schpo  YCPYQTADQF LERFSMRYTH RWRNLGIFVG YVFFNIFAVL LLFYVFRVMK
Snq2sacce  YCIYEVGDNY LTHISSKYSY LWRNFGIFWI YIFFNIIAMV CVYYLFHVRQ
Pdrbsacce  YCMYTVGDDF LAQENMSYHH RWRNFGFEWV FVCFNIAAMF VGFYLTYIKK
Consensus  .C........ L......Y.. RWRN.GIF.. ...FNI.A.. ....L..V.K

           1601                           1630
Pdr5sacce  K...N.GKLS KK........ ..........
Cdr1canal  G...NREKKN KK........ ..........
Bfr1schpo  ...LRSTWLG KKITGTG... ..........
Snq2sacce  SSFLSPVSIL NKIKNIRKKK Q.........
Pdrbsacce  IWPSVIDGIK KCIPSMRRSK TSHNPNEQSV
Consensus  .......... KK........ ..........
```

Residues listed in the consensus sequence are present in at least 75% of the aligned sequences. Proteins listed subsequently in italics are at least 90% identical to the paired transporters listed in parenthesis and are therefore not included in the alignment: *Ydr1sacce* (Bfr1schpo). Residues indicated in boldface type are also conserved in at least one other family of the ABC transporter superfamily.

Database accession numbers

	SWISSPROT	PIR	EMBL/GENBANK
Bfr1schpo	P41820		X82891; G609264
Cdr1canal	P43071		X77589; G454277
Pdr5sacce	P33302	S34702; A49730	X74113; G395259
Pdrbsacce	P40550		Z47047; G763333
Snq2sacce	P32568	S30918	X66732; G295839
Ydr1sacce			D26548

References

[1] Prasad, R. et al. (1995) Curr. Genet. 27, 320–329.

[2] Nagao, K. et al. (1995) J. Bacteriol. 177, 1536–1543.

[3] **Higgins, C.F. (1992) Annu. Rev. Cell Biol. 8, 67–113.**

Summary

Transporters of the cystic fibrosis transmembrane conductance regulator family, the example of which is the human cystic fibrosis transmembrane conductance regulator [1,2] (Cftrhomsa) act as cAMP-dependent chloride channels [1]. In humans, mutations of the CFTR gene leading to defects in channel function cause cystic fibrosis [1]. This disorder of exocrine gland function is the most common genetic disease in the Caucasian population, affecting 1 in 2000–2500 live births [3,4]. Members of this family have only been found in vertebrates.

Statistical analysis of multiple amino acid sequence comparisons places the cystic fibrosis transmembrane conductance regulator family in the multidrug resistance subdivision of the ATP binding cassette (ABC) superfamily [5]. The CFTR proteins are the only ABC transporters which function principally as channels, rather than as ATP-dependent active transporters. Transporters of this family consist of a single polypeptide chain made up of five domains. Two homologous two-domain polypeptides, each made up of a transmembrane domain followed by an ATP binding domain, are separated by a central "R domain" which contains many charged residues and phosphorylation sites [6]. This domain is unique to this family. The most common cystic fibrosis-causing mutation in the Caucasian population – ΔF508 [7] – occurs in the N-terminal ATP binding domain. Each transmembrane domain is predicted to contain six membrane-spanning helices by the hydropathy of its amino acid sequences, so the functional transporters contain 12 such helices. Proteins are glycosylated.

Many residues, including several long sequence motifs, are well conserved within the cystic fibrosis transmembrane conductance regulator family, including motifs unique to the family, signature motifs of the ABC superfamily, and motifs necessary for function by the criterion of site-directed mutagenesis.

Nomenclature, biological sources and substrates

CODE	DESCRIPTION [SYNONYMS]	ORGANISM [COMMON NAMES]	SUBSTRATE(S)
Cftrbosta	Cystic fibrosis transmembrane conductance regulator [CFTR, cAMP-dependent Cl⁻ channel]	Bos taurus [cow]	Cl⁻
Cftrhomsa	Cystic fibrosis transmembrane conductance regulator [CFTR, cAMP-dependent Cl⁻ channel]	Homo sapiens [human]	Cl⁻
Cftrmusmu	Cystic fibrosis transmembrane conductance regulator [CFTR, cAMP-dependent Cl⁻ channel]	Mus musculus [mouse]	Cl⁻
Cftrorycu	Cystic fibrosis transmembrane conductance regulator [CFTR, cAMP-dependent Cl⁻ channel]	Oryctolagus cuniculus [rabbit]	Cl⁻
Cftrsquac	Cystic fibrosis transmembrane conductance regulator [CFTR, cAMP-dependent Cl⁻ channel]	Squalus acanthias [dogfish]	Cl⁻
Cftrxenla	Cystic fibrosis transmembrane conductance regulator [CFTR, cAMP-dependent Cl⁻ channel]	Xenopus laevis [toad]	Cl⁻

Phylogenetic tree

Proteins listed subsequently in italics are at least 90% identical to the paired transporters listed in parenthesis and are therefore not included in the phylogenetic tree: *Cftrbosta* (Cftrhomsa); *Cftrorycu* (Cftrmusmu).

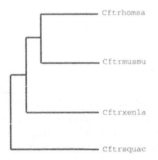

Proposed orientation of CFTR [2] in the membrane

The model is based on predictions of membrane-spanning regions and α-helical content. The N-terminus of the protein is illustrated on the inside and is folded 12 times through the membrane. The predicted membrane-spanning helices

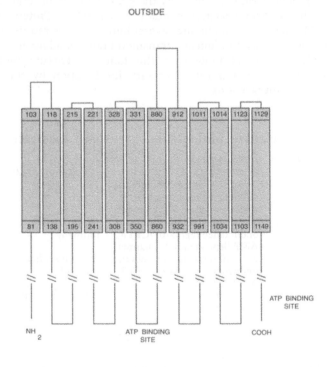

are portrayed as rectangles. The numbers corresponding to the first and last residue of each membrane-spanning helix are boxed. More than half of the residues are conserved in at least 75% of the members of the cystic fibrosis transmembrane conductance regulator family and, therefore, are not mapped onto the model.

Physical and genetic characteristics

	AMINO ACIDS	MOL. WT	EXPRESSION SITES	CHROMOSOMAL LOCUS
Cftrbosta	1481	167 758		
Cftrhomsa	1480	168 173	lung	7q31–q32
Cftrmusmu	1476	167 852	lung	
Cftrorycu	1450	164 629	heart ventricle	
Cftrsquac	1492	169 384	rectal gland	
Cftrxenla	1485	168 895		

Multiple amino acid sequence alignments

```
          1                                                  50
Cftrhomsa MQRSPLEKAS VVSKLFFSWT RPILRKGYRQ RLELSDIYQI PSVDSADNLS
Cftrmusmu MQKSPLEKAS FISKLFFSWT TPILRKGYRH HLELSDIYQA PSADSADHLS
Cftrxenla MQKTPLEKAS IFSQIFFSWT KPILWKGYRQ RLELSDIYQI HPGDSADNLS
Cftrsquac MQRSPIEKAN AFSKLFFRWP RPILKKGYRQ KLELSDIYQI PSSDSADELS
Consensus MQ.SPLEKAS ..SKLFFSWT .PIL.KGYRQ .LELSDIYQI PS.DSAD.LS


          51                                                 100
Cftrhomsa EKLEREWDRE LA.SKKNPKL INALRRCFFW RFMFYGIFLY LGEVTKAVQP
Cftrmusmu EKLEREWDRE QA.SKKNPQL IHALRRCFFW RFLFYGILLY LGEVTKAVQP
Cftrxenla ERLEREWDRE VATSKKNPKL INALRRCFFW KFLFYGILLY LGEVTKAVQP
Cftrsquac EMLEREWDRE LATSKKNPKL VNALRRCFFW RFLFYGILLY FVEFTKAVQP
Consensus E.LEREWDRE .A.SKKNPKL INALRRCFFW RFLFYGILLY LGEVTKSVQP


          101                                                150
Cftrhomsa LLLGRIIASY DPDNKEERSI AIYLGIGLCL LFIVRTLLLH PAIFGLHHIG
Cftrmusmu VLLGRIIASY DPENKVERSI AIYLGIGLCL LFIVRTLLLH PAIFGLHRIG
Cftrxenla LLLGRIIASY DRDNEHERSI AYYLAIGLCL LFVVRMLLLH PAIFGLHHIG
Cftrsquac LCLGRIIASY NAKNTYEREI AYYLALGLCL LFVVRTLFLH PAVFGLQHLG
Consensus LLLGRIIASY D..N..ERSI A.YL.IGLCL LF.VRTLLLH PAIFGLHHIG


          151                                                200
Cftrhomsa MQMRIAMFSL IYKKTLKLSS RVLDKISIGQ LVSLLSNNLN KFDEGLALAH
Cftrmusmu MQMRTAMFSL IYKKTLKLSS RVLDKISIGQ LVSLLSNNLN KFDEGLALAH
Cftrxenla MQMRIAMFSL IYKKTLKLSS KVLDKISTGQ LVSLLSNNLN KFDEGLALAH
Cftrsquac MQMRIALFSL IYKKILKMSS RVLDKIDTGQ LVSLLSNNLN KFDEGVAVAH
Consensus MQMRIAMFSL IYKKTLKLDD RVLDKISTGQ LVSLLSNNLN KFDEGLALAH


          201                                                250
Cftrhomsa FVWIAPLQVA LLMGLIWELL QASAFCGLGF LIVLALFQAG LGRMMMKYRD
Cftrmusmu FIWIAPLQVT LLMGLLWDLL QFSAFCGLGL LIILVIFQAI LGKMMVKYRD
Cftrxenla FVWIAPLQVL LLMGLLWDLL QASAFCGLGF LIILSLFQAR LGRMMMKYKD
Cftrsquac FVWIAPVQVV LLMGLIWNEL TEFVFCGLGF LIMLALFQAW LGKKMMQYRD
Consensus FVWIAPLQV  LLMGL.W.LL Q.SAFCGLGF LI.L.LFQA. LG.MMMKYRD
```

137

```
               251                                                  300
Cftrhomsa QRAGKISERL VITSEMIENI QSVKAYCWEE AMEKMIENLR QTELKLTRKA
Cftrmusmu QRAAKINERL VITSEIIDNI YSVKAYCWES AMEKMIENLR EVELKMTRKA
Cftrxenla KRAGKINERL VITSQIIENI QSVKAYCWEN AMEKIIETIR ETELKLTRKA
Cftrsquac KRAGKINERL AITSEIIDNI QSVKVYCWED AMEKIIDDIR QVELKLTRKV
Consensus .RAGKINERL VITSEII.NI QSVKAYCWE. AMEK.IE..R ..ELKLTRKA

               301                                                  350
Cftrhomsa AYVRYFNSSA FFFSGFFVVF LSVLPYALIK GIILRKIFTT ISFCIVLRMA
Cftrmusmu AYMRFFTSSA FFFSGFFVVF LSVLPYTVIN GIVLRKIFTT ISFCIVLRMS
Cftrxenla AYVRYFNSSA FFFSGFFVVF LSIVPHLLLD GISLRKIFTT ISFSIVLRMA
Cftrsquac AYCRYFSSSA FFFSGFFVVF LSVVPYAFIH TIKLRRIFTT ISYNIVLRMT
Consensus AY.RYF.SSA FFFSGFFVVF LSV.PY..I. GI.LRKIFTT ISF.IVLRM.

               351                                                  400
Cftrhomsa VTRQFPWAVQ TWYDSLGAIN KIQDFLQKQE YKTLEYNLTT TEVVMENVTA
Cftrmusmu VTRQFPTAVQ IWYDSFGMIR KIQDFLQKQE YKVLEYNLMT TGIIMENVTA
Cftrxenla VTRQFPWAVQ TWYDSLGVIN KIQEFLQKEE YKSLEYNLTT TEVAMENVSA
Cftrsquac VTRQFPSAIQ TWYDSLGAIR KIQDFLHKDE HKTVEYNLTT KEVEMVNVTA
Consenuss VTRQFP.AVQ TWYDSLG.I. KIQDFLQK.E YK.LEYNLTT TEV.MENVTA

               401                                                  450
Cftrhomsa FWEEGFGELF EKAKQNNNNR KTSNGDDSLF FSNFSLLGTP VLKDINFKIE
Cftrmusmu FWEEGFGELL QKAQQSNGDR KHSSDENNVS FSHLCLVGNP VLKNINLNIE
Cftrxenla SWDEGIGEFF EKAKLEVNGG NISNEDPSAF FSNFSLHVAP VLRNINFKIE
Cftrsquac SWDEGIGELF EKVKQNDSER KMANGDDGLF FSNFSLHVTP VLKNISFKLE
Consensus .W.EG.GELF EKAKQ....R K.SN.D...F FSNFSL...P VLKNINFKIE

               451                                                  500
Cftrhomsa RGQLLAVAGS TGAGKTSLLM MIMGELEPSE GKIKHSGRIS FCSQFSWIMP
Cftrmusmu KGEMLAITGS TGLGKTSLLM LILGELEASE GIIKHSGRVS FCSQFSWIMP
Cftrxenla KGQLLAIAGS TGAGKTSLLM MIMGELEPSA GKIKHSGRIS FSPQVSWIMP
Cftrsquac KGELLAIAGS TGSGKSSLLM MIMGELEPSD GKIKHSGRIS YSPQVPWIMP
Consensus KG.LLAIAGS TG.GKTSLLM MIMGELEPS. GKIKHSGRIS F..Q.SWIMP

               501                                                  550
Cftrhomsa GTIKENIIFG VSYDEYRYRS VIKACQLEED ISKFAEKDNI VLGEGGITLS
Cftrmusmu GTIKENIIFG VSYDEYRYKS VVKACQLQQD ITKFAEQDNT VLGEGGVTLS
Cftrxenla GTIKENIVFG VSYDQYRYLS VIKACQLEED ISKFPEKDNT VLGEGGITLS
Cftrsquac GTIKDNIIFG LSYDEYRYTS VVNACQLEED ITVFPNKDKT VLGDGGITLS
Consensus GTIKENIIFG VSYDEYRY S V KACQLEED I KF EKDNT VLGEGGITLS

               551                                                  600
Cftrhomsa GGQRARISLA RAVYKDADLY LLDSPFGYLD VLTEKEIFES CVCKLMANKT
Cftrmusmu GGQRARISLA RAVYKDADLY LLDSPFGYLD VFTEEQVFES CVCKLMANKT
Cftrxenla GGQRARISLA RAVYKDADLY LLDSPFSYLD LFTEKEIFES CVCKLMANKT
Cftrsquac GGQRARISLA RALYKDADLY LLDSPFSHLD VTTEKDIFES CLCKLMVNKT
Consensus GGQRARISLA RAVYKDADLY LLDSPF.YLD V.TE..IFES CVCKLMANKT

               601                                                  650
Cftrhomsa RILVTSKMEH LKKADKILIL HEGSSYFYGT FSELQNLQPD FSSKLMGCDS
Cftrmusmu RILVTSKMEH LRKADKILIL HQGTSYFYGT FSELQSLRPS FSSKLMGYDT
Cftrxenla RILVTSKVEQ LKKADKVLIL HEGSCYFYGT FSELEDQRPE FSSHLIG...
Cftrsquac RILVTSKLEH LKKADKILLL HEGHCYFYGT FSELQGEKPD FSSQLLGSVH
Consensus RILVTSK.EH LKKDAKILIL HEG..YFYGT FSELQ..P.. FSS.L.G...
```

```
           651                                                  700
Cftrhomsa  FDQFSAERRN SILTETLHRF SL...EGDAP VSWTETKKQS FKQ.TGEFGE
Cftrmusmu  FDQFTEERRS SILTETLRRF SV...DDSSA PWS..KPKQS FRQ.TGEVGE
Cftrxenla  FDHFNAERRN SIITETLRRC SI...DSDPS AVRNEVKNKS FKQ.VADFTE
Cftrsquac  FDSFSAERRN SILTETFRRC SVSSGDGAGL GSYSETRKAS FKQPPPEFNE
Consensus  FD.F.AERRN SILTETLRR. S....D.... ....E..K.S FKQ...EF.E

           701                                                  750
Cftrhomsa  KRKNS.ILNP INSIRKFSIV QKTPLQMNGI EED..SDEPL ERRLSLVPDS
Cftrmusmu  KRKNS.ILNS FSSVRKISIV QKTPL...CI DGE..SDDLQ EKRLSLVPDS
Cftrxenla  KRKSS.IINP RKSSRKFSLM QKSQPQMSGI EEEDMPAEQG ERKLSLVPES
Cftrsquac  KRKSSLIVNP ITSNKKFSLV QTAMSYPQTN GMEDATSEPG ERHFSLIPEN
Consensus  KRK.S.I.NP ..S.RKFS.V QK......I ..E....E   ER LSLVP.S

           751                                                  800
Cftrhomsa  EQGEAILPRI SVISTGPTLQ ARRRQSVLNL MTH.SVNQGQ NIHRKTTAST
Cftrmusmu  EQGEAALPRS NMIATGPTFP GRRRQSVLDL MTF.TPNSGS SNLQRTRTSI
Cftrxenla  EQGEASLPRS NFLNTGPTFQ GRRRQSVLNL MTRTSISQGS NAFATRNASV
Cftrsquac  ELGEPTKPRS NIFKSELPFQ AHRRQSVLAL MTHSSTS..P NKIHARRSAV
Consensus  EQGEA.LPRS N...TGPTFQ .RRRQSVL.L MT..S...G. N.......S.

           801                                                  850
Cftrhomsa  RKVSLAPQAN L.T.ELDIYS RRLSQETGLE ISEEINEEDL KECFFDDMES
Cftrmusmu  RKISLVPQIS L.N.EVDVYS RRLSQDSTLN ITEEINEEDL KECFLDDVIK
Cftrxenla  RKMSVNSYSN S.SFDLDIYN RRLSQDSILE VSEEINEEDL KECFLDDTDS
Cftrsquac  RKMSMLSQTN FASSEIDIYS RRLSEDGSFE ISEEINEEDL KECFADEEEI
Consensus  RK.S...Q.N ....E DIYS RRLSQD..LE ISEEINEEDL KECF.DD...

           851                                                  900
Cftrhomsa  IPAVTTWNTY LRYITVHKSL IFVLIWCLVI FLAEVAASLV VLWLLGNTP.
Cftrmusmu  IPPVTTWNTY LRYFTLHKGL LLVLIWCVLV FLVEVAASLF VLWLLKNNP.
Cftrxenla  QSPTTTWNTY LRFLTAHKNF IFILVFCLVI FFVEVAASSA WLWIIKRNAP
Cftrsquac  QNVTTTWSTY LRYVTTNRNL VFVLILCLVI FLAEVAASLA GLWIISGLAI
Consensus  ....TTWNTY LRY.T.HK.L .FVLI.CLVI FL.EVAASL. .LW.......

           901                                                  950
Cftrhomsa  ...LQDKGNS THSRNNS.YA VIITSTSSYY VFYIYVGVAD TLLAMGFFRG
Cftrmusmu  ...VNSGNNG TKISNSS.YV VIITSTSFYY IFYIYVGVAD TLLALSLFRG
Cftrxenla  AINMTSNENV SEVSD.T.LS VIVTHTSFYY VFYIYVGVAD SLLALGIFRG
Cftrsquac  NTGSQTNDTS TDLSHLSVFS KFITNGSHYY IFYIYVGLAD SFLALGVIRG
Consensus  ........N  T..S..S... VIIT.TS.YY .FYIYVGVAD .LLALG.FRG

           951                                                 1000
Cftrhomsa  LPLVHTLITV SKILHHKMLH SVLQAPMSTL NTLKAGGILN RFSKDIAILD
Cftrmusmu  LPLVHTLITA SKILHRKMLH SILHAPMSTI SKLKAGGILN RFSKDIAILD
Cftrxenla  LPLVHSLISV SKVLHKKMLH AILHAPMSTF NTMRAGRILN RFSKDTAILD
Cftrsquac  LPLVHTLVTV SKDLHKQMLH SVLQGPMTAF NKMKAGRILN RFIKDTAIID
Consensus  LPLVHTLITV SK.LH.KMLH S.L.APMST. N...AG.ILN RFSKD.AILD

          1001                                                 1050
Cftrhomsa  DLLPLTIFDF IQLLLIVIGA IAVVAVLQPY IFVATVPVIV AFIMLRAYFL
Cftrmusmu  DFLPLTIFDF IQLVFIVIGA IIVVSALQPY IFLATVPGLV VFILLRAYFL
Cftrxenla  DILPLSIFDL TQLVLIVIGA ITVVSLLEPY IFLATVPVIV AFILLRSYFL
Cftrsquac  DMLPLTVFDF VQLILIVVGA ICVVSVLQPY TLLAAIPVAV IFIMLRAYFL
Consensus  D.LPLTIFDF .QL.LIVIGA I.VVS.LQPY IFLATVPV.V .FI.LRAYFL
```

```
              1051                                                    1100
Cftrhomsa QTSQQLKQLE SEGRSPIFTH LVTSLKGLWT LRAFGRQPYF ETLFHKALNL
Cftrmusmu HTAQQLKQLE SEGRSPIFTH LVTSLKGLWT LRAFRRQTYF ETLFHKALNL
Cftrxenla HTSQQLKQLE SKARSPIFAH LITSLKGLWT LRAFGRQPYF ETLFHKALNL
Cftrsquac RTSQQLKQLE SEARSPIFSH LITSLRGLWT VRAFGRQSYF ETLFHKALNL
Consensus .TSQQLKQLE SE.RSPIF.H L.TSLKGLWT LRAFGRQ.YF ETLFHKALNL
```

```
              1101                                                    1150
Cftrhomsa HTANWFLYLS TLRWFQMRIE MIFVIFFIAV TFISILTTGE GEGRVGIILT
Cftrmusmu HTANWFMYLA TLRWFQMRID MIFVLFFIVV TFISILTTGE GEGTAGIILT
Cftrxenla HTANWFLYLS TLRWFQMTIE MIFVIFFIAV SFISIATSGA GEEKVGIVLT
Cftrsquac HTANWFLYLS TLRWFQMRID IVFVLFFIAV TFIAIATHDV GEGQVGIILT
Consensus HTANWFLYLS TLRWFQMRI. MIFVLFFIAV TFISI.T.G. GEG.VGIILT
```

```
              1151                                                    1200
Cftrhomsa LAMNIMSTLQ WAVNSSIDVD SLMRSVSRVF KFIDMPTEGK P.TKSTKPYK
Cftrmusmu LAMNIMSTLQ WAVNSSIDTD SLMRSVSRVF KFIDIQTEES MYTQIIKELP
Cftrxenla LAMNIMNTLQ WAVNASIDVD SLMRSVSRIF RFIDLPVEEL INENKNKE..
Cftrsquac LAMNITSTLQ WAVNSSIDVD GLMRSVSRVF KYIDIPPEGS ETKNRHNA..
Consensus LAMNIMSTLQ WAVNSSIDVD SLMRSVSRVF KFID.P.E.. ......K...
```

```
              1201                                                    1250
Cftrhomsa NGQLSKVMII ENSHVKKDDI WPSGGQMTVK DLTAKYTEGG NAILENISFS
Cftrmusmu REGSSDVLVI KNEHVKKSDI WPSGGEMVVK DLTVKYMDDG NAVLENISFS
Cftrxenla .EQLSEVLIY ENDYVKKTQV WPSGGQMTVK NLSANYIDGG NTVLENISFS
Cftrsquac .NNPSDVLVI ENKHLTKE.. WPSGGQMMVN NLTAKYTSDG RAVLQDLSFS
Consensus ....S.VL.I EN.HVKK... WPSGGQM.VK .LTAKY...G NAVLENISFS
```

```
              1251                                                    1300
Cftrhomsa ISPGQRVGLL GRTGSGKSTL LSAFLRLLNT EGEIQIDGVS WDSITLQQWR
Cftrmusmu ISPGQRVGLL GRTGSGKSTL LSAFLRMLNI KGDIEIDGVS WNSVTLQEWR
Cftrxenla LSPGQRVGLL GRTGSGKSTL LSAFLRLLST QGDIQIDGVS WQTIPLQKWR
Cftrsquac VNAGQRVGLL GRTGAGKSTL LSALLRLLST EGEIQIDGIS WNSVSLQKWR
Consensus .SPGQRVGLL GRTGSGKSTL LSAFLRLL.T .G.IQIDGVS W.S..LQ.WR
```

```
              1301                                                    1350
Cftrhomsa KAFGVIPQKV FIFSGTFRKN LDPYEQWSDQ EIWKVADEVG LRSVIEQFPG
Cftrmusmu KAFGVITQKV FIFSGTFRQN LDPNGKWKDE EIWKVADEVG LKSVIEQFPG
Cftrxenla KAFGVIPQKV FIFSGSIRKN LDPYGKWSDE ELLKVTEEVG LKLIIDQFPG
Cftrsquac KAFGVIPQKV FVFSGTFRKN LDPYEQWSDE EIWKVTEEVG LKSMIEQFPD
Consensus KAFGVIPQKV FIFSGTFRKN LDPY..WSDE EIWKV..EVG LKS.IEQFPG
```

```
              1351                                                    1400
Cftrhomsa KLDFVLVDGG CVLSHGHKQL MCLARSVLSK AKILLLDEPS AHLDPVTYQI
Cftrmusmu QLNFTLVDGG YVLSHGHKQL MCLARSVLSK AKIILLDEPS AHLDPITYQV
Cftrxenla QLDFVLLDGG CVLSHGHKQL VCLARSVLSK AKILLLDEPS AHLDPITFQI
Cftrsquac KLNFVLVDGG YILSNGHKQL MCLARSILSK AKILLLDEPT AHLDPVTFQI
Consensus .L.FVLVDGG .VLSHGHKQL MCLARSVLSK AKILLLDEPS AHLDP.T.QI
```

```
              1401                                                    1450
Cftrhomsa IRRTLKQAFA DCTVILCEHR IEAMLECQQF LVIEENKVRQ YDSIQKLLNE
Cftrmusmu IRRVLKQAFA GCTVILCEHR IEAMLDCQRF LVIEESNVWQ YDSLQALLSE
Cftrxenla IRKTLKHAFA DCTVILSEHR LEAMLECQRF LVIEDNTVRQ YDSIQKLVNE
Cftrsquac IRKTLKHTFS NCTVILSEHR VEALLECQQF LVIEGCSVKQ FDALQKLLTE
Consensus IR.TLK.AFA .CTVIL.HER .EAMLECQ.F LVIE...V.Q YDS.QKLL.E
```

```
           1451                              1499
Cftrhomsa RSLFRQAISP SDRVKLFP.. HRNSSKCKSK PQIAALKEET EEEVQDTRL
Cftrmusmu KSIFQQAISS SEKMRFFQ.. GRHSSKHKPR TQITALKEET EEEVQETRL
Cftrxenla KSFFKQAISH SDRLKLFPLH RRNSSKRKSR PQISALQEET EEEVQDTRL
Cftrsquac ASLFKQVFGH LDRAKLFTAH RRNSSKRKTR PKISALQEEA EEDLQETRL
Consensus .S.F.QAIS. SDR.KLF... RNSSSK.K.R PQI.AL.EET EEEVQ.TRL
```

Proteins listed subsequently in italics are at least 90% identical to the paired transporters listed in parenthesis and are therefore not included in the alignment: *Cftrbosta* (Cftrhomsa); *Cftrorycu* (Cftrmusmu). Residues listed in the consensus sequence are present in at least 75% of the aligned transporter sequences. Residues indicated in boldface type are also conserved in at least one other family of the ABC transporter superfamily.

Database accession numbers

	SWISSPROT	PIR	EMBL/GENBANK
Cftrbosta	P35071	A39323	M76128; G163742
Cftrhomsa	P13569	A30300; DVHUCF	M28668; G180332
Cftrmusmu	P26361	A39901; A40303	L26098; G915270
Cftrorycu	Q00554	E39323	U40227; G1100985
Cftrsquac	P26362	A39322	M83785; G213870
Cftrxenla	P26363	S23756	X65256; G64623

References
[1] Mcintosh, I. and Cutting, G.R. (1992) FASEB J. 6, 2775–2782.
[2] **Riordan, J.R. et al. (1989) Science 245, 1066–1073.**
[3] **Boat, T.F. et al. (1989) In The Metabolic Basis of Inherited Disease (Scriver, C.L. et al., eds). McGraw-Hill, New York, pp. 2649–2680.**
[4] Quinton, P.M. (1990) FASEB J. 4, 2709–2717.
[5] **Higgins, C.F. (1992) Annu. Rev. Cell Biol. 8, 67–113.**
[6] Cheng, S.H. et al. (1991) Cell 66, 1027–1036.
[7] Cystic Fibrosis Genetic Analysis Consortium (1990) Am. J. Hum. Genet. 47, 354–359.

Summary

Transporters of the P-glycoprotein family, examples of which are the human multidrug resistance (MDR) protein MDR1[1] (Mdr1homsa) and bacterial haemolysin[2] and cyclolysin[3] secretion proteins (e.g. Hlybescco, Cyabborpe), act as ATP-dependent efflux pumps. Many of these transporters mediate drug resistance, often to many structurally dissimilar drugs. The overexpression of MDR proteins in cancer cells is a common cause of treatment failure[4]. Other family members mediate the secretion of many different molecules including peptides and glucans. Some family members, such as human P-glycoprotein, are also associated with chloride channel activity[5,6]; the sulfonylurea receptor, SUR[7], in some mammals acts as a mediator of insulin release. Members of this family are widely distributed throughout many taxa, including bacteria, plants and mammals.

Statistical analysis of multiple amino acid sequence comparisons places the P-glycoprotein transporter family in the multidrug resistance subdivision of the ATP binding cassette (ABC) superfamily[8]. Proteins in this superfamily use the energy of ATP hydrolysis to pump substrates across cell membranes. Many transporters of this family (e.g. MDR1[1]) consist of a single polypeptide chain made up of four domains. The N- and C-terminal halves of the protein are homologous, and each half is made up of a transmembrane domain followed by an ATP binding domain. Others (e.g. the antigen peptide transporter TAP[9]) consist of a single transmembrane domain followed by an ATP binding domain. In these cases the transporter functions as a dimer: in TAP, this is a heterodimer of the homologous proteins TAP1 and TAP2 (also known as RING4 and RING11). Each transmembrane domain is predicted to contain six membrane-spanning helices by the hydropathy of the amino acid sequences, so the functional transporters contain 12 such helices. Many members of this family are glycosylated.

Many residues, including several long sequence motifs, are well conserved within the P-glycoprotein transporter family, including motifs unique to the family, signature motifs of the ABC superfamily, and motifs necessary for function by the criterion of site-directed mutagenesis.

Nomenclature, biological sources and substrates

CODE	DESCRIPTION [SYNONYMS]	ORGANISM [COMMON NAMES]	SUBSTRATE(S) [RESISTANCE][a]
Aprdpseae	Alkaline protease secretion ATP binding protein [APRD]	*Pseudomonas aeruginosa* [gram-negative bacterium]	Alkaline protease
Atm1sacce	Mitochondrial transporter ATM1 precursor [ATM1, MDY, YMR301C, YM9952.03C]	*Saccharomyces cerevisiae* [yeast]	Unknown
Chvaagrtu	β-(1→2)Glucan export ATP binding protein [CHVA, attachment protein]	*Agrobacterium tumefaciens* [gram-positive bacterium]	β-(1→2)Glucans
Comastrpn	Transport ATP binding protein [COMA]	*Streptococcus pneumoniae* [gram-positive bacterium]	Competence factor ?

CODE	DESCRIPTION [SYNONYMS]	ORGANISM [COMMON NAMES]	SUBSTRATES [RESISTANCE][a]
Cvabescco	Colicin V secretion ATP binding protein [CVAB]	Escherichia coli [gram-negative bacterium]	Colicin V
Cyabborpe	Cyclolysin secretion ATP binding protein [CYAB]	Bordetella pertussis [gram-negative bacterium]	Cyclolysin
Cydcescco	Transport ATP binding protein [CYDC, MDRA, MDRH, SURB]	Escherichia coli [gram-negative bacterium]	Unknown
Cydchaein	Transport ATP binding protein [CYDC, HI1156]	Haemophilus influenzae [gram-negative bacterium]	Unknown
Cyddescco	Transport ATP binding protein [CYDD, HTRD]	Escherichia coli [gram-negative bacterium]	Unknown
Cyddhaein	Transport ATP binding protein [CYDD, HI1157]	Haemophilus influenzae [gram-negative bacterium]	
Hemsentfa	Hemolysin secretion protein [CYLB]	Enterococcus faecalis [gram-positive bacterium]	Hemolysin, bacteriocin cytolysin B
Hetaanasp	Heterocyst differentiation ATP binding protein [HETA]	Anabaena sp. [alga]	Unknown
Hly2escco	Hemolysin secretion ATP binding protein [HLYB]	Escherichia coli [gram-negative bacterium]	Hemolysin A
Hlybactac	Leukotoxin secretion ATP binding protein [LKTB, AALTB]	Actinobacillus actinomycetemcomitans [gram-positive bacterium]	Leukotoxin (hemolysin)
Hlybactpl	Hemolysin secretion ATP binding protein [CLYI-B, APXIB, APPB, CLYIB]	Actinobacillus pleuropneumoniae [gram-positive bacterium]	Leukotoxin (hemolysin)
Hlybescco	Haemolysin secretion ATP binding protein [HLYB]	Escherichia coli [gram-negative bacterium]	Hemolysin A
Hlybpasha	Leukotoxin secretion ATP binding protein [LKTB]	Pasteurella haemolytica [gram-negative bacterium]	Leukotoxin (hemolysin)
Hlybprovu	Leukotoxin secretion ATP binding protein [HLYB]	Proteus vulgaris [gram-negative bacterium]	Hemolysin A
Hmt1schpo	Heavy metal tolerance protein precursor [HMT1]	Schizosaccharomyces pombe [yeast]	Metal-bound phytochelatins
Lcn3lacla	Lacticin 481/lactococcin transport ATP binding protein [LCNDR3]	Lactococcus lactis [gram-positive bacterium]	Lacticin, 481/lactococcin
Lcnclacla	Lactococcin A transport ATP binding protein [LCNC]	Lactococcus lactis [gram-positive bacterium]	Lactococcin A
Mdlescco	Multidrug resistance-like ATP binding protein [MDL]	Escherichia coli [gram-negative bacterium]	Unknown
Mdl1sacce	ATP-dependent permease [MDL1, YLR188W, L9470.3]	Saccharomyces cerevisiae [yeast]	Unknown
Mdl2sacce	ATP-dependent permease [MDL2, SSH1]	Saccharomyces cerevisiae [yeast]	Unknown
Mdrleita	Multidrug resistance protein [PGPA]	Leishmania tarentolae [trypanosome]	Methotrexate
Mdrplafa	Multidrug resistance protein [Chloroquine resistance protein, MDR1]	Plasmodium falciparum [trypanosome]	Chloroquine
Mdr1caeel	Multidrug resistance protein [P-glycoprotein A, PGP1]	Caenorhabditis elegans [nematode]	Multiple drugs

CODE	DESCRIPTION [SYNONYMS]	ORGANISM [COMMON NAMES]	SUBSTRATE(S) [RESISTANCE][a]
Mdr1crigr	Multidrug resistance protein 1 [P-glycoprotein 1, PGP1]	Cricetulus griseus [hamster]	[Multiple drugs]
Mdr1homsa	Multidrug resistance protein 1 [P-glycoprotein 1, PGY, MDR1]	Homo sapiens [human]	[Multiple drugs]
Mdr1leien	Multidrug resistance protein 1 [P-glycoprotein 1, PGY1]	Leishmania enriettii [trypanosome]	[Vinblastine, puromycin]
Mdr1musmu	Multidrug resistance protein 1 [P-glycoprotein 1, MDR1, MDR1B, PGY1]	Mus musculus [mouse]	[Multiple drugs]
Mdr1ratno	Multidrug resistance protein 1 [P-glycoprotein 1, PGY1, MDR1, MDR1B]	Rattus norvegicus [rat]	[Multiple drugs]
Mdr2crigr	Multidrug resistance protein 2 [P-glycoprotein 2, PGP2]	Cricetulus griseus [hamster]	[Multiple drugs]
Mdr2musmu	Multidrug resistance protein 2 [P-glycoprotein 2, PGY2]	Mus musculus [mouse]	[Multiple drugs]
Mdr3caeel	Multidrug resistance protein 3 [P-glycoprotein C, PGP3]	Caenorhabditis elegans [nematode]	[Multiple drugs]
Mdr3crigr	Multidrug resistance protein 3 [P-glycoprotein 3, PGP3]	Cricetulus griseus [hamster]	[Multiple drugs]
Mdr3homsa	Multidrug resistance protein 3 [P-glycoprotein 3, PGY3, MDR3]	Homo sapiens [human]	[Multiple drugs]
Mdr3musmu	Multidrug resistance protein 3 [P-glycoprotein 3, PGY3, MDR3, MDR1A]	Mus musculus [mouse]	[Multiple drugs]
Mdr4drome	Multidrug resistance protein homolog 50 [P-glycoprotein 50, MDR50, MDR49]	Drosophila melanogaster [fruit fly]	[Colchicine]
Mdr5drome	Multidrug resistance protein homolog 65 [P-glycoprotein 65, MDR65]	Drosophila melanogaster [fruit fly]	Unknown
Msbaescco	Probable transport ATP binding protein [MSBA]	Escherichia coli [gram-negative bacterium]	Unknown
Msbahaein	Probable transport ATP binding protein [MSBA, MSH1, HI0060]	Haemophilus influenzae [gram-negative bacterium]	Unknown
Mt2ratno	Mt2 protein	Rattus norvegicus [rat]	Peptides?
Natabacsu	ATP binding transport protein [NATA]	Bacillus subtilis [gram-positive bacterium]	Unknown
Ndvarhime	β-(1→2)Glucan export ATP binding protein [NDVA]	Rhizobium meliloti [gram-negative bacterium]	β-(1→2)Glucans
Nistlacla	Nisin transport ATP binding protein [NIST]	Lactococcus lactis [gram-positive bacterium]	Nisin
Peddpedac	Pediocin PA-1 transport ATP binding protein [PEDD]	Pediococcus acidilactici [gram-positive bacterium]	Pedoicin PA-1
Pglyarath	P-glycoprotein [ATPGP1]	Arabidopsis thaliana [mouse-ear cress]	Unknown

CODE	DESCRIPTION [SYNONYMS]	ORGANISM [COMMON NAMES]	SUBSTRATE(S) [RESISTANCE][a]
Pgp1ara	P-glycoprotein	Arabidopsis thaliana [mouse-ear cress]	Unknown
Pmd1schpo	Leptomycin B resistance protein [PMD1]	Schizosaccharomyces pombe [yeast]	Leptomycin B Mating factor?
Prtderwch	Proteases secretion ATP binding protein [PRTD]	Erwinia chrysanthemi [gram-negative bacterium]	Proteases A, B, C, G
Rt3bactpl	RTX toxin-III operon protein [APXIIIB, RTXB]	Actinobacillus pleuropneumoniae [gram-positive bacterium]	RTX toxin-III
Spabbacsu	Subtilin transport ATP binding protein [SPAB, SPAY, SPAT]	Bacillus subtilis [gram-positive bacterium]	Subtilin
Ste6sacce	Mating Factor A secretion protein [Multidrug resistance protein homolog, STE6, YKL209C]	Saccharomyces cerevisiae [yeast]	Mating factor A
Surcricr	Sulfonylurea receptor [SUR]	Cricetus cricetus [hamster]	Insulin
Surratno	Sulfonylurea receptor [SUR]	Rattus norvegicus [rat]	Insulin
Syrdpsesy	ATP binding protein [SYRD]	Pseudomonas syringae [gram-negative bacterium]	Syringomycin
Tap1homsa	Antigen peptide transporter 1 [Peptide supply factor 1, TAP1, PSF1, RING4, Y3]	Homo sapiens [human]	Peptides
Tap1musmu	Antigen peptide transporter 1 [Histocompatibility antigen modifier 1, TAP1, HAM1]	Mus musculus [mouse]	Peptides
Tap2homsa	Antigen peptide transporter 2 [Peptide supply factor 1, TAP2, PSF2, RING11, Y1]	Homo sapiens [human]	Peptides
Tap2musmu	Antigen peptide transporter 2 [Histocompatibility antigen modifier 2, TAP2, HAM2]	Mus musculus [mouse]	Peptides
Tap2ratno	Antigen peptide transporter 2 [TAP2, MTP2]	Rattus norvegicus [rat]	Peptides

[a] Presumed substrates; protein confers resistance to specified compounds.

145

Phylogenetic tree

Proteins listed subsequently in italics are at least 90% identical to the paired transporters listed in parenthesis and are therefore not included in the phylogenetic tree: *Hlybprovu, Hly2escco* (Hlybescco); *Mdr1ratno* (Mdr1musmu); *Tap2ratno* (Tap2musmu).

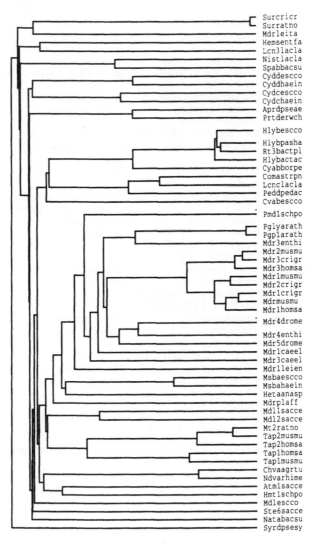

Proposed orientation of MDR1 [1] in the membrane

The model is based on predictions of membrane-spanning regions and α-helical content. The N-terminus of the protein is illustrated on the inside and is folded 12 times through the membrane. The predicted membrane-spanning helices are portrayed as rectangles. The numbers corresponding to the first and last residue of each membrane-spanning helix are boxed. Residues that are conserved in more than 75% of the aligned transporters (see below) are shown.

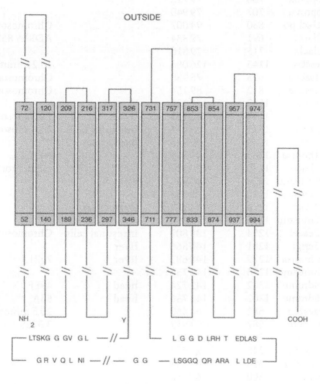

Physical and genetic characteristics

	AMINO ACIDS	MOL. WT	EXPRESSION SITES	CHROMOSOMAL LOCUS
Aprdpseae	593	63 670		63 minutes
Atm1sacce	690	77 522		Chromosome 13
Chvaagrtu	588	64 651		
Comastrpn	717	80 350		comAB locus
Cvabescco	698	78 245		
Cyabborpe	712	77 969		
Cydcescco	573	62 920		19.5 minutes
Cydchaein	576	64 831		66.849

	AMINO ACIDS	MOL. WT	EXPRESSION SITES	CHROMOSOMAL LOCUS
Cyddescco	588	64956		19.3 minutes
Cyddhaein	586	65645		66.943
Hemsentfa	714	82051		
Hetaanasp	607	67789		
Hly2escco	707	79463		
Hlybactac	707	79578		
Hlybactpl	707	79663		apxI operon
Hlybescco	707	79672		Plasmid pHly152
Hlybpasha	708	79712		
Hlybprovu	707	79940		
Hmt1schpo	830	94007		Chromosome 3
Lcn3lacla	691	79834		ADRIA 85LO30 op.
Lcnclacla	715	79810		
Mdlescco	1143	126083		10.2 minutes
Mdl1sacce	695	75950		Chromosome 12
Mdl2sacce	812	89754		Chromosome 16
Mdrleita	1548	172235	H circle	
Mdrplafa	1419	162251		Chromosome 5
Mdr1caeel	1321	145074	intestinal cells	Chromosome 4
Mdr1crigr	1276	140925	liver, ovary	
Mdr1homsa	1280	141504		7q21.1
Mdr1leien	1280	139728		extrachromosomal circle
Mdr1musmu	1276	140993		
Mdr1ratno	1277	141386		
Mdr2crigr	1276	141057	liver, ovary	
Mdr2musmu	1276	140332		
Mdr3caeel	1254	138807	intestinal cells	Chromosome 10
Mdr3crigr	1281	140866	liver	
Mdr3homsa	1279	140682	liver	7q21.1
Mdr3musmu	1276	140754		
Mdr4drome	1302	142724	head	49EF
Mdr5drome	1302	143736	head	65A
Msbaescco	582	64460		20.5 minutes
Msbahaein	587	64912		3.3225
Mt2ratno	703	77811		
Natabacsu	246	27878		
Ndvarhime	616	67238		ndvA locus
Nistlacla	600	69210		
Peddpedac	724	81651		Plasmid pSRQ11
Pglyarath	1286	140571		
Pgp1arath	1233	135313		
Pmd1schpo	1362	149652		Plasmid pDB248'
Prtderwch	575	61617		prt locus
Rt3bactpl	711	80405		
Spabbacsu	614	71188		spa region
Ste6sacce	1290	144765		Chromosome 11, left arm
Surcricr	1581	177015	pancreatic islets	
Surratno	1580	176750	pancreatic islets	
Syrdpsesy	566	63195		
Tap1homsa	748	80964		6p21.3
Tap1musmu	577	63450		MHC class II region
Tap2homsa	686	75663		6p21.3
Tap2musmu	702	77444		MHC class II region
Tap2ratno	703	77712		MHC class II region

Multiple amino acid sequence alignments

```
           1                                                  50
Surcricr   .........P LAFCGTENHS AAYRVDQGVL NN.GCFVDAL NVVPHVFLLF
Surratno   .........P LAFCGTENHS AAYRVDQGVL NN.GCFVDAL NVVPHVFLLF
Mdrleita   MVDNGHVTIA MADLGTVVEI AQVRCQQEAQ RKFAEQLDEL WGGEPAYTPT
Hmt1schpo  ......... ......... ......... ......M VLRYNSPRLN
Consensus  ......... ......... ......... ......... .........
```

```
           51                                                 100
Surcricr   IT.....FPI LFIGWGSQSS KVHIHHSTWL HFPGHNLRWI LTFILLFVLV
Surratno   IT.....FPI LFIGWGSQSS KVHIHHSTWL HFPGHNLRWI LTFILLFVLV
Mdrleita   VEDQASWFQQ LYYGWIGD.. ...YIYKAAAG NITEADLPPP TRSTRTYHIG
Hmt1schpo  ILELVLLYVG FFSIGSLNLL QKRKATSDPY RRKNRFGKEP IGIISWWILG
Consensus  ......... ......... ......... ......... .........
```

```
           101                                                150
Surcricr   CEIAEGILSD GVTESRHLHL YMPAGMAFMA ...AITSVVY YHNIETSNFP
Surratno   CEIAEGILSD GVTESRHLHL YMPAGMAFMA ...AITSVVY YHNIETSNFP
Mdrleita   RKLSRQAHAD .IDASRRWQG YIGCEVVYKS EAEAKGVLRW VGHLQQSDYP
Mdl2sacce  ......... ......... ......... ....MKTY VLLYGKLIMT
Mt2ratno   ......... ......... ......... ......MAL SHPRPWASLL
Tap2musmu  ......... ......... ......... ......MAL SYLRPWVSLL
Tap2homsa  ......... ......... ......... ......MRL PDLRPWTSLL
Tap1homsa  ......... ...MASSRCP APRGCRCLPG ASLAWLGTVL LLLADWVLLR
Hmt1schpo  IALTYVVDIS NLVIYALAVP NWWPCKTTVV CLILFLLFWI IVLISCADSK
Consensus  ......... ......... ......... ......... .........
```

```
           151                                                200
Surcricr   KLLIALLIYW TLAFITKTIK FVKFYDHAIG FSQLRFCLTG LLVILYGMLL
Surratno   KLLIALLIYW TLAFITKTIK FVKFYDHAIG FSQLRFCLTG LLVILYGMLL
Mdrleita   RSLVAGVEWR ......... ....MP PRHRRLAVLG SAAALHN...
Hemsentfa  ......... ...MKR LKYVAQGEHS ECALACITML LNYYGNQSTL
Lcn3lacla  ......... MKIVLQNNEQ DCLLACYSMI LGYFGRDVAI
Hlybescco  ......... ....MDSCH KIDYGLYALE ILAQYHNVSV
Hlybpasha  ......... .MEAN...HQ RNDLGLVALT MLAQYHNISL
Rt3bactpl  ......... .MESQMPFNE KIDYGLHALV ILAQYHNVAV
Hlybactac  ......... ...MDSQK NTNLALQALE VLAQYHNISI
Hlybactpl  ......... ...MDFYR EEDYGLYALT ILAQYHNIAV
Cyabborpe  ......... MTSPVAQCAS VPDSGLLCLV MLARYHGLAA
Comastrpn  ......... .MKFGK RH..YRPQVD QMDCGVASLA MVFGYYGSYY
Lcnclacla  ......... .MKFKK KN..YTSQVD EMDCGCAALS MILKSYGTEK
Peddpedac  ......... .MWTQK WHKYYTAQVD ENDCGLAALN MILKYYGSDY
Cvabescco  ..MTNRNFRQ IINLLDLRWQ RRVPVIHQTE TAECGLACLA MICGHFGKNI
Mdl1sacce  ......... ......... ......MIVR MIRLCKGPKL
Mdl2sacce  TMILNTGRFE EWYKVCIIAL KEKEIYVPSS PIAMLNGRLP LLRLGICRNM
Mt2ratno   LVDLALLGLL QSSLGTLLPP GLPGLWLEGT LRLGV..... .....LWGLL
Tap2musmu  LADMALLGLL QGSLGNLLPQ GLPGLWIEGT LRLGV..... .....LWGLL
Tap2homsa  LVDAALLWLL QGPLGTLLPQ GLPGLWLEGT LRLGG..... .....LWGLL
Tap1homsa  TALPRIFSLL VPTALPLLRV WAVGLSRWAV LWLGACGVLR ATVGSKSENA
Atm1sacce  ......... ..·........ ....MLLLPR CPVIGRIVRS KFRSGLIRNH
Hmt1schpo  ALPKNADSIL KAYRLSVLYV WAIDIVFETI FIVYSPHPNE TFQGIVLADH
Consensus  ......... ......... ......... ......... .........
```

```
           201                                                      250
Surcricr   LVEVNVIRVR RYIFFKTPRE VKPPEDLQDL GVRFLQPFVN LLSKGTYWWM
Surratno   LVEVNVIRVR RYVFFKTPRE VKPPEDLQDL GVRFLQPFVN LLSKGTYWWM
Mdrleita   ....GVVHGE RLFWPHEDNY LCSCEPVEQL YVK........ ..SK......
Hemsentfa  VELREKYGVP KGGLTIKNIR TVFDEYGFDV STFKSSFSN. ....YLDLPT
Lcn3lacla  HELYSGEMIP PDGLSVSYLK NINMKHQVSM HVYKTDKKNS PNKIFYPKML
Hlybescco  NPEEIKHRFD TDGTGLGLTS WLLAAKSLEL KVKQVKKTID RLNF..IFLP
Hlybpasha  NPEEIKHKFD LDGKGLSLTA WLLAAKSLAL KAKHIKKEIS RLHL..VNLP
Rt3bactpl  NPEEVKHKFD LDGKGLDLVA WLLAAKSLEL KMKRVKKSIE RLPF..IHLP
Hlybactac  NPEEIKHKFD IDGHGLNQTK WLLAAKSLGL KVRTANKTVD RLPF..LHLP
Hlybactpl  NPEELKHKFD LEGKGLDLTA WLLAAKSLEL KAKQVKKAID RLAF..IALP
Cyabborpe  DPEQLRHEF. .AEQAFCSET IQLAARRVGL KVRRHRPAPA RLPR..APLP
Comastrpn  FLAHLRELAK TTMDGTTALG LVKVAEEIGF ETRAIKADMT LFDLPDLTFP
Lcnclacla  SLASLRLLAG TTIEGTSALG IKKAAEILEF SVQALRTDAS LFEMKNAPYP
Peddpedac  MLAHLRQLAK TTADGTTVLG LVKAAKHLNL NAEAVRADMD ALTASQLPLP
Cvabescco  DLIYLRRKFN LSARGATLAG INGIAEQLGM ATRALSLELD ELRV..LKTP
Pmdlschpo  .......MSL HSKKSTSTVK DNEHSLDLSI KSIPSNEKNF STEKSENEAS
Pglyarath  ................. ..........MDNDGGAP ..PPPPTLVV
Pgplarath  ................. ..........MDNDGGAP ..PPPPTLVV
Mdr2musmu  ..........M DLEAARNGTA RR...LDGDF ..ELGSISNQ
Mdr3crigr  ..........M DLEAARNGTA RRPGTVEGDF ..ELGSISNQ
Mdr3homsa  ..........M DLEAAKNGTA WRPTSAEGDF ..ELGISSKQ
Mdr1musmu  ..........M EFEENLK... ...GRADKNF ..SKMGKKSK
Mdr2crigr  ..........M EFEEDFS... ...ARADKDF ..LKMGRKSK
Mdr1crigr  ..........M EFEEDFS... ...GRKDKNF ..LKMGRKSK
Mdr3musmu  ..........M ELEEDLK... ...GRADKNF ..SKMGKKSK
Mdr1homsa  ..........M DLEGDRNG.. ...GAKKKNF ..FKLNNKSE
Mdr4drome  ................MV KKEESRLPQA ..GDFQLKE.
Mdr5drome  ................ME RDEVSTSSSE ..GKSQEEAP
Mdr1caeel  .MLRNGSLRQ SLRTLDSFSL APEDVLKTAI KTVEDYEGDN IDSNGEIKIT
Mdr3caeel  ................MK KTKVNPEDDI TLGKFTPKPS
Mdrplafa   ................ ..........MGK
Mdl1sacce  L.RSQFASAS ALYSTKSLFK PPMYQKAEIN LIIPHR.... KHFLLRSIRL
Mdl2sacce  LSRPRLAKLP SI.RFRSLVT PSSSQLIPLS RLCLRSPAVA KSLILQSFRC
Mt2ratno   KVGGLLRLVG TFLPLLCLTN PLFFSLRALV GSTMSTSVVR VASASWGWLL
Tap2musmu  KVGELLGLVG TLLPLLCLAT PLFFSLRALV GGTASTSVVR VASASWGWLL
Tap2homsa  KLRGLLGFVG TLLLPLCLAT PLTVSLRALV AGASRAPPAR VASAPWSWLL
Tap1homsa  GAQGWLAALK PLAAALGLAL PGLALFRELI SWGAPGSADS TRLLHWGSHP
Ndvarhime  ................ ..........MKI
Atmlsacce  .SPVIFTV.. ........SK LSTQRPLLFN S..AVNLWNQ AQKDITHKKS
Hmtlschpo  VARLVLCVFA TAIYLTYRRK RHTHDPLDFE ERQLTEESNV NENAISQNPS
Consensus  ................ ..........
```

```
           251                                                      300
Surcricr   NAFIKTAHKK PIDLRAIAKL PIAMRALTNY QRLCVAFDAQ ARKDTQSPQG
Surratno   NAFIKTAHKK PIDLRAIGKL PIAMRALTNY QRLCLAFDAQ ARKDTQSQQG
Mdrleita   .......... .......... .......... ......YNLI PPRPPPSPDL
Hemsentfa  PVISYWNNQH FVVIEKIKKK KVLILDPASN KRWIDISEFK KNF......S
Lcn3lacla  PVIIQWNDNH FVVVTKIYRK NVTLIDPAIG KVKYNYNDFM KKF......S
Nistlacla  .......... MDEVKEFTSK QFFYTLLTLP STLKLIFQLE
Spabbacsu  .......... MEVKEQLKLK ELLFIMKQMP KTFKLIFTLE
Cyddescco  .......... .......... MNKSRQKELT RWLKQQSVIS
Cyddhaein  .......... .......... MNKLRQKYLQ KWLRAQQEPI
Cydcescco  .......... .......... .....MRALL PYLA.....L
```

```
Cydchaein  ........... ........... ........... .....MRTLL PFIR.....L
Hlybescco  ALVWREDGRH FILT.KISKE VNRYLIFDLE QRNPRV..LE QSEFEALYQG
Hlybpasha  ALVWQDNGKH FLLV.KVDTD NNRYLTYNLE QDAPQI..LS QDEFEACYQG
Rt3bactpl  ALIWRDDGQH VILM.KIDTQ TNRYLIFDLE ERNPKV..LS AAEFHEIFQG
Hlybactac  ALAWRDDGEH FILL.KIDQE TDRYLIFDLI QKNPIV..LD KNEFEERYQS
Hlybactpl  ALVWREDGKH FILT.KIDNE AKKYLIFDLE THNPRI..LK QTEFESLYQG
Cyabborpe  AIALDRQGGY FVLVPRFEPG ADQAVLIQRP GQAPAR..LG QAEFEALWAG
Comastrpn  FVAHVLKEGK LLHYYVVTGQ DKDSIHIADP DPGVKLTKLP RERFEEEWTG
Lcnclacla  FIAHVIKDQK YPHYYVITGA NKNSVFIADP DPTIKMTKLS KEAFLSEWTG
Peddpedac  VIVHVFKKNK LPHYYVVYQV TENDLIIGDP DPTVKTTKIS KSQFAKEWTQ
Cvabescco  CILHWDFSHF VVL...VSVK RNRYVL...H DPARGIRYIS REEMSRYFTG
Lepbschpo  ESHVVDVVKD PFEQYTPEEQ EILYKQINDT PA..KLSGYP RILSYADKWD
Pmdlschpo  ESHVVDVVKD PFEQYTPEEQ EILYKQINDT PA..KLSGYP RILSYADKWD
Pglyarath  ........... ....EEPKKA EI..RGVAFK ELFRFADGLD
Pgplarath  ........... ....EEPKKA EI..RGVAFK ELFRFADGLD
Mdr2musmu  ........... ....GREKKK KV..NLIGLL TLFRYSDWQD
Mdr3crigr  ........... ....GRNKKK KV..NLIGPL TLFRYSDWQD
Mdr3homsa  ........... ....KRKKTK TV..KMIGVL TLFRYSDWQD
Mdr1musmu  ........... ....KEKKEK .K..PAVGVF GMFRYADWLD
Mdr2crigr  ........... ....KEKKEK EN..PNVGIF GMFRYADWLD
Mdr1crigr  ........... ....KEKKEK .K..PVVSVF TMFRYAGWLD
Mdr3musmu  ........... ....KEKKEK .K..PAVSVL TMFRYAGWLD
Mdr1homsa  ........... ....KDKKEK .K..PTVSVF SMFRYSNWLD
Mdr4drome  ........... ....GSVVD AT..RKYSYF DLFRYSTRCE
Mdr5drome  ........... ....MAEGLE PT..EPIAFL KLFRFSTYGE
Mdr1caeel  ........... ....RDAKEE VV..NKVSIP QLYRYTTTLE
Mdr3caeel  ........... ....PQDSYQ G......NFF DVFRDADYKD
Mdr1leien  GIAGKDGSTR DCSGYGSQGP LFSAEEEVKG TVVRETVGPI EIFRYADATD
Msbaescco  ........... ....MHNDKDL STWQTFRRLW PTIAPFKA..
Msbahaein  ........... ...M QEQKLQENDF STLQTFKRLW PMIKPFKA..
Hetaanasp  ........... ..MPKS PHKLFKANSF WKENNL..IL REIKHFRKIA
Mdrplafa   EQKEKKDGNL SIKEEVEKEL NKKSTAELFR KIKNEKISFF LPFKCLPAQH
Mdllsacce  QSDIAQGKKS TKPTLKLSNA .........N SKSSGFKDIK RLFVLSKP.E
Mdl2sacce  NSSKTVPETS LPSASPISKG SARSAHAKEQ SKTDDYKDII RLFMLAKR.D
Mt2ratno   ADYGAV.ALS LAVWAVLSPA ...GAQEKEP GQENNRALMI RLLRLXKP.D
Tap2musmu  AGYGAV.ALS WAVWAVLSPA ...GVQEKEP GQE.NRTLMK RLLKLSRP.D
Tap2homsa  VGYGAA.GLS WSLWAVLSPP ...GAQEKEQ DQVNNKVLMW RLLKLSRP.D
Tap1homsa  TAFVVSYAAA LPAAALWHKL ...GSLWVPG GQGGSGNPVR RLLGCLGS.E
Tap1musmu  ........... ...........MLC RMLGFLGP.K
Chvaagrtu  ........... .....MT LFQVYTRALR YL...TVHKW
Ndvarhime  ILAVGSRRNA LPHRAVAAPI PIPERGETVS LFQVYARALQ YL...AVHKF
Atm1sacce  VEQF...... SSAPKVKTQV KKTSKAPTLS ELKILKDLFR YIWPKGNNKV
Hmt1schpo  TVQLGVSAST SNFGTLKSTS KKPSDKSWAE YFRSFSTLLP YLWPTKDYRL
Ste6sacce  ........... ....MN FLSFKTTKHY HIFRYVNIRN
Syrdpsesy  ........... ........ ....MKTKQE
Consensus  ........... ........... ........... ...........
```

```
           301                                          350
Surcricr   ARAIWRA.LC HAFGRRLILS STFRILADLL GFAGPLCIFG IVDHLGKENH
Surratno   ARAIWRA.LC HAFGRRLVLS STFRILADLL GFAGPLCIFG IVDHLGKENH
Mdrleita   LRTLFKVHWY HVWA........QILPKLL SDVTALMLPV LLEYFVK...
Hemsentfa  NILIY..AHK KKTKKEGKRK QFFLKSFIFT KFKRYFFSLI ILSFVSQLLL
Lcn3lacla  GYIITLSPNS SFTKKKRISE IIFPLKKIFK NRNTFLYIFS L..FISQIVA
Nistlacla  KRYAIYLIVL NAITAFVPLA SLFIYQDLIN SVLGSGRH.. ..........
```

```
Spabbacsu RSLFLKLIRF SIITGILPIV SLYISQELIN SLVTIRKEVS I.........
Cyddescco QRWLNISRLL GFVSGILIIA QAWFMARILQ HMIMEN.IPR EALLLPFTLL
Cyddhaein KKLMRANIVL ATLSSFILVA QTYFLATLLD KLIMQN.VPR DELIPYFLGL
Cydcescco YKRHKWMLSL GIVLAIVTLL ASIGLLTLSG WFLSASAVAG VAGLYSFNYM
Cydchaein FKFAKFPLIL GLVLMILGLG SSMGLLTVSG WFLAATAIAG LGTL..FNFF
Aprdpseae ........... ...MARLGSS VTNEIKQALA ASRGALRSVA AFSGVINLLM
Prtderwch ........... ...MNASSE RDRSLFGVLR QFRRSFWSVG IFSAVINVLM
Hlybescco HIILITSRSS VTG.KLA.KF DFTWFIPAII KYRRIFIETL VVSVFLQLFA
Hlybpasha QLILVTSRAS VVG.QLA.KF DFTWFIPAVI KYRKIFLETL IVSIFLQIFA
Rt3bactpl GMILITSRAS IMG.QLA.KF DFTWFIPAVI KYRKIFVETI IVSIFLQLFA
Hlybactac KVILIASRAS IVG.NLA.KF DFTWFIPAVI KYRKIFIETL IVSIFLQIFA
Hlybactpl KLILVASRAS IVG.KLA.KF DFTWFIPAVI KYRKIFIETL IVSIFLQIFA
Cyabborpe ELLLCACAAS PTQ.ALA.RF DFSWFIPALV KHRHLIGEVL LISLVLQFIS
Comastrpn VTLFMAPSPD YKPHKEQ.KN GLLSFIPILV KQRGLIANIV LATLLVTVIN
Lcnclacla ISLFLSTTPS YHPTKEK.AS SLLSFIPIIT RQKKVILNIV IASFIVTLIN
Peddpedac IAIIIAPTVK YKPIKES.RH TLIDLVPLLI KQKRLIGLII TAAAITTLIS
Cvabescco VALEVWPGSE FQSETLQTRI SLRSLINSIY GIKRTLAKIF CLSVVIEAIN
Pmdlschpo IMLQLAGTIT GIGAGLGMPL MSLVSGQLAQ AFTDLASGKG ASS.......
Pglyarath YVLMGIGSVG AFVHGCSLPL FLRFFADLVN SFGSNSNNVE ..........
Pgplarath YVLMGIGSVG AFVHGCSLPL FLRFFADLVN SFGSNSNNVE ..........
Mdr2musmu KLFMFLGTLM AIAHGSGLPL MMIVFGEMTD KFVDNTGNFS LPVNF.SLSM
Mdr3crigr KLFMLLGTIM AIAHGSGLPL MMIVFGEMTD KFVNNAGNFS LPVNF.SLSM
Mdr3homsa KLFMSLGTIM AIAHGSGLPL MMIVFGEMTD KFVDTAGNFS FPVNF.SLSL
Mdr1musmu KLCMILGTLA AIIHGTLLPL LMLVFGNMTD SFTKAE..AS ILPSITNQSG
Mdr2crigr KLYMVLGTLA AVLHGTSLPL LMLVFGNMTD SFTKAE..TS IWPNMTNQSE
Mdr1crigr RLYMLVGTLA AIIHGVALPL MMLVFGDMTD SFASVGNIPT ..NATNNATQ
Mdr3musmu RLYMLVGTLA AIIHGVALPL MMLIFGDMTD SFASVGNVSK ..NST.NMSE
Mdr1homsa KLYMVVGTLA AIIHGAGLPL MMLVFGEMTD IFANAGNLED LMSNITNRSD
Mdr4drome RFLLVVSLLV ATAASAFIPY FMIIYGEFTS LLVDRTVGVG TSSPAFALPM
Mdr5drome IGWLFFGFIM CCIKALTLPA VVIIYSEFTS MLVDRAMQFG TSSNVHALPL
Mdr1caeel KLLLFIGTLV AVITGAGLPL MSILQGKVSQ AFINEQIVIN ..........
Mdr3caeel YILFSGGLIL SAVNGALVPF NSLIFEGIAN ALMEGESQYQ NGTINMPW..
Mdr1leien RVLMIAGTAF AVACGAGMPV FSFIFGRIAM DLMSGVGSAE ..........
Msbaescco ..GLIVAGVA LILNAASDTF MLSLLKPL.. ....LDDGFG KTDRSV....
Msbahaein ..GLIVSGVA LVFNALADSG LIYLLKPL.. ....LDDGFG KANHSF....
Hetaanasp ILAVIFSFLA ASFEGVSIGF LLSFLQKLTS PNDPIQTGIS WVDMILAADA
Mdrplafa  RKLLFISFVC AVLSGGTLPF FISVFGVILK NM................
Mdl1sacce SKYIGLALLL ILISSSVSMA VPSVIGKLLD LASESDGEDE EGSKSNKLYG
Mdl2sacce WKLLLTAILL LTISCSIGMS IPKVIGIVLD TLKTSSGSDF FDLKI.PIFS
Mt2ratno  LPFLIVAFIF LAMAVWWEMF IPHYSGRVID ILGGDFDPDA FASAIF....
Tap2musmu LPFLIAAFFF LVVAVWGETL IPRYSGRVID ILGGDFDPDA FASAIF....
Tap2homsa LPLLVAAFFF LVLAVLGETL IPHYSGRVID ILGGDFDPHA FASAIF....
Tap1homsa TRRLSLFLVL VVLSSLGEMA IPFFTGRLTD WILQDGSADT FTRNLT....
Tap1musmu KRRLYLVLVL LILSCLGEMA IPFFTGRITD WILQDKTVPS FTRNIW....
Chvaagrtu RVAVVVIANV ILAA..ITIA EPVLFGRIID AISSGTNVTP I.......LI
Ndvarhime RVGAIVIANI VLAA..ITIA EPILFGRIID AISSQKDVAP M.......LL
Atmlsacce RIRVLIALGL LISAKILNVQ VPFFFKQTID SMNIAWDDPT VALPAAIGLT
Hmt1schpo QFQIFICIVL LFLGRAVNIL APRQLGVLTE KLTKHSEK.. IPWSDVILFV
Ste6sacce DYRLLMIMII GTVATGLVPA ITSILTGRVF DLLSVFVANG SHQGLY....
Syrdpsesy KKARPGSIMR LLWSSHPWLT FFTLLLTGLIS GVASIAVVNV INQAIHEETF
Consensus ........... .......... .......... ..........
```

```
          351                                             400
Surcricr  VFQPKTQFLG VYFVSSQEFL GNAYVLAVLL FLALLLQRTF LQASYYVAIE
Surratno  VFQPKTQFLG VYFVSSQEFL GNAYVLAVLL FLALLLQRTF LQASYYVAIE
Mdrleita  .......... .YLNADNATW GWGLGLALTI FLTNVIQSCS AHKYDHISIR
Hemsentfa LLIPIATKYS IDNIRSFQEI PTYVLLLILT SFFSVL.YVV QYLKSSVVAE
Lcn3lacla LWFSIILR.. .DILNKSHDI TYSFIMMISL VLFQTL.SLL ..MKLGAQKN
Nistlacla .......... ......LINIIIIY FIVQVITTVL GQLESYVSGK
Spabbacsu .......... ......VITIFLTY LGVSFFSELI SQISEFYNGK
Cyddescco V......... ....LTFVL RAWVV..... .WLRERVGYH
Cyddhaein I......... ....IGFGM RAIIL..... .WAREKIGFQ
Cydcescco L......... ....PAAGV RGAAITRTAG RYFERLVSHD
Cydchaein Y......... ....PSASV RGLAIGRTVM RYFEKIVTHD
Aprdpseae LVPSLYMLQV YDRVLSSANE VTLLMLTLMA LGVFVFMGAL EALRSFVLVR
Prtderwch LAPSVYMLQV YDRVLASGNG ITLLMLTLLM AGLCAFMGAL EWVRSLLVVR
Hlybescco LITPLFFQVV MDKVLVHRGF STLNVITVAL SVVVVFEIIL SGLRTYIFAH
Hlybpasha LITPLFFQVV MDKVLVHRGF STLNIITVAL AIVIIFEIVL SGLRTYVFSH
Rt3bactpl LITPLFFQVV MDKVLVHRGF STLNVITVAL SVVVIFEIVL SGLRTYIFSH
Hlybactac LITPLFFQVV MDKVLVHRGF STLNVITVAL AIVVLFEIIL GGLRTYVFAH
Hlybactpl LITPLFFQVV MDKVLVHRGF STLNVITVAL AIVVLFEIVL NGLRTYIFAH
Cyabborpe LLTPLFFQVV MDKVLVNNAM ETLNVITVGF LAAILFEALL TGIRTYLFAH
Comastrpn IVGSYYLQSI IDTYVPDQMR STLGIISIGL VIVYILQQIL SYAQEYLLLV
Lcnclacla ILGSYYLQSM IDSYIPNALM GTLGIISVGL LLTYIIQQVL EFAKAFLLNV
Peddpedac IAGAYFFQLI IDTYLPHLMT NRLSLVAIGL IVAYAFQAII NYIQSFFTIV
Cvabescco LLMPVGTQLV MDHAIPAGDR GLLTLISAAL MFFILLKAAT STLRAWSSLV
Pmdlschpo .......... ....FQ HTVDHFCLYF IYIAIGVFGC SYIYTVTFII
Pglyarath .......... ....KMM EEVLKYALYF LVVGAAIWAS SWAEISCWMW
Pgplarath .......... ....KMM EEVLKYALYF LVVGAAIWAS SWAEISCWMW
Mdr2musmu LNPGRI.... .LEE...... .EMTRYAYYY SGLGGGVLVA AYIQVSFWTL
Mdr3crigr INPGRI.... .LEE...... .EMTRYAYYY SGLGGGVLVA AYIQVSFWTL
Mdr3homsa LNPGKI.... .LEE...... .EMTRYAYYY SGLGGAGVLVA AYIQVSFWTL
Mdr1musmu PNSTLIISNS SLEE...... .EMAIYAYYY TGIGAGVLIV AYIQVSLWCL
Mdr2crigr INNTEVIS.G SLEE...... .DMATYAYYY TGIGAGVLIV AYIQVSFWCL
Mdr1crigr VNASDIF..G KLEE...... .EMTTYAYYY TGIGAGVLIV AYIQVSFWCL
Mdr3musmu ADKRAMF..A KLEE...... .EMTTYAYYY TGIGAGVLIV AYIQVSFWCL
Mdr1homsa INDTGFF..M NLEE...... .DMTRYAYYY SGIGAGVLVA AYIQVSFWCL
Mdr4drome FGGGQQLTNA SKEENNQAII DDATAFGIGS LVGSVAMFLL ITLAIDLANR
Mdr5drome FGGGKTLTNA SREENNEALY DDSISYGILL TIASVVMFIS GIFSVDVFNM
Mdr1caeel .NNGSTFLPT GQNYTKTDFE HDVMNVVWSY AAMTVGMWAA GQIIVTCYLY
Mdr3caeel .......... ....FS SEIKMFCLRY FYLGVALFLC SYFANSCLYT
Mdr1leien .......... ... EKAAKTSLIM VYVGIAMLIA CAGHVMCWTV
Msbaescco .......... ..LVWMPLVV IGLMILRGIT SYVSSYCISW
Msbahaein .......... ..LKMMAFVV VGMIILRGIT NFISNYCLAW
Hetaanasp WP........ ....I PPIYRISLLI LLSTWMRATF NYFGGVYTES
Mdrplafa  .......... ...NLG DDINPIILSL VSIGLVQFIL SMISSYCMDV
Mdl1sacce F......... ....TKKQFFTAL GAVFIIGAVA NASRIIILKV
Mdl2sacce L......... ....PLYEFLSFF TVALLIGCAA NFGRFILLRI
Mt2ratno  .......... ....FMCLF SVGSSLSAGC RGGSFLF...
Tap2musmu .......... ....FMCLF SVGSSFSAGC RGGSFLF...
Tap2homsa .......... ....FMCLF SFGSSLSAGC RGGCFTY...
Tap1homsa .......... ....LMSIL TIASAVLEFV GDGIYNN...
Tap1musmu .......... ....LMSIL TIASTALEFA SDGIYNI...
Chvaagrtu LW........ ...... AGFGVFNTVA YVAVAREADR
Ndvarhime LW........ ...... AGFGVFNTIA FVLVSREADR
Atm1sacce ILCYGVARFG S......... VLFGELRNAV FAKVAQNAIR
```

```
Hmt1schpo IYRFLQGNMG .................. ......... .VIGSLRSFL WVPVSQYAYR
Mdlescco  VPPKVVGIVV DGVTEQHFTT GQILMWIATM VLIAVVVYLL RYVWRVLLFG
Ste6sacce ................ ... SQLVQRSMAV MALGAASVPV MWLSLTSWMH
Syrdpsesy QRQSL.... ............ .......FWF VGLSVVALLF RNGASLFPAY
Consensus ................ ..........  ..................... .
```

```
                  401                                        450
Surcricr  TGINLRGAIQ TKIYNKIMHM STSNLSMGEM TAGQICNLVA IDTNQLMWFF
Surratno  TGINLRGAIQ TKIYNKIMHL STSNLSMGEM TAGQICNLVA IDTNQLMWFF
Mdrleita  TAALFETSSM ALLFEKCFTV SRRSLQRPDM SVGRIMNMVG NDVDNIGSLN
Hemsentfa FQYEFDFKLM FSYIDKLFSM PLMYF..SNR STGELVFRAN LNIYIRQILS
Lcn3lacla TNLLYESKIS RQIFKGIFSR PLLYF..RNN SVGTIIEKIN LRTGIRDGIL
Nistlacla FDMRLSYSIN MRLMRTTSSL ELSDY..EQA DMYNIIEKVT QDSTYKPFQL
Spabbacsu FQLNIGYKLN YKVMKKSSNL ALKDF..ENP EIYDKLERVT KEISYKPYQI
Cyddescco AGQHIRFAIR RQVLDRLQQA GPAWI..QGK PAGSWATLVL EQIDDMHDYY
Cyddhaein SGQLLRNHIR QKILDKIHLV GPATI..NQK PAGSWASIML EQVENLHNFY
Cydcescco ATFRVLQHLR IYTFSKLLPL SPAGL..ARY RQGELLNRVV ADVDTLDHLY
Cydchaein ATFRILSKLR VQVFEKIIPL SPAVL..NRY RNSDLLNRLV SDVDTLDSLY
Aprdpseae VSERFDGQLH GRIYAAAFER ......NLRA GGQEASQALH DLTTLRQFIT
Prtderwch LGTRIDLALN QDVFNAAFAR ......NLEA GDGRAGLALT DLTLLRQFIT
Hlybescco STSRIDVELG AKLFRHLLAL .PISYFESR. RVGDTVARVR ELDQIRNFLT
Hlybpasha STSRIDVELG AKLFRHLLSL .PISYFENR. RVGDTVARVR ELDQIRNFLT
Rt3bactpl STSRIDVELG AKLFRHLLAL .PISYFENR. RVGDTVARVR ELDQIRNFLT
Hlybactac STSRIDVELG ARLFRHLLAL .PISYFEAR. RVGDTVARVR ELDQIRNFLT
Hlybactpl STSRIDVELG ARLFRHLLAL .PISYFENR. RVGDTVARVR ELDQIRNFLT
Cyabborpe TSSKLDVELG ARLYAHLLRL .PLAYFQAR. RVGDSVARVR ELEHIRAFLT
Comastrpn LGQRLSIDVI LSYIKHVFHL .PMSFFATR. RTGEIVSRFT DANSIIDALA
Lcnclacla LSQRLAIDVI LSYIRHIFQL .PMSFFSTR. RTGEITSRFS DASSILDAIA
Peddpedac LGQRLMIDIV LKYVHHLFDL .PMNFFTTR. HVGEMTSRFS DASKIIDALG
Cvabescco MSTLINVQWQ SGLFDHLLRL .PLAFFERR. KLGDIQSRFD SLDTLRATFT
Pmd1schpo AGERIARRIR QDYLHAILSQ .NIGYFD.RL GAGEITTRIT TDTNFIQDGL
Pglyarath SGERQTTKMR IKYLEAALNQ .DIQFFDTEV RTSDVVFAIN TDAVMVQDAI
Pgplarath SGERQTTKMR IKYLEAALNQ .DIQFFDTEV RTSDVVFAIN TDAVMVQDAI
Mdr2musmu AAGRQIKKIR QKFFHAILRQ .EMGWFDIK. GTTELNTRLT DDVSKISEGI
Mdr3crigr AAGRQIKKIR QNFFHAILRQ .EMGWFDIK. GTTELNTRLT DDISKISEGI
Mdr3homsa AAGRQIRKIR QKFFHAILRQ .EIGWFDIN. DTTELNTRLT DDISKISEGI
Mdr1musmu AAGRQIHKIR QKFFHAIMNQ .EIGWFDVH. DVGELNTRLT DDVSKINDGI
Mdr2crigr AAGRQINKIR QKFFHAIMNQ .EIGWFDVH. DIGELNTRLT DDVSKINDGI
Mdr1crigr AAGRQIHKIR QKFFHAIMNQ .EIGWFDVH. DVGELNTRLT DDVSKINEGI
Mdr3musmu AAGRQIHKIR QKFFHAIMNQ .EIGWFDVH. DVGELNTRLT DDVSKINEGI
Mdr1homsa AAGRQIHKIR KQFFHAIMRQ .EIGWFDVH. DVGELNTRLT DDVSKINEVI
Mdr4drome IALNQIDRIR KLFLEAMLRQ .DIAWYDTS. SGSNFASKMT EDLDKLKEGI
Mdr5drome VALRQVTRMR IKLFSSVIRQ .DIGWHDLA. SKQNFTQSMV DDVEKIRDGI
Mdr1caeel VAEQMNNRLR REFVKSILRQ .EISWFDTNH S.GTLATKLF DNLERVKEGT
Mdr3caeel LCERRLHCIR KKYLKSVLRQ .DAKWFD.ET TIGGLTQKMS SGIEKIKDGI
Mdr1leien AACRQVARIR LLFFRAVLRQ .DIGWHDEH. SPGALTARMT GDTRVIQNGI
Msbaescco VSGKVVMTMR RRLFGHMMGM .PVSFFD.KQ STGTLLSRIT YDSEQVASSS
Msbahaein VSGKVVMTMR RRLFKHLMFM .PVSFFD.QN STGRLLSRIT YDSQMIASSS
Hetaanasp AQLNLADRLH KQIFEQLQAL .RLSYFA.QT RSGELINTIT TEIERIKQGF
Mdrplafa  ITSKILKTLK LEYLRSVFYQ .DGQFHD.NN PGSKLRSDLD FYLEQVSSGI
Mdl1sacce TGERLVARLR TRTMKAALDQ .DATFLD.TN RVGDLISRLS SDASIVAKSV
Mdl2sacce LSERVVARLR ANVIKKTLHQ .DAEFFD.NH KVGDLISRLG SDAYVVSRSM
Mt2ratno  AESRINLRIR EQLFSSLLRQ .DLAFFQ.ET KTGELNSRLS SDTSLMSQWL
Tap2musmu TMSRINLRIR EQLFSSLLRQ .DLGFFQ.ET KTGELNSRLS SDTSLMSRWL
```

```
Tap2homsa TMSRINLRIR EQLFSSLLRQ .DLGFFQ.ET KTGELNSRLS SDTTLMSNWL
Tap1homsa TMGHVHSHLQ GEVFGAVLRQ .ETEFFQ.QN QTGNIMSRVT EDTSTLSDSL
Tap1musmu TMGHMHGRVH REVFRAVLRQ .ETGFFL.KN PAGSITSRVT EDTANVCESI
Chvaagrtu LAHGRRATLL TEAFGRIISM .PLSWHHLRG TSNALHTLLR ASETLFGLW.
Ndvarhime LAHGRRASLL TEAFGRIVSM .PLSWHSQRG TSNALHTLLR ACETLFGLW.
Atm1sacce .......TVS LQTFQHLMKL .DLGWHLSRQ TGGLTRAMDR GTKGISQVLT
Hmt1schpo .......AIS TKALRHVLNL .SYDFHLNKR AGEVLTALTK GS.SLNTFAE
Mdlescco  ASYQLAVELR EDYYRQLSRQ .HPEFY.LRH RTGDLMARAT NDVDRVVFAA
Ste6sacce IGERQGFRIR SQILEAYLEE KPMEWYDNNE KLLGDFTQIN RCVEELRSSS
Syrdpsesy ASMRIMTRLR IALCRKILGT PLEE..VDRR GAPNVLTLLT SDIPQLNATL
Consensus ..........  ..........  ..........  ..........  ..........
```

```
          451                                             500
Surcricr  FLCPNLWTMP VQII..VGVI LLYYILGVSA LIGAAVIILL APVQYFVATK
Surratno  FLCPNLWAMP VQII..VGVI LLYYILGVSA LIGAAVIILL APVQYFVATK
Mdrleita  WYVMYFWSAP LQLV..LCLL LLIRLVGWLR VPGMAVLFVT LPLQAVISKH
Hemsentfa QKVITTLIDS LFLG..IYLF LMV.....NY SILLTIIALV LISLIAFLSI
Lcn3lacla LKIFPSLLNF FTVF..IVII YLG.....TI SFTLTLF.LV IMNLLYMIFS
Nistlacla FNAIIVELSS FISL..LSSL FFI..GTWNI GVAILLLIVP VLSLVLFLRV
Spabbacsu IQAIITMTTS FVTL..LSSI AFL..MSWNP KVSLLLLVIP VISLFYFLKI
Cyddescco ARYLPQMALA VSVP..LLIV VAI....FPS NWAAALILLG TAPLIPLFMA
Cyddhaein ARFLPQQSLS AIVP..VVIF IAV....FPL NWAAGLILMI TAPLVPLFMI
Cydcescco LRVISPLVGA FVVI..MVVT IGLSFLDFTL AFTLGGIMLL TLFLMPPLFY
Cydchaein LRLLAPFFTA VFVI..IAMM IGLSFINIPL ALGLGLFLLI LLMIIPTVFY
Aprdpseae GQALFAFFDA PWFP..VYLL V.I....FLF DPWLGLLSLV GALALMALAW
Prtderwch GNALFAFFDV PWFP..LFLL V.L....FLL HPWLGMLALG GTVVPGGVGL
Hlybescco GQALTSVLDL LFSL..IFFA V.M....WYY SPKLTLVILF SLPCYAAWSV
Hlybpasha GQALTSVLDL LFSF..IFFA V.M....WYY SPKLTLVILG SLPCYILWSI
Rt3bactpl GQALTSVLDL LFSF..IFFA V.M....WYY SPKLTIVILL SLPCYIAWSI
Hlybactac GQALTSILDL LFSF..IFFA V.M....WYY SPKLTLVVLG SLPCYVIWSV
Hlybactpl GQALTSVLDL MFSF..IFFA V.M....WYY SPKLTLVILG SLPFYMGWSI
Cyabborpe GNAVTVLLDV VFSV..VFIA V.M....FFY SVKLTLVVLA ALPCYFLLSL
Comastrpn STILSIFLDV STVV..IISL V.L....FSQ NTNLFFMTLL ALPIYTVIIF
Lcnclacla STILSLFLDL TIVV..MTGL I.L....GLQ NMQLFLLVLL AIPLYIVVII
Peddpedac STTLTLFLDM WILL..AVGL F.L....AYQ NINLFLCSLV VVPIYISIVW
Cvabescco TSVIGFIMDS IMVV..GVCV M.M....LLY GGYLTWIVLC FTTIYIFIRL
Pmd1schpo GEKVGLVFFA IATF..VSGF V.I...AFIR HWKFTLILSS MFPAICGGIG
Pglyarath SEKLGNFIHY MATF..VSGF I.V...GFTA VWQLALVTLA VVPLIAVIGG
Pgp1arath SEKLGNFIHY MATF..VSGF I.V...GFTA VWQLALVTLA VVPLIAVIGG
Mdr2musmu GDKVGMFFQA IATF..FAGF I.V...GFIR GWKLTLVIMA ISPILGLSTA
Mdr3crigr GDKVGMFFQA VATF..FAGF I.V...GFIR GWKLTLVIMA ISPILGLSAA
Mdr3homsa GDKVGMFFQA VATF..FAGF I.V...GFIR GWKLTLVIMA ISPILGLSAA
Mdr1musmu GDKIGMFFQS ITTF..LAGF I.I...GFIS GWKLTLVILA VSPLIGLSSA
Mdr2crigr GDKIGMFFQS IATF..LAAF I.V...GFIS GWKLTLVILA VSPLIGLSSA
Mdr1crigr GDKIGMFFQA MATF..FGGF I.I...GFTR GWKLTLVILA ISPVLGLSAG
Mdr3musmu GDKIGMFFQA MATF..FGGF I.I...GFTR GWKLTLVILA ISPVLGLSAG
Mdr1homsa GDKIGMFFQA MATF..FTGF I.V...GFTR GWKLTLVILA ISPVLGLSAA
Mdr4drome GEKIVIVVFL IMTF..VIGI V.S...AFVY GWKLTLVVLS CVPFIIAATS
Mdr5drome SEKVGHFVYL VVGF..IITV A.I...SFSY GWKLTLAVSS YIPLVILLNY
Mdr1caeel GDKIGMAFQY LSQF..ITGF I.V...AFTH SWQLTLVMLA VTPIQALCGF
Mdr3caeel GDKVGVLVGG VATF..ISGV S.I...GFYM CWQLTLVMMI TVPLQLGSMY
Mdr1leien NDKLSQGIMN GSMG..VIGY I.A...GFVF SWELTLMMIG MMPFIIVMAA
Msbaescco SGALITVVRE GASI..IGLF I.M...MFYY SWQLSIILIV LAPIVSIAIR
Msbahaein SGSLITIVRE GAYI..ISLF A.V...MFYT SWELTIVLFI IGPIIAVLIR
```

155

```
Hetaanasp SGLAFVLTRI MT.V..CVYF V.V...MFSI SWQLSIISVL IFLLLAVGLS
Mdrplafa  GTKFITIFTY ASSF..LGLY I.W...SLIK NARLTLCITC VFPLIYVCGV
Mdl1sacce TQNVSDGTRA IIQG..FVGF G.M...MSFL SWKLTCVMMI LAPPLGAMAL
Mdl2sacce TQKVSDGVKA LICG..VVGV G.M...MCCL SPQLSILLLF FTPPVLFSAS
Mt2ratno  SLNANILLRS LVKV..VGLY Y.F...MLQV SPRLTFLSLL DLPLTIAAEK
Tap2musmu PFNANILLRS LVKV..VGLY F.F...MLQV SPRLTFLSLL DLPLTIAAEK
Tap2homsa PLNANVLLRS LVKV..VGLY G.F...MLSI SPRLTLLSLL HMPFTIAAEK
Tap1homsa SENLSLFLWY LVRG..LCLL G.I...MLWG SVSLTMVTLI TLPLLFLLPK
Tap1musmu SDTLSLLLWY LGRA..LCLL V.F...MFWG SPYLTLVTLI NLPLLFLLPK
Chvaagrtu ...LEFMRTH LATF..VALV L.LIPTAMAM DLRLSFVLIG LGIVYWFIGK
Ndvarhime ...LEFMRQH LATA..VALM L.LIPTAFAM DVRLSLILVV LGAAYVMISK
Atm1sacce AMVFHIIPIS FEIS..VVCG I.LT...YQF GASFAAITFS TMLLYSIFTI
Hmt1schpo QVVFQIGPVL LDLG..VAMV Y.FF...IKF DIYFTLIVLI MTLCYCYVTV
Mdlescco  GEGVLTLVDS LVMA..CAVL I.MMSTQI.. SWQLTLFSLL PMPVMAIMIK
Ste6sacce AEASAITFQN LVAI..CALL G....TSFYY SWSLTLIILC SSPIITFFAV
Syrdpsesy LIMPTILVES AVFLFGIAYL AYLSWVVFAI TISLMILGVA MYLLFFMGGM
Consensus ........... ........ . ........... ........ . ..........
```

```
          501                                               550
Surcricr  L......SQA QRTTLEHSNE RLKQTNEMLR GMKLLKLYAW ESIFCSRVEV
Surratno  L......SQA QRTTLEYSNE RLKQTNEMLR GIKLLKLYAW ENIFCSRVEK
Mdrleita  V......QDV SERMASVVDL RIKRTNELLS GVRIVKFMGW EPVFLARIQD
Hemsentfa INSHTIKRFV DKEIMEQG.N VQRIITEAIE GIETIKSANA KKSFLLNWKN
Lcn3lacla FSLISIKRQA NIQYTQQTID FTSVVQEDLN QIEQIKAQAN EKECVKRWTK
Nistlacla GQLEFLIQWQ RAS.SERETW YIVYLLTHDF SFKEIKLNNI SNYFIHKFGK
Spabbacsu GQEEFFIHWK RAG.KERKSW YISYILTHDF SFKELKLYNL KDYLLNKYWD
Cyddescco LVGMGAADAN RRN.FLALAR LSGHFLDRLR GMETLRIFGR GEAEIESIRS
Cyddhaein IVGIAAADNS QKN.MDTLSR LSAQFLDRLR GLETLRLFNR TSEQTEHIEN
Cydcescco RAGKSTGQ.. .NL.THLRGQ YRQQLTAWLQ GQAELTIFGA SDRYRTQLEN
Cydchaein RLGQQFGE.. .RL.IQARAT YRTQFLEFIQ AQAELLLFNA EDKLKEKMLV
Aprdpseae FNERATRAPL AKA.GELSIK SGQLASNNLR NAEVIEAMGM LGSMRGRWER
Prtderwch AEPASDQSTA GGS.NQQSQQ ATHLADAQLR NADVIEAMGM LGNLRRRWLA
Hlybescco FISPILRRRL DDK.FSRNAD NQSFLVESVT AINTIKAMAV SPQMTNIWDK
Hlybpasha FISPILRRRL DEK.FARSAD NQAFLVESVT SINMIKAMAV APQMTDTWDK
Rt3bactpl FISPILRRRL DEK.FARNAD NQSFLVESVS AIDTIKALAV TPQMTNIWDK
Hlybactac FISPILRRRL DDK.FARNAD NQSFLVESVT AINTIKAMAI SPQMTNIWDK
Hlybactpl FISPILRRRL DEK.FARGAH NQSFLVESVT AINTIKALAV TPQMTNTWDK
Cyabborpe VLTPVLRRRL HVK.FNRGAE NQAFLVETVS GIDTVKSLAV EPQWQRNWDR
Comastrpn AFMKPFEKMN RDT.MEANAV LSSSIIEDIN GIETIKSLTS ESQRYQKIDK
Lcnclacla IFTPLFEKQN HEV.MQTNAV LNSSIIEDIN GIETIKALAS EQERYQKIDY
Peddpedac LFKKTFNRLN QDT.MESNAV LNSAIIESLS GIETIKSLTG EATTKKKIDT
Cvabescco VTYGNYRQIS EEC.LVREAR AASYFMETLY GIATVKIQGM VGIRGAHWLN
Pmd1schpo LGVPFITKNT KGQ.IAVVAE SSTFVEEVFS NIRNAFAFGT QDILAKLYNK
Pglyarath IHTTTLSKLS NKS.QESLSQ AGNIVEQTVV QIRVVMAFVG ESRASQAYSS
Pgp1arath IHTTTLSKLS NKS.QESLSQ AGNIVEQ... .......PLNK NKKTHHHQQT
Mdr2musmu VWAKILSTFS DKE.LAAYAK AGAVAEEAPG AIRTVIAFGG QNKELERYQK
Mdr3crigr VWAKILSTFS DKE.LAAYAK AGAVAEEALG AIRTVIAFGG QNKELERYQK
Mdr3homsa VWAKILSAFS DKE.LAAYAK AGAVAEEALG AIRTVIAFGG QNKELERYQK
Mdr1musmu LWAKVLTSFT NKE.LQAYAK AGAVAEEVLA AIRTVIAFGG QQKELERYNK
Mdr2crigr MWAKVLTSFT NKE.LQAYAK AGAVAEEVLA AIRTVIAFGG QNKELERYNK
Mdr1crigr IWAKVLTSFT DKE.LQAYAK AGAVAEEVLA AIRTVIAFGG QKKELERYNN
Mdr3musmu IWAKILSSFT DKE.LHAYAK AGAVAEEVLA AIRTVIAFGG QKKELERYNN
Mdr1homsa VWAKILSSFT DKE.LLAYAK AGAVAEEVLA AIRTVIAFGG QKKELERYNK
Mdr4drome VVARLQGSLA EKE.LKSYSD AANVVEEVFS GIRTVFAFSG QEKEKERFGK
```

```
Mdr5drome YVAKFQGKLT ARE.QESYAG AGNLAEEILS SIRTVVSFGG EKSEVQRYEN
Mdr1caeel AIAKSMSTFA IRE.TLRYAK AGKVVEETIS SIRTVVSLNG LRYELERYST
Mdr3caeel LSAKHLNRAT KNE.MSAYSN AGGMANEVIA GIRTVMAFNA QPFEINRYAH
Mdr1leien IIGSIVSKIT ESS.RKYFAK AGSLATEVME NIRTVQAFGR EDYELERFTK
Msbaescco VVSKRFRNIS KNM.QNTMGQ VTTSAEQMLK GHKEVLIFGG QEVETKRFDK
Msbahaein LVSKIFRRLS KNL.QDSMGE LTSATEQMLK GHKVVLSFGG QHVEEVHFNH
Hetaanasp TLNKRVRETS FGI.SHANAQ FTAVAVEFIN GIRTIQAFGT QEFERQRFYK
Mdrplafa  ICNKKVKLNK KTS.LLYNNN TMSIIEEALM GIRTVASYCG EKTILNKFNL
Mdl1sacce IYGRKIRNLS RQL.QTSVGG LTKVAEEQLN ATRTIQAYGG EKNEVRRYAK
Mdl2sacce VFGKQIRNTS KDL.QEATGQ LTRVAEEQLS GIKTVQSFVA EGNELSRYNV
Mt2ratno  VYNPRHQAVL KEI.QDAVAK AGQVVREAVG GLQTVRSFGA EEQEFRRYKE
Tap2musmu VYNPRHQAVL KEI.QDAVAK AGQVVREAVG GLQTVRSFGA EEQEVSHYKE
Tap2homsa VYNTRHQEVL REI.QDAVAR AGQVVREAVG GLQTVRSFGA EEHEVCRYKE
Tap1homsa KVGKWYQLLE VQV.RESLAK SSQVAIEALS AMPTVRSFAN EEGEAQKFRE
Tap1musmu KLGKVHQSLA VKV.QESLAK STQVALEALS AMPTVRSFAN EEGEAQKFRQ
Chvaagrtu WVMGRTKDGQ ASV.EEHYHS VFAHVSDSIS NVSVLHSYNR IEAETKALKS
Ndvarhime VVMSRTKEGQ AAV.EGHYHT VFSHVSDSIS NVSVVHSYNR IEAETRELKK
Atm1sacce KTTAWRTHFR RDA.NKADNK AASVALDSLI NFEAVKYFNN EKYLADKYNG
Hmt1schpo KITSWRTEAR RKM.VNTWRE SYAVQNDAIM NFETVKNFDA DDFENERYGH
Mdlescco  RNGDALHERF KLA.QAAFSS LNDRTQESLT SIRMIKAFGL EDRQSALFAA
Ste6sacce VFSRMIHVYS EKE.NSETSK AAQLLTWSMN AAQLVRLYCT QRLERKKFKE
Syrdpsesy KFTHKVRDEF TAFNEYTHAL VFGLKELKLN GIRR.RWFSR SAIQESSVRV
Consensus .......... .......... .......... .......... ..........
```

```
          551                                                600
Surcricr  TRRKEMTSLR AFAVYTSISI FMNTAIP... IAAVL..... ITFVGHVSFF
Surratno  TRRKEMTSLR AFAVYTSISI FMNTAIP... IAAVL..... ITFVGHVSFF
Mdrleita  ARSRELRCLR DVHVANVFFM FVNDATPTLV IAVVF..... ILY..HVS..
Hemsentfa MFTSQLLITK NKNRYIAIFG ILPEIIQSVM PALFLIIGIK LII.......
Lcn3lacla KSAQTIFSYN KILNIDGITS AFNQGFNYIC VILMMIFGIY LNQ.......
Nistlacla LKKGFINQDL AIARKKTYFN IFLDFILNLI NILTIFAMIL SVR.......
Spabbacsu IKKSFIEQDT KILRKKTLLN LIYEIAVQLV GAVIIFIAIM SAF.......
Cyddescco ASEDFRQRTM EVLRLAFLSS GILEFFTSLS IALVAVYFGF SYLGELDFGH
Cyddhaein ATEDFRETTM DVLKLAFLSS AVLEFFTSIS IALMAVYFGF SYLGQIEFGT
Cydcescco TEIQWLEAQR RQSELTALSQ AIMLLIGALA VILM.....L WMAS.GGVGG
Cydchaein TEKTWQEDQA KEAKLSGFST ALVLFLNGLL ISGM.....L WFASNADFGT
Aprdpseae LHQAFLDQQS LASERAARIN ALSKYLRIAL QSLVLGLGAW LAV.......
Prtderwch RHYRFISLQN LASERAAAVG GASKYSRIAL QSLMLGLGAL LAI.......
Hlybescco QLAGYVAAGF KVTVLATIGQ QGIQLIQKTV MIINLWLG.. ......AHLV
Hlybpasha QLASYVSSSF RVTVLATIGQ QGVQLIQKTV MVINLWLG.. ......AHLV
Rt3bactpl QLASYVSADF RVTVLATIGQ QGVQLIQKTV MIINLWLG.. ......AHLV
Hlybactac QLASYVAVSF KVTVLATIGQ QGIQLIQKAV MVINLWLG.. ......AHLV
Hlybactpl QLASYVSAGF RVTTLATIGQ QGVQFIQKVV MVITLWLG.. ......AHLV
Cyabborpe QLAGYVAAGL SVANVAMLAN TGVTLISRL. ....LRWESC ....GWAHRG
Comastrpn EFVDYLKKSF TYSRAESQQK ALKKVAHLLL NVGILWMG.. ......AVLV
Lcnclacla EFASYLKKAF TLQKSEAIQG LIKAIIQLTL SVTILWFG.. ......ATLV
Peddpedac LFSDLLHKNL AYQKADQGQQ AIKAATKLIL TIVILWWG.. ......TFFV
Cvabescco MKIDAINSGI KLTRMDLLFG GINTFVTACD QIVILWLG.. ......AGLV
Pmd1schpo YLITAQRFGI NKAIAMGLMV GWMFFVAYGV YGLAFWEGGR LL........
Pglyarath ALKIAQKLGY KTGLAKGMGL GATYFVVFCC YALLLWYGGY LV........
Pgplarath NIKPNDNKSD EQRRAKS... .......... .......... ..........
Mdr2musmu HLENAKKIGI KKAISANISM GIAFLLIYAS YALAFWYGST LVI.......
Mdr3crigr HLENAKKIGI KKAISANISM GIAFLLIYAS YALAFWYGST LVI.......
Mdr3homsa HLENAKEIGI KKAISANISM GIAFLLIYAS YALAFWYGST LVI.......
```

```
Mdr1musmu NLEEAKNVGI KKAITASISI GIAYLLVYAS YALAFWYGTS LVL.......
Mdr2crigr NLEEAKNVGI KKAVTANISI GIAYLLVYAS YALAFWYGTS LVL.......
Mdr1crigr NLEEAKRLGI KKAITANISM GAAFLLIYAS YALAFWYGTS LVI.......
Mdr3musmu NLEEAKRLGI KKAITANISM GAAFLLIYAS YALAFWYGTS LVI.......
Mdr1homsa NLEEAKRIGI KKAITANISI GAAFLLIYAS YALAFWYGTT LVL.......
Mdr4drome LLIPAENTGR KKGLYSGMGN ALSWLIIYLC MALAIWYGVT LILDE..RDL
Mdr5drome FLVPARKASQ WKGAFSGLSD AVLKSMLYLS CAGAFWYGVN LIIDD..RNV
Mdr1caeel AVEEAKKAGV LKGLFLGISF GAMQASNFIS FALAFYIGVG WV.......
Mdr3caeel QLNEARRMGI RKAIILAICT AFPLMLMFTC MAVAFWYGAT LA.......
Mdr1leien AVLYAQGRGI RKELASNLSA AVIMALMYVS YTVAFFFGSY LVEWG..RR.
Msbaescco VSNRMRLQGM KMVSASSISD PIIQLIASLA LAFVLYAASF PSV.......
Msbahaein VSNDMRRKSM KMVTANSISD PVVQVIASLA LATVLYLATT PLI.......
Hetaanasp ASTNQLNAAI KVVLAWTLVK PIAEGIA... .TTVLISLIV ISF.......
Mdrplafa  SETFYSKYIL KANFVEALHI GLINGLILVS YAFGFWYGTR IIINSATNQY
Mdl1sacce EVRNVFHIGL KEAVTSGLFF GSTGLVGNTA MLSLLLVGTS MIQ.......
Mdl2sacce AIRDIFQVGK TAAFTNAKFF TTTSLLGDLS FLTVLAYGSY LVL.......
Mt2ratno  ALERCRQLWW RRDLEKSLYL VIQRVMALGM QVLILNVGVQ QIL.......
Tap2musmu ALERCRQLWW RRDLEKDVYL VIRRVMALGM QVLILNCGVQ QIL.......
Tap2homsa ALEQCRQLYW RRDLERALYL LVRRVLHLGV QMLMLSCGLQ QMQ.......
Tap1homsa KLQEIKTLNQ KEAVAYAVNS WTTSISGMLL KVGILYIGGQ LVT.......
Tap1musmu KLEEMKTLNK KEALAYVAEV WTTSVSGMLL KVGILYLGGQ LVI.......
Chvaagrtu FTEKLLSAQY PVLDWWAFAS ALNRTASTVS MMIILVIGTV LVK.......
Ndvarhime FTQRLLSAQY PVLDWWALAS GLNRIASTIS MMAILVIGTV LVQ.......
Atm1sacce SLMNYRDSQI KVSQSLAFLN SGQNLIFTTA LTAMMYMGCT GVI.......
Hmt1schpo AVDIYLKQER KVLFSLNFLN IVQGGIFTFS LAIACLLSAY RVT.......
Mdlescco  DAEDTGKKKL RVARIDARFD PTIYIAIGMA NLLAIGGGSW MVV.......
Ste6sacce IILNCNTFFI KSCFFVAANA GILRFLTLTM FVQGFWFGSA MI........
Syrdpsesy AKYNYIERLW FTAAENVGQL TLSLLVGCLL FAAPMF............
Consensus .......... .......... .......... .......... ..........

          601                                                 650
Surcricr  KESDLSPSVA FASLSLFHIL VTPLFLLSSV VRSTVKALVS VQKLSEF...
Surratno  KESDFSPSVA FASLSLFHIL VTPLFLLSSV VRSTVKALVS VQKLSEF...
Mdrleita  .GKVLKPEVV FPTIALLNTM RVSFFMIPII ISSILQCFVS AKRVTAF...
Hemsentfa .NNSLSLGSL IGFVSIVTMV MKPILSLVSS YNDFLLLNVY FQKLSEV...
Lcn3lacla .GNLVSIPDL IIFQSGISLF VSAVNQIQDV MFEISRLSIY GNKISDL...
Nistlacla .AGKLLIGNL VSLIQAISKI NTYSQTMIQN IYIIYNTSLF MEQLFEF...
Spabbacsu .AGKIMVGNV MSYIRSVSLV QNHSQSIMTS IYSIYNSNLY MNQLYEF...
Cyddescco YDTGVTLAAG FLALILAPEF FQPLRDLGTF YHAKAQAVGA ADSLKTF...
Cyddhaein YNAPLTLFTG FFCLILAPEF YQPLRDLGTY YHDRAAGIGA ADAIVDF...
Cydcescco NAQPGALIAL FVFCALAA.. FEALAPVTGA FQHLGQVIAS AVRISDL...
Cydchaein DEYHTAYIAL FTFAALAA.. FEIIMPLGAA FLHIGQVIAA AERVTEI...
Aprdpseae .EGRITPGMM IAGSILMGRA LGPIDQLIGV WKQWGAARDA YRRLSGL...
Prtderwch .DGKITPGMM IAGSILVGRV LSPIDQLIGV WKQWSSARIA WQRLTRL...
Hlybescco ISGDLSIGQL IAFNMLAGQI VAPVIRLAQI WQDFQQVGIS VTRLGDV...
Hlybpasha ISGDLSIGQL IAFNMLSGQV IAPVIRLAQL WQDFQQVGIS VTRLGDV...
Rt3bactpl ISGDLSIGQL ITFNMLSGQV IAPVVRLAQL WQDFQQVGIS ITRLGDV...
Hlybactac ISGDLSIGQL IAFNMLAGQI ISPVIRLAQI WQDFQQVGIS VTRLGDV...
Hlybactpl ISGDLSIGQL IAFNMLSGQV IAPVIRLAQL WQDFQQVGIS VTRLGDV...
Cyabborpe GRARMTVGEL VAFNMLSGHV TQPVIRLAQL WNDFQQTGVS MQRLGDI...
Comastrpn MDGKMSLGQL ITYNTLLVYF TNPLENIINL QTKLQTAQVA NNRLNEV...
Lcnclacla ISQKITLGQL ITFNALLSYF TNPITNIINL QTKLQKARVA NERLNEV...
Peddpedac MRHQLSLGQL LTYNALLAYF LTPLENIINL QPKLQAARVA NNRLNEV...
Cvabescco IDNQMTIGMF VAFSSFRGQF SERVASLTSF LLQLRIMSLH NERIADI...
```

```
Pmd1schpo HAGDLDVSKL IGCFFAVLIA SYSLANISPK MQSFVSCASA AKKIFDT...
Pglyarath RHHLTNGGLA IATMFAVMIG GLALGQSAPS MAAFAKAKVA AAKIFRI...
Pgp1arath ........... .........C LQALGQSAPS MAAFAKAKVA AAKIFRI...
Mdr2musmu .SKEYTIGNA MTVFFSILIG AFSVGQAAPC IDAFANARGA AYVIFDI...
Mdr3crigr .SKEYTIGNA MTVFFSILIG AFSVGQAAPC IDAFANARGA AYVIFDI...
Mdr3homsa .SKEYTIGNA MTVFFSILIG AFSVGQAAPC IDAFANARGA AYVIFDI...
Mdr1musmu .SNEYSIGEV LTVFFSILLG TFSIGHLAPN IEAFANARGA AFEIFKI...
Mdr2crigr .SNEYSVGQV LTVFFSILFG TFSIGHIAPN IEVFANARGA AYEIFKI...
Mdr1crigr .SKEYSIGQV LTVFFAVLIG AFSIGQASPN IEAFANARGA AYEIFNI...
Mdr3musmu .SKEYSIGQV LTVFFSVLIG AFSVGQASPN IEAFANARGA AYEVFKI...
Mdr1homsa .SGEYSIGQV LTVFFSVLIG AFSVGQASPS IEAFANARGA AYEIFKI...
Mdr4drome PDRVYTPAVL VIVLFAVIMG AQNLGFASPH VEAIAVATAA GQTLFNI...
Mdr5drome ENKEYTPAIL MIAFFGIIVG ADNIARTAPF LESFASARGC ATNLFKV...
Mdr1caeel HDGSLNFGDM LTTFSSVMMG SMALGLAGPQ LAVLGTAQGA ASGIYEV...
Mdr3caeel AAGAVSSGAV FAVFWAVLIG TRRLGEAAPH LGAITGARLA IHDIFKV...
Mdr1leien .....DMADI ISTFLAVLMG SFGLGFVAPS RTAFTESRAA AYEIFKA...
Msbaescco M.DSLTAGTI TVVFSSMIAL MRPLKSLTNV NAQFQRGMAA CQTLFTI...
Msbahaein AEDNLSAGSF TVVFSSMLAM MRPLKSLTAV NAQFQSGMAA CQTLFAI...
Hetaanasp ATFTLPVASL LTFFFVLVRV IPNIQDINGT VAFLSTLQGS SENIKNI...
Mdrplafa  PNNDFNGASV ISILLGVLIS MFMLTIILPN ITEYMKALEA TNSLYEI...
Mdl1sacce .SGSMTVGEL SSFMMYAVYT GSSLFGLSSF YSELMKGAGA AARVFEL...
Mdl2sacce .QSQLSIGDL TAFMLYTEYT GNAVFGLSTF YSEIMQGAGA ASRLFEL...
Mt2ratno  .AGEVTRGGL LSFLLYQEEV GHHVQNLVYM YGDMLSNVGA AEKVFSY...
Tap2musmu .AGEVTRGGL LSFLLYQEEV GQYVRNLVYM YGDMLSNVGA AEKVFSY...
Tap2homsa .DGELTQGSL LSFMIYQESV GSYVQTLVYI YGDMLSNVGA AEKVFSY...
Tap1homsa .SGAVSSGNL VTFVLYQMQF TQAVEVLLSI YPRVQKAVGS SEKIFEY...
Tap1musmu .RGTVSSGNL VSFVLYQLQF TQAVQVLLSL YPSMQKAVGS SEKIFEY...
Chvaagrtu .NGELRVGDV IAFIGFANLL IGRLDQMRQF VTQIFEARAK LEDFFVL...
Ndvarhime .RGELGVGEV IAFIGFANLL IGRLDQMKAF ATQIFEARAK LEDFFQL...
Atm1sacce .GGNLTVGDL VLINQLVFQL SVPLNFLGSV YRDLKQSLID METLFKL...
Hmt1schpo .FGFNTVGDF VILLTYMIQL QQPLNFFGTL YRSLQNSIID TERLLEI...
Mdlescco  .QGSLTLGQL TSFMMYLGLM IWPMLALAWM FNIVERGSAA YSRIRAM...
Ste6sacce KKGKLNINDV ITCFHSCIML GSTLNNTLHQ IVVLQKGGVA MEKIMTL...
Syrdpsesy ..AVIDAKTM SASVLAVLYI MGPLVMLVSA MPMLAQGRIA CTRLADFGFS
Consensus ......... ......... ......... ......... .........
```

```
          651                                            700
Surcricr  LSSAEIREEQ CAPREPAPQG QAGKYQAVPL KVVNRKRPAR EEVRDLLGPL
Surratno  LSSAEIREEQ CAPREPAPQG QAGKYQAVPL KVVNRKRPAR EEVRDLLGPL
Mdrleita  IECPDTHSQV QDIASIDVPD AAAIFKGASI HTYLPVKLPR CKSR..LTAM
Hemsentfa LTYE...... .EKNDFNNKK GNVG.KEEFI YRVNNVYYTI SVFEKNI...
Lcn3lacla LI.E...... .NPQRIDNIE KHSN.NAIIL KDISYSY... .ELNNYI...
Nistlacla LKRESVVHKK ...IEDTEIC NQHI.GTVKV INLSYVYPNS NA..F.A...
Spabbacsu L..ELKEEKS ...QGHKKPI VEPI.HSVVF QNVSFIYPNQ GE..Q.T...
Cyddescco METPLAHPQR GEAELA.... ..ST.DPVTI EAEELFITSP EGKTL.A...
Cyddhaein LESDYLTVHQ NEKTIS.... ..LE.SAVEI SAENLVVLST QGSAL.T...
Cydcescco TDQK.PEVTF PDTQTR.... ..VA.DRVSL TLRDVQFTYP EQSQQ.A...
Cydchaein IEQK.PLVEF NGNEEF.... ..ET.KVRLI SAKNLNFSYP EQETL.V...
Aprdpseae LD.......E FPARERRMEL PEPR.GHLLL ESLDA..APP GSEAR.T...
Prtderwch IA.......A YPPRPAAMAL PAPE.GHLSV EQVSL..RTA QGNTR.....
Hlybescco ..LNSPTESY ...HGKL.TL PEIN.GDITF RNIRFRYKPD SPV...I...
Hlybpasha ..LNSPTEQY ...QGKL.SL PEIK.GDISF KNIRFRYKPD APT...I...
Rt3bactpl ..LNSPTENY ...QGKL.SL PEIF.GDIAF KHIRFRYKPD API...I...
Hlybactac ..LNSPTENN ...TASV.SL PEIQ.GEISF RNIKFRYKPD SPM...I...
```

```
Hlybactpl ..LNSPTESY ...QGKL.AL PEIK.GDITF RNIRFRYKPD APV...I...
Cyabborpe ..LNCRTEVA ...GDKA.QL PALR.GSIEL DRVSFRYRPD AAD...A...
Comastrpn YLVASEFEEK ...KTVE.DL SLMK.GDMTF KQVHYKY.GY GRD...V...
Lcnclacla YLVPSEFEEK ...KT...EL SLSH.FNLNM SDISYQY.GF GRK...V...
Peddpedac YLVESEFSKS ...REIT.AL EQLN.GDIEV NHVSFNY.GY CSN...I...
Cvabescco ALHEKEEKKP ...EIEI.VA DMGP.ISLET NGLSYRYDSQ SAP...I...
Pmdlschpo IDRVSPI.NA FTPTGDV.VK D.IK.GEIEL KNIRFVYPTR PEV.L.V...
Pglyarath IDHKPTI.ER NSESGVE.LD S.VT.GLVEL KNVDFSYPSR PDV.K.I...
Pgplarath IDHKPTI.ER NSESGVE.LD S.VT.GLVEL KNVDFSYPSR PDV.K.I...
Mdr2musmu IDNNPKI.DS FSERGHK.PD N.IK.GNLEF SDVHFSYPSR ANI.K.I...
Mdr3crigr IDNNPKI.DS FSERGHK.PD S.IK.GNLDF SDVHFSYPSR ANI.K.I...
Mdr3homsa IDNNPKI.DS FSERGHK.PD S.IK.GNLEF NDVHFSYPSR ANV.K.I...
Mdr1musmu IDNEPSI.DS FSTKGYK.PD S.IM.GNLEF KNVHFNYPSR SEV.Q.I...
Mdr2crigr IDNEPSI.DS FSTQGHK.PD S.VM.GNLEF KNVHFSYPSR SGI.K.I...
Mdr1crigr IDNKPSI.DS FSKNGYK.PD N.IK.GNLEF KNIHFSYPSR KDV.Q.I...
Mdr3musmu IDNKPSI.DS FSKSGHK.PD N.IQ.GNLEF KNIHFSYPSR KEV.Q.I...
Mdr1homsa IDNKPSI.DS YSKSGHK.PD N.IK.GNLEF RNVHFSYPSR KEV.K.I...
Mdr4drome IDRPSQV.DP MDEKGNR.PE NTA..GHIRF EGIRFRYPAR PDV.E.I...
Mdr5drome IDLTSKI.DP LSTDGKL.LN YGLR.GDVEF QDVFFRYPSR PEV.I.V...
Mdr1caeel LDRKPVI.DS SSKAGRK..D MKIK.GDITV ENVHFTYPSR PDV.P.I...
Mdr3caeel IDHEPEI.KC TSSEGKI.PE K..IQ.GKLTF DGIEFTYPTR PEL.K.I...
Mdr1leien IDRVPPV.D. .IDAGGV.PV PGFK.ESIEF RNVRFAYPTR PGM.I.L...
Msbaescco LDSEQ...EK ..DEGKR.VI ERAT.GDVEF RNVTFTYPGR .DV.P.A...
Msbahaein LDLEP...EK ..DDGAY.KA EPAK.GELEF KNVSFAYQGK .DE.L.A...
Hetaanasp LQTNN...KP YLKNGKL.HF QGLK.RSIDL VSVDFGYTA. .DN.L.V...
Mdrplafa  INRKPLV.EN NDDGETL.PN IK....KIEF KNVRFHYDTR KDV.E.I...
Mdl1sacce NDRKPLI.RP ..TIGKD.PV SLAQ.KPIVF KNVSFTYPTR PKH.Q.I...
Mdl2sacce TDRKPSI.SP ..TVG.H.KY KPDR.GVIEF KDVSFSYPTR PSV.Q.I...
Mt2ratno  LDRRPNL.PN ..PGTLA.PP RL.E.GRVEF QDVSFSYPSR PEK.P.V...
Tap2musmu LDRKPNL.PQ ..PGILA.PP WL.E.GRVEF QDVSFSYPRR PEK.P.V...
Tap2homsa MDRQPNL.PS ..PGTLA.PT TL.Q.GVVKF QDVSFAYPNR PDR.P.V...
Tap1homsa LDRTPRC.PP ..SGLLT.PL HL.E.GLVQF QDVSFAYPNR PDV.L.V...
Tap1musmu LDRTPCS.PL ..SGSLA.PS NM.K.GLVEF QDVSFAYPNQ PKV.Q.V...
Chvaagrtu EDAVKEREEP GDARELS... .NVS.GTVEF RNINFGFANT K...Q.G...
Ndvarhime EDSVQDREEP ADAGELK... .GVV.GEVEF RDISFDFANS A...Q.G...
Atm1sacce ..RKNEVKIK NAERPLM.LP ENVP.YDITF ENVTFGYHPD R...K.I...
Hmt1schpo FEEKPTVVEK PNAPDLK.VT QG....KVIF SHVSFAYDPR K...P.V...
Mdlescco  LAERPVVND. ....GSE.PV PEGR.GELDV NIHQFTYP.Q TDH.P.A...
Ste6sacce LKDGSKRNPL NKTVAHQFPL DYAT.SDLTF ANVSFSYPSR PSE.A.V...
Natabacsu .......... ..........MITL TDCSRRFQDK KKVVK.A...
Syrdpsesy INEPHPEPET SDADNVLLLD HKKSWGSIQL KNVHMNYKDP QSSSGFA...
Consensus .......... .......... .......... ......Y... ..........
```

```
                701                                         750
Surcricr  QRLA...... .........  PSMDGDADNF CVQIIGGFFT WTPDGIPT..
Surratno  QRLT...... .........  PSTDGDADNF CVQIIGGFFT WTPDGIPT..
Mdrleita  QRSTLWFRRR GVPETEWYEV DSPDASASSL AVHSTTVHMG STQTVITDSD
Hemsentfa .......... .......... .......... .......... ..........
Lcn3lacla .......... .......... .......... .......... ..........
Nistlacla .......... .......... .......... .......... ..........
Spabbacsu .......... .......... .......... .......... ..........
Cyddescco .......... .......... .......... .......... ..........
Cyddhaein .......... .......... .......... .......... ..........
Cydcescco .......... .......... .......... .......... ..........
```

```
Cydchaein .........................................................
Aprdpseae .........................................................
Prtderwch .........................................................
Hlybescco .........................................................
Hlybpasha .........................................................
Rt3bactpl .........................................................
Hlybactac .........................................................
Hlybactpl .........................................................
Cyabborpe .........................................................
Comastrpn .........................................................
Lcnclacla .........................................................
Peddpedac .........................................................
Cvabescco .........................................................
Pmdlschpo .........................................................
Pglyarath .........................................................
Pgplarath .........................................................
Mdr2musmu .........................................................
Mdr3crigr .........................................................
Mdr3homsa .........................................................
Mdr1musmu .........................................................
Mdr2crigr .........................................................
Mdr1crigr .........................................................
Mdr3musmu .........................................................
Mdr1homsa .........................................................
Mdr4drome .........................................................
Mdr5drome .........................................................
Mdr1caeel .........................................................
Mdr3caeel .........................................................
Mdr1leien .........................................................
Msbaescco .........................................................
Msbahaein .........................................................
Hetaanasp .........................................................
Mdrplafa  .........................................................
Mdl1sacce .........................................................
Mdl2sacce .........................................................
Mt2ratno  .........................................................
Tap2musmu .........................................................
Tap2homsa .........................................................
Tap1homsa .........................................................
Tap1musmu .........................................................
Chvaagrtu .........................................................
Ndvarhime .........................................................
Atm1sacce .........................................................
Hmt1schpo .........................................................
Mdlescco  .........................................................
Ste6sacce .........................................................
Natabacsu .........................................................
Syrdpsesy .........................................................
Consensus .........................................................

          751                                                  800
Surcricr  ......................LSNI TIRIPRGQLT MIVGQVGCGK
Surratno  ......................LSNI TIRIPRGQLT MIVGQVGCGK
Mdrleita  GAAGEDEKGE VEEGDREYYQ LVSKELLRNV SLTIPKGKLT MVIGSTGSGK
```

```
Hemsentfa ........................ ...... LNGI SFDIRKGDKV AIVGRSGSGK
Lcn3lacla ........................ ...... FNNI NFSIKKGEKI AIVGKSGSGK
Nistlacla ........................ ...... LKNI NLSFEKGELT AIVGKNGSGK
Spabbacsu ........................ ...... LKHI NVSLHKGERV AIVGPNGSGK
Cyddescco ........................ ...... .GPL NFTLPAGQRA VLVGRSGSGK
Cyddhaein ........................ ...... .KPL NFQIPANHNV ALVGQSGAGK
Cydcescco ........................ ...... LKGI SLQVNAGEHI AILGRTGCGK
Cydchaein ........................ ...... LKNL TLDLEQGKKI AILGKTGSGK
Aprdpseae ........................ ...... LRGL TLAIPAGSVV GVIGPSGSGK
Prtderwch ........................ ...... LQNI HFSLQAGETL VILGASGSGK
Hlybescco ........................ ...... LDNI NLSIKQGEVI GIVGRSGSGK
Hlybpasha ........................ ...... LNNV NLEIRQGEVI GIVGRSGSGK
Rt3bactpl ........................ ...... LDDV NLSVKQGEVI GIVGRSGSGK
Hlybactac ........................ ...... LNNI NLDISQGEVI GIVGRSGSGK
Hlybactpl ........................ ...... LNDV NLSIQQGEVI GIVGRSGSGK
Cyabborpe ........................ ...... LRNV SLRIAPGEVV GVVGRSGSGK
Comastrpn ........................ ...... LSDI NLTVPQGSKV AFVGISGSGK
Lcnclacla ........................ ...... LSEI ELSIKENEKL TIVGMSGSGK
Peddpedac ........................ ...... LEDV SLTIPHHQKI TIVGMSGSGK
Cvabescco ........................ ...... FSAL SLSVAPGESV AITGASGAGK
Pmd1schpo ........................ ...... LDNF SLVCPSGKIT ALVGASGSGK
Pglyarath ........................ ...... LNNF CLSVPAGKTI ALVGSSGSGK
Pgp1arath ........................ ...... LNNF CLSVPAGKTI ALVGSSGSGK
Mdr2musmu ........................ ...... LKGL NLKVKSGQTV ALVGNSGCGK
Mdr3crigr ........................ ...... LKGL NLKVQSGQTV ALVGNSGCGK
Mdr3homsa ........................ ...... LKGL NLKVQSGQTV ALVGSSGCGK
Mdr1musmu ........................ ...... LKGL NLKVKSGQTV ALVGNSGCGK
Mdr2crigr ........................ ...... LKGL NLKVQSGQTV ALVGKSGCGK
Mdr1crigr ........................ ...... LKGL NLKVQSGQTV ALVGNSGCGK
Mdr3musmu ........................ ...... LKGL NLKVKSGQTV ALVGNSGCGK
Mdr1homsa ........................ ...... LKGL NLKVQSGQTV ALVGNSGCGK
Mdr4drome ........................ ...... LKGL TVDVLPGQTV AFVGASGCGK
Mdr5drome ........................ ...... HRGL NIRIRAGQTV ALVGSSGCGK
Mdr1caeel ........................ ...... LRGM NLRVNAGQTV ALVGSSGCGK
Mdr3caeel ........................ ...... LKGV SFEVNPGETV ALVGHSGCGK
Mdr1leien ........................ ...... FRDL SLKIKCGQKV AFSGASGCGK
Msbaescco ........................ ...... LRNI NLKIPAGKTV ALVGRSGSGK
Msbahaein ........................ ...... LNNI SFSVPAGKTV ALVGRSGSGK
Hetaanasp ........................ ...... LNNI TLTIERGKTT ALVGASGAGK
Mdrplafa  ........................ ...... YKDL SFTLKEGKTY AFVGESGCGK
Mdl1sacce ........................ ...... FKDL NITIKPGEHV CAVGPSGSGK
Mdl2sacce ........................ ...... FKNL NFKIAPGSSV CIVGPSGRGK
Mt2ratno  ........................ ...... LQGL TFTLHPGKVT ALVGPNGSGK
Tap2musmu ........................ ...... LQGL TFTLHPGTVT ALVGPNGSGK
Tap2homsa ........................ ...... LKGL TFTLRPGEVT ALVGPNGSGK
Tap1homsa ........................ ...... LQGL TFTLRPGEVT ALVGPNGSGK
Tap1musmu ........................ ...... LQGL TFTLHPGTVT ALVGPNGSGK
Chvaagrtu ........................ ...... VHDV SFTAKAGETV AIVGPTGAGK
Ndvarhime ........................ ...... VRNV SFKAKAGQTI AIVGPTGAGK
Atm1sacce ........................ ...... LKNA SFTIPAGWKT AIVGSSGSGK
Hmt1schpo ........................ ...... LSDI NFVAQPGKVI ALVGESGGGK
Mdlescco  ........................ ...... LENV NFALKPGQML GICGPNGSGK
Ste6sacce ........................ ...... LKNV SLNFSAGQFT FIVGKSGSGK
Natabacsu ........................ ...... VRDV SLTIEKGEVV GILGENGAGK
```

```
Syrdpsesy ........... .......... ......LGPI DLTIHSGELV YIVGGNGCGK
Consensus .......... .......... ......L... .......G... ..VG..G.GK


         801                                            850
Surcricr   SSLLLATLGE MQKVSGAVFW NSNLPDSEGE DPSSPERETA AGSDIRSRGP
Surratno   SSLLLATLGE MQKVSGAVFW NS.LPDSEGE DPSNPERETA ADSDARSRGP
Mdrleita   STLLGALMGE YSVESGELW. .......... ....AER.... ........S
Hemsentfa  STLLKLLAGL LQPSNG.... ......EIL Y.....EGYP LSNNSNNRRN
Lcn3lacla  STLFNILLGL .......... ......IS Y.....EGEV TYGYENLRQI
Nistlacla  STLVKIISGL YQPTM..... ......G IIQYDKMRSS LMPEEFYQKN
Spabbacsu  KTFIKLLTGL YEVHE..... ......G DILINGINIK ELDMDSYMNQ
Cyddescco  SSLLNALS.G FLSYQ..... ......GS LRI.NGIELR DLSPESWRKH
Cyddhaein  TSLMNVIL.G FLPYE..... ......GS LKI.NGQELR ESNLADWRKH
Cydcescco  STLLQQLTRA WDPQQ..... ......GE ILL.NDSPIA SLNEAALRQT
Cydchaein  SSLLQLLVRN YDANQ..... ......GE LLL.AEKPIS AYSEETLRHQ
Aprdpseae  SSLARVVLGI WPTLHG.... ......SVR L...DGAEIR QYERETLGPR
Prtderwch  SSLARLLVGA QSPTQG.... ......KVR L...DGADLN QVDKNTFGPT
Hlybescco  STLTKLIQRF YIPEN..... ......GQ VLI.DGHDLA LADPNWLRRQ
Hlybpasha  STLTKLLQRF YIPEN..... ......GQ VLI.DGHDLA LADPNWLRRQ
Rt3bactpl  STLTKLLQRF YIPEN..... ......GQ VLI.DGHDLA LADPNWLRRQ
Hlybactac  STLTKLIQRF YIPEQ..... ......GQ VLI.DGHDLA LADPNWLRRQ
Hlybactpl  STLTKLIQVF YIPEN..... ......GQ VLI.DGHDLA LADPNWLRRQ
Cyabborpe  STLTRLIQRM FVADR..... ......GR VLI.DGHDIG IVDSASLRRQ
Comastrpn  TTLAKMMVNF YDPSQ..... ......GE ISL.GGVNLN QIDKKALRQY
Lcnclacla  STLVKLLVNF FQPTS..... ......GT ITL.GGIDLQ QFDKHQLRRL
Peddpedac  TTLAKLLVGF FEPQEQ.... ......HGE IQI.NHHNIS DISRTILRQY
Cvabescco  TTLMKVLCGL FEPDS..... ......GR VLI.NGIDIR QIGINNYHRM
Pmdlschpo  STIIGLVERF YDPIG..... ......GQ VFL.DGKDLR TLNVASLRNQ
Pglyarath  STVVSLIERF YDPNS..... ......GQ VLL.DGQDLK TLKLRWLRQQ
Pgplarath  STVVSLIERF YDPNS..... ......GQ VLL.DGQDLK TLKLRWLRQQ
Mdr2musmu  STTVQLLQRL YDPTE..... ......GK ISI.DGQDIR NFNVRCLREI
Mdr3crigr  TTTLQLLQRL YDPTE..... ......GT ISI.DGQDIR NFNVRYLREI
Mdr3homsa  STTVQLIQRL YDPDE..... ......GT INI.DGQDIR NFNVNYLREI
Mdr1musmu  STTVQLMQRL YDPLE..... ......GV VSI.DGQDIR TINVRYLREI
Mdr2crigr  STTVQLLQRL YDPTE..... ......GV VSI.DGQDIR TINVRYLREI
Mdr1crigr  STTVQLLQRL YDPTE..... ......GV VSI.DGQDIR TINVRYLREI
Mdr3musmu  STTVQLMQRL YDPLD..... ......GM VSI.DGQDIR TINVRYLREI
Mdr1homsa  STTVQLMQRL YDPTE..... ......GM VSV.DGQDIR TINVRFLREI
Mdr4drome  STLIQLMQRF YDPEA..... ......GS VKL.DGRDLR TLNVGWLRSQ
Mdr5drome  STCVQLLQRF YDPVF..... ......GS VLL.DDLDIR KYNIQWLRSN
Mdr1caeel  STIISLLLRY YDVLK..... ......GK ITI.DGVDVR DINLEFLRKN
Mdr3caeel  STSIGLLMRF YNQCA..... ......GM IKL.DGIPIQ EYNIRWLRST
Mdr1leien  SSVIGLIQRF YDPIG..... ......GA VLV.DGVRMR ELCLREWRDQ
Msbaescco  STIASLITRF YDIDE..... ......GE ILM.DGHDLR EYTLASLRNQ
Msbahaein  STIANLVTRF YDIEQ..... ......GE ILL.DGVNIQ DYRLSNLREN
Hetaanasp  TTLADLIPRF YDPTE..... ......GQ ILV.DGLDVQ YFEINSLRRK
Mdrplafa   STILKLIERL YDPTE..... ......GD IIVNDSHNLK DINLKWWRSK
Mdl1sacce  STIASLLLRY YDVNS..... ......GS IEF.GDEDIR NFNLRKYRRL
Mdl2sacce  STIALLLLRY YNPTT..... ......GT ITI.DNQDIS KLNCKSLRRH
Mt2ratno   STVAALLQNL YQPTG..... ......GQ LLL.DGEPLV QYDHHYLHRQ
Tap2musmu  STVAALLQNL YQPTG..... ......GQ LLL.DGEPLT EYDHHYLHRQ
Tap2homsa  STVAALLQNL YQPTG..... ......GQ VLL.DEKPIS QYEHCYLHSQ
Tap1homsa  STVAALLQNL YQPTG..... ......GQ LLL.DGKPLP QYEHRYLHRQ
Tap1musmu  STVAALLQNL YQPTG..... ......GQ LLL.DGQRLV QYDHHYLHTQ
```

```
Chvaagrtu TTLINLLQRV YDPDS..... ........GQ ILI.DGTDIS TVTKNSLRNS
Ndvarhime TTLVNLLQRV HEPKH..... ........GQ ILI.DGVDIA TVTRKSLRRS
Atm1sacce STILKLVFRF YDPES..... ........GR ILI.NGRDIK EYDIDALRKV
Hmt1schpo STIMRILLRF FDVNS..... ........GS ITI.DDQDIR NVTLSSLRSS
Mdlescco  STLLSLIQRH FDVSP..... ........GD IRFHD.IPLT KLQLDSWRTG
Ste6sacce STLSNLLLRF YD..GY.... ........NGS ISI.NGHNIQ TIDQKLLIEN
Natabacsu TTMLRMIASL LEPSQ..... ........GV ITVDGFDTVK QPAEVKQRIG
Syrdpsesy STLAKVFCGL YIPQEGQ... ...........LLLDGAAVT DDSRGDYRDL
Consensus ST...L.... .......... ......G... ..........R..

          851                                          900
Surcricr  VAYASQKPWL LNATVEENIT FESP..... .......... FNKQRYKMVI
Surratno  VAYASQKPWL LNATVEENIT FESP..... .......... FNKQRYKMVI
Mdrleita  IAYVPQQAWI MNATLRGNIL FFDE..... .......... ERAEDLQDVI
Hemsentfa IFYVNQNAHI FNETIEKNIS LEFKPNS.. .........  .SINEKKRLK
Lcn3lacla IGVVSQNMNL RKGSLIENIV ..SNNNS.. .........  .EELDIQKIN
Nistlacla ISVLFQDFVK YELTIRENIG LSD...... .......... LSSQWEDEKI
Spabbacsu IAALFQDFMK YEMTLKENIG FGQ...... .......... IDKLHQTNKM
Cyddescco LSSVGQNPQL PAATLRDNVL LARP..... .......... .DA.SEQELQ
Cyddhaein IAWVGQNPLL LQGTIKENLL LGDV..... .......... .QA.NDEEIN
Cydcescco ISVVPQRVHL FSATLRDNLL LASP..... .......... .GS.SDEALS
Cydchaein ICFLTQRVHV FSDTLRQNLQ FASA..... .......... .DKISDEQMI
Aprdpseae IGYLPQDIEL FAGTVAENIA .......... ........R FGEVQADKVV
Prtderwch IGYLPQDVQL FKGSLAENIA .......... ........R FGDADPEKVV
Hlybescco VGVVLQDNVL LNRSIIDNIS L.AN..... .......... .PGMSVEKVI
Hlybpasha IGVVLQDNVL LNRSIRENIA L.SD..... .......... .PGMPMERVI
Rt3bactpl IGVVLQDNVL LNRSIRDNIA L.TD..... .......... .PSMSMERVI
Hlybactac VGVVLQDNVL LNRSIRENIA L.TN..... .......... .PGMPMEKVI
Hlybactpl VGVVLQDNVL LGRSIRDNIA L.AD..... .......... .PGMPMEKIV
Cyabborpe LGVVLQESTL FNRSVRDNIA L.TR..... .......... .PGASMHEVV
Comastrpn INYLPQQPYV FNGTILENLL LGAK..... .......... .EGTTQEDIL
Lcnclacla INYLPQQPYI FTGSILDNLL LGAN..... .......... .ENASQEEIL
Peddpedac INYVPQEPFI FSGSVLENLL LGSR..... .......... .PGVTQQMID
Cvabescco IACVMQDDRL FSGSIRENIC GFAE..... .......... ..EMDEEWMV
Pmd1schpo ISLVQQEPVL FATTVFENIT YGLPDTIKGT LSKEELERRV YDAAKLAN..
Pglyarath IGLVSQEPAL FATSIKENIL LG..RP..DA .DQVEIE... .EAARVAN..
Pgplarath IGLVSQEPAL FATSIKENIL LG..RP..DA .DQVEIE... .EAARVAN..
Mdr2musmu IGVVSQEPVL FSTTIAENIR YGRGNV..TM ...DEIE... .KAVKEAN..
Mdr3crigr IGVVSQEPVL FSTTIAENIR YGRGNV..TM ...EEIK... .KAVKEAN..
Mdr3homsa IGVVSQEPVL FSTTIAENIC YGRGNV..TM ...DEIK... .KAVKEAN..
Mdr1musmu IGVVSQEPVL FATTIAENIR YGREDV..TM ...DEIE... .KAVKEAN..
Mdr2crigr IGVVSQEPVL FATTIAENIR YGRENV..TM ...DEIE... .KAVKEAN..
Mdr1crigr IGVVSQEPVL FATTIAENIR YGRENV..TM ...DEIE... .KAVKEAN..
Mdr3musmu IGVVSQEPVL FATTIAENIR YGREDV..TM ...DEIE... .KAVKEAN..
Mdr1homsa IGVVSQEPVL FATTIAENIR YGRENV..TM ...DEIE... .KAVKEAN..
Mdr4drome IGVVGQEPVL FATTIGENIR YGRPSA..TQ ...ADIE... .KAARAAN..
Mdr5drome IAVVGQEPVL FLGTIAQNIS YGKPGA..TQ ...KEIE... .AAATQAG..
Mdr1caeel VAVVSQEPAL FNCTIEENIS LGKEGI..TR ...EEMV... .AACKMAN..
Mdr3caeel IGIVQQEPII FVATVAENIR MGDVLI..T. ...DQDIE... .EACKMAN..
Mdr1leien IGIVSQEPNL FAGTMMENVR MGKPNA.... ......TDEEV VEACRQAN..
Msbaescco VALVSQNVHL FNDTVANNIA YARTE....Q YSREQIE... .EAARMAY..
Msbahaein CAVVSQQVHL FNDTIANNIA YAAQD....K YSREEII... .AAAKAAY..
Hetaanasp MAVVSQDTFI FNTSIRDNIA YGTSG....A .SEAEIR... .EVARLAN..
Mdrplafa  IGVVSQDPLL FSNSIKNNIK YSLYSLKDLE AMENYYEENT NDTYENKNFS
```

```
Mdl1sacce IGYVQQEPLL FNGTILDNIL YCIPPEI... .......... ..AEQDDRIR
Mdl2sacce IGIVQQEPVL MSGTIRDNIT YGLT..Y... .......... ..TPTKEEIR
Mt2ratno  VVLVGQEPVL FSGSVKDNIA YGL.R..... .......... ..DCEDAQVM
Tap2musmu VVLVGQEPVL FSGSVKDNIA YGL.R..... .......... ..DCEDAQVM
Tap2homsa VVSVGQEPVL FSGSVRNNIA YGL.Q..... .......... ..SCEDDKVM
Tap1homsa VAAVGQEPQV FGRSLQENIA YGLTQ..... .......... ..KPTMEEIT
Tap1musmu VAAVGQEPLL FGRSFRENIA YGLNR..... .......... ..TPTMEEIT
Chvaagrtu IATVFQDAGL LNRSIRENIR LGRE...... .......... ..TATDAEVV
Ndvarhime IATVFQDAGL MNRSIGENIR LGRE...... .......... ..DASLDEVM
Atml1sacce IGVVPQDTPL FNDTIWENVK FGRI...... .......... ..DATDEEVI
Hmt1schpo IGVVPQDSTL FNDTILYNIK YAKP...... .......... ..SATNEEIY
Mdlescco  LAVVSQTPFL FSDTVANNIA LG........ .......... CPNATQQEIE
Ste6sacce ITVVEQRCTL FNDTLRKNIL LGSTDSVRNA DCSTNENRHL IKDACQMALL
Natabacsu VLFGGETGLY DRMTAKENLQ YF........ .......... ...GRLYGLN
Syrdpsesy FSAVFSDFHL FNRLIGPDEK .......... .......... .EHPSTDQAQ
Consensus ...V.Q...L .......NI.. .......... .......... ..........
```

```
          901                                              950
Surcricr  EACSLQPDID I......... ..............LPHGDQTQI
Surratno  EACSLQPDID I......... ..............LPHGDQTQI
Mdrleita  RCCQLEADLA Q......... ...............FCGGLDTEI
Hemsentfa GSMSKSKMDE VLL....... ...........GIPQYEKTIV
Lcn3lacla DVLKDVNMLE LVD....... ...........SLPQKIFSQL
Nistlacla IKVLDNLGLD FLKTNNQY.. ...............VLDTQL
Spabbacsu HEVLDIVRAD FLKSHSSY.. ...............QFDTQL
Cyddescco AALDNAWVSE FLP....... ...........LLPQGVDTPV
Cyddhaein QALMRSQAKE FTD....... ...........KL..GLHHEI
Cydcescco EILRRVGLEK LLE....... ...........DA..GLNSWL
Cydchaein EMLHQVGLSK LLE....... ...........QEGKGLNLWL
Aprdpseae EAARLAGVHE LV........ ...........L RLPQGYDTVL
Prtderwch AAAKLAGVHE LI........ ...........L SLPNGYDTEL
Hlybescco YAAKLAGAHD FIS....... ...........ELREGYNTIV
Hlybpasha YAAKLAGAHD FIS....... ...........ELREGYNTIV
Rt3bactpl YAAKLAGAHD FIS....... ...........ELREGYNTIV
Hlybactac AAAKLAGAHD FIS....... ...........ELREGYNTVV
Hlybactpl HAAKLAGAHE FIS....... ...........ELREGYNTIV
Cyabborpe AAARLAGAHE FIC....... ...........QLPEGYDTML
Comastrpn RAVELAEIRE DIE....... ...........RMPLNYQTEL
Lcnclacla KAVELAEIRA DIE....... ...........QMQLGYQTEL
Peddpedac QACSFAEIKT DIE....... ...........NLPQGYHTRL
Cvabescco ECARASHIHD VIM....... ...........NMPMGYETLI
Pmd1schpo .......AYD FI.MT..... ...........LPEQFSTNV
Pglyarath .......AHS FI.IK..... ...........LPDGFDTQV
Pgp1arath .......AHS FI.IK..... ...........LPDGFDTQV
Mdr2musmu .......AYD FI.MK..... ...........LPQKFDTLV
Mdr3crigr .......AYE FI.MK..... ...........LPQKFDTLV
Mdr3homsa .......AYE FI.MK..... ...........LPQKFDTLV
Mdr1musmu .......AYD FI.MK..... ...........LPHQFDTLV
Mdr2crigr .......AYD FI.MK..... ...........LPHKFDTLV
Mdr1crigr .......AYD FI.MK..... ...........LPHKFDTLV
Mdr3musmu .......AYD FI.MK..... ...........LPHQFDTLV
Mdr1homsa .......AYD FI.MK..... ...........LPHKFDTLV
Mdr4drome .......CHD FI.TR..... ...........LPKGYDTQV
Mdr5drome .......AHE FI.TN..... ...........LPESYRSMI
```

```
Mdr1caeel .......AEK FIKT..... ......... ..........LPNGYNTLV
Mdr3caeel .......AHE FIC.K..... .......... ..........LSDRYDTVI
Mdr1leien .......IHD TI.MA..... ......... ..........LPDRYDTPV
Msbaescco .......AMD FIN.K..... .......... ..........MDNGLDTVI
Msbahaein .......ALE FIE.K..... .......... ..........LPQVFDTVI
Hetaanasp .......ALQ FIE.E..... ......... ..........MPEGFDTKL
Mdrplafa  LISNSMTSNE LLEMKKEYQT IKDSDVVDVS KKVLIHDFVS SLPDKYDTLV
Mdl1sacce RAIGKANCTK FLA........ ............ ........NFPDGLQTMV
Mdl2sacce SVAKQCFCHN FIT........ ............ ........KFPNTYDTVI
Mt2ratno  AAAQAACADD FIG........ ............ ........EMTNGINTEI
Tap2musmu AAAQAACADD FIG........ ............ ........EMTNGINTEI
Tap2homsa AAAQAAHADD FIQ........ ............ ........EMEHGIYTDV
Tap1homsa AAAVKSGAHS FIS........ ............ ........GLPQGYDTEV
Tap1musmu AVAVESGAHD FIS........ ............ ........GFPQGYDTEV
Chvaagrtu EAAAAAAATD FID........ ............ ........SRINGYLTQV
Ndvarhime AAAEAAAASD FIE........ ............ ........DRLNGYDTVV
Atm1sacce TVVEKAQLAP LIK........ ............ ........KLPQGFDTIV
Hmt1schpo AAAKAAQIHD RIL........ ............ ........QFPDGYNSRV
Mdlescco  HVARLASVHD DI........ ............ ......L RLPQGYDTEV
Ste6sacce D........ ............ ........RFIL DLPDGLETLI
Natabacsu RHEIKARIED LSK........ ............ ......RFGMRDYM
Syrdpsesy TYLSTLGLED KVKIE..... ......... ..........
Consensus ..... ......... ......... ..........
```

```
          951                                              1000
Surcricr  G...ERGINL SGGQRQRISV ARALYQQTNV VFLDDPFSAL DVHLSDHLMQ
Surratno  G...ERGINL SGGQRPGISV ARALYQHTNV VFLDDPFSAL DVHLSDHLMQ
Mdrleita  G...EMGVNL SGGQKARVSL ARAVYANRDV YLLDDPLSAL DAHVGQRIVQ
Hemsentfa ...SENGSNF SGGQRQKIAL ARAFYSNVNT LLLDEPTSAM DNI.SEFEVF
Lcn3lacla ...FENGKNL SGGQIQRLLI AKSLLNNNKF IFWDEPFSSL DNQ.NRIHIY
Nistlacla GNWFQEGHQL SGGQWQKIAL ARTFFKKASI YILDEPSAAL DPV.AEKEIF
Spabbacsu GLWFDEGRQL SGGQWQKIAL ARAYFREASL YILDEPSSAL DPI.AEKETF
Cyddescco G...DQAARL SVGQAQRVAV ARALLNPCSL LLLDEPAASL DAH.SEQRVM
Cyddhaein K...DGGLGI SVGQAQRLAI ARALLRKGDL LLLDEPTASL DAQ.SENLVL
Cydcescco G...EGGRQL SGGELRRLAI ARALLHDAPL VLLDEPTEGL DAT.TESQIL
Cydchaein G...DGGRPL SGGEQRRLGL ARILLNNASI LLLDEPTEGL DRE.TERQIL
Aprdpseae G...VGGAGL SGGQRQRIAL ARALYGAPTL VVLDEPNSNL DDS.GEQALL
Prtderwch G...DGGGGL SGGQRQRIGL ARAMYGDPCL LILDEPNASL DSE.GDQALM
H1ybescco G...EQGAGL SGGQRQRIAI ARALVNNPKI LIFDEATSAL DYE.SEHVIM
H1ybpasha G...EQGAGL SGGQRQRIAI ARALVNNPKI LIFDEATSAL DYE.SEHIIM
Rt3bactpl G...ELGAGL SGGQRQRIAI ARALVNNPRI LIFDEATSAL DYE.SEHIIM
H1ybactac G...EQGAGL SGGQRQRIAI ARALVNNPRI LIFDEATSAL DYE.SENIIM
H1ybactpl G...EQGAGL SGGQPNRIAI ARALVNNPKI LIFDEATSAL DYESEHIIMR
Cyabborpe G...ENGVGL SGGQRQRIGI ARALIHRPRV LILDEATSAL DYE.SEHIIQ
Comastrpn T...SDGAGI SGGQRQRIAL ARALLTDAPV LILDEATSSL DIL.TEKRIV
Lcnclacla S...SDASSL SGGQKQRIAL ARALLSPAKI LILDEATSNL DMI.TEKKIL
Peddpedac S...ESGFNL SGGQKQRLSI ARALLSPAQC FIFDESTSNL DTI.TEHKIV
Cvabescco G...ELGEGL SGGQKQRIFI ARALYRKPGI LFMDEATSAL DSE.SEHFVN
Pmd1schpo G...QRGFLM SGGQKQRIAI ARAVISDPKI LLLDEATSAL DSK.SEVLVQ
Pglyarath G...ERGLQL SGGQKQRIAI ARAMLKNPAI LLLDEATSAL DSE.SEKLVQ
Pgp1arath G...ERGLQL SGGQKQRIAI ARAMLKNPAI LLLDEATSAL DSE.SEKLVQ
Mdr2musmu G...DRGAQL SGGQKQRIAI ARALVRNPKI LLLDEATSAL DTE.SEAEVQ
Mdr3crigr G...ERGAQL SGGQKQRIAI ARALVRNPKI LLLDEATSAL DTE.SEAEVQ
Mdr3homsa G...ERGAQL SGGQKQRIAI ARALVRNPKI LLLDEATSAL DTE.SEAEVQ
```

```
Mdr1musmu G...ERGAQL SGGQKQRIAI ARALVRNPKI LLLDEATSAL DTE.SEAVVQ
Mdr2crigr G...ERGAQL SGGQKQRIAI ARALVRNPKI LLLDEATSAL DTE.SEAVVQ
Mdr1crigr G...ERGAQL SGGQKQRIAI ARALVRNPKI LLLDEATSAL DTE.SEAVVQ
Mdr3musmu G...ERGAQL SGGQKQRIAI ARALVRNPKI LLLDEATSAL DTE.SEAVVQ
Mdr1homsa G...ERGAQL SGGQKQRIAI ARALVRNPKI LLLDEATSAL DTE.SEAVVQ
Mdr4drome G...EKGAQI SGGQKQRIAI ARALVRQPQV LLLDEATSAL DPT.SEKRVQ
Mdr5drome G...ERGSQL SGGQKQRIAI ARALIQNPKI LLLDEATSAL DYQ.SEKQVQ
Mdr1caeel G...DRGTQL SGGQKQRIAI ARALVRNPKI LLLDEATSAL DAE.SEGIVQ
Mdr3caeel G...AGAVQL SGGQKQRVAI ARAIVRKPQI LLLDEATSAL DTE.SERMVQ
Mdr1leien G...PVGSLL SGGQKQRIAI ARALVKRPPI LLLDEATSAL DRK.SEMEVQ
Msbaescco G...ENGVLL SGGQRQRIAI ARALLRDSPI LILDEATSAL DTE.SERAIQ
Msbahaein G...ENGTSL SGGQRQRLAI ARALLRNSPV LILDEATSAL DTE.SERAIQ
Hetaanasp G...DRGVRL SGGQRQRIAI ARALLRDPEI LILDEATSAL DSV.SERLIQ
Mdrplafa  G...SNASKL SGGQKQRISI ARAIMRNPKI LILDEATSSL DNK.SEYLVQ
Mdl1sacce G...ARGAQL SGGQKQRIAL ARAFLLDPAV LILDEATSAL DSQ.SEEIVA
Mdl2sacce G...PHGTLL SGGQKQRIAI ARALIKKPTI LILDEATSAL DVE.SEGAIN
Mt2ratno  G...ERGSQL AVGQKQRLAI ARALVRNPRV LILDEATSAL DA.....ECE
Tap2musmu G...EKGGQL AVGQKQRLAI ARALVRNPRV LILDEATSAL DA.....QCE
Tap2homsa G...EKGSQL AAGQKQRLAI ARALVRDPRV LILDEATSAL DV.....QCE
Tap1homsa D...EAGSQL SGGQRQAVAL ARALIRKPCV LILDDATSAL DAN.SQLQVE
Tap1musmu G...ETGNQL SGGQRQAVAL ARALIRKPLL LILDDATSAL DAG.NQLRVQ
Chvaagrtu G...ERGNRL SGGERQRIAI ARAILKNAPI LVLDEATSAL DVE.TEARVK
Ndvarhime G...ERGNRL SGGERQRVAI ARAILKNAPI LVLDEATSAL DVE.TEARVK
Atm1sacce G...ERGLMI SGGEKQRLAI ARVLLKNARI MFFDEATSAL DTH.TEQALL
Hmt1schpo G...ERGLKL SGGEKQRVAV ARAILKDPSI ILLDEATSAL DTN.TERQIQ
Mdlescco  G...ERGVML SGGQKQRISI ARALLVNAEI LILDDALSAV DGR.TEHQIL
Ste6sacce G...TGGVTL SGGQQQRVAI ARAFIRDTPI LFLDEAVSAL DIV.HRNLLM
Natabacsu N...RRVGGF SKGMRQKVAI ARALIHDPDI ILFDEPTTGL DIT.SSNIFR
Syrdpsesy GLGYSTTTAL SYGQQKRLAL VCAYLEDRPI YLLDEWAADQ DPPFKRFFYE
Consensus G.....G..L SGGQ.QR... ARA....... L.LDE..SAL D....E....
```

```
          1001                                        1050
Surcricr  AGILELLRDD KRTVVLVTHK LQYLPHADWI ..................
Surratno  AGILELLRDD KRTVVLVTHK LQYLPHADWI ..................
Mdrleita  DVILGRLRG. .KTRVLATHQ IHLLPLADYI ..................
Hemsentfa SNLLDE.... KRTVITVAHR ISTVKNFDKI ..................
Lcn3lacla KNVLENPDYK SQTIIMISHH LDVLKYVDRV ..................
Nistlacla DYF..VALSE NNISIFISHS LNAARKANKI ..................
Spabbacsu DTF..FSLSK DKIGIFISHR LVAAKLADRI ..................
Cyddescco EALNA..ASL RQTTLMVTHQ LEDLADWDVI ..................
Cyddhaein QALNE..ASQ HQTTLMITHR IEDLKQCDQI ..................
Cydcescco ELLAE..MMR EKTVLMVTHR LRGLSRFQQI ..................
Cydchaein RLILQ..HAE NKTLIIVTHR LSSIEQFDKI ..................
Aprdpseae AAIQALKAR. GCTVLLITHR AGVLGCADRL ..................
Prtderwch QAIVALQKR. GATVVLITHR PALTTLAQKI ..................
Hlybescco RNMHK..ICK GRTVIIIAHR LSTVKNADRI ..................
Hlybpasha QNMQK..ICQ GRTVILIAHR LSTVKNADRI ..................
Rt3bactpl QNMQK..ICH GRTVIIIAHR LSTVKNADRI ..................
Hlybactac HNMHK..ICQ NRTVLIIAHR LSTVKNAASI ..................
Hlybactpl NMHQI...CK GRTVIIIAHR LSTVKNAASI ..................
Cyabborpe RNMRD..ICD GRTVIIIAHR LSAVRCADRI ..................
Comastrpn DNLIA..L.. DKTLIFIAHR LTIAERTEKV ..................
Lcnclacla KNLLP..L.. DKTIIFIAHR LSVAEMSHRI ..................
Peddpedac SKLLF..M.K DKTIIFVAHR LNIASQTDKV ..................
```

```
Cvabescco VAIKN..M.. NITRVIIAHR ETTLRTVDRV ..................
Pmd1schpo KALDN..ASR SRTTIVIAHR LSTIRNADNI ..................
Pglyarath EALDR..FMI GRTTLIIAHR LSTIRKADLV ..................
Pgp1arath EALDR..FMI GRTTLIIAHR LSTIRKADLV ..................
Mdr2musmu AALDK..ARE GRTTIVIAHR LSTIRNADVI ..................
Mdr3crigr AALDK..ARE GRTTIVIAHR LSTVRNADVI ..................
Mdr3homsa AALDK..ARE GRTTIVIAHR LSTVRNADVI ..................
Mdr1musmu AALDK..ARE GRTTIVIAHR LSTVRNADVI ..................
Mdr2crigr AALDK..ARE GRTTIVIAHR LSTVRNADVI ..................
Mdr1crigr AALDK..ARE GRTTIVIAHR LSTVRNADII ..................
Mdr3musmu AALDK..ARE GRTTIVIAHR LSTVRNADVI ..................
Mdr1homsa VALDK..ARK GRTTIVIAHR LSTVRNADVI ..................
Mdr4drome SALEL..ASQ GPTTLVVAHR LSTITNADKI ..................
Mdr5drome QALDL..ASK GRTTIVVSHR LSAIRGADKI ..................
Mdr1caeel QALDK..AAK GRTTIIIAHR LSTIRNADLI ..................
Mdr3caeel TALDK..ASE GRTTLCIAHR LSTIRNA... ..................
Mdr1leien AALDQLIQRG GTTVVVIAHR LATIRDMDRI YYVKH.........
Msbaescco AALDEL..QK NRTSLVIAHR LSTIEKADEI ..................
Msbahaein SALEEL..KK DRTVVVIAHR LSTIENADEI ..................
Hetaanasp ESIEKL..SV GRTVIAIAHR LSTIAKADKV ..................
Mdrplafa  KTINNLKGNE NRITIIIAHR LSTIRYANTI FVLSNRERSD NNNNNNNDDN
Mdl1sacce KNL.QRRVER GFTTISIAHR LSTIKHSTRV I...........
Mdl2sacce YTFGQLMKSK SMTIVSIAHR LSTIRRSENV I...........
Mt2ratno  QALQTWRSQE DRTMLVIAHR LHTVQNADQV L...........
Tap2musmu QALQNWRSQG DRTMLVIAHR LHTVQNADQV L...........
Tap2homsa QALQDWNSRG DRTVLVIAHR LQTVQRAHQI L...........
Tap1homsa QLLYESPERY SRSVLLITQH LSLVEQADHI L...........
Tap1musmu RLLYESPKRA SRTVLLITQQ LSLAEQAHHI L...........
Chvaagrtu AAV..DALRK NRTTFIIAHR LSTVRDADLV ..................
Ndvarhime DAI..DALRK DRTTFIIAHR LSTVREADLV ..................
Atm1sacce RTIRDNFTSG SRTSVYIAHR LRTIADADKI ..................
Hmt1schpo AAL..NRLAS GRTAIVIAHR LSTITNADLI ..................
Mdlescco  HNLRQW..GQ GRTVIISAHR LSALTEASEI ..................
Ste6sacce KAIRHW..RK GKTTIILTHE LSQIESDDYL Y.................
Natabacsu EFIQQLK.RE QKTILFSSHI MEEVQA.....................
Syrdpsesy ELLPDLK.RR GKTILIITHD DQYFQLADRI ..................
Consensus .......... ..T.....HR L......D.. ..................
```

```
          1051                                          1100
Surcricr  .......... IAMKDGTIQR EGTLKDFQRS ..ECQLFEHW KTLMNRQDQE
Surratno  .......... IAMKDGTIQR EGTLKDFQRS ..ECQLFEHW KTLMNRQDQE
Mdrleita  .......... VVLQHGSIVF AGDFAAFSAT ALEETLRGEL KGSKDVESCS
Hemsentfa .......... ILMDNGEIVC IGKHEDLIEN SELYRSLYYK KQQFGGTK..
Lcn3lacla .......... IYIDDKKIM. IDKHNNLLLN .DSYNSFVNE ..........
Nistlacla .......... VVMKDGQVED VGSHDVLLRR CQ..YYQELY YSEQYEDNDE
Spabbacsu .......... IVMDKGEIVG IGTHEELLKT CP..LYKKMD ESENYMNPLE
Cyddescco .......... WVMQDGRIIE QGRYAELSVA GGPFATLLAH RQEEI.....
Cyddhaein .......... FVMQRGEIVQ QGKFTELQ.H EGFFAELLAQ RQQDIQ....
Cydcescco .......... IVMDNGQIIE QGTHAELLAR QGRYY.QFKQ GL........
Cydchaein .......... CVIDNGRLIE EGDYNSLITK ENGFFKRLIE RV........
Aprdpseae .......... LALNAGQLHL YGERDQVLAA LNNQRAASAS QQRADYRVAG
Prtderwch .......... LILHEGQQQR MGLARDVLTE LQQRSAANQA RMNPTAAMPQ
Hlybescco .......... IVMEKGKIVE QGKHKELLS. EPESLYSYLY QLQSD.....
Hlybpasha .......... IVMEKGEIVE QGKHHELLQ. NSNGLYSYLH QLQLN.....
```

```
Rt3bactpl .......... IVMEKGHIVE QGKHNQLLE. NENGLYYYLN QLQSN.....
Hlybactac .......... IVMDKGEIIE QGKHQELLK. DEKGLYSYLH QLQVN.....
Hlybactpl .......... IVMEKGQIVE QGKHKELLR. DPNGLYHYLH QLQSE.....
Cyabborpe .......... VVMEGGEVAE CGSHETLL.. AAGGLYARLQ ALQAGEAG..
Comastrpn .......... VVLDQGKIVE EGKHADLL.. AQGGFYAHLV NS........
Lcnclacla .......... IVVDQGKVIE SGSHVDLL.. AQNGFYEQLY HN........
Peddpedac .......... VVLDHGKIVE QGSHRQLL.. NYNGYYARLI HNQE......
Cvabescco .......... ISI.......................................
Pmdlschpo .......... VVVNAGKIVE QGSHNELLDL ..NGAYARLV EAQKLSGGEK
Pglyarath .......... AVLQQGSVSE IGTHDELFSK GENGVYAKLI KMQE...AAH
Pgplarath .......... AVLQQGSVSE IGTHDELFSK GENGVYAKLI KMQE...AAH
Mdr2musmu .......... AGFEDGVIVE QGSHSELMK. .KEGIYFRLV NMQT...AGS
Mdr3crigr .......... AGFEDGVIVE QGSHSELMQ. .KEGVYFKLV NMQT...SGS
Mdr3homsa .......... AGFEDGVIVE QGSHSELMK. .KEGVYFKLV NMQT...SGS
Mdr1musmu .......... AGFDGGVIVE QGNHDELMR. .EKGIYFKLV MTQT...RGN
Mdr2crigr .......... AGFDGGVIVE QGNHEELMK. .EKGIYCRLV MMQT...RGN
Mdr1crigr .......... AGFDGGVIVE QGNHEELMR. .EKGIYFKLV MTQT...AGN
Mdr3musmu .......... AGFDGGVIVE QGNHDELMR. .EKGIYFKLV MTQT...AGN
Mdr1homsa .......... AGFDDGVIVE KGNHDELMK. .EKGIYFKLV TMQT...AGN
Mdr4drome .......... VFLKDGVVAE QGTHEELME. .RRGLYCELV SITQ...RKE
Mdr5drome .......... VFIHDGKVLE EGSHDDLMA. .LEGAYYNMV RAGD...INM
Mdr1caeel .......... ISCKNGQVVE VGDHRALMA. .QQGLYYDLV TAQT...FTD
Mdr3caeel .......... ..........STHDELISK .DDGIYASMV KAQEIERAKE
Mdr1leien .......... DGAEGSRITE SGTFDELLEL ..DGEFAAVA KMQGVLAGDA
Msbaescco .......... VVVEDGVIVE RGTHNDLLE. .HRGVYAQLH KMQFGQ....
Msbahaein .......... LVIDHGEIRE RGNHKTLLE. .QNGAYKQLH SMQFTG....
Hetaanasp .......... VVMEQGRIVE QGNYQELLE. .QRGKLWKYH QMQHESGQTN
Mdrplafa NNNNNNNNNK INNEGSYIIE QGTHDSLMK. NKNGIYHLMI NNQKISSNKS
Mdl1sacce .......... VLGKHGSVVE TGSFRDLIAI PNSELNALLA EQQDEEGKGG
Mdl2sacce .......... VLGHDGSVVE MGKFKELYAN PTSALSQLLN EKAAPGPSDQ
Mt2ratno .......... VL.KQGQLVE ...HDQLRDE QDVYAHLVQQ RLEA......
Tap2musmu .......... VL.KQGRLVE ...HDQLRDG QDVYAHLVQQ RLEA......
Tap2homsa .......... VL.QEGKLQK ...LAQL...................
Tap1homsa .......... FL.EGGAIRE GGTHQQLMEK KGCYWAMVQA PADAPE....
Tap1musmu .......... FL.REGSVGE QGTHLQLMKR GGCYRAMVEA LAAPAD....
Chvaagrtu .......... LFLDQGRIIE KGTFDELTQR GGRFTSLLRT SGLLTEDEGQ
Ndvarhime .......... IFMDQGRVVE MGGFHELSQS NGRFAALLRA SGILTDEDVR
Atm1sacce .......... IVLDNGRVRE EGKHLELLAM PGSLYRELWT IQEDLDHLEN
Hmt1schpo .......... LCISNGRIVE TGTHEELIKR DGGRYKKMW. FQQAMGKTSA
Mdlescco .......... IVMQHDISPS VAILMCWHNK AAGIAICIAI N.........
Ste6sacce .......... .LMKEGEVVE SGTQSELLAD PTTTFSTWYH LQNDYSDAKT
Natabacsu .....LCDSV IMIHSGEVIY RGALESLYES ERSEDLNYIF MSKLVRGIS.
Syrdpsesy .......... IKLADGCIVS DVKCAVEGKR A.................
Consensus ................G.... .G....L...................
```

```
          1101                                              1150
Surcricr LEKETVMERK ASEPSQGLPR AMSSRDGLLL DEEEEEEAA ESEEDDNLSS
Surratno LEKETVMERK APEPSQGLPR AMSSRDGLLL DEDEEEEEAA ESEEDDNLSS
Mdrleita SDVDTESATA ETAPYVAKAK GLNAEQETSL AGGEDPLR.S DVEAGRLMTT
Spabbacsu EEGSKWKEAL YQG........................................
Aprdpseae YGAPQVVAAP RQGGVE......................................
Pmdlschpo DQEMVEEELE DAPREIPITS FGDDDEDNDM ASLEAPMMSH NTDTDTLNNK
Pglyarath ETAMSNARKS SARPSSARNS VSSPIMTRNS SYGRSPYSRR LSDFSTSDFS
Pgplarath ETAMSNARKS SARPSSARNS VSSPIMTRNS SYGRSPYSRR LSDFSTSDFS
```

```
Mdr2musmu QILSEEFEV. .......... .ELSDEKAAG DVAPNGWKAR IFR.NSTKKS
Mdr3crigr QILSQEFEV. .......... .ELSEEKAAD GMTPNGWKSH IFR.NSTKKS
Mdr3homsa QIQSEEF... .......... .ELNDEKAAT RMAPNGWKSR LFR.HSTQKN
Mdr1musmu EIEPGNNAY. .......... .GSQSDTDAS ELTSEESKSP LIR.RSIYRS
Mdr2crigr EVELGSEAD. .......... .GSQSDTIAS ELTSEEFKSP SVR.KSTCRS
Mdr1crigr EIELGNEVG. .......... .ESKNEIDNL DMSSKDSASS LIRRRSTRRS
Mdr3musmu EIELGNEAC. .......... .KSKDEIDNL DMSSKDSGSS LIRRRSTRKS
Mdr1homsa EVELENAAD. .......... .ESKSEIDAL EMSSNDSRSS LIRKRSTRRS
Mdr4drome ATEADEGAVA GRPLQKSQNL SDEETDDDEE DEEEDEEPEL QTSGSSRDSG
Mdr5drome PDEVEKEDSI EDTKQKSLAL FEKSFETSPL NLEKGQKNSV QFEEPIIKAL
Mdr1caeel AVDSAAEGKF SRENSVARQT SEHEGLSRQA SEMDDIMNRV RSSTIGSITN
Mdr3caeel DTTLDDEEDE KTHRSFHRDS VTSDEERELQ QSLARDSTRL RQSMISTTTQ
Mdr1leien KSGASVRDAK KASGHLGVIL DEADLAQLDE DVPRTARQNV PIDELAKW..
Hetaanasp S......... .......... .......... .......... ..........
Mdrplafa  SNNGNDNGSD NKSSAYKDSD TGNDADNMNS LSIHENENIS NNRRNCK....
Mdl1sacce VIDLDNSVAR EV........ .......... .......... ..........
Mdl2sacce QLQIEKVIEK EDLNESKEHD DQKKDDNDDN DNNHDNDSNN QSPETKDNNS
Chvaagrtu QPRPKAIAS. .......... .......... .......... ..........
Ndvarhime KSLTAA.... .......... .......... .......... ..........
Atm1sacce ELKDQQEL.. .......... .......... .......... ..........
Hmt1schpo ETH....... .......... .......... .......... ..........
Ste6sacce IVDTETEEKS IHTVESFNSQ LETPKLGSCL SNLGYDETDQ LSFYEAIYQK
Consensus .......... .......... .......... .......... ..........


               1151                                         1200
Surcricr  V......... .......... .......... .......... ...LHQRAKI
Surratno  V......... .......... .......... .......... ...LHQRAKI
Mdrleita  E......... .......... .......... .......... ...EKATGKV
Pmd1schpo LNEKDNVVFE DKTLQHVASE IVPNLPPADV GELNEEPKKS KKSKKNNHEI
Pglyarath LSIDA..... .......... .......... SSYPNYRNE. ....KLAFKD
Pgp1arath LSIDA..... .......... .......... SSYPNYRNE. ....KLAFKD
Mdr2musmu LKS....... .......... .......... PHQNRLDEE. ....TNELDA
Mdr3crigr LKSSR..... .......... .......... AHHHRLDVD. ....ADELDA
Mdr3homsa LKNSQ..... .......... .......... MCQKSLDVE. ....TDGLEA
Mdr1musmu VHRKQ..... .......... .......... DQERRLSMK. .....EAVDE
Mdr2crigr ICGSQ..... .......... .......... DQERRVSVK. .....EAQDE
Mdr1crigr IRGPH..... .......... .......... DQDRKLSTK. .....EALDE
Mdr3musmu ICGPH..... .......... .......... DQDRKLSTK. .....EALDE
Mdr1homsa VRGSQ..... .......... .......... AQDRKLSTK. .....EALDE
Mdr4drome FRAST..... .......... .......... RRKRRSQRR. ....KKKKDK
Mdr5drome IKDTN..... .......... .......... AQSAEAPPE. ....KPNFFR
Mdr1caeel GPVID..... .......... .......... EKEERIGKDA LSRLKQELEE
Mdr3caeel VPEWE..... .......... .......... IENAR..... ....EEMIE
Mdr1leien .......... .......... .......... .......... .........E
Mdrplafa  .......... .......... .......... NTAENEKEEK VPFFKRMFRR
Mdl2sacce DDIEKCVCRT SPERRGKGG. .......... .......... ..........
Mdlescco  .......... .......... .......... NWRRRSTTLR KIARRPSMRS
Ste6sacce RSN....... .......... .VRTRRVKVE EENIGYALKQ QKNTESSTGP
Consensus .......... .......... .......... .......... ..........


               1201                                         1250
Surcricr  PWRACTKYLS SAGILLLSLL VFSQL.LKHM VLVAIDYWLA KWTDSALVLS
Surratno  PWRACTKYLS SAGILLLSLL VFSQL.LKHM VLVAIDYWLA KWTDSALVLS
Mdrleita  PWSTYVAYLK SCGGLEAWGC LLATFALTEC VTAASSVWLS IWSTGSLMWS
```

```
Pmd1schpo NSLTALWFIH SFVRTMIEII CLLIGILASM ICGAAYPVQA AVFARFLNIF
Pglyarath Q...ANSFWR LAKMNSPEWK YALLGSVGSV ICGSLSAFFA YVLSAVLSVY
Pgp1arath Q...ANSFWR LAKMNSPEWK YALLGSVGSV ICGSLSAFFA YVLSAVLSVY
Mdr2musmu N.VPPVSFLK VLKLNKTEWP YFVVGTVCAI ANGALQPAFS IILSEMIAIF
Mdr3crigr N.VPPVSFLK VLKLNKTEWP YFVVGTVCAI VNGALQPAIS IILSEMIAIF
Mdr3homsa N.VPPVSFLK VLKLNKTEWP YFVVGTVCAI ANGGLQPAFS VIFSEIIAIF
Mdr1musmu D.VPLVSFWR ILNLNLSEWP YLLVGVLCAV INGCIQPVFA IVFSRIVGVF
Mdr2crigr D.VPLVSFWG ILKLNITEWP YLVVGVLCAV INGCMQPVFS IVFSGIIGVF
Mdr1crigr D.VPPISFWR ILKLNSSEWP YFVVGIFCAI VNGALQPAFS IIFSKVVGVF
Mdr3musmu D.VPPASFWR ILKLNSTEWP YFVVGIFCAI INGGLQPAFS VIFSKVVGVF
Mdr1homsa S.IPPVSFWR IMKLNLTEWP YFVVGVFCAI INGGLQPAFA IIFSKIIGVF
Mdr4drome EVVSKVSFTQ LMKLNSPEWR FIVVGGIASV MHGATFPLWG LFFGDFFGIL
Mdr5drome ......TFSR ILQLAKQEWC YLILGTISAV AVGFLYPAFA VIFGEFYAAL
Mdr1caeel NNAQKTNLFE ILYHARPHAL SLFIGMSTAT IGGFIYPTYS VFFTSFMNVF
Mdr3caeel EGAMEASLFD IFKYASPEMR NIIISLVFTL IRGFTWPAFS IVYGQLFKIL
Mdr1leien VKHAKVGFLR LMRMNKDKAW AVALGILSSV VIGSARPASS IVMGHMLRVL
Mdrplafa  KKKAPNNLRI IYKEIFSYKK DVTIIFFSIL VAGGLYPVFA LLYARYVSTL
Mdlescco  FSQLWPTLKR LLAYGSPWRK PLGIAVLMMW VAAAAEVSGP LLISYFIDNM
Ste6sacce QLLSIIQIIK RMIKSIRYKK ILILGLLCSL IAGATNPVFS YTFSFLLEGI
Consensus .......... .......... .......... .......... ..........
```

```
          1251                                              1300
Surcricr  PAARNCSLSQ ECDLDQSVYA MVFTLLCSLG IVLCLVTSVT VEWTGLKVAK
Surratno  PAARNCSLSQ ECALDQSVYA MVFTVLCSLG IALCLVTSVT VEWTGLKVAK
Mdr1eita  ADTYLYVY.. .......... LFIVFLEIFG SPLRFFLCYY LIRIG...SR
Pmd1schpo TD...LSSTD FL.HKVNVFA VYWLILAIVQ FFAYAISNFA MTYAMEAVLQ
Pglyarath Y....NPDHE YMIKQIDKYC YLLIGLSSAA LVFNTLQHSF WDIVGENLTK
Pgp1arath Y....NPDHE YMIKQIDKYC YLLIGLSSAA LVFNTLQHSF WDIVGENLTK
Mdr2musmu GP...GDDA. VKQQKCNMFS LVFLGLGVLS FFTFFLQGFT FGKAGEILTT
Mdr3crigr GP...GDDA. VKQQKCNLFS LVFLGLGVLS FFTFFLQGFT FGKAGEILTT
Mdr3homsa GP...GDDA. VKQQKCNIFS LIFLFLGIIS FFTFFLQGFT FGKAGEILTR
Mdr1musmu SR...DDDHE TKRQNCNLFS LFFLVMGLIS FVTYFFQGFT FGKAGEILTK
Mdr2crigr TR...DDDPK TKQQNCNLFS LFFLVMGMIC FVTYFFQGFT FGKAGEILTK
Mdr1crigr TR...NTDDE TKRHDSNLFS LLFLILGVIS FITFFLQGFT FGKAGEILTK
Mdr3musmu TN...GGPPE TQRQNSNLFS LLFLILGIIS FITFFLQGFT FGKAGEILTK
Mdr1homsa TR...IDDPE TKRQNSNLFS LLFLALGIIS FITFFLQGFT FGKAGEILTK
Mdr4drome S....DGDDD VVRAEVLKIS MIFVGIGLMA GLGNMLQTYM FTTAGVKMTT
Mdr5drome A....EKDPE DALRRTAVLS WACLGLAFLT GLVCFLQTYL FNYAGIWLTT
Mdr1caeel A.....GNPA DFLSQGHFWA LMFLVLAAAQ GICSFLMTFF MGIASESLTR
Mdr3caeel SA...GGDDV SI..KALLNS LWFILLAFTG GISTLISGSL LGKAGETMSG
Mdr1leien GEYSATKDVE ALRSGTNLYA PLFIVFAVAN FSGWILHGF. YGYAGEHLTT
Mdrplafa  F......DFA NLEYNSNKYS IYILLIAIAM FISETLKNYY NNKIGEKVEK
Mdlescco  .....VAKNN LPLKVVAGLA AAYVGLQLFA AGLHYAQSLL FNRAAVGVVQ
Ste6sacce VP..STDGKT GSSHYLAKWS LLVLGVAAAD GIFNFAKGFL LDCCSEYWVM
Consensus .......... .......... .......... .......... ..........
```

```
          1301                                              1350
Surcricr  RLHRSLLNRI ILAPMRFFE. .TTPLGSILN RFSSDCNTID QHIPSTLECL
Surratno  RLHRSLLNRI ILAPMRFFE. .TTPLGSILN RFSSDCNTID QHIPSTLECL
Mdr1eita  NMHRDLLESI GVARMSFFD. .TTPVGRVLN RFTKDMSILD NTLNDGYLYL
Pmd1schpo RIRYHLFRTL LRQDVEFFDR SENTVGAITT SLSTKIQSLE GLSGPTLGTF
Pglyarath RVREKMLSAV LKNEMAWFDQ EENESARIAA RLALDANNVR SAIGDRISVI
Pgp1arath RVREKMLSAV LKNEMAWFDQ EENESARIAA RLALDANNVR SAIGDRISVI
Mdr2musmu RLRSMAFKAM LRQDMSWFDD HKNSTGALST RLATDAAQVQ GATGTKLALI
```

```
Mdr3crigr RLRSMAFKAM LRQDMSWFDD YKNSTGALST RLATDRAQVQ GATGTRLALI
Mdr3homsa RLRSMAFKAM LRQDMSWFDD HKNSTGALST RLATDAAQVQ GATGTRLALI
Mdr1musmu RVRYMVFKSM LRQDISWFDD HKNSTGSLTT RLASDASSVK GAMGARLAVV
Mdr2crigr RLRYMVFKSM LRQDISWFDD HRNSTGALTT RLASDAANVK GAMSSRLAGI
Mdr1crigr RLRYMVFKSM LRQDVSWFDN PKNTTGALTT RLANDAGQVK GATGARLAVI
Mdr3musmu RLRYMVFKSM LRQDVSWFDD PKNTTGALTT RLANDAAQVK GATGSRLAVI
Mdr1homsa RLRYMVFRSM LRQDVSWFDD PKNTTGALTT RLANDAAQVK GAIGSRLAVI
Mdr4drome RLRKRAFGTI IGQDIAYFDD ERNSVGALCS RLASDCSNVQ GATGARVGTM
Mdr5drome RMRAMTFNAM VNQEVGWFDD ENNSVGALSA RLSGEAVDIQ GAIGYPLSGM
Mdr1caeel DLRNKLFRNV LSQHIGFFDS PQNASGKIST RLATDVPNLR TAIDFRFSTV
Mdr3caeel RLRMDVFRNI MQQDASYFDD SRHNVGSLTS RLATDAPNVQ AAIDQRLAEV
Mdr1leien KIRVLLFRQI MRQDINFFDI PGRDAGTLAG MLSGDCEAVH QLWGPSIGLK
Mdrplafa  TMKRRLFENI LYQEMSFFDQ DKNTPGVLSA HINRDVHLLK TGLVNNIVIF
Mdlescco  QLRTDVMDAA LRQPLSEFD. .TQPVGQVIS RVTNDTEVIR DLYVTVVATV
Ste6sacce DLRNEVMEKL TRKNMDWFSG ENNKASEISA LVLNDLRDLR SLVSEFLSAM
Consensus ........................................................
```

```
                     1351                                   1400
Surcricr  SRSTLLCVSA LTVISYVTPV FLVALLPLAV VCYFIQKYFR VASRDLQQLD
Surratno  SRSTLLCVSA LAVISYVTPV FLVALLPLAV VCYFIQKYFR VASRDLQQLD
Mdrleita  LEYFFSMCST VIIMVVVQPF VLVAIVPCVY SYYKLMQVYN ASNRETRRIK
Pmd1schpo FQILTNIISV TILSLATGWK LGLVTLSTSP VIITAGYYRV RALDQVQEKL
Pglyarath VQNTALMLVA CTAGFVLQWR LALVLVAVFP VVVAATVLQK MFMTGFSGDL
Pgp1arath VQNTALMLVA CTAGFVLQWR LALVLVAVFP VVVAATVLQK MFMTGFSGDL
Mdr2musmu AQNTANLGTG IIISFIYGWQ LTLLLLSVVP FIAVAGIVEM KMLAGNAKRD
Mdr3crigr AQNTANLGTG IIISFIYGWQ LTLLLLSVVP FIAVSGIVEM KMLAGNAKRD
Mdr3homsa AQNIANLGTG IIISFIYGWQ LTLLLLAVVP IIAVSGIVEM KLLAGNAKRD
Mdr1musmu TQNVANLGTG VILSLVYGWQ LTLLLVVIIP LIVLGGIIEM KLLSGQALKD
Mdr2crigr TQNVANLGTG IIISLVYGWQ LTLLLVVIAP LIILSGMMEM KVLSGQALKD
Mdr1crigr TQNIANLGTG IIISLIYGWQ LTLLLLAIVP IIAIAGVVEM KMLSGQALKD
Mdr3musmu FQNIANLGTG IIISLIYGWQ LTLLLLAIVP IIAIAGVVEM KMLSGQALKD
Mdr1homsa TQNIANLGTG IIISFIYGWQ LTLLLLAIVP IIAIAGVVEM KMLSGQALKD
Mdr5drome IQALSNFISS VSVAMYYNWK LALLCLANCP IIVGSVILEA KMMSNAVVRE
Mdr1caeel ITTLVSMVAG IGLAFFYGWQ MALLIIAILP IVAFGQYLRG RRFTGKNVKS
Mdr3caeel LTGIVSLFCG VGVAFYYGWN MAPIGLATEL LLVVVQSSVA QYLKFRGQRD
Mdr1leien VQTMCIIASG LVVGFIYQWK LALVALACMP LMIGCSLTRR LMINGYTKSR
Mdrplafa  SHFIMLFLVS MVMSFYFCPI VAAVLTFIYF INMRVFAVRA RLTKSKEIEK
Mdlescco  LRSAALVGAM LVAMFSLDWR MALVAIMIFP VVLVVMVIYQ RYSTPIVRRV
Ste6sacce TSFVTVSTIG LIWALVSGWK LSLVCISMFP LIIIFSAIYG GILQKCETDY
Consensus ........................................................
```

```
                     1401                                   1450
Surcricr  DTTQLPLVSH FAETVEGLTT IRAFRYEARF QQKLLEYTDS NNIASLFLTA
Surratno  DTTQLPLLSH FAETVEGLTT IRAFRYEARF QQKLLEYTDS NNIASLFLTA
Mdrleita  SIAHSPVFTL LEESLQGQRT IATYGKLHLV LQEALGRLDV VYSALYMQNV
Pmd1schpo SAA............ ......YKE SAAFACESTS AIRTVASLNR EENVFAEYCD
Pglyarath EAA............ ......HAK GTQLAGEAIA NVRTVAAFNS EAKIVRLYTA
Pgp1arath EAA............ ......HAK GTQLAGEAIA NVRTVAAFNS EAKIVRLYTA
Mdr2musmu KKE............ ......MEA AGKIATEAIE NIRTVVSLTQ ERKFESMYVE
Mdr3crigr KKA............ ......LEA AGKIATEAIE NIRTVVSLTQ ERKFESMYVE
Mdr3homsa KKE............ ......LEA AGKIATEAIE NIRTVVSLTQ ERKFESMYVE
Mdr1musmu KKQ............ ......LEI SGKIATEAIE NFRTIVSLTR EQKFETMYAQ
Mdr2crigr KKE............ ......LEV SGKIATEAIE NFRTVVSLTR EQKFENMYAQ
Mdr1crigr KKE............ ......LEG SGKIATEAIE NFRTVVSLTR EQKFENMYAQ
```

```
Mdr3musmu KKE...... .......LEG SGKIATEAIE NFRTVVSLTR EQKFETMYAQ
Mdr1homsa KKE...... .......LEG AGKIATEAIE NFRTVVSLTQ EQKFEHMYAQ
Mdr4drome KAS...... .......IEE ASQVAVEAIT NIRTVNGLCL ERQVLDQYVQ
Mdr5drome KQV...... .......IEE ACRIATE6IT NIRTVAGLRR EADVIREYTE
Mdr1caeel ASE...... .......FAD SGKIAIEAIE NVRTVQALAR EDTFYENFCE
Mdr3caeel MDS...... .......AIE ASRLVTESIS NWKTVQALTK QEYMYDAFTA
Mdr1leien EGD...... .......TDD T..IVTEALS NVRTVTSLNM KEDCVEAFQA
Mdrplafa  KENMSSGVFA FSSDDEMFKD PSFLIQEAFY NMHTVINYGL EDYFCNLIEK
Mdlescco  RA........ ......YLAD INDGFKQIIN GMSVIQQFRQ QARFGERMGE
Ste6sacce KTS...... .......VAQ LENCLYQIVT NIKTIKCLQA EFHFQLTYHD
Consensus .......... .......... .......... .......... ..........

          1451                                            1500
Surcricr  ANRWLEVCME YIGACVVLIA AATSISNSLH RELSA..GLV GLGLTYALMV
Surratno  ANRWLEVRME YIGACVVLIA AATSISNSLH RELSA..GLV GLGLTYALMV
Mdrleita  SNRWLGVRLE FLSCVVTFMV AFIGVIGKME GASSQNIGLI SLSLTMSMTL
Pmd1schpo S.....LIKP GRESAIASLK SGLFFSAAQG VTFLINALTF WYGSTLMRKG
Pglyarath N.....LEPP LKRCFWKGQI AGSGYGVAQF CLYASYALGL WYASWLVKHG
Pgp1arath N.....LEPP LKRCFWKGQI AGSGYGVAQF CLYASYALGL WYASWLVKHG
Mdr2musmu K.....LHGP YRNSVRKAHI YGITFSISQA FMYFSYAGCF RFGSYLIVNG
Mdr3crigr K.....LHEP YRNSVQMAHI YGITFSISQA FMYFSYAGCF RFGAYLIVNG
Mdr3homsa K.....LYGP YRNSVQAHI YGITFSISQA FMYFSYAGCF RFGAYLIVNG
Mdr1musmu S.....LQVP YRNAMKKAHV FGITFSFTQA MMYFSYAACF RFGAYLVAQQ
Mdr2crigr S.....LQIP YRNALKKAHV FGITFSFTQA MMYFSYAACF RFGAYLVAHQ
Mdr1crigr S.....LQIP YRNALKKAHV FGITFSFTQA MMYFSYAACF RFGAYLVARE
Mdr3musmu S.....LQIP YRNAMKKAHV FGITFFFTQA MMYFSYAGCF RFGAYLVTQQ
Mdr1homsa S.....LQVP YRNSLRKAHI FGITFSFTQA MMYFSYAGCF RFGAYLVAHK
Mdr4drome Q.....IDRV DVACRRKVRF RGLVFALGQA APFLAYGISM YYGGILVAEE
Mdr5drome E.....IQRV EVLIRQKLRW RGVLNSTMQA SAFFAYAVAL CYGGVLVSEG
Mdr1caeel K.....LDIP HKEAIKEAFI QGLSYGCASS VLYLLNTCAY RMGLALIITD
Mdr3caeel A.....SKSP HRRAIVRGLW QSLSFALAGS FVMWNFAIAY MFGLWLISNN
Mdr1leien A.....LREE APRSVRKGII AGGIYGITQF IFYGVYALCF WYGSKLIDKG
Mdrplafa  A.....IDYK NKGQKRRIIV NAALWGFSQS AQLFINSFAY WFGSFLIKRG
Mdlescco  A.....SRSH YMARMQTLRL DGFLLR.... .PLLSLFSSL ILCGLLMLFG
Ste6sacce L.....KIKM QQIASKRAIA TGFGISMTNM IVMCIQAIIY YYGLKLVM..
Consensus .......... .......... .......... .......... ..........

          1501                                            1550
Surcricr  SNYLNWMVRN LADMEIQLGA VKR........ .......... ..IHALLKTE
Surratno  SNYLNWMVRN LADMEIQLGA VKG........ .......... ..IHTLLKTE
Mdrleita  TETLNWLVRQ VAMVEANMNS VERVLHYTQE VEHEHVPEMG ELVAQLVRSE
Pmd1schpo ..EYNIVQFY TCFIAIVFGI QQAGQFFGYS ADVTKAKAAA GEIKYLSESK
Pglyarath ISDFS..KTI RVFMVLMVSA NGAAETLTLA PDFIKGGQAM RSVFELLDRK
Pgp1arath ISDFS..KTI RVFMVLMVSA NGAAETLTLA PDFIKGGQAM RSVFELLDRK
Mdr2musmu HMRFK..DVI LVFSAIVLGA VALGHASSFA PDYAKAKLSA AYLFSLFERQ
Mdr3crigr HMRFR..DVI LVFSAIVFGA VALGHASSFA PDYAKAKLSA AHLFSLFERQ
Mdr3homsa HMRFR..DVI LVFSAIVFGA VALGHASSFA PDYAKAKLSA AHLFMLFERQ
Mdr1musmu LMTFE..NVM LVFSAVVFGA MAAGNTSSFA PDYAKAKVSA SHIIRIIEKT
Mdr2crigr IMTFE..NVM LVFSAVVFGA IAAGNASSFA PDYAKAKVSA SHIIRIMEKI
Mdr1crigr LMTFE..NVL LVFSAIVFGA MAVGQVSSFA PDYAKAKVSA SHIIMIIEKV
Mdr3musmu LMTFE..NVL LVFSAIVFGA MAVGQVSSFA PDYAKATVSA SHIIRIIEKT
Mdr1homsa LMSFE..DVL LVFSAVVFGA MAVGQVSSFA PDYAKAKISA AHIIMIIEKT
Mdr4drome RMNYE..DII KVAEALIFGS WMLGQALAYA PNVNDAILSA GRLMDLFKRT
Mdr5drome QLPFQ..DII KVSETLLYGS MMLAQSLAFT PAFSAALIAG HRLFQILDRK
```

```
Mdr1caeel PPTMQPMRVL RVMYAITIST STLGFATSYF PEYAKATFAG GIIFGMLRKI
Mdr3caeel WST..PYTVF QVIEALNMAS MSVMLAASYF PEYVRARISA GIMFTMIRQK
Mdr1leien EAEFK..DVM IASMSILFGA QNAGEAGAFA TKLADAEASA KRVFSVIDRV
Mdrplafa  ..TILVDDFM KSLFTFIFTG SYAGKLMSLK GDSENAKLSF EKYYPLMIRK
Mdlescco  FSASGTIEVG VLYAFISYLG RLNEPLIELT TQQAMLQQA. .....VVAGE
Ste6sacce IHEYTSKEMF TTFTLLLFTI MSCTSLVSQI PDISRGQRAA SWIYRILDEK
Consensus ......... ......... ......... ......... .........
```

```
          1551                                            1600
Surcricr  ....AESYEG LLAPSLIPKN WP....DQGK IQIQNLSVRY DSSLKPVLKH
Surratno  ....AESYEG LLAPSLIPKN WP....DQGK IQIQNLSVRY DSSLKPVLKH
Mdrleita  SGRGANVTET VVIESAGAAS SALHPVQAGS LVLEGVQMRY REGLPLVLRG
Pmd1schpo PKI....DTW STEGKKVESL QSAA.....I EFRQVEFSYP TRRHIKVLRG
Pglyarath TEIEPDD... PDTTPVPDRL R.GE.....V ELKHIDFSYP SRPDIQIFRD
Pgplarath TEIEPDD... PDTTPVPDRL R.GE.....V ELKHIDFSYP SRPDIQIFRD
Mdr2musmu PLI....DSY SGEGLWPDKF E.GS.....V TFNEVVFNYP TRANVPVLQG
Mdr3crigr PLI....DSY SGEGLWPDKF E.GS.....V TFNEVVFNYP TRANMPVLQG
Mdr3homsa PLI....DSY SEEGLKPDKF E.GN.....I TFNEVVFNYP TRANVPVLQG
Mdr1musmu PEI....DSY STEGLKPTLL E.GN.....V KFNGVQFNYP TRPNIPVLQG
Mdr2crigr PSI....DSY STRGLKPNWL E.GN.....V KFNEVVFNYP TRPDIPVLQG
Mdr1crigr PSI....DSY STGGLKPNTL E.GN.....V KFNEVVFNYP TRPDIPVLQG
Mdr3musmu PEI....DSY STQGLKPNML E.GN.....V QFSGFVFNYP TRPSIPVLQG
Mdr1homsa PLI....DSY STEGLMPNTL E.GN.....V TFGEVVFNYP TRPDIPVLQG
Mdr4drome STQPNPPQSP YNTVEKSE...GD.....I VYENVGFEYP TRKGTPILQG
Mdr4enthi .......... .......... .......... .......... ..........
Mdr5drome PKIQSPMGTI KNTLAKQLNL F.EG.....V RYRGIQFRYP TRPDAKILNG
Mdr1caeel SKI....DSL SLAGEK.KKL Y.GK.....V IFKNVRFAYP ERPEIEILKG
Mdr3caeel SVIDNRGLTG DTPTIK.....GN.....I NMRGVYFAYP NRRRQLVLDG
Mdr1leien PDVDIEQAGN KDLG...... EGCD....I EYRNVQFIYS ARPKQVVLAS
Mdrplafa  SNIDVRDDG. ..GIRINKNL IKGK.....V DIKDVNFRYI SRPNVPIYKN
Mdlescco  RVFELMDGPR QQYGNDDRPF TSGT.....I EVDNVSFAY. .RDDNLVLNN
Ste6sacce HNTLEVENNN ARTVGIAGHT YHGKEKKPIV SIQNLTFAYP SAPTAFVYKN
Consensus ......... ......... ......... ......... .........
```

```
          1601                                            1650
Surcricr  VNTLISPGQK IGICGRTGSG KSSFSLAFFR MVDMFEGRII IDG.......
Surratno  VNALISPGQK IGICGRTGSG KSSFSLAFFR MVDMFEGRII IDG.......
Mdrleita  VSFQIAPREK VGIVGRTGSG KSTLLLTFMR MVEVCGGVIH VNG.......
Pmd1schpo LNLTVKPGQF VAFVGSSGCG KSTTIGLIER FYDCDNGAVL VD.GV.....
Pglyarath LSLRARAGKT LALVGPSGCG KSSVISLIQR FYEPSSGRVM ID.GK.....
Pgplarath LSLRARAGKT LALVGPSGCG KSSVISLIQR FYEPSSGRVM ID.GK.....
Mdr2musmu LSLEVKKGQT LALVGSSGCG KSTVVQLLER FYDPMAGSVL LD.GQ.....
Mdr3crigr LSLEVKKGQT LALVGSSGCG KSTVVQLLER FYDPMAGTVL LD.GQ.....
Mdr3homsa LSLEVKKGQT LALVGSSGCG KSTVVQLLER FYDPLAGTVL LD.GQ.....
Mdr1musmu LSLEVKKGQT LALVGSSGCG KSTVVQLLER TKYDPMAGSVF LD.GK.....
Mdr2crigr LSLEVKKGQT LALVGSSGCG KSTVVQLLER FYDPNAGTVF LD.GK.....
Mdr1crigr LNLEVKKGQT VLALVGSSGCG KSTVVQLLER FYDPMAGTVF LD.GK.....
Mdr3musmu LSLEVKKGQT LALVGSSGCG KSTVVQLLER FYDPMAGSVF LD.GK.....
Mdr1homsa LSLEVKKGQT LALVGSSGCG KSTVVQLLER FYDPLAGKVL LD.GK.....
Mdr4drome LNLTIKKSTT VALVGPSGSG KSTCVQLLLR YYDPVSGSVN LS.GV.....
Mdr5drome LDLEVLKGQT VALVGHSGCG KSTCVQLLQR YYDPDEGTIH IDHDD.....
Mdr1caeel LSFSVEPGQT LALVGPSGCG KSTVVALLER FYDTLGGEIF ID.GS.....
Mdr3caeel FNMSANFGQT VALVGPSGCG KSTTIQLIER YYDALCGSVK ID.DS.....
Mdr1leien VNMRFGDATS NGLIGQTGCG KSTVIQMLAR FYERRSGLIS VN.GR.....
```

```
Mdrplafa  LSFTCDSKKT TAIVGETGSG KSTFMNLLLR FYDLKNDHII LKNDMTNFQD
Mdlescco  INLSVPSRNF VALVGHTGSG KSTLASLLMG YYPLTEGEIR LDGRPLSSLS
Ste6sacce MNFDMFCGQT LGIIGESGTG KSTLVLLLTK LYNCEVGKIK IDGT......
Consensus ......... ......... ......... ......... .........

          1651                                             1700
Surcricr  ......... ......... ......... ......... IDIAKLPLHT
Surratno  ......... ......... ......... ......... IDIAKLPLHT
Mdrleita  ......... ......... ......... ......... REMSAYGLRD
Pmd1schpo ......... ......... ......... ......... .NVRDYNIND
Pglyarath ......... ......... ......... ......... .DIRKYNLKA
Pgp1arath ......... ......... ......... ......... .DIRKYNLKA
Mdr2musmu ......... ......... ......... ......... .EAKKLNVQW
Mdr3crigr ......... ......... ......... ......... .EAKKLNIQW
Mdr3homsa ......... ......... ......... ......... .EAKKLNVQW
Mdr1musmu ......... ......... ......... ......... .EIKQLNVQW
Mdr2crigr ......... ......... ......... ......... .EIKQLNVQW
Mdr1crigr ......... ......... ......... ......... .EVNQLNVQW
Mdr3musmu ......... ......... ......... ......... .EIKQLNVQW
Mdr1homsa ......... ......... ......... ......... .EIKRLNVQW
Mdr4drome ......... ......... ......... ......... .PSTEFPLDT
Mdr5drome ......... ......... ......... ......... .IQHDLTLDG
Mdr1caeel ......... ......... ......... ......... .EIKTLNPEH
Mdr3caeel ......... ......... ......... ......... .DIRDLSVKH
Mdr1leien ......... ......... ......... ......... .DLSSLDIAE
Mdrplafa  YQNNNNNSLV LKNVNEFSNQ SGSAEDYTVF NNNGEILLDD INICDYNLRD
Mdlescco  HTRG...... ......... ......... ......... .........
Consensus ......... ......... ......... ......... .........

          1701                                             1750
Surcricr  LRSRLSIILQ DPVLFSGTIR FNL....DPE KKCSDSTLWE ALEIAQLKLV
Surratno  LRSRLSIILQ DPVLFSGTIR FNL....DPE KKCSDSTLWE ALEIAQLKLV
Mdrleita  VRRHFSMIPQ DPVLFDGTVR QNV....DPF LEASSAEVWA ALELVGLRER
Pmd1schpo YRKQIALVSQ EPTLYQGTVR ENIVLGAS.. KDVSEEEMIE ACKKANIHEF
Pglyarath IRKHIAIVPQ EPCLFGTTIY ENIAYGHEC. .ATEAEIIQ AATLASAHKF
Pgp1arath IRKHIAIVPQ EPCLFGTTIY ENIAYGHEC. .ATEAEIIQ AATLASAHKF
Mdr2musmu LRAQLGIVSQ EPILFDCSIA ENIAYGDNSR .VVPHDEIVR AAKEANIHPF
Mdr3crigr LRAQLGIVSQ EPVLFDCSIA ENIAYGDNSR .VVSQDEIVR AAKAANIHPF
Mdr3homsa LRAQLGIVSQ EPILFDCSIA ENIAYGDNSR .VVSQDEIVS AAKAANIHPF
Mdr1musmu LRAHLGIVSQ EPILFDCSIA ENIAYGDNSR .AVSHEEIVR AAKEANIHQF
Mdr2crigr LRAHLGIVSQ EPILFDCSIA ENIAYGDNSR .VVSQDEIER AAKEANIHQF
Mdr1crigr LRAHLGIVSQ EPAILFDCSIA ENIAYGDNSR .VVSQDEIER AAKEANIHQF
Mdr3musmu LRAQLGIVSQ EPILFDCSIA ENIAYGDNSR .VVSYEEIVR AAKEANIHQF
Mdr1homsa LRAHLGIVSQ EPILFDCSIA ENIAYGDNSR .VVSQDEIVR AAKEANIHAF
Mdr4drome LRSKLGLVSQ EPVLFDRTIA ENIAYGNNFR DDVSMQEIIE AAKKSNIHNF
Mdr5drome VRTKLGIVSQ EPTLFERSIA ENIAYGDN.R RSVSMVEIIA AAKSANAHSF
Mdr1caeel TRSQIAIVSQ EPTLFDCSIA ENIIYGLD.P SSVTMAQVEE AARLANIHNF
Mdr3caeel LRDNIALVGQ EPTLFNLTIR ENITYGLEN. .ITQDQVEK AATLANIHTF
Mdr1leien WRRNISIVLQ EPNLFSGTVR ENIRYA...R EGATDEEVEE AARLAHIHHE
Mdrplafa  LRNLFSIVSQ EPMLFNMSIY ENIKFG...R EDATLEDVKR VSKFAAIDEF
Mdlescco  ..QGVAMVQQ DPVVLADTFL ANVTLGRD.. ...ISEERVWQ ALETVQLADV
Ste6sacce LRKEISVVEQ KPLLFNGTIR DNLTYGLQ.. DEILEIEMYD ALKYVGIHDF
Consensus ......... ......... ......... ......... .........
```

```
                    1751                                              1800
Surcricr    VKALPGGLDA IITEGGENFS QGQRQLFCLA RAFVRKTSIF I.MDEATASI
Surratno    VKALPGGLDA IITEGGENFS QGQRQLFCLA RAFVRKTSIF I.MDEATASI
Mdrleita    VASESEGIDS RVLEGGSNYS VGQRQLMCMA RALLKRGSGF ILMDEATANI
Pmd1schpo   ILGLPNGYNT LCGQKGSSLS GGQKQRIAIA RALIRNPKIL L.LDEATSAL
Pglyarath   ISALPEGYKT YVGERGVQLS GGQKQRIAIA RALVRKAEIM L.LDEATSAL
Pgplarath   ISALPEGYKT YVGERGVQLS GGQKQRIAIA RALVRKAEIM L.LDEATSAL
Mdr2musmu   IETLPQKYNT RVGDKGTQLS GGQKQRIAIA RALIRQPRVL L.LDEATSAL
Mdr3crigr   IETLPQKYKT RVGDKGTQLS GGQKQRLAIR RALIRQPRVL L.LDEATSAL
Mdr3homsa   IETLPHKYET RVGDKGTQLS GGQKQRIAIA RALIRQPQIL L.LDEATSAL
Mdr1musmu   IDSLPDKYNT RVGDKGTQLS GGQKQRIAIA RALVRQPHIL L.LDEATSAL
Mdr2crigr   IESLPDKYNT RVGDKGTQLS GGQKQRIAIA RALVRQPHIL L.LDEATSAL
Mdr1crigr   IESLPDKYNT RVGDKGTQLS GGQKQRIAIA RALVRQPHIL L.LDEATSAL
Mdr3musmu   IDSLPDKYNT RVGDKGTQLS GGQKQRIAIA RALVRQPHIL L.LDEATSAL
Mdr1homsa   IESLPNKYST KVGDKGTQLS GGQKQRIAIA RALVRQPHIL L.LDEATSAL
Mdr4drome   ISALPQGYDT RLG.KTSQLS GGQKQRIAIA RALVRNPKIL I.LDEATSAL
Mdr5drome   IISLPNGYDT RMGARGTQLS GGQKQRIAIA RALVRNPKIL L.LDEATSAL
Mdr1caeel   IAELPEGFET RVGDRGTQLS GGQKQRIAIA RALVRNPKIL L.LDEATSAL
Mdr3caeel   VMGLPDGYDT SVGASGGRLS GGQKQRVAIA RAIVRDPKIL L.LDEATSAL
Mdr1leien   IIKWTDGYDT EVGYKGRALS GGQKQRIAIA RGLLRRPRLL L.LDEATSAL
Mdrplafa    IESLPNKYDT NVGPYGKSLS GGQKQRIAIA RALLREPKIL L.LDEATSSL
Mdlescco    ARSMSDGIYT PLGEQGNNLS VGQKQLLALA RVLVETPQIL I.LDEATASI
Ste6sacce   VISSPQGLDT RID..TTLLS GGQAQRLCIA RALLRKSKIL I.LDECTSAL
Consensus   .......... .......... .......... .......... ..........

                    1801                                              1850
Surcricr    DMATENILQK VV..MTAFAD RTVVTIAHRV HTILSADLVM VLKRGAILEF
Surratno    DMATENILQK VV..MTAFAD RTVVTIAHRV HTILSADLVM VLKRGAILEF
Mdrleita    DPALDRQIQA TV..MSAFSA YTVITIAHRL HTVAQYDKII VMDHGVVAEM
Pmd1schpo   DSHSEKVVQE ALN..AASQG RTTVAIAHRL SSIQDADCIF VFDGGV....
Pglyarath   DAESERSVQE ALD..QACSG RTSIVVAHRL STIRNAHVIA VIDDGK....
Pgplarath   DAESERSVQE ALD..QACSG RTSIVVAHRL STIRNAHVIA VIDDGK....
Mdr2musmu   DTESEKVVQE ALD..KAREG RTCIVIAHRL STIQNADLIV VIENGK....
Mdr3crigr   DTESEKVVQE ALD..KAREG RTCIVIAHRL STIQNADLIV VIQNGK....
Mdr3homsa   DTESEKVVQE ALD..KAREG RTCIVIAHRL STIQNADLIV VFQNGR....
Mdr1musmu   DTESEKVVQE ALD..KAREG RTCIVIAHRL STIQNADLIV VIENGK....
Mdr2crigr   DTESEKVVQE ALD..KAREG RTCIVIAHRL STIQNADLIV VIQNGK....
Mdr1crigr   DTESEKVVQE ALD..KAREG RTCIVIAHRL STIQNADLIV VIQNGK....
Mdr3musmu   DTESEKVVQE ALD..KAREG RTCIVIAHRL STIQNADLIV VIQNGK....
Mdr1homsa   DTESEKVVQE ALD..KAREG RTCIVIAHRL STIQNADLIV VFQNGR....
Mdr4drome   DLESEKVVQQ ALD..EARSG RTCLTIAHRL TTVRNADLIC VLKRGV....
Mdr5drome   DLQSEQLVQQ ALD..TACSG RTCIVIAHRL STVQNADVIC VIQNGQ....
Mdr1caeel   DTESEKVVQE ALD..RAREG RTCIVIAHRL NTVMNADCIA VVSNGT....
Mdr3caeel   DTESEKIVQE ALD..KARLG RTCVVIAHRL STIQNADKII VCRNGK....
Mdr1leien   DSVTEAKVQE GIEAFQAKYK VTTVSIAHRL TTIRHCDQII LLDSGC....
Mdrplafa    DSNSEKLIEK TIVDIKDKAD KTIITIAHRI ASIKRSDKIV VFNNPDRNGT
Mdlescco    DSGTEQAIQH ALA..AVREH TTLVVIAHRL STIVDAATIL VLHRGQ....
Ste6sacce   DSVSSSIINE IVKKGPPALL TMVITHSEQM MRSCNSIAVL KDGKVVERGN
Consensus   .......... .......... .......... .......... ..........

                    1851                                              1900
Surcricr    DKPETLLSQK DSVFASFV.. ...RADK.... .......... ..........
Surratno    DKPEKLLSQK DSVFASFV.. ...RADK.... .......... ..........
Mdrleita    GSPRELVMNH QSMFHSMVES LGSRGSKDFY ELLMGRRIVQ PAVLSD....
```

```
Pmd1schpo .TCEAGTHAE LVKQ..R... ..GRYYELVV EQGLNKA............
Pglyarath .VAEQGSHSH LLKNHPD... ..GIYARMIQ LQRFTHTQVI GMTSGSSSRV
Pgplarath .VAEQGSHSH LLKNHPD... ..GIYARMIQ LQRFTHTQVI GMTSGSSSRV
Mdr2musmu .VKEHGTHQQ LLAQ..K... ..GIYFSMVN IQAGTQNL...........
Mdr3crigr .VKEHGTHQQ LLAQ..K... ..GIYFSMVN IQAGAQNS...........
Mdr3homsa .VKEHGTHQQ LLAQ..K... ..GIYFSMVS VQAGTQNL...........
Mdr1musmu .VKEHGTHQQ LLAQ..K... ..GIYFSM.. VQAGAKRS...........
Mdr2crigr .VKEHGTHQQ LLAQ..K... ..GIYFSM.. VQAGAKRL...........
Mdr1crigr .VKEHGTHQQ LLAQ..K... ..GIYFSMVS VQAGAKR............
Mdr3musmu .VKEHGTHQQ LLAQ..K... ..GIYFSMVS VQAGAKRS...........
Mdr1homsa .VKEHGTHQQ LLAQ..K... ..GIYFSMVS VQAGTKRQ...........
Mdr5drome .VVEQGNHMQ LISQ..G... ..GIYAKLHK TQKDH.............
Mdr1caeel .IIEKGTHTQ LMSE..K... ..GAYYKLTQ KQMTEKK...........
Mdr3caeel .AIEEGTHQT LLAR..R... ..GLYYRLVE KQSS..............
Mdr1leien .IIEQGSHEE LMALGGEYKT RYDLYMSALS ................
Mdrplafa  FVQSHGTHDE LLSAQD.... ..GIYKKYVK LAK...............
Mdlescco  .AVEQGTHQQ LLRPRDVLAD VSTATCGRRA GSQRA.............
Ste6sacce FDTLYNNRGE LFQIVSNQSS ................
Consensus ................
```

```
          1901
Pglyarath KEDDA
Pgplarath KEDDA
Consensus .....
```

Proteins listed subsequently in italics are at least 90% identical to the paired transporters listed in parenthesis and are therefore not included in the alignment: *Hlybprovu*, *Hly2escco* (Hlybescco); *Mdr1ratno* (Mdr1musmu); *Tap2ratno* (Tap2musmu). Residues listed in the consensus sequence are present in at least 75% of the aligned transporter sequences. Residues indicated in boldface type are also conserved in at least one other family of the ABC transporter superfamily.

Database accession numbers

	SWISSPROT	PIR	EMBL/GENBANK
Aprdpseae	Q03024	S26696	X64558; G45280
Atm1sacce	P40416		X82612; G575393
Chvaagrtu	P18768	A32810; VXAGCA	M24198; G142224
Comastrpn	Q03727	A39203	M36180; G520738
Cvabescco	P22520	S12272; IKEC5B	X57524; G41176
Cyabborpe	P18770	S02386; BVBRCB	X14199; G39733
Cydcescco	P23886	PS0228; B36888	L10383; G145165
Cydchaein	P45081		L45792; G1006503
Cyddescco	P29018	B40632	L21749; G347240
Cyddhaein	P45082		L45793; G1006505
Hemsentfa		A41464	M38052
Hetaanasp	P22638	A35391	M31722; G142020
Hly2escco	P10089	B24433; LEECB	M10133; G146380
Hlybactac	P23702	S12601; A61378	X53955; G38647
Hlybactpl	P26760	A40366; S18855	M65808; G141821
Hlybescco	P08716	S10057	M14107; G150683
Hlbypasha	P16532	A32051; S29517	M20730; G150495
Hlybprovu	P11599	S05477; LEEBBV	X12852; G45904
Hmt1schpo	Q02592	S25198	Z14055; G4972
Lcn3lacla	P37608		U04057; G433323

	SWISSPROT	PIR	EMBL/GENBANK
Lcnclacla	Q00564	B43943	M90969; G387688
Mdlescco	P30751		L08627; G146801
Mdl1sacce	P33310	S42681	U17246; G577195
Mdl2sacce	P33311	S42682	L16959; G311095
Mdrleita	P21441	A34207; DVLNS	X17154; G9659
Mdrplafa	P13568	A32547; DVZQF	M29154; G160399
Mdr1caeel	P34712	S27337	X65054; G6809
Mdr1crigr	P21448	A38696; DVHY1C	M60040; G191165
Mdr1homsa	P08183	A25059; DVHU1	M14758; G307180
Mdr1leien	Q06034		L08091; G159370
Mdr1musmu	P06795	A33719; DVMS1	M14757; G387426
Mdr1ratno	P43245		M81855
Mdr2crigr	P21449	B27126; DVHY2C	M60041; G191167
Mdr2musmu	P21440	A30409; DVMS2	J03398; G387428
Mdr3caeel	P34713	S27338	X65055; G6811
Mdr3crigr	P23174		M60042; G191169
Mdr3homsa	P21439	JS0051; DVHU3	M23234; G307181
Mdr3musmu	P21447	A34175; DVMS1A	M30697; G387429
Mdr4drome	Q00449	A41249	M59076; G157871
Mdr5drome	Q00748	B41249	M59077; G157875
Msbaescco	P27299	S21588; S27998	Z11796; G42023
Msbahaein	P44407		L44704; G1005423
Mt2ratno		S25577	
Natabacsu	P46903	U30873; G973330	
Ndvarhime	P18767	A31094; VXZRNA	M20726; G152269
Nistlacla	Q03203	S38790	X68307; G44044
Peddpedac	P36497	D48941	M83924; G150568
Pglyarath		A42150	X61370
Pgp1arath		S21957	X61370
Pmd1schpo	P36619	S20548	D10695; G218550
Prtderwch	P23596	S12525	M60395; G148484
Rt3bactpl	Q04473		L12145; G470686
Spabbacsu	P33116	B43935	M86869; G143715
Ste6sacce	P12866	S05789; DVBYS6	X15428; G4564
Surcricr	Q09427		L40623; G1311522
Surratno	Q09429		L40624; G1311534
Syrdpsesy	P33951	S27646; S37347	M97223; G151562
Tap1homsa	Q03518	A41538 S27333	X66401; G34636
Tap1musmu	P21958	A37779	M55637; G199305
Tap2homsa	Q03519	A40224; S27334	X66401; G34638
Tap2musmu	P36371	A44135	M90459; G199434
Tap2ratno	P36372	S19603	X63854; G56719

References

1 Chen, C-J. et al. (1986) Cell 47, 381–389.
2 Felmlee, T. et al. (1985) J. Bacteriol. 163, 94–105.
3 Glaser, P. et al. (1988) EMBO J. 7, 3997–4004.
4 **Endicott, J.A. and Ling, V. (1989) Annu. Rev. Biochem. 58, 137–171.**
5 Valverde, M.A. et al. (1992) Nature 355, 830–833.
6 Gill, D.R. et al. (1992) Cell 71, 23–32.
7 Aguilar-Bryan, L. et al. (1995) Science 268, 423–426.
8 **Higgins, C.F. (1992) Annu. Rev. Cell Biol. 8, 67–113.**
9 Beck, S. et al. (1992) J. Mol. Biol. 228, 433–441.

Summary

Transporters of the peroxisomal membrane family, examples of which are the human adrenoleukodystrophy protein (Aldhomsa)[1,2], and the rat 70 kDa peroxisomal membrane protein[3] (Pmp7ratno), mediate import of large molecules – proteins or long chain fatty acids – through the peroxisomal membrane. Mutations in human ALDP give rise to the serious genetic disease adrenoleukodystrophy[4]. As this gene is located on the X-chromosome, the disease only occurs in males; it affects about 1 in 20 000 live male births. Mutations in human 70 kDa peroxisomal membrane protein have been implicated in the Zellweger syndrome, an inherited defect of peroxisome assembly[5]. Members of the peroxisomal membrane transporter family have only been found in mammals and fungi.

Statistical analysis of multiple amino acid sequence comparisons places the peroxisomal membrane transporter family within the multidrug resistance subdivision of the ATP binding cassette (ABC) superfamily[6]. Proteins in this superfamily use the energy of ATP hydrolysis to pump substrates across cell membranes. Members of this family consist of a single transmembrane domain followed by an ATP binding domain; the functional transporter is expected to be a dimer and may be glycosylated. Unusually, the transmembrane domains of proteins in this family are predicted to contain five membrane-spanning helices by the hydropathy of the amino acid sequences, so the functional transporters will contain ten such helices. The ATP binding domain has been found experimentally to be exposed to the cytosol[3], leaving the N-terminus extracellular, similar to the topology of the transmembrane subunits of the histidine transporter from *Salmonella typhimurium*[7].

Many residues, including several long sequence motifs, are well conserved within the peroxisomal membrane transporter family, including motifs unique to the family, signature motifs of the ABC superfamily, and motifs necessary for function by the criterion of site-directed mutagenesis. These sequence motifs occur throughout the protein, but are most common within the ATP binding domains.

Nomenclature, biological sources and substrates

CODE	DESCRIPTION [SYNONYMS]	ORGANISM [COMMON NAMES]	SUBSTRATE(S)
Aldhomsa	Adrenoleukodystrophy protein [VLCFA-CoA, synthetase, ALDP]	*Homo sapiens* [human]	VLCFA-CoA synthetase?
Aldmusmu	Adrenoleukodystrophy protein [VLCFA-CoA synthetase homolog, ALDGH]	*Mus musculus* [mouse]	VLCFA-CoA synthetase?
Aldrmusmu	ALDR	*Mus musculus* [mouse]	VLCFA-CoA synthetase?
Pat2sacce	Peroxisomal long chain fatty acid import protein 2 [PAT2, PXA1, PAL1, SSH2, YPL147W, LPI1W]	*Saccharomyces cerevisiae* [yeast]	Long chain fatty acids
Pmp7homsa	70 kDa peroxisomal membrane protein [PXMP1, PMP70]	*Homo sapiens* [human]	Peroxisomal matrix enzymes

CODE	DESCRIPTION [SYNONYMS]	ORGANISM [COMMON NAMES]	SUBSTRATE(S)
Pmp7musmu	70 kDa Peroxisomal membrane protein [PMP70]	*Mus musculus* [mouse]	Peroxisomal matrix enzymes
Pmp7ratno	70 kDa Peroxisomal membrane protein [PMP70]	*Rattus norvegicus* [rat]	Peroxisomal matrix enzymes

Phylogenetic tree

Proteins listed subsequently in italics are at least 90% identical to the paired transporters listed in parenthesis and are therefore not included in the phylogenetic tree: *Aldmusmu* (Aldhomsa); *Pmp7musmu, Pmp7ratno* (Pmp7homsa).

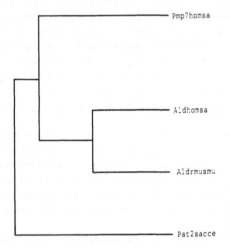

Proposed orientation of ALD [1] in the membrane

The model is based on predictions of membrane-spanning regions and α-helical content. The N-terminus of the protein is illustrated on the outside and is folded five times through the membrane. The predicted membrane-spanning helices are portrayed as rectangles. The numbers corresponding to the first and last residue of each membrane-spanning helix are boxed. Residues that are conserved in more than 75% of the aligned transporters (see below) are shown.

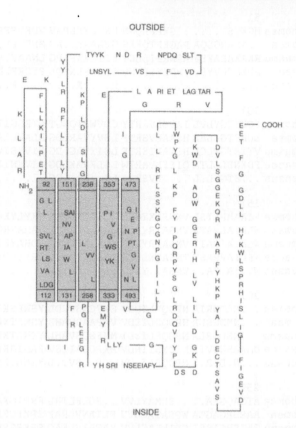

Physical and genetic characterics

	AMINO ACIDS	MOL. WT	EXPRESSION SITES	CHROMOSOMAL LOCUS
Aldhomsa	745	82 908	liver	Xq28
Aldmusmu	736	81 858	liver	
Pmp7homsa	659	75 475	liver	1p22–p21
Pmp7musmu	660	75 482	liver	
Pmp7ratno	659	75 315	liver	
Pat2sacce	758	86 955		Chromosome 16
Aldrmusmu	741	83 483		

Multiple amino acid sequence alignments

```
           1                                                50
Pmp7homsa .....MAAFS KYLTARNSSL AGAAFL.... ......LLCL LHKRRRALGL
Aldhomsa  .....MPVLS RPRPWRGNTL KRTAVLLALA AYGAHKVYPL VRQCL.....
Aldrmusmu MIHMLNAAAY RVKWTRSGAA KRAACLVA.A AYALKTLYPI IGKRLKQPGH
Pat2sacce ........MQ LDSGARIMYI PEVELVDRQS PDDNKFMNAT DKKKRKRIFI
Consensus .......... .....R.... ...A.L.... .......... ..K.......
```

```
          51                                                   100
Pmp7homsa HGKKS...... ..GKPPLQNN EKEGKKERAV VDKVFFSRLI QILKIMVPRT
Aldhomsa  ..APARGLQA PAGEPTQEAS GVAAAK..AG MNRVFLQRLL WLLRLLFPRV
Aldrmusmu RKAKAEAYSP AENREILHCT EIICKKPAPG LNAAFFKQLL ELRKILFPKL
Pat2sacce PPKDNDVYE. .HDKFLFKNV ELERAKNSQL FYSKFLNQMN VLSKILIPTV
Consensus .......... .......... E....K.... ....F...L. .L.KIL.P..

          101                                                  150
Pmp7homsa FCKETGYLVL IAVMLVSRTY CDVWMIQNGT LIESGIIGRS RKDFKRYLLN
Aldhomsa  LCRETGLLAL HSAALVSRTF LSVYVARLDG RLARCIARKD PRAFGWQLLQ
Aldrmusmu VTTETGWLCL HSVALISRTF LSIYVAGLDG KIVKSIVEKK PRTFIIKLIK
Pat2sacce FDKNFLLLTA QIFFLVMRTW LSLFVAKLDG QIVKNIIAGR GRSFLWDLGC
Consensus ...ETG.L.L ....LVSRT. LS..VA.LDG .I...I.... .R.F...L..

          151                                                  200
Pmp7homsa FIAAMPLISL VNNFLKYGLN ELKLCFRVRL TKYLYEEYLQ A.FTYYKMG.
Aldhomsa  WLLIALPATF VNSAIRYLEG QLALSFRSRL VAHAYRLYFS Q.QTYYRVS.
Aldrmusmu WLMIAIPATF VNSAIRYLEC KLALAFRTRL VDHAYETYFA N.QTYYKVI.
Pat2sacce WFLIAVPASY TNSAIKLLQR KLSLNFRVNL TRYIHDMYLD KRLTFYKLIF
Consensus W..IA.PA.. VNSAI.YL.. .L.L.FR.RL ....Y..Y.. ...TYYK...

          201                                                  250
Pmp7homsa ...NLDNRIA NPDQLLTQDV EKFCNSVVDL YSNLSKPFLD IVLYIFKLTS
Aldhomsa  ...NMDGRLR NPDQSLTEDV VAFAASVAHL YSNLTKPLLD VAVTSYTLLR
Aldrmusmu ...NMDGRLA NPDQSLTEDI MMFSQSVAHL YSNLTKPILD VILTSYTLIR
Pat2sacce DAKASNSVIK NIDNSITNDV AKFCDATCSV FANIAKPVID LIFFSVYLRD
Consensus ...N.D.R.. NPDQSLT.DV ..F..SV..L YSNL.KP.LD ....S..L..

          251                                                  300
Pmp7homsa AIGAQGPA.. ..SMMAYLVV ..SGLFLTRL RRPIGKMTIT EQKYEGEYRY
Aldhomsa  AARSRGAGTA WPSAIAGLVV FLTANVLRAF SPKFGELVAE EARRKGELRY
Aldrmusmu TATSRGASPI GPTLLAGLVV YATAKVLKAC SPKFGSLVAE EAHRKGYLRY
Pat2sacce NLGTVGVAGI ......FVNY FITGFILRKY TPPLGKLAGE RSASDGDYYN
Consensus .....G.... .......LVV ......L... .P..G.L..E E....G..RY

          301                                                  350
Pmp7homsa VNSRLITNSE EIAFYNGNKR EKQTVHSVFR KLVEHLHNFI LFRFSMGFID
Aldhomsa  MHSRVVANSE EIAFYGGHEV ELALLQRSYQ DLASQINLIL LERLWYVMLE
Aldrmusmu VHSRIIANVE EIAFYRGHKV EMKQLQKCYK ALAYQMNLIL SKRLWYIMIE
Pat2sacce YHLNMINNSE EIAFYQGTAV ERTKVKELYD VLMEKMLLVD KVKFGYNMLE
Consensus .HSR.I.NSE EIAFY.G..V E......Y. .L.....L... .R..Y.M.E

          351                                                  400
Pmp7homsa SIIAKYLATV VGYLVVSRPF L.......... .....DLSHP RHLKSTHSEL
Aldhomsa  QFLMKYVWSA SGLLMVAVPI ITATGYSESD AEAVKKAALE KKEEELVSER
Aldrmusmu QFLMKYVWSS CGLIMVAIPI ITATGFADGD LEDGPKQA.. .....MVSDR
Pat2sacce DYVLKYTWSG LGYVFASIPI VMSTLATGIN SEE....... .......KN
Consensus ....KY.WS. .G...V..PI .......... .......... ..........

          401                                                  450
Pmp7homsa LEDYYQSGRM LLRMSQALGR IVLAGREMTR LAGFTARITE LMQVLKDLNH
Aldhomsa  TEAFTIARNL LTAAADAIER IMSSYKEVTE LAGYTARVHE MFQVFEDVQR
Aldrmusmu TEAFTTARNL LASGADAIER IMSSYKEITE LAGYTARVYN MFWVFDEVKR
Pat2sacce MKEFIVNKRL MLSLADAGSR LMHSIKDISQ LTGYTNRIFT LLSVLHRVHS
Consensus .E........ L.....A..R I.....E.T. LAG.TAR... ...V......
```

```
            451                                          500
Pmp7homsa GKYE....RT MVSQQEK.... ...GIEGVQV IPLIPGAGEI IIADNIIKFD
Aldhomsa  CHFK....RP RELEDAQAGS GTIGRSGVRV EGPLKIRGQV VDVEQGIICE
Aldrmusmu GIYK....RT VT.QEPENHS KRGGNLELPL SDTLAIKGTV IDVDHGIICE
Pat2sacce LNFNYGAVPS ILSIRTEDAS RNSNLLPTTD NSQDAIRGTI QRNFNGIRLE
Consensus ........R. .......... ...G...... ....I.G.. .....GI..E

            501                                          550
Pmp7homsa HVPLATPN.. ...GDVLIRD LNFEVR.... .......... ..........
Aldhomsa  NIPIVTPS.. ...GEVVVAS LNIRVE.... .......... ..........
Aldrmusmu NVPIITPA.. ...GEVVASR LNFKVE.... .......... ..........
Pat2sacce NIDVIIPSVR ASEGIKLINK LTFQIPLHID PITSKSNSIQ DLSKANDIKL
Consensus N.P..TP... ...G.V.... LN..V..... .......... ..........

            551                                          600
Pmp7homsa .....SGANV LICGPNGCGK SSLFRVLGEL WPLF..GGRL TKPERGKLFY
Aldhomsa  .....EGMHL LITGPNGCGK SSLFRILGGL WPTY..GGVL YKPPPQRMFY
Aldrmusmu .....EGMHL LITGPNGCGK SSLFRILSGL WPVY..EGVL YKPPPQHMFY
Pat2sacce PFLQGSGSSL LILGPNGCGK SSIQRIIAEI WPVYNKNGLL SIPSENNIFF
Consensus ......G..L LI.GPNGCGK SSLFR.L..L WP.Y...G.L .KP.....FY

            601                                          650
Pmp7homsa VPQRPYMT.L GTLRDQVIYP DGREDQKRKG ISDLVLKEYL DNVQLGHILE
Aldhomsa  IPQRPYMS.V GSLRDQVIYP DSVEDMQRKG YSEQDLEAIL DVVHLHHILQ
Aldrmusmu IPQRPYMS.L GSLRDQVIYP DSADDMREKG YTDQDLERIL HSVHLYHIVQ
Pat2sacce IPQKPYFSRG GTLRDQIIYP MSSDEFFDRG FRDKELVQIL VEVKLDYLLK
Consensus IPQRPYMS.. G.LRDQVIYP DS..D...KG ..D..L..IL ...V.L.HI..

            651                                          700
Pmp7homsa REGG...WDS VQDWMDVLSG GEKQRMAMAR LFYHKPQFAI LDECTSAVSV
Aldhomsa  REGG...WEA MCDWKDVLSG GEKQRIGMAR MFYHRPKYAL LDECTSAVSI
Aldrmusmu REGG...WDA VMDWKDVLSG GEKQRMGMAR MFYHRPKYAL LDECTSAVSI
Pat2sacce RGVGLTYLDA IADWKDLLSG GEKQRVNFAR IMFHKPLYVV LDEATNAISV
Consensus REGG...WDA ..DWKDVLSG GEKQR..MAR .FYHKP.YA. LDECTSAVS.

            701                                          750
Pmp7homsa DVEGYIYSHC RKVGITLFTV SHRKSLWKHH EYYLHMDGRG NYEFKQ....
Aldhomsa  DVEGKIFQAA KDAGIALLSI THRPSLWKYH THLLQFDGEG GWKFEKLDSA
Aldrmusmu DVEGKIFQAA IGAGISLLSI THRPSLWKYH THLLQFDGEG GWRFEQLDTA
Pat2sacce DMEDYLFNLL KRYRFNFISI SQRPTLIKYH EMLLEIGENR DGKWQLQAVG
Consensus DVEG.IF... ...GI.L.SI .HRPSLWKYH ..LL..DG.G ...F......

            751                                          800
Pmp7homsa ....ITEDTV EFGS...... .......... .......... ..........
Aldhomsa  ARLSLTEEKQ RLEQQLAGIP KMQRRLQELC QILGEAVAPA HVPAPSPQGP
Aldrmusmu IRLTLSEEKQ KLESQLAGIP KMQQRLNELC KILGEDSVLK TIQTPEKTS.
Pat2sacce TDEAITSIDN EIEEELERKLE RVKGWEDERT KLREKLEII. ..........
Consensus .....TE... ..E....... .......... .......... ..........

            801
Pmp7homsa ........
Aldhomsa  GGLQGAST
Aldrmusmu ........
Pat2sacce ........
Consensus ........
```

Proteins listed subsequently in italics are at least 90% identical to the paired transporters listed in parenthesis and are therefore not included in the alignment: *Aldmusmu* (Aldhomsa); *Pmp7musmu, Pmp7ratno* (Pmp7homsa). Residues listed in the consensus sequence are present in at least 75% of the aligned transporter sequences. Residues indicated in boldface type are also conserved in at least one other family of the ABC transporter superfamily.

Database accession numbers

	SWISSPROT	PIR	EMBL/GENBANK
Aldhomsa	P33897	S30059	Z21876; G38591
Aldmusmu	P48410		Z33637; G520955
Aldrmusmu			Z48670
Pat2sacce	P41909		U17065; G619668
Pmp7homsa	P28288	S20313	M81182; G190129
Pmp7musmu			L28836
Pmp7ratno	P16970	JS0371; A35723	D90038; G220862

References
1 Mosser, J. et al. (1993) Nature 361, 726–730.
2 Aubourg, P. et al. (1993) Biochimie 75, 293–302.
3 Kamijo, K. et al. (1990) J. Biol. Chem. 265, 4534–4540.
4 **Moser, H.W. and Moser, A. (1989) In The Metabolic Basis of Inherited Disease (Scriver, C.L. et al., eds). McGraw-Hill, New York, pp. 1511–1532.**
5 Gartner, J. et al. (1992) Nature Genet. 1, 16–23.
6 **Higgins, C.F. (1992) Annu. Rev. Cell Biol. 8, 67–113.**
7 Higgins, C.F. et al. (1982) Nature 298, 723–727.

Part 4

ABC-2 Transporters

Summary

Members of the ABC-2 nodulation protein family, the example of which is the NODJ nodulation protein from *Azorhizobium*[1] (Nodjazoca) mediate export of modified oligosaccharides – probably modified β-1,4-linked *N*-acetylglucosamine oligosaccharides. With the ABC transporter protein NODI[1] they comprise a membrane transport complex involved in the nodulation process. Members of the family have been found only in gram-negative bacteria.

Statistical analysis of multiple amino acid sequence comparisons indicates that the ABC-2 nodulation protein family, with the polysaccharide export family, comprise a superfamily which has been termed ABC-2-type transporters[2]. These proteins are not homologous to any other families of ATP binding cassette (ABC) transporters, although their transport mechanism is believed to involve bound ATP and to be similar to that of other ABC transporters[2]. These two families share several amino acid sequence motifs[2,3]. They are both predicted to contain six membrane-spanning helices by the hydropathy of their amino acid sequences.

Many residues, including several long sequence motifs, are well conserved within the ABC-2 nodulation protein family, including motifs unique to the family, signature motifs of the ABC-2 superfamily, and motifs necessary for function by the criterion of site-directed mutagenesis.

Nomenclature, biological sources and substrates

CODE	DESCRIPTION [SYNONYMS]	ORGANISM [COMMON NAMES]	SUBSTRATE(S)
Nodjazoca	Nodulation protein J [NODJ]	*Azorhizobium caulinodans* [gram-negative bacterium]	Modified oligosaccharide
Nodjbraja	Nodulation protein J [NODJ]	*Bradyrhizobium japonicum* [gram-negative bacterium]	Modified oligosaccharide
Nodjrhiga	Nodulation protein J [NODJ]	*Rhizobium galegae* [gram-negative bacterium]	Modified oligosaccharide
Nodjrhile	Nodulation protein J [NODJ]	*Rhizobium leguminosarum* [gram-negative bacterium]	Modified oligosaccharide
Nodjrhilt	Nodulation protein J [NODJ]	*Rhizobium leguminosarum* [gram-negative bacterium]	Modified oligosaccharide

Phylogenetic tree

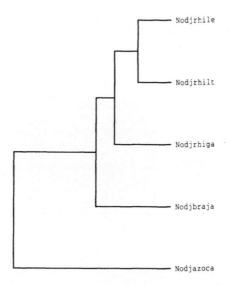

Nodjrhile

Nodjrhilt

Nodjrhiga

Nodjbraja

Nodjazoca

Proposed orientation of NODJ [1] in the membrane

The model is based on predictions of membrane-spanning regions and α-helical content. The N-terminus of the protein is illustrated on the inside and is folded six times through the membrane. The predicted membrane-spanning helices are portrayed as rectangles. The numbers corresponding to the first and last residue of each membrane-spanning helix are boxed. Residues that are conserved in more than 75% of the aligned transporters (see below) are shown.

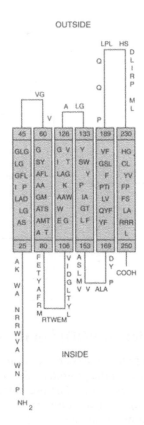

Physical and genetic characteristics

	AMINO ACIDS	MOL. WT	CHROMOSOMAL LOCUS
Nodjazoca	254	27 701	nod locus 1
Nodjbraja	262	28 194	nod gene cluster
Nodjrhiga	168	17 825	nod box
Nodjrhile	259	27 698	Plasmid pRL1JI
Nodjrhilt	262	28 033	nodIJ region

Multiple amino acid sequence alignments

```
          1                                                  50
Nodjrhile ...MGVATLP AGGLNWLAVW RRNYLAWKKA ALASILGNLA DPVIYLFGLG
Nodjrhilt MSGDSVTALP GGSLNWIAVW RRNYIAWKKA ALASLLGHLA EPLIYLFGLG
Nodjrhiga MGRVSTEALP SGLLNWVAVW RRNFLAWKKV APASLLGNLA DPMIYIFGLG
Nodjbraja MDDGYASVMP ANAYNWTAVW RRNYLAWRKV ALASLLGNLA DPITNLFGLG
Nodjazoca ........MR ERMVTWAAVF QRNAMSWRRE MAASVLGSVI DPLIMLFGLG
Consensus .........P ....NW.AVW RRN..AW.K. A.AS.LG.LA DP.I.LFGLG
```

```
          51                                                              100
Nodjrhile AGLGVMVGRV DGVSYTAFLA AGMIATSAMT AATFETIYAA FGRMQGQRTW
Nodjrhilt AGLGVMVGRV GGVSYTAFLA AGMVATSAMT AATFETIYAA FGRMEGQRTW
Nodjrhiga SGLGVMLGNV GGVSYSAFLA AGMVATSAMT ASTFETIYAT FARMRDHRTW
Nodjbraja FGLGLIVGRV EGTSYIAFLA AGMVAISAMT SATFETLYAA FARMDVKRTW
Nodjazoca VGLGKIVDSV DGRSYAEFLA CGLILTSAMS ASNYEMLYGT YSRIYVTGTL
Consensus .GLG..VG.V .G.SY.AFLA AGM.ATSAMT A.TFET.YA. F.RM...RTW

          101                                                             150
Nodjrhile EAMLYTQLTQ GDIVVGEMAW AATKASLAGT GIGIVAAMLG YTHWLALLYA
Nodjrhilt EAMLYTQLRL GDIVLGEMAW AATKAALAGA GIGVVAAALG YTQWLSLLYA
Nodjrhiga EAMLYTKLTL GDIVLGEMAW AATKASLAGT AIGIVTATLA YSEWDSLIYV
Nodjbraja EGILFTQLTL GDIVLGELVW AASKSVLAGT AIGIVAATLG YASWTSVLCA
Nodjazoca KSMRYAPICV SDYLIGEVLW AAYEGVVAGT IVAVCTAFLG YIPGWSVIYI
Consensus E.MLYT.L.. GDIV.GE..W AA.K..LAGT .IG.V.A.LG Y..W.S..Y.

          151                                                             200
Nodjrhile LPVIAITGLA FASLGMVVTA LAPSYDYFIF YQTLVITPML FLSGAVFPVD
Nodjrhilt LPVIALTGLA FASLGMVVTA LAPSYDYFIF YQTLVITPIL FLSGAVFPVD
Nodjrhiga FPVIALTGLA FASLSMVVAA LAPSYDYLVF YQSLVITPML VLSGSVFPVE
Nodjbraja IPTIALTGLV FASLAMVVIS LAPTYDYFVF YQSLVLTPMV FLCGAVFPTS
Nodjazoca LPDILFVALI FSSTSLLVAA ISRGYALFAF YQSIAIAPLV FLSGVIVPRF
Consensus .P.IA.TGL. FASL.MVV.A LAP.YDYF.F YQ.LVITP.. FLSG.VFP..

          201                                                             250
Nodjrhile QLPVAFQQIA AFLPLAHSID LIRPTMLGQP IANVCLHIGV LCIYIVVPFL
Nodjrhilt QLPIVFQTAA RFLPLSHSID LIRPIMLGHP VVDVCQHVGA LCIYIVIPFF
Nodjrhiga QLSPMLQRIT HLLPLAHSID LIRPAMLGHP VPDITLHLGA LCLYIVLPFF
Nodjbraja QMPDSFQHFA GLLPLAHSVD LIRPVMLERG ADNAALHVGA LCVYAVLPFF
Nodjazoca TGNDVISGMI HFSPLYRAVN DVRNVVYEGR GTQVGPLLLL SLLYASVMVF
Consensus Q.....Q.. ..LPL.HS.D LIRP.ML... ......H.G. LC.Y.V.PFF

          251       263
Nodjrhile VSTALLRRRL MR.
Nodjrhilt LSTALLRRRL LR.
Nodjrhiga VSIALLRRRL TQ.
Nodjbraja ASIALFRRRL LR.
Nodjazoca ISAKVICVRL DD.
Consensus .S.AL.RRRL ...
```

Residues listed in the consensus sequence are present in at least 75% of the aligned transporter sequences. Residues indicated in boldface type are also conserved in the other family of the ABC-2 transporter superfamily.

Database accession numbers

	SWISSPROT	PIR	EMBL/GENBANK
Nodjazoca	Q07757	S35008	L18897; G310298
Nodjbraja	P26025	S27497	J03685; G152118
Nodjrhiga			X87578
Nodjrhile	P06755	B24400; S10231	Y00548; G46213
Nodjrhilt	P24144	S08617	X51411; G46244

References
[1] Geelen, D. et al. (1993) Mol. Microbiol. 9, 145–154.
[2] **Reizer, J. et al. (1992) Protein Sci. 1, 1326–1332.**
[3] Vazquez, M. et al. (1993) Mol. Microbiol. 8, 369–377.

Summary

Transporters of the ABC-2 polysaccharide exporter family [1], examples of which are the BEXB capsular polysaccharide exporters of *Haemophilus influenzae* [2] (Bexbhaein) and the polysialic acid transporter of *Escherichia coli*, KPSM [3] (Kpm1escco), mediate polysaccharide export. KPSM comprises a membrane transport complex for the transport of polysialic acid across the membrane with KPST; BEXB comprises a transport complex with BEXA. These transporters are only found in gram-negative bacteria.

Statistical analysis of multiple amino acid sequence comparisons indicates that the ABC-2 polysaccharide exporter family, with the nodulation transporter family, comprise a superfamily which has been termed ABC-2-type transporters [4-6]. These proteins are not homologous to any other families of ATP binding cassette (ABC) transporters, although their transport mechanism is believed to involve bound ATP and to be similar to that of other ABC transporters [2]. Members of the ABC-2 polysaccharide exporter family are predicted to contain six membrane-spanning helices by the hydropathy of their amino acid sequences.

Many residues, including several long sequence motifs, are well conserved within the ABC-2 polysaccharide exporter family, including motifs unique to the family and signature motifs of the ABC-2 superfamily [4-6].

Nomenclature, biological sources and substrates

CODE	DESCRIPTION [SYNONYMS]	ORGANISM [COMMON NAMES]	SUBSTRATE(S)
Bex1haein	Capsule polysaccharide export inner-membrane protein [BEXB]	*Haemophilus influenzae* [gram-negative bacterium]	Capsule polysaccharide
Bex2haein	Capsule polysaccharide export inner-membrane protein [BEXB]	*Haemophilus influenzae* [gram-negative bacterium]	Capsule polysaccharide
Bex3haein	Capsule polysaccharide export inner-membrane protein [BEXB]	*Haemophilus influenzae* [gram-negative bacterium]	Capsule polysaccharide
Ctrcneime	Capsule polysaccharide export inner-membrane protein [BEXB, CTRC]	*Neisseria meningitidis* [gram-negative bacterium]	Capsule polysaccharide
Kpm1escco	Polysialic acid transport protein [KPSM]	*Escherichia coli* [gram-negative bacterium]	Polysialic acid
Kpm2escco	Polysialic acid transport protein [KPSM]	*Escherichia coli* [gram-negative bacterium]	Polysialic acid
Vexbsalti	Vi polysaccharide export protein [VEXB]	*Salmonella typhi* [gram-negative bacterium]	Vi polysaccharide

Phylogenetic tree

Proteins listed subsequently in italics are at least 90% identical to the paired transporters listed in parenthesis and are therefore not included in the phylogenetic tree: *Kpm2escco* (Kpm1escco); *Bex2haein, Bex3haein* (Bex1haein).

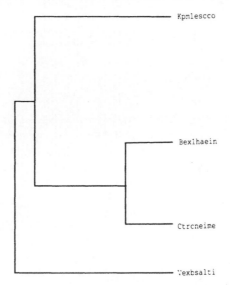

Proposed orientation of KPM1 [3] in the membrane

The model is based on predictions of membrane-spanning regions and α-helical content. The N-terminus of the protein is illustrated on the inside and is folded six times through the membrane. The predicted membrane-spanning helices are portrayed as rectangles. The numbers corresponding to the first and last residue of each membrane-spanning helix are boxed. Residues that are conserved in more than 75% of the aligned transporters (see below) are shown.

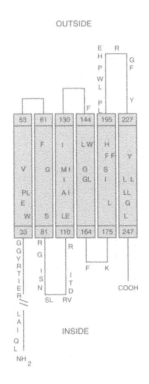

Physical and genetic characteristics

	AMINO ACIDS	MOL. WT	CHROMOSOMAL LOCUS
Bex1haein	265	30 181	capsulation locus
Bex2haein	265	30 108	capsulation locus
Bex3haein	265	30 195	capsulation locus
Ctrcneime	265	30 168	capsulation locus
Kpm1escco	258	29 557	polysialic acid gene cluster
Kpm2escco	258	29 561	K5 antigen gene cluster
Vexbsalti	264	30 429	Plasmid pGBM124

Multiple amino acid sequence alignments

```
          1                                                  50
Kpm1escco ........MA RSGFEVQKVT VEALFLREIR TRFGKFRLGY LWAILEPSAH
Bex1haein .MQYGDKTTF KQSLAIQGRV INALLMREII TRYGRQNIGF FWLFVEPLLM
Ctrcneime .MKALHKTSF WESLAIQRRV IGALLMREII TRYGRNNIGF LWLFVEPLLM
Vexbsalti MNILKNNSYY FMKLITVCEL IILLMSRDIK TRYNGNLLNY MMVLAVPLVW
Consensus .......... ...L..Q... I.AL..REI. TRYG....G. .W...EPL..

          51                                                 100
Kpm1escco LLILLGILGY VMHRTMPDIS FPVFLLNGLI PFFIFSSISK RSIGAIEANQ
Bex1haein TFFIVMMWKF IRADKFSTLN MIAFVMTGYP MAMMWRNASN RAIGSISANL
Ctrcneime TFVIVLMWKF LKADRYSTLN IVAFAITGYP MLMMWRNASK RAVGSISSNA
Vexbsalti ISITVISFQY LNRSVPISTD DISFVIAGIL PYLLFRYTIT ATMRTHSFST
Consensus ....V..... .......... ...F...G.. .......... S.R..G.IS.N.
```

```
           101                                                    150
Kpm1escco GLFNYRPVKP IDTIIARALL ETLIYVAVYI LLMLIVWMTG EYFEITNFLQ
Bex1haein SLLYHRNVRV LDTIFTRVLL EVAGASIAQI LFMAILVMID WIDAPHDVFY
Ctrcneime SLLYHRNVRV LDTILARMIL EIAGATIAQI VIMAVLIAIG WIEMPADMFY
Vexbsalti SLAVVSQVKK RHVIFSLAAI EFVNAVIIYI IISLINFLIF SRWEAQKPFL
Consensus SL...R.V.. .DTI..R..L E...A.I..I ..M.I...I .........F.

           151                                                    200
Kpm1escco LVLTWSLLII LSCGVGLIFM VVGKTFPEMQ KVLPILLKPL YFISCIMFPL
Bex1haein MLIAWFLMAM FAFGLGLIIC AIAQQFDVFG KIWGTLSFVL LPISGAFFFV
Ctrcneime MLMAWLLMAF FAIGLGLVIC SIAFNFEPFG KIWGTLTFVM MPLSGAFFFV
Vexbsalti IFEGMVIAWL LGLSFGYFCD ALSERFPLVY KAVPVMLRPM FLISAVFYTA
Consensus ....W.L.. ...G.GL... .....F.... K....L..... ..IS..FF..

           201                                                    250
Kpm1escco HSIPKQYWSY LLWNPLVHVV ELSREAVMPG YISEGVSLNY LAMFTLVTLF
Bex1haein HNLPAQAQSI ALWFPMIHGT EMFRHGYFGD TVVTYESIGF LVVSDLALLL
Ctrcneime HNLPPKVQEY ALMIPMVHGT EMFRAGYFGS DVITYENPWY IVLCNLVLLL
Vexbsalti NELPYSLLSI FSWNPLLHAN EIVREGMFEG YHSLYLEPFY PLAFSATLFL
Consensus H.LP...... .LW.P..H.. E..R.G.F.. ....Y....Y .....L.LLL

           251      266
Kpm1escco IGLALYRTRE EAMLTS
Bex1haein LGLVMVKNFS KGVEPQ
Ctrcneime FGLAMVSKFS KGVEPQ
Vexbsalti AGLIFHLICD TENH..
Consensus .GL..........
```

Residues listed in the consensus sequence are present in at least 75% of the aligned transporter sequences. Proteins listed subsequently in italics are at least 90% identical to the paired transporters listed in parenthesis and are therefore not included in the alignment: *Kpm2escco* (Kpm1escco); *Bex2haein, Bex3haein* (Bex1haein). Residues indicated in boldface type are also conserved in the other family of the ABC-2 transporter superfamily.

Database accession numbers

	SWISSPROT	PIR	EMBL/GENBANK
Bex1haein	P19390		M33787; G148869
Bex2haein	P19391		M33788; G148871
Bex3haein	P22235	S12234; BWHIXB	X54987; G45299
Ctrcneime	P32015	S15222	M57677; 50252
Kpm1escco	P23889	A42469	M57382; G146565
Kpm2escco	P24584	S12236	X53819; G41878
Vexbsalti	P43109		D14156; G426450

References
1 Frosch, M. et al. (1991) Mol. Microbiol. 5, 1251–1263.
2 Kroll, J.S. and Moxon, E.R. (1990) J. Bacteriol. 172, 1374–1379.
3 Pavelka, M.S. et al. (1991) J. Bacteriol. 173, 4603–4610.
4 Geelen, D. et al. (1993) Mol. Microbiol. 9, 145–154.
5 **Reizer, J. et al. (1992) Protein Sci. 1, 1326–1332.**
6 Vazquez, M. et al. (1993) Mol. Microbiol. 8, 369–377.

Summary

Most transporters of the ABC-2-associated (cytoplasmic) protein family, examples of which are the ATP binding protein NODI in *Azorhizobium caulinodans*[1] (Nodiazoca), and BEXA[2] in *Haemophilus influenzae* (Bexahaein) and CTRD[3] in *Neisseria meningitidis* (Ctrdneime) export oligo- or polysaccharides, spermidine, putrescine, teichoic acid and sulfate; the substrates of a few members are unknown. One member of the family, the TNRB2 protein[4] of *Streptomyces longisporoflavus* (Tnrbstrlo), confers resistance to the antibiotic tetronasin. NODI is involved in the nodulation process by an unknown mechanism. These transporters are only found in prokaryotes, but are found in both gram-positive and gram-negative bacteria.

Statistical analysis of multiple amino acid sequence comparisons places the ABC-2-associated (cytoplasmic) protein family in the ATP binding cassette (ABC) superfamily[5]. Proteins in this transporter superfamily use the energy of ATP hydrolysis to pump substrates across cell membranes. The family is characterized by their cytoplasmic ATP binding domains[5], which exist as separate chains and are described in the following tables. In many, but not all, of these proteins the ATP binding domains are associated with transmembrane proteins of the ABC-2 superfamily[6]: NODI is associated with NODJ, KPST with KPSM and BEXA with BEXB. In other proteins in this family, such as ABCA from *Aeromonas salmonicida*[7] (Abcaaersa), the associated transmembrane protein is unknown. Transmembrane proteins in the ABC-2 superfamily, which includes the nodulation protein family and the polysaccharide export family, are predicted to contain six membrane-spanning helices by the hydropathy of their amino acid sequences.

A small number of amino acids, scattered throughout the whole length of the sequences, are conserved within the ABC-2 family, including motifs unique to the family, signature motifs of the ABC superfamily, and motifs necessary for function by the criterion of site-directed mutagenesis.

Nomenclature, biological sources and substrates

CODE	DESCRIPTION [SYNONYMS]	ORGANISM [COMMON NAMES]	SUBSTRATE(S) [RESISTANCE][a]
Abcaaersa	Abca protein [ABCA]	*Aeromonas salmonicida* [gram-negative bacterium]	Unknown
Bexahaein	ATP binding protein [BEXA]	*Haemophilus influenzae* [gram-negative bacterium]	Capsular polysaccharide
Ctrdneime	Capsule polysaccharide export ATP binding protein C [CTRD]	*Neisseria meningitidis* [gram-negative bacterium]	Capsular polysaccharide
Glyustrli	tRNA-GlyU	*Streptomyces lividans* [gram-negative bacterium]	Unknown
Kst1escco	Polysialic acid transport ATP binding protein [KPST]	*Escherichia coli* [gram-negative bacterium]	Polysialic acid
Kst5escco	Polysialic acid transport ATP binding protein [KPST]	*Escherichia coli* [gram-negative bacterium]	Polysialic acid
Natabacsu	ATP binding transport protein [NATA]	*Bacillus subtilis* [gram-positive bacterium]	Unknown

CODE	DESCRIPTION [SYNONYMS]	ORGANISM [COMMON NAMES]	SUBSTRATE(S)
Nodescco	Nodulation protein homolog	Escherichia coli [gram-negative bacterium]	
Nodiazoca	Nodulation ATP binding protein I [NODI]	Azorhizobium caulinodans [gram-negative bacterium]	Modified oligosaccharide
Nodibraja	Nodulation ATP binding protein I [NODI]	Bradyrhizobium japonicum [gram-negative bacterium]	Modified oligosaccharide
Nodirhiga	Nodulation ATP binding protein I [NODI]	Rhizobium galegae [gram-negative bacterium]	Modified oligosaccharide
Nodirhilo	Nodulation ATP binding protein I [NODI]	Rhizobium loti [gram-negative bacterium]	Modified oligosaccharide
Nodirhile	Nodulation ATP binding protein I [NODI]	Rhizobium leguminosarum [gram-negative bacterium]	Modified oligosaccharide
Nodirhime	Nodulation protein [NODI]	Rhizobium meliloti [gram-negative bacterium]	Modified oligosaccharide
Nosfpsest	Copper transport ATP binding protein [NOSF]	Pseudomonas stutzeri [gram-negative bacterium]	Copper
Potgescco	Putrescine transport ATP binding protein [POTG]	Escherichia coli [gram-negative bacterium]	Putrescine
Sppaescco	Spermidine/putrescine transport protein A	Escherichia coli [gram-negative bacterium]	Spermidine putrescine
Sulfsynsp	Sulfate transport protein	Synechococcus sp. [cyanobacterium]	Sulfate
Taghbacsu	Teichoic acid translocation ATP binding protein [TAGH]	Bacillus subtilis [gram-negative bacterium]	Teichoic acid
Tnrbstrlo	TnrB2 protein	Streptomyces longisporoflavus [gram-negative bacterium]	[Tetronasin]
Vexcsalti	Vi polysaccharide export ATP binding protein [VEXC]	Salmonella typhi [gram-negative bacterium]	Vi polysaccharide

[a] Presumed substrates; protein confers resistance to specified compounds.

Phylogenetic tree

Proteins listed subsequently in italics are at least 90% identical to the paired transporters listed in parenthesis and are therefore not included in the phylogenetic tree: *Nodirhime* (Nodirhilo).

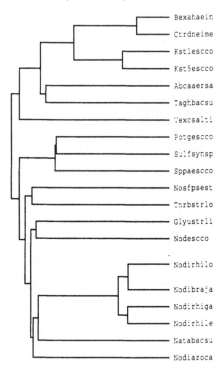

Physical and genetic characteristics

	AMINO ACIDS	MOL. WT	CHROMOSOMAL LOCUS
Abcaaersa	308	34 015	
Bexahaein	217	24 746	cap
Ctrdneime	216	24 595	
Glyustrli	326	34 954	
Kst1escco	219	24 939	kps gene cluster
Kst5escco	224	25 481	
Natabacsu	246	27 878	
Nodescco	308	34 647	2.4–4.1 minutes
Nodiazoca	320	35 310	nod operon
Nodibraja	306	34 127	nod gene cluster
Nodirhiga	347	38 435	nod box
Nodirhilo	339	37 264	
Nodirhile	311	34 300	Plasmid pRL1JI
Nodirhime	339	37 264	
Nosfpsest	308	33 777	
Potgescco	404	44 784	19 minutes
Sppaescco	378	43 028	15 minutes
Sulfsynsp	344	38 476	

	AMINO ACIDS	MOL. WT	CHROMOSOMAL LOCUS
Taghbacsu	527	59 243	310°
Tnrbstrlo	300	32 278	
Vexcsalti	246	27 368	Plasmid pGBM124

Multiple amino acid sequence alignments

```
            1                                                50
Bexahaein   ..............................................MIR
Ctrdneime   ..............................................MIS
Kstlescco   ..............................................MIK
Kst5escco   ..............................................MIK
Abcaaersa   ..........................................MSEPVLA
Taghbacsu   ...........................................MKLKVS
Vexcsalti   ...............................................MT
Potgescco   ..MPEGRTTP AGNSHHYGAL AHIQCRRAAV NDAIPRPQAK TRKALTPLLE
Sulfsynsp   ........................................MPKDKAVGIQ
Sppaescco   .......................MGQSKKLNK QPSSLSPLVQ
Nosfpsest   .........................................MNAVE
Tnrbstrlo   ..................................................
Glyustrli   ................................M QTNEHEHVIE
Nodescco    .........................................MTIALE
Nodirhilo   .......MK RKLGPEELRR LETPAIERES HGQTSAKSSV PDSASTVAVD
Nodibraja   ..........................................MNMSNMAID
Nodirhiga   MGENMEREML RPKTIAMDQN SASARSNPER EIKTGRLEPA SNSAPTMAID
Nodirhile   ...............................MDSP SGSLSPVAID
Natabacsu   ..............................................MIT
Nodiazoca   ..........................MMLMRE SPDDRHYTVS
Consensus   ..................................................

            51                                               100
Bexahaein   VNNVCKKYH. ..............TNSGW KTVLKNINFE LQKGEKIGIL
Ctrdneime   VEHVSKQYQ. ............MRGGM RTVLDDINFS LQKGEKVGIL
Kstlescco   IENLTKSYR. ..............TPTGR HYVFKNLNII FPKGYNIALI
Kst5escco   IENLTKSYR. ..............TPTGR HYVFKDLNIE IPSGKSVAFI
Abcaaersa   VSGVNKSFPI YRSPWQALWH ALNPKADVKV FQALRDIELT VYRGETIGIV
Taghbacsu   FRNVSKQYHL YKKQSDKIKG LFFP.AKDNG FFAVRNVSFD VYEGETIGFV
Vexcsalti   LKITDFSFRR DAFVFGLLGC TRYFESDKGP RVVLDKTDFV MGYHEHIGIL
Potgescco   IRNLTKSYD. ..............GQ... .HAVDDVSLT IYKGEIFALL
Sulfsynsp   VSQVSKQFG. ..............SF... .QAVKDVDLT VETGSLVALL
Sppaescco   LAGIRKCFD. ..............GK... .EVIPQLDLT INNGEFLTLL
Nosfpsest   IQGVSQRYG. ..............SM... .TVLHDLNLN LGEGEVLGLF
Tnrbstrlo   MAGLHKSFG. ..............RT... .HALDGLDLA VDSGEVHGFL
Glyustrli   VTDLRRVYG. ..............G...G FEAVRGIDFS VRRGEVFALL
Nodescco    LQQLKKTYP. ..............G...G VQALRGIDLQ VEAGDFYALL
Nodirhilo   FAGVTKSYG. ..............NK... .IVVDELSFS VASGECFGLL
Nodibraja   LVGVRKSFG. ..............DK... .VIVNDLSFS VARGECFGLL
Nodirhiga   LQAVTMIYR. ..............DK... .TVVDSLSFG VRAGECFGLL
Nodirhile   LAGVSKSYG. ..............GK... .IVVNDLSFT IAAGECFGLL
Natabacsu   LTDCSRRFQ. ..............DKKKV VKAVRDVSLT IEKGEVVGIL
Nodiazoca   ATGVWK.... ..............KRGG IDVLRGLDMY VRRGERYGIM
Consensus   .....K.... ..............................GE.....
```

```
          101                                                    150
Bexahaein GRNGAGKSTL IRLMSGVEPP TSGTIERSM. .......... .SISWPLAFS
Ctrdneime GRNGAGKSTL VRLISGVEPP TSGEIKRTM. .......... .SISWPLAFS
Kst1escco GQNGAGKSTL LRIIGGIDRP DSGNIITEH. .......... .KISWPVGLA
Kst5escco GRNGAGKSTL LRMIGGIDRP DSGKIITNK. .......... .TISWPVGLA
Abcaaersa GHNGAGKSTL LQLITGVMQP DCGQITRTG. .......... .RVVGLLELG
Taghbacsu GINGSGKSTM SNLLAKIIPP TSGEIEMNG. .......... .Q.PSLIAIA
Vexcsalti AAPGSGKTTL TRLLCGLDAP DEGDFIGLR. .......... .GDALPLGAN
Potgescco GASGCGKSTL LRMLAGFEQP SAGQIMLDGV DLSQVPPYLR PINMMF..QS
Sulfsynsp GPSGSGKSTL LRLIAGLEQP DSGRIFLTGR DATNESVRDR QIGFVF..QH
Sppaescco GPSGCGKTTV LRLIAGLETV DSGRIMLDNE DITHVPAENR YVNTVF..QS
Nosfpsest GHNGAGKTTS MKLILGLLSP SEGQVKVLGR AP..NDPQVR RQLGYL.PEN
Tnrbstrlo GPNGAGKSTT IRVLLGLLRA DSGAPSSSAR TPWKDAVALH RRLAYV.PGD
Glyustrli GTNGAGKTST VELLEGLAAP AGGRVRVLGH DPYTERAAVR PRTGVM.LQE
Nodescco  GPNGAGKSTT IGIISSLVNK TSGRVSVFGY DLEKDVVNAK RQLGLV.PQE
Nodirhilo GPNGAGKSTI ARMLLGMTCP DAGAITVLGV PVPARARLAR RGIGVV.PQF
Nodibraja GPNGAGKSTI ARMLLGMISP DRGKITVLDE PVPSRARAAR VAVGVV.PQF
Nodirhiga GPNGAGKSTI TRMLLGMATP SAGKISVLGL PVPGKARLAR ASIGVV.SQF
Nodirhile GPNGAGKSTI TRMILGMTSP SVGKITVLGA QEPGQVRLAR AKIGIV.SQF
Natabacsu GENGAGKTTM LRMIASLLEP SQGVITVDGF DTVKQPAEVK QRIGVLFGGE
Nodiazoca GTNGAGKSSL INIILGLTPP DRGTVTVFGK DMRRQGHLAR ARIGVV.PQD
Consensus G.NGAGKST. .....G...P ...G..... .......... ..........


          151                                                    200
Bexahaein GAFQGSLTGM DNLRFICRLY DV.DPDYVTR ..FTKEFSEL GDYLYEPVKK
Ctrdneime GAFQGSLTGM DNLRFICRIY NV.DIDYVKA ..FTEEFSEL GQYLYEPVKR
Kst1escco GGFQGSLTGR ENVKFVARLY AK.RDELNER VDFVEEFSEL GKYFDMPIKT
Kst5escco GGFQGSLTGR ENVKFVARLY AK.QEELKEK IEFVEEFAEL GKYFDMPIKT
Abcaaersa SGFNPEFTGR ENIFFNGAIL GMSQREMDDR LERILSFAAI GDFIDQPVKN
Taghbacsu AGLNNQLTGR DNVRLKCLMM GLTNKEIDDM YDSIVEFAEI GDFINQPVKN
Vexcsalti SFILPGLTGE ENARMMASLY GLDGDEFS.. .HFCYQLTQL EQCYTDRVSE
Potgescco YALFPHMTVE QNIAFGLKQD KLPKAEIASR VNEMLGLVHM QEFAKRKPHQ
Sulfsynsp YALFKHLTVR KNIAFGLELR KHTKEKVRAR VEELLELVQL TGLGDRYPSQ
Sppaescco YALFPHMTVF ENVAFGLRMQ KTPAAEITPR VMEALRMVQL ETFAQRKPHQ
Nosfpsest VTFYPQLSGR ETLRHFARLK GAALTQ.... VDELLEQVGL AHAADRRVKT
Tnrbstrlo VELWPNLTGG EAIDLLSRLR GGLDRQ...R RDELIERFDL ..DPTKKGRA
Glyustrli GGFPSELTVA ETARMWAGCV SGARPPAEV. ....LALVGL EAKSGTRVKQ
Nodescco  FNFNPFETVQ QIVVNQAGYY GVERKEAYIR SEKYLKQLDL WGKRNERARM
Nodirhilo DNLDQELTVR ENLLVFGRYF GMSTRQSEAV IPSLLEFARL ERKANARVSE
Nodibraja DNLEPEFTVR ENLLVFGRYF GMSARTIEAV VPSLLEFARL ESKADVRVSL
Nodirhiga DNLDMEFTVR ENLLVFGRYF QMSTRAIEKL IPSLLEFAQL EAKADVRVSD
Nodirhile DNLDLEFTVR ENLLVYGRYF RMSTREIETV IPSLLEFARL ESKANTRVAD
Natabacsu TGLYDRMTAK ENLQYFGRLY GLNRHEIKAR IEDLSKRFGM RDYMNRRVGG
Nodiazoca DCLEQNMTPY ENVMLYGRLC RMTASEARKR ADQIFEKFSM MDCANRPVRL
Consensus .......T.. .N........ .......... ..........L ..........


          201                                                    250
Bexahaein YSSGMKARLA FALSLSVEFD CYLIDEVIAV GDSRFAEKCK YELFE.KRK.
Ctrdneime YSSGMKARLA FALSLAVEFD CYLIDEVIAV GDSRFADKCK YELFE.KRK.
Kst1escco YSSGMRSRLA FGLSMAFKFD YYLIDEITAV GDAKFKKKCS DIFDK.IRE.
Kst5escco YSSGMRSRLG FGLSMAFKFD YYIVDEVTAV GDARFKEKCA QLFKE.RHK.
Abcaaersa YSSGMMVRLA FSVIINTDPD VLIIDEALAV GDDAFQRKCY ARLKQLQSQ.
Taghbacsu YSSGMKSRLG FAISVHIDPD ILIIDEALSV GDQTFYQKCV DRINEFKKQ.
Vexcsalti YSVTMKTHLA FAINLLLPCR LYIADGKLYT GDNATQLRMQ AAL.ACQLQ.
```

```
Potgescco LSGGQRQRVA LARSLAKRPK LLLLDEPMGA LDKKLRDRMQ LEVVDILERV
Sulfsynsp LSGGQRQRVA LARALAVQPQ VLLLDEPFGA LDAKVRKDLR SWLRKLHDEV
Sppaescco LSGGQQQRVA IARAVVNKPR LLLLDESLSA LDYKLRKQMQ NELKALQRKL
Nosfpsest YSKGMRQRLG LAQALLGEPR LLLLDEPTVG LDPIATQDLY LLIDRLRQR.
Tnrbstrlo YSKGNRQKVP IVAALASDAE LLLLDEPTAG LDPLMEVVFQ DVILRAKAA.
Glyustrli LSGGQRRRLD LALALLGDPE VLFLDEPSTG LDAEGRRDTW ELVGALRDQ.
Nodescco  LSGGMKRRLM IARALMHEPK LLILDEPTAG VDIELRRSMW GFLKDLNDK.
Nodirhilo LSGGMKRCLT MARALINDPQ LIVMDEPTTG LDPHARHLIW ERLRALLAR.
Nodibraja LSGGMKRRLT LARALINDPH LLVMDEPTTG LDPHARHLIW ERLRALLAR.
Nodirhiga LSGGMKRRLT LARALVNDPQ LLILDEPTTG LDPPARHQIW ERLRSLLIR.
Nodirhile LSGGMKRRLT LAGALINDPQ LLILDEPTTG LDPHARHLIW ERLRSLLAR.
Natabacsu FSKGMRQKVA IARALIHDPD IILFDEPTTG LDITSSNIFR EFIQQLKRE.
Nodiazoca LSGGMRRLTM VARALVNDPY VIILDEATVG LDAKSRSALW QQIEASNAT.
Consensus .S.G...... .A........ ....DE.... .D........ ........
```

```
              251                                          300
Bexahaein DRSIILVSHS PSAMKSYCDN AVVLENGIMH HF.EDMDKAY QYYNETQK..
Ctrdneime DRSIILVSHS HSAMKQYCDN AMVLEKGHMY QF.EDMDKAY EYYNSLP...
Kst1escco KSHLIMVSHS ERALKEYCDV AIYLNKEGQG KFYKNVTEAI ADYK..KDL.
Kst5escco ESSFLMVSHS LNSLKEFCDV AIVFKNSYII GYYENVQSGI DEYKMYQDLD
Abcaaersa GVTILLVSHA AGSVIELCDR AVLLDRGEV. LLQGEPKAVV HNYHKLLH..
Taghbacsu GKTIFFVSHS IGQIEKMCDR VAWMHYGEL. RMFDETKTVV KEYKAFIDWF
Vexcsalti QKGLIVLTHN PRLIKEHCHA FGVLLHGKI. TMCEDLAQAT ALFEQYQSNQ
Potgescco GVTCVMVTHD QEEAMTMAGR IAIMNRGKFV QIGEPEEIYE HP.TTRYSAE
Sulfsynsp HVTTVFVTHD QEEAMEVADQ IVVMNHGKVE QIGSPAEIYD NP.ATPFVMS
Sppaescco GITFVFVTHD QEEALTMSDR IVVMRDGRIE QDGTPREIYE EP.KNLFVAG
Nosfpsest GTSIILCSHV LPGVEAHINR AAILAKGCLQ AVGSLSQLRA E...AGLPVR
Tnrbstrlo GKTVLLSSHI LAQVEKLCDR VSIIRKGRNV QSGTLTEMRH L...TRTTVE
Glyustrli GTTVLLTTHY LEEAEHLADR LAIMHDGR.I AATGTPAEVT AAQPSHISFE
Nodescco  GTTIILTTHY LEEAEMLCRN IGIIQHGELV ENTSMKALLA KLKSETFILD
Nodirhilo GKTIILTTHF MEEAERLCDR LCVLEKGRNI AEGGPQALID EHIGCQVM..
Nodibraja GKTILLTTHF MEEAERLCDR LCVLESGCKI AEGKPDALID EHIGCNVI..
Nodirhiga GKTILLTTHM MDEAERMCDR LCVLEGGRMI AEGPPLSLIE DIIGCPVI..
Nodirhile GKTILLTTHI MEEAERLCDR LCVLEAGRKI AEGRPHALIE EQIGCPVI..
Natabacsu QKTILFSSHI MEEVQALCDS VIMIHSGEVI YRGALESLYE SERSEDLN..
Nodiazoca GATVLVISHL AEDLERMTDR IACICAGTVR AEWETAALLK ALGALKIL..
Consensus ........H. ........D. ......G... .......... ........
```

```
              301                                          350
Bexahaein .......... .......... .......... .......... ........
Ctrdneime .......... .......... .......... .......... ........
Kst1escco .......... .......... .......... .......... ........
Kst5escco IE........ .......... .......... .......... ........
Abcaaersa ..MEGDERAR FRYHLRQTGR GD..SYISDE STSEPKIKSA PGILSVDLQP
Taghbacsu NKLSKKEKET YKKEQTEERK KEDPEAFARF RKKKKKPKSL ANAIQIAILS
Vexcsalti ATIQTEDYSF DI........ .......... .......... ........
Potgescco FIGSVNVFEG VLKERQEDGL VLDSPGLVHP LKVDADASVV DNVPVHVALR
Sulfsynsp FIGPVNVL.. ......PNSSH IFQAGGL... .......... DTPHPEVFLR
Sppaescco FIGEINMFNA TVIERLDEQR V.RANVEGRE CNIYVNFAVE PGQKLHVLLR
Nosfpsest IRAS.GISER DSWLQRWTDA GHSARGLSES SIEVVAVNGH KLVLLRQLLG
Tnrbstrlo AETERAVTGL DAM......A GVHAVRVTER RVH.FAVDGA HLDAAIHRLG
Glyustrli LPDGYFVGDL PPLAELGVSD HETDGRTVK. LRTRELQRAA TGLLVWAAQA
Nodescco  LAPK...SPL PKLDGYQYRL VDTATLEVEV LREQGINSVF TQL....SEQ
Nodirhilo ...EIYGGDP HELLSL.... VKPHSQRIEV SGETLYCYAP DPDQVRTQLQ
```

```
Nodibraja ...EIYGGDL DQLREL.... IRPYARHIEV SGETLFCYAR CPDEISVHLR
Nodirhiga ...EVYGGNP DELSLI.... VRPHVDRIET SGETLFCYTV NSDQVRAKLR
Nodirhile ...EIYGGDP QELSLL.... IRPNARRLEI SGETLFCYTP DPEQVRAQLR
Natabacsu ...YIF.... ..MSKL.... VRGIS..... .........
Nodiazoca ...EIDHSVS PRGKEL.... FCNHGLHVHE NDGRLSAVHP SSDFALEAVL
Consensus ......... ......... ......... ......... .........

          351                                            400
Bexahaein ......... ......... ......... ......... .........
Ctrdneime ......... ......... ......... ......... .........
Kst1escco ......... ......... ......... ......... .........
Kst5escco ......... ......... ......... ......... .........
Abcaaersa QSTVWYESKG AVLSDVH.IE SF....... ......... .........
Taghbacsu ILTVFMAGTM FFNAPLRTIA SFGAIPQNEV KNHHGDAKGK SEERLTAINK
Vexcsalti ......... ......... ......... ......... .........
Potgescco PEKIMLCEEP PANGCNFAVG EVIHIAYLGD LSVYHVRLKS GQ.MISAQLQ
Sulfsynsp PHDIEIAIDP IPETVPARID RIVHLGWEVQ A...EVRLED GQ.VLVAHLP
Sppaescco PEDLRVEEIN DDNHAEGLIG YVRERNYKGM TLESVVELEN GKMVMVSEFF
Nosfpsest EGEPEDIEIH QPSLEDL.YR YYMERAGDVR AQEGRL.... .........
Tnrbstrlo EFGIRSLTSH PPTLEELMLR HYGDELAAGA GGNGAAR... .........
Glyustrli AVELRRLDVR SASLEEAFLS IAKQVSARSD GTATDTTEKE YAA......
Nodescco  GIQVLSMRNK ANRLEELFVS L...VNEKQG DRA...... .........
Nodirhilo GRAGLRLLLR PANLEDVF.. ..LRLTGREM EE....... .........
Nodibraja GRTDLRVLQR PPNLEDVF.. ..LRLTGREM EK....... .........
Nodirhiga EFPSLRLLER PANLEDVF.. ..LRLTGREM EK....... .........
Nodirhile AYSNLRLLER PPNLEDVF.. ..LRLTGREM EK....... .........
Natabacsu ......... ......... ......... ......... .........
Nodiazoca KKEAVPTTTR FATLDDIIRF PGFPWTGVGG VNGEK.... .........
Consensus ......... ......... ......... ......... .........

          401                                            450
Taghbacsu QGFIANEKAA AYKDQGLKQK ADVTLPFGTK VTVAAKGKQA AKIKFDGHSY
Vexcsalti ......... ......... ......... ......... .........
Potgescco NAHRHRKGLP T......WGD EVRLCWEVDS CVVLTV.... .........
Sulfsynsp RDRYRDLQLE PEQQVFVRPK QAR.SFPLNY SI....... .........
Sppaescco NEDDPDFDHS LDQKMAINWV ESWEVVLADE EHK...... .........
Consensus ......... ......... ......... ......... .........

          451                                            500
Taghbacsu YVKQSAVATN MKHAELHATA FTSYVSQNAA SSYEYFLKFL GDSSTSIQSK

          501                                            550
Taghbacsu LNGYTEGNKA DGRKTLNFDY EKISYVLEND KATELIFHNI SPINPASLSL

          551                    587
Taghbacsu SDSDVLYDSS KKRFLVNTDD QVFAVDNEEH TLTLMLK
```

Residues listed in the consensus sequence are present in at least 75% of the aligned transporter sequences. Proteins listed subsequently in italics are at least 90% identical to the paired transporters listed in parenthesis and are therefore not included in the alignment: *Nodirhime* (Nodirhilo). Residues indicated in boldface type are also conserved in at least one other family of the ABC transporter superfamily.

Database accession numbers

	SWISSPROT	PIR	EMBL/GENBANK
Abcaaersa	Q07698	A36918	L11870; G304013
Bexahaein	P10640	A28781, BVHIXA	M19995; G148867
Ctrdneime	P32016	S15223	M57677; G150253
Glyustrli			X65556
Kst1escco	P23888	S13590	M57381; G146567
Kst5escco	P24586	S12237	X53819; G41879
Natabacsu	P46903		U30873; G973330
Nodescco		S45204	
Nodiazoca	Q07756	S35007	L18897; G310297
Nodibraja	P26050	S27496	J03685; G152117
Nodirhiga			X87578
Nodirhilo	P23703		X55705; G581510
Nodirhile	P08720		Y00548; G46214
Nodirhime		S13590	
Nosfpsest	P19844	S13584	X53676; G45849
Potgescco	P31134	B45313	M93239; G147335
Sppaescco		A40840	M64519
Sulfsynsp		GRYCS7	
Taghbacsu	P42954		U13832; G755153
Tnrbstrlo		S34187	X73633
Vexcsalti	P43110		D14156; G475034

References

[1] Geelen, D. et al. (1993) Mol. Microbiol. 9, 145–154.
[2] Kroll, J.S. et al. (1988) Cell 53, 347–356.
[3] Frosch, M. et al. (1991) Mol. Microbiol. 5, 1251–1263.
[4] Linton, K.J. et al. (1994) Mol. Microbiol. 11, 777–785.
[5] **Higgins, C.F. (1992) Annu. Rev. Cell Biol. 8, 67–113.**
[6] **Reizer, J. et al. (1992) Protein Sci. 1, 1326–1332.**
[7] Chu, S. and Trust, T.J. (1993) J. Bacteriol. 175, 3105–3114.

Part 5

ABC Binding Protein-Dependent Transporters: Transmembrane Elements

Summary

Transporters of the ABC-associated binding protein-dependent maltose transporter family, the example of which is the MALG maltose permease of *Escherichia coli* (Malgescco), mediate uptake of maltose, starch degradation products and glycerol-3-phosphate. Members of the family are found in both gram-negative and gram-positive bacteria.

Members of the ABC-associated binding protein-dependent maltose transporter family are associated with cytoplasmic elements of the ATP binding cassette (ABC) superfamily[1]. Proteins in this superfamily use the energy of ATP hydrolysis to pump substrates across cell membranes. Two cytoplasmic chains containing the ATP binding domains associate with two transmembrane domains of the ABC-associated binding protein-dependent maltose transporter family to form the active complex.

Statistical analysis of multiple amino acid sequence comparisons suggests that the ABC-associated binding protein-dependent maltose transporter family is most closely related to the ABC-associated binding protein-dependent peptide transporter family. Members of the ABC-associated binding protein-dependent maltose transporter family are predicted to form six membrane-spanning helices by the hydropathy of their amino acid sequences and activities of reporter gene fusions[2]. Several amino acid sequence motifs are highly conserved in the ABC-associated binding protein-dependent maltose transporter family that are necessary for function by the criterion of site-directed mutagenesis.

Nomenclature, biological sources and substrates

CODE	DESCRIPTION [SYNONYMS]	ORGANISM [COMMON NAMES]	SUBSTRATE(S)
Amyctheth	Starch degradation products transport system permease [AMYC]	*Thermoanaerobacter thermosulfurogenes* [gram-positive bacterium]	Starch degradation products
Cymgkleox	Starch utilization protein [CYMG]	*Klebsiella oxytoca* [gram-negative bacterium]	Starch degradation products
Maldstrpn	Maltodextrin transport system permease [MALD]	*Streptococcus pneumoniae* [gram-positive bacterium]	Maltodextrin
Malgescco	Maltodextrin transport system permease [MALG]	*Escherichia coli* [gram-negative bacterium]	Maltose
Malgsalty	Maltodextrin transport system permease [MALG]	*Salmonella typhimurium* [gram-negative bacterium]	Maltose
Malgentae	Maltodextrin transport system permease [MALG]	*Enterobacter aerogenes* [gram-negative bacterium]	Maltodextrin
Ugpeescco	*sn*-Glycerol-3-phosphate transport system permease [UGPE]	*Escherichia coli* [gram-negative bacterium]	*sn*-Glycerol-3-phosphate

Phylogenetic tree

Proteins listed subsequently in italics are at least 90% identical to the paired transporters listed in parenthesis and therefore are not included in the phylogenetic tree: *Malgsalty, Malgentae* (Malgescco).

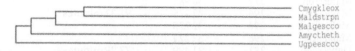

```
                                    Cmygkleox
                                    Maldstrpn
                                    Malgescco
                                    Amyctheth
                                    Ugpeescco
```

Proposed orientation of MALG in the membrane

The model is based on predictions of membrane-spanning regions and α-helical content. The N-terminus of the protein is illustrated on the inside and is folded six times through the membrane[2]. The predicted membrane-spanning helices are portrayed as rectangles. The numbers corresponding to the first and last residue of each membrane-spanning helix are boxed. Residues that are conserved in more than 75% of the aligned transporters (see below) are shown. Consensus residues indicated by an asterisk are not conserved in MALG.

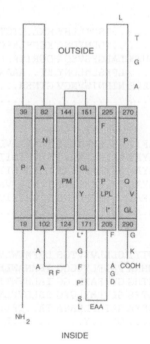

Physical and genetic characteristics

	AMINO ACIDS	MOL. WT	CHROMOSOMAL LOCUS
Amyctheth	274	30 994	
Cymgkleox	277	30 884	
Maldstrpn	277	30 983	
Malgescco	296	32 225	91.32 minutes
Malgsalty	296	32 225	
Malgentae	296	32 221	
Ugpeescco	281	31 499	77.29 minutes

Multiple amino acid sequence alignments

```
          1                                                   50
Cmygkleox MRNIISKIMT ILVY...LFL LLNALVVLGP VIWTVMSSLK PGNNLFSSGF
Maldstrpn MILITQRRLT QSYY...LYL IGLSIVIIYP LLITIMSAFK AGNVSAFKLD
Malgescco MAMVQPKSQK ARLFITHLLL LLFIAAIMFP LLMVVAISLR QGNFATGSLI
Amyctheth .....MRKVH VQKYLLTFLG IVLSLLWISP FYIILVNSFK TKLELFTNTL
Ugpeescco ...MIENRPW LTIFSHTMLI LGIA.VILFP LYVAFVAATL DKQAVYAAPM
Consensus .......... .......... .......... ....P..... ..........

          51                                                  100
Cmygkleox TE........ ..ISFTLEHY HNL.LTGT.P YLKWYKNTFI LATCNMLISL
Maldstrpn TN........ ..IDLNFDNF KGPSLKPC.T VLGTF.NTLI IALITMAVQT
Malgescco PEQISWDHWK LALGFSVEQA DGRITPPPFP VLLWLWNSVK VAGISAIGIV
Amyctheth S........ LPKSLMLDNY KT..AAANLN LSEAFSNTLI ITVFSILIIA
Ugpeescco TLIPGTHLLE NIHNIWVNGV GTNSA....P FWRMLLNSFV MAFSITLGKI
Consensus .......... .......... .......... ....N... .A........

          101                                                 150
Cmygkleox VVVTITAFIF SRYRFKAKKK ILMSILVLQM FPAFLSMTAI YILLSKMN..
Maldstrpn SIIVLAGYAY SRYNFLARKQ SLVFFLIIQM VPTMAALTAF FVMALMLN..
Malgescco ALSTTCAYAF ARMRFPGKAT LLKGMLIFQM FPAVLSLVAL YALFDRLGEY
Amyctheth IFSSMTAYAL QRVKRKSSVI IYMIFTVAML IPFQSVMIPL VAEFGKFH..
Ugpeescco TVSMLSAFAI VWFRFPLRNL FFWMIFITLM LPVEVRIFPT VEVIANLQ..
Consensus ......A.A. .R..F..... ........M .P........ ..........

          151                                                 200
Cmygkleox ....LIDTYI GLLLVYVTGS LPFMTWLVKG YFDAIPTSLD EAAKIDGAGH
Maldstrpn ....ALNHNW FLIFLYVGGG IPMNAWLMKG YFDTVPMSLD ESAKLDGAGH
Malgescco IPFIGLNTHG GVIFAYL.GG IALHVWTIKG YFETIDSSLE EAAALDGATP
Amyctheth .....FLTRS GLVFMYLGFG SSLGVFLYYG ALKGIPTSLD EAALIDGCSR
Ugpeescco ....MLDSYA GLTLPLMASA TA..TFLFRQ FFMTLPDELV EAARIDGASP
Consensus .......... GL...Y.... .....L..G .F...P.SL. EAA..DGA..

          201                                                 250
Cmygkleox LTIFFEIILP LAKPILVFVA LVSFTGPWMD FILPTLILRS EDKMTLAIGI
Maldstrpn FRRFWQIVLP LVRPMVAVQA LWAFMGPFGD YILSSFLLRE KEYFTVAVGL
Malgescco WQAFRLVLLP LSVPILAVVF ILSFIAAITE VPVASLLLRD VNSYTLAVGM
Amyctheth FRIYWNIILP LLNPTTITLA VLDIMWIWND YLLPSLVINK VGSRTLPLMI
Ugpeescco MRFFCDIVFP LSKTNLAALF VITFIYGWNQ YLWPLLIITD VDLGTTVAGI
Consensus ...F..I.LP L..P...... ...F...... ....L.... ....T...G.
```

```
             251                                             299
Cmygkleox FSWIS.SNSA ENFTLFAAGA LLVAVPITLL FIVTQKHITT GLVSGAVKE
Maldstrpn QTFVN.NAKN LKIAYFSAGA ILIALPICIL FFFLQKNFVS GLTSGGDKG
Malgescco QQYLN.PQNY L.WGDFAAAA VMSALPITIV FLLAQRWLVN GLTAGGVKG
Amyctheth FYFF..SQYT KQWNLGMAGL TIAILPVVIF YFLAQRKLVT AIIAGAVKQ
Ugpeescco KGMIATGEGT TEWNSVMVAM LLTLIPPVVI VLVMQRAFVR GLVDSEK..
Consensus ...............................A.. .....P.... ....Q...V. GL..G..K.
```

Proteins listed subsequently in italics are at least 90% identical to the paired transporters listed in parenthesis and therefore are not included in the alignments: *Malgsalty, Malgentae* (Malgescco). Residues listed in the consensus sequence are present in at least 75% of the aligned transporter sequences. Residues indicated by boldface type are also conserved in the ABC-associated binding protein-dependent peptide transporter family.

Database accession numbers

	SWISSPROT	*PIR*	*EMBL/GENBANK*
Amyctheth	P37729	S37705; S17297	M57692; M54654
Cymgkleox		S55409	C86014
Maldstrpn	Q04699	S32571	L08611
Malgescco	P07622	A24361	U00006; X02871
Malgsalty	P26468	A60175; S20605	X54292; M33921
Malgentae	P18814	S05333	
Ugpeescco	P10906	S03782	X13141; U00039

References

1 Higgins, C.F. (1992) Annu. Rev. Cell Biol. 8, 67–113.
2 Dassa, E. and Muir, S. (1993) Mol. Microbiol. 7, 29–38.

ABC-associated binding protein-dependent peptide transporter family

Summary

Transporters of the ABC-associated binding protein-dependent peptide transporter family, the example of which is the DPPC dipeptide transporter of *Escherichia coli* (Dppcescco), mediate the uptake of oligopeptides. One member of the family, NIKC, mediates the uptake of nickel. Members of the family are found in both gram-negative and gram-positive bacteria.

Members of the ABC-associated binding protein-dependent peptide transporter family are associated with cytoplasmic elements of the ATP binding cassette (ABC) superfamily[1]. Proteins in this superfamily use the energy of ATP hydrolysis to pump substrates across cell membranes. Two cytoplasmic chains containing the ATP binding domains associate with two transmembrane domains of the ABC-associated binding protein-dependent peptide transporter family to form the active complex.

Statistical analysis of multiple amino acid sequence comparisons suggests that the ABC-associated binding protein-dependent peptide transporter family is most closely related to the ABC-associated binding protein-dependent maltose transport family. Members of the ABC-associated binding protein-dependent peptide transporter family are predicted to form six membrane-spanning helices by the hydropathy of their amino acid sequences, activities of reporter gene fusions, reaction with peptide specific antibodies, and susceptibility to proteolysis[2].

Several amino acid sequence motifs are highly conserved in the ABC-associated binding protein-dependent peptide transporter family.

Nomenclature, biological sources and substrates

CODE	DESCRIPTION [SYNONYMS]	ORGANISM [COMMON NAMES]	SUBSTRATE(S)
Amidstrpn	Oligopeptide transport system permease [AMID]	*Streptococcus pneumoniae* [gram-positive bacterium]	Oligopeptides
Appcbacsu	Oligopeptide transport system permease [APPC]	*Bacillus subtilis* [gram-positive bacterium]	Oligopeptides
Dppcbacsu	Dipeptide transport system permease [DPPC]	*Bacillus subtilis* [gram-positive bacterium]	Dipeptides
Dppcescco	Dipeptide transport system permease [DPPC]	*Escherichia coli* [gram-negative bacterium]	Dipeptides
Dppchaein	Dipeptide transport system permease [DPPC]	*Haemophilus influenzae* [gram-negative bacterium]	Dipeptides
Nikcescco	Nickel transport system permease [NIKC]	*Escherichia coli* [gram-negative bacterium]	Nickel
Oppcbacsu	Oligopeptide transport system permease [OPPC, SPOOKc]	*Bacillus subtilis* [gram-positive bacterium]	Oligopeptides
Oppchaein	Oligopeptide transport system permease [OPPC, HI1122]	*Haemophilus influenzae* [gram-negative bacterium]	Oligopeptides
Oppclacla	Oligopeptide transport system permease [OPPC]	*Lactobacillus lactis* [gram-positive bacterium]	Oligopeptides
Oppcsalty	Oligopeptide transport system permease [OPPC]	*Salmonella typhimurium* [gram-negative bacterium]	Oligopeptides

CODE	DESCRIPTION [SYNONYMS]	ORGANISM [COMMON NAMES]	SUBSTRATE(S)
Sapcescco	Peptide transport system permease [SAPC]	Escherichia coli [gram-negative bacterium]	Oligopeptides
Sapchaein	Peptide transport system permease [SAPC, HI1640]	Haemophilus influenzae [gram-negative bacterium]	Peptides
Sapcsalty	Peptide transport system permease [SAPC]	Salmonella typhimurium [gram-negative bacterium]	Peptides

Phylogenetic tree

Proteins listed subsequently in italics are at least 90% identical to the paired transporters listed in parenthesis and therefore are not included in the phylogenetic tree: *Sapcescco* (Sapcsalty).

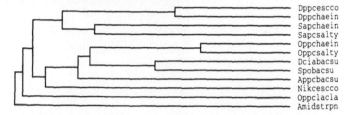

Proposed orientation of DPPC in the membrane

The model is based on predictions of membrane-spanning regions and α-helical content. The N-terminus of the protein is illustrated on the inside and is folded six times through the membrane [2]. The predicted membrane-spanning helices are portrayed as rectangles. The numbers corresponding to the first and last residue of each membrane-spanning helix are boxed. Residues that are conserved in more than 75% of the aligned transporters (see below) are shown. Consensus residues indicated by an asterisk are not conserved in DPPC.

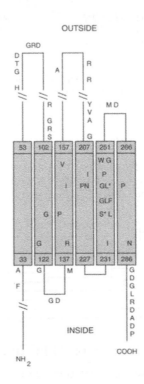

OUTSIDE

INSIDE

NH₂

COOH

Physical and genetic characteristics

	AMINO ACIDS	MOL. WT	CHROMOSOMAL LOCUS
Amidstrpn	308	34 634	
Appcbacsu	303	33 420	104°
Dppcbacsu	320	35 836	113°
Dppcescco	300	32 308	79.75 minutes
Dppchaein	295	31 840	
Nikcescco	277	30 362	77.87 minutes
Oppcbacsu	305	33 621	104°
Oppchaein	311	34 631	
Oppclacla	294	32 835	
Oppcsalty	302	33 090	
Sapcescco	296	31 548	
Sapchaein	295	33 014	
Sapcsalty	296	31 548	

Multiple amino acid sequence alignments

```
          1                                                50
Sapchaein ......................................M QNKEPDEFRE STSIFQIWLR
Sapcsalty ...................... ..........MPYDSVYS EKRPPGTLRT A......WRK
Oppchaein .........M TDYRTQPINQ KNADFVEQVA DRIEEMQLEG RSLWQDAKRR
Oppcsalty ...........MMLSK KNSETLENFS EKLE...VEG RSLWQDARRR
Dppcbacsu MNLPVQTDER QPEQHNQVPD EWFVLNQEKN READSVKRPS LSYTQDAWRR
```

```
Oppcbacsu ........... ....MQNIPK NMFEPAAANA GDAEKISKKS LSLWKDAMLR
Appcbacsu ............ ......MS ELQTTPSPEI RLKENISKKP ETMTKIFWEK
Dppcescco ................ .....MSQVT ENKVISAPVP MTPLQEFWHY
Dppchaein ............. ........... MSDTPLTFAP KTPLQEFWFY
Nikcescco ................ ........ ........... ......MNFF
Oppclacla ................ ........... .MTEKKHKNS LSLVHSIKEE
Amidstrpn ........... ....MSTIDK EKFQFVKRDD FASETIDAPA YSYWKSVFKQ
Consensus ................ ......... ........... ...........
```

```
           51                                             100
Sapchaein FRQNTIALFS FYLLIALIFT A.LFASYLAP YA.DNRQFIG QELMPPSWVD
Sapcsalty FYSDAPAMVG LYGCAGLALL C.IFGGWIAP YG.IDQQFLG YQLLPPSWSR
Oppchaein FFRNKAAVAS LIILAFIIIF I.TVAPWFFP FT.....YED TDWNMMSAAP
Oppcsalty FMHNRAAVAS LIVLFLIALF V.TVAPMLSQ FT.....YFD TDWGMMSSAP
Dppcbacsu LKKNKLAMAG LFILLFLFVM A.VIGPFLSP HS.....VVR Q..SLTEQNL
Oppcbacsu FRSNKLAMVG LIIIVLIILM A.IFAPMFSR YD.....YST T..NLLNADK
Appcbacsu FSKNKLAILG AVILFIIIMS A.VFAPLIAP YPQEQQSLLD KYKAP.....
Dppcescco FKRNKGAVVG LVYVVIVLFI A.IFANWIAP YNPAEQ.FRD ALLAPPAWQE
Dppchaein FKQNRGALIG LIFILIVALI S.ILAPYIAP FDPTEQ.NRT ALLLPPAWYE
Nikcescco LSSRWSVRLA LIIIALLALI A.LTSQWWLP YDP....... QAIDLPSRLL
Oppclacla LKKDKLAMIS TIFLVAVFLI VYIYSMFLKQ SNYVDVNIMD QYLA......
Amidstrpn FMKKKSTVVM LGILVAIILI SFIY.PMFSK FDFND...VS KVNDFSVRYI
Consensus F.....A... .......... .......... ..........
```

```
           101                                            150
Sapchaein RGKIAFFFGT DDLGRDILSR LIMGTRYTLG SALLVVFSVA IIGGALGIIA
Sapcsalty YGEVSFFLGT DDLGRDVLSR LLSGAAPTVG GAFIVTLAAT LCGLVLGVVA
Oppchaein TMEGYHFFGT DASGRDLLVR TAIGGRISLL VGIAGAFISV TIGTIYGAIS
Oppcsalty DMASGHYFGT DSSGRDLLVR VAIGGRISLM VGIAAALVAV IVGTLYGSLS
Dppcbacsu PPSADHWFGT DELGRDVFTR TWYGARISLF VGVMAALIDF LIGVIYGGVA
Oppcbacsu PPSKDHWFGT DDLGRDIFVR TWVGARISIF IGVAAAVLDL LIGVIWGSIS
Appcbacsu ..GLEHLMGT DKFGRDIFSR ILYGARVSLL VGFASVVGSI LIGTVLGALA
Dppcescco GGSMAHLLGT DDVGRDVLSR LMYGARLSLL VGCLVVVLSL IMGVILGLIA
Dppchaein GGNPAYLLGT DDIGRDILSR IIYGTRISVF AGFIIVLLSC AFGTSLGLIS
Nikcescco SPDAQHWLGT DHLGRDIFSR LMAATRVSLG SVMACLLLVL TLGLVIGGSA
Oppclacla PLTNGHLLGT DNGGRDIIMM LMISARNSFN IAFAVTLITL VVGNILGVIT
Amidstrpn KPNAEHWFGT DSNGKSLFDG VWFGARNSIL.ISVIATVINL VIGVFVGGIW
Consensus .....H..GT D..GRD...R ...G.R.S.. .......... ..G...G...
```

```
           151                                            200
Sapchaein GLLKGIKARF VGHIFDAFLS LPILLIAVVI STLMEP.SLW NAMFATLLAI
Sapcsalty GATHGLRSAV LNHILDTLLS IPSLLLAIIV VAFAGP.HLS HAMFAVWLAL
Oppchaein GYVGGKTDML MMRFLEILSS FPFMFFVILL VTLFGQ.NIF LIFIAIGAIA
Oppcsalty GYLGGKIDSV MMRLLEILNS FPFMFFVILL VTFFGQ.NIL LIFVAIGMVS
Dppcbacsu GYKGGRIDSI MMRIIEVLYG LPYLLVVILL MVLMGP.GLG TIIVALTVTG
Oppcbacsu GFRGGRTDEI MMRIADILWA VPSLLMVILL MVVLPK.GLF TIIIAMTITG
Appcbacsu GYFRGIVDAV IMRVVDIVLS IPDIFLLITL VTIFKP.GVD KLILIFCLTG
Dppcescco GYFGGLVDNI IMRVVDIMLA LPSLLLALVL VAIFGP.SIG NAALALTFVA
Dppchaein GYYGGVLDTI IIRLIDIMLA IPNLLLTIVV VSILEP.SLA NATLAIAVVS
Nikcescco GLIGGRVDQA TMRVADMFMT FPTSILSFFM VGVLGT.GLT NVIIAIALSH
Oppclacla GYFGGRFDLI FMRFTDFVMI LPSMMIIIVF VTIIPRFNSW SLIGIISIFS
Amidstrpn G.ISKSVDRV MMEVYNVISN IPPLLIVIVL TYSIGA.GFW NLIFAMSVTT
Consensus G...G..D.. .MR....... .P.....I.. V........ ...A.....
```

```
            201                                              250
Sapchaein LPYFIHTIYR AIQKELEKDY VVMLKLEGIS NQALLKSTIL PNITVIYIQE
Sapcsalty LPRMVRSVYS MVHDELEKEY VIAARLDGAT TLNILWFAIL PNITAGLVTE
Oppchaein WLGLARIVRG QTLSLKNKEF VEAAIVCGVP RRQIILKHII PNVLGLVAVY
Oppcsalty WLDMARIVRG QTLSLKRKEF IEAAQVGGVS TASIVIRHIV PNVLGVVVVY
Dppcbacsu WVGMARIVRG QVLQIKNYEY VLASKTFGAK TFRIIRKNLL RNTMGAIIVQ
Oppcbacsu WINMARIVRG QVLQLKNQEY VLASQTLGAK TSRLLFKHIV PNAMGSILVT
Appcbacsu WTTTARLVRG EFLSLRSREY VLAAKTIGTK THKIIFSHIL PNALGPIIVS
Dppcescco LPHYVRLTRA AVLVEVNRDY VTASRVAGAG AMRQMFINIF PNCLAPLIVQ
Dppchaein IPSYVRLTRA AMMNEKNRDY VTSSKVAGAG ILRLMFIVIL PNCLAPLIVQ
Nikcescco WAWYARMVRS LVISLRQREF VLASRLSGAG HVRVFVDHLA GAVIPSLLVL
Oppclacla WIGTTRLIRA RTMTEVNRDY VRASKTSGTS DFKIMFREIW PNLSTLVIAE
Amidstrpn WIGIAFMIRV QILRYRDLEY NLASRTLGTP TLKIVAKNIM PQLVSVIVTT
Consensus .....R..R. .........Y V.A....G.. .........I. PN........

            251                                              300
Sapchaein VARAFVIAVL DISALSFISL GAQRPTPEWG AMIKDSLELL YLAP..WTVL
Sapcsalty ITRALSMAIL DIAALGFLDL GAQLPSPEWG AMLGDALELI YVAP..WTVM
Oppchaein ASLEVPGLIL FESFLSFLGL GTQEPMSSWG ALLSDG.AAQ MEVSPWLL.I
Oppcsalty ASLLVPSMIL FESFLSFLGL GTQEPLSSWG ALLSDG.ANS MEVSPWLL.L
Dppcbacsu MTLTVPAAIF AESFLSFLGL GIQAPFASWG VMANDGLPTI LSGHWWRL.F
Oppcbacsu MTLTVPTAIF TEAFLSYLGL GVPAPLASWG TMASDGLPA. LTYYPWRL.F
Appcbacsu ATLKVGSVIL AESALSYLGF GIQPPIASWG NMLQDAQNFT VMIQAWWYPL
Dppcescco ASLGFSNAIL DMAALGFLGM GAQPPTPEWG TMLSDVLQFA Q..SAWWVVT
Dppchaein MTMGISNAIL ELATLGFLGI GAKPPTPELG TMLSEARGFM Q..AANWLVT
Nikcescco ATLDIGHMML HVAGMSFLGL GVTAPTAEWG VMINDARQYI WTQP..LQMF
Oppclacla ATLVFAGNIG LETGLSFLGF GLPAGTPSLG TMINEATNPE TMTDKPWTWV
Amidstrpn MTQMLPSFIS YEAFLSFFGL GLPITVPSLG RLISDYSQNV TTNA..YLFW
Consensus ........I. ....LSFLGL G...P...WG .M..D.... ..........

            301                            334
Sapchaein LPGFAIIFTI LLSIIFSNGL TKAINQHQE. ....
Sapcsalty LPGAAITLSV LLVNLLGDGI RRAIIAGVE. ....
Oppchaein FPAFFLCLTL FCFNFIGDGL RDALDPKDR. ....
Oppcsalty FPAGFLVVTL FCFNFIGDGL RDALDPKDR. ....
Dppcbacsu FPAFFISSTM YAFNVLGDGL QDALDPKLRR ....
Oppcbacsu FPAGFICITM FGFNVVGDGL RDALDPKLRK ....
Appcbacsu FPGLFILMTV LCFNFVGDGL RDALDPKNIK ....
Dppcescco FPGLAILLTV LAFNLMGDGL RDALDPKLKQ ....
Dppchaein IPGLVILSLV LAFNLMGDGL RDALDPKLKQ ....
Nikcescco WPGLALFISV MAFNLVGDAL RDHLDPHLVT EHAH
Oppclacla PATVVILIVV LAIIFIGNAL RRVADQRQAT R...
Amidstrpn IPLTTLVLVS LSLFVVGQNL ADASDPRTHR ....
Consensus .P........ ...N..GDGL RDA.DP.... ....
```

Proteins listed subsequently in italics are at least 90% identical to the paired transporters listed in parenthesis and therefore are not included in the alignments: *Sapcescco* (Sapcsalty). Residues listed in the consensus sequence are present in at least 75% of the aligned transporter sequences. Residues indicated by boldface type are also conserved in the ABC-associated binding protein-dependent maltose transport family.

Database accession numbers

	SWISSPROT	PIR	EMBL/GENBANK
Amidstrpn	P18794	S11151	X17337
Appcbacsu	P42063		U20909
Dppcbacsu	P26904	S16649	X56678
Dppcescco	P37315		L08399; U00039
Dppchaein			U17295
Nikcescco	P33592	S39596	X73143; U00039
Oppcbacsu	P24139	S15232	X56347; M57689
Oppchaein	P45053		U32792
Oppclacla	Q07743		L18760
Oppcsalty	P08006	C29333	X05491
Sapcescco			X97282
Sapchaein	P45287		U32837
Sapcsalty	P36669	S39587	X74212

References

1 Higgins, C.F. (1992) Annu. Rev. Cell Biol. 8, 67–113.
2 Pearce, S.R. et al. (1992) Mol. Microbiol. 6, 47–57.

ABC-associated binding protein-dependent iron transporter family

Summary

Transporters of the ABC-associated binding protein-dependent iron transporter family, the example of which is the BTUC vitamin B12 transport protein of *Escherichia coli* (Btucescco), mediate uptake of iron. Members of the family are found in both gram-negative and gram-positive bacteria.

Members of the ABC-associated binding protein-dependent iron transporter family are associated with cytoplasmic elements of the ATP binding cassette (ABC) superfamily [1]. Proteins in this superfamily use the energy of ATP hydrolysis to pump substrates across cell membranes. Two cytoplasmic chains containing the ATP binding domains associate with two trans-membrane domains of the ABC-associated binding protein-dependent iron transporter family to form the active complex.

Statistical analysis reveals no apparent relationship between the amino acid sequences of the ABC-associated binding protein-dependent iron transporter family and any other family of transporters. They are predicted to form nine membrane-spanning helices by the hydropathy of their amino acid sequences.

Several amino acid sequence motifs are highly conserved in the ABC-associated binding protein-dependent iron transporter family.

Nomenclature, biological sources and substrates

CODE	DESCRIPTION [SYNONYMS]	ORGANISM [COMMON NAMES]	SUBSTRATE(S)
Btucescco	Vitamin B12 transport system permease [BTUC]	*Escherichia coli* [gram-negative bacterium]	Vitamin B12
Cbrberwch	Ferri-siderophore permease [CBRB]	*Erwinia chrysanthemi* [gram-negative bacterium]	Ferri-siderophores
Cbrcerwch	Ferri-siderophore permease [CBRC]	*Erwinia chrysanthemi* [gram-negative bacterium]	Ferri-siderophores
Fatcviban	Ferric anguibactin transport system permease [FATC]	*Vibrio anguillarum* [gram-negative bacterium]	Ferric anguibactin
Fatdviban	Ferric anguibactin transport system permease [FATD]	*Vibrio anguillarum* [gram-negative bacterium]	Ferric anguibactin
Feccescco	Ferric dicitrate transport system permease [FECC]	*Escherichia coli* [gram-negative bacterium]	Ferric dicitrate
Fecdescco	Ferric dicitrate transport system permease [FECD]	*Escherichia coli* [gram-negative bacterium]	Ferric dicitrate
Fepdescco	Ferric enterobactin transport system [FEPD]	*Escherichia coli* [gram-negative bacterium]	Ferric enterobactin
Fepgescco	Ferric enterobactin transport system [FEPG]	*Escherichia coli* [gram-negative bacterium]	Ferric enterobactin
Feubbacsu	Iron uptake protein [FEUB]	*Bacillus subtilis* [gram-positive bacterium]	Iron
Fhubbacsu	Ferrichrome transport protein [FHUB]	*Bacillus subtilis* [gram-positive bacterium]	Ferrichrome
Fhubescco	Ferrichrome transport protein [FHUB]	*Escherichia coli* [gram-negative bacterium]	Ferrichrome
Fxuamycsm	Ferric exochelin uptake system [FXUA]	*Mycobacterium smegmatus* [gram-positive bacterium]	Ferric exochelin
Hempyeren	Hemin uptake system [HEMP, HEMU]	*Yersinia enterocolitica* [gram-negative bacterium]	Hemin

Phylogenetic tree

Proposed orientation of BTUC in the membrane

The model is based on predictions of membrane-spanning regions and α-helical content. The N-terminus of the protein is illustrated on the inside and is folded nine times through the membrane. The predicted membrane-spanning helices are portrayed as rectangles. The numbers corresponding to the first and last residue of each membrane-spanning helix are boxed. Residues that are conserved in more than 75% of the aligned transporters (see below) are shown. Consensus residues indicated by an asterisk are not conserved in BTUC.

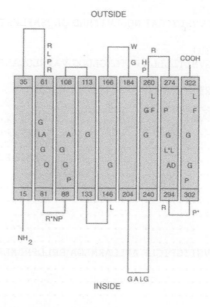

Physical and genetic characteristics

	AMINO ACIDS	MOL. WT	CHROMOSOMAL LOCUS
Btucescco	326	35 007	38.56 minutes
Cbrberwch	340	34 832	
Cbrcerwch	349	36 777	
Fatcviban	317	34 929	

	AMINO ACIDS	MOL. WT	CHROMOSOMAL LOCUS
Fatdviban	314	33 922	
Feccescco	332	34 892	97.19 minutes
Fecdescco	318	34 131	97.17 minutes
Fepdescco	334	33 871	13.37 minutes
Fepgescco	330	34 993	13.35 minutes
Feubbacsu	334	35 896	15°
Fhubbacsu	384	40 720	286°
Fhubescco	659	70 335	3.68 minutes
Fxuamycsm	344	35 170	
Hempyeren	334	35 528	

Multiple amino acid sequence alignments

```
          1                                                  50
Fhubescco MSKRIALFPA LLLALLVIVA TALTWMNFSQ ALPRSQWAQA AWSPDIDVIE

          51                                                 100
Fhubescco QMIFHYSLLP RLAISLLVGA GLGLVGVLFQ QVLRNPLAEP TTLGVATGAQ

          101                                                150
Fhubescco LGITVTTLWA IPGAMASQFA AQAGACVVGL IVFGVAWGKR LSPVTLILAG

          151                                                200
Fhubescco LVVSLYCGAI NQLLVIFHHD QLQSMFLWST GTLTQTDWGG VERLWPQLLG

          201                                                250
Fhubescco GVMLTLLLLR PLTLMGLDDG VARNLGLALS LARLAALSLA IVISALLVNA

          251                                                300
Fxuamycsm .......... .......... .......... .......... ..........
Fepgescco .......... .......... .......... .......... ..........
Cbrcerwch .......... .......... .......... .......... ..........
Feubbacsu .......... .......... .......... .......... ..........
Fhubbacsu .......... .......... ...MHFHFC SKHSIKSAEK SDILKQQLII
Fecdescco .......... .......... .......... .......... ..........
Hempyeren .......... .......... .......... .......... ..........
Cbrberwch .......... .......... .......... .......... ..........
Fepdescco .......... .......... .......... .......... ..........
Feccescco .......... .......... .......... .......... ..........
Btucescco .......... .......... .......... .......... ..........
Fhubescco VGIIGFIGLF APLLAKMLGA RRLLPRLMLA SLIGALILWL SDQIILWLTR
Fatdviban .......... .......... .......... .......... ..........
Fatcviban .......... .......... .......... .......... ..........
Consensus .......... .......... .......... .......... ..........

          301                                                350
Fxuamycsm ........M RGSRRQRVGA HVVACGDRGR RTAGGCLRGP RVLGDRDRPS
Fepgescco .......... ........ .MIYVSRRLL ITCLLLVSAC VVAGIWGLRS
Cbrcerwch .......MDK RGGMDHVLVW RQGRFSRQIN LTTVGRVSLA LLLVLAVMVA
Feubbacsu .......... .......... ....MYSKQ WTRIILITSP FA.IALSLLL
Fhubbacsu IISNRKEVRQ LSQHKNIRTA SEEIQWTSRT YGAVIVLIAG LCLLCLGAFL
Fecdescco .......... .......... .......... .......... ..MKIAL VIFITLALAG
```

```
Hempyeren .................. ............. .....MNCRI HPRLMLSILL MILIILALG.
Cbrberwch ................ ........MS HAVIPTGRRI APGQVLAGGG VCLLALAVLS
Fepdescco ................ .....MS GSVAVTRAIA VPGLLLL.... .LIIATALS
Feccescco .................. .....MTAIK HPVLLWGLPV AALIIIFWLS
Btucescco .................. .MLTLARQQQ RQNIRWLLCL SVLMLLALLL
Fhubescco VWMEVSTGSV IALIGAPLLL WLLPRLRSIS APDMKVNDRV AAERQHVLAF
Fatdviban .................. ............. .......... MTFRMILAFF
Fatcviban .................. ............. .......... .........
Consensus .................. ............. .......... .........
```

```
            351                                                400
Fxuamycsm ....TPIT.. PVDVLRVLVG TNTTFDR..V VVLEW...RM PRMLMALLIG
Fepgescco ....GAVTLE TSQVFAALMG DAPRSMT..M VVTEW...RL PRVLMALLIG
Cbrcerwch SLGLGKLMLS PWEVLRALWS SQPEGAA..L IVQQL...RL PRVVLAALVG
Feubbacsu SILYGAKHLS TDIVFTSLIH FDPGNTDHQI IW.HS...RI PRAAGALLIG
Fhubbacsu SISLGAADIH LRTVWEAIFH YQPTKTSHQI IH.DL...RL PRTAAAALVG
Fecdescco CALLSLHMGV IPVPWRALLT DWQAGHEHYY VLMEY...RL PRLLLALFVG
Hempyeren ....SANMGA LTLSFRTLWH AS.LDDAMWH IWLNI...RL PRVLLAVVVG
Cbrberwch LMV.GPVWIA PSQVLGALWH PDPLNVSHIL V.TST...RL SRTLIAIVVG
Fepdescco LLI.GAKSLP ASVVLEAFS. GTCQSADCTI V.LDA...RL PRTLAGLLAG
Feccescco LFCYSAIPVS GADATRALLP GHTPTLPEAL V.QNL...RL PRSLVAVLIG
Btucescco SLCAGEQWIS PGDWFTP... .......RGEL FVWQI...RL PRTLAVLLVG
Fhubescco ALAGGVLLLM AVVALSFGRD AHGWTWASGA LLEDLMPWRW PRIMAALFAG
Fatdviban TLCATSLFFG ANQIEWSLLP TFNEKAWLPI I.....ASRL PRLVALILTG
Fatcviban .MTSLNLNFR VSVVLVILLS IAFIFINSGF DLEYIIPRRL IKLSAIIIGG
Consensus .................. ............. .......RL PR.......G
```

```
            401                                                450
Fxuamycsm AALGVSGAIF QALTRNPLGS PDIIGLNAG. AYTGALVALA GLGTGGQHGG
Fepgescco AALGVSGAIF QSLMRNPLGS PDVMGFNTG. AWSGVLVAMV LFG.....QD
Cbrcerwch GALAVSGLIL QAMIRNPLAS PDILGITSG. ASAAAVFYLS FLAATL...G
Feubbacsu AALAVSGALM QGITRNYLAS PSIMGVSDG. SAFIITLCMV LLPQSSSI..
Fhubbacsu ALLAVSGAIM QGMTRNPLAE PSIMGVTSG. SAFAVSIAFA FFPGLSAM..
Fecdescco AALAVAGVLI QGIVRNPLAS PDILGVNHA. ASLASVGALL LMP.SLPVMV
Hempyeren CALAVSGAIM QGLFRNPLAD PGLLGISSG. AAL.CVGLII VMPFSLPPLV
Cbrberwch AGLAVAGALM QVLTRNPLAS PGLFGINAG. AMF....FLI VCVSLFPKVA
Fepdescco GALGLAGALM QTLTRNPLAD PGLLGVNAG. ASF....AIV LGAALFGYSS
Feccescco ASLALAGTLL QTLTHNPMAS PSLLGINSG. AAL....AMA LTSALSPTPI
Btucescco AALAISGAVM QALFENPLAE PGLLGVSNG. AGVGLIAAVL LGQGLTP...
Fhubescco VMLAVAGCII QRLTGNPMAS PEVLGISSG. AAFGVVLMLF LVPG....NA
Fatdviban SGLAMCGVIL QHIVRNRFVE PGTTGSLDA. AKLGILVSIV MLPSSDKLER
Fatcviban SCVAISAVIF QALARNRILT PSIMGYESIY LVWQALLLLF VGTSGSAVLG
Consensus ..LA..G... Q...RNP... P...G...G. A.................
```

```
            451                                                500
Fxuamycsm Y.YAVAGGAL VGGLITAAAV YALS..YRNG LAGYRLIVVG IGVGAVLSSV
Fepgescco L.TAIALSAM VGGIVTSLLV WLLA..WRNG IDTFRLIIIG IGVRAMLVAF
Cbrcerwch A.HYLPLAAM IGAATAALAV YWLA..WQAG VSPQRLVLTG VGVSALLMAA
Feubbacsu ...EMMIYSF IGSALGAVLV FGLAAMMPNG FTPVQLAIIG TVTSMLLSSL
Fhubbacsu ...GLVLWSF AGAGLGASTV MGIGMFSRGG LTPVKLALAG TAVTYFFTGI
Fecdescco L....PLLAF AGGMA.GLIL LKMLAKTHQP M...KLALTG VALSA.CWAS
Hempyeren ALYSHMVGAF IGSLAISAII FTLSRWGHGN LS..RLLLAG IAINALCGAA
Cbrberwch MSVWL.WSAF AGAAVAGCLV WLIGTMGKGS LNPLRMVLAG AAITAMF.AA
Fepdescco AQEQL.AMAF AGALVASLIV AFTGSQGGGQ LSPVRLTLAG VALAAVL.EG
```

```
Feccescco AGYSLSFIAA CGGGVSWLLV MTAGGGFRHT HDRNKLILAG IALSAFC.MG
Btucescco .NWALGLCAI RGALIITLIL LRFA...RRH LSTSRLLLAG VALGIICSAL
Fhubescco FGWLLPAGSL GAAVT...LL IIMIAAGRGG FSPHRMLLAG MALSTAFTML
Fatdviban MFFAVLFC.. FAAGLVYIAI IRKVKFSNTA LVPVIGLMFG SVLSALAE..
Fatcviban VVGNFVVSAV LILLYSFVIQ FWVLKRFQHD M..HQVLLIG FVLTMVLTTV
Consensus .......... .G........ ...... .......... .....L...G ..........
```

```
              501                                              550
Fxuamycsm NQWIVIKLDH HTAVTASVWQ QGTLNGLTWS QVVPMTVCLA VVTVALLAMG
Fepgescco NTWLLLKASL ETALTAGLWN AGSLNGLTWA KTSPSAPIII LMLIAAALLV
Cbrcerwch TTFMLVFSPL TTTLSAYVWL TGSVYGASWR ETRELGGWLL LIAPWLVLLA
Feubbacsu SAAMSIYF.. QISQDLSFWY SARLHQMSPD FLKLAAPFFL IGIIMAISLS
Fhubbacsu STAIAIRF.. DVAQDISFWY AGGVAGVKWS GVQLLLIAGA VGLTLAFFIA
Fecdescco LTDYLMLSRP QDVNNALLWL TGSLWGRDWS FVKIAIPLMI LFLPLSLSFC
Hempyeren VGVLTYISDD QQLRQFSLWS MGSLGQAQWS TLMVAASLIL PACVLGLLQA
Cbrberwch FSQAMLVVNQ EGLDTVLFWL AGSVADRELA DGAAADGLLL AALVGALLLS
Fepdescco LTSGIALLNP DVYDQLRFWQ AGSLDIRNLH TLKVVLIPVL IAGATALLLS
Feccescco LTRITLLLAE DHAYGIFYWL AGGVSHARWQ DVWQLLPVVV TAVPVVLLLA
Btucescco MTWAIYFSTS VDLRQLMYWM MGGFGGVDWR QSWLMLALIP VLLWIC.CQS
Fhubescco L.MMLQASGD PRMAQVLTWI SGSTYNATDA QVWRTGIVMV ILLAITPLCR
Fatdviban ....FYAYQN NILQSMSGWL MGDFSKVV.Q EHYEIIFLIL PITLLTYLYA
Fatcviban AQFIQIRISP GEFSIFQGLS YTSFERAKPS TLLFAGTVLS ILALFANKWV
Consensus .......... .........W. .G........ .......... ..........
```

```
              551                                              600
Fxuamycsm PQLPVLQMGD DAAGGLGVNP ERVRLSYLVA GVALVALACA AAGPISFVAL
Fepgescco RRMRLLEMGD DTACALGVRL ERSRLLMMLV AVVLTAAATA LAGPISFIAL
Cbrcerwch RQVRVQQLDD GLAQGIGVGV EWLAVALLLL SVRLAGAAIA WGAAMAFVGL
Feubbacsu KKVTAVSLGD DISKSLGQKK KTIKIMAMLS VIILTGSAVA LAGKIAFVGL
Fhubbacsu RSVTVLSLGD DLAKGLGQYT SAVKLVGMLI VVILTGAAVS IAGTIAFIGL
Fecdescco RDLDLLALGD ARATTLGVSV PHTRFWALLL AVAMTSTGVA ACGPISFIGL
Hempyeren RQLNLLQLGD EEAHYLGVNV KQAKLRLLLL SAILIGAAVA VSGVIGFIGL
Cbrberwch GQVNVLNAGE AIARGLGHGT GRIRLLMSLL VVALAGGAVA MAGSIGFVGL
Fepdescco RALNSLSLGS DTATALGSRV ARTQLIGLLA ITVLCGSATA IVGPIAFIGL
Feccescco NQLNLLNLSD STAHTLGVNL TRLRLVINML VLLLVGACVS VAGPVAFIGL
Btucescco RPMNMLALGE ISARQLGLPL WFWRNVLVAA TGWMVGVSVA LAGAIGFIGL
Fhubescco RWLTILPLGG DTARAVGMAL TPTRIALLLL AACLTATATM TIGPLSFVGL
Fatdviban HRFTVMGMGE DIASNLGISY AMTAALGLIL VSITVAVTVV TVGAIHFVGL
Fatcviban SELDVIGLGR DQAMSLGLND AHYIPKYFSV IAILVAISTS LIGPTAFMGV
Consensus ........G. ..A..LG... .......... .......... ..G...F.GL
```

```
              601                                              650
Fxuamycsm AAPQLARRLT AS.PGVALVP AAAM.GAVLL LASDLVAQHL FTANELPVGA
Fepgescco VAPHIARRIS GT.ARWGLTQ AALC.GALLL LAADLCAQQL FMPYQLPVGV
Cbrcerwch IAPHIRKRLV AP.GFAGQAA MAFLSGAGLV MVADLCGRTL FLPLDLPAGI
Feubbacsu VVPHITRFLV GS.DYSRLIP CSCILGGIFL TLCDLASRFI NYPFETPIEV
Fhubbacsu IIPHITRFLV GV.DYRWIIP CSAVLGAVLL VFADIAARLV NAPFETPVGA
Fecdescco VVPHMMRSIT GG.RHRRLLP VSALTGALLL VVADLLARII HPPLELPVGV
Hempyeren VVPHLIRMRI GA.DHRWLLP GAALGGACLL LTADTLARTL VAPAEMPVGL
Cbrberwch IVPHMARKLL PA.DHRWLLP GCALLGACLL LLADILARVV IVPQEVPVGV
Fepdescco MMPHMARWLV GA.DHRWSLP VTLLATPALL LFADIIGR.V IVPGELRVSV
Feccescco LVPHLARFWA GF.DQRNVLP VSMLLGATLM LLADVLARAL AFPGDLPAGA
Btucescco VIPHILRLCG LT.DHRVLLP GCALAGASAL LLADIVARLA LAAAELPIGV
Fhubescco MAPHIARMMG FR.RTMPHIV ISALVGGLLL VFADWCGRMV LFPFQIPAGL
```

```
Fatdviban VIPNLVALKY GD.HLKNTLP IVALGGASLL IFCDVISRVV LFPFEVPVGL
Fatcviban FIANIAYSIT GSPQYRHTLP VACTIAIVMF LTAQLMVEHF F.NYKTTVSI
Consensus ..PH..R... .........P .....G..LL ..AD...R.. ..P...P.G.
```

```
               651                 673
Fxuamycsm VTVSLGGIYL VYLLVTQARR ...
Fepgescco VTVSLGGIYL IVLLIQESRK K..
Cbrcerwch FVSALGAPFF LYLLIKQRH. ...
Feubbacsu VTSIIGVPFF LYLIKRKGGE QNG
Fhubbacsu LTSLIGVPFF FYLARRERRG L..
Fecdescco LTAIIGAPWF VWLLVRMR.. ...
Hempyeren ITSLLGGPYF LWLILRQREQ RSG
Cbrberwch MTALFGAPFF IFLLRRGGRY G..
Fepdescco VSAFIGAPVL IFLVRRKTRG GA.
Feccescco VLALIGSPCF VWLVRRRG.. ...
Btucescco VTATLGAPVF IWLLLKAGR. ...
Fhubescco LSTFIGAPYF IYLLRKQSR. ...
Fatdviban TASAVGGVMF LAFLLKGAKA ...
Fatcviban LVNVLCGGYF LIITMRARSQ L..
Consensus .....G...F ..L........ ...
```

Residues listed in the consensus sequence are present in at least 75% of the aligned transporter sequences.

Database accession numbers

	SWISSPROT	PIR	EMBL/GENBANK
Btucescco	P06609	A24498; S04777	M14031
Cbrberwch		S54821	X87208
Cbrcerwch		S54822	X87208
Fatcviban	P37737	B41671	M74068
Fatdviban	P37738	A41671	M74068
Feccescco	P15030	JS0113	M26397; U14003
Fecdescco	P15029	JS0114	M26397; U14003
Fepdescco	P23876	S16296; S16305	X57471; X59402
Fepgescco	P23877	S16297	X57471
Feubbacsu	P40410		L19954
Fhubbacsu			X93092
Fhubescco	P06972	S07318; S45222	X05810; D26562
Fxuamycsm			U10425
Hempyeren			X77867

References
[1] Higgins, C.F. (1992) Annu. Rev. Cell Biol. 8, 67–113.

Part 6

ABC Binding Protein-Dependent Transporters: Cytoplasmic Elements

Binding protein-dependent monosaccharide transporter family

Summary

Transporters of the binding protein-dependent monosaccharide transporter family, examples of which are the L-arabinose transport ATP binding protein ARAG from *Escherichia coli*[1] (Aragescco) and the ribose transport ATP binding protein, also from *E. coli*[2] (Rbsaescco), mediate import of monosaccharides. These transporters are found mostly in gram-negative bacteria; one member of the family is found in the gram-positive species *Bacillus subtilis*[3].

Statistical analysis of multiple amino acid sequence comparisons places the binding protein-dependent monosaccharide transporter family in the ATP binding cassette (ABC) superfamily[4]. Proteins in this superfamily use the energy of ATP hydrolysis to pump substrates across cell membranes. In transporters of this family the two ATP binding domains form one chain, separate from any transmembrane domains. The family is characterized by the cytoplasmic ATP binding domains[4], which are described in the following tables. Each transporter is associated with two transmembrane subunits. In the arabinose transport system, ARAG is associated with two copies of the ARAH transmembrane protein[1]; in contrast, in the ribose transport system, RBSA is associated with two homologous transmembrane proteins, RBSC and RBSD[5]. Unusually, the transmembrane proteins of the ribose transport system are predicted to contain eight membrane-spanning helices by the hydropathy of their amino acid sequences[5].

Quite a large number of amino acids are conserved within the binding protein-dependent monosaccharide transporter family, including several long sequence motifs unique to the family, signature motifs of the ABC superfamily, and motifs necessary for function by the criterion of site-directed mutagenesis.

Nomenclature, biological sources and substrates

CODE	DESCRIPTION [SYNONYMS]	ORGANISM [COMMON NAMES]	SUBSTRATE(S)
Aragescco	L-Arabinose transport ATP binding protein [ARAG]	*Escherichia coli* [gram-negative bacterium]	L-Arabinose
Mglaescco	Galactoside transport ATP binding protein [MGLA]	*Escherichia coli* [gram-negative bacterium]	Galactosides
Mglahaein	Galactoside transport ATP binding protein [MGLA, HI0823]	*Haemophilus influenzae* [gram-negative bacterium]	Galactosides
Mglatrepa	Methylgalactoside transport ATP binding protein [MGLA]	*Treponema pallidum* [gram-negative bacterium]	Methyl-galactosides
Rbsabacsu	Ribose transport ATP binding protein [RBSA]	*Bacillus subtilis* [gram-positive bacterium]	Ribose
Rbsaescco	Ribose transport ATP binding protein [RBSA]	*Escherichia coli* [gram-negative bacterium]	Ribose
Rbsahaein	Ribose transport ATP binding protein [RBSA, HI0502]	*Haemophilus influenzae* [gram-negative bacterium]	Ribose
Xylgescco	D-Xylose transport ATP binding protein [XYLG]	*Escherichia coli* [gram-negative bacterium]	D-Xylose
Xylghaein	D-Xylose transport ATP binding protein [XYLG, HI1110]	*Haemophilus influenzae* [gram-negative bacterium]	D-Xylose

Phylogenetic tree

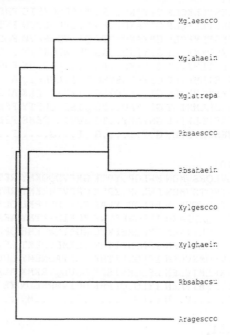

Physical and genetic characteristics

	AMINO ACIDS	MOL. WT	CHROMOSOMAL LOCUS
Aragescco	504	55 018	42.77 minutes
Mglaescco	506	56 401	48.14 minutes
Mglahaein	506	56 567	47.757
Mglatrepa	496	55 157	
Rbsabacsu	453	49 911	
Rbsaescco	501	55 041	84.69 minutes
Rbsahaein	493	54 157	
Xylgescco	513	56 470	80.36 minutes
Xylghaein	503	55 679	64.177

Multiple amino acid sequence alignments

```
          1                                                50
Mglaescco MVSSTTPSSG EYLLEMSGIN KSFPGVKALD NVNLKVRPHS IHALMGENGA
Mglahaein MTAQTQCQDS QVLLTMTNVC KSFPGVKALD NANLTVRSHS VHALMGENGA
Mglatrepa .........M CDVLTIRDLS KSFARNRVLN GVNFRMGKGA VVGLMGENGA
Rbsaescco .........M EALLQLKGID KAFPGVKALS GAALNVYPGR VMALVGENGA
Rbsahaein .........M ETLLKISGVD KSFPGVKALN NACLSVYAGR VMALMGENGA
Xylgescco .........M PYLLEMKNIT KTFGSVKAID NVCLRLNAGE IVSLCGENGS
Xylghaein .......... MALLEMKHIT KKFGDVTALH NISIELEAGE ILSLCGENGS
Rbsabacsu .......... .MQIEMKDIH KTFGKNQVLS GVSFQLMPGE VHALMGENGA
Aragescco ......MQQS TPYLSFRGIG KTFPGVKALT DISFDCYAGQ VHALMGENGA
Consensus .......... ...L...... K.F..V.AL. ........G. ...L.GENGA
```

223

```
           51                                                  100
Mglaescco GKSTLLKCLF GIYQKD..SG TILFQGKEID FHSAKEALEN GISMVHQELN
Mglahaein GKSTLLKCLF GIYAKD..EG EILFLGEPVN FKTSKEALEN GISMVHQELN
Mglatrepa GKSTLMKCLF GMYAKD..TG QILVDGSPVD FQSPKEALEN GVAMVHQELN
Rbsaescco GKSTMMKVLT GIYTRD..AG TLLWLGKETT FTGPKSSQEA GIGIIHQELN
Rbsahaein GKSTLMKVLT GIYSKD..AG TIEYLNRSVN FNGPKASQEA GISIIHQELN
Xylgescco GKSTLMKVLC GIYPHGSYEG EIIFAGEEIQ ASHIRDTERK GIAIIHQELA
Xylghaein GKSTLMKILC GIYPCGDYSG DIYFSESELK ARNIRDTEEK GISIIHQELT
Rbsabacsu GKSRLMNILT GLHKAD..KG QISINGNETY FSNPKEAEQH GIAFIHQELN
Aragescco GKSTLLKILS GNYAPT..TG SVVINGQEMS FSDTTAALNA GVAIIYQELH
Consensus GKSTL.K.L. G.Y......G .I...G.... F......... GI...HQEL.

           101                                                 150
Mglaescco LVLQRSVMDN MWLGR.YPTK GMFVDQDKMY RETKAIFDEL DIDIDPRARV
Mglahaein LVRQTSVMDN LWLGR.YPLK GPFVDHAKMY RDTKAIFDEL DIDVDPKEKV
Mglatrepa QCLDRTVMDN LFLGR.YPAR FGIVDEKRML DDSLTLFASL KMDVNPRAVM
Rbsaescco LIPQLTIAEN IFLGREFVNR FGKIDWKTMY AEADKLLAKL NLRFKSDKLV
Rbsahaein LVGNLTIAEN IFLGREFKTS WGAINWQKMH QEADKLLARL GVTHSSKQLC
Xylgescco LVKELTVLEN IFLGNEITHN .GIMDYDLMT LRCQKLLAQV SLSISPDTRV
Xylghaein LVKNMSVLEN IFLGNEITHK .GLTADNEMY LRCKNLLQQV QLDADPNTRV
Rbsabacsu IWPEMTVLEN LFIGKEISSK LGVLQTRKMK ALAKEQFDKL SVSLSLDQEA
Aragescco LVPEMTVAEN IYLG.QLPHK GGIVNRSLLN YEAGLQLKHL GMDIDPDTPL
Consensus L.....V..N ..LG...... .G......M. .........L ..........

           151                                                 200
Mglaescco GTLSVSQMQM IEIAKGFSYN AKIVIMDEPT SSLTEKEVNH LFTIIRKLKE
Mglahaein AKLSVSQMQM IEIAKAFSYN AKIVIMDEPT SSLSEKEVEH LFKIIDKLKQ
Mglatrepa RSMSVSQRQM VEIAKAMSYN AKIIVLDEPT SSLTEREIVR LFAIIRDLSK
Rbsaescco GDLSIGDQQM VEIAKVLSFE SKVIIMDEPT DALTDTETES LFRVIRELKS
Rbsahaein AELSIGEQQM VEIAKALSFE SKVIIMDEPT DALTDTETEA LFNVIRELKA
Xylgescco GDLGLGQQQL VEIAKALNKQ VRLLILDEPT ASLTEQETSI LLDIIRDLQQ
Xylghaein GELGLGQQQL VEIAKALNKQ VRLLILDEPT ASLTEKETEI LLNLIKDLKA
Rbsabacsu GECSVGQQQM IEIAKALMTN AEVIIMDEPT AALTEREISK LFEVITALKK
Aragescco KYLSIGQWQM VEIAKALARN AKIIAFDEPT SSLSAREIDN LFRVIRELRK
Consensus ...S..Q.QM .EIAKA.... ....I.DEPT ..L...E... LF..I..L..

           201                                                 250
Mglaescco RGCGIVYISH KMEEIFQLCD EVTVLRDGQW IAT.EPLAGL TMDKIIAMMV
Mglahaein RGCGIIYISH KMDEIFKICD EITILRDGKW INT.VNVKES TMEQIVGMMV
Mglatrepa KGVAFIYISH KMDEIFQICS EVIVLRDGVL TLS.QSIGEV EMSDLITAMV
Rbsaescco QGRGIVYISH RMKEIFEICD DVTVFRDGQF IAE.REVASL TEDSLIEMMV
Rbsahaein ENRGIVYISH RLKEIFQICD DVTVLRDGQF IGE.RVMAEI TEDDLIEMMV
Xylgescco HGIACIYISH KLNEVKAISD TICVIRDGQH IGT.RDAAGM SEDDIITMMV
Xylghaein HNIACIYISH KLNEVKAISD KICVIRDGEH VGT.KDASTM TEDDIITMMV
Rbsabacsu NGVSIVYISH RMEEIFAICD RITIMRDGKT VDT.TNISET DFDEVVKKMV
Aragescco EGRVILYVSH RMEEIFALSD AITVFKDGRY VKTFTDMQQV DHDALVQAMV
Consensus .G....YISH ...EIF.I.D ...V.RDG.. ......... ..D.....MV

           251                                                 300
Mglaescco GRSLNQRFPD KENKPGEVIL EVRNLTSLRQ .....PSIRD VSFDLHKGEI
Mglahaein GRELTQRFPE KTNVPKEVIL QVENLTAKNQ .....PSIQD VSFELRKGEI
Mglatrepa GRTLDKRFPD ADNTVGDDYL EIRGLSTRYA .....PQLRD ISLSVKRGEI
Rbsaescco GRKLEDQYPH LDKAPGDIRL KVDNLCG... .....PGVND VSFTLRKGEI
```

```
Rbsahaein GRRLDEQYPH LSQEKGECVL DVKHVSG... .....SGIDD VSFKLHAGEI
Xylgescco GRELTALYPN EPHTTGDEIL RIEHLTAWHP VNRHIKRVND VSFSLKRGEI
Xylghaein GREITSLYPH EPHEIKDEIL RVENLSAWHP INTHIKRVDN VSFSLHEGEI
Rbsabacsu GRELTERYPK RTPSLGDKVF EVKNASVK.. .....GSFED VSFYVRSGEI
Aragescco GRDIGDIYGW QPRSYGEERL RLDAVKA... ....PGVRTP ISLAVRSGEI
Consensus GR.L....P. .....G...L .......... .......D VSF....GEI


          301                                                  350
Mglaescco LGIAGLVGAK RTDIVETLFG I.REKSAGTI TLHGKQINNH NANEAINHGF
Mglahaein LGIAGLVGAK RTDIVEAIFG V.RELIEGTI KLHGKTVKNH TALEAINNGF
Mglatrepa FGLYGLVGAG RSELLEAIFG L.RTIADGEI SLAGKKIRLK SSRDAMKLNF
Rbsaescco LGVSGLMGAG RTELMKVLYG ALP.RTSGYV TLDGHEVVTR SPQDGLANGI
Rbsahaein VGVSGLMGAG RTELGKLLYG ALP.KTAGKV RLKNQEIENR SPQDGLDNGI
Xylgescco LGIAGLVGAG RTETIQCLFG VWPGQWEGKI YIDGKQVDIR NCQQAIAQGI
Xylghaein LGVAGLVGSG RTDMVQCLFG SYEGKFEGNI FINQKQVNIK NCAQAIEHKI
Rbsabacsu VGVSGLMGAG RTEMMRALFG VDRLDT.GEI WIAGKKTAIK NPQEAVKKVS
Aragescco VGLFGLVGAG RSELMKGMFG G.TQITAGQV YIDQQPIDIR KPSHAIAAGM
Consensus .G..GL.GAG RT.......G .......G.. ..........  ....A.....


          351                                                  400
Mglaescco ALVTEERRST GIYAYLDIGF NSLISNIRNY KNKVGLLDNS RMKSDTQWVI
Mglahaein ALVTEERRST GIYSNLSIEF NSLISNMKSY LTPWKLLSTK KMKSDTQWVI
Mglatrepa AFVPEERKLN GMFAKGSIEY NTTIANLPAY K.RYGLLSKK KLQEAAEREI
Rbsaescco VYISEDRKRD GLVLGMSVKE NMSLTALRYF SRAGGSLKHA DEQQAVSDFI
Rbsahaein VYISEDRKGD GLVLGMSVKE NMSLTSLDHF SQKGG.IRHQ AEKMAVDDFI
Xylgescco AMVPEDRKRD GIVPVMAVGK NITLAALNKF TGGISQLDDA AEQKCILESI
Xylghaein VMVPEDRKKH GIVSIMGVGK NITLSSLKSY CFGKMVVNEA KEEQIIGSAI
Rbsabacsu ALLQRIARMK G.......... .......... .......... ..........
Aragescco MLCPEDRKAE GIIPVHSVRD NINISARRKH VLGGCVINNG WEENNADHHI
Consensus ....E.R... G......... N......... .......... .........I


          401                                                  450
Mglaescco DSMRVKTPGH RTQIGSLSGG NQ.QKVIIGR WLLTQPEILM LDEPTRGIDV
Mglahaein DSMNVKTPSH RTTIGSLSGG NQ.QKVIIGR WLLTQPEILM LDEPTRGIDI
Mglatrepa KAMRVKCVSP SELISALSGG NQ.QKVIIGK WLERDPDVLL LDEPTRGIDV
Rbsaescco RLFNVKTPSM EQAIGLLSGG NQ.QKVAIAR GLMTRPKVLI LDEPTRGVDV
Rbsahaein LMFNIKTPNR DQQVGLLSGG NQ.QKVAIAR GLMTRPNVLI LDEPTRGVDV
Xylgescco QQLKVKTSSP DLAIGRLSGG NQ.QKAILAR CLLLNPRILI LDEPTRGIDI
Xylghaein KRLKVKTFSP DLPIGRLSGG NQ.QKAILAK CLSLNPKILI LDEPTRGIDV
Rbsabacsu ...SCSTASP ETHARHLSGG KPGKKVVIAK WIGIGPKVLI LDEPTRGVDV
Aragescco RSLNIKTPGA EQLIMNLSGG NQ.QKAILGR WLSEEMKVIL LDEPTRGIDV
Consensus .....KT... ...I..LSGG NQ.QK..... .L...P..L. LDEPTRG.DV


          451                                                  500
Mglaescco GAKFEIYQLI AELAKKGKGI IIISSEMPEL LGITDRILVM SNGLVSGIVD
Mglahaein GAKFEIYQLI QELAKKDKGI IMISSEMPEL LGVTDRILVM SNGKLAGIVE
Mglatrepa GAKYEIYQLI IRMAREGKTI IVVSSEMPEL LGITNRIAVM SNYRLAGIVD
Rbsaescco GAKKEIYQLI NQFKADGLSI ILVSSEMPEV LGMSDRIIVM HEGHLSGEFT
Rbsahaein GAKKEIYQLI NEFKKEGLSI LMISSDMPEV LGMSDRVLVM REGKISAEFS
Xylgescco GAKYEIYKLI NQLVQQGIAV IVISSELPEV LGLSDRVLVM HEGKLKANLI
Xylghaein GAKYEIYKLI NQLAQEGIAI IVISSELPEV LGISDRVLVM HQGKLKASLI
Rbsabacsu GAKREIYTLM NELTERGVAI IMVSSELPEI LGMSDRIIVV HEGRISGEIH
Aragescco GAKHEIYNVI YALAAQGVAV LFASSDLPEV LGVADRIVVM REGEIAGELL
Consensus GAK.EIY.LI ......G..I II..SSE.PE. LG..DR..VM ..G......
```

```
              501                    525
Mglaescco TKTTTQNEIL RLASLHL... .....
Mglahaein SAKTSQEEIL QLAAKYL... .....
Mglatrepa TKSTDQEALL RLSARYL... .....
Rbsaescco REQATQEVLM AAAVGKLNRV NQE..
Rbsahaein REEATQEKLL AAAIGK... .....
Xylgescco NHNLTQEQVM EAALRSEHHV EKQSV
Xylghaein NTALTQEQVM ETALKE.... .....
Rbsabacsu AREATQERIM TLATGGR... .....
Aragescco HEQADERQAL SLAMPKVSQA VA...
Consensus .....QE... ..A....... .....
```

Residues listed in the consensus sequence are present in at least 75% of the aligned transporter sequences. Residues indicated in boldface type are also conserved in at least one other family of the ABC transporter superfamily.

Database accession numbers

	SWISSPROT	PIR	EMBL/GENBANK
Aragescco	P08531	S01074	X06091; G40946
Mglaescco	P23199	B37277	M59444; G146854
Mglahaein	P44884		L45461; G1005849
Mglatrepa			U45323
Rbsabacsu	P36947		Z25798; G397497
Rbsaescco	P04983	B26304	M13169; G147513
Rbsahaein	P44735		L45143; G1003888
Xylgescco	P37388		U00039; G466705
Xylghaein	P45046		L45746; G1006414

References

[1] Scripture, J.B. et al. (1987) J. Mol. Biol. 197, 37–46.
[2] Buckel, S.D. et al. (1986) J. Biol. Chem. 261, 7659–7662.
[3] Woodson, K. and Devine, K.M. (1994) Microbiology 140, 1829–1838.
[4] **Higgins, C.F. et al. (1992) Annu. Rev. Cell Biol. 8, 67–113.**
[5] Burland, V.D. et al. (1993) Genomics 16, 551–561.

Summary

The majority of transporters of the binding protein-dependent peptide trans-porter family, examples of which are the oligopeptide transport ATP binding protein OPP[1], found in both gram-positive and gram-negative bacteria (e.g. Oppdbacsu, Oppdhaein), and the glutamine[2] and histidine[3] transporters of *Escherichia coli* (Glnqescco, Hispescco), mediate import of amino acids and small peptides. Some members of this family import mono- or oligosaccharides (e.g. Lackagrra)[4], metals (e.g. Nikdescco)[5] or nitrate (e.g. Nrtsynsp)[6]. A few mediate antibiotic resistance – Sap transporters mediate export of, and confer resistance to, antimicrobial peptides – and the substrates of a few are unknown[7]. These transporters are only found in unicellular organisms; most are found in bacteria (both gram-positive and gram-negative), but a few are found in simple eukaryotes.

Statistical analysis of multiple amino acid sequence comparisons places the binding protein-dependent peptide transporter family, in the ATP binding cassette (ABC) superfamily[8]. Proteins in this superfamily use the energy of ATP hydrolysis to pump substrates across cell membranes. In transporters of this family each ATP binding domain forms one chain, with a single ATP binding motif. Two ATP binding domains and two transmembrane domains form the active complex. The family is characterized by the cytoplasmic ATP binding domains[8], which are described in the following tables. The ATP binding domains of each transporter system are associated with two transmem-brane subunits. Most transport systems in this family contain two homologous ATP binding domains (e.g. OPPD and OPPF[1]); others, such as the histidine trans-porter, only contain one ATP binding component (HISP[3] in this example) which presumably functions as a homodimer. In most transporters in this family, each domain of the associated transmembrane proteins is predicted to contain six membrane-spanning helices by the hydropathy of their amino acid sequences. In the case of the transmembrane domains OPPB and OPPC from the oligo-peptide transport system of *Salmonella typhimurium*, this has been confirmed by β-lactamase fusions[9]. Two exceptions to the six-transmembrane-helix rule occur in this family. The transmembrane domains of HISQ and HISM, associated with HISP, each have five transmembrane helices[10] and the MALK protein of the *E. coli* maltose/maltodextrin transport ATP binding protein has been shown experimentally to have eight such helices[11]. The two N-terminal helices can be deleted without loss of transport function[12].

Many residues and short sequence motifs are conserved within the binding protein-dependent peptide transporter family, including motifs unique to the family, signature motifs of the ABC superfamily, and motifs necessary for function by the criterion of site-directed mutagenesis.

Nomenclature, biological sources and substrates

CODE	DESCRIPTION [SYNONYMS]	ORGANISM [COMMON NAMES]	SUBSTRATE(S) [RESISTANCE][a]
Abchaein	ATP binding protein [ABC]	*Haemophilus influenzae* [gram-negative bacterium]	Unknown
Abcxantsp	Probable ATP dependent transporter [YCF16]	*Antithamnion* sp. [red alga]	Unknown

CODE	DESCRIPTION [SYNONYMS]	ORGANISM [COMMON NAMES]	SUBSTRATE(S) [RESISTANCE][a]
Abcxcyapa	Probable ATP dependent transporter [YCF1]	*Cyanophora paradoxa* [flagellate protozoan]	Unknown
Abcxgalsu	Probable ATP dependent transporter [YCF16]	*Galdieria sulphuraria* [alga]	Unknown
Abcxodosi	Probable ATP dependent transporter [YCF16]	*Odontella sinensis* [diatom]	Unknown
Amiestrpn	Oligopeptide transport ATP binding protein [AMIE]	*Streptococcus pneumoniae* [gram-positive bacterium]	Oligopeptides
Amifstrpn	Oligopeptide transport ATP binding protein [AMIE]	*Streptococcus pneumoniae* [gram-positive bacterium]	Oligopeptides
Appdbacsu	Oligopeptide transport ATP binding protein [APPD]	*Bacillus subtilis* [gram-positive bacterium]	Oligopeptides
Appfbacsu	Oligopeptide transport ATP binding protein [APPF]	*Bacillus subtilis* [gram-positive bacterium]	Oligopeptides
Artpescco	Arginine transport ATP binding protein [ARTP]	*Escherichia coli* [gram-negative bacterium]	Arginine
Artphaein	Arginine transport ATP binding protein [ARTP, HI1180]	*Haemophilus influenzae* [gram-negative bacterium]	Arginine
Brafpseae	High-affinity branched-chain amino acid transport ATP binding protein [BRAF]	*Pseudomonas aeruginosa* [gram-negative bacterium]	Branched amino acids
Bztrhoca	BztABCD transporter ATP binding protein	*Rhodobacter capsulatus* [gram-negative bacterium]	Gln/Glu/Asn/Asp
Cysaescco	Sulfate transport ATP binding protein [CYSA]	*Escherichia coli* [gram-negative bacterium]	Sulfate, thiosulfate
Cysasynsp	Sulfate transport ATP binding protein [CYSA]	*Synechococcus* sp. [cyanobacterium]	Sulphur-containing compounds
Devaanasp	DevA protein	*Anabaena* sp. [alga]	Unknown
Dppdbacsu	Dipeptide transport ATP binding protein [DPPD, DCIAD]	*Bacillus subtilis* [gram-positive bacterium]	Dipeptides
Dppdescco	Dipeptide transport ATP binding protein [DPPD]	*Escherichia coli* [gram-negative bacterium]	Dipeptides
Dppdhaein	Dipeptide transport ATP binding protein [DPPD, HI1185]	*Haemophilus influenzae* [gram-negative bacterium]	Dipeptides
Dppfescco	Dipeptide transport ATP binding protein [DPPD, DPPE]	*Escherichia coli* [gram-negative bacterium]	Dipeptides
Dppfhaein	Dipeptide transport ATP binding protein [DPPF, HI1184]	*Haemophilus influenzae* [gram-negative bacterium]	Dipeptides
Drrastrpe	Daunorubicin resistance ATP binding protein [DRRA]	*Streptomyces peucetius* [gram-negative bacterium]	[Daunorubicin, doxorubicin]
Feceescco	Fe^{3+}-Dicitrate transport ATP binding protein [FECE]	*Escherichia coli* [gram-negative bacterium]	Fe^{3+}-Citrate
Fecehaein	Fe^{3+}-Dicitrate transport ATP binding protein [FECE homolog, FECE, HI0361]	*Haemophilus influenzae* [gram-negative bacterium]	Fe^{3+}-Citrate
Ftseescco	Cell division ATP binding protein [FTSE]	*Escherichia coli* [gram-negative bacterium]	Unknown

CODE	DESCRIPTION [SYNONYMS]	ORGANISM [COMMON NAMES]	SUBSTRATE(S) [RESISTANCE][a]
Ftsehaein	Cell division ATP binding protein [FTSE, HI0769]	Haemophilus influenzae [gram-negative bacterium]	Unknown
Glnqbacst	Glutamine transport ATP binding protein [GLNQ]	Bacillus stearothermophilus [gram-positive bacterium]	Glutamine
Glnqescco	Glutamine transport ATP binding protein [GLNQ]	Escherichia coli [gram-negative bacterium]	Glutamine
Gltlescco	Glutamate/aspartate transport ATP binding protein [GLTL]	Escherichia coli [gram-negative bacterium]	Glutamate, aspartate
Gltlhaein	Glutamate/aspartate transport ATP binding protein [GLTL, HI1078]	Haemophilus influenzae [gram-negative bacterium]	Glutamate, aspartate
Gluacorgl	Glutamate transport ATP binding protein [GLUA]	Corynebacterium glutamicum [gram-positive bacterium]	Glutamate
Hispescco	Histidine transport ATP binding protein [HISP]	Escherichia coli [gram-negative bacterium]	Histidine
Hispsalty	Histidine transport ATP binding protein [HISP]	Salmonella typhimurium [gram-negative bacterium]	Histidine
Lackagrra	Lactose transport ATP binding protein [LACK]	Agrobacterium radiobacter [gram-positive bacterium]	Lactose
Livgescco	High-affinity branched-chain amino acid transport ATP binding protein [LIV1 protein, [LIVG]	Escherichia coli [gram-negative bacterium]	Branched amino acids .
Livgsalty	High-affinity branched-chain amino acid transport ATP binding protein [LIV1 protein G, LIVG]	Salmonella typhimurium [gram-negative bacterium]	Branched amino acids
Malkentae	Maltose/maltodextrin transport ATP binding protein [MALK]	Enterobacter aerogenes [gram-negative bacterium]	Maltose, maltodextrin
Malkescco	Maltose/maltodextrin transport ATP binding protein [MALK]	Escherichia coli [gram-negative bacterium]	Maltose, maltodextrin
Malksalty	Maltose/maltodextrin transport ATP binding protein [MALK]	Salmonella typhimurium [gram-negative bacterium]	Maltose, maltodextrin
Mbpxmarpo	Probable transport protein [MBPX]	Marchantia polymorpha [liverwort]	Unknown
Mklmycle	Possible ribonucleotide transport ATP binding protein [MKL]	Mycobacterium leprae [gram-positive bacterium]	Ribonucleotides?
Modcescco	Molybdenum transport ATP binding protein [MODC, CHLD, NARD]	Escherichia coli [gram-negative bacterium]	Molybdenum
Modchaein	Molybdenum transport ATP binding protein [MODC, HI1691]	Haemophilus influenzae [gram-negative bacterium]	Molybdenum
Modcrhoca	Molybdenum transport ATP binding protein [MODC, MOLD]	Rhodobacter capsulatus [gram-negative bacterium]	Molybdenum
Moddazovi	Molybdenum transport ATP binding protein [MODD]	Azotobacter vinelandii [gram-negative bacterium]	Molybdenum

CODE	DESCRIPTION [SYNONYMS]	ORGANISM [COMMON NAMES]	SUBSTRATE(S) [RESISTANCE][a]
Msmkstrmu	Multiple sugar-binding transport ATP binding protein [MSMK]	Streptococcus mutans [gram-positive bacterium]	Melibiose, raffinose, somaltotriose
Nasdklepn	Nitrate transport protein [NASD]	Klebsiella pneumoniae [gram-negative bacterium]	Nitrate
Nikdescco	Nickel transport ATP binding protein [NIKD]	Escherichia coli [gram-negative bacterium]	Nickel
Nikeescco	Nickel transport ATP binding protein [NIKE]	Escherichia coli [gram-negative bacterium]	Nickel
Nocpagrtu	Nopaline permease ATP binding protein [NOCP]	Agrobacterium tumefaciens [gram-negative bacterium]	Nopaline
Nrtcsynsp	Nitrate transport ATP binding protein [NRTC]	Synechococcus sp. [cyanobacterium]	Nitrate
Nrtdsynsp	Nitrate transport ATP binding protein [NRTD]	Synechococcus sp. [cyanobacterium]	Nitrate
Occpagrtu	Octopine permease ATP binding protein P [OCCP]	Agrobacterium tumefaciens [gram-positive bacterium]	Octopine
Oppdbacsu	Oligopeptide transport ATP binding protein [OPPD, SPO0KD]	Bacillus subtilis [gram-positive bacterium]	Oligopeptides
Oppdhaein	Oligopeptide transport ATP binding protein [OPPD, HI1121]	Haemophilus influenzae [gram-negative bacterium]	Oligopeptides
Oppdlacla	Oligopeptide transport ATP binding protein [OPPD]	Lactococcus lactis [gram-positive bacterium]	Oligopeptides
Oppdmycge	Oligopeptide transport ATP binding protein [OPPD, MG079]	Mycoplasma genitalium [gram-negative bacterium]	Oligopeptides
Oppdsalty	Oligopeptide transport ATP binding protein [OPPD]	Salmonella typhimurium [gram-negative bacterium]	Oligopeptides
Oppfbacsu	Oligopeptide transport ATP binding protein [OPPF, SPO0KE]	Bacillus subtilis [gram-positive bacterium]	Oligopeptides
Oppfhaein	Oligopeptide transport ATP binding protein [OPPF, HI1120]	Haemophilus influenzae [gram-negative bacterium]	Oligopeptides
Oppflacla	Oligopeptide transport ATP binding protein [OPPF]	Lactococcus lactis [gram-positive bacterium]	Oligopeptides
Oppfsalty	Oligopeptide transport ATP binding protein [OPPF]	Salmonella typhimurium [gram-negative bacterium]	Oligopeptides
Opuabacsu	Glycine betaine transport ATP binding protein [OPUAA]	Bacillus subtilis [gram-positive bacterium]	Glycine betaine
P29mycge	Probable ABC transporter ATP binding protein P29 [MG290]	Mycoplasma genitalium [gram-negative bacterium]	Unknown
Pebccamje	Probable ABC transporter ATP binding protein [PEB1C]	Campylobacter jejuni [gram-negative bacterium]	Amino acid?
Phncescco	Phosphonates transport ATP binding protein [PHNC]	Escherichia coli [gram-negative bacterium]	Alkylphosphonates
Phnkescco	Phosphonates transport ATP binding protein [PHNK]	Escherichia coli [gram-negative bacterium]	Alkylphosphonates

CODE	DESCRIPTION [SYNONYMS]	ORGANISM [COMMON NAMES]	SUBSTRATE(S) [RESISTANCE][a]
Phnlescco	Phosphonates transport ATP binding protein [PHNL]	Escherichia coli [gram-negative bacterium]	Alkylphosphonates
Potaescco	Spermidine/putrescine transport ATP binding protein [POTA]	Escherichia coli [gram-negative bacterium]	Spermidine, putrescine
Potahaein	Spermidine/putrescine transport ATP binding protein [POTA, HI1347]	Haemophilus influenzae [gram-negative bacterium]	Spermidine, putrescine
Provescco	Glycine betaine/L-proline transport ATP binding protein [PROV]	Escherichia coli [gram-negative bacterium]	Glycine betaine, L-proline
Provsalty	Glycine betaine/L-proline transport ATP binding protein [PROV]	Salmonella typhimurium [gram-negative bacterium]	Glycine betaine, L-proline
Pstbescco	Phosphate transport ATP binding protein [PSTB, PHOT]	Escherichia coli [gram-negative bacterium]	Phosphate
Sapdhaein	Peptide transport system ATP binding protein [SAPD, HI1641]	Haemophilus influenzae [gram-negative bacterium]	[Antimicrobial peptides]
Sapdsalty	Peptide transport system ATP binding protein [SAPD]	Salmonella typhimurium [gram-negative bacterium]	[Antimicrobial peptides]
Sapfescco	Peptide transport system ATP binding protein [SAPF]	Escherichia coli [gram-negative bacterium]	[Antimicrobial peptides]
Sapfhaein	Peptide transport system ATP binding protein [SAPF, HI1642]	Haemophilus influenzae [gram-negative bacterium]	[Antimicrobial peptides]
Sapfsalty	Peptide transport system ATP binding protein [SAPF]	Salmonella typhimurium [gram-negative bacterium]	[Antimicrobial peptides]
Sfucserma	Fe^{3+}-transport ATP binding protein [SFUC]	Serratia marcescens [gram-negative bacterium]	Fe^{3+}
Ugpcescco	sn-Glycerol-3-phosphate transport ATP binding protein [UGPC]	Escherichia coli [gram-negative bacterium]	sn-Glycerol-3-phosphate

[a] Presumed substrates; protein confers resistance to specified compounds.

Phylogenetic tree

Proteins listed subsequently in italics are at least 90% identical to the paired transporters listed in parenthesis and are therefore not included in the phylogenetic tree: *Hispsalty* (Hispescco); *Livgsalty* (Livgescco); *Malksalty, Malkentae* (Malkescco); *Sapfsalty, Sapfhaein* (Sapfescco); *Provsalty* (Provescco).

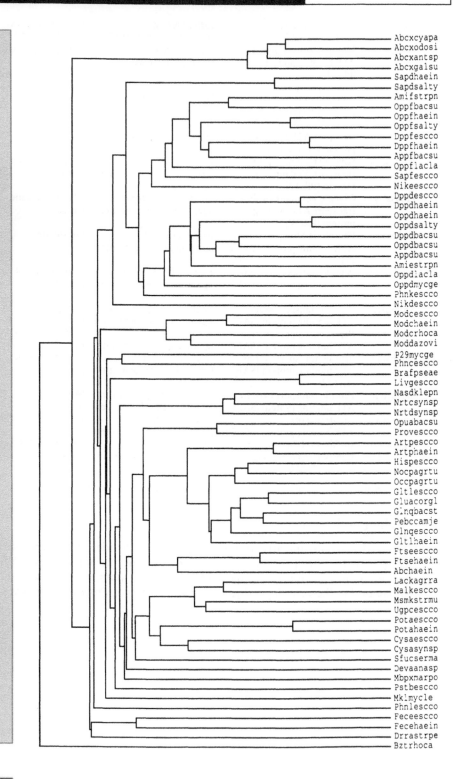

Abcxcyapa
Abcxodosi
Abcxantsp
Abcxgalsu
Sapdhaein
Sapdsalty
Amifstrpn
Oppfbacsu
Oppfhaein
Oppfsalty
Dppfescco
Dppfhaein
Appfbacsu
Oppflacla
Sapfescco
Nikeescco
Dppdescco
Dppdhaein
Oppdhaein
Oppdsalty
Dppdbacsu
Oppdbacsu
Appdbacsu
Amiestrpn
Oppdlacla
Oppdmycge
Phnkescco
Nikdescco
Modcescco
Modchaein
Modcrhoca
Moddazovi
P29mycge
Phncescco
Brafpseae
Livgescco
Nasdklepn
Nrtcsynsp
Nrtdsynsp
Opuabacsu
Provescco
Artpescco
Artphaein
Hispescco
Nocpagrtu
Occpagrtu
Gltlescco
Gluacorgl
Ginqbacst
Pebccamje
Glnqescco
Gltlhaein
Ftseescco
Ftsehaein
Abchaein
Lackagrra
Malkescco
Msmkstrmu
Ugpcescco
Potaescco
Potahaein
Cysaescco
Cysasynsp
Sfucserma
Devaanasp
Mbpxmarpo
Pstbescco
Mklmycle
Phnlescco
Feceescco
Fecehaein
Drrastrpe
Bztrhoca

Physical and genetic characteristics

	AMINO ACIDS	MOL. WT	CHROMOSOMAL LOCUS
Abchaein	345	37 877	35.757
Abcxantsp	251	28 205	ATP synthase operon
Abcxcyapa	259	28 995	
Abcxgalsu	257	29 072	ATP synthase operon
Abcxodosi	251	28 316	
Amiestrpn	355	39 546	ami locus
Amifstrpn	308	34 743	ami locus
Appdbacsu	328	36 311	app operon
Appfbacsu	329	37 112	app operon
Artpescco	242	26 844	19.47 minutes
Artphaein	243	27 138	68.119
Brafpseae	255	28 284	bra operon
Bztrhoca	263	30 012	
Cysaescco	365	41 054	cysTWAM gene cluster
Cysasynsp	344	38 476	
Devaanasp	244	26 723	
Dppdbacsu	335	36 681	dciA operon
Dppdescco	327	35 844	79.73 minutes
Dppdhaein	330	36 328	68.309
Dppfescco	334	37 560	
Dppfhaein	327	36 917	68.255
Drrastrpe	330	35 700	drrAB locus
Feceescco	255	28 190	97.15 minutes
Fecehaein	306	34 251	45.418
Ftseescco	222	24 439	77.57 minutes
Ftsehaein	218	24 349	
Glnqbacst	242	27 436	glnQH operon
Glnqescco	240	26 731	18.23 minutes
Gltlescco	241	26 661	
Gltlhaein	257	28 894	
Gluacorgl	242	26 485	glu gene cluster
Hispescco	257	28 667	52.21 minutes
Hispsalty	258	28 771	histidine transport operon
Lackagrra	363	39 324	lac operon
Livgescco	255	28 427	77.38 minutes
Livgsalty	255	28 452	liv cluster
Malkentae	294	32 247	malB region
Malkescco	371	40 974	91.41 minutes
Malksalty	369	40 755	malB region
Mbpxmarpo	370	42 799	
Mklmycle	347	37 583	cosmid B1790
Modcescco	352	39 144	17.21 minutes
Modchaein	351	39 582	96.211
Modcrhoca	363	38 545	
Moddazovi	380	41 573	
Msmkstrmu	377	41 964	msm operon
Nasdklepn	261	28 725	nasFEDCBA operon
Nikdescco	253	26 503	77.89 minutes
Nikeescco	268	29 619	77.9 minutes
Nocpagrtu	257	28 188	Plasmid PTIC58
Nrtcsynsp	659	72 346	
Nrtdsynsp	274	30 365	
Occpagrtu	262	28 954	Plasmid pTiA6
Oppdbacsu	336	37 196	opp operon
Oppdhaein	323	35 720	64.847

	AMINO ACIDS	MOL. WT	CHROMOSOMAL LOCUS
Oppdlacla	338	37 349	
Oppdmycge	402	45 494	
Oppdsalty	335	36 864	opp operon
Oppfbacsu	307	34 912	
Oppfhaein	332	36 760	64.793
Oppflacla	319	35 976	opp operon
Oppfsalty	334	37 214	opp operon
Opuabacsu	418	46 468	upstream of amyE
P29mycge	245	28 006	
Pebccamje	242	27 262	
Phncescco	262	29 430	93.11 minutes
Phnkescco	252	27 831	92.98 minutes
Phnlescco	226	24 705	92.96 minutes
Potaescco	378	43 028	25.46 minutes
Potahaein	381	43 429	77.723
Provescco	400	44 162	60.38 minutes
Provsalty	400	44 210	proU locus
Pstbescco	257	29 027	84.1 minutes
Sapdhaein	349	39 478	
Sapdsalty	330	37 611	
Sapfescco	268	30 570	29.05 minutes
Sapfhaein	269	30 294	93.27
Sapfsalty	268	30 671	Plasmid pEG6162
Sfucserma	345	36 692	sfu region
Ugpcescco	356	39 523	77.27 minutes

Multiple amino acid sequence alignments

```
                 1                                                50
Abcxcyapa ................. .MSTEKTKIL EVKNLKAQV. ...........
Abcxodosi ................. .MNTN.YPIL EIKNLKACI. ...........
Abcxantsp ................. ...MNNRILL NIKNLDVTI. ...........
Abcxgalsu ................. ...MKHKSLL QIKNLHVKL. ...........
Sapdhaein ................. .....MALL DICNLNIEIQ T..........
Sapdsalty ................. ....MPLL DIRNLTIEFK T..........
Amifstrpn ................. ....MSEKLV EIKDLEISFG .........E
Oppfbacsu ................. .MNELTEKLL EIKHLKQHFV .........T
Oppfhaein ............... .M TVSNNKELLL EVNHLGVSFK IKNDKSLFFA
Oppfsalty ............... .MN AVIEQRKVLL EIADLKVHFD IKEGKQWFWQ
Dppfescco ............. .MST QEATLQQPLL QAIDLKKHYP VK..KGMFA.
Dppfhaein ............. .MT NEVKENTPLL NAIGLKKYYP VK..KGLFA.
Appfbacsu .............. MTAANQETIL ELRDVKKYFP IR..SGLFQR
Oppflacla .............. .....MSEIL NLKDLKVYYP IR..SGFFNR
Sapfescco .............. ....MIETLL EVRNLSKTFR YRTGW....F
Nikeescco ............. .....MTLL NISGLSHHYA ....HGGFNG
Dppdescco ............. .....MALL NVDKLSVHFG D.........
Dppdhaein ............. .....MALL DVKELSVHFG D.........
Oppdhaein ............ .....MNPLL DVKNLYVRFK T.........
Oppdsalty .......... .MSLSETAT QAPQPANVLL EVNDLRVTFA T.........
Dppdbacsu ............. .....MEKVL SVQNLHVSFT T.........
Oppdbacsu ............. ..MIRVTRLL EVKDLAISFK T.........
Appdbacsu ............. ....MSTLL EVNNLKTYFF R.........
Amiestrpn ............. .MTKEKNVIL TARDIVVEFD V.........
```

```
Oppdlacla ................. ...MESENIL EAKQVSVAFR I........
Oppdmycge ........MA LKRSNFFVDK DQQLKDNLIL DITDLHVNFK V........
Phnkescco .............. ....MNQPLL SVNNLTHLY. ..........
Nikdescco .............. .....MPQQI ELRNIAL... ..........
Modcescco ................. ...MLELNFS Q........T
Modchaein ................. ...MLQINVK K........Q
Modcrhoca ................. ...MISARFS G........R
Moddazovi ......... MNTSFEPGTK GETGALSGGL PADGIRARFR V.......D
P29mycge ............ ...MENKPIL SFEKVSIIY. ..........
Phncescco ............ ...MQTII RVEKLAKTF. ..........
Brafpseae ............ ...MSRPIL EVSGLTMRF. ..........
Livgescco ............ ...MSQPLL SVNGLMMRF. ..........
Nasdklepn ............ ...MKPLI QVQGVTSVLA P.......P
Nrtcsynsp ............ ...MSVFL AVDHVHQVFD L.......P
Nrtdsynsp ............ ...MTAILPS TAATVNTGFL HFDCVGKTFP T.......P
Opuabacsu MSVDEKPIKI KVEKVSKIFG KQTKKAVQML ANGKTKKEIL K.......A
Provescco .....MAIKL EIKNLYKIFG EHPQRAFKYI EQGLSKEQIL E........K
Artpescco ................ ...MSI QLNGINCFY. ..........
Artphaein ................ ...MAI RVKNLNFFY. ..........
Hispescco ................ ....MSENKL NVIDLHKRY. ..........
Nocpagrtu ................ ...MDATQPTL VAEDVHKNF. ..........
Occpagrtu ................ .MPNPVRPAV QLKDIRKNF. ..........
Gltlescco ................ ...MI TLKNVSKWY. ..........
Gluacorgl ................ ...MI KMTGVQKYF. ..........
Glnqbacst ................ ...MI YFHQVNKYY. ..........
Pebccamje ................ ...MI ELKNVNKYY. ..........
Glnqescco ................ ...MI EFKNVSKHF. ..........
Gltlhaein ................ ...MSML KVSNIQKNF. ..........
Ftseescco ................ ...MI RFEHVSKAYL ..........
Ftsehaein ................ ...MI KFSNVSKAYH ..........
Abchaein ................ ...MI KLNNIXKIFE L.......P
Lackagrra ................ ...MAEV RLTDIRKSY. .......G
Malkescco ................ ...MASV QLQNVTKAW. .......G
Msmkstrmu ................ ...MVEL NLNHIYKKY. .......P
Ugpcescco ................ ...MAGL KLQAVTKSW. .......D
Potaescco ............ ..MGQSKKLN KQPSSLSPLV QLAGIRKCF. .......D
Potahaein ............ MKLGRKPKLV ENQLQNKPII ELRSIKKSY. .......G
Cysaescco ................ .......MSI EIANIKKSF. .......G
Cysasynsp ................ .MPKDKAVGI QVSQVSKQF. .......G
Sfucserma ................ ...MSTL ELHGIGKSY. .......N
Devaanasp ................ ...MRQEAVI AIKSLNHYYG K.......G
Mbpxmarpo ................ ...MSI LIYKVSKSL. ..........
Pstbescco ............ ...M SMVETAPSKI QVRNLNFYY. ..........
Mklmycle .......... ...MAAIGG DGRMPMGVAI EVKGLTKSF. ..........
Phnlescco ................ ...MI NVQNVSKTFI LH.......QQ
Feceescco ................ .............M TLRTENLTVS
Fecehaein ................ ....MDSFST SIWVNDVTVR
Drrastrpe ................ ....MNTQPT RAIETSGLVK
Bztrhoca ................ ..........MSDT
Consensus ................ ..........
```

```
          51                                                    100
Abcxcyapa DGTE..ILKG VNLTINSGEI HAIMGPN..G SGKSTFSKIL AG...HPAYQ
Abcxodosi NENE..ILKD LNLKIHKGEI HAIMGPN..G SGKSTFSKVL AG...HPAYN
```

```
Abcxantsp GETQ..ILNS LNLSIKPGEI HAIMGKN..G SGKSTLAKVI AG...HPSYK
Abcxgalsu ANTEEYILNG INLNVNQGEI HAIMGPN..G SGKSTLSKVI AG...HSLYK
Sapdhaein SNGRIKIVDG VNLSLNEGEI SGLVGES..G SGKSLIAKVI CNAIKEN.WI
Sapdsalty SEGWVKAVDR VSMTLSEGEI RGLVGES..G SGKSLIAKAI CGVAKDN.WR
Amifstrpn GSKKFVAVKN ANFFINKGET FSLVGES..G SGKTTIGRAI IGLNDT....
Oppfbacsu PRGTVKAVDD LSFDIYKGET LGLVGES..G CGKSTTGRSI IRLYEA....
Oppfhaein KPQTLKAVKD VSFKLYAGET LGVVGES..G CGKSTLARAI IGLVEA....
Oppfsalty PPKTLKAVDG VTLRLYEGT LGVVGES..G CGKSTFARAI IGLVKA....
Dppfescco PERLVKALDG VSFNLERGKT LAVVGES..G CGKSTLGRLL TMIEMP....
Dppfhaein KPQQVKALDG VSFQLERGKT LAVVGES..G CGKSTLGRLL TMIEEP....
Appfbacsu KVGDVKAVDG VSFSLKKGET LGIVGES..G CGKSTAGRTM IRLYKP....
Oppflacla VTDNVLAVDG VDLTIHEGET VGLVGES..G SGKSTIGKTI VGLEQM....
Sapfescco RRQTVEAVKP LSFTLREGQT LAIIGEN..G SGKSTLAKML AGMIEP....
Nikeescco KHQHQAVLNN VSLTLKSGET VALLGGT..G CGKSTLARLL VGLESP....
Dppdescco ESAPFRAVDR ISYSVKQGEV VGIVGES..G SGKSVSSLAI MGLID.YPGR
Dppdhaein KKTPFKAVDR ISYQVAQGEV LGIVGES..G SGKSVSSLAI MGLID.HPGR
Oppdhaein PDGVVTAVND LNFTLNAGST LGIVGES..G SGKSQTAFAL MGLLAANGE.
Oppdsalty PDGDVTAVND LNFTLRAGET LGIVGES..G SGKSQTAFAL MGLLATNGR.
Dppdbacsu YGGTVQAVRG VSFDLYKGET FAIVGES..G CGKSVTSQSI MGLLPPYSAK
Oppdbacsu YGGEVQAIRG VNFHLDKGET LAIVGES..G SGKSVTSQAI MKLIPMPPGY
Appdbacsu KKEPIPAVDG VDFHISKGET VALVGES..G SGKSITSLSI MGLVQSSGGK
Amiestrpn RDKVLTAIRG VSLELVEGEV LALVGES..G SGKSVLTKTF RGMLEENG.R
Oppdlacla AGKFQKAIYD IDLSLKRGEV LAIVGES..G SGKSTFATAV MGLHNPNQTQ
Oppdmycge KDGILHAVRG IDLKVERGSI VGIVGES..G SGKSVSVKSI IGFNDNAQTK
Phnkescco ..APGKGFSD VSFDLWPGEV LGIVGES..G SGKTTLLKSI SARLTPQQ..
Nikdescco .QAAQPLVHG VSLTLQRGRV LALVGGS..G SGKSLTCAAT LGILPA.GVR
Modcescco LGNHCL...T INETLPANGI TAIFGVS..G AGKTSLINAI SGL.....TR
Modchaein LGQLAL...Q ANIQVPDQGV TAIFGLS..G SGKTSLINLV SGL.....IQ
Modcrhoca QGDFTL...D AAFDVPGQGV TALFGPS..G CGKTTVLRCM AGL.....TR
Moddazovi YAGFAL...D VDLTLPGHGV TALFGHS..G SGKTTLLRCV AGL.....ER
P29mycge ..KKAPLLQN ISFKVMAKEN VCLLGKS..G VGKSSLL....NSVT..NTK
Phncescco ..NQHQALHA VDLNIHHGEM VALLGPS..G SGKSTLLRHL SGLIT..GDK
Brafpseae ..GGLLAVNG VNLKVEEKQV VSMIGPN..G AGKTTVFNCL TGFY.....Q
Livgescco ..GGLLAVNN VNLELYPQEI VSLIGPN..G AGKTTVFNCL TGFY.....K
Nasdklepn ..AQFLALQN VSFDIYEGET ISLIGHS..G CGKSTLLNLI AGI.....AL
Nrtcsynsp GGGQYIALKD VSLNIRPGEF ISLIGHS..G CGKSTLLNLI AGL.....AQ
Nrtdsynsp .RGPYVAIED VNLSVQQGEF ICVIGHS..G CGKSTLLNLV SGF.....SQ
Opuabacsu T.GSTVGVNQ ADFEVYDGEI FVIMGLS..G SGKSTLVRML NRL.....IE
Provescco T.GLSLGVKD ASLAIEEGEI FVIMGLS..G SGKSTMVRLL NRL.....IE
Artpescco ..GAHQALFD ITLDCPQGET LVLLGPS..G AGKSSLLRVL NLL.....EM
Artphaein ..GSSQTLFD INLEAEEGDT VVLLGPS..G AGKSTLIRTL NLL.....EV
Hispescco ..GEHEVLKG VSLQANAGDV ISIIGSS..G SGKSTFLRCI NFL.....EK
Nocpagrtu ..GTLEILKG ISLTANKGDV VSIIGSS..G SGKSTFLRCM NFL.....ET
Occpagrtu ..GNLEVLHG VSLSANEGEV ISILGSS..G SGKSTLLRCV NML.....EV
Gltlescco ..GHFQVLTD CSTEVKKGEV VVVCGPS..G SGKSTLIKTV NGL.....EP
Gluacorgl ..GDFHALTD IDLEIPRGQV VVVLGPS..G SGKSTLCRTI NRL.....ET
Glnqbacst ..GDFHVLKD INLTIHQGEV VVIIGPS..G SGKSTLVRCI NRL.....ET
Pebccamje ..GTHHVLKI FNLSVKEGEK LVIIGPS..G SGKSTTIRCM NGL.....EE
Glnqescco ..GPTQVLHN IDLNIAQGEV VVIIGPS..G SGKSTLLRCI NKL.....EE
Gltlhaein ..NGNHVLKG IDFEINKGEV VAILGPS..G SGKTTFLRCL NLL.....ER
Ftseescco G.GRQ.ALQG VTFHMQPGEM AFLTGHS..G AGKSTLLKLI CGI.....ER
Ftsehaein G.ATQPALQG LNFHLPVGSM TYLVGHS..G AGKSTLLKLI MGM.....EK
Abchaein X.KKLTALDN VSLNIEKGQI CGVIGAS..G AGKSTLIRCV NLL.....EK
Lackagrra S.LE..VIKG VNLEVSSGEF VVFVGPS..G CGKSTLLRMI AGL.....ED
```

```
Malkescco E.VV..VSKD INLDIHEGEF VVFVGPS..G CGKSTLLRMI AGL.....ET
Msmkstrmu N.SSHYSVED FDLDIKNKEF IVFVGPS..G CGKSTTLRMV AGL.....ED
Ugpcescco G.KTQ.VIKP LTLDVADGEF IVMVGPS..G CGKSTLLRMV AGL.....ER
Potaescco G.KE..VIPQ LDLTINNGEF LTLLGPS..G CGKTTVLRLI AGL.....ET
Potahaein S.NT..IIND FNLTINNGEF VTILGPS..G CGKTTVLRLL AGL.....EE
Cysaescco R.TQ..VLND ISLDIPSGQM VALLGPS..G SGKTTLLRII AGL.....EH
Cysasynsp S.FQ..AVKD VDLTVETGSL VALLGPS..G SGKSTLLRLI AGL.....EQ
Sfucserma A.IR..VLEH IDLQVAAGSR TAIVGPS..G SGKTTLLRII AGF.....EI
Devaanasp A.LKRQILFD INLEIYPGEI VIMTGPS..G SGKTTLLSLI GGL.....RS
Mbpxmarpo ..GNLKILDR VSLYVPKFSL IALLGPS..G SGKSSLLRII AGL.....DN
Pstbescco ..GKFHALKN INLDIAKNQV TAFIGPS..G CGKSTLLRTF NKMFELYPEQ
Mklmycle  ..GSSRIWED VTLDIPAGEV SVLLGPS..G TGKSVFLKSL IGLL.....R
Phnlescco NGVRLPVLNR ASLTVNAGEC VVLHGHS..G SGKSTLLRSL YA........
Feceescco YGTD.KVLND VSLSLPTGKI TALIGPN..G CGKSTLLNCF SRLL.....M
Fecehaein YNNGHTAIHN MTFSLNSGTI CALVGVN..G SGKSTLFKSI MGLV.....K
Drrastrpe VYNGTRAVDG LDLNVPAGLV YGILGPN..G AGKSTTIRML ATLL.....R
Bztrhoca  SFVRTEMLAP RPAPVSQVGA IKWMRENLFS GPLNTALTVF GLLATVWLVQ
Abcxcyapa VTGGEILFKN KNLL..... ........E LEPEERARAG VFLAFQYPIE
Abcxodosi ILSGDILFKG SSIL..... ........N LDPEERSHMG IFLAFQYPIE
Abcxantsp ITNGQILFEN QDVT..... ........E IEPEDRSHLG IFLAFQYPVE
Abcxgalsu VVKGEIIFQD QNLL..... ........N YTIEDRANLG IFLAFQYPLE
Sapdhaein ITADRFRFHD VELLKL.... ......SPNK RRKLVGKEIS MIFQNPLSCL
Sapdsalty VTADRMRFDD IDLLRL.... ......SSRE RRKLVGHNVS MIFQEPQSCL
Amifstrpn .SNGDIIFDG QKINGK.... ......KSRE QAAELIRRIQ MIFQDPAASL
Oppfbacsu .TDGEVLFNG ENVHGR.... ......KSRK KLLEFNRKMQ MIFQDPYASL
Oppfhaein .SEGEILWLG KHLRKQ.... ......SAKQ WKET.RKDIQ MIFQDPLASL
Oppfsalty .TDGKVAWLG KDLLGM.... ......KADE WREV.RSDIQ MIFQDPLASL
Dppfescco .TGGELYYQG QDLLK..... ......HDPQ AQKLRRQKIQ IVFQNPYGSL
Dppfhaein .TKGELYYKG HNFLE..... ......NDSE TKALRRKKIQ IVFQNPYASL
Appfbacsu .TEGQILFKG QDISNL.... ......SEEK LRKSVRKNIQ MVFQDPFASL
Oppflacla .TSGQLIYKG QDVSKK.... ......KIRN QLK.YNKDVQ MIFQDAFSSL
Sapfescco .TSGELLIDD HPLHFG.... ......DYS. ...FRSQRIR MIFQDPSTSL
Nikeescco .AQGNISWRG EPLAKL.... ......N.RA QAKAFRRDIQ MVFQDSISAV
Dppdescco VMAEKLEFNG QDLQRI.... ......SEKE RRNLVGAEVA MIFQDPMTSL
Dppdhaein VSAESLQFEN TDLLTL.... ......ESKA KRQLIGADVA MIFQDPMTSL
Oppdhaein .VEGSAIFEG KELVNL.... ......PNAE LNKIRAEQIS MIFQDPMTSL
Oppdsalty .IGGSATFNG REILNL.... ......PERE LNTLRAEQIS MIFQDPMTSL
Dppdbacsu VTDGRILFKN KDLCRL.... ......SDKE MRGIRGADIS MIFQDPMTAL
Oppdbacsu FKRGEILFEG KDLVPL.... ......SEKE MQNVRGKEIG MIFQDPMTSL
Appdbacsu IMDGSIKLED KDLTSF.... ......TEND YCKIRGNEVS MIFQEPMTSL
Amiestrpn IAEGSIDYRG QDLTALS... ......SHKD WEQIRGAKIA TIFQDPMTSL
Oppdlacla IT.GSILLDD EEVIG.K... ......TGDS MASIRGSKVG MIFQNPLTAL
Oppdmycge ..AKLMNFKN VDITKLK... ......KH.Q WKYYRGTYVS YISQDPLFSL
Phnkescco ...GEIHYEN RSLYAM.... ......SEAD RRRLLRTEWG VVHQHPLDGL
Nikdescco QTAGEILADG KPVSP..... ..... .CALRGIKIA TIMQNPRSAF
Modcescco PQKGRIVLNG RVLND..... .....AEKGI CLTPEKRRVG YVFQD..ARL
Modchaein PDEGFICLND RTLVD..... .....MESQE SLPTHLRKIG YVFQD..ARL
Modcrhoca LPGGHLVVNG VTWQE..... .....GRQ. ITPPHRRAVG YVFQE..ASL
Moddazovi AAEARLEING ELWQD..... .....SAAGV FLPTHRRALG YVFQE..ASL
P29mycge  IVKSGLVYFD GVASNK.... KEYKKL... .....KKQCS YLDQI..PNL
Phncescco SVGSHIELLG RTVQREGRLA RDIRKS.... ....RAHTG YIFQQ..FNL
Brafpseae PTGGLIRLDG EEIQGLPG.. ..........HKIARKGVV RTFQN..VRL
Livgescco PTGGTILLRD QHLEGLPG.. ..........QQIARMGVV RTFQH..VRL
Consensus ...G...... .......... .......... .......FQ.....L
```

```
           101                                                    150
Nasdklepn PTEGGLLCDN REIAG.... ........... ....PGPERA VVFQN..HSL
Nrtcsynsp PSSGGIILEG RQVTE.... ........... ....PGPDRM VVFQN..YSL
Nrtdsynsp PTSGGVYLDG QPIQE.... ........... ....PGPDRM VVFQN..YSL
Opuabacsu PTAGNIYIDG DMIT...... ....NMSKDQ LREVRRKKIS MVFQK..FAL
Provescco PTRGQVLIDG VDIA...... ....KISDAE LREVRRKKIA MVFQS..FAL
Artpescco PRSGTLNIAG NHFDFTKT.P S......DKA IRDLRR.NVG MVFQQ..YNL
Artphaein PKSGELSIAN NEFNLSNAMA N......PKA IQQLRQ.DVG MVFQQ..YHL
Hispescco PSEGSIVVNG QTINLVRDKD GQLKVADKNQ LRLLRT.RLT MVFQH..FNL
Nocpagrtu PNKGRIAVGQ EEVVVKTDAA GRLIGVDRKK IERMRM.QLG MVFQS..FNL
Occpagrtu PNAGSVAIMG EEIALEHRAG RLARPKDLKQ VNRLRE.RAA MVFQG..FNL
Gltlescco VQQGEITVDG IVVN.....D K......KTD LAKLRS.RVG MVFQH..FEL
Gluacorgl IEEGTIEIDG KVLP.....E E......GKG LANLRA.DVG MVFQS..FNL
Glnqbacst ISSGELIVDN VKVN.....D K......HID INQLRR.NIG MVFQH..FNL
Pebccamje VSSGEVVVNN LVLN.....H K......N.K IEICRK.YCA MVFQH..FNL
Glnqescco ITSGDLIVDG LKVN.....D P......KVD ERLIRQ.EAG MVFQQ..FYL
Gltlhaein PEQGILEFTD GSLK.....I DFSQKISKAD ELKLRR.RSS MVFQQ..YNL
Ftseescco PSAGKIWFSG HDIT...... ....RLKNRE VPFLRR.QIG MIFQD..HHL
Ftsehaein ANAGQIWFNG HDIT...... ....RLSKYE IPFLRR.QIG MVHQD..YRL
Abchaein  PTSGSVIVDG VELT...... ....KLSDRE LVLARR.QIG MIFQH..FNL
Lackagrra ISSGELTIGG TVMN............D.... .VDPSKRGIA MVFQT..YAL
Malkescco ITSGDLFIGE KRMN............D.... .TPPAERGVG MVFQS..YAL
Msmkstrmu ITKGELKIDG EVVN............D.... .KAPKDRDIA MVFQN..YAL
Ugpcescco VTEGDIWIND QRVT............E.... .MEPKDRGIA MVFQN..YAL
Potaescco VDSGRIMLDN EDIT............H.... .VPAENRYVN TVFQS..YAL
Potahaein LDSGSIILDG EDIT............N.... .VPAEKRHIN TVFQS..YAL
Cysaescco QTSGHIRFHG TDVS............R.... .LHARDRKVG FVFQH..YAL
Cysasynsp PDSGRIFLTG RDAT............N.... .ESVRDRQIG FVFQH..YAL
Sfucserma PDGGQILLQG QAMG............NGSG WVPAHLRGIG FVPQD..GAL
Devaanasp VQEGNLQFLG VELS...... ....GASQNK LVQIR.RSIG YIFQA..HNL
Mbpxmarpo CDYGNIWLHG IDVTN............ISTQYRRMS FVFQH..YAL
Pstbescco RAEGEILLDG DNI....... .....LTNSQ DIALLRAKVG MVFQK..PTP
Mklmycle  PERGSILIDG TDIIECSAKE LYEI....... .....RTLFG VLFQD..GAL
Phnlescco .....NYLPD EGQIQIKHGD EWVDLVTAPA RKVVEIRKTT VGWVSQFLRV
Feceescco PQSGTVFLGD NPINMLSS.. ........... .RQLARRLSL LPQHHLTPE.
Fecehaein PQQGEIKLCD LPI....S.. ........... .QALKRNLVA YVPQSEEVDW
Drrastrpe PDGGTARVFG HDVT..SE.. ........... .PDTVRRRIS VTGQYASVDE
Bztrhoca  AAAPWLLHGV WNANSLTECR AIIAERWGPE ATGACWAVIR VRWNQFLFGF
Consensus ...G...... ........... .......... ........FQ .....L
```

```
           151                                                    200
Abcxcyapa IAGVSNIDFL RLAYNNRRKE E.................................
Abcxodosi IPGVSNEDFL RLAYNSKQKF L.................................
Abcxantsp IPGVTNADFL RIAYNAKRAF D.................................
Abcxgalsu IAGVNNIDFL RLAYNSKLKF N.................................
Sapdhaein DPSRKIGKQL IQNIPNWTFK NKWWKWFGW. ................
Sapdsalty DPSERVGRQL MQNIPAWTYK GRWWQRLGW. ................
Amifstrpn NERATVDYII SEGLYN.....................
Oppfbacsu NPRMTVADII AEGLDI.....................
Oppfhaein NPRMNIGEII AEPLKI.....................
Oppfsalty NPRMTIGEII AEPLRT.....................
Dppfescco NPRKKVGQIL EEPLLI.....................
Dppfhaein NPRKKIGSIL EEPLII.....................
Appfbacsu NPRKTLRSII KEPFNT.....................
```

```
Oppflacla NPRKTIYDII AEPIRN.... ........................
Sapfescco NPRQRISQIL DFPLRL.... ........................
Nikeescco NPRKTVREIL REPMR..... ........................
Dppdescco NPCYTVGFQI MEAIKV.... ........................
Dppdhaein NPAYTVGFQI MEALKT.... ........................
Oppdhaein NPYMKIGEQL MEVLQL.... ........................
Oppdsalty NPYMRVGEQL MEVLML.... ........................
Dppdbacsu NPTLTVGDQL GEALLR.... ........................
Oppdbacsu NPTMKVGKQI TEVLFK.... ........................
Appdbacsu NPVLTIGEQI TEVLIY.... ........................
Amiestrpn DPIKTIGSQI TEVIVK.... ........................
Oppdlacla NPLMKIGQQI KEMLAV.... ........................
Oppdmycge NPTMTIGKQV KEAIYVASKR RYFQAKSDLK FALSNKEIDK KTYKSKLKEI
Phnkescco RRQVSAGGNI GERLMATGAR HY........ ........................
Nikdescco NPLHTMHTHA RET....... ........................
Modcescco FPHYKVRGNL RYGMS..... ........................
Modchaein FPHYTVKGNL RYGMK..... ........................
Modcrhoca FTHLSVRENL VYGLR..... ........................
Moddazovi FPHLSVRRNL EYGMK..... ........................
P29mycge  IDTDYVYEAI LRSAKQKLTW L.......QK LI......... ........
Phncescco VNRLSVLENV LIGALGSTPF W.......RT CF......... ........
Brafpseae FKEMTAVENL LVAQHRHLNT NFLAGLFKTP AF......... ........
Livgescco FREMTVIENL LVAQHQQLKT GLFSGLLKTP SF......... ........
Nasdklepn LPWLTCFDNV ALAVDQVFR. R.......... ........
Nrtcsynsp LPWRTVRQNI ALAVDSVL.. H.......... ........
Nrtdsynsp LPWKSARDNI ALAVKAA.R. P.......... ........
Opuabacsu FPHRTILENT EYG....LEL Q.......... ........
Provescco MPHMTVLDNT AFG....MEL A.......... ........
Artpescco CAHLTVQQNL IEA...PCRV L.......... ........
Artphaein WPHLTVIENL IEA...PMKV R.......... ........
Hispescco WSHMTVLENV MEA...PIQV L.......... ........
Nocpagrtu WGHMTVLQNV MEG...PLHV L.......... ........
Occpagrtu WSHQTILQNV MEA...PVHV Q.......... ........
Gltlescco FPHLSIIENL TLA...QVKV L.......... ........
Gluacorgl FPHLTIKDNV TLA...PIKV R.......... ........
Glnqbacst YPHMTVLQNI TLA...PMKV L.......... ........
Pebccamje YPHMTVLQNL TLA...PMKL Q.......... ........
Glnqescco FPHLTALENV MFG...PLRV R.......... ........
Gltlhaein FPHRSALENV MEG...MVVV Q.......... ........
Ftseescco LMDRTVYDNV AI....PLII A.......... ........
Ftsehaein LTDRTVVENV AL....PLII A.......... ........
Abchaein  LSSRTVFENV AL....PLEL E.......... ........
Lackagrra YPHMTVRENM GFALRF.... A.......... ........
Malkescco YPHLSVAENM SFGLKP.... A.......... ........
Msmkstrmu YPHMSVYDNM AFGLKL.... R.......... ........
Ugpcescco YPHMSVEENM AWGLKI.... R.......... ........
Potaescco FPHMTVFENV AFGLRM.... Q.......... ........
Potahaein FPHMTIFENV AFGLRM.... Q.......... ........
Cysaescco FRHMTVFDNI AFGLTVLPRR E.......... ........
Cysasynsp FKHLTVRKNI AFGLEL.... R.......... ........
Sfucserma FPHFTVAGNI GFGLK..... .......... ........
Devaanasp LGFLTARQNV QMAVELNEHI .......... ........
Mbpxmarpo FKHMTVYENI SFG....LRL R.......... ........
Pstbescco FP.MSIYDNI AFGVRLFEKL S.......... ........
```

239

```
Mklmycle  FGSMNLYDNT AFPLREHTK..................................
Phnlescco  IPRISALEVV MQPL.......................................
Feceescco  ..GITVQELV SYGRNPWLSL W...............................
Fecehaein  QFPVSVYDVV MMGRYGYMNF L...............................
Drrastrpe  GLTGT.ENLV MMGR.....L Q...............................
Bztrhoca   YPVDQYWRLF VTFAGLFLAL APVLFDALPR KLIWGTLLYP LAAFWLL....
Consensus  .....................................................
```

```
           201                                               250
Abcxcyapa  ...GLTELDP LTFYSIVKEK LNVVKMD.PH FLNRNVNEGF SGGEKKRNEI
Abcxodosi  ...NKDEVDP ISFFTIINEK LKLVDMS.PV FLSRNVNEGF SGGEKKRNEI
Abcxantsp  ...NKEELDP LSFFSFIENK ISNIDLN.ST FLSRNVNEGF SGGEKKKNEI
Abcxgalsu  ...QRNPVDP LKFLEIVYPK LKLVGLD.ES FLYRKVNEGF SGGEKKKNEI
Sapdhaein  ...........KKRRAIEL LHRVGIKDHR DIMASYPNEL TEGEGQKVMI
Sapdsalty  ...........RKRRAIEL LHRVGIKDHK EPMRSFPYEL TDGECQKVMI
Amifstrpn  ...HRLFKDE EERKEKVQSI IREVGLL..A EHLTRYPHEF SGGQRQRIGI
Oppfbacsu  ...HKLAKTK KERMQRVHEL LETVGLN..K EHANRYPHEF SGGQRQRIGI
Oppfhaein  ...YQPHLSA AEVKEKVQAM MLKVGLL..P NLINRYPHEF SGGQCQRIGI
Oppfsalty  ...YHPKLSR QDVRDRVKAM MLKVGLL..P NLINRYPHEF SGGQCQRIGI
Dppfescco  ....NTSLSK EQRREKALSM MAKVGLK..T EHYDRYPHMF SGGQRQRIAI
Dppfhaein  ....NTKLSA KERREKVLSM MEKVGLR..A EFYDRYPHMF SGGQRQRIAI
Appfbacsu  ....HNMYTM RERNEKVEEL LARVGLH..P SFAGRYPHEF SGGQRQRIGI
Oppflacla  ...FEKIDAN TENK.RIHEL LDIVGLP..K QALEQYPFQF SGGQQQRIGI
Sapfescco  ....NTDLEP EQRRKQIIET MRMVGLL..P DHVSYYPHML APGQKQRLGL
Nikeescco  ...HLLSLKK SEQLARARQM LKAVDLD..D SVLDKRPPQL SGGQLQRVCL
Dppdescco  ....HQGGNK STRRQRAIDL LNQVGIPDPA SRLDVYPHQL SGGMSQRVMI
Dppdhaein  ....HEGGTK KARKDRTLEL LKLVGIPDPE SRIDVYPHQL SGGMSQRVMI
Oppdhaein  ....HKGYDK QTAFAESVKM LDAVKMPEAK KRMGMYPHEF SGGMRQRVMI
Oppdsalty  ....HKGMSK AEAFEESVRM LDAVKMPEAR KRMKMYPHEF SGGMRQRVMI
Dppdbacsu  ....HKKMSK KAARKEVLSM LSLVGIPDPG ERLKQYPHQF SGGMRQRIVI
Oppdbacsu  ....HEKISK EAAKKRAVEL LELVGIPMPE KRVNQFPHEF SGGMRQRVVI
Appdbacsu  ....HKNMKK KEARQRAVEL LQMVGFSRAE QIMKEYPHRL SGGMRQRVMI
Amiestrpn  ....HQGKTA KEAKELAIDY MNKVGIPDAD RRFNEYPFQY SGGMRQRIVI
Oppdlacla  ....HDVYPE NQYESRIFQL LEQVGIPNPK RVVNQFPHQL SGGMRQRVMI
Oppdmycge  KQTYQQKIKP INVEKKTLEI LQFIGINDAK KRLKAFPSEF SGGMRQRIVI
Phnkescco  ..........GDIRATAQKW LEEVEIP..A NRIDDLPTTF SGGMQQRLQI
Nikdescco  ....CLALGK PADDATLTAA IEAVGLENAA RVLKLYPFEM SGGMLQRMMI
Modcescco  ......KSMV D....QFDKL VALLGIE..P .LLDRLPGSL SGGEKQRVAI
Modchaein  ......NVSQ D....DFNYI VDLLGIT..H .LLKRYPLTL SGGEKQRVAI
Modcrhoca  ......RARG PLR.ISEAEV TQLLGID..P .LLRRPTATL SGGERQRVAI
Moddazovi  ......RVDA ASRQVSWERV LELLGIG..H .LLERLPGRL SGGERQRVGI
P29mycge   ......CFEP KWIKDKILAI LKEVNLN..D YVSCII.KDL SAGQKQRVEI
Phncescco  ......SWFT GEQKQRALQA LTRVGMV..H FAHQRV.STL SGGQQQRVAI
Brafpseae  ......RRSE REAMEYAAHW LEEVNLT..E FA.NRSAGTL AYGQQRRLEI
Livgescco  ......RRAQ SEALDRAATW LERIGLL..E HA.NRQASNL AYGDQRRLEI
Nasdklepn  ......SMSK GERKEWIEHN LERVQMG..H .ALHKRPGEI SGGMKQRVGI
Nrtcsynsp  ......DRNR TERRTIIEET IDLVGLR..A .AADKYPHEI SGGMKQRVAI
Nrtdsynsp  ......HLST SEQRQVVDHH LELVGLT..E .AQHKRPDQL SGGMKQRVAI
Opuabacsu  ......GVDK QERQQKALES LKLVGLE..G FE.HQYPDQL SGGMQQRVGL
Provescco  ......GINA EERREKALDA LRQVGLE..N YA.HSYPDEL SGGMRQRVGL
Artpescco  ......GLSK DQALASAEKL LERLRLK..P YS.DRYPLHL SGGQQQRVAI
Artphaein  ......GVSE NEAKTDAMEL LKRLRLE..Q LA.DRFPLHL SGGQQQRVAI
Hispescco  ......GLSK QEARERAVKY LAKVGID..E RAQGKYPVHL SGGQQQRVSI
Nocpagrtu  ......KQPK GEVRDRAMDF LDKVGIA..N K.HAAYPSQL SGGQQQRVSI
```

```
Occpagrtu ......GRDR KACRDEAEAL LERVGIA..S K.RDAYPSEL SGGQQQRAAI
Gltlescco ......KRDK APAREKALKL LERVGLS..A HA.NKFPAQL SGGQQQRVAI
Gluacorgl ......KMKK SEAEKLAMSL LERVGIA..N QA.DKYPAQL SGGQQQRVAI
Glnqbacst ......RIPE KEAKETAMYY LEKVGIP..D KA.NAYPSEL SGGQQQRVAI
Pebccamje ......KKSK KEAEETAFKY LKVVGLL..D KA.NVYPATL SGGQQQRVAI
Glnqescco ......GANK EEAEKLAREL LAKVGLA..E RA.HHYPSEL SGGQQQRVAI
Gltlhaein ......KQDK AQAREKALSL LEKVGLK..N KA.DLFPSQL SGGQQQRVGI
Ftseescco ......GASG DDIRRRVSAA LDKVGLL..D KA.KNFPIQL SGGEQQRVGI
Ftsehaein ......GMHP KDANTRAMAS LDRVGLR..N KA.HYLPPQI SGGEQQRVDI
Abchaein  .....SESK AKIQEKITAL LDLVGLS..E KR.DAYPSNL SGGQKQRVAI
Lackagrra ......GMAK DEIERRVNAA AKILELD... ALMDRKPKAL SGGQRQRVAI
Malkescco ......GAKK EVINQRVNQV AEVLQLA... HLLDRKPKAL SGGQRQRVAI
Msmkstrmu ......HYSK EAIDKRVKEA AQILGLT... EFLERKPADL SGGQRQRVAM
Ugpcescco ......GMGK QQIAERVKEA ARILELD... GLLKRRPREL SGGQRQRVAM
Potaescco ......KTPA AEITPRVMEA LRMVQLE... TFAQRKPHQL SGGQQQRVAI
Potahaein ......KVPN EEIKPRVLEA LRMVQLE... EMADRKPTQL SGGQQQRIAI
Cysaescco ......RPNA AAIKAKVTKL LEMVQLA... HLADRYPAHV SGGQKQRVAL
Cysasynsp ......KHTK EKVRARVEEL LELVQLT... GLGDRYPSQL SGGQRQRVAL
Sfucserma .......GGK REKQRRIEAL MEMVALD..R RLAALWPHEL SGGQQQRVAL
Devaanasp ........SQ EEAIAKAEAM LTAVGLE..N RV.DYYPENL SGGQKQRVAI
Mbpxmarpo ......GFSA QKITNKVNDL LNCLRIA..D ISFE.YPAQL SGGQKQRVAL
Pstbescco ......RADM DERVQWALTK AALWNET..K DKLHQSGYSL SGGQQQRLCI
Mklmycle  .......KKE SEIRDIVMEK LQLVGLG..G DE.KKFPGEI SGGMRNVPGL
Phnlescco ...LDTGVPR EACAAKAARL LTRLNVP..E RLWHLAPSTF SGGEQQRVNI
Feceescco ......GRLS AEDNARVNVA MNQTRI...N HLAVRRLTEL SGGQRQRAFL
Fecehaein ......RIPK AIDKQKVQEA MQRVNI...E HLAHRQIGEL SGGQKKRVFL
Drrastrpe ......GYSW ARARERAAEL IDGFGL...G DARDRLLKTY SGGMRRRLDI
Bztrhoca  .............WGGPI WGPVSVLAGF AILGLLFTAL APKLGVPVSA
Consensus .............................P... SGG..QR..I
```

                251   300
```
Abcxcyapa LQMALLNPSL AILDETDSGL DIDALRIVAE GVNQLSNK...ENSIILITH
Abcxodosi LQMILLDSEL SILDETDSGL DIDALKIISK GINTFMNQ...NKAIILITH
Abcxantsp LQMSLLNSKL AILDETDSGL DIDALKTIAK QINSLKTQ...ENSIILITH
Abcxgalsu LQMSLLDAKL AILDETDSGL DIDALQDISN AIKSILKMSQ FKQSIIIITH
Sapdhaein AMAVANQPRL LIADEPTNAL ESTTALQVFR .L.LSSMNQN QGTTILLTSN
Sapdsalty AIALANQPRL LIADEPTNSM EPTTQAQIFR .L.LTRLNQN SNTTILLISH
Amifstrpn ARALVMQPDF VIADEPISAL DVSVRAQVLN .L.LKKFQKE LGLTYLFIAH
Oppfbacsu ARALAVDPEF IIADEPISAL DVSIQAQVVN .L.MKELQKE KGLTYLFIAH
Oppfhaein ARALIIEPKM IICDEPVSAL DVSIQAQVVN .L.LKSLQKE MGLSLIFIAH
Oppfsalty ARALILEPKL IICDDAVSAL DVSIQAQVVN .L.LQQLQRE MGLSLIFIAH
Dppfescco ARGLMLDPDV VIADEPVSAL DVSVRAQVLN .L.MMDLQQE LGLSYVFISH
Dppfhaein ARGLMLDPDV VVADEPVSAL DVSVRAQVLN .L.MMDLQDE LGLSYVFISH
Appfbacsu ARALTLNPEL IIADEPVSAL DVSIQPQVIN .L.MEELQEE FNLTYLFISH
Oppflacla ARAVATNPKL IVADEPVSAL DLSVQAQVLN .F.MKLIQKD LGIAFLFISH
Sapfescco ARALILRPKV IIADEALASL DMSMRSQLIN .L.MLELQEK QGISYIYVTQ
Nikeescco ARALAVEPKL LILDEAVSNL DLVLQAGVIR .L.LKKLQQQ FGTACLFITH
Dppdescco AMAIACRPKL LIADEPTTAL DVTIQAQIIE .L.LLELQQK ENMALVLITH
Dppdhaein AMAIACRPKL LIADEPTTAL DVTIQAQIME .L.LLELQKK ECMSLILITH
Oppdhaein AMALLCRPKL LIADEPTTAL DVTVQAQIMT .L.LNELKRE FNTAIIMITH
Oppdsalty AMALLCRPKL LIADEPTTAL DVTVQAQIMT .L.LNELKRE FNTAIIMITH
Dppdbacsu AMALICEPDI LIADEPTTAL DVTIQAQILE .L.FKEIQRK TDVSVILITH
Oppdbacsu AMALAANPKL LIADEPTTAL DVTIQAQILE .L.MKDLQKK IDTSIIFITH
Appdbacsu AIALSCNPKL LIADEPTTAL DVTIQAQVLE .L.MKDLCQK FNTSILLITH
```

```
Amiestrpn AIALACRPDV LICDEPTTAL DVTIQAQIID .L.LKSLQNE YHFTTIFITH
Oppdlacla AIAIANDPDL IIADEPTTAL DVTIQAQILD .L.ILEIQKK KNAGVILITH
Oppdmycge AIAVATEPDL IIADEPTTAL DVTIQAKVLT .L.IKQLRDL LNITIIFISH
Phnkescco ARNLVTHPKL VFMDEPTGGL DVSVQARLLD .L.LRGLVVE LNLAVVIVTH
Nikdescco AMAVLCESPF IIADEPTTDL DVVAQGGASS .ICWKHYAKQ CGNAA..GDH
Modcescco GRALLTAPEL LLLLDEPLASL DIPRKRELLP .Y.LQRLTRE INIPMLYVSH
Modchaein GRALLTDPDI LLMDEPLSAL DVPRKRELMQ .Y.LERLSKE INIPILYVTH
Modcrhoca GRALLSQPEL LLMDEPLSAL DRISRDEILP .Y.LERLHAS LQMPVILVSH
Moddazovi ARALLTSPRL LLMDEPLAAL DLKRKNEILP .Y.LERLHDE LDIPMLFVSH
P29mycge  AKLFFKSPKL LLVDEPTTGL DPLTASKIMD .L.ITDFVKR EKITLVFVTH
Phncescco ARALMQQAKV ILADEPIASL DPESARIVMD .T.LRDINQN DGITVVVTLH
Brafpseae ARCMMTRPRI LMLDEPAAGL NPKETDDLKA .L.IAKLRSE HNVTVLLIEH
Livgescco ARCMVTQPEI LMLDEPAAGL NPKETKELDE .L.IAELRNH HNTTILLIEH
Nasdklepn ARALAMKPKV LLLLDEPFGAL DALTRAHLQD .A.VMQIQQS LNTTIVMITH
Nrtcsynsp ARGLAIRPKL LLLLDEPFGAL DALTRGNLQE .Q.LMRICQE AGVTAVMVTH
Nrtdsynsp ARALSIRPEV LILDEPFGAL DAITKEELQE .E.LLNIWEE ARPTVLMITH
Opuabacsu ARALTNDPDI LLMDEAFSAL DPLIRKDMQD .E.LLDLHDN VGKTIIFITH
Provescco ARALAINPDI LLMDEAFSAL DPLIRTEMQD .E.LVKLQAK HQRTIVFISH
Artpescco ARALMMEPQV LLFDEPTAAL DPEITAQIVS .I.IRELAET N.ITQVIVTH
Artphaein ARALMMKPQV LLFDEPTAAL DPEITAQVVD .I.IKELQET G.ITQVIVTH
Hispescco ARALAMEPEV LLFDEPTSAL DPELVGELLR .I.MQQLAEE GK.TMVVVTH
Nocpagrtu ARALAMQPSA LLFDEPTSAL DPELVGEVLK .V.IRKLAEE GR.TMVVVTH
Occpagrtu ARALAMRPDV MLFDEPTSAL DPELVGEVLK .V.MRDLAAE GR.TMLIVTH
Glt1escco ARALCMDPIA MLFDEPTSAL DPEMINEVLD .V.MVELANE G.MTMMVVTH
Gluacorgl ACALAMNPKI MLFDEPTSAL DPEMVNEVLD .V.MASLAKE G.MTMVCVTH
Glnqbacst ARGLAMKPKI MLFDEPTSAL DPETIGEVLD .V.MKQLAKE G.MTMVVVTH
Pebccamje ARSLCTKKPY ILFDEPTSAL DPETIQEVLD .V.MKEISHQ SNTTMVVVTH
Glnqescco ARALAVKPKM MLFDEPTSAL DPELRHEVLK .V.MQDLAEE G.MTMVIVTH
Glt1haein ARALAVKPDI ILLDEPTSAL DPELVGEVLQ .T.LKMLAQE GW.TMIIVTH
Ftseescco ARAVVNKPAV LLADEPTGNL DDALSEGILR .L.FEEFNR. VGVTVLMATH
Ftsehaein ARAIVHKPQL LLADEPTGNL DDELSLGIFN .L.FEEFNR. LGMTVLIATH
Abchaein  ARALASDPKV LLCDEATSAL DPATTQSILK .L.LKEINRT LGITILLITH
Lackagrra GRAIVRQPDV FLFDEPLSNL DAELRVHMRV .E.IARLHKE LNATIVYVTH
Malkescco GRTLVAEPSV FLLDEPLSNL DAALRVQMRI .E.ISRLHKR LGRTMIYVTH
Msmkstrmu GRAIVRDAKV FLMDEPLSNL DAKLRVSMRA .E.IAKIHRR IGATTIYVTH
Ugpcescco GRAIVRDPAV FLFDEPLSNL DAKLRVQMRL .E.LQQLHRR LKTTSLYVTH
Potaescco ARAVVNKPRL LLLDESLSAL DYKLRKQMQN .E.LKALQRK LGITFVFVTH
Potahaein ARAVVNKPKV LLLDESLSAL DYKLRKQMQQ .E.LKMLQRQ LGITFIFVTH
Cysaescco ARALAVEPQI LLLDEPFGAL DAQVRKELRR .W.LRQLHEE LKFTSVFVTH
Cysasynsp ARALAVQPQV LLLDEPFGAL DAKVRKDLRS .W.LRKLHDE VHVTTVFVTH
Sfucserma ARALSQQPRL MLLDEPFSAL DTGLRAATRK .A.VAELLTE AKVASILVTH
Devaanasp ARALVNNPPL VLADEPTAAL DKQSGRDVVE .I.MQRLAKD QGTSILLVTH
Mbpxmarpo ARSLAIQPDF LLLDEPFGAL DGELRRHLSK .W.LKRYLQD NKITTIMVTH
Pstbescco ARGIAIRPEV LLLDEPCSAL DPISTGRIEE .L.ITELKQD ..YTVVIVTH
Mklmycle  ARALVLDPQI ILCDEPDSGL DPVRTAYLSQ .L.IMDINAQ IDATILIVTH
Phnlescco ARGFIVDYPI LLLDEPTASL DAKNSAAVVE ...LIREAKT RGAAIVGIFH
Feceescco AMVLAQNTPV VLLDEPTTYL DINHQVDLMR ...LMGELRT QGKTVVAVLH
Fecehaein ARALAQQSPI ILLDEPFTGV DVKTENAIVD ...LLQQLRE EGHLILVSTH
Drrastrpe AASIVVTPDL LFLDEPTTGL DPRSRNQVWD ...IVRALVD AGTTVLLTTQ
Bztrhoca  GIGLVVAALF WLY..AAAPI EAALQSALPL ALPEVDSDQF GGFLLALVIG
Consensus A......P.. ...DEP...L D......... ........... .........H
```

301   350
Abcxcyapa YQRLLDYIVP DYIHVMQNGR ILKTGGAELA KELEIKGYDW LNELEMVKK.

```
Abcxodosi YQRLLDYVQP NYVHVMQNGK IIKTGTADLA KELESKGYEW LK........
Abcxantsp YQRLLDYIKP DYIHVMQKGE IIYTGGSDTA MKLEKYGYDY LNK.......
Abcxgalsu YQRILNYIQP DYIHVMYKGK IIKTGDASLA NQLESQGYEW LASE......
Sapdhaein DIKSISEW.C DQISVLYCGQ N........T E...SAPTEI L.IESPHHPY
Sapdsalty DLQMLSQW.A DKINVLYCGQ T........V E...TAPSKD L.VTMPHHPY
Amifstrpn DLSVVRFI.S DRIAVIYKGV I........V E...VAETEE L.FNNPIHPY
Oppfbacsu DLSMVKYI.S DRIGVMYFGK L........V E...LAPADE L.YENPLHPY
Oppfhaein DLAVVKHI.S DRVLVMYLGN A........M E...LGSDVE V.YNDTKHPY
Oppfsalty DLAVVKHI.S DRVLVMYLGH A........V E...LGTYDE V.YHNPLHPY
Dppfescco DLSVVEHI.A DEVMVMYLGR C........V E...KGTKDQ I.FNNPRHPY
Dppfhaein DLSVVEHI.A DEVMVMYLGR C........I E...KGTTEQ I.FSNPQHPY
Appfbacsu DLSVVRHI.S DRVGVMYLGK M........M E...LTGKHE L.YDNPLHPY
Oppflacla DLGVVRHM.T DNIAVMHNGR I........V E...KGTRRD I.FDEPQHIY
Sapfescco HIGMMKHI.S DQVLVMHQGE V........V E...RGSTAD V.LASPLHEL
Nikeescco DLRLVERF.C QRVMVMDNGQ I........V ETQVVGEKLT F.SSDAGRVL
Dppdescco DLALVAEA.A HKIIVMYAGQ V........V E...TGDAHA I.FHAPRHPY
Dppdhaein DLALVAEA.A ERIIVMYAGQ I........V E...EGTAKD I.FREPKHPY
Oppdhaein DLGVVAGI.C DQVMVMYAGR T........M E...YGTAEQ I.FYHPTHPY
Oppdsalty DLGVVAGI.C DKVLVMYAGR T........M E...YGKARD V.FYQPVHPY
Dppdbacsu DLGVVAQV.A DRVAVMYAGK M........A E...IGTRKD I.FYQPQHPY
Oppdbacsu DLGVVANV.A DRVAVMYAGQ I........V E...TGTVDE I.FYDPRHPY
Appdbacsu DLGVVSEA.A DRVIVMYCGQ V........V E...NATVDD L.FLEPLHPY
Amiestrpn DLGVVASI.A DKVAVMYAGE I........V E...YGTVEE V.FYDPRHPY
Oppdlacla DLGVVAEV.A DTVAVMYAGQ L........V E...KTSVEE L.FQNPKHPY
Oppdmycge NISLIANF.C DFVYVMYAGK I........V E...QGLVEE I.FTNPLHPY
Phnkescco DLGVARLL.A DRLLVMKQGQ V........V E...SGLTDR V.LDDPHHPY
Nikdescco DMGVVARL.A DDVAVMSDGK I........V E...QGDVET L.FNAPKHTV
Modcescco SLDEILHL.A DRVMVLENGQ VKAFGALEEV WGSSVMNPWL P.KEQQSSIL
Modchaein SLDELLRL.A DRVVLMENGI VKAYDRVEKI WNSPIFAPWK G.ESEQSSVL
Modcrhoca DLSEVERL.A DTLVLMEAGR VRAAGPIAAM QADPNL.PLI H.RPDLAAVI
Moddazovi LPDEVARL.A DHVVLLDQGR VTAQGSLQDI MARLDL.PTA F.HEDAGVVI
P29mycge  DIDLALKY.S TRIIALKNH. .......... ALVLDRLTEK L.TKEQLYKI
Phncescco QVDYALRY.C ERIVALRQG. .......... HVFYDGSSQQ F.DNERFDHL
Brafpseae DMKLVMSI.S DHIVVINQGA P......... ..LADGTPEQ I.RDNP..DV
Livgescco DMKLVMGI.S DRIYVVNQGT P......... ..LANGTPEQ I.RNNP..DV
Nasdklepn DVDEAVLL.S DRVLMMTNGP AATVGEILDV NLPRPANRVQ L.ADDSRYHH
Nrtcsynsp DVDEALLL.S DRVVMLTNGP AAQIGQILEV DFPRPRQRLE M.METPHYYD
Nrtdsynsp DIDEALFL.A DRVVMMTNGP AATIGEVLEI PFDRPREREA V.VEDPRYAQ
Opuabacsu DLDEALRI.G DRIVLMKD.. ........G. NIVQIGTPEE I.LMNPSNEY
Provescco DLDEAMRI.G DRIAIMQN.. ........G. EVVQVGTPDE I.LNNPANDY
Artpescco EVEVARKT.A SRVVVYMEN. ........G. HIVEQGDASC ..FTEPQTEA
Artphaein EVNVAQKV.A TKVVVYMEQ. ........G. KIVEMGSADC ..FENPKTEQ
Hispescco EMGFARHV.S THVIFLHQ.. ........G. KIEEEGAPEQ L.FGNPQSPR
Nocpagrtu EMGFARDV.S SKVLFLEK.. ........G. QIEEQGTPQE V.FQNPTSPR
Occpagrtu EMDFARDV.S SRTVFLHQ.. ........G. VIAEEGPSSE M.FAHPRTDR
Gltlescco EMGFARKV.A NRVIFMDE.. ........G. KIVEDSPKDA F.FDDPKSDR
Gluacorgl EMGFARA.A DRVLFMAD.. ........G. LIVEDTEPDS F.FTNPKSDR
Glnqbacst EMGFAREV.A DRIVFMDQ.. ........G. RILEEAPPEE F.FSNPKEER
Pebccamje EMGFAKEV.A DRIIFMED.. ........G. AIVEENIPSE F.FSNPKTER
Glnqescco EIGFAEKV.A SRLIFIDK.. ........G. RIAEDGNPQV L.IKNPPSQR
Gltlhaein EMQFAKDV.A DRVILMAD.. ........G. HIVEQNTADK F.FTCPQHER
Ftseescco DINLISRR.S YRMLTLSD.. ........G. HLHGGVGHE. ..........
Ftsehaein DINLIQQK.P KPCLVLEQ.. ........G. YLRY...... ..........
Abchaein  EMEVVKQI.C DQVAVIDQ.. ........G. RLVEQGTVGE I.FANPKTEL
```

243

```
Lackagrra DQVEAMTL.A DKIVVMRG.. ........G. IVEQVGAPLA L.YDDPDNMF
Malkescco DQVEAMTL.A DKIVVLDA.. ........G. RVAQVGKPLE L.YHYPADRF
Msmkstrmu DQTEAMTL.A DRIVIMSSTK NEDGSGTIG. RVEQVGTPQE L.YNRPANKF
Ugpcescco DQVEAMTL.A QRVMVMNG.. ........G. VAEQIGTPVE V.YEKPASLF
Potaescco DQEEALTM.S DRIVVMRD.. ........G. RIEQDGTPRE I.YEEPKNLF
Potahaein DQEEAITM.S DRIVLLRK.. ........G. KIAQDGSPRE I.YEDPANLF
Cysaescco DQEEATEV.A DRVVVMSQ.. ........G. NIEQADAPDQ V.WREPATRF
Cysasynsp DQEEAMEV.A DQIVVMNH.. ........G. KVEQIGSPAE I.YDNPATPF
Sfucserma DQSEALSF.A DQVAVMRS.. ........G. RLAQVGAPQD L.YLRPVDEP
Devaanasp D.NRILDI.A DRIVEMEDG. ........ILARDSQTA I.VSYDSGAW
Mbpxmarpo DQKEAISM.A DEIVILKE.. ........G. RLLQQGKPKN L.YDQPINFF
Pstbescco NMQQAARC.S DHTAFM.... ....YLG ELIEFSNTDD L.FTKPAKKQ
Mklmycle  NVNIARTV.P DNMGMLFRKH LVMFGPREVL LTSDEPVVRQ F.LNGRRIGP
Phnlescco DEAVRNDV.A DRLHPMGASS .......... .......... ..........
Feceescco DLNQASRY.C DQLVVMANGH VMAQGTPE.. .......... ..........
Fecehaein NLGSVPDF.C DQ.VVMINRT VIAAGKTE.. .......... ....DTFNQH
Drrastrpe YLDEADQL.A DRIAVIDHGR VIAEGTTGEL KSSLGSNVLR LRLHDAQSRA
Bztrhoca  VTAIVVSLPL GILLALGRQS DMLIVKSLSV GIIEFVRGVP LITLLFTASL
Consensus .......... D......... .......... .......... ..........
```

```
          351                                                  400
Sapdhaein TQALINA.VP DFTQPLGFKT KLGTLEGTAP ILEQMPI.GC RLGPRCPFAQ
Sapdsalty TQALIRA.IP DFGSAMPHKS RLNTLPGAIP LLEQLPI.GC RLGPRCPYAQ
Amifstrpn TQALLSA.VP IPDPILERKK VLKVYEGSQH .......... ..........
Oppfbacsu TKSLLSA.IP LPDPDYERNR C.SEYDPSVH .......... ..........
Oppfhaein TKALMSA.VP IPDPKLERNK SIELLEGDLP SPINPP.SGC VFRTRCLKAD
Oppfsalty TKALMSA.VP IPDPDLERNK KIQLLEGELP SPINPP.SGC VFRTRCPIAG
Dppfescco TQALLSA.TP ..RLNPDDRR ERIKLSGELP SPLNPP.PGC AFNARCRRRF
Dppfhaein TKALLSA.TP ..RLSPNLRR ERIKLTGELP SPINPP.KGC AFNPRCWKAT
Appfbacsu TQALLSS.VP VTRKRGSVKR ERIVLKGELP SPANPP.KGC VFHTRCPVAK
Oppflacla TKRLLSA.IP SIDVTRRAEN RKNRLKVEQD FEDKKA.NFY DKDGHALPLK
Sapfescco TKRLIAG.HF GEALTADAWR KDR....... .......... ..........
Nikeescco QNAVLPA.FP VRRRTTEKV. .......... .......... ..........
Dppdescco TQALLRA.LP ..EFA.QDKE RLASLPGVVP GKYDRP.NGC LLNPRCPYAT
Dppdhaein TQALLRS.LP ..EFA.EGKS RLESLQGVVP GKYDRP.TGC LLNPRCPYAT
Oppdhaein SIGLMDA.IP ..RLDGNE.E HLVTIPGNPP NLLHLP.KGC PFSPRCQFAT
Oppdsalty SIGLLNA.VP ..RLDSEG.A EMLTIPGNPP NLLRLP.KGC PFQPRCPHAM
Dppdbacsu TKGLLGS.VP ..RLDLNG.A ELTPIDGTPP DLFSPP.PGC PFAARCPNRM
Oppdbacsu TWGLLAS.MP ..TLESSGEE ELTAIPGTPP DLTNPP.KGD AFALRSSYAM
Appdbacsu TEGLLTS.IP ..VID.GEID KLNAIKGSVP TPDNLP.PGC RFAPRCPKAM
Amiestrpn TWSLLSS.LP ..QLA.DDKG DLYSIPGTPP SLYTDL.KGD AFALRSDYAM
Oppdlacla TRSLLRS.NP ..S.AETVSD DLYVIPGSVP SLSKIEYDKD LFLARVPWMK
Oppdmycge TWALISS.IP ..E.QKDKNK PLTSIPGVIP NMLTPP.KGD AFASRNQYAL
Phnkescco TQLLVSS.VL ..QN...... .......... .......... ..........
Nikdescco TRTLVSAHLA LYGMDLAS.. .......... .......... ..........
Modcescco KVTVLEHHSA LRDDRLALGD QHLWVNKLDE PLQAALRIRI QASDVSLVLQ
Modchaein ALPVHLHNPP YKMTALSLGE QVLWIHQVPA NVGERVRVCI YSSDVSITLQ
Modcrhoca EGVVIALDPA YGLSTLQVPG GRIVVPGNLG PIGARRRLRV PATDVSLGRH
Moddazovi ESVVAEHDDH YHLTRLAFPG GAVLVARRPE APGQRLRLRV HARDVSLANS
P29mycge  YDN....... .......... .......... .......... ..........
Phncescco YRSINRVEEN AKAA...... .......... .......... ..........
Brafpseae IKAYLGEA.. .......... .......... .......... ..........
Livgescco IRAYLGEA.. .......... .......... .......... ..........
Nasdklepn LRQQILHFLY E.....KQPK AA........ .......... ..........
```

```
Nrtcsynsp LRNELINFLQ QQRRAKRRAK AAAPAPAVAA SQQKTVRLGF LPGNDCAPLA
Nrtdsynsp LRTEALDFLY RRFAHDDD.. .......... .......... ..........
Opuabacsu VEKFVEDVDL SKVLTAGHIM KRAETVRIDK ....GPRVAL TLMKNLGISS
Provescco VRTFFRGVDI SQVFSAKDIA RRTPNGLIRK TPGFGPRSAL KLLQDEDREY
Artpescco FKNYLSH... .......... .......... .......... ..........
Artphaein FKHYLSH... .......... .......... .......... ..........
Hispescco LQRFLKGSLK .......... .......... .......... ..........
Nocpagrtu CRAFLSSVL. .......... .......... .......... ..........
Occpagrtu FRQFLRRDGG TSH....... .......... .......... ..........
Gltlescco AKDFL..AKI LH........ .......... .......... ..........
Gluacorgl AKDFL..GKI LAH....... .......... .......... ..........
Glnqbacst AKVFL..SRI LNH....... .......... .......... ..........
Pebccamje ARLFL..GKI LKN....... .......... .......... ..........
Glnqescco LQEFL..QHV S......... .......... .......... ..........
Gltlhaein TKQFLLQAKI PLELDYYI.. .......... .......... ..........
Abchaein  AQEFIRSTFH ISLPDEYLEN LTDTPKHSKA YPIIKFEFTG RSVDAPLLSQ
Lackagrra VAGFIGSPRM NFLPAVVIGQ A.E...GGQV TVALKARPDT QLTVACATPP
Malkescco VAGFIGSPKM NFLPVKVTAT AID...QVQV ELPMPNRQQV WLPVES.RDV
Msmkstrmu VAGFIGSPAM NFFD..VTIK DGHLVSKDGL TIAVTEGQLK MLESKGFK..
Ugpcescco VASFIGSPAM NLLTGRVNNE GTHFELDGGI ELPLNGG... ...YRQYA..
Potaescco VAGFIG...EI NMFNATVIER LDE....... QRVRANVEGR ECNIYVNFAV
Potahaein VARFIG..EI NVFEATVIER KSE....... QVVLANVEGR ICDIYTDMPV
Cysaescco VLEFMG..EV NRLQGTIRGG QFH....... VGAHRWPLG. .......YTP
Cysasynsp VMSFIG..PV NVLPNS..SH IFQ....... AGGLDTP... ..........
Sfucserma TASFLGETLV ..........LTAEL AHGWADCALG RIAVD...DR
Devaanasp NETP...... .......... .......... .......... ..........
Mbpxmarpo VGIFLG.... LLIEIPKLNE SITLKNIPSK TPQNLKKFAF DPIWVKIFAN
Pstbescco TEDYITGRYG .......... .......... .......... ..........
Mklmycle  IGMSEEKDES TMAEEAALLE AGHYAGGAEE VEGVPPQITV TPGMPKRKAV
Feceescco ..EVMTPGLL .......... .......... RTVFSVEAEI HPEPVSGRPM
Fecehaein NLEIVFGGVL .......... .......... RHIKLLGENL HNDE.DKRSV
Drrastrpe EAERLLSAEL GVTIHRDSDP TALSARIDDP RQGMRALAEL SRTHLEVRSF
Bztrhoca  LLQYFLPPGT NFDLILRVVI LVTLFAAAYI AEVIRGGLAA LPRGQYEAAD
Consensus .......... .......... .......... .......... ..........

              401                                           450
Sapdhaein KKCM.EKPRR LKIKQHEFSC HYPINLREKN FKEKTTATPF ILNCKGNE..
Sapdsalty RECI.ITPRL TGAKNHLYAC HFPLNMERE. .......... ..........
Amifstrpn DY.ETDKPSM VEIRPGHYVW ANQAELARYQ KGLN...... ..........
Oppfbacsu QLKDGETMEF REVKPGHFVM CTEAEFKAFS .......... ..........
Oppfhaein ENCAKQKPPF TSQNNSHFVA CLKVL..... .......... ..........
Oppfsalty PECAQTRPVL EG.SFRHAVS CLKVDPL... .......... ..........
Dppfescco GPCTQLQPQL KDY.GGQLVA CFAVDQDENP QR........ ..........
Dppfhaein EKCRENQPHL EQHTDGKLIA CFHID..... .......... ..........
Appfbacsu PICKEQIPEF KEAAPSHFVA CHLYS..... .......... ..........
Oppflacla KLSESHWAAL PKGGENVESN Y......... .......... ..........
Dppdescco DRCRAEEP.A LNMLADGRQS K..CHYPLDD AGRPTL.... ..........
Dppdhaein EYCRQVEP.Q LHHI.GSRKV K..CHTPLNE QGNPVEYQGA ..........
Oppdhaein EQC.QIAP.K LTTFNHGQLR N..CWLSAEK FNL....... ..........
Oppdsalty EIC.NNAP.P LEAFSPGRLR A..CFKPVEE LL........ ..........
Dppdbacsu VVCDRVYP.G QTIRSDSHTV N..CWLQDQR AEHAVLSGDA KD........
Oppdbacsu KIDFEQEP.P MFKVSDTHYV K..SWLLHPD A.PKVEPPEA ..........
Appdbacsu DKCWTNQP.S LLTHKSGRTV R..CFLYEEE GAEQS..... ..........
Amiestrpn QIDFEQKA.P QFSVSETHWA K..TWLLHED APKVEKPAVI ANLHDKIREK
```

```
Oppdlacla EEAQKVISEK MTEISSNHFV RGQAWKKFEF PDQKLKGGEK .........
Oppdmycge AIDFEYHPP. ..FFEVTKTH KAATWLLHPQ APKVEPPQAV IDNITLTKKA
Modcescco PPQQTSIRNV LRAKVVNSYD DNG......Q VEVELEVGGK TLWARISPWA
Modchaein KPEQTSIRNI LRGKITQIEI QDS......R VDLAVLVEGH KIWASISKWA
Modcrhoca APTDTTILNA LPAVILG.AE AAEGYQITVR LALGASGEGA SLLARVSRKS
Moddazovi RIEDSSITNV LPATVREVVE ADTPAHVLVR LE....AEGT PLIARITRRS
Nrtcsynsp IAQELGLFQD LGLSVELQSF LTWEALEDSI RLGQLEGALM MAAQPLAMTM
Opuabacsu IYAVDKQKKL LGVIYASDAK KAAESDLSLQ DILNTEFTTV PENTYLTEIF
Provescco GYVIERGNKF VGAVSIDSLK TALTQQQGLD AALIDAPLAV DAQTPLSELL
Abchaein  ASKKFGVELS ILTSQIDYAG GVKFGYTIAE VEGDEDAITQ TKVYLMENNV
Lackagrra QGGDAVTVGV RPEHFLPA.. ...GSGDTQL TAHVD..... .VVEHLGNTS
Malkescco QVGANMSLGI RPEHLLPS.. ...DIADVIL EGEVQ..... .VVEQLGNET
Msmkstrmu ..NKNLIFGI RPEDISSSLL VQETYPDATV DAEVV..... .VSELLGSET
Ugpcescco ..GRKMTLGI RPEHIALS.. ...SQAEGGV PMVMD..... .TLEILGADN
Potaescco EPGQKLHVLL RPEDLRVEEI NDDNHAEGLI GYVRERNYKG MTLESVVELE
Potahaein EKDQKLQVLL RPEDIVIEEL DENEHSKAII GHIIDRTYKG MTLESTVEFD
Cysaescco AYQGPVDLFL RPWEVDIS.. RRTSLDSPLP VQVLEASPKG HYTQLVVQPL
Cysasynsp ....HPEVFL RPHDIEIA.. .....IDPIP ETVPARIDRI VHLGWEVQAE
Sfucserma QRSGPARIML RPEQIQIGLS DPAQRGQAVI TG........ ..IDFAGFVS
Mbpxmarpo RSINKYRFFL RPYEFCIKSE MDLEATPVQI KTIIYKRTFV QLDLFVTSFL
Mklmycle  ARRQARVRAM LPTLPKGAQA AILDDLEGAH NYQAHEFGD. ..........
Feceescco CLMR...... .......... .......... .......... ..........
Fecehaein TVLTDDEKAV VFYGETKQDP PAPTTQNCHF EDCPYKSAVK NKRD......
Drrastrpe SLGQSSLDEV FLALTGHPAD DRSTEEAAEE EKVA...... ..........
Bztrhoca  ALGLDYWQAQ RLIIMPQALK ISIPGIVSSF IGLFKDTTLV AFVGLFDPLK
Consensus .......... .......... .......... .......... ..........
```

```
          451                                              500
Amiestrpn MGFAHLAD.. .......... .......... .......... ..........
Oppdmycge LQFKDQ.... .......... .......... .......... ..........
Modcescco RDELAIKPGL WLYAQIKSVS ITA....... .......... ..........
Modchaein QNELRFAIGQ DVYVQIKAVS VM........ .......... ..........
Modcrhoca FDLLGFQPGE QVVARLKAMA LSAPAQTGG. .......... ..........
Moddazovi CDQLGIAPGR RMWAQIKAVA LLG....... .......... ..........
Nrtcsynsp GLGGHRPFAI ATPLTVSRNG GAIALSRRYL NAGVRSLEDL CQFLAATPQR
Opuabacsu DVVSDANIPI AVVDEKQRMK GIVVRGALIG ALAGNNEYIN AEGTNEQTQD
Provescco SHVGQAPCAV PVVDEDQQYV GIISKGMLLR AL.......D REGVNNG...
Abchaein  RVEVLGYVQ. .......... .......... .......... ..........
Lackagrra YVYAHTVPGE QIIIEQEERR HGGRYGDEIA VGISAKTSFL FDASGRRIR.
Malkescco QIHIQIPSIR QNLVYRQNDV VLVEEGATFA IGLPPERCHL FREDGTACRR
Msmkstrmu MLYLKL..GQ TEFAARVDAR DFHEPGEKVS LTFNVAKGHF FDAETEAAIR
Ugpcescco LAHGRW..GE QKLVVRLAHQ ERPTAGSTLW LHLAENQLHL FDGETGQRV.
Potaescco ...NGKMVMV SEFFNEDDPD FDHSLDQKMA INWVESWEVV LADEEHK...
Potahaein ..HNGMRVLV SEFFNEDDPH MDHSIGQRVG ITWHEGWEVV LNDEDNQ...
Cysaescco GWYNEPLTVV MH......GD DAPQRGERLF VGLQHARLYN GDERIETRDE
Cysasynsp VRLEDGQVLV AHLPRDRYRD LQLEPEQQVF VRPKQARSFP LNYSI.....
Sfucserma TLNLQMAATG AQLEIKTVSR EGLRPGAQVT LNVMGQAHIF AG........
Devaanasp .......... .......... .......... .......... ..........
Mbpxmarpo WNLTIPIGYQ SFRNLHIESF MQTLYIKPRL QVFLRAYPIL TNIKKN....
Bztrhoca  GISNVVRSDM AWKGTYWEPY IFVALIFFLF NFSMSRYSMY LERKLKRDHR
Consensus .......... .......... .......... .......... ..........
```

```
          501                                              550
Nrtcsynsp LRLAIPDPIA MPALLLRYWL ASAGLNPEQD VELVGMSPYE MVEALKAGDI
```

```
Opuabacsu PSAQEVK... ........ ........ ........ ........
Malkescco LHKEPGV... ........ ........ ........ ........
Cysaescco ELALAQSA.. ........ ........ ........ ........
Consensus ........ ........ ........ ........ ........

          551                                              600
Nrtcsynsp DGFAAGEMRI ALAVQAGAAY VLATDLDIWA GHPEKVLGLP EAWLQVNPET

          601                                              650
Nrtcsynsp AIALCSALLK AGELCDDPRQ RDRIVEVLQQ PQYLGSAAGT VLQRYFDFGL

          651                                              700
Nrtcsynsp GDEPTQILRF NQFHVDQANY PNPLEGTWLL TQLCRWGLTP LPKNRQELLD

          701                                              750
Nrtcsynsp RVYRRDIYEA AIAAVGFPLI TPSQRGFELF DAVPFDPDSP LRYLEQFEIK

          751       764
Nrtcsynsp APIQVAPIPL ATSA
```

Proteins listed subsequently in italics are at least 90% identical to the paired transporters listed in parenthesis and are therefore not included in the alignment: *Hispsalty* (Hispescco); *Livgsalty* (Livgescco); *Malksalty, Malkentae* (Malkescco); *Sapfsalty, Sapfhaein* (Sapfescco); *Provsalty* (Provescco). Residues listed in the consensus sequence are present in at least 75% of the aligned transporter sequences. Residues indicated in boldface type are also conserved in at least one other family of the ABC transporter superfamily.

Database accession numbers

	SWISSPROT	PIR	EMBL/GENBANK
Abchaein	P44785		L45262; G1005459
Abcxantsp	Q02856	S37635	X63382; G14178
Abcxcyapa	P48255		U30821; G1016162
Abcxgalsu	P35020	S39521	X67814; G429179
Abcxodosi	Q00830	S21682	X60752; G11945
Amiestrpn	P18765	S11152	X17337; G47346
Amifstrpn	P18766	S11153	X17337; G47347
Appdbacsu	P42064		U20909; G677943
Appfbacsu	P42065		U20909; G677944
Artpescco	P30858	S31694	X86160; G769790
Artphaein	P45092		L45815; G1006549
Brafpseae	P21629	D36125	D90223; G216866
Bztrhoca			U37407
Cysaescco	P16676	C35402; QRECSA	M32101; G145661
Cysasynsp	P14788	A30301; GRYCS7	J04512; G142152
Devaanasp		A55541	
Dppdbacsu	P26905	S16650	X56678; G48807
Dppdescco	P37314		L08399; G349228
Dppdhaein	P45095		L45820; G1006559
Dppfescco	P37313		L08399; G349229
Dppfhaein	P45094		L45819; G1006557
Drrastrpe	P32010	S27707	M73758; G153230
Feceescco	P15031	JS0115; QRECM3	M26397; G145928
Fecehaein	P44662		L45003; G1003608
Ftseescco	P10115	S03131; CEECFE	X04398; G41499
Ftsehaein	P44871		L45407; G1005744

	SWISSPROT	PIR	EMBL/GENBANK
Glnqbacst	P27675	A42478	M61017; G142988
Glnqescco	P10346	S03183; QRECGQ	X14180; G581098
Gltlescco	P41076		U10981; G624632
Gltlhaein	P45022		L45715; G1006354
Gluacorgl	P48243		X81191; G732701
Hispescco	P07109	A27835	Y00455; G41705
Hispsalty	P02915	A03412; QREBPT	V01373; G47734
Lackagrra	Q01937	34734	X66596
Livgescco	P22730	F37074	J05516; G146635
Livgsalty	P30293	JH0670	D12589; G217074
Malkentae	P18813	S05328	
Malkescco	P02914	A03411; MMECMK	U00006; G409797
Malksalty	P19566	S05329; S20602	X54292; G47772
Mbpxmarpo	P10091	S01592; BVLVMX	X04465; G11666
Mklmycle	P30769	S31144	Z14314; G581333
Modcescco	P09833	B26871; BVECHD	U27192; G973216
Modchaein	P45321		L46321; G1008029
Modcrhoca	Q08381	C36914	L06254; G310274
Moddazovi	P37732	S31045	X69077; G49180
Msmkstrmu	Q00752	E42400	M77351; G153741
Nasdklepn	P39459		L27431; G473439
Nikdescco	P33593	S39597	X73143; G404848
Nikeescco	P33594	S39598	X73143; G404849
Nocpagrtu	P35116	G42600	
Nrtcsynsp	P38045	S30893	X61625; G48971
Nrtdsynsp	P38046	S30894; S36604	X61625; G48972
Occpagrtu	P35117	C41044; C42600	M80607; G154771
Oppdbacsu	P24136	S15233; D38447	X56347; G580898
Oppdhaein	P45052		L45757; G1006435
Oppdlacla	Q07733		U09553; G495177
Oppdmycge	P47325		U39688; G1045756
Oppdsalty	P04285	A03413; QREBOT	X05491; G47805
Oppfbacsu	P24137	S15234; E38447	X56347; G580899
Oppfhaein	P45051		L45756; G1006433
Oppflacla	Q07734		L18760; G308851
Oppfsalty	P08007	D29333; QREBOF	X05491; G47806
Opuabacsu	P46920		U17292; G984803
P29mycge	P47532		U39709; G1045987
Pebccamje	P45677		L13662; G388564
Phncescco	P16677	D35718	D90227; G216591
Phnkescco	P16678	C35719	D90227; G216600
Phnlescco	P16679	D35719	D90227; G216601
Potaescco	P23858	A40840	M64519; G147326
Potahaein	P45171		L45980; G1007359
Provescco	P14175	JS0128; BVECPV	M24856; G147373
Provsalty	P17328	S05374; QREBVT	X52693; G47831
Pstbescco	P07655	Q00616; BVECZB	X02723; G42398
Sapdhaein	P45288		L46271; G1007931
Sapdsalty	P36636	S39588	X74212; G414211
Sapfescco	P36637		U08190; G470683
Sapfhaein	P45289		L46272; G1007933
Sapfsalty	P36638	S39589	X74212; G414212
Sfucserma	P21410	C35108; QRSEUC	M33815; G152862
Ugpcescco	P10907	S03783; QRECUC	X13141; G43249

References

[1] Perego, M. et al. (1991) Mol. Microbiol. 5, 173–185.
[2] Nohno, T. et al. (1986) Mol. Gen. Genet. 205, 260–269.

[3] Kraft, R. and Leinwand, L.A. (1987) Nucleic Acids Res. 15, 8568.

[4] Williams, S.G. et al. (1992) Mol. Microbiol. 6, 1755–1768.

[5] Sofia, H.J. et al. (1994) Nucleic Acids Res. 22, 2576–2586.

[6] Omata, T. et al. (1993) Mol. Gen. Genet. 236, 193–202.

[7] Parra-Lopez, C. et al. (1993) EMBO J. 12, 4053–4062.

[8] **Higgins, C.F. (1992) Annu. Rev. Cell Biol. 8, 67–113.**

[9] Pearce, S.R. et al. (1992) Mol. Microbiol. 6, 47–57.

[10] Higgins, C.F. et al. (1992) Nature 298, 723–727.

[11] Froshauer, S. et al. (1988) Mol. Biol. 200, 501–511.

[12] Ehrmann, M. et al. (1990) Proc. Natl Acad. Sci. USA 87, 7574–7578.

Part 7

Other ABC-Associated (Cytoplasmic) Proteins

Summary

Most transporters of the heme exporter family, the example of which is the heme exporter CYCV from *Bradyrhizobium japonicum*[1] (Ccmabraja), mediate export of heme into the periplasm for the biogenesis of C-type cytochromes. The COB transporters[2], which transport cobalt for the synthesis of vitamin B12 (cobalamin), and the NIK transporters[3] which transport nickel (e.g. Nikeescco), are also members of this family. These transporters are found only in gram-negative bacteria.

Statistical analysis of multiple amino acid sequence comparisons places the heme exporter family in the ATP binding cassette (ABC) superfamily[4]. Proteins in this transporter superfamily use the energy of ATP hydrolysis to pump substrates across cell membranes. Transporters of the heme exporter family exist as four or five separate chains – two separate ATP binding domains and two separate transmembrane domains. The family is characterized by the cytoplasmic ATP binding domains[4], which are described in the following tables. The associated transmembrane domains are predicted to contain six membrane-spanning helices by the hydropathy of their amino acid sequences.

A relatively small number of amino acids, scattered throughout the whole length of the sequences, are conserved within the family, almost all of which are conserved in other families of the ABC superfamily.

Nomenclature, biological sources and substrates

CODE	DESCRIPTION [SYNONYMS]	ORGANISM [COMMON NAMES]	SUBSTRATE(S)
Cbiosalty	Cobalt transport ATP binding protein [CBIO]	*Salmonella typhimurium* [gram-negative bacterium]	Co^{2+}
Ccmabraja	Heme exporter protein A [Cytochrome C type biogenesis ATP binding protein A, CYCV]	*Bradyrhizobium japonicum* [gram-negative bacterium]	Heme
Ccmaescco	Heme exporter protein A [Cytochrome C type biogenesis ATP binding protein A, CCMA]	*Escherichia coli* [gram-negative bacterium]	Heme
Ccmahaein	Heme exporter protein A [Cytochrome C type biogenesis ATP binding protein A, CCMA, HI1089]	*Haemophilus influenzae* [gram-negative bacterium]	Heme
Ccmarhoca	Heme exporter protein A [Cytochrome C type biogenesis ATP binding protein A, HELA]	*Rhodobacter capsulatus* [gram-negative bacterium]	Heme
Nikeescco	Nickel transport ATP binding protein [NikE]	*Escherichia coli* [gram-negative bacterium]	Ni^{2+}

Phylogenetic tree

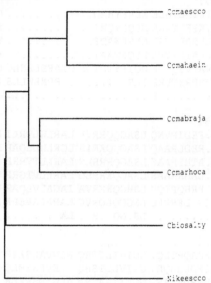

Physical and genetic characteristics

	AMINO ACIDS	MOL. WT	CHROMOSOMAL LOCUS
Cbiosalty	271	30 147	cob operon
Ccmabraja	200	21 132	cyc
Ccmaescco	205	22 865	49.42 minutes
Ccmahaein	212	24 009	
Ccmarhoca	214	22 168	hel
Nikeescco	268	29 619	77.9 minutes

Multiple amino acid sequence alignments

```
          1                                                  50
Ccmaescco ...........MLEARELLC ERDERTLFSG LSFTLNAGEW VQITGSNGAG
Ccmahaein ......MFEQ HKLSLQNLSC QRGERVLFRA LTCDFNSGDF VQIEGHNGIG
Ccmabraja .......... MQLSGRRVIC VRGGREVFAG LDFEAVSGEA VAVVGRNGSG
Ccmarhoca .........M TLLAVDQLTV SRGGLAVLEG VSFSLAAGHA LVLRGPNGIG
Cbiosalty ...........MLATSDLWF RYQNEPVLKG LNMDFSLSPV TGLVGANGCG
Nikeescco MTLLNISGLS HHYAHGGFNG KHQHQAVLNN VSLTLKSGET VALLGGTGCG
Consensus ................L............................G. .....G.NG.G


          51                                                 100
Ccmaescco KTTLLRLLTG LSRPDAGEVL WQGQPLHQVR DSYHQ..... .NLLWIGHQP
Ccmahaein KTSLLRILAG LVRPLEGEVR WDSEAISKQR EQYHQ..... .NLLYLGHLS
Ccmabraja KTSLLRLIAG LLIPAGGTIA LDGG...DAE LTLPE..... .QCHYLGHRD
Ccmarhoca KTTLLRTLAG LQPPLAGRVS MP.........PE..... .GIAYAAHAD
Cbiosalty KSTLFMNLSG LLRPQKGAVL WQGKPLDYSK RGL........ ...LARRQ
Nikeescco KSTLARLLVG LESPAQGNIS WRGEPLAKLN RAQAKAFRRD IQMVFQDSIS
Consensus K..L.R.L.G L..P..G....................................
```

```
          101                                                    150
Ccmaescco GIKTRLTALE NLHFYHRD.. .......... GDTAQC.... LEALAQAGLA
Ccmahaein GVKPELTAWE NLQFYQRI.. .......... SQAEQNTDML WDLLEKVGLL
Ccmabraja ALKPALSVAE NLSFWADF.. .......... LGGERLDA.. HESLATVGLD
Ccmarhoca GLKATLSVRE NLQFWAAI.. .......... HATDTVET.. ..ALARMNLN
Cbiosalty QVATVFQDPE QQIFYTDIDS DIAFSLRNLG GPEAEITRRV DEALTLVDAQ
Nikeescco AVNPRKTVRE ILR....... ..EPMRHLLS LKKSEQLARA RQMLKAVDLD
Consensus .........E .......... .......... .......... ...L......

          151                                                    200
Ccmaescco G.FEDIPVNQ LSAGQQRRVA LARLWLTRAT LWILDEPFTA IDVNGVDRLT
Ccmahaein G.REDLPAAQ LSAGQQKRIA LGRLWLSQAP LWILDEPFTA IDKKGVEILT
Ccmabraja H.ATHLPAAF LSAGQRRRLS LARLLTVRRP IWLLDEPTTA LDVAGQDMFG
Ccmarhoca A.LEHRAAAS LSAGQKRRLG LARLLVTGRP VWVLDEPTVS LDAASVALFA
Cbiosalty H.FRHQPIQC LSHGQKKRVA IAGALVLQAR YLLLDEPTAG LDPAGRTQMI
Nikeescco DSVLDKRPPQ LSGGQLQRVC LARALAVEPK LLILDEAVSN LDLVLQAGVI
Consensus .......... LS.GQ..R.. LA........ ...LDEP... .D........

          201                                                    250
Ccmaescco QRMAQHTEQ. GGIVILTTHQ PLNVAESKIR RISLTQTRAA ..........
Ccmahaein ALFDEHAQR. GGIVLLTSHQ ..EVPSSHLQ KLNLAAYKAE ..........
Ccmabraja GLMRDHLAR. GGLIIAATHM ALGIDSRELR IGGVA..... ..........
Ccmarhoca EAVRAHLAA. GGAALMATHI DLGLS..EAR VLDLAPFKAR PPEAGGHRGA
Cbiosalty AIIRRIVAQ. GNHVIISSHD IDLIYEISDA VYVLRQGQIL THGAPGEVFA
Nikeescco RLLKKLQQQF GTACLFITHD LRLVERFCQR VMVMDNGQIV ETQVVGEKLT
Consensus .......... G.......H. .......... .......... ..........

          251                                                    296
Ccmaescco .......... .......... .......... .......... ......
Ccmahaein .......... .......... .......... .......... ......
Ccmabraja .......... .......... .......... .......... ......
Ccmarhoca FDHGFDGAFL .......... .......... .......... ......
Cbiosalty CTEAMEHAGL TQPWLVKLHT QLGLPLCKTE TEFFHRMQKC AFREAS
Nikeescco FSSDAGRVLQ NAVLPAFPVR RRTTEKV... .......... ......
Consensus .......... .......... .......... .......... ......
```

Residues listed in the consensus sequence are present in at least 75% of the aligned transporter sequences. Residues indicated in boldface type are also conserved in at least one other family of the ABC transporter superfamily.

Database accession numbers

	SWISSPROT	PIR	EMBL/GENBANK
Cbiosalty		Q05596	L12006; G154435
Ccmabraja	P30963	A39741	M60874; G152074
Ccmaescco	P33931		U00008; G405926
Ccmahaein	P45032		L45726; G1006376
Ccmarhoca	P29959	S23663	X63462; G46024
Nikeescco	P33594	S39598	X73143; G404849

References
1 Ramseier, T.M. et al. (1991) J. Biol. Chem. 266, 7793–7803.
2 Roth, J.R. et al. (1993) J. Bacteriol. 175, 3303–3316.
3 Navarro, C. et al. (1993) Mol. Microbiol. 9, 1181–1191.
4 **Higgins, C.F. (1992) Annu. Rev. Cell Biol. 8, 67–113.**

Summary

Transporters of the macrolide-streptogramin-tylosin resistance family, the examples of which are the erythromycin resistance protein MSRA[1] of *Staphylococcus epidermidis* (Mrsastaep) and the tylosin resistance protein TLRC[2] of *Streptomyces fradiae* (Tlrcstrfr), confer resistance to antibiotics by acting as ATP-dependent efflux pumps. These transporters are found mostly in gram-positive bacteria from the genera *Staphylococcus* and *Streptococcus*. Several members of this family are plasmid-encoded.

Statistical analysis of multiple amino acid sequence comparisons places the macrolide-streptogramin-tylosin resistance family, in the ATP binding cassette (ABC) superfamily[3]. Proteins in this superfamily use the energy of ATP hydrolysis to pump substrates across cell membranes. In transporters of the macrolide-streptogramin-tylosin resistance family the two ATP binding domains form one chain, separate from any transmembrane domains. The MsrA protein is characterized by a long "Q-linker" domain between the two ATP binding domains. Unusually, it is not known to associate with a specific transmembrane protein or complex. Instead it is believed to interact with other transmembrane resistance proteins to alter their specificity[4]. Other members of this family are associated with specific transmembrane proteins. The macrolide-streptogramin-tylosin resistance family is characterized by the cytoplasmic ATP binding domains[3], which are described in the following tables. Where present, the associated transmembrane domains are predicted to contain six membrane-spanning helices by the hydropathy of their amino acid sequences.

Several amino acid sequence motifs are highly conserved in the macrolide-streptogramin-tylosin resistance family, including motifs unique to the family, signature motifs of the ABC superfamily, and motifs necessary for function by the criterion of site-directed mutagenesis.

Nomenclature, biological sources and substrates

CODE	DESCRIPTION [SYNONYMS]	ORGANISM [COMMON NAMES]	RESISTANCE[a]
Abcsacer	ABC tranporter [ertx]	*Saccharopolyspora erythraea* [gram-negative bacterium]	Unknown
Lmrcstrli	LmrC protein	*Streptomyces lincolnensis* [gram-negative bacterium]	[Linocomycin]
Msrastaep	Erythromycin resistance ATP binding protein [MSRA]	*Staphylococcus epidermidis* [gram-positive bacterium]	[Antibiotics]
Srmbstram	SrmB protein polyketide synthase	*Streptomyces ambofaciens* [gram-negative bacterium]	[Streptogramin B]
Tlrcstrfr	Tylosin resistance ATP binding protein [TLRC]	*Streptomyces fradiae* [gram-negative bacterium]	[Tylosin]
Vgastaau	VgA protein	*Staphylococcus aureus* [gram-positive bacterium]	[Virginiamycin A-like antibiotics]

[a] Presumed substrates; protein confers resistance to specified compounds.

Phylogenetic tree

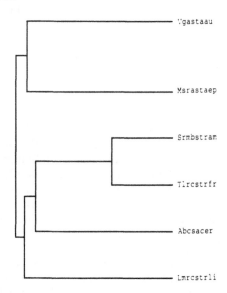

Physical and genetic characteristics

	AMINO ACIDS	MOL. WT	CHROMOSOMAL LOCUS
Abcsacer	481	52 832	
Lmrcstrli	579	62 803	
Msrastaep	488	55 912	Plasmid PUL5050
Srmbstram	550	60 146	
Tlrcstrfr	548	59 129	
Vgastaau	522	60 184	

Multiple amino acid sequence alignments

```
           1                                                  50
Vgastaau   ....MKIMLE GLNIKHYVQD RLLLNINRLK IYQNDRIGLI GKNGSGKTTL
Msrastaep  ...MEQYTIK FNQINHKLTD LRSLNIDHLY AYQFEKIALI GGNGTGKTTL
Srmbstram  ...MSIAQYA LHDITKRYHD CVVLDRVGFS IKPGEKVGVI GDNGSGKSTL
Tlrcstrfr  MRTSPSSQLS LHGVTKRYDD RVVLSQVSLA ISPGEKAGII GDNGAGKSTL
Abcsacer   ....MVNLIN LESVSKSYGV RPLLDEVSLG VGASDRIGVV GLNGGGKTTL
Lmrcstrli  ..MADASIVC TNLSFSWPDE TPVFDGLSFA LGDG.RCGLV GPNGAGKSTL
Consensus  .......... .......... .......L... .......G.. G.NG.GK.TL

           51                                                 100
Vgastaau   LHILYKKIVP EEGIVKQFSH C......... .......... ..........
Msrastaep  LNMIAQKTKP ESGTVETNGE I......... .......... ..........
Srmbstram  LKILAGRVEP DNGALTVVAP GGVGYLAQTL ELPLDATVQD AVDLALSDLR
Tlrcstrfr  LRLLAGEERP DAGEVTVIAP GGVGYLPQTL GLPPRATVQD AIDLAMTELR
Abcsacer   LEVLSGSVDR DSGRVSHSRD LRMAVVTQRT ELPEGSTVRN AV........
Lmrcstrli  LRLAVGELTP TAGSIT...A QDMSVPAESL PL.IDGTVDE A.......WR
Consensus  L.......P ..G...... .......... .......... ..........
```

```
          101                                                   150
Vgastaau  ..................................  ........EL IPQLKL....
Msrastaep ..................................  ........QY FEQLNMDVEN
Srmbstram ELE.AAMREA EAELGESDEN GSERELSAGL QRYAALVEQY QARGGYEADV
Tlrcstrfr VLE.AELRRT EAALAEA... ATDEALQDAL TAYARLTEQY EVRDGYGADA
Abcsacer  .LD.P..... .......... .......... HGF TAEHEWAADA
Lmrcstrli SLHRAALHAI ESGDVDEAHF TT........ .......... .VGDHWDIEE
Consensus .......... .......... .......... .......... ..........


          151                                                   200
Vgastaau  ...............IE STKSGGEVTR NYIRQALDKN PELLLADEPT
Msrastaep DFNTLDGSLM SELHIPMHTT DSMSGGEKAK YKLANVISNY SPILLLDEPT
Srmbstram RVEVALHGLG LPSLDRDRKL GTLSGGERSR LALAATLASS PELLLLDEPT
Tlrcstrfr RVDAALHGLG LPGLPRDRRL GTLSGGERSR LALAATLASQ PELLLLDEPT
Abcsacer  KVRSVLTGLG MTSLGLDTPV ADFSGGERRR VALAA.LVRE LDLLVLDEPT
Lmrcstrli RTTIVLDRLG LGDVSLDRPL RSLSGGQVLA IGLAAQLLKR PDVLILDEPT
Consensus ........L. .......... ...SGGE... ..LA...... ...L.LDEPT


          201                                                   250
Vgastaau  TNLDNNYIEK LEQDLKNWHG AFIIVSHDRA FLDNLCTTIW EI.DEGRITE
Msrastaep NHLDKIGKDY LNNILKYYYG TLIIVSHDRA LIDQIADTIW DIQEDGTIRV
Srmbstram NDLDDRAMEW LEDHLAGHRG TVIAVTHDRV FLDRLTTTIL EV.DSGSVTR
Tlrcstrfr NDLDDRAVHW LEEHLSGHRG TVVTVTHDRV FLDRLTATVL EV.DGRGVSR
Abcsacer  NHLDVEGVRW LADHLLQRRC ALVIVTHDRW FLDTVCNRTW EV.VQGRVEQ
Lmrcstrli NNLDLAARQR LYQVVEEWKG ALLVVSHDRE LLDRV.DTIA EL.QASELRL
Consensus N.LD...... L...L....G ....V.HDR. .LD....T.. E.........


          251                                                   300
Vgastaau  YKGNYSNYVE QKELERHREE LEYEKYEKEK KRLEKAINIK EQKAQ...RA
Msrastaep FKGNYTQYQN QYEQEQLEQQ RKYEQYISEK QRLSQASKAK RNQAQQMAQA
Srmbstram YGNGYEGYLT AKAVERE... RRLREYEEWR AELDRNRGLI TSNVARMDGI
Tlrcstrfr HGDGYAGYLA AKAAERR... RRQQQYDEWR AELDRNRRLA EANVARLDGI
Abcsacer  YEGGYRLGLR PRRAGPG... WR........ ...SRPRRSA RTSPAR....
Lmrcstrli YGGNFTAYTE AVELEQENVQ RAVLRADRSC AATSARRRRA QERAQRRASN
Consensus ....Y..Y. ....E..... .......... .......... ..........


          301                                                   350
Vgastaau  TKKPKNLSLS EGKIKGAKPY FAGKQKKLRK TVKSLETRLE K.LESVEKRN
Msrastaep SSKQKNKSIA PDRLSASKE. KGTVEKAAQK QAKHIEKRME .HLEEVEKPQ
Srmbstram PRKMSLSVFG HGAYRRRGRD HGAMVRI..R NAKQRVAQLT E..NPVHAPA
Tlrcstrfr PRKMGKAAFG HGAFRARGRD HGAMSRV..R NAKERVERLT A..NPVAPPA
Abcsacer  .......... ....SSRGCS GGAKART..S KPRFRVEAAE ALIADVPPPR
Lmrcstrli AKRNKVRRGC PVSTRAPLQR QAQESAG..R AASVHQDRVS QAKAKLDEAS
Consensus .......... .......... .......... .......... .........P.


          351                                                   400
Vgastaau  ELPPLKMDL. VNL...ESVK NRTIIRGEDV SGTIEGRVLW KAKSFS..IR
Msrastaep SYHEFNFPQN KIY...DIHN NYPII.AQNL TLVKGSQKLL TQVRFQ..IP
Srmbstram DPLSFAARID T....AGPEA EEAVAELTDV ..RVAGR..L AVDSLT..IR
Tlrcstrfr DRLSLTARIA T....ADGPG EAPAAELDGV ..VVGSR..L RVPKLR...LG
Abcsacer  DTVEL...VS F....AKRRL GKTVLELENV DLRIADR..V LLEDLTWLIG
Lmrcstrli QGMREEARLA ITLPQTSVPA GRTVLTCHEA NVRYGERTLF TGSGVDLGIR
Consensus .......... .......... .......... .......R... .........I.
```

```
           401                                              450
Vgastaau   GGDKMAIIGS NGTGKTTFIK KIV...HGNP G.ISLSPSVK IGYFSQKIDT
Msrastaep  YGKNIALVGA NGVGKTTLLE AIY...HQIE G.IDCSPKVQ MAYYRQLAYE
Srmbstram  PGERLLVTGP NGAGKSTLLR VLSGELEPDG GSVRVG..CR VGHLRQDETP
Tlrcstrfr  AAERLLITGP NGAGKSTLLS VLAGELSPDA GAVSVP..GR VGHLRQEETP
Abcsacer   PGDRIGLVGV NGSGKTTLLR LLAGERDADA GRRIEGKTVR LAHLTQELHD
Lmrcstrli  GPERIALLGP NGSGKSTLLK LIAGELEPSS GTVTAP.TDR VSYLSQRLDL
Consensus  ........G. NG.GK.TLL. ..........G G............ .....Q....

           451                                              500
Vgastaau   LELDKSILEN VQ........ SSSQQNETLI RTILARMHFF RDDVYKPISV
Msrastaep  DMRDVSLLQY LM........ DETDSSESFS RAILNNLGL. NEALERSCNV
Srmbstram  WAPGLTVLRA FAQ....... GREGYLEDHA EKLLSLGLFS PSDLRRRVKD
Tlrcstrfr  WPAKLTVLEA FAH....... NRPGDRDEQA DRRLSLGLFE PEALRLRVGE
Abcsacer   LPGDWRVLEA IEDVAERVTL DKYELTASQL GERFGFG... KGRQWTPVSD
Lmrcstrli  LDLDASVLDN LRRFA..... ..PHLQDGEV RYRLAQFLFR GDRVHRTAGW
Consensus  .......L.. .......... .........L .......... ..........

           501                                              550
Vgastaau   LSGGERVKVA LTKVFLS..E VNTLVLDEPT NFLDMEAIEA FESLLKEYNG
Msrastaep  LSGGERTKLS LAVLFST..K ANMLILDEPT NFLDIKTLEA LEMFMNKYPG
Srmbstram  LSYGQRRRIE IARLVSD..P MDLLLLDEPT NHLTPVLVEE LEQALADYRG
Tlrcstrfr  LSYGQRRRIE LARLVSE..P VGLLLLDEPT NHLSPALVEE LEEALTGYGG
Abcsacer   LSGGERRSVQ LARLLMA..E PNVLVLDEPT NDLDIDTLQQ LEDLLDTWPG
Lmrcstrli  LSGGERLRAT LACVLSTDPA PQLLLLDEPT NNLDLNSAAQ LENALNAFQG
Consensus  LS.G.R.... LA........ ...L.LDEPT N.L....... LE..L....G

           551                                              600
Vgastaau   SIIFVSHDRK FIEKV..ATR IMTIDNKEIK IFDGTYEQFK QAEKPTRNIK
Msrastaep  IILFTSHDTR FVKHV..SDK KWELTGQSIH DIT...............
Srmbstram  AVVVVTHDRR MRSRF..TGA RLTMGDGRIA EFSAG.............
Tlrcstrfr  ALVLVTHDRR MRSRF..TGS HLELREGVVS GAR...............
Abcsacer   TLVVVSHDRY LVERVCDTST RCSATGG...................
Lmrcstrli  AFVVVSHDQA FL.RAIGVSR WLRLADGTLE EIAEADDAWP HRISEAAAAR
Consensus  ......HD.. .......... .......... .......... ..........

           601                                    647
Vgastaau   EDKKLLLETK ITEVLSRLSI EPSE.....E LEQEFQNLIN EKRNLDK
Msrastaep  .................................................
Srmbstram  .................................................
Tlrcstrfr  .................................................
Abcsacer   .................................................
Lmrcstrli  VVRGVNPVAR VRNWPYAVGI RPVC......................
Consensus  .................................................
```

Residues listed in the consensus sequence are present in at least 75% of the aligned transporter sequences. Residues indicated in boldface type are also conserved in at least one other family of the ABC transporter superfamily.

Database accession numbers

	SWISSPROT	PIR	EMBL/GENBANK
Abcsacer		S47441	X80735
Lmrcstrli		S44975	X79146
Msrastaep	P23212	S11158; YESAEE	X52085; G47001

	SWISSPROT	PIR	EMBL/GENBANK
Srmbstram		S25202	X63451
Tlrcstrfr	P25256	JQ1142	M57437; G153508
Vgastaau		JC1204	M90056

References

[1] Ross, J.I. et al. (1990) Mol. Microbiol. 4, 1207–1214.
[2] Rosteck, P.R. et al. (1991) Gene 102, 27–32.
[3] **Higgins, C.F. (1992) Annu. Rev. Cell Biol. 8, 67–113.**
[4] Ross, J.I. et al. (1995) Gene 153, 93–98.

Part 8

H⁺-Dependent Symporters

Summary

Transporters of the H⁺/sugar-symporter-uniporter family, the example of which is the GLUT1 facilitative glucose transporter of humans (Gtr1homsa), mediate either symport (H⁺-coupled substrate uptake) or uniport (facilitative uptake) of structurally dissimilar sugars, including mono- and disaccharides, aldohexoses and aldopentoses, and carboxylated compounds [1-6]. In addition, they also serve as ion and water channels [7,8]. Possible transport of nicotinamide by the GLUT1 glucose transporter also has been reported [9]. Some members of the family are inhibited by forskolin, cytochalasin-B, or both, while others are insensitive to both antibiotics [5]. Members of the H⁺/sugar-symporter-uniporter family have a broad biological distribution that includes bacteria, plants and humans.

Statistical analysis of multiple amino acid sequence comparisons places the H⁺/sugar-symporter-uniporter family in the uniporter-symporter-antiporter (USA) superfamily, also known as the major facilitator superfamily (MFS) [10,11]. Members of the H⁺/sugar-symporter-uniporter family are predicted to contain 12 membrane-spanning helices by the hydropathy of their amino acid sequences [5], reaction with peptide-specific antibodies [12] and glycosylation-scanning mutagenesis [13]. Eukaryotic proteins are glycosylated [7] and may exist as oligomers [14]. There is considerable similarity between the sequences of the N- and C-terminal halves of these proteins, implying they arose through gene duplication of an ancestral six-helix protein [5,10].

Several amino acid sequence motifs are highly conserved in the H⁺/sugar-symporter-uniporter family, including motifs unique to the family, signature motifs of the USA/MFS superfamily, and motifs necessary for function by the criterion of site-directed mutagenesis [3,5,7,10].

Nomenclature, biological sources and substrates

CODE	DESCRIPTION [SYNONYMS]	ORGANISM [COMMON NAMES]	SUBSTRATE(S)
Araeescco	Arabinose-H⁺ symporter [ARAE]	Escherichia coli [gram-negative bacterium]	H⁺/arabinose
Araekleox	Arabinose-H⁺ symporter [ARAE]	Klebsiella oxytoca [gram-negative bacterium]	H⁺/arabinose
Gal2sacce	Facilitative galactose transporter [GAL2, IMP1]	Saccharomyces cerevisiae [yeast]	Galactose
Galpescco	Galactose-H⁺ symporter [GALP]	Escherichia coli [gram-negative bacterium]	H⁺/galactose
Glcpsynsp	Glucose transporter [GLCP, GTR]	Synechocystis sp. strain PCC 6803 [cyanobacterium]	Glucose
Glfzymmo	Glucose facilitated diffusion protein [GLF]	Zymomonas mobilis [gram-negative bacterium]	Glucose
Gtr1leido	Membrane transporter 1 [D1]	Leishmania donovani [trypanosome]	Glucose
Gtr1bosta	Facilitative glucose transporter type 1 [GTR1, GLUT1]	Bos taurus [cow]	Aldopentoses, aldohexoses
Gtr1galga	Facilitative glucose transporter type 1 [GTR1, GLUT1]	Gallus gallus [chicken]	Aldopentoses, aldohexoses

CODE	DESCRIPTION [SYNONYMS]	ORGANISM [COMMON NAMES]	SUBSTRATE(S)
Gtr1homsa	Facilitative glucose transporter type 1 [GTR1, GLUT1, SLC2a1]	*Homo sapiens* [human]	Aldopentoses, aldohexoses
Gtr1musmu	Facilitative glucose transporter type 1 [GTR1, GLUT1, GT1]	*Mus musculus* [mouse]	Aldopentoses, aldohexoses
Gtr1orycu	Facilitative glucose transporter type 1 [GTR1, GLUT1]	*Oryctolagus cuniculus* [rabbit]	Aldopentoses, aldohexoses
Gtr1ratno	Facilitative glucose transporter type 1 [GTR1, GLUT1]	*Rattus norvegicus* [rat]	Aldopentoses, aldohexoses
Gtr1sussc	Facilitative glucose transporter type 1 [GTR1, GLUT1]	*Sus scrofa* [pig]	Aldopentoses, aldohexoses
Gtr2galga	Facilitative glucose transporter type 2 [GTR2, GLUT2]	*Gallus gallus* [chicken]	Glucose
Gtr2homsa	Facilitative glucose transporter type 2 [GTR2, GLUT2, SLC2a2]	*Homo sapiens* [human]	Glucose
Gtr2leido	Membrane transporter 2 [D2]	*Leishmania donovani* [trypanosome]	Glucose
Gtr2musmu	Facilitative glucose transporter type 2 [GTR2, GLUT2]	*Mus musculus* [mouse]	Glucose
Gtr2ratno	Facilitative glucose transporter type 2 [GTR2, GLUT2]	*Rattus norvegicus* [rat]	Glucose
Gtr2sacsp	Facilitative glucose transporter type 2 [GTR2, GLUT2]	*Saccharum* sp. [Sugar cane]	Glucose
Gtr3canfa	Facilitative glucose transporter type 3 [GTR3, GLUT3]	*Canis familaris* [dog]	Glucose
Gtr3galga	Facilitative glucose transporter type 3 [CEF-GT3, GTR3, GLUT3]	*Gallus gallus* [chicken]	Glucose
Gtr3homsa	Facilitative glucose transporter type 3 [GTR3, GLUT3, SLC2a3]	*Homo sapiens* [human]	Glucose
Gtr3musmu	Facilitative glucose transporter type 3 [GTR3, GLUT3]	*Mus musculus* [mouse]	Glucose
Gtr3oviar	Facilitative glucose transporter type 3 [GTR3, GLUT3]	*Ovis aries* [sheep]	Glucose
Gtr3ratno	Facilitative glucose transporter type 3 [GTR3, GLUT3]	*Rattus norvegicus* [rat]	Glucose
Gtr4homsa	Insulin responsive facilitative glucose transporter type 4 [GTR4, GLUT4, SLC2a4]	*Homo sapiens* [human]	Glucose
Gtr4musmu	Insulin responsive facilitative glucose transporter type 4 [GTR4, GLUT4]	*Mus musculus* [mouse]	Glucose
Gtr4ratno	Insulin responsive facilitative glucose transporter type 4 [GTR4, GLUT4, GT2]	*Rattus norvegicus* [rat]	Glucose
Gtr5homsa	Facilitative glucose transporter type 5 [GTR5, GLUT5, SLC2a5]	*Homo sapiens* [human]	Fructose
Gtr5orycu	Facilitative glucose transporter type 5 [GTR5, GLUT5]	*Oryctolagus cuniculus* [rabbit]	Fructose
Gtr5ratno	Facilitative glucose transporter type 5 [GTR5, GLUT5]	*Rattus norvegicus* [rat]	Fructose
Gtr7ratno	Facilitative glucose transporter type 7 [GTR7, GLUT7]	*Rattus norvegicus* [rat]	Glucose
Gtrkluma	Glucose transporter [GTR, KHT2]	*Kluyveromyces marxianus* [yeast]	Glucose

CODE	DESCRIPTION [SYNONYMS]	ORGANISM [COMMON NAMES]	SUBSTRATE(S)
Hex6ricco	Hexose carrier protein [HEX6]	*Ricinus communis* [castor oil plant]	Hexoses
Hgt1klula	High-affinity glucose transporter [HGT1]	*Kluyveromyces lactis* [yeast]	Glucose
Hup1chlke	Hexose-H⁺ cotransporter [HUP1]	*Chlorella kessleri* [alga]	H⁺/hexose
Hup2chlke	Hexose transporter [HUP2]	*Chlorella kessleri* [alga]	Hexoses
Hxt0sacce	Hexose transporter [HXT10, YFL011W]	*Saccharomyces cerevisiae* [yeast]	Glucose
Hxt1sacce	High-affinity glucose transporter [HXT1, YHR094C]	*Saccharomyces cerevisiae* [yeast]	Glucose, mannose
Hxt2sacce	High-affinity glucose transporter [HXT2]	*Saccharomyces cerevisiae* [yeast]	Glucose
Hxt3sacce	High-affinity glucose transporter [HXT3]	*Saccharomyces cerevisiae* [yeast]	Glucose
Hxt4sacce	Low-affinity glucose transporter [HXT4, RAG1, LGT1, YHR092C]	*Saccharomyces cerevisiae* [yeast]	Glucose
Hxt5sacce	Probable glucose transporter [HXT5]	*Saccharomyces cerevisiae* [yeast]	Glucose
Hxt6sacce	Hexose transporter [HXT6]	*Saccharomyces cerevisiae* [yeast]	Glucose
Hxt7sacce	Hexose transporter [HXT7]	*Saccharomyces cerevisiae* [yeast]	Glucose
Hxt8sacce	Hexose transporter [HXT8, HRA569, YJL214W]	*Saccharomyces cerevisiae* [yeast]	Glucose
Hxtasacce	Low-affinity hexose transporter [HXT11, LGT3]	*Saccharomyces cerevisiae* [yeast]	Glucose
Hxtchlke	Hexose transport protein homolog	*Chlorella kessleri* [alga]	Hexoses
Hxtcsacce	Hexose transporter [HXT13, HXT8, YEL069C]	*Saccharomyces cerevisiae* [yeast]	Glucose
Hxtdsacce	Hexose transporter [HXT14, HXT9, NO345]	*Saccharomyces cerevisiae* [yeast]	Glucose
Itr1sacce	Myo-inositol transporter 1 [ITR1]	*Saccharomyces cerevisiae* [yeast]	Myo-inositol
Itr2sacce	Myo-inositol transporter 2 [ITR2]	*Saccharomyces cerevisiae* [yeast]	Myo-inositol
Lacpklula	Lactose permease [LACP, LAC12]	*Kluyveromyces lactis* [yeast]	Lactose
Ma3tsacce	Maltose permease [Mal3T, MA3T, MAL31, YBR298C, YBR2116]	*Saccharomyces cerevisiae* [yeast]	Maltose
Ma6tsacce	Maltose permease [MAL6T, MAL61]	*Saccharomyces cerevisiae* [yeast]	Maltose
Qayneucr	Quinate permease [QA-y]	*Neurospora crassa* [mold]	Quinate
Qutdemeni	Quinate permease [QUTD]	*Emericella nidulans* [mold]	Quinate
Rag1klula	Low-affinity glucose transporter [RAG1]	*Kluyveromyces lactis* [yeast]	Glucose
Sgt1schma	Glucose transporter [SGT1, SGTP1]	*Schistosoma mansoni* [fluke]	Glucose
Sgt2schma	Glucose transporter [SGT2, SGTP2]	*Schistosoma mansoni* [fluke]	Glucose
Sgt4schma	Glucose transporter [SGT4, SGTP4]	*Schistosoma mansoni* [fluke]	Glucose
Snf3sacce	High-affinity glucose transporter [SNF3]	*Saccharomyces cerevisiae* [yeast]	Glucose, mannose, fructose

CODE	DESCRIPTION [SYNONYMS]	ORGANISM [COMMON NAMES]	SUBSTRATE(S)
Stl1sacce	Sugar transporter [STL1]	Saccharomyces cerevisiae [yeast]	Glucose
Stp1arath	Glucose-H⁺ symporter [STP1]	Arabidopsis thaliana [mouse-ear cress]	H⁺/glucose
Stp4arath	Monosaccharide transporter [STP4]	Arabidopsis thaliana [mouse-ear cress]	Monosaccharides
Sugricco	Sugar carrier protein [RCSTC]	Ricinus communis [castor oil plant]	Hexoses
Tgtptaeso	Facilitative glucose transporter [TGTP1]	Taenia solium [tapeworm]	Glucose
Xyleescco	Xylose-H⁺ symporter [XYLE]	Escherichia coli [gram-negative bacterium]	H⁺/xylose

Cotransported ions are listed for known symporters.

Phylogenetic tree

Proteins listed subsequently in italics are at least 90% identical to the paired transporters listed in parenthesis and therefore are not included in the phylogenetic tree: *Araekleox* (Araeescco); *Gtr1musmu, Gtr1orycu, Gtr1sussc, Gtr1bosta, Gtr1ratno* (Gtr1homsa); *Gtr2musmu* (Gtr2ratno); *Gtr3ratno* (Gtr3musmu); *Gtr4ratno, Gtr4musmu* (Gtr4homsa); *Hxt6sacce* (Hxt7sacce).

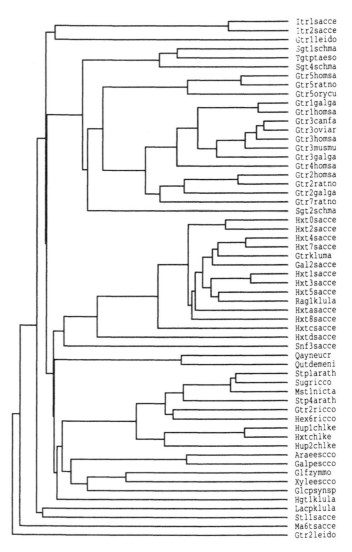

Proposed orientation of human GLUT1 in the membrane

The model is based on predictions of membrane-spanning regions and α-helical content [5,12,13]. The N-terminus of the protein is illustrated on the inside and is folded 12 times through the membrane. The predicted membrane-spanning helices are portrayed as rectangles. The numbers corresponding to the first and last residue of each membrane-spanning helix are boxed. Residues that are conserved in more than 75% of the aligned transporters (see below) are shown. Consensus residues indicated by an asterisk are not conserved in GLUT1.

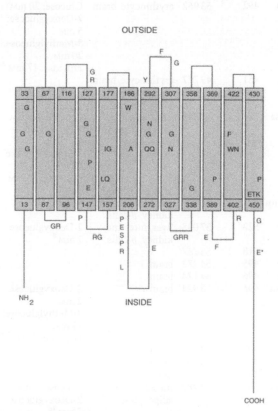

Physical and genetic characteristics

	AMINO ACIDS	MOL. WT	EXPRESSION SITES	K_m	CHROMOSOMAL LOCUS
Araeescco	472	51684		Arabinose: 140–320 μM [15]	64.17 minutes
Araekleox	472	51732			
Gal2sacce	574	63738			Chromosome 12
Galpescco	464	50982		Galactose: 42–49 μM [15]	66.48 minutes
Glcpsynsp	468	49747			
Glfzymmo	473	50200			
Gtr1bosta	492	54131	erythrocyte, brain		

	AMINO ACIDS	MOL. WT	EXPRESSION SITES	K_m	CHROMOSOMAL LOCUS
Gtr1galga	490	54 086	erythrocyte, brain		
Gtr1homsa	492	54 117	erythrocyte, brain	Glucose: 17 mM 2-Deoxyglucose: 7 mM 3-Methylglucose: 18 mM [2,16]	1p35–p31.3
Gtr1leido	547	58 787			
Gtr1musmu	492	53 991	erythrocyte, brain		
Gtr1orycu	492	54 097	erythrocyte, brain		
Gtr1ratno	492	53 962	erythrocyte, brain	Glucose: 20 mM 2-Deoxyglucose: 5 mM 3-Methylglucose: 20 mM Fructose: 17 mM [2,16]	
Gtr1sussc	451	49 777	erythrocyte, brain		
Gtr2galga	533	57 699	liver, intestine, kidney, β-cells		
Gtr2homsa	524	57 489	liver, intestine, kidney, β-cells	Glucose: 66 mM 2-Deoxyglucose: 17 mM 3-Methylglucose: 42 mM Fructose: 36 mM [2,16]	3q26.1–q26.3
Gtr2leido	558	60 215	promastigotes		
Gtr2musmu	523	57 075	liver, intestine, kidney, β-cells		
Gtr2ratno	522	57 085	liver, intestine, kidney, β-cells	2-Deoxyglucose: 7 mM [2]	
Gtr2ricco	518	55 737			
Gtr3canfa	495	54 282	brain		
Gtr3galga	496	54 174	brain		
Gtr3homsa	496	53 924	brain	2-Deoxyglucose: 2 mM 3-Methylglucose: 11 mM Fructose: 6 mM [2,16]	12p13.3
Gtr3musmu	493	53 478	brain		
Gtr3oviar	494	54 194	brain		
Gtr3ratno	493	53 580	brain		
Gtr4homsa	509	54 787	muscle, adipocytes	Glucose: 5 mM 2-Deoxyglucose: 5 mM [16]	17p13
Gtr4musmu	510	54 951	muscle, adipocytes		
Gtr4ratno	509	54 895	muscle, adipocytes	Glucose: 2 mM 3-Methylglucose: 3 mM [16]	
Gtr5homsa	501	54 974	intestine	Fructose: 6 mM [2]	1p31
Gtr5orycu	486	53 854	intestine		
Gtr5ratno	502	55 543	intestine		
Gtr7ratno	528	58 393	liver microsomes		
Gtrkluma	566	62 727			
Hex6ricco	510	55 594			
Hgt1klula	551	60 761			
Hup1chlke	533	57 463			
Hup2chlke	540	58 343			

	AMINO ACIDS	MOL. WT	EXPRESSION SITES	K_m	CHROMOSOMAL LOCUS
Hxt0sacce	546	60662			Chromosome 6
Hxt1sacce	570	63261			Chromosome 8
Hxt2sacce	541	59840			Chromosome 13
Hxt3sacce	567	62557			Chromosome 4
Hxt4sacce	576	63910			Chromosome 8
Hxt5sacce	592	66251			Chromosome 8
Hxt6sacce	570	62719			Chromosome 4
Hxt7sacce	570	62735			Chromosome 4
Hxt8sacce	569	63492			Chromosome 10
Hxtasacce	567	62857			Chromosome 15
Hxtchlke	534	57771			
Hxtcsacce	564	62734			Chromosome 5
Hxtdsacce	515	58153			Chromosome 14
Itr1sacce	584	63605	Myoinositol: 100 μM [6]		Chromosome 4
Itr2sacce	612	67041	Myoinositol: 400 μM [6]		Chromosome 14
Lacpklula	587	65383	Lactose: 1 mM [6]		
Ma3tsacce	614	68262			
Ma6tsacce	614	68225			
Qayneucr	537	60103			
Qutdemeni	533	59476			
Rag1klula	567	63123			
Sgt1schma	521	56815			
Sgt2schma	489	54060			
Sgt4schma	505	55009			
Snf3sacce	884	96718			Chromosome 4
Stl1sacce	536	59943			Chromosome 4
Stp1arath	522	57596			
Stp4arath	514	57095			
Sugricco	523	57768			
Tgtptaeso	510	55191			
Xyleescco	491	53608	Xylose: 70–170 μM [15]		91.29 minutes

Multiple amino acid sequence alignments

```
          1                                                  50
Itr1sacce ........... .....MGIHI PYLTSKTSQS NVGDAVGNAD SVEFN.....
Itr2sacce MAEMKNSTAA SSRWTKSRLS HFFPSYTNSS GMGAASTDQS STQGEELHHR
Hxt4sacce .......... .......... .......... ...MS..E EAAYQEDTAV
Hxt7sacce .......... .......... .......... ...MS..Q DAAIAEQTPV
Gtrkluma  .......... .......... .......... ..MSELE TGTAAHGTPV
Gal2sacce .......... .......... .......... ...... MAVEENNVPV
Hxt1sacce .......... .......... .......... ...... MNSTPDLISP
Hxt3sacce .......... .......... .......... ...... MNSTPDLISP
Hxt5sacce .......... .......... .......... .MSELE NAHQGPLEGS ATVSTNSNSY
Rag1klula .......... .......... .......... ..MSNQ MTDSTSAGSG
Hxtasacce .......... .......... .......... ...... ......MSGV
Hxt8sacce .......... .......... .......... ...... MTDRKTNLPE
Snf3sacce .......... .....MDPNS NSSSETLRQE KQGFLDKALQ RVKGIALRRN


          51                                                 100
Itr1sacce ...SEHDSPS KRGKIHIESH EIQ...RAPA SDDEDRIQIK PVNDEDDTSV
Itr2sacce KHCEEDNDGQ KPKKSPVSTS TMQIKSRQDE DEDDGRIVIK PVNDEDDTSV
```

```
Sgt1schma .......................... .............. ..M
Gtr5homsa .......................... ......... .MEQQDQ
Gtr5ratno .......................... ......... ..MEKEDQ
Gtr4homsa .......................... .........MP SGFQQIGSED
Gtr2galga .......................... ........ .....MDGKS
Hxt0sacce .........M VSSSVSILGT SAKASTSLSR K...DEIKLT PETREASLDI
Hxt2sacce .....MSEFA TSRVESGSQQ TSIHSTPIVQ KLETDESPIQ TKSEYTNAEL
Hxt4sacce QNTPADALSP VES..DSNSA LSTPSNKAER DDMKDFDENH EESNNY.VEI
Hxt7sacce EH.....LSA VDS..ASHSV LSTPSNKAER DEIKAYGEG. EEHEPV.VEI
Gtrkluma  EN........ .KS..VSSSQ ASTPTNVGSR DDLKVDDDNH ..SVDA.IEL
Gal2sacce VSQQPQAGED VISSLSKDSH LSAQSQKYSN DELKAGESGP EGSQSVPIEI
Hxt1sacce QKSNSSNSYE LESGRSKAMN TPE......G KNESFHDNLS ESQVQPAVAP
Hxt3sacce QKSSENSNAD LPSNSSQVMN MPE......E K..GVQDDF. QAEADQVLTN
Hxt5sacce NEKSGNSTAP GTAGYNDNLA QAKPVSSYIS HEGPPKDELE ELQKEVDKQL
Rag1klula TEHSVDTNTA LKAGSPNDLK ........VS HE....EDLN DLEKTAEETL
Hxtasacce NNTSANDLST TESNSNSVAN AP....SVKT EHNDSKNSLN LDATEPPIDL
Hxt8sacce EPIFEEAEDD GCPSIENSSH LSVPTVEENK DFSEYNGEEA EE.....VVV
Hxtcsacce ....MSSAQS SIDSDGDVRD ADIHVAPPVE KEWSDGFDDN EVINGDNVEP
Hxtdsacce .......... ......MGR DYNVTIKYLD DKEENIEGQA
Snf3sacce NSNKDHTTDD TTGSIRTPTS LQRQNSDRQS NMTSVFTDDI STIDDNSILF
Qayneucr  .......... .......... .......... .MTLLALKED
Qutdemeni .......... .......... .......... .MSILALVED
Stp1arath .......... .......... ....MPA GGFVV..GDG
Sugricco  .......... .......... ....MPA VGGIPPSGGN
Mst1nicta .......... .......... .....MAG GGGIGP..GN
Stp4arath .......... .......... ......M AGGFVSQTPG
Gtr2ricco .......... .......... ....MAGGG MAALGVKTER
Hex6ricco .......... .......... ...... MAAGLAITSE
Hup1chlke .......... .......... ...MAGGGVVV VSGRGLSTGD
Hxtchlke  .......... .......... ...MAGGAIV ASGGASRSSE
Hup2chlke .......... .......... ...MAGGGPV ASTTTNRASQ
Araeescco .......... .......... ...... MVTINTESAL
Galpescco .......... .......... ...... ....MPD
Glcpsynsp .......... .......... ...... ...MNPSS
Hgt1klula .......... .......... .MSLKNWLL LRDIQYEGTF
Lacpklula .......... .....MADHSSS SSSLQKKPIN TIEHKDTLGN
Ma6tsacce .MKGLSSLIN RKKDRNDSHL DEIENGVNAT EFNSIEMEEQ GKKSDFDLSH
Gtr2leido .......... .....MTLKK RSSAPELPTS LDEDEEEDSP
Consensus .......... .......... .......... ..........
```

```
          101                                         150
Itr1sacce MITFNQSLSP FIITLTFVAS IS.GFMFGYD TGYIS............
Itr2sacce IITFNQSISP FIITLTFVAS IS.GFMFGYD TGYIS............
Gtr1leido .......... MRASVMLCAA LG.GFLFGYD TGVIN............
Sgt1schma GVASNNGITG KLVLTVLITC VGSSFLIGYN LGVLN............
Tgtptaeso ....MKGISG PLVLAIFTTC FGSSFLLGYN LGVAN............
Sgt4schma .MGSGKKFTK SLSLSVLLAC LGSSFTIGYN LGVLN............
Gtr5homsa S.MKEGRLTL VLALATLIAA FGSSFQYGYN VAAVN............
Gtr5ratno E..KTGKLTL VLALATFLAA FGSSFQYGYN VAAVN............
Gtr5orycu EKKKEGRLTL VLALRTLIAA FGSSFQYAYN VSVCN............
Gtr1galga MESGS.KMTA RLMLAVGGAV LG.SLQFGYN TGVIN............
Gtr1homsa MEPSSKKLTG RLMLAVGGAV LG.SLQFGYN TGVIN............
Gtr3musmu ..MGTTKVTP SLVFAVTVAT IG.SFQFGYN TGVIN............
```

```
Gtr3canfa  ..MGTQKVTV SLIFALSIAT IG.SFQFGYN TGVIN.............
Gtr3oviar  ..MGTTKVTT PLIFAISIAT IG.SFQFGYN TGVIN.............
Gtr3homsa  ..MGTQKVTP ALIFAITVAT IG.SFQFGYN TGVIN.............
Gtr3galga  .MADKKKITA SLIYAVSVAA IG.SLQFGYN TGVIN.............
Gtr4homsa  GEPPQQRVTG TLVLAVFSAV LG.SLQFGYN IGVIN.............
Gtr2homsa  ..MTEDKVTG TLVFTVITAV LG.SFQFGYD IGVIN.............
Gtr2ratno  ..MSEDKITG TLAFTVFTAV LG.SFQFGYD IGVIN.............
Gtr2galga  KMQAEKHLTG TLVLSVFTAV LG.FFQYGYS LGVIN.............
Gtr7ratno  ..MSDDSLTA TLSLFVFTAV LG.SFQFGYD IGVIN.............
Sgt2schma  ......MRQL KFFLPYCIIT LGSSFPFGYH TGVIN.............
Hxt0sacce  PYKPIIAYWT VMG.LCLMIA FG.GFIFGWD TGTIS.............
Hxt2sacce  PAKPIAAYWT VIC.LCLMIA FG.GFVFGWD TGTIS.............
Hxt4sacce  PKKPASAYVT VSI.CCLMVA FG.GFVFGWD TGTIS.............
Hxt7sacce  PKRPASAYVT VSI.MCIMIA FG.GFVFGWD TGTIS.............
Gtrkluma   PKKPRSAYIT VSI.LCLMVA FG.GFVFGWD TGTIS.............
Gal2sacce  PKKPMSEYVT VSL.LCLCVR FG.GFMFGWD TSTIS.............
Hxt1sacce  PNTGKGVYVT VSI.CCVMVA FG.GFIFGWD TGTIS.............
Hxt3sacce  PNTGKGAYVT VSI.CCVMVA FG.GFVFGWD TGTIS.............
Hxt5sacce  EKKSKSDLLF VSV.CCLMVA FG.GFVFGWD TGTIS.............
Rag1klula  QQKPAKEYIF VSL.CCVMVA FG.GFVFGWD TGTIS.............
Hxtasacce  PQKPLSAYTT VAI.LCLMIA FG.GFIFGWD TGTIS.............
Hxt8sacce  PEKPASAYAT VSI.MCLCMA FG.GFMSGWD TGTIS.............
Hxtcsacce  PKRGLIGYLV IYL.LCYPIS FG.GFLPGWD SGITA.............
Hxtdsacce  AKISHNASLH IPVLLCLVIS LG.GFIFGWD IGTIG.............
Snf3sacce  SEPPQKQSMM MSICVGVFVA VG.GFLFGYD TGLIN.............
Qayneucr   RPTPKAVYNW RVYTCAAIAS FA.SCMIGYD SAFIG.............
Qutdemeni  RPTPREVYNW RVYLLAAVAS FT.SCMIGYD SAFIG.............
Stp1arath  QKAYPGKLTP FVLFTCVVAA MG.GLIFGYD IGISG.............
Sugricco   RKVYPGNLTL YVTVTCVVAA MG.GLIFGYD IGISG.............
Mst1nicta  GKEYPGNLTL YVTVTCIVAA MG.GLIFGYD IGISG.............
Stp4arath  VRNYNYKLTP KVFVTCFIGA FG.GLIFGYD LGISG.............
Gtr2ricco  AAQYKGRMTL AVAMTCLVAA VG.GAIFGYD IGISG.............
Hex6ricco  GGQYNGRMTS FVALSCMMAA MG.GVIFGYD IGVSG.............
Hup1chlke  Y...RGGLTV YVVMVAFMAA CG.GLLLGYD NGVTG.............
Hxtchlke   Y...QGGLTA YVLLVALVAA CG.GMLLGYD NGVTG.............
Hup2chlke  YGYARGGLNW YIFIVALTAG SG.GLLFGYD IGVTG.............
Araeescco  TPRSLRDTRR MNMFVSVAAA VA.GLLFGLD IGVIA.............
Galpescco  AKKQGRSNKA MTFFVCFLAA LA.GLLFGLD IGVIA.............
Glfzymmo   ..MSSESSQG LVTRLALIAA IG.GLLFGYD SAVIA.............
Xyleescco  ..MNTQYNSS YIFSITLVAT LG.GLLFGYD TAVIS.............
Glcpsynsp  SPSQSTANVK FVLLISGVAA LG.GFLFGFD TAVIN.............
Hgt1klula  YKKFPHVYNI YV..IGFIAC IS.GLMFGFD IASMS.............
Lacpklula  DRDHKEALNS DNDNTSGLKI NGVPIEDARE EVLLP.............
Stl1sacce  ...........    ....MKDLKL SNFKG.............
Ma6tsacce  LEYGPGSLIP NDNNEEVPDL LDEAMQDAKE ADESE.............
Gtr2leido  QPLSNTPFFS MKNLIVATPI ILTPLLYGYN LGFVGPYSTM YGYASNCQLY
Consensus  ................ .G.....G.. .G...............
```

```
           151                                              200
Itr1sacce  ....SALISI GT............ ...........    ...........
Itr2sacce  ....SALISI NR............ ...........    ...........
Gtr1leido  ....AALFQM KD............ ...........    ...........
Sgt1schma  ....LPRRNI EIYFNETVVP .NTPE...... ..LDS...........
Tgtptaeso  ....LPGDNI KKFLVNYYKP DNSSA...... ..LNA...........
```

```
Sgt4schma ....LPGENI KEFLSRTMLG KNASEAENTA NLVTP............
Gtr5homsa ....SPALLM QQFYNETYYG RTG...... ..........
Gtr5ratno ....SPSEFM QQFYNDTYYD RNK...... ..........
Gtr5orycu ....SPSELM TEFYNDTYYD RTG...... ..........
Gtr1galga ....RPQKVI EDFYNHTWLY RYE...... ..........
Gtr1homsa ....APQKVI EEFYNQTWVH RYG...... ..........
Gtr3musmu ....APETIL KDFLNYTLEE RLE...... ..........
Gtr3canfa ....APETII KDFLNYTLEE KSE...... ..........
Gtr3oviar ....APEAII KDFLNYTLEE RSE...... ..........
Gtr3homsa ....APEKII KEFINKTLTD KGN...... ..........
Gtr3galga ....APEKII QAFYNRTLSQ RSG...... ..........
Gtr4homsa ....APQKVI EQSYNETWLG RQGPE..... ..........
Gtr2homsa ....APQQVI ISHYRHVLGV PLDDRKAINN YVINSTDELP TISYSMNPKP
Gtr2ratno ....APQEVI ISHYRHVLGV PLDDRRATIN YDINGTDT.P LIVTPAHTTP
Gtr2galga ....APQKVI EAHYGRMLGA IPMVRHATNT SRDNATITVT IPGTEAWGSS
Gtr7ratno ....APQEVI ISHYRHVLGV PLDDRRATIN YDINGTDT.P LIVTPAHTTP
Sgt2schma ....APADLI KSFINTTLAA RSV...... ..........
Hxt0sacce ....GFIN.Q TDFKRRF..G ELQ..R.DGS F..........
Hxt2sacce ....GFVN.Q TDFKRRF..G QMK..S.DGT Y..........
Hxt4sacce ....GFVA.Q TDFIRRF..G M.K..HHDGT Y..........
Hxt7sacce ....GFIN.Q TDFIRRF..G M.K..HKDGT N..........
Gtrkluma  ....GFVN.Q TDFIRRF..G Q.E..KADGS H..........
Gal2sacce ....GFVV.Q TDFLRRF..G M.K..HKDGT H..........
Hxt1sacce ....GFVA.Q TDFLRRF..G M.K..HHDGS H..........
Hxt3sacce ....GFVA.Q TDFLRRF..G M.K..HKDGS Y..........
Hxt5sacce ....GFVR.Q TDFIRRF..G S.T..RANGT T..........
Rag1klula ....GFVN.Q TDFLRRF..G Q.E..KADGS H..........
Hxtasacce ....GFVN.L SDFIRRF..G Q.K..NDKGT Y..........
Hxt8sacce ....GFVN.Q TDFLRRF..G NYS..HSKNT Y..........
Hxtcsacce ....GFIN.M DNFKMNF..G SYK..HSTGE Y..........
Hxtdsacce ....GMTN.M VSFQEKF..G TTNIIHDDET I..........
Snf3sacce ....SITS.M .....NY..V KSHVAPNHDS F..........
Qayneucr  ....TTLA.L PSFTKEF... ..........D F..........
Qutdemeni ....TTLS.L QSFQNEF... ..........N W..........
Stp1arath ....GVTS.M PSFLKRFFPS VY.RKQQEDA S..........
Sugricco  ....GVTS.M DSFLKKFFPS VY.RKKKADE S..........
Mst1nicta ....GVTS.M DSFLSRFFPS VF.RKQKADD S..........
Stp4arath ....GVTS.M EPFLEEFFPY VY.KKMKSAH ..........
Gtr2ricco ....GVTS.M DPFLEKFFPV VF.HR.KNSG G..........
Hex6ricco ....GVTS.M DPFLKKFFPD VY.RKMKEDT E..........
Hup1chlke ....GVVS.L EAFE.KFFPD VWAKKQEVHE ..........
Hxtchlke  ....GVAS.M EQFERKFFPD VYEKKQQIVE ..........
Hup2chlke ....GVTS.M PEFLQKFFPS IYDRTQQPSD S..........
Araeescco ....GALP.. ......FITD HFV...... ..........
Galpescco ....GALP.. ......FIAD EFQ...... ..........
Glfzymmo  ....AIGT.. ......PVDI HFIAPRHLSA T..........
Xyleescco ....GTVE.. ......SLNT VFVAPQNLSE S..........
Glcpsynsp ....GAVA.. ......ALQK HF........ Q.........
Hgt1klula ....SMIG.T DVY...... ..........
Lacpklula ....GYLSK. QYY..KLYGL ...CFITYLC AT...MQGYD GALMGSIYTE
Stl1sacce ....KFISRT SHW..GLTGK KLRYFITIAS MTGFSLFGYD QGLMASLITG
Ma6tsacce ....RGMPLM TALKTYPKAA AWSLLVSTTL IQEGYDTAIL GAFYALPVFQ
Gtr2leido SAKKSCETLT AAKCRWFNAS TYVSNTTYGE VCGWADRTTC FLKYSDEAGC
Consensus ......... .......... .......... ..........
```

```
          201                                              250
Itr1sacce ........... ..DLDHKVLT ..YGEKEIVT AATSLGALIT SIFAGTAADI
Itr2sacce ........... ..DLDNKVLT ..YGEKELIT AATSLGALIT SVGAGTAADV
Gtr1leido ........... ..HFGFSEHS ..W.QYALIV AIAIAGAFVG AFISGFISAA
Sgt1schma ............ ....SFFYTHVS TIFVVAAAIG AFSCGWVADG
Tgtptaeso ............ ..NFLYGQVT SVLVICAAIA AFTCGWVADG
Sgt4schma ............ ..SFLYAQVS TAFVVAGAIG AFSCGAIADC
Gtr5homsa ........... ..EFMEDFPL ..TLLWSVTV SMFPFGGFIG SLLVGPLVNK
Gtr5ratno ........... ..ENIESFTL ..TLLWSLTV SMFPFGGFIG SLMVGFLVNN
Gtr5orycu ........... ..ELIDEFPL ..TLLWSVTV SMFPSGGFAG SLLVGPLVNK
Gtr1galga ........... ..EPISPATL ..TTLWSLSV AIFSVGGMIG SFSVGLFVNR
Gtr1homsa ........... ..ESILPTTL ..TTLWSLSV AIFSVGGMIG SFSVGLFVNR
Gtr3musmu ........... ..DLPSEGLL ..TALWSLCV AIFSVGGMIG SFSVGLFVNR
Gtr3canfa ........... ..NLPTEVLL ..TSLWSLSV AIFSVGGMIG SFSVGLFVNR
Gtr3oviar ........... ..TPPSSVLL ..TSLWSLSV AIFSVGGMIG SFSVGLFVNR
Gtr3homsa ........... ..APPSEVLL ..TSLWSLSV AIFSVGGMIG SFSVGLFVNR
Gtr3galga ........... ..ETISPELL ..TSLWSLSV AIFSVGGMIG SFSVSLFFNR
Gtr4homsa ........... GPSSIPPGTL ..TTLWALSV AIFSVGGMIS SFLIGIISQW
Gtr2homsa TPWAE..... EETVAAAQLI ..TMLWSLSV SSFAVGGMTA SFFGGWLGDT
Gtr2ratno DAW.E..... EETEGSAHIV ..TMLWSLSV SSFAVGGMVA SFFGGWLGDK
Gtr2galga EGTLAPSAGF EDPTVSPHIL ..TMYWSLSV SMFAVGGMVS SFTVGWIGDR
Gtr7ratno DAWE...... EETEGSAHIV ..TMLWSLSV SSFAVGGMVA SFFGGWLGDK
Sgt2schma ........... ...TCDERFI ..DLLWSLCV TSFLLGGFFG GLIGGVLANK
Hxt0sacce ........... ..QLS...... ..DVRTGLIV GIFNIGCALG GLTLGRLGDI
Hxt2sacce ........... ..YLS...... ..DVRTGLIV GIFNIGCAFG GLTLGRLGDM
Hxt4sacce ........... ..YLS...... ..KVRTGLIV SIFNIGCAIG GIILAKLGDM
Hxt7sacce ........... ..YLS...... ..KVRTGLIV SIFNIGCAIG GIILSKLGDM
Gtrkluma  ........... ..YLS...... ..NVRTGLIV SIFNIGCAIG GIILSKLGDM
Gal2sacce ........... ..YLS...... ..NVRTGLIV AIFNIGCAFG GIILSKGGDM
Hxt1sacce ........... ..YLS...... ..KVRTGLIV SIFNIGCAIG GIVLAKLGDM
Hxt3sacce ........... ..YLS...... ..KVRTGLIV SIFNIGCAIG GIILAKLGDM
Hxt5sacce ........... ..YLS...... ..DVRTGLMV SIFNIGCAIG GIVLSKLGDM
Rag1klula ........... ..YLS...... ..NVRTGLIV SIFNIGCAVG GIVLSNIGDR
Hxtasacce ........... ..YLS...... ..KVRMGLIV SIFNIGCAIG GIVLSKVGDI
Hxt8sacce ........... ..YLS...... ..NVRTGLIV SIFNVGSAIG CLFLSKLGDI
Hxtcsacce ........... ..YLS...... ..NVRMGLLV AMFSIGCAIG GLIFARLADT
Hxtdsacce ........... ..FVSTKKLT ..DLQIGLII SIFNISCGVG ALTLSKIGDW
Snf3sacce ........... ...T...... ..AQQMSILV SFLSLGTFFG ALTAPFISDS
Qayneucr  ........... ..ASYTPGAL ..ALLQSNIV SVYQAGAFFG CLFAYATSYF
Qutdemeni ........... ..ESLNTD.. ...LISANIV SLYQRGAFFG ALFAYPIGHF
Stp1arath ........... ..TNQYCQYD ..SPTLTMFT SSLYLAALIS SLVASTVTRK
Sugricco  ........... ..SNQYCQYD ..SQTLTMFT SSLYLAALIA SLVASTITRK
Mst1nicta ........... ..TNQYCKFD ..SQTLTMFT SSLYLAALLS SLVASTVTRK
Stp4arath ........... ..ENEYCRFD ..SQLLTLFT SSLYVAALVS SLFASTITRV
Gtr2ricco ........... ..KNNYCKYD ..NQGLAAFT SSLYLAGLVA SLVASPVTRN
Hex6ricco ........... ..ISNYCKFD ..SQLLTSFT SSLYVAGLVA SFFASSVTRA
Hup1chlke ........... ..DSPYCTYD ..NAKLQLFV SSLFLAGLVS CLFASWITRN
Hxtchlke  ........... ..TSPYCTYD ..NPKLQLFV SSLFLAGLIS CIFSAWITRN
Hup2chlke ........... ..KDPYCTYD ..DQKLQLFT SSFFLAGMFV SFFAGSVVRR
Araeescco ........... ..L......T ..SRLQEWVV SSMMLGAAIG ALFNGWLSFR
Galpescco ........... ..I......T ..SHTQEWVV SSMMFGAAVG AVGSGWLSFK
Glfzymmo  ........... ..A......A ..ASLSGMVV VAVLVGCVTG SLLSGWIGIR
Xyleescco ........... ..A......A ..NSLLGFCV ASALIGCIIG GALGGYCSNR
Glcpsynsp ........... ..T......D ..SLLTGLSV SLALLGSALG AFGAGPIADR
```

273

```
Hgt1klula .......... ..KDYFSNPD ..SLTYGGIT ASMAGGSFLG SLISPNFSDA
Lacpklula DAYLKYY... ....HLDINS ..SSGTGLVF SIFNVGQICG AFFVPL.MDW
Stl1sacce KQFNYEFPAT KENGDHDRHA ..TVVQGATT SCYELGCFAG SLFVMFCGER
Ma6tsacce KKYGSLNSNT GDYEISVSWQ ..IGLCLCYM AGEIVGLQVT GPSVDYMGNR
Gtr2leido LSDSACKWSY SANTCGNQVG YSSIQSGVFA GSLVIGSTMG ALMGGYLTKR
Consensus ......... ......... ......... ....G.... ..........
```

```
                 251                                          300
Itr1sacce FGRKRCLMGS NLMFVIGAI. ......... ...LQV SAHTFWQMAV
Itr2sacce FGRRPCLMFS NLMFLIGAI. ......... ...LQI TAHKFWQMAA
Gtr1leido FGRRPCIAVA DALFVIGSV. ......... ...LMG AAPNVEVVLV
Sgt1schma LGRRNGLILN NVIGIIGGV. ......... ..IVGPCV LVKQPALLYV
Tgtptaeso LGRKRSLMVN NGIGIVGSV. ......... ..ISSVCV VANQPALLYV
Sgt4schma LGRRNGLIVN SLLAIIGGI. ......... ..LVGPCV AYSQPALLFV
Gtr5homsa FGRKGALLFN NIFSIVPAI. ......... ..LMGCSR VATSFELIII
Gtr5ratno LGRKGALLFN NIFSILPAI. ......... ..LMGCSK IAKSFEIIIA
Gtr5orycu FGRKGALLFN NIFSIVPAI. ......... ..LMGCSK VARSFELIII
Gtr1galga FGRRNSMLMS NILAFLAAV. ......... ..LMGFSK MALSFEMLIL
Gtr1homsa FGRRNSMLMM NLLAFVSAV. ......... ..LMGFSK LGKSFEMLIL
Gtr3musmu FGRRNSMLLV NLLAIIAGC. ......... ..LMGFAK IAESVEMLIL
Gtr3canfa FGRRNSMLMV NLLAVAGGC. ......... ..LMGFCK IAQSVEMLIL
Gtr3oviar FGRRNSMLIV NLLAIAGGC. ......... ..LMGFCK IAESVEMLIL
Gtr3homsa FGRRNSMLIV NLLAVTGGC. ......... ..FMGLCK VAKSVEMLIL
Gtr3galga FGRRNSMLLV NVLAFAGGA. ......... ..LMALSK IAKAVEMLII
Gtr4homsa LGRKRAMLVN NVLAVLGGS. ......... ..LMGLAN AAASYEMLIL
Gtr2homsa LGRIKAMLVA NILSLVGAL. ......... ..LMGFSK LGPSHILIIA
Gtr2ratno LGRIKAMLAA NSLSLTGAL. ......... ..LMGCSK FGPAHALIIA
Gtr2galga LGRVKAMLVV NVLSIAGNL. ......... ..LMGLAK MGPSHILIIA
Gtr7ratno LGRIKAMLAA NSLSLTGAL. ......... ..LMGCSK FGPSHALIIA
Sgt2schma LGRKNSLFLL SIPTVIGSL. ......... ..LMMFSK MAQSFEMIII
Hxt0sacce YGRKIGLMC. VILVYVVGI. ......... ...VIQIA SSDKWYQYFI
Hxt2sacce YGRRIGLMC. VVLVYIVGI. ......... ...VIQIA SSDKWYQYFI
Hxt4sacce YGRKMGLIV. VVVIYIIGI. ......... ...IIQIA SINKWYQYFI
Hxt7sacce YGRKVGLIV. VVVIYIIGI. ......... ...IIQIA SINKWYQYFI
Gtrkluma  YGRRIGLMI. VVLIYVVGI. ......... ...IIQIA SIDKWYQYFI
Gal2sacce YGRKKGLSI. VVSVYIVGI. ......... ...IIQIA SINKWYQYFI
Hxt1sacce YGRRIGLIV. VVVIYTIGI. ......... ...IIQIA SINKWYQYFI
Hxt3sacce YGRKMGLIV. VVVIYIIGI. ......... ...IIQIA SINKWYQYFI
Hxt5sacce YGRKIGLMT. VVVIYSIGI. ......... ...IIQIA SIDKWYQYFI
Rag1klula WGRRIGLIT. VIIIYVIGI. ......... ...IIQIA SVDKWYQYFI
Hxtasacce YGRRIGLIT. VTAIYVVGI. ......... ...LIQIT SINKWYQYFI
Hxt8sacce YGRCMGLII. VIVVYMVGI. ......... ...VIQIA SIDKWYQYFI
Hxtcsacce LGRRLAIVI. VVLVYMVGA. ......... ...IIQIS SNHKWYQYFV
Hxtdsacce IGRKGGIWF. ALVVYCIGI. ......... ...TIQIL SYGRWYFLTL
Snf3sacce YGRKPTIIFS TIFIFSIGN. ......... ...SLQVG AGGITL.LIV
Qayneucr  LGRRKSLIAF SVVFIIGA.. ......... .AIMLAADG QGRGIDPIIA
Qutdemeni WGRRWGLMFS ALIFFLGA.. ......... .GMMLGANG .DRGLGLIYG
Stp1arath FGRRLSMLFG GILFC.AG.. ......... ...ALING FAKHVWMLIV
Sugricco  FGRKLSMLFG GVLFC.AG.. ......... ...AIING AAKAVWMLIL
Mst1nicta LGRRLSMLCG GVLFC.AG.. ......... ...ALING FAQNVAMLIV
Stp4arath FGRKWSMFLG GFTFF.IG.. ......... ...SAFNG FAQNIAMLLI
Gtr2ricco YGRKASIVCG GVSFL.IG.. ......... ...AALNV AAVNLAMLIL
Hex6ricco FGRKPSILLG GXVFL.AX.. ......... ...AALGG AAVNVYMLIF
Hup1chlke WGRKVTMGIG GAFFV.AG.. ......... ....GLVNA FAQDMAMLIV
```

```
Hxtchlke  WGRKASMGIG GIFFIAAG.. .......... .....GLVNA FAQDIAMLIV
Hup2chlke WGRKPTMLIA SVLFLAGA.. .......... .....GL.NA GAQDLAMLVI
Araeescco LGRKYSLMAG AILFVLGSIG SAF....... .......... .ATSVEMLIA
Galpescco LGRKKSLMIG AILFVAGSLF SAA....... .......... .APNVEVLIL
Glfzymmo  FGRRGGLLMS SICFVAAGFG AALTEKLFGT GGSA...... ....LQIFCF
Xyleescco FGRRDSLKIA AVLFFISGVG SAWPELGFTS INPDNTVPVY LAGYVPEFVI
Glcpsynsp HGRIKTMILA AVLFTLSSIG SGLPFTIWD. .......... .......FIF
Hgt1klula FGRKVSLHIC AALWIIG... .......... .....AILQC AAQDQAMLIV
Lacpklula KGRKPAILIG CLGVVIGAII S......... .....SLTT. ...TKSALIG
Stl1sacce IGRKPLILMG SVITIIGAVI S......... .....TCAFR GYWALGQFII
Ma6tsacce YTLIMALFFL AAFIFI.... .......... ........LY FCKSLGMIAV
Gtr2leido LDYCKSFLFI GLLSVIGNV. .......... ....LTHVAT GLFHYWVLFV
Consensus .GR....... .......... .......... .......... ..........

          301                                                350
Itr1sacce GRLIMGFGVG IGSLIAPLFI SEIAPKMIRG RLTVINSLWL TGGQLVAYGC
Itr2sacce GRLIMGFGVG IGSLISPLFI SEIAPKMIRG RLTVINSLWL TGGQLIAYGC
Gtr1leido SRVIVGLAIG ISSATIPVYL AEVTSPKHRG ATIVLNNLFL TGGQFVAAGF
Sgt1schma GRFVIGINSG ITIGIASLYL TEVAPRDLRG GIGACHQLAV TVGIAFSYFI
Tgtptaeso GRAISGLNSG LSIGIAAMFL TEIAPRHLRG MIGACNQLAI TIGIVISYVL
Sgt4schma GRVFNGFNFG ISMGIAPMYL TEIAPLSLRG GIGSLHQLAL TIGILVSYLM
Gtr5homsa SRLLVGICAG VSSNVVPMYL GELAPKNLRG ALGVVPQLFI TVGILVAQIF
Gtr5ratno SRLLVGICAG ISSNVVPMYL GELAPKNLRG ALGVVPQLFI TVGILVAQLF
Gtr5orycu SRLLVGICAG VSSNVVPMYL GELAPKNLRG ALGVESQLFI TLGILVAQIF
Gtr1galga GRFIIGLYSG LTTGFVPMYV GEVSPTALRG ALGTFHQLGI VLGILIAQVF
Gtr1homsa GRFIIGVYCG LTTGFVPMYV GEVSPTAFRG ALGTLHQLGI VVGILIAQVF
Gtr3musmu GRLLIGIFCG LCTGFVPMYI GEVSPTALRG AFGTLNQLGI VVGILVAQIF
Gtr3canfa GRLIIGLFCG LCTGFVPMYI GEISPTALRG AFGTLNQLGI VIGILVAQIF
Gtr3oviar GRLIIGLFCG LCTGFVPMYI GEISPTALRG AFGTLNQLGI VIGILVAQIF
Gtr3homsa GRLVIGLFCG LCTGFVPMYI GEISPTALRG AFGTLNQLGI VVGILVAQIF
Gtr3galga GRFIIGLFCG LCTGFVPMYI SEVSPTSLRG AFGTLNQLGI VVGILVAQIF
Gtr4homsa GRFLIGAYSG LTSGLVPMYV GEIAPTHLRG ALGTLNQLAI VIGILIAQVL
Gtr2homsa GRSISGLYCG LISGLVPMYI GEIAPTALRG ALGTFHQLAI VTGILISQII
Gtr2ratno GRSVSGLYCG LISGLVPMYI GEIAPTTLRG ALGTLHQLAL VTGILISQIA
Gtr2galga GRAITGLYCG LSSGLVPMYV SEVSPTALRG ALGTLHQLAI VTGILISQVL
Gtr7ratno GRSVSGLYCG LISGLVPMYI GEISPHTLRG AAGTLLQLGI TVGIIISQIL
Sgt2schma GRFTIGIACG AHTVVGPMFL SEIAPVNFRG AAGTFNQFVI VSAILLSQVL
Hxt0sacce GRIVSGMGVG GVAVLSPTLI SEISPKHLRG TCVSFYQLMI TLGIFLGYCT
Hxt2sacce GRIISGMGVG GIAVLSPTLI SETAPKHIRG TCVSFYQLMI TLGIFLGYCT
Hxt4sacce GRIISGLGVG GIAVLSPMLI SEVSPKHIRG TLVSCYQLMI TLGIFLGYCT
Hxt7sacce GRIISGLGVG GIAVLSPMLI SEVSPKHLRG TLVSCYQLMI TAGIFLGYCT
Gtrkluma  GRIISGLGVG GISVLSPMLI SETAPKHIRG TLVSFYQLMI TFGIFLGYCT
Gal2sacce GRIISGLGVG GIAVLCPMLI SEIAPKHLRG TLVSCYQLMI TAGIFLGYCT
Hxt1sacce GRIISGLGVG GITVLSPMLI SEVAPSEMRG TLVSCYQVMI TLGIFLGYCT
Hxt3sacce GRIISGLGVG GIAVLSPMLI SEVAPKEMRG TLVSCYQLMI TLGIFLGYCT
Hxt5sacce GRIISGLGVG GITVLAPMLI SEVSPKQLRG TLVSCYQLMI TFGIFLGYCT
Rag1klula GRIISGLGVG GITVLSPMLI SETAPKHLRG TLVSCYQLMI TFGIFLGYCT
Hxtasacce GRIISGLGVG GIAVLSPMLI SEVAPKQIRG TLVQLYQLMC TMGIFLGYCT
Hxt8sacce GRIIAGIGAG SISVLAPMLI SETAPKHIRG TLLACWQLMV TFAIFLGYCT
Hxtcsacce GKIIYGLGAG GCSVLCPMLL SEIAPTDLRG GLVSLYQLNM TFGIFLGYCS
Hxtdsacce GRAVTGIGVG VTTVLVPMFL SENSPLKIRG SMVSTYQLIV TFGILMGNIL
Snf3sacce GRVISGIGIG AISAVVPLYQ AEATHKSLRG AIISTYQWAI TWGLLVSSAV
Qayneucr  GRVLAGIGVG GASNMVPIYI SELAPPAVRG RLVGIYELGW QIGGLVGFWI
Qutdemeni GRVLAGIGVG AGSNICPIYI SEMAPSAIRG RLVGVYELGW QIGGVVGFWI
```

```
Stp1arath GRILLGFGIG FANQAVPLYL SEMAPYKYRG ALNIGFQLSI TIGILVAEVL
Sugricco   GRILLGFGIG FANQSVPLYL SEMAPYKYRG ALNIGFQLSI TIGILVANVL
Mst1nicta GRILLGFGIG FANQSVPLYL SEMAPYKYRG ALNLGFQLSI TIGILVANVL
Stp4arath GRILLGFGVG FANQSVPVYL SEMAPPNLRG AFNNGFQVAI IFGIVVATII
Gtr2ricco GRIMLGVGIG FGNQAVPLYL SEMAPAHLRG GLNIMFQLAT TLGIFTANLI
Hex6ricco GRVLLGVGVG FANQAVPLYL SEMAPPRYRG AINNGFQFSV GIGALSANLI
Hup1chlke GRVLLGFGVG LGSQVVPQYL SEVAPFSHRG MLNIGYQLFV TIGILIAGLV
Hxtchlke   GRVLLGFGVG LGSQVVPQYL SEVAPFSHRG MLNIGYQLFV TIGILIAGLV
Hup2chlke GRVLLGFGVG GGNNAVPLYL SECAPPKYRG GLNMMFQLAV TIGIIVAQLV
Araeescco ARVVLGIAVG IASYTAPLYL SEMASENVRG KMISMYQLMV TLGIVLAFLS
Galpescco SRVLLGLAVG VASYTAPLYL SEIAPEKIRG SMISMYQLMI TIGILGAYLS
Glfzymmo   FRFLAGLGIG VVSTLTPTYI AEIRPPDKRG QMVSGQQMAI VTGALTGYIF
Xyleescco YRIIGGIGVG LASMLSPMYI AELAPAHIRG KLVSFNQFAI IFGQLLVYCV
Glcpsynsp WRVLGGIGVG AASVIAPAYI AEVSPAHLRG RLGSLQQLAI VSGIFIALLS
Hgt1klula GRVISGMGIG FGSSAAPVYC SEISPPKIRG TISGLFQFSV TVGIMVLFYI
Lacpklula GRWFVAFFAT IANAAAPTYC AEVAPAHLRG KVAGLYNTLW SVGSIVAAFS
Stl1sacce GRVVTGVGTG LNTSTIPVWQ SEMSKAENRG LLVNLEGSTI AFGTMIAYWI
Ma6tsacce GQALCGMPWG CFQCLTVSYA SEICPLALRY YLTTYSNLCW TFGQLFAAGI
Gtr2leido ARIVLGFPLG WQSITSSHYT DKFAPANHAK TLGTLFQVSV STGIFVTSFF
Consensus GR...G...G ......P... .E..P...RG ......QL.. ..GI......
```

```
                351                                          400
Itr1sacce GA...GLNYV .........N NGWRILVGLS LIPTAVQ.FT CLCFLPDTPR
Itr2sacce GA...GLNHV .........K NGWRILVGLS LIPTVLQ.FS FFCFLPDTPR
Gtr1leido TAIMVVFTSK .........N IGWRVAIGIG ALPAVVQAFC LLFFLPESPR
Sgt1schma TFTF...... sgt......LLNTL NLWPLAVALG AVPAAISLVT LP.FCPESPR
Tgtptaeso TLSH...... .......LLNTP TLWPVAMGVG AIPAVIALII SP.FTVESPR
Sgt4schma TLTY...... .......TLNTP TLWPISVAVG SVPALIALIL LP.YCPESPR
Gtr5homsa GLRN...... .......LLANV DGWPILLGLT GVPAALQLLL LP.FFPESPR
Gtr5ratno GLRS...... .......VLASE EGWPILLGLT GVPAGLQLLL LP.FFPESPR
Gtr5orycu GLRS...... .......I.RQQ KGWPILLGLT GGPAAAA..C PP.FFPESPR
Gtr1galga GLDL...... .......IMGND SLWPLLLGFI FVPALLQCII LP.FAPESPR
Gtr1homsa GLDS...... .......IMGNK DLWPLLLSII FIPALLQCIV LP.FCPESPR
Gtr3musmu GLDF...... .......ILGSE ELWPGLLGLT IIPAILQSAA LP.FCPESPR
Gtr3canfa GLKV...... .......IMGTE ELWPLLLGFT IIPAVLQSAA LP.FCPESPR
Gtr3oviar GLKV...... .......ILGTE DLWPLLLGFT ILPAIIQCAA LP.FCPESPR
Gtr3homsa GLEF...... .......ILGSE ELWPLLLGFT ILPAILQSAA LP.FCPESPR
Gtr3galga GLEG...... .......IMGTE ALWPLLLGFT IVPAVLQCVA LL.FCPESPR
Gtr4homsa GLES...... .......LLGTA SLWPLLLGLT VLPALLQLVL LP.FCPESPR
Gtr2homsa GLEF...... .......ILGNY DLWHILLGLS GVRAILQSLL LF.FCPESPR
Gtr2ratno GLSF...... .......ILGNQ DYWHILLGLS AVPALLQCLL LL.FCPESPR
Gtr2galga GLDF...... .......LLGND ELWPLLLGLS GVAALLQFFL LL.LCPESPR
Gtr7ratno GLDN...... .......SSGNV NTWPHLLSLS RIPAALQPAI LP.FPPESPP
Sgt2schma SLPE...... .......VMGTT ELWPYLLALC TVSSVIHILL LF.TCPESPT
Hxt0sacce NYGTKKY... .S......NS IQWRVPLGLC FAWAIFMVIG MV.MVPESPR
Hxt2sacce NYGTKDY... .S......NS VQWRVPLGLN FAFAIFMIAG ML.MVPESPR
Hxt4sacce NYGTKTY... .T......NS VQWRVPLGLG FAWALFMIGG MT.FVPESPR
Hxt7sacce NFGTKNY... .S......NS VQWRVPLGLC FAWALFMIGG MT.FVPESPR
Gtrkluma  NYGTKTY... .S......NS VQWRVPLGLC FAWAIFMITG ML.FVPESPR
Gal2sacce NYGTKSY... .S......NS VQWRVPLGLC FAWSLFMIGA LT.LVPESPR
Hxt1sacce NFGTKNY... .S......NS VQWRVPLGLC FAWALFMIGG MM.FVPESPR
Hxt3sacce NFGTKNY... .S......NS VQWRVPLGLC FAWALFMIGG MT.FVPESPR
Hxt5sacce NFGTKNY... .S......NS VQWRVPLGLC FAWSIFMIVG MT.FVPESPR
Rag1klula NYGTKNY... .S......NS VQWRVPLGLC FAWAIFMVLG MM.FVPESAR
```

276

```
Hxtasacce NYGTKNY... .H......NA TQWRVGLGLC FAWTTFMVSG MM.FVPESPR
Hxt8sacce NYGTKTY... .S......NS VQWRVPLGLC FAWAIIMIGG MT.FVPESPR
Hxtcsacce VYGTRKY... .D......NT AQWRVPLGLC FLWALIIIIG ML.LVPESPR
Hxtdsacce NFICERCYKD PT....,.QN IAWQLPLFLG YIWAIIIGMS LV.YVPESPQ
Snf3sacce SQGTHA.... RN......DA SSYRIPIGLQ YVWSSFLAIG MF.FLPESPR
Qayneucr  NYGVNTTMAP TR....... SQWLIPFAVQ LIPAGLLFLG SF.WIPESPR
Qutdemeni NYGVDETLAP SH....... KQWIIPFAVQ LIPAGLLIIG AL.LIRESPR
Stplarath NYFFAKIKGG W........ .GWRLSLGGA VVPALIITIG SL.VLPDTPN
Sugricco  NYFFAKIKGG W........ .GWRLSLGGA MVPALIITVG SL.VLPDTPN
Mst1nicta NYFFAKIH.. W........ .GWRLSLGGA MVPALIITIG SL.FLPETPN
Stp4arath NYFTAQMKGN I........ .GWRISLGLA CVPAVMIMIG AL.ILPDTPN
Gtr2ricco NYGTQNIK.P W........ .GWRLSLGLA AAPALLMTLA GL.FLPETPN
Hex6ricco NYGTEKIEGG W........ .GWRISLAMA AVPAAILTFG AL.FLPETPN
Hup1chlke NYAVRDWEN. ......... .GWRLSLGLA AAPGAILFLG SL.VLPESPN
Hxtchlke  NYGVRNWDN. ......... .GWRLSLGLA AVPGLILLLG AI.VLPESPN
Hup2chlke NYGTQTMNN. ......... .GWRLSLGLA GVPAIILLIG SL.LLPETPN
Araeescco DTAFSYSG.. ......... .NWRAMLGVL ALPAVLLIIL VV.FLPNSPR
Galpescco DTAFSYTG.. ......... .AWRWMLGVI IIPAILLLIG VF.FLPDSPR
Glfzymmo  TWLLAHFGSI D.....WVNA SGWCWSPASE GLIGIAFLLL LL.TAPDTPH
Xyleescco NYFIARSGDA S.....WLNT DGWRYMFASE CIPALLFLML LY.TVPESPR
Glcpsynsp NWFIALMAGG SAQNPWLFGA AAWRWMFWTE LIPALLYGVC AF.LIPESPR
Hgt1klula GYGCHFIDGA ......... AAFRITWGLQ MVPGLILMVG VF.FIPESPR
Lacpklula TYGTNK.... ...NFPNSS KAFKIPLYLQ MMFPGLVCI. FGWLIPESPR
Stl1sacce DFGL...... ...SYTNSS VQWRFPVSMQ IVFALFLLA. FMIKLPESPR
Ma6tsacce MKNSQN.... ...KYANSE LGYKLPFALQ WIWPLPLAVG IF.LAPESPW
Gtr2leido GLVLGNTIQ. ...YDAASNA NTMGRMQGLV SVSTLLSIFV VFLPLITKDG
Consensus .......... ........... ..W....... ...A...... ....PESPR
```

```
          401                                            450
Itr1sacce YYVMK.GDLA RATEVLKRS. .....YTDTS EEIIERKVEE LVTLNQSIPG
Itr2sacce YYVMK.GDLK RAKMVLKRS. .....YVNTE DEIIDQKVEE LSSLNQSIPG
Gtr1leido WLLSK.GHAD RAKAVADKF. .....EVDLC EF...QEGDE LPS.......
Sgt1schma FLYMKKHKEA EARKAFLQLN VKE.NVDTFI GELREEIEVA KNQPVFK...
Tgtptaeso WLYLKKKDEK AAREAFARIN GSE.NVDMFI AEMREELEVA QNQPEFK...
Sgt4schma FLFIKKGKEA KARKAFQRLN CID.DINETF NEMKREMHEA EKRPKFK...
Gtr5homsa YLLIQKKDEA AAKKALQTLR GWD.SVDREV AEIRQEDEAE KAAGFIS...
Gtr5ratno YLLIQKKNES AAEKALQTLR GWK.DVDMEM EEIRKEDEAE KAAGFIS...
Gtr5orycu YLLIG.QEPR CRQKALQSLR GWD.SVDREL EEIRREDEAA RAAGLVS...
Gtr1galga FLLINRNEEN KAKSVLKKLR GTT.DVSSDL QEMKEESRQM MREKKVT...
Gtr1homsa FLLINRNEEN RAKSVLKKLR GTA.DVTHDL QEMKEESRQM MREKKVT...
Gtr3musmu FLLINKKEED QATEILQRLW GTS.DVVQEI QEMKDESVRM SQEKQVT...
Gtr3canfa FLLINRKEEE NAKEILQRLW GTQ.DVSQDI QEMKDESARM AQEKQVT...
Gtr3oviar FLLINRKEEE KAKEILQRLW GTE.DVAQDI QEMKDESMRM SQEKQVT...
Gtr3homsa FLLINRKEEE NAKQILQRLW GTQ.DVSQDI QEMKDESARM SQEKQVT...
Gtr3galga FLLINKMEEE KAQTVLQKLR GTQ.DVSQDI SEMKEESAKM SQEKKAT...
Gtr4homsa YLYIIQNLEG PARKSKRLT GWA.DVSGVL AELKDEKRKL ERERPLS...
Gtr2homsa YLYIKLDEEV KAKQSLKRLR GYD.DVTKDI NEMRKEREEA SSEQKVS...
Gtr2ratno YLYLNLEEEV RAKKSLKRLR GTE.DITKDI NEMRKEKEEA STEQKVS...
Gtr2galga YLYIKLGKVE EAKKSLKRLR GNC.DPMKEI AEMEKEKQEA ASEKRVS...
Gtr7ratno WLTIDIDDEG NAKRILQSLQ GYD.EVSHEL QEIKDESQKE EAETFLT...
Sgt2schma YLYIIKGDRR RSENALVYLR GQDCDVHAEL ELLKLETEQS STHKS.N...
Hxt0sacce YLVEK.GKYE EARRSLAKSN KVTVTDPGVV FEFDTIVANM ELERAVGNAS
Hxt2sacce FLVEK.GRYE DAKRSLAKSN KVTIEDPSIV AEMDTIMANV ETERLAGNAS
Hxt4sacce YLVEV.GKIE EAKRSIALSN KVSADDPAVM AEVEVVQATV EAEKLAGNAS
```

277

```
Hxt7sacce YLAEV.GKIE EAKRSIAVSN KVAVDDPSVL AEVEAVLAGV EAEKLAGNAS
Gtrkluma  FLVEK.DRID EAKRSIAKSN KVSYEDPAVQ AEVDLICAGV EAERLAGSAS
Gal2sacce YLCEV.NKVE DAKRSIAKSN KVSPEDPAVQ AELDLIMAGI EAEKLAGNAS
Hxt1sacce YLVEA.GRID EARASLAKVN KCPPDHPYIQ YELETIEASV EEMRAAGTAS
Hxt3sacce YLVEA.GQID EARASLSKVN KVAPDHPFIQ QELEVIEASV EEARAAGSAS
Hxt5sacce YLVEV.GKIE EAKRSLARAN KTTEDSPLVT LEMENYQSSI EAERLAGSAS
Rag1klula FLVET.DQIE EARKSLAKTN KVSIDDPVVK YELLKIQSSI ELEKAAGNAS
Hxtasacce YLIEV.GKDE EAKRSLSKSN KVSVDDPALL AEYDTIKAGI ELEKLAGNAS
Hxt8sacce FLVQV.GKIE QAKASFAKSN KLSVDDPAVV AEIDLLVAGV EAEEAMGTAS
Hxtcsacce YLIEC.ERHE EARASIAKIN KVSPEDPWVL KQADEINAGV LAQRELGEAS
Hxtdsacce YLAKIKNDVP SAKYSFARMN GIPATDSMVI EFIDDLLENN YNNEETNNES
Snf3sacce YYV.LKDKLD EAAKSLSFLR GVPVHDSGLL EELVEIKATY DYEASFGSSN
Qayneucr  WLYANGKREE .AMKVLCWIR NLEPTDRYIV QEVSFIDADL ERYTRQVGNG
Qutdemeni WLFLRGNREK .GIETLAWIR NLPADHIYMV EEINMIEQSL EQQRVKIGLG
Stplarath SMIERGQHEE .AKTKLRRIR GVD....DVS QEFDDLVAAS KESQ..SIEH
Sugricco  SMIERGQHEE .ARAHLKRVR GVE....DVD EEFTDLVHAS EDSK..KVEH
Mst1nicta SMIERGNHDE .AKARLKRIR GID....DVD EEFNDLVVAS EASR..KIEN
Stp4arath SLIERGYTEE .AKEMLQSIR GTN....EVD EEFQDLIDAS EESK..QVKH
Gtr2ricco SLIERGRVEE .GRRVLERIR GTA....DVD AEFTDMVEAS ELAN..TIEH
Hex6ricco SLIQRSNDHE RAKLMLQRVR GTT....DVQ AELDDLIKAS IISR..TIQH
Hup1chlke FLVEKGKTEK .GREVLQKLR GTS....EVD AEFADIVAAV EIARPITMRQ
Hxtchlke  FLVEKGRTDQ .GRRILEKLR GTS....HVE AEFADIVAAV EIARPITMRQ
Hup2chlke SLIERGHRRR .GRAVLARLR RTE....AVD TEFEDICAAA EESTRYTLRQ
Araeescco WLAEKGRHIE AEEVLRMLRD TSE....KAR EELNEIRESL KL..KQGGWA
Galpescco WFAAKRRFVD AERVLLRLRD TSA....EAK RELDEIRESL QV..KQSGWA
Glfzymmo  WLVMKGRHSE ASKILARL.E PQA....DPN LTIQKIKAGF DKAMDKSSAG
Xyleescco WLMSRGKQEQ AEGILRKI.M GNT....LAT QAVQEIKHSL DHGR.KTGGR
Glcpsynsp YLVAQGQGEK AAAILWKV.E GGD....VPS R.IEEIQATV SLDHKPRFSD
Hgt1klula WLANHDRWEE TSLIVANIVA NGDVNNEQVR FQLEEIKEQV IIDSAAK.NF
Lacpklula WLVGVGREEE AREFIIKYHL NGDRTHPLLD MEMAEIIESF HGTDLSNPLE
Stllsacce WLISQSRTEE ARYLVGTLDD ADPN.....D EEVITEVAML HDAVNRTKHE
Ma6tsacce WLV.KKGRID QARRSLERIL SGKGPEKELL VSMELDKIKT TIEKEQKMSD
Gtr2leido YSKSRRGDYE GENSEDASRK AAEE...... .......... ..........
Consensus .L........ .......... .......... ..E....... ..........
```

```
          451                                           500
Itr1sacce KNVPEKVWNT IKELHTVPSN LRALIIGCGL QAIQQFTGWN SLMYFSGTIF
Itr2sacce KNPITKFWNM VKELHTVPSN FRALIIGCGL QAIQQFTGWN SLMYFSGTIF
Gtrlleido ......VRID YRPLMARDMR FR.VVLSSGL QIIQQFSGIN TIMYYSSVIL
Sgt1schma .......... FTQLFTQRDL RMPVLIACLI QVLQQLSGIN AVITYSSLML
Tgtptaeso .......... FTELFRRRDL RMPVIIAVLI QVMQQLSGIN AVVANSSEML
Sgt4schma .......... FFRLFTQRDL RMPVLIACII QVFQQLSGIN AVITYSSTML
Gtr5homsa .......... VLKLFRMRSL RWQLLSIIVL MGGQQLSGVN AIYYYADQIY
Gtr5ratno .......... VWKLFRMQSL RWQLISTIVL MAGQQLSGVN AIYYYADQIY
Gtr5orycu .......... VRALCAMRGL AWQLISVVPL MW.QQLSGVN AIYYY.DQIY
Gtr1galga .......... IMELFRSPMY RQPILIAIVL QLSQQLSGIN AVFYYSTSIF
Gtr1homsa .......... ILELFRSPAY RQPILIAVVL QLSQQLSGIN AVFYYSTSIF
Gtr3musmu .......... VLELFRSPNY VQPLLISIVL QLSQQLSGIN AVFYYSTGIF
Gtr3canfa .......... VLELFRSRSY RQPIIISIML QLSQQLSGIN AVFYYSTGIF
Gtr3oviar .......... VLELFRAPNY RQPIIISIML QLSQQLSGIN AVFYYSTGIF
Gtr3homsa .......... VLELFRVSSY RQPIIISIVL QLSQQLSGIN AVFYYSTGIF
Gtr3galga .......... VLELFRSPNY RQPIIISITL QLSQQLSGIN AVFYYSTGIF
Gtr4homsa .......... LLQLLGSRTH RQPLIIAVVL QLSQQLSGIN AVFYYSTSIF
Gtr2homsa .......... IIQLFTNSSY RQPILVALML HVAQQFSGIN GIFYYSTSIF
```

```
Gtr2ratno ..........  VIQLFTDPNY RQPIVVALML HLAQQFSGIN GIFYYSTSIF
Gtr2galga ..........  IGQLFSSSKY RQAVIVALMV QISQQFSGIN AIFYYSTNIF
Gtr7ratno ..........  LIELLRSRDT RWPSLLIAFL MQSQQTSGVN GIFYYHQHIY
Sgt2schma ..........  VCDLLRIPYL RWGLIVALVP HIGQQFSGIN GILYYFVSLF
Hxt0sacce WH........ .ELFSNKGAI LPRVIMGIVI QSLQQLTGCN YFFYYGTTIF
Hxt2sacce WG........ .ELFSNKGAI LPRVIMGIMI QSLQQLTGNN YFFYYGTTIF
Hxt4sacce WG........ .EIFSTKTKV FQRLIMGAMI QSLQQLTGDN YFFYYGTTVF
Hxt7sacce WG........ .ELFSSKTKV LQRLIMGAMI QSLQQLTGDN YFFYYGTTIF
Gtrkluma  IK........ .ELFSTKTKV FQRLIMGMLI QSFQQLTGNN YFFYYGTTIF
Gal2sacce WG........ .ELFSTKTKV FQRLLMGVFV QMFQQLTGNN YFFYYGTVIF
Hxt1sacce WG........ .ELFTGKPAM FQRTMMGIMI QSLQQLTGDN YFFYYGTIVF
Hxt3sacce WG........ .ELFTGKPAM FKRTMMGIMI QSLQQLTGDN YFFYYGTTVF
Hxt5sacce WG........ .ELVTGKPQM FRRTLMGMMI QSLQQLTGDN YFFYYGTTIF
Rag1klula WG........ .ELITGKPSM FRRTLMGIMI QSLQQLTGDN YFFYYGTTIF
Hxtasacce WS........ .ELLSTKTKV FQRVLMGVMI QSLQQLTGDN YFFYYGTTIF
Hxt8sacce WK........ .ELFSRKTKV FQRLTMTVMI NSLQQLTGDN YFFYYGTTIF
Hxtcsacce WK........ .ELFSVKTKV LQRLITGILV QTFLQLTGEN YFFFYGTTIF
Hxtdsacce KKQSLVKRNT FEFIMGKPKL WLRLIIGMMI MAFQQLSGIN YFFYYGTSVF
Snf3sacce FIDCFISS.. ....KSRPKQ TLRMFTGIAL QAFQQFSGIN FIFYYGVNFF
Qayneucr  F......WKP F.LSLKQRKV QWRFFLGGML FFWQNGSGIN AINYYSPTVF
Qutdemeni F......WKP FKAAWTNKRI LYRLFLGSML FLWQNGSGIN AINYYSPRVF
Stp1arath P......WRN LL....RRKY RPHLTMAVMI PFFQQLTGIN VIMFYAPVLF
Sugricco  P......WRN LL....QRKY RPHLSMAIAI PFFQQLTGIN VIMFYAPVLF
Mst1nicta P......WRN LL....QRKY RPHLTMAIMI PFFQQLTGIN VIMFYAPVLF
Stp4arath P......WKN IM....LPRY RPQLIMTCFI PFFQQLTGIN VITFYAPVLF
Gtr2ricco P......FRN IL....EPRN RPQLVMAVCM PAFQILTGIN SILFYAPVLF
Hex6ricco P......FKN IM....RRKY RPQLVMAVAI PFFQQVTGIN VIAFYAPILF
Hup1chlke S......WAS LF....TRRY MPQLLTSFVI QFFQQFTGIN AIIFYVPVLF
Hxtchlke  S......WRS LF....TRRY MPQLLTSFVI QFFQQFTGIN AIIFYVPVLF
Hup2chlke S......WAA LF....SRQY SPMLIVTSLI AMLQQLTGIN AIMFYVPVLF
Araeescco L......FK. .I....NRNV RRAVFLGMLL QAMQQFTGMN IIMYYAPRIF
Galpescco L......FK. .E....NSNF RRAVFLGVLL QVMQQFTGMN VIMYYAPKIF
Glfzymmo  L......FA. .F....G... ITVVFAGVSV AAFQQLVGIN AVLYYAPQMF
Xyleescco L......LM. .F....G... VGVIVIGVML SIFQQFVGIN VVLYYAPEVF
Glcpsynsp L......LS. .R....RGGL LPIVWIGMGL SALQQFVGIN VIFYYSSVLW
Hgt1klula G......YKD LF....RKKT LPKTIVGVSA QMWQQLCGMN VMMYYIVYIF
Lacpklula MLDVRSLF.R .....TRSDR Y.RAMLVILM AWFGQFSGNN VCSYYLPTML
Stl1sacce KHSLSSLFSR .....GRSQN LQRALIAAST QFFQQFTGCN AAIYYSTVLF
Ma6tsacce EGTYWD.... .....CVKDGI NRRRTRIACL CWIGQCSCGA SLIGYSTYFY
Gtr2leido .......... ...YTMTQM IGPILNGVAM GCVTQLTGIN ANMNFAPTIM
Consensus .......... .......... ..........  ...QQ..G.N ....Y....F
```

```
                   501                                            550
Itr1sacce ETVGFKN... SSAVSIIVSG TNFIFTLVAF F.SIDKIGRR TILLIGLPGM
Itr2sacce ETVGFKN... SSAVSIIVSG TNFVFTLIAF F.CIDKIGRR YILLIGLPGM
Gtr1leido YDAGFRDAIM PVVLSIPLAF MNALFTAVAI F.TVDRFGRR RMLLISVFGC
Sgt1schma ELAGIPDVYL QY.CVFAIGV LNVIVTVVSL P.LIERAGRR TLLLWPTVSL
Tgtptaeso KSAKVSPDML EY.FVVGLGL LNVICTIVAL P.LLEKAGRR TLLLWPSLVV
Sgt4schma KTAGIPLVYI QF.CVVAVPA INVLMTVLSV Y.LIERAGRR TLLLWPTVLL
Gtr5homsa LSAGVPEEHV QY.VTAGTGA VNVVMTFCAV F.VVELLGRR LLLLLGFSIC
Gtr5ratno LSAGVKSNDV QY.VTAGTGA VNVFMTMVTV F.VVELWGRR NLLLIGFSTC
Gtr5orycu LSP..LDTDT QY.YTAATGA VNVLMTVCTV F.VVESWA.R LLLLLGFSPL
Gtr1galga EKSGV..EQP VY.ATIGSGV VNTAFTVVSL F.VVERAGRR TLHLIGLAGM
Gtr1homsa EKAGV..QQP VY.ATIGSGI VNTAFTVVSL F.VVERAGRR TLHLIGLAGM
```

279

```
Gtr3musmu KDAGV..QEP IY.ATIGAGV VNTIFTVVSL F.LVERAGRR TLHMIGLGGM
Gtr3canfa KDAGV..EEP IY.ATIGAGV VNTIFTVVSL F.LVERAGRR TLHMIGLGGM
Gtr3oviar KDAGV..QEP VY.ATIGAGV VNTIFTVVSV F.LVERAGRR TLHLIGLGGM
Gtr3homsa KDAGV..QEP IY.ATIGAGV VNTIFTVVSL F.LVERAGRR TLHMIGLGGM
Gtr3galga ERAGI..TQP VY.ATIGAGV VNTVFTVVSL F.LVERAGRR TLHLVGLGGM
Gtr4homsa ETAGV..GQP AY.ATIGAGV VNTVFTLVSV L.LVERAGRR TLHLLGLAGM
Gtr2homsa QTAGI..SKP VY.ATIGVGA VNMVFTAVSV F.LVEKAGRR SLFLIGMSGM
Gtr2ratno QTAGI..SQP VY.ATIGVGA INMIFTAVSV L.LVEKAGRR TLFLAGMIGM
Gtr2galga QRAGV..GQP VY.ATIGVGV VNTVFTVISV F.LVEKAGRR SLFLAGLMGM
Gtr7ratno KQAGA..QDP AY.VTLGSGS VNFLTTVVSL I.VVEKAGRR TLFLAGMIGM
Sgt2schma ISNGLTKQVA SY.ANLGTGV TILIGAFASI F.VIDRKGRR PLLMFGTSVC
Hxt0sacce NAVGM...QD SFETSIVLGA VNFASTFVAL Y.IVDKFGRR KCLLWGSASM
Hxt2sacce NAVGM...KD SFQTSIVLGI VNFASTFVAL Y.TVDKFGRR KCLLGGSASM
Hxt4sacce TAVGL...ED SFETSIVLGI VNFASTFVGI F.LVERYGRR RCLLWGAASM
Hxt7sacce KAVGL...SD SFETSIVLGI VNFASTFVGI Y.VVERYGRR TCLLWGAASM
Gtrkluma  NSVGM...DD SFETSIVLGI VNFASTFVAI Y.VVDKFGRR KCLLWGAAAM
Gal2sacce KSVGL...DD SFETSIVIGV VNFASTFFSL W.TVENLGRR KCLLLGAATM
Hxt1sacce QAVGL...SD SFETSIVFGV VNFFSTCCSL Y.TVDRFGRR NCLMWGAVGM
Hxt3sacce NAVGM...SD SFETSIVFGV VNFFSTCCSL Y.TVDRFGRR NCLLYGAIGM
Hxt5sacce QAVGL...ED SFETAIVLGV VNFVSTFFSL Y.TVDRFGRR NCLLWGCVGM
Rag1klula QSVGM...DD SFETSIVLGI VNFASTFFAL Y.TVDHFGRR NCLLYGCVGM
Hxtasacce KSVGL...KD SFQTSIIIGV VNFFSSFIAV Y.TIERFGRR TCLLWGAASM
Hxt8sacce KSVGM...ND SFETSIVLGI VNFASCFFSL Y.SVDKLGRR RCLLLGAATM
Hxtcsacce KSVGL...TD GFETSIVLGT VNFFSTIIAV M.VVDKIGRR KCLLFGAAGM
Hxtdsacce KGVGI...KD PYITSIILSS VNFLSTILGI Y.YVEKWGHK TCLLYGSTNL
Snf3sacce NKTGV...SN SYLVSFITYA VNVVFNVPGL F.FVEFFGRR KVLVVGGVIM
Qayneucr  RSIGITGTDT GFLTTGIFGV VKMVLTIIWL LWLVDLVGRR RILFIGAAGG
Qutdemeni KSIGVSGGNT SLLTTGIFGV VTAVITFVWL LYLIDHFGRR NLLLVGAAGG
Stp1arath NTIGFT.TDA SLMSAVVTGS VNVGATLVSI Y.GVDRWGRR FLFLEGGTQM
Sugricco  DTIGFG.SDA ALMSAVITGL VNVFATMVSI Y.GVDKWGRR FLFLEGGVQM
Mst1nicta KTIGFG.ADA SLMSAVITGG VNVLATVVSI Y.YVDKLGRR FLFLEGGIQM
Stp4arath QTLGFG.SKA SLLSAMVTGI IELLCTFVSV F.TVDRFGRR ILFLQGGIQM
Gtr2ricco QSMGFG.GNA SLYSSVLTGA VLFSSTLISI G.TVDRLGRR KLLISGGIQM
Hex6ricco RTIGLE.ESA SLLSSIVTGL VGSASTFISM L.IVDKLGRR ALFIFGGVQM
Hup1chlke SSLGSA.NSA ALLNTVVVGA VNVGSTLIAV M.FSDKFGRR FLLIEGGIQC
Hxtchlke  SSLGSA.SSA ALLNTVVVGA VNVGSTMIAV L.LSDKFGRR FLLIEGGITC
Hup2chlke SSFGTA.RHA ALLNTVIIGA VNVAATFVSI F.SVDKFGRR GLFLEGGIQM
Araeescco KMAGFTTTEQ QMIATLVVGL TFMFATFIAV F.TVDKAGRK PALKIGFSVM
Galpescco ELAGYTNTTE QMWGTVIVGL TNVLATFIAI G.LVDRWGRK PTLTLGFLVM
Glfzymmo  QNLGFG.ADT ALLQTISIGV VNFIFTMIAS R.VVDRFGRK PLLIWGALGM
Xyleescco KTLGAS.TDI ALLQTIIVGV INLTFTVLAI M.TVDKFGRK PLQIIGALGM
Glcpsynsp RSVGFT.EEK SLLITVITGF INILTTIVAI A.FVDKFGRK PLLLMGSIGM
Hgt1klula NMAGYT.GNT NLVASSIQYV LNVVMTIPAL F.LIDKFGRR PVLIIGGIFM
Lacpklula RNVGMKSVSL NVLMNGVYSI VTWISSICGA F.FIDKIGRR EGFLGSISGA
Stl1sacce NKTIKLDYRL SMIIGGVFAT IYALSTIGSF F.LIEKLGRR KLFLLGATGQ
Ma6tsacce EKAGVS.TDT AFTFSIIQYC LGIAATFVS. WWASKYCGRF DLYAFGLAFQ
Gtr2leido SNLGL....Q PLVGNIIVMA WNMLATF.CV IPLSRRFSMR TLFLFCGFVG
Consensus ...G..... ........G. .N........ .........GRR .....G....
```

```
             551                                          600
Itr1sacce TMALVVCSIA F......HFL GIKFDGAVAV VVSSGFSSWG IVIIVFIIVF
Itr2sacce TVALVICAIA F......HFL GIKFNGADAV VASDGFSSWG IVIIVFIIVY
Gtr1leido LVLLVVIAII G......FFI GTRI...... ....SYSVGG GLFLALLAVF
```

```
Sgt1schma ALSLLLLTIF VN......... .....LADSG PQSTKN.AMG IISIILILIY
Tgtptaeso AIILLLLVIF VN......... .....IANYG GVVNKT.PFV LVSAVLVFIY
Sgt4schma AFSLLCLTIS VN......... .....IASST KDPTTARTAG IISAVLIILY
Gtr5homsa LIACCVLTAA LA......... ..LQD.. TVSW....MP YISIVCVISY
Gtr5ratno LTACIVLTVA LA......... ..LQN.. TISW....MP YVSIVCVIVY
Gtr5orycu APTCCVLTAA LA......... ..LQD.. TVSW....MP YISIVCIIVY
Gtr1galga AGCAILMTIA LT......... ..LLD.. QMPW....MS YLSIVAIFGF
Gtr1homsa AGCAILMTIA LA......... ..LLE.. QLPW....MS YLSIVAIFGF
Gtr3musmu AVCSVFMTIS LL......... ..LKD.. DYEA....MS FVCIVAILIY
Gtr3canfa AVCSILMTIS LL......... ..LKD.. NYNW....MS FVCIGAILVF
Gtr3oviar AFCSILMTIS LL......... ..LKD.. NYSW....MS FICIGAILVF
Gtr3homsa AFCSTLMTVS LL......... ..LKD.. NYNG....MS FVCIGAILVF
Gtr3galga AVCAAVMTIA LA......... ..LKE.. KW......IR YISIVATFGF
Gtr4homsa CGCAILMTVA LL......... ..LLE.. RVPA....MS YVSIVAIFGF
Gtr2homsa FVCAIFMSVG LV......... ..LLN.. KFSW....MS YVSMIAIFLF
Gtr2ratno FFCAVFMSLG LV......... ..LLD.. KFTW....MS YVSMTAIFLF
Gtr2galga LISAVAMTVG LV......... ..LLS.. QFAW....MS YVSMVAIFLF
Gtr7ratno FFCAVFMSLV LV......... ..LLD.. KFTW....MS YVSMTAIFLF
Sgt2schma LFSLLLFTLT LI......... ..IKQVT EINK....LT ILSIVLTYTF
Hxt0sacce AICFVIFATV GVTRLWP... ...QGKD QP..SSQSAG NVMIVFTCFF
Hxt2sacce AICFVIFSTV GVTSLYP... ...NGKD QP..SSKAAG NVMIVFTCLF
Hxt4sacce TACMVVFASV GVTRLWP... ...NGKK NG..SSKGAG NCMIVFTCFY
Hxt7sacce TACMVVYASV GVTRLWP... ...NGQD QP..SSKGAG NCMIVFACFY
Gtrkluma  TACMVVFASV GVTRLWP... ...DGAN HPETASKGAG NCMIVFACFY
Gal2sacce MACMVIYASV GVTRLYP... ...HGKS QP..SSKGAG NCMIVFTCFY
Hxt1sacce VCCYVVYASV GVTRLWP... ...NGQD QP..SSKGAG NCMIVFACFY
Hxt3sacce VCCYVVYASV GVTRLWP... ...NGEG NG..SSKGAG NCMIVFACFY
Hxt5sacce ICCYVVYASV GVTRLWP... ...NGQD QP..SSKGAG NCMIVFACFY
Rag1klula VACYVVYASV GVTRLWP... ...DGPD HPDISSKGAG NCMIVFACFY
Hxtasacce LCCFAVFASV GVTKLWP... ...QGSS HQDITSQGAG NCMIVFTMFF
Hxt8sacce TACMVIYASV GVTRLYP... ...NGKS EP..SSKGAG NCTIVFTCFY
Hxtcsacce MACMVIFASI GVKCLYP... ...HGQD GP..SSKGAG NAMIVFTCFY
Hxtdsacce LFYMMTYATV GT...FG... ...RETD FSNI.......VLIIVTCCF
Snf3sacce TIANFIVAIV GCS......... ...LKTVAAA KVMIAFICLF
Qayneucr  SLCMWFIGAY IKI........ ...ADPGSNK AEDAKLTSGG IAAIFFFYLW
Qutdemeni SVCLWIVGGY IKI........ ...AKPE.NN PEGTQLDSGG IAAIFFFYLW
Stp1arath LICQAVVAAC IG......... ...AKFGVDG TPGELPKWYA IVVVTFICIY
Sugricco  LICQAIVAAC IG......... ...AKFGVDG APGDLPQWYA VVVVLFICIY
Mst1nicta LICQIAVSIC IA......... ...IKFGVNG TPGDLPKWYA IVVVIFICVY
Stp4arath LVSQIAIGAM IG......... ...VKFGVAG T.GNIGKSDA NLIVALICIY
Gtr2ricco IVCQVIVAVI LG......... ...AKFGAD. ..KQLSRSYS IAVVVVICLF
Hex6ricco FVAQIMVGSI MA......... ...AELGDH. ..GGIGKGYA YIVLILICIY
Hup1chlke CLAMLTTGVV LA......... ...IEFAKYG T.DPLPKAVA SGILAVICIF
Hxtchlke  CLAMLAAGIT LG......... ...VEFGQYG T.EDLPHPVS AGVLAVICIF
Hup2chlke FIGQVVTAAV LG......... ...VELNKYG T.N.LPSSTA AGVLVVICVY
Araeescco ALGTLVLGYC L.......... ...MQFD N.GTASSGLS WLSVGMTMMC
Galpescco AAGMGVLG.T M......... .....MHI...GIHSPSAQ YFAIAMLLMF
Glfzymmo  AAMMAVLGCC I......... ......WFKVGG VLPLASVLLY
Xyleescco AIGMFSLGTA F......... ......YTQAPG IVALLSMLFY
Glcpsynsp TITLGILSVV FG......... ...GATVVNG Q.PTLTGAAG IIALVTANLY
Hgt1klula FTWLFSVAGI LATYSVPAPG GVNGDDTVTI QIPSENTSAA NGVIASSYLF
Lacpklula ALALTGLSIC TARY......... .....EKTKKKSAS NGALVFIYLF
Stl1sacce AVSFT...IT FACL....... ..........VKENKENAR GAAVGL.FLF
Ma6tsacce AIMFFIIGGL GC......... ..... ....SDTHGAK MGSGALLMVV
```

```
Gtr2leido SLCCVFLGGI PVY.......  ........P GVTKSDKAIS GIAITGIAIF
Consensus .......... .......... .......... .......... ..........
```

```
           601                                              650
Itr1sacce AAFYALGIGT VPWQ.QSELF PQNVRGIGTS YATATNWAGS LVIASTFLT.
Itr2sacce AAFYALGIGT VPWQ.QSELF PQNVRGVGTS YATATNWAGS LVIASTFLT.
Gtr1leido LALYAPGIGC IPWVIMGEIF PTHLRTSAAS VATMANWGAN VLVSQVFPI.
Sgt1schma ICSFALGLGP VPALIVSEIF RQGPRAAAYS LSQSIQWLSN LIVLCSYPV.
Tgtptaeso VAAFAMGLGP MPALIVAEIF RQGPRAAAYS LSQSIQWACN LIVVASFPS.
Sgt4schma ICGFALGLGP IPGVIVAEIF RQEPRAAAYS LSQGVNLLCN LLVLFSYPS.
Gtr5homsa VIGHALGPSP IPALLITEIF LQSSRPSAFM VGGSVHWLSN FTVGLIFPF.
Gtr5ratno VIGHAVGPSP IPALFITEIF LQSSRPSAYM IGGSVHWLSN FIVGLIFPF.
Gtr5orycu VIGHAIGP.A IRSLY.TEIF LQSGRPPTWW ..GQVHWLSN FTVGLVFPL.
Gtr1galga VAFFEIGPGP IPWFIVAELF SQGPRPAAFA VAGLSNWTSN FIVGMGFQY.
Gtr1homsa VAFFEVGPGP IPWFIVAELF SQGPRPAAIA VAGFSNWTSN FIVGMCFQY.
Gtr3musmu VAFFEIGPGP IPWFIVAELF SQGPRPAAIA VAGCCNWTSN FLVGMLFPS.
Gtr3canfa VAFFEIGPGP IPWFIVAELF SQGPRPAAMA VAGCSNWTSN FLVGLLFPS.
Gtr3oviar VAFFEIGPGP IPWFIVAELF GQGPRPAAMA VAGCSNWTSN FLVGLLFPS.
Gtr3homsa VAFFEIGPGP IPWFIVAELF SQGPRPAAMA VAGCSNWTSN FLVGLLFPS.
Gtr3galga VALFEIGPGP IPWFIVAELF SQGPRPAAMA VAGCSNWTSN FLVGMLFPY.
Gtr4homsa VAFFEIGPGP IPWFIVAELF SQGPRPAAMA VAGFSNWTSN FIIGMGFQY.
Gtr2homsa VSFFEIGPGP IPWFMVAEFF SQGPRPAALA IAAFSNWTCN FIVALCFQY.
Gtr2ratno VSFFEIGPGP IPWFMVAEFF SQGPRPTALA LAAFSNWVCN FIIALCFQY.
Gtr2galga VIFFEVGPGP IPWFIVAELF SQGPRPAAIA VAGFCNWACN FIVGMCFQY.
Gtr7ratno VSFFEIGPIP IPFFGVREWF TQIWRPGAIV CVATLDWVPN FKKGICFQS.
Sgt2schma LFGFSVS... IPWFLVSELF TQENRDAAVS IAAATNWLCN AIVALIFPQ.
Hxt0sacce IFSFAITWAP IAYVIVAETY PLRVKNRAMA IAVGANWMWG FLIGFFTPF.
Hxt2sacce IFFFAISWAP IAYVIVAESY PLRVKNRAMA IAVGANWIWG FLIGFFTPF.
Hxt4sacce LFCFATTWAP IPFVVNSETF PLRVKSKCMA IAQACNWIWG FLIGFFTPF.
Hxt7sacce IFCFATTWAP IPYVVVSETF PLRVKSKAMS IATAANWLWG FLIGFFTPF.
Gtrkluma  IFCFATSWAP IAYVVVAESY PLRVKAKCMA IATASNWIWG FLNGFFTPF.
Gal2sacce IFCYATTWAP VAWVITAESF PLRVKSKCMA LASASNWVVG FLIAFFTPF.
Hxt1sacce IFCFATTWAP IAYVVISECF PLRVKSKCMS IASAANWIWG FLISFFTPF.
Hxt3sacce IFCFATTWAP IAYVVISETF PLRVKSKAMS IATAANWLWG FLIGFFTPF.
Hxt5sacce IFCFATTWAP VAYVLISESY PLRVRGKAMS IASACNWIWG FLISFFTPF.
Rag1klula IFCFATTWAP IAYVVISESY PLRVKGKAMA IASASNWIWG FLIGFFTPF.
Hxtasacce IFSFATTWAG GCYVIVSETF PLRVKSRGMA IATAANWMWG FLISFFTPF.
Hxt8sacce IFCFSCTWGP VCYVIISETF PLRVRSKCMS VATAANLLWG FLIGFFTPF.
Hxtcsacce IFCFATTWAP VAYIVVAESF PSKVKSRAMS ISTACNWLWQ FLIGFFTPF.
Hxtdsacce IFWFAITLGP VTFVLVSELF PLRTRAISMA ICTFINWMFN FLISLLTPM.
Snf3sacce IAAFSATWGG VVWVISAELY PLGVRSKCTA ICAAANWLVN FICALITPY.
Qayneucr  TAFYTPSWNG TPWVINSEMF DQNTRSLGQA SAAANNWFWN FIISRF....
Qutdemeni TAFYTPSWNG TPWVINSEMF DPTVRSLAQA CAAASNWLWN FLISRF....
Stp1arath VAGFAWSWGP LGWLVPSEIF PLEIRSAAQS ITVSVNMIFT FIIAQIFLT.
Sugricco  VSGFAWSWGP LGWLVPSEIF PLEIRSAAQS VNVSVNMFFT FVVAQVFLI.
Mst1nicta VAGFAWSWGP LGWLVPSEIF PLEIRSAAQS INVSVNMIFT FIVAQVFLT.
Stp4arath VAGFAWSWGP LGWLVPSEIS PLEIRSAAQA INVSVNMFFT FLVAQLFLT.
Gtr2ricco VLAFGWSWGP LGWTVPSEIF PLETRSAGQS ITVAVNLLFT FAIAQAFLS.
Hex6ricco VAGFGWSWGP LGWLVPSEIF PLEIRSAGQS IVVAVSFLFT FVVAQTFLS.
Hup1chlke ISGFAWSWGP MGWLIPSEIF TLETRPAGTA VAVVGNFLFS FVIGQAFVS.
Hxtchlke  IAGFAWSWGP MGWLIPSEIF TLETRPAGTA VAVMGNFLFS FVIGQAFVS.
Hup2chlke VAAFAWSWGP LGWLVPSEIQ TLETRGAGMS MAVIVNFLFS FVIGQAFLS.
Araeescco IAGYAMSAAP VVWILCSEIQ PLKCRDFGIT CSTTTNWVSN MIIGATFLT.
Galpescco IVGFAMSAGP LIWVLCSEIQ PLKGRDFGIT CSTATNWIAN MIVGATFLT.
```

```
Glfzymmo    IAVFGMSWGP VCWVVLSEMF PSSIKGAAMP IAVTGQWLAN ILVNFLFKV.
Xyleescco   VAAFAMSWGP VCWVLLSEIF PNAIRGKALA IAVAAQWLAN YFVSWTFPM.
Glcpsynsp   VFSFGFSWGP IVWVLLGEMF NNKIRAAALS VAAGVQWIAN FIISTTF...
Hgtlklula   VCFFAPTWGI GIWIYCSEIF NNMERAKGSA LSAATNWAFN FALAMFVPS.
Lacpklula   GGIFSFAFTP MQSMYSTEVS TNLTRSKAQL LNFVVSGVAQ FVNQFATPK.
Stllsacce   ITFFGLSLLS LPWIYPPEIA SMKVRASTNA FSTCTNWLCN FAVVMFTPI.
Ma6tsacce   AFFYNLGIAP VVFCLVSEMP SSRLRTKTII LARNAYNVIQ VVVTVLIMY.
Gtr2leido   IALYEMGVGP CFYVLAVDVF PESFRPIGSS ITVGVMFIFN LIINICYPIA
Consensus   .........P .......E.F ....R..... .....NW... F.........
```

```
            651                                              700
Itr1sacce   ...M...... LQNITPAGTF AFFAGLSCLS TIFCYFCYP. ..ELSGLELE
Itr2sacce   ...M...... LQNITPTGTF SFFAGVACLS TIFCYFCYP. ..ELSGLELE
Gtr1leido   ...L...... MGAIGVGGTF TIISGLMALG CIFVYFFAV. ..ETKGLTLE
Sgt1schma   ...I...... .QKNIGGYSF LPFLVVVVIC WIFFFLFMP. ..ETKNRTFD
Tgtptaeso   ...L...... .NELLKGYVY LPYLVVVAVC WVVFFLFMP. ..ETKNRTFD
Sgt4schma   ...I...... .NDAIGGYSF LPFLVIVIIC WIFFFLYMI. ..ETKNRTCD
Gtr5homsa   ...I...... .QEGLGPYSF IVFAVICLLT TIYIFLIVP. ..ETKAKTFI
Gtr5ratno   ...I...... .QVGLGPYSF IIFAIICLLT TIYIFMVVP. ..ETKGRTFV
Gtr5orycu   ...I...... .QWA.GLYSF IIFGVACLST TVYTFLIVP. ..ETKGKSFI
Gtr1galga   ...I...... .AQLCGSYVF IIFTVLLVLF FIFTYFKVP. ..ETKGRTFD
Gtr1homsa   ...V...... .EQLCGPYVF IIFTVLLVLF FIFTYFKVP. ..ETKGRTFD
Gtr3musmu   ...A...... .AAYLGAYVF IIFAAFLIFF LIFTFFKVP. ..ETKGRTFE
Gtr3canfa   ...A...... .AFYLGAYVF IIFTGFLIVF LVFTFFKVP. ..ETRGRTFE
Gtr3oviar   ...A...... .TFYLGAYVF IVFTVFLVIF WVFTFFKVP. ..ETRGRTFE
Gtr3homsa   ...A...... .AHYLGAYVF IIFTGFLITF LAFTFFKVP. ..ETRGRTFE
Gtr3galga   ...A...... .EKLCGPYVF LIFLVFLLIF FIFTYFKVP. ..ETKGRTFE
Gtr4homsa   ...V...... .AEAMGPYVF LLFAVLLLGF FIFTFLRVP. ..ETRGRTFD
Gtr2homsa   ...I...... .ADFCGPYVF FLFAGVLLAF TLFTFFKVP. ..ETKGKSFE
Gtr2ratno   ...I...... .ADFLGPYVF FLFAGVVLVF TLFTFFKVP. ..ETKGKSFD
Gtr2galga   ...I...... .ADLCGPYVF VVFAVLLLVF FLFAYLKVP. ..ETKGKSFE
Gtr7ratno   ...L...... .RDFKGPYHF WAFHGVVIVW YGNYWFKVP. ..ETKGKSFD
Sgt2schma   ...L...... .VIYIGIYAF IPFICALLVV LIFVGLYLP. ..ETKGKTPA
Hxt0sacce   ...I...... TRSIGFSYGY VFMGCL.IFS YFYVFFFVC. ..ETKGLTLE
Hxt2sacce   ...I...... TSAIGFSYGY VFMGCL.VFS FFYVFFFVC. ..ETKGLTLE
Hxt4sacce   ...I...... SGAIDFYYGY VFMGCL.VFS YFYVFFFVP. ..ETKGLTLE
Hxt7sacce   ...I...... TGAINFYYGY VFMGCL.VFM FFYVLLVVP. ..ETKGLTLE
Gtrkluma    ...I...... TSAIHFYYGY VFMGCL.VAM FFYVFFFVP. ..ETKGLTLE
Gal2sacce   ...I...... TSAINFYYGY VFMGCL.VAM FFYVFFFVP. ..ETKGLSLE
Hxt1sacce   ...I...... TGAINFYYGY VFMGCM.VFA YFYVFFFVP. ..ETKGLSLE
Hxt3sacce   ...I...... TGAINFYYGY VFMGCM.VFA YFYVFFFVP. ..ETKGLTLE
Hxt5sacce   ...I...... TSAINFYYGY VFMGCM.VFA YFYVFFFVP. ..ETKGLTLE
Rag1klula   ...I...... TSAIHFYYGY VFMGCM.VFA FFYVYFFVP. ..ETKGLTLE
Hxtasacce   ...I...... TGAINFYYGY VFLGCL.VFA YFYVFFFVP. ..ETKGLTLE
Hxt8sacce   ...I...... TSAINFYYGY VFMGCL.AFS YFYVFFFVP. ..ETKGLTLE
Hxtcsacce   ...I...... TGSIHFYYGY VFVGCL.VAM FLYVFFFLP. ..ETIGLSLE
Hxtdsacce   ...I...... VSKIDFKLGY IFAACL.LAL IIFSWILVP. ..ETRKKNEQ
Snf3sacce   ...IVDTGSH TSSLGAKIFF IWGSLN.AMG VIVVYLTVY. ..ETKGLTLE
Qayneucr    ......TPQM FIKMEYGV.Y FFFASLMLLS IVFIYFFLP. ..VTKSIPLE
Qutdemeni   ......TPQM FTSMGYGV.Y FFFASLMILS IVFVFFLIP. ..ETKGVPLE
Stp1arath   .........M LCHLKFGL.F LVFAFFVVVM SIFVYIFLP. ..ETKGIPIE
Sugricco    .........M LCHLKFGL.F IFFSFFVLIM SIFVYYFLP. ..ETKGIPIE
Mst1nicta   .........M LCHLKFGL.F LFFAFFVVIM TVFIYFFLP. ..ETKNIPIE
Stp4arath   .........M LCHMKFGL.F FFFAFFVVIM TIFIYLMLP. ..ETKNVPIE
```

```
Gtr2ricco ........L LCAFKFGI.F LFFAGWITVM TVFVCVFLP. ..ETKGVPIE
Hex6ricco ........M LCHFKSGI.F FFFGGWVVVM TAFVHFLLP. ..ETKKVPIE
Hup1chlke ........M LCAMEYGV.F LFFAGWLVIM VLCAIFLLP. ..ETKGVPIE
Hxtchlke  ........M LCAMKFGV.F LFFAGWLVIM VLCAIFLLP. ..ETKGVPIE
Hup2chlke ........M MCAMRWGV.F LFFAGWVVIM TFFVYFCLP. ..ETKGVPVE
Araeescco ........L LDSIGAAGTF WLYTALNIAF VGITFWLIP. ..ETKNVTLE
Galpescco ........M LNTLGNANTF WVYAALNVLF ILLTLWLVP. ..ETKHVSLE
Glfzymmo  ...ADGSPAL NQTFNHGFSY LVFAALSILG GLIVARFVP. ..ETKGRSLD
Xyleescco ...MDKNSWL VAHFHNGFSY WIYGCMGVLA ALFMWKFVP. ..ETKGKTLE
Glcpsynsp .......PPL LDTVGLGPAY GLYATSAAIS IFFIWFFVK. ..ETKGKTLE
Hgt1klula ...AFKNISW K.......TY IIFGVFSVAL TIQTFFMFP. ..ETKGKTLE
Lacpklula ...AMKNIKY .......WFY VFYVFFDIFE FIVIYFFFV. ..ETKGRSLE
Stl1sacce ...FIGQSGW .......GCY LFFAVMNYLY IPVIFFFYP. ..ETAGRSLE
Ma6tsacce ...QLNSEKW NWGAKSGF.. .FWGGFCLAT LAWAVVDLP. ..ETAGRTFI
Gtr2leido TEGISGGPSG NPNKGQAVAF IFFGCIGVVA CVIEYFFLQP WVEPEAKMTD
Consensus .......... .......... .......... .......P. ..ETKG...E
```

```
             701                                           750
Itr1sacce EVQTILKDGF NIKASKALAK KRKQQVARV. ..HELKYEPT QEIIEDI...
Itr2sacce EVQTILKDGF NIKASKALAK KRKQQVAEGA AHHKLKFEPT QEIVES....
Gtr1leido QIDNMFRKRA GLPPRFHEEG ESGESGAGYR EDGDLGRLAT EDVCDLSSLG
Sgt1schma EVARDLAFGN IVVGKRTTAL EDRNLTVFTK QGNNEGPASE SLLYPRSDND
Tgtptaeso EVARDLAFGS IVVGKRTAAL QA...PVFTK EDEEAATA.. ...LRRSDEE
Sgt4schma SNARDLATAK VVACQRPSRL TYKNEEPFYS DE........ ..........
Gtr5homsa EINQIFTKMN KVSEVY.... PEKE..ELKE LPPVTSEQ.. ..........
Gtr5ratno EINQIFAKKN KVSDVY.... PEKEEKELND LPPATREQ.. ..........
Gtr5orycu EIIRRFIRMN KV.EVS.... PDRE..ELKD FPPDVSE... ..........
Gtr1galga EIAYRFRQGG ASQS...... DKTPDE.FHS LGADSQV... ..........
Gtr1homsa EIASGFRQGG ASQS...... DKTPEELFHP LGADSQV... ..........
Gtr3musmu DIARAFEGQA HSGKGP.... ...AGVELNS MQPVKETPGN A.........
Gtr3canfa EITRAFEGQG QDANRA.... EKGPIVEMNS MQPVKETATV .........
Gtr3oviar EITRAFEGQV QTGTRG.... EKGPIMEMNS IQPTKDTNA. ..........
Gtr3homsa DITRAFEGQA HGADRS.... GKDGVMEMNS IEPAKETTTN V.........
Gtr3galga DISRGFEEQV ETSSPSSPPI EKNPMVEMNS IEPDKEVA.. ..........
Gtr4homsa QISAAFH... ..RTPSLLEQ EVKPSTELEY LGPDEND... ..........
Gtr2homsa EIAAEFQKKS GSAHR..... .PKAAVEMKF LGATETV... ..........
Gtr2ratno EIAAEFRKKS GSAPP..... .RKATVQMEF LGSSETV... ..........
Gtr2galga EIAAAFRRKK LPA....... ..KSMTELED LRGGEEA... ..........
Gtr7ratno EIAAEFRKKH GGRPP..... .KLRWITANF IIASDQVKKM KND.......
Sgt2schma SIEDYFMRVC GFRGTEAHEN PTFTDIIDDT TQY....... ..........
Hxt0sacce EVNEMYEERI KPWKSGGWIP SSR.RTPQPT SSTPLVIVDS K.........
Hxt2sacce EVNEMYVEGV KPWKSGSWIS KEK.RVSEE. .......... ..........
Hxt4sacce EVNTLWEEGV LPWKSPSWVP PNK.RGTDYN ADDLMHDDQP FYKKMFGKK.
Hxt7sacce EVNTMWEEGV LPWKSASWVP PSR.RGANYD AEEMTHDDKP LYKRMFSTK.
Gtrkluma  EVQEMWEEGV LPWKSSSWVP SSR.RNAGYD VDALQHDEKP WYKAML....
Gal2sacce EIQELWEEGV LPWKSEGWIP SSR.RGNNYD LEDLQHDDKP WYKAMLE...
Hxt1sacce EVNDMYAEGV LPWKSASWVP VSK.RGADYN ADDLMHDDQP FYKSLFSRK.
Hxt3sacce EVNDMYAEGV LPWKSASWVP TSQ.RGANYD ADALMHDDQP FYKKMFGKK.
Hxt5sacce EVNEMYEENV LPWKSTKWII ᴾSR.RTTDYD LDATRNDPRP FYKRMFTKEK
Rag1klula EVNEMYSEGV LPWKSSSWVP SSR.RGAEYD VDALQHDDKP WYKAML....
Hxtasacce EVNTMWLEGV PAWKSASWVP PER.RTADYD ADAIDHDDRP IYKRFFSS..
Hxt8sacce EVDEMWMDGV LPWKSESWVP ASR.RDGDYD NEKLQHDEKP FYKRMF....
Hxtcsacce EIQLLYEEGI KPWKSASWVP PSR.RGISSE ESKTEKKDWK KFLKFSKNSD
Hxtdsacce EINKIFEPE. .......... .......... .......... ..........
```

```
Snf3sacce EIDELYIKSS TGVVSPKFNK DIRERALKFQ YDPLQRLEDG KNTFVAKRNN
Qayneucr  AMDRLFEIKP VQNANKNLMA ELNFDRNPER EESSSLDDKD RVTQTENAV.
Qutdemeni SMETLFDKKP VWHAHSQLIR EL...RENEE AFRADMGASG KGGVTKEYVE
Stp1arath EMGQVWRSH. ..WYWSRFVE DGEYG....N .ALEMGKNSN QAGTKHV...
Sugricco  EMGQVWKQH. ..WYWSRYVV DEDYP....N GGLEMGKEGR IP..KNV...
Mst1nicta EMVIVWKEH. ..WFWSKFMT EVDYPGTRNG TSVEMSKGS. .AGYKIV...
Stp4arath EMNRVWKAH. ..WFWGKFIP DEA....VNM GAAEMQQKSV .........
Gtr2ricco EMVLLWRKH. ..WFWKKVMP VDMPLEDGWG AAPASNNHK. .........
Hex6ricco KMDIVWRDH. ..WFWKKIIG EEAAEENNKM EAA........
Hup1chlke RVQALYARH. ..WFWNRVMG PAAAEVIAED EKRVAAASAI IKEEELSKAM
Hxtchlke  RVQALYARH. ..WFWKKVMG PAAQEIIAED EKRVAASQAI MKEERISQTM
Hup2chlke TVPTMFARH. ..WLWGRVMG EKGRALVAAD EARKAGTVAF KVESGSEDGK
Araeescco HIERKLMAG. ..EKLRNIGV .........
Galpescco HIERNLMKG. ..RKLREIGA HD........
Glfzymmo  EIEEMWRSQ. ...K.......
Xyleescco ELEALWEPE. ..TKKTQQTA TL........
Glcpsynsp QM........
Hgt1klula EIDQMWVDNI PAWRTANYIP QLPIVKDEEG NKLGLLGNPQ HLEDVHSNEK
Lacpklula ELEVVFEAPN PRKASVDQAF LAQVRATLVQ RNDVRVANAQ NLKEQEPLKS
Stl1sacce EIDIIF.... ......AKAY EDGTQPWRVA NHLPKLSLQE VEDHANALGS
Ma6tsacce EINELFRLGV PARKFKSTKV DPFAAAKAAA AEINVKDPKE DLETSVVDEG
Gtr2leido DLDGAAVPEG KHD.......
Consensus .........
```

```
          751                                          800
Tgtptaeso DAKVDA....
Snf3sacce FDDETPRNDF RNTISGEIDH SPNQKEVHSI PERVDIPTST EILESPNKSS
Qutdemeni EA........
Hup1chlke K.........
Hxtchlke  K.........
Hup2chlke PASDQ.....
Hgt1klula GLLDRSDSAS NSN.......
Lacpklula DADHVEKLSE AESV......
Stl1sacce YDDEMEKEDF GEDR......
Ma6tsacce RSTPSVVNK.
Gtr2leido .........
Consensus .........
```

```
          801                                          850
Gtr1leido AAIKAAPHEP K.........
Snf3sacce GMTVPVSPSL QDVPIPQTTE PAEIRTKYVD LGNGLGLNTY NRGPPSLSSD
```

```
          851                                          900
Snf3sacce SSEDYTEDEI GGPSSQGDQS NRSTMNDIND YMARLIHSTS TASNTTDKFS
```

```
          901                                          950
Snf3sacce GNQSTLRYHT ASSHSDTTEE DSNLMDLGNG LALNAYNRGP PSILMNSSDE
```

```
          951                                         1000
Snf3sacce EANGGETSDN LNTAQDLAGM KERMAQFAQS YIDKRGGLEP ETQSNILSTS
```

```
          1001                    1034
Snf3sacce LSVMADTNEH NNEILHSSEE NATNQPVNEN NDLK
```

Proteins listed subsequently in italics are at least 90% identical to the paired transporters listed in parenthesis and therefore are not included in the alignments. *Araekleox* (Araeescco); *Gtr1musmu, Gtr1orycu, Gtr1sussc, Gtr1bosta, Gtr1ratno* (Gtr1homsa); *Gtr2musmu* (Gtr2ratno); *Gtr3ratno* (Gtr3musmu); *Gtr4ratno, Gtr4musmu* (Gtr4homsa); *Hxt6sacce* (Hxt7sacce). Residues listed in the consensus sequence are present in at least 75% of the aligned transporter sequences. Residues indicated by boldface type are also conserved in at least one other family of the USA/MFS superfamily.

Database accession numbers

	SWISSPROT	PIR	EMBL/GENBANK
Araeescco	P09830	B26430	J03732
Araekleox	P45598		X79598
Gal2sacce	P13181	A33865; JQ0383	M68547; M81879
Galpescco	P37021		U28377
Glcpsynsp	P15729	S06973; S10014	X15988; X16472
Glfzymmo	P21906	A37855	M60615
Gtr1bosta	P27674		M60448
Gtr1galga	P46896		L07300
Gtr1homsa	P11166	A27217	K03195; M20653
Gtr1leido	Q01440	A48442	M85072
Gtr1musmu	P17809	A30310; S09705	M22998; M23384
Gtr1orycu	P13355	A30797	M21747
Gtr1ratno	P11167	A25949	M13979; M22061
Gtr1sussc	P20303	S04223	X17058
Gtr2galga		S37476	Z22932
Gtr2homsa	P11168	A31318	J03810
Gtr2leido	Q01441		M85073
Gtr2musmu	P14246	S06920; S05319	X16986; X15684
Gtr2ratno	P12336	A31556	J03145
Gtr2ricco			L21753
Gtr3canfa			L35267
Gtr3galga	P28568	A41264	M37785
Gtr3homsa	P11169	A31986	M20681
Gtr3musmu	P32037	A41751	M75135; X61093
Gtr3oviar			L39214
Gtr3ratno	Q07647	S38981	D13962
Gtr4homsa	P14672	A33801; A49158	M20747; M91463
Gtr4musmu	P14142	B30310	M23383
Gtr4ratno	P19357	S03349; A32101	X14771; J04524
Gtr5homsa	P22732	A36629	M55531
Gtr5orycu	P46408		D26482
Gtr5ratno	P43427		L05195; D13871
Gtr7ratno	Q00712	S24344	X66031
Gtrkluma		S51081	Z47080
Hex6ricco			L08188
Hgt1klula			U22525
Hup1chlke	P15686	S07096	Y07520
Hup2chlke			X66855
Hxt0sacce	P43581		D50617; Z46255
Hxt1sacce	P32465	A39728; S38798	L07079; M82963
Hxt2sacce	P23585	S12200	M33270
Hxt3sacce	P32466	S31294	L07080; S52309
Hxt4sacce	P32467	S31314; S39817	M81960; X67321
Hxt5sacce	P38695	S43742; S46726	X77961; U00060
Hxt6sacce	P39003	S43185	Z31691
Hxt7sacce	P39004	S43186	Z31692

	SWISSPROT	PIR	EMBL/GENBANK
Hxt8sacce	P40886	S45159	Z34098; Z49489
Hxtasacce	P40885	S45153	Z34098; X82621
Hxtchlke		S38435	X75440
Hxtcsacce	P39924		U18795
Hxtdsacce	P42833		Z46259
Itr1sacce	P30605	A40538	D90352
Itr2sacce	P30606	B40538	D90353
Lacpklula	P07921	A31776	X06997
Ma3tsacce	P38156	S46182	Z36167
Ma6tsacce	P15685	S07686	M27823
Qayneucr	P11636	S04254; G31277	X14603
Qutdemeni	P15325	S08498	X13525
Rag1klula	P18631	S11295	X53752
Sgt1schma		A53153	L25065
Sgt2schma		B53153	L25066
Sgt4schma		C53153	L25067
Snf3sacce	P10870	A31928	J03246
Stl1sacce	P39932		L07492
Stp1arath	P23586	S12042	X55350
Stp4arath		S25009	X66857
Sugricco			L08196
Tgtptaeso			U39197
Xyleescco	P09098	A26430; A27418	J02812; X0663

References

1. **Olson, A.L. and Pessin, J.E. (1996) Annu. Rev. Nutr. 16, 235–256.**
2. **Thorens, B. (1996) Am. J. Physiol. 270, G541–G553.**
3. **Baldwin, S.A. (1993) Biochim. Biophys. Acta 1154, 17–50.**
4. **Mueckler, M. (1993) Eur. J. Biochem. 219, 713–725.**
5. **Henderson, P.J.F. (1993) Curr. Opin. Cell Biol. 5, 708–721.**
6. **Bisson, L.F. et al. (1993) CRC Crit. Rev. Biochem. Mol. Biol. 28, 259–308.**
7. **Fischbarg, J. and Vera, J.C. (1995) Am. J. Physiol. 268, C1077–C1089.**
8. Boorer, K.J. et al. (1994) J. Biol. Chem. 269, 20417–20424.
9. Sofue, M. et al. (1992) Biochem. J. 288, 669–674.
10. **Griffith, J. et al. (1992) Curr. Opin. Cell Biol. 4, 684–695.**
11. **Marger, M.D. and Saier, M. (1993) Trends Biochem. Sci. 18, 13–20.**
12. Davies, A. et al. (1987) J. Biol. Chem. 262, 9347–9352.
13. Hresko, R. et al. (1994) J. Biol. Chem. 269, 20482–20488.
14. Herbert, D. and Carruthers, A. (1992) J. Biol. Chem. 267, 23829–23838.
15. **Henderson, P.J.F. (1992) Int. Rev. Cytol. 137, 149–208.**
16. Gould, G. et al. (1991) Biochemistry 30, 5139–5145.

Summary

Transporters of the H⁺/rhamnose symporter family, the example of which is the RHAT rhamnose-H⁺ symporter of *Escherichia coli* (Rhatescco), mediate symport (H⁺-coupled substrate uptake) of rhamnose. The two known members of the family are found only in gram-negative bacteria.

Statistical analysis reveals no apparent relationship between the amino acid sequences of the H⁺/rhamnose symporter family and any other family of transporters. They are predicted to contain ten membrane-spanning helices by the hydropathy of their amino acid sequences and activities of reporter gene fusions [1].

The amino acid sequences of the two known members of the H⁺/rhamnose symporter family are more than 90% identical.

Nomenclature, biological sources and substrates

CODE	DESCRIPTION [SYNONYMS]	ORGANISM [COMMON NAMES]	SUBSTRATE(S)
Rhatescco	Rhamnose-H⁺ symporter [RHAT]	*Escherichia coli* [gram-negative bacterium]	H⁺/L-rhamnose
Rhatsalty	Rhamnose-H⁺ symporter [RHAT]	*Salmonella typhimurium* [gram-negative bacterium]	H⁺/L-rhamnose

Cotransported ions are listed.

Proposed orientation of RHAT in the membrane

The model is based on predictions of membrane-spanning regions and α-helical content. The N-terminus of the protein is illustrated on the inside and is folded ten times through the membrane [2]. The predicted membrane-spanning helices are portrayed as rectangles. The numbers corresponding to the first and last residue of each membrane-spanning helix are boxed.

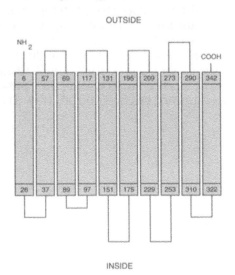

Physical and genetic characteristics

	AMINO ACIDS	MOL. WT	K_m	CHROMOSOMAL LOCUS
Rhatescco	344	37319	Rhamnose: 28 μM [2]	88.25 minutes
Rhatsalty	344	37390		

Multiple amino acid sequence alignments

```
            1                                                  50
Rhatescco   MSNAITMGIF WHLIGAASAA CFYAPFKKVK KWSWETMWSV GGIVSWIILP

            51                                                 100
Rhatescco   WAISALLLPN FWAYYSSFSL STRLPVFLFG AMWGIGNINY GLTMRYLGMS

            101                                                150
Rhatescco   MGIGIAIGIT LIVGTLMTPI INGNFDVLIS TEGGRMTLLG VLVALIGVGI

            151                                                200
Rhatescco   VTRAGQLKER KMGIKAEEFN LKKGLVLAVM CGIFSAGMSF AMNAAKPMHE

            201                                                250
Rhatescco   AAAALGVDPL YVALPSYVVI MGGGAIINLG FCFIRLAKVK DLSLKADFSL

            251                                                300
Rhatescco   AKSLIIHNVL LSTLGGLMWY LQFFFYAWGH ARIPAQYDYI SWMLHMSFYV

            301                                           350
Rhatescco   LCGGIVGLVL KEWNNAGRRP VTVLSLGCVV IIVAANIVGI GMAN
```

Proteins listed subsequently in italics are at least 90% identical to the paired transporters listed in parenthesis and therefore are not included in the alignments: *Rhatsalty* (Rhatescco).

Database accession numbers

	SWISSPROT	PIR	EMBL/GENBANK
Rhatescco	P27125	B42436; S26145	M85158; X60699
Rhatsalty	P27135	A42436	M85157

References

[1] Tate, C. and Henderson, P.J.F. (1993) J. Biol. Chem. 268, 26850–26857.
[2] Muir, J. et al. (1993) Biochem. J. 290, 833–842.

Summary

Transporters of the H⁺/amino acid symporter family, the example of which is the PHEP phenylalanine transporter of *Escherichia coli* (Phepescco), mediate proton-dependent uptake of one or more amino acids[1]. Members of the family occur in both gram-positive and gram-negative bacteria and in various fungi.

Statistical analysis reveals no apparent relationship between the amino acid sequences of the H⁺/amino acid symporter family and any other family of transporters. They are predicted to form 12 membrane-spanning helices by the hydropathy of their amino acid sequences and the activities of reporter gene fusions[2]. Eukaryotic transporters of the H⁺/amino acid symporter family may be glycosylated.

Several amino acid sequence motifs are highly conserved in the H⁺/amino acid symporter family.

Nomenclature, biological sources and substrates

CODE	DESCRIPTION [SYNONYMS]	ORGANISM [COMMON NAMES]	SUBSTRATE(S)
Alp1sacce	High-affinity basic amino acid transporter [ALP1]	*Saccharomyces cerevisiae* [yeast]	Lysine, arginine
Anspsalty	L-Asparagine transporter [ANSP]	*Salmonella typhimurium* [gram-negative bacterium]	L-Asparagine
Aropescco	Aromatic amino acid transporter [AROP]	*Escherichia coli* [gram-negative bacterium]	Aromatic amino acids
Aropcorgl	Aromatic amino acid transporter [AROP]	*Corynebacterium glutamicum* [gram-positive bacterium]	Aromatic amino acids
Can1sacce	High-affinity arginine transporter [CAN1, YELO63c]	*Saccharomyces cerevisiae* [yeast]	Arginine
Can1canal	High-affinity basic amino acid transporter [CAN1]	*Candida albicans* [yeast]	Lysine, arginine
Cycaescco	Serine-alanine, -glycine transporter [CYCA, DAGA]	*Escherichia coli* [gram-negative bacterium]	Serine, alanine, glycine
Dip5sacce	Dicarboxylic amino acid transporter [DIP5]	*Saccharomyces cerevisiae* [yeast]	Dicarboxylic acids
Gabpbacsu	GABA transporter [4-amino butyrate transporter, GABP]	*Bacillus subtilis* [gram-positive bacterium]	4-amino butyrate
Gabpescco	GABA transporter [4-amino butyrate transporter, GABP]	*Escherichia coli* [gram-negative bacterium]	4-amino butyrate
Gap1sacce	General amino acid transporter [GAP1, YKR039w]	*Saccharomyces cerevisiae* [yeast]	Many amino acids
Hip1sacce	High-affinity histidine transporter [HIP1, G7572]	*Saccharomyces cerevisiae* [yeast]	Histidine
Hutmbacsu	Putative histidine transporter [HUTM, EE57d]	*Bacillus subtilis* [gram-positive bacterium]	Histidine
Ina1triha	General amino acid transporter [INA1, INDA1]	*Trichoderma harzianum* [fungus]	Many amino acids
Isp5schpo	Putative amino acid transporter [ISP5]	*Schizosaccharomyces pombe* [yeast]	Many amino acids
Lyp1sacce	High-affinity lysine transporter [LYP1]	*Saccharomyces cerevisiae* [yeast]	Lysine
Lyspescco	Lysine transporter [LYSP, CADR]	*Escherichia coli* [gram-negative bacterium]	Lysine

CODE	DESCRIPTION [SYNONYMS]	ORGANISM [COMMON NAMES]	SUBSTRATE(S)
Pap1sacce	General amino acid transporter [PAP1]	*Saccharomyces cerevisiae* [yeast]	Many amino acids
Phepescco	Phenylalanine transporter [PHEP]	*Escherichia coli* [gram-negative bacterium]	Phenylalanine
Proysalty	Proline transporter [PROY]	*Salmonella typhimurium* [gram-negative bacterium]	Proline
Put4sacce	Proline transporter [PUT4]	*Saccharomyces cerevisiae* [yeast]	Proline, 4-amino butyrate
Putxemeni	High-affinity proline transporter [PUTX, PRNB]	*Emericella nidulans* [mold]	Proline
Roccbacsu	Amino acid transporter [ROCC, IPA78d]	*Bacillus subtilis* [gram-positive bacterium]	Arginine, ornithine
Rocebacsu	Amino acid transporter [ROCE]	*Bacillus subtilis* [gram-positive bacterium]	Arginine, ornithine
Tat2sacce	High-affinity tryptophan transporter [TAT2, SCM2, LTG3, TAP2]	*Saccharomyces cerevisiae* [yeast]	Tryptophan
Val1sacce	Valine-tyrosine-tryptophan transporter [VAL1, VAP1, TAT1, TAP1, YBR069c, YBR0710]	*Saccharomyces cerevisiae* [yeast]	Valine, alanine, tyrosine, tryptophan

Phylogenetic tree

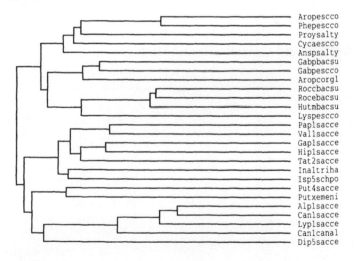

Proposed orientation of PHEP in the membrane

The model is based on predictions of membrane-spanning regions and α-helical content. The N-terminus of the protein is illustrated on the inside and is folded 12 times through the membrane [2]. The predicted membrane-spanning helices are portrayed as rectangles. The numbers corresponding to the first and last residue of each membrane-spanning helix are boxed. Residues that are conserved in more than 75% of the aligned transporters (see below) are shown. Consensus residues indicated by an asterisk are not conserved in PHEP.

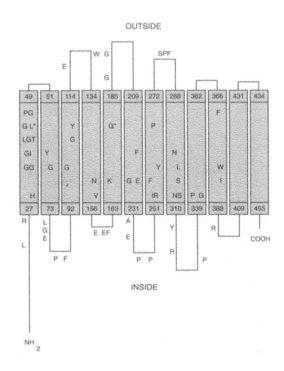

Physical and genetic characteristics

	AMINO ACIDS	MOL. WT	CHROMOSOMAL LOCUS
Alp1sacce	574	64066	
Anspsalty	497	54004	
Aropescco	456	49809	2.58 minutes
Aropcorgl	463	49268	
Can1sacce	590	65785	Chromosome 5
Can1canal	571	63317	
Cycaescco	470	51659	95.4–95.42 minutes
Dip5sacce	608	68097	Chromosome 16
Gabpbacsu	469	51084	52°
Gabpescco	466	51080	60.17 minutes
Gap1sacce	602	65655	Chromosome 11
Hip1sacce	603	66006	Chromosome 7
Hutmbacsu	475	51581	335°
Ina1triha	573	62850	
Isp5schpo	543	60074	
Lypsacce	611	68118	Chromosome 14
Lyspescco	488	53471	48.34 minutes
Pap1sacce	566	62707	Chromosome 11
Phepescco	458	50667	12.96 minutes
Proysalty	292	31824	
Put4sacce	627	68786	Chromosome 15
Putxemeni	570	63101	
Roccbacsu	470	51730	327°
Rocebacsu	467	51634	354°
Tat2sacce	592	65404	Chromosome 15
Val1sacce	619	68757	Chromosome 2

Multiple amino acid sequence alignments

```
          1                                                50
Pap1sacce ............. ....MEKSA EFEVTD..SA LYNNFNTSTT ASLTP.....
Val1sacce ............. ....MDDSV SFIAKEASPA QYSHSLHERT HSEKQKRDFT
Gap1sacce ............. ... MSNTSSYEKN NPDNLKHN.. .GITIDSEFL
Hip1sacce ............. ...MPR NPLKKEYWAD VVDGFKPATS PAFENEKEST
Tat2sacce ............. ...... MTEDFISSVK RSNEELKERK
Ina1triha ............ ................ ................ .......M
Put4sacce ...MVNILPF HKNNRHSAGV VTCADDVSGD GSGGDTKKEE NVVQVTESPS
Alp1sacce ............ ....... ......... .MDETVNIQM
Can1sacce ............ ....MTNSKE DADIEEKHMY NEPVTTLFHD
Lyp1sacce MGRFSNIITS NKWDEKQNNI GEQSMQELPE DQIEHEMEAI DPSNKTTPYS
Can1canal ............ ................ ...........M
Dip5sacce ............ ...... MKMPLKKMFT STSPRNSSSL


          51                                               100
Aropescco ........... ................ ................ .......
Phepescco ........... ................ ...........MKNASTV
Proysalty ........... ............MCTPRG LTPPWFFIVL
Cycaescco ........... ................ ..........MVDQVKV
Anspsalty ........... ................ .....MKTQ TTHAAEQHAA
Gabpbacsu ........... ................ ................ .......
Gabpescco ........... ................ ................ .......
Aropcorgl ........... ................ ................ .......
Roccbacsu ........... ................ ................ .......
Rocebacsu ........... ................ ................ .......
Hutmbacsu ........... ................ ................ .......
Lyspescco ........... ................ ................ .......
Pap1sacce ......EIKE HSEESR.... NGLVHR.... FVDSFRRAES QRLEEDNDLE
Val1sacce ITEKQDEVSG QTAEPRRTDS KSILQRKCKE FFDSFKRQLP P..DRNSELE
Gap1sacce TQEPITIPSN GSAVSIDETG SGSKWQDFKD ...SFKRVKP IEVDPNLSEA
Hip1sacce TFVTELTSKT DSAFPLSSKD SPGINQTTND .ITSSDRFRR NE......DT
Tat2sacce SNFGFVEYKS KQLTSSSSHN SNSSHHDDDN .QHGKRNIFQ RCVDSFKSPL
Ina1triha SKEESGHVTP EKGDNVVDYQ ASTTVLPSEG PERDANWFTR NGLNV..DSF
Isp5schpo ........MP AMKRKKLDME SSRWFPKGET C...FQRWYR SFLPP..EDG
Put4sacce SGSRNNHRSD NEKDDAIRME KISKNQSASS NGTIREDLIM DV..DLEKSP
Putxemeni ........... ....MSPPS AKSMEEGRTP SVQYGYGDPK
Alp1sacce SKEGQYEINS SSIIKEEEFV DEQYSGENVT KAITTERKVE DDDAAKETE.
Can1sacce VEASQTHHRR GSIPLKDEKS KELYPLRSFP TRVNGEDTFS MEDGIGDED.
Lyp1sacce IDEKQYNTKK KHGSLQGGAI AD...VNSIT NSLTRLQVVS HEPDIDEDE.
Can1canal PEDYEKYRMG SSNESHQKSV QPISSSISKS NKKTKHQTDF VQDSDIIEA.
Dip5sacce DSDHDAYYSK QNPDNFPVKE QEIYNIDLEE NNVSSRSSTS TSPSARDDSF
Consensus ........... ................ ................ .......


          101                                              150
Aropescco .MMEGQQHGE QLKRGLKNRH IQLIALGGSI GTGLFLGSAS VIQSAGP.GI
Phepescco SEDTASNQEP TLHRGLHNRH IQLIALGGAI GTGLFLGIGP AIQMAGP.AV
Proysalty IGQKLMESNN KLKRGLSTRH IRFMALGSAI GTGLFYGSAD AIKMAGP.SV
Cycaescco VADDQAPAEQ SLRRNLTNRH IQLIAIGGAI GTGLFMGSGK TISLAGP.SI
Anspsalty KRRWLNAHEE GYHKAMGNRQ VQMIAIGGAI GTGLFLGAGA RLQMAGP.AL
Gabpbacsu ..MNQS..QS GLKKELKTRH MTMISIAGVI GAGLFVGSGS VIHSTGP.GA
Gabpescco ..MGQSSQPH ELGGGLKSRH VTMLSIAGVI GASLFVGSSV AIAEAGP.AV
```

```
Aropcorgl ....MAKSNE GLGTGLRTRH LTMMGLGSAI GAGLFLGTGV GIRAAGP.AV
Roccbacsu ....MQNHKN ELQRSMKSRH LFMIALGGVI GTGLFLGSGF TISQAGPLGA
Rocebacsu .MNTNQDNGN QLQRTMKSRH LFMISLGGVI GTGFFLGTGF TINQAGPLGA
Hutmbacsu .MNLQENSSQ QLKRTMKSRH LFMISLGGVI GTGLFLSTGY TLHQAGPGGT
Lyspescco VSETKTTEAP GLRRELKARH LTMIAIGGSI GTGLFVASGA TISQAGPGGA
Paplsacce DGTKSMKSNN HLKKSMKSRH VVMMSLGTGI GTGLLVANAK GLSLAGPGSL
Vallsacce SQEK.....N NLTKSIKSRH LVMISLGTGI GTGLLVGNGQ VLGTAGPAGL
Gaplsacce EKVAIITAQT PLKHHLKNRH LQMIAIGGAI GTGLLVGSGT ALRTGGPASL
Hiplsacce EQEDI..NNT NLSKDLSVRH LLTLAVGGAI GTGLYVNTGA ALSTGGPASL
Tat2sacce DGS...FDTS NLKRTLKPRH LIMIAIGGSI GTGLFVGSGK AIAEGGPLGV
Inaltriha KKKHYGPGMV ELERPMKARH LHMIAIGGSI GAGFFVGSGG ALAKGGPGSL
Isp5schpo KPQ....... KLKRTLTARH IQMIGIGGAI GTGVWVGSKN TLREGGAASV
Put4sacce SVDGDSEPHK .LKQGLQSRH VQLIALGGAI GTGLLVGTSS TLHTCGPAGL
Putxemeni TLEGEIEEHT ATKRGLSSRQ LQLLAIGGCI GTGLFVGTST VLTQTGPAPL
Alplsacce ..SSPQ.ERR EVKRKLKQRH IGMIALGGTI GTGLIIGIGP PLAHAGPVGA
Canlsacce ..EGEV.QNA EVKRELKQRH IGMIALGGTI GTGLFIGLST PLTNAGPVGA
Lyplsacce ..EEAHYEDK HVKRALKQRH IGMIALGGTI GTGLFVGIST PLSNAGPVGS
Canlcanal ..SSINDEFG EVKRDLKARH VSMIAIGGTI GTGLFISTGS LLHTTGPVMS
Dip5sacce AVPDGKDENT RLRKDLKARH ISMIAIGGSL GTGLLIGTGT ALLTGGPVAM
Consensus .......... .L......RH ......GG.I GTGLL.G... ......GP...


          151                                                  200
Aropescco ILGYAIAGFI AFLIMRQLGE MVVEEPVA.. ...GSFSHFA YKYWGSFAGF
Phepescco LLGYGVAGII AFLIMRQLGE MVVEEPVS.. ...GSFAHFA YKYWGPFAGF
Proysalty LLAYIIGGVA AYIIMRALGE MSVHNPAA.. ...SSFSRYA QENLGPLAGY
Cycaescco IFVYMIIGFM LFFVMRAMGE LLLSNLEY.. ...KSFSDFA SDLLGPWAGY
Anspsalty ALVYLICGIF SFFILRALGE LVLHRPSS.. ...GSFVSYA REFLGEKAAY
Gabpbacsu VVSYALAGLL VIFIMRMLGE MSAVNPTS.. ...GSFSQYA HDAIGPWAGF
Gabpescco LLAYLFAGLL VVMIMRMLAE MAVATPDT.. ...GSFSTYA DKAIGRWAGY
Aropcorgl LLAYIIAGAI VVLVMQMLGE MAAARPAS.. ...GSFSRYG EDAFGHWAGF
Roccbacsu IAAYIIGGFL MYLVMLCLGE LAVAMPVA.. ...GSFQAYA TKFLGQSTGF
Rocebacsu VLSYLVGGFI MFLTMLCLGE LAVAFPVS.. ...GSFQTYA TKFISPAFGF
Hutmbacsu ILAYVIGGLM MYLVMQCLGE LSVANAVT.. ...GSFQKYA TTFIGPSTGF
Lyspescco LLSYMLIGLM VYFLMTSLGE LAAYMPVS.. ...GSFATYG QNYVEEGFGF
Paplsacce VIGYVMVSFV TYFMVQAAGE MGVTYPTLP. ...GNFNAYN SIFISKSFGF
Vallsacce VLGYGIASIM LYCIIQAAGE LGLCYAGLT. ...GNYTRYP SILVDPSLGF
Gaplsacce LIGWGSTGTM IYAMVMALGE LAVIFP.IS. ...GGFTTYA TRFIDESFGY
Hiplsacce VIDWVIISTC LFTVINSLGE LSAAFP.VV. ...GGFNVYS MRFIEPSFAF
Tat2sacce VIGWAIAGSQ IIGTIHGLGE ITVRFP.VV. ...GAFANYG TRFLDPSISF
Inaltriha FVDFLIIGIM MFNVVYALGE LAIMYP.VS. ...GSFYTYS ARFIDPAWGF
Isp5schpo LICYSLVGSM VLMTVYSLGE LAVAFP.IN. ...GSFHTYG TRFIHPSWGF
Put4sacce FISYIIISAV IYPIMCALGE MVCFLPGDGS DSAGSTANLV TRYVDPSLGF
Putxemeni LMSYIVMASI VWFVMNVLGE MTTYLPIRG. ...VSVPYLI GRFTEPSIGF
Alplsacce LISYLFMGTV IYSVTQSLGE MATFIPVTS. ....SFSVFA QRFLSPALGA
Canlsacce LISYLFMGSL AYSVTQSLGE MATFIPVTS. ....SFTVFS QRFLSPAFGA
Lyplsacce LIAYIFMGTI VYFVTQSLGE MATFIPVTS. ....SITVFS KRFLSPAFGV
Canlcanal LISFLFVTTI CFSVTQSLGE MATYIPISG. ....SFAQFV TRWVSKSCGA
Dip5sacce LIAYAFVGLL VFYTMACLGE MASYIPLDG. .....FTSYA SRYVDPALGF
Consensus ...Y...G.. ........LGE .....P.... .....F.... .........G.


          201                                                  250
Aropescco ASGWNYWVLY VLVAMAELTA VGKYIQFW.. YPEIPT...W VSAAVFFVVI
Phepescco LSGWNYWVMF VLVGMAELTA AGIYMQYW.. FPDVPT...W IWAAAFFIII
```

```
Proysalty ITGWTYCFEI LIVAIADVTA FGIYMGVW.. FPAVPH...W IWVLSVVLII
Cycaescco FTGWTYWFCW VVTGMADVVA ITAYAQFW.. FPDLSD...W VASLAVIVLL
Anspsalty VAGWMYFINW AMTGIVDITA VALYMHYWGA FGDVPQ...W VFALGALTIV
Gabpbacsu TIGWLYWFFW VIVIAIEAIA GAGIIQYW.. FHDIPL...W LTSLILTIVL
Gabpescco TIGWLYWWFW VLVIPLEANI AAMILHSW.. VPGIPI...W LFSLVITLAL
Aropcorgl SLGWLYWFML IMVMGAEMTG AAAIMGAW.. F.GVEP...W IPSLVCVVFF
Roccbacsu MIGWLYWFSW ANTVGLELTS AGILMQRW.. LPSVPI...W IWCLVFGIVI
Rocebacsu AFGWLYWLGW AVTCAIEFLS AGQLMQRW.. FPHIDV...W IWCLVFAALM
Hutmbacsu MVGIMYWINW VVTVGSEFTA SGILMQRW.. FPDSSV...W MWSAIFAALL
Lyspescco ALGWNYWYNW AVTIAVDLVA AQLVMNWW.. FPDTPG...W IWSALFLGVI
Paplsacce ATTWLFCIQW LTVLPLELIT SSMTVKYW.. ..N.DTINAD VFIVIFYVFL
Vallsacce AVSVVYTIQW LTVLPLQLVT AAMTVKYW.. ....TSVNAD IFVAVVFVFV
Gaplsacce ANNFNYMLQW LVVLPLEIVS ASITVNFW.. ..GTDPKYRD GFVALFWLAI
Hiplsacce AVNLNYLAQW LVLLPLELVA ASITIKYW.. ..N.DKINSD AWVAIFYATI
Tat2sacce VVSTIYVLQW FFVLPLEIIA AAMTVQYW.. ..N.SSIDPV IWVAIFYAVI
Inaltriha AMGWNYVLQW AAVLPLELTV CGITISYW.. ..NSE.ITTA AWISLFLGVI
Isp5schpo TLGWNYLASF LATYPLELIT ASICLQFW.. ...IN.INSG IWITVFIALL
Put4sacce ATGWNYFYCY VILVAAECTA ASGVVEYWTT AVPKGV.... .WITIFLCVV
Putxemeni ASGYNYWYSF AMLLACEVST M.ALLSFLSC WNPDNVGHCL GLIIEYWNPP
Alplsacce TNGYMYWLSW CFTFALELSV LGKVIQYWT. .EAVPLA... AWIVIFWCLL
Canlsacce ANGYMYWFSW AITFALELSV VGQVIQFWT. .YKVPLA... AWISIFWVII
Lyplsacce SNGYMYWFNW AITYAVEVSV IGQVIEYWT. .DKVPLA... AWIAIFWVII
Canlcanal ANGWLYWFSW AVTFGLELSV VGQVIQFWT. .DAVPLA... AWISIFFVIL
Dip5sacce AIGYTYLFKY FILPPNQLTA AALVIQYWIS RDRVNPG... VWITIFLVVI
Consensus ..G..Y.... ......E... ......W... .......... ..........
```

```
                251                                            300
Aropescco NAINLTNVKV FGEMEFWFAI IKVIAVVAMI I..FGGW.LL FSGNGGPQAT
Phepescco NAVNLVNVRL YGETEFWFAL IKVLAIIGMI G..FGLW.LL FSGHGGEKAS
Proysalty CAINLMSVKV FGELEFWFSF FKVATIIIMI VAGIGII.VW GIGNGGQPTG
Cycaescco LTLNLATVKM FGEMEFWFAM IKIVAIVSLI VVGLVMVAMH FQSPTGVEAS
Anspsalty GTMNMIGVKW FAEMEFWFAL IKVLAIVIFL VVG.TIFLGT GQPLEGNATG
Gabpbacsu TLTNVYSVKS FGEFEYWFSL IKVVTIIAFL IVGFAFIFGF A...PGSEPV
Gabpescco TGSNLLSVKN YGEFEFWLAL CKVIAILAFI FLGAVAISGF Y...PYAEVS
Aropcorgl AVVNLVAVRG FGEFEYWFAF IKVAVIIAFL IIGIALIFGW L...PGSTFV
Roccbacsu FLINALSVRS FAEMEFWFSS IKVAAIILFI VIGGAAVFGL IDFKGGQETP
Rocebacsu FILNAITTKA FAESEFWFSG IKILIILLFI ILGGAAMFGL IDLKGGEQAP
Hutmbacsu FICNAFSVKL FAETEFWFSS VKIVTIILFI ILGGAAMFGL ISLNGTADAP
Lyspescco FLLNYISVRG FGEAEYWFSL IKVTTVIVFI IVGVLMIIGI ..FKGAQPAG
Paplsacce LFIHFFGVKA YGETEFIFNS CKILMVAGFI ILSVVINCGG AGVD....GY
Vallsacce IIINLFGSRG YAEAEFIFNS CKILMVIGFV ILAIIINCGG AGDR....RY
Gaplsacce VIINMFGVKG YGEAEFVFSF IKVITVVGFI ILGIILNCGG GPTG....GY
Hiplsacce ALANMLDVKS FGETEFVLSM IKILSIIGFT ILGIVLSCGG GPHG....GY
Tat2sacce VSINLFGVRG FGEAEFAFST IKAITVCGFI ILCVVLICGG GPDH....EF
Inaltriha IIINLFGALG YAEEEFWASC FKLAATVIFM IIAFVLVLGG GPKDGRYHEY
Isp5schpo CFVNMFGVRG YGEVEFFVSS LKVMAMVGFI ICGIVIDCGG VRTDHR..GY
Put4sacce VILNFSAVKV YGESEFWFAS IKILCIVGLI ILSFILFWGG GPNHDR....
Putxemeni VSVGLWIAIV LVESEFWFAG LKILAIIGLI ILGVVLFFGG GPNHER....
Alplsacce TSMNMFPVKY YGEFEFCIAS IKVIALLGFI IFSFCVVCGA GQSDGP....
Canlsacce TIMNLFPVKY YGEFEFWVAS IKVLAIIGFL IYCFCMVCGA G.VTGP....
Lyplsacce TLMNFFPVKV YGEFEFWVAS VKVLAIMGYL IYALIIVCGG SH.QGP....
Canlcanal TIFNFFPVKF YGEVEFWIAS IKIIAVFGWI IYAFIMVCGA GK.TGP....
Dip5sacce VAINVVGVKF FGEFEFWLSS FKVMVMLGLI LLLFIIMLGG GPNHDR....
Consensus ...N...V.. ..E.EF.... .K........ .......... ..G.......
```

```
              301                                                    350
Aropescco .VSNLWDQGG FLPHA.... ......FTGL VMMMAIIMFS FG.GLELVGI
Phepescco .IDNLWRYGG FFATG.... ......WNGL ILSLAVIMFS FG.GLELIGI
Proysalty .IHNLWSNGG FFSNG.... ......WLGM IMSLQMVMFA YG.GIEIIGI
Cycaescco .FAHLWNDGG WFPKG.... ......LSGF FAGFQIAVFA FV.GIELVGT
Anspsalty .FHLITDNGG FFPHG.... ......LLPA LVLIQGVVFA FA.SIELVGT
Gabpbacsu GFSNLTGKGG FFPEG.... ......ISSV LLGIVVVIFS FM.GTEIVAI
Gabpescco GISRLWDSGG FMPNG.... ......FGAV LSAMLITMFS FM.GAEIVTI
Aropcorgl GTSNFIGDHG FMPNG.... ......ISGV AAGLLAVAFA FG.GIEIVTI
Roccbacsu FLSNFMTDRG LFPNG.... ......VLAV MFTLVMVNFS FQ.GTELVGI
Rocebacsu FLTHFYED.G LFPNG.... ......IKAM LITMITVNFA FQ.GTELIGV
Hutmbacsu MLSNFTDHGG LFPNG.... ......FLAV FIAMISVSFA FS.GTELIGV
Lyspescco W.SNWTIGEA PFAGG.... ......FAAM IGVAMIVGFS FQ.GTELIGI
Pap1sacce IGGKYWRDPG SFAEGSGATR ......FKGI CYILVSAYFS FG.GIELFVL
Val1sacce IGAEYWHNPG PFAHG.... ......FKGV CTVFCYAAFS YG.GIEVLLL
Gap1sacce IGGKYWHDPG AFAGDTPGAK ......FKGV CSVFVTAAFS FA.GSELVGL
Hip1sacce IGGKYWHDPG AFVGHSSGTQ ......FKGL CSVFVTAAFS YS.GIEMTAV
Tat2sacce IGAKYWHDPG CLANG.... ......FPGV LSVLVVASYS LG.GIEMTCL
Ina1triha WGARYWYDPG AFKNG.... ......FKGF CSVFVTAAFS FS.GTELVGL
Isp5schpo IGATI.FRKN AFIHG.... ......FHGF CSVFSTAAFS YA.GTEYIGI
Put4sacce LGFRYWQHPG AFAHHLTGGS LGN...FTDI YTGIIKGAFA FILGPELVCM
Putxemeni LGFRYWQDPG AFNPYLVPGD TGK...FLGF WTALIKSGFS FIFSPELITT
Alp1sacce IGFRYWRNPG AWGPGIISSN KNEGRFL.GW VSSLINAAFT YQ.GTELVGI
Can1sacce VGFRYWRNPG AWGPGIISKD KNEGRFL.GW VSSLINAAFT FQ.GTELVGI
Lyp1sacce IGFRYWRNPG AWGPGIISSD KSEGRFL.GW VSSLINAAFT YQ.GTELVGI
Can1canal VGFRYWRNGY AWGDGILV.. NNNGKYVAAF VSGLINSIFT FQ.GSELVAV
Dip5sacce LGFRYWRDPG AFKEYSTAIT GGKGKFV.SF VAVFVYSLFS YT.GIELTGI
Consensus .........G ....G..... ...............F. ...G.E....

              351                                                    400
Aropescco TAAEADNPEQ SIPKATNQVI YRILIFYIGS LAVLLSLMPW TRVTA.....
Phepescco TAAEARDPEK SIPKAVNQVV YRILLFYIGS LVVLLALYPW VEVKS.....
Proysalty TAGEAKDPEK SIPRAINSVP MRIWYF.... ..........  ..MSA.....
Cycaescco TAAETKDPEK SLPRAINSIP IRIIMFYVFA LIVIMSVTPW SSVVP.....
Anspsalty AAGECKDPQK MVPKAINSVI WRIGLFYVGS VVLLVLLLPW NAYQA.....
Gabpbacsu AAGETSNPIE SVTKATRSVV WRIIVFYVGS IAIVVALLPW ...NSA...N
Gabpescco AAAESDTPEK HIVRATNSVI WRISIFYLCS IFVVVALIPW ...NMP...G
Aropcorgl AAAESDKPRE AISLAVRAVI WRISVFYLGS VLVITFLMPY ESINGA...D
Roccbacsu AAGESESPEK TLPKSIRNVI WRTLFFFVLA MFVLVAILPY KTAGVI...E
Rocebacsu AAGESEDPEK TIPRSIKQTV WRTLVFFVLS IIVIAGMIPW KQAGVV...E
Hutmbacsu TAGESANPQK DIPRSIRNVA WRTVIFFIGA VFILSGLISW KDAGVI...E
Lyspescco AAGESEDPAK NIPRAVRQVF WRILLFYVFA ILIISLIIPY TDPSLL...R
Pap1sacce SINEQSNPRK STPVAAKRSV YRILIIYLLT MILIGFNVPH NNDQLMG...
Val1sacce SAAEQENPTK SIPNACKKVV YRILLIYMLT TILVCFLVPY NSDELLG..S
Gap1sacce AASESVEPRK SVPKAAKQVF WRITLFYILS LLMIGLLVPY NDKSLI...G
Hip1sacce SAAESKNPRE TIPKAAKRTF WLITASYVTI LTLIGCLVPS NDPRLLN..G
Tat2sacce ASGET..DPK GLPSAIKQVF WRILFFFLIS LTLVGFLVPY TNQNLL...G
Ina1triha AAAESTNPTK NMPGAIKQVF WRITIFYILG LFFVGLLINS DDPALLS..S
Isp5schpo AASETKNPAK AFPKAVKQVF IRVSLFYILA LFVVSLLISG RDERLTT..L
Put4sacce TSAECADQRR NIAKASRRFV WRLIFFYVLG TLAISVIVPY NDPTLVNALA
Putxemeni AAGEVEAPRR NIPKATKRFI YRVFTFYILG SLVIGVTVAY NDPTLEAGVE
Alp1sacce TAGEAANPRK ALPRAIKKVV VRILVFYILS LFFIGLLVPY NDPKL....D
Can1sacce TAGEAANPRK SVPRAIKKVV FRILTFYIGS LLFIGLLVPY NDPKL....T
Lyp1sacce TAGEAANPRK TVPRAINKVV FRIVLFYIMS LFFIGLLVPY NDSRL....S
```

```
Can1canal TAGEA..SPR ALRSAIRKVM FRILVFYVLC MLFMGLLVPY NDPKL....T
Dip5sacce VCSEAENPRK SVPKAIKLTV YRIIVFYLCT VFLLGMCVAY NDPRLLST.K
Consensus .A.E...P.. ..P....... .RI..FY... ........P. ..........
```

```
          401                                                 450
Aropescco ......DTSP FVLIFHELGD TFVANALNIV VLTAALSVYN S...CVYCNS
Phepescco ......NSSP FVMIFHNLDS NVVASALNFV ILVASLSVYN S...GVYSNS
Proysalty ......RCSS LCLSIRGIRS AQTAVHLC.. .......... ..........
Cycaescco ......EKSP FVELFVLVGL PAAASVINFV VLTSAASSAN S...GVFSTS
Anspsalty ......GQSP FVTFFSKLGV PYIGSIMNIV VLTAALSSLN S...GLYCTG
Gabpbacsu ....ILE.SP FVAVLEHIGV PAAAQIMNFI VLTAVLSCLN S...GLYTTS
Gabpescco ....LKAVGS YRSVLELLNI PHAKLIMDCV ILLSVTSCLN S...ALYTAS
Aropcorgl ....TAAESP FTQILAMANI PGTVGFMEAI IVLALLSAFN A...QIYATS
Roccbacsu .......SP FVAVLDQIGI PFSADIMNFV ILTAILSVAN S...GLYAAS
Rocebacsu .......SP FVAVFEQIGI PYAADIMNFV ILIALLSVAN S...GLYAST
Hutmbacsu .......SP FVAVFAEIGI PYAADIMNFV ILTALLSVAN S...GLYAST
Lyspescco NDVKDISVSP FTLVFQHAGL LSAAAVMNAV ILTAVLSAGN S...GMYAST
Pap1sacce SGGSATHASP YVLAASIHKV RVIPHIINAV ILISVISVAN S...ALYAAP
Val1sacce SDSSGSHASP FVIAVASHGV KVVPHFINAV ILISVISVAN S...SLYSGP
Gap1sacce ASSVDAAASP FVIAIKTHGI KGLPSVVNVV ILIAVLSVGN S...AIYACS
Hip1sacce SSSVDAASSP LVIAIENGGI KGLPSLMNAI ILIAVVSVAN S...AVYACS
Tat2sacce GSSVD..NSP FVIAIKLHHI KALPSIVNAV ILISVLSVGN S...CIFASS
Ina1triha AAYADSKASP FVLVGKYAGL KGFDHFMNLV ILASVLSIGV S...GVYGGS
Isp5schpo SATA...ASP FILALMDAKI RGLPHVLNAV ILISVLTAAN G...ITYTGS
Put4sacce QGKPGAGSSP FVIGIQNAGI KVLPHIINGC ILTSAWSAAN AFM...FAST
Putxemeni SGGSGAGASP FVVAIKTL.. .VLEGSTMSS MLPSGSLPGH PVTHGCYAGS
Alp1sacce SDGIFVSSSP FMISIENSGT KVLPDIFNAV VLITILSAGN S...NVYIGS
Can1sacce QSTSYVSTSP FIIAIENSGT KVLPHIFNAV ILTTIISAAN S...NIYVGS
Lyp1sacce ASSAVIASSP FVISIQNAGT YALPDIFNAV VLITVVSAAN S...NVYVGS
Can1canal QDGGFTRNSP FLIAMENSGT KVLPHIFNAV IVTTIISAGN S...NIYSGS
Dip5sacce GKSMSAAASP FVVAIQNSGI EVLPHIFNAC VLVFVFSACN S...DLYVSS
Consensus .......SP F......... ......N.. .L....S..N S.....Y...
```

```
          451                                                 500
Aropescco RMLFGLAQQG NAPKALASVD KRGVPVNTIL VSALVTALCV LINYLAPE..
Phepescco RMLFGLSVQG NAPKFLTRVS RRGVPINSLM LSGAITSLVV LINYLLPQ..
Proysalty .......... .......... .......... .......... ..........
Cycaescco RMLFGLAQEG VAPKAFAKLS KRAVPAKGLT FSCICLLGGV VMLYVNPSVI
Anspsalty RILRSMSMGG SAPKFMAKMS RQHVPYAGIL ATLVVYVVGV FLNYLVPS..
Gabpbacsu RMLYSLAERN EAPRRFMKLS KKGVPVQAIV AGTFFSYIAV VMNYFSPD..
Gabpescco RMLYSLSRRG DAPAVMGKIN RSKTPYVAVL LSTGAAFLTV VVNYYAPA..
Aropcorgl RLVFSMANRQ DAPRVFSKLS TSHVPTNAVL LSMFFAFVSV GLQYWNPA..
Roccbacsu RMMWSLSSNQ MGPSFLTRLT KKGVPMNALL ITLGISGCSL LTSVMAAE..
Rocebacsu RILYAMANEG QAFKALGKTN QRGVPMYSLI VTMAVACLSL LTKFAQAE..
Hutmbacsu RMMWSLANEN MISSRFKKVT SKGIPLNALM ISMAVSCLSL VSSIVAPG..
Lyspescco RMLYTLACDG KAPRIFAKLS RGGVPRNALY ATTVIAGLCF LTSMFGNQ..
Pap1sacce RLMCSLAQQG YAPKFLNYID REGRPLRALV VCSLVGVVGF VA..CSPQEE
Val1sacce RLLLSLAEQG VLPKCLAYVD RNGRPLLCFF VSLVFGCIGF VA..TSDAEE
Gap1sacce RTMVALAEQR FLPEIFSYVD RKGRPLVGIA VTSAFGLIAF VA..ASKKEG
Hip1sacce RCMVAMAHIG NLPKFLNRVD KRGRPMNAIL LTLFFGLLSF VA..ASDKQA
Tat2sacce RTLCSMAHQG LIPWWFGYID RAGRPLVGIM ANSLFGLLAF LV..KSGSMS
Ina1triha RTLTALAQQG YAPKLFTYID KSGRPLPSVI FLILFGFIAY VS..LDATGP
Isp5schpo RTLHSMAEQG HAPKWFKYVD REGRPLLAMA FVLCFGALGY IC..ESAQSD
Put4sacce RSLLTMAQTG QAPKCLGRIN KWGVPYVAVG VSFLCSCLAY LN..VSSSTA
```

```
Putxemeni EKLYSLAGEG QAPKIFTRTN RTGVPYVAVL ATWTIGLLSF LN..LSSSGQ
Alp1sacce RVLYSLSKNS LAPRFLSNVT RGGVPYFSVL STSVFGFLAF LE..VSAGSG
Can1sacce RILFGLSKNK LAPKFLSRTT KGGVPYIAVF VTAAFGALAY ME..TSTGGD
Lyp1sacce RVLYSLARTG NAPKQFGYVT RQGVPYLGVV CTAALGLLAF LV..VNNNAN
Can1canal RILYGLAQAG VAPKFFLRTN KGGVPFFAVA FTAAFGALGY LA..CSSQGN
Dip5sacce RNLYALAIDG KAPKIFAKTS RWGVPYNALI LSVLFCGLAY MN..VSSGSA
Consensus R......... ..P....... ..G.P..... .......... ..........
```

```
          501                                               550
Aropescco SAFGLLMALV VSALVINWAM ISLAHM.... .......... ..KFRRA.KQ
Phepescco KAFGLLMALV VATLLLNWIM ICLAHL.... .......... ..RFRAA.MR
Proysalty ......... .......... .......... .......... ..........
Cycaescco GAFTMITTVS AILFMFVWTI ILCSYL.... .......... ..VYRKQ.RP
Anspsalty RVFEIVLNFA SLGIIASWAF IMVCQM.... .......... ..RLRQAIKE
Gabpbacsu TVFLFLVNSS GAIALLVYLV IAVSQL.... .......... ..KMRKKLEK
Gabpescco KVFKFLIDSS GAIALLVYLV IAVSQL.... .......... ..RMRKIL.R
Aropcorgl GLLDFLLNAV GGCLIVVWAM ITLSQL.... .......... ..KLRKEL.Q
Roccbacsu TVYLWCISIS GMVTVVAWMS ICASQF.... .......... ..FFRRRFLA
Rocebacsu TVYMVLLSLA GMSAQVGWIT ISLSQL.... .......... ..MFRRKYIR
Hutmbacsu TVYVVMVAIA GFAGVVVWMS IALSQL.... .......... ..LFRKRFLK
Lyspescco TVYLWLLNTS GMTGFIAWLG IAISHY.... .......... ..RFRRGYVL
Pap1sacce QAFTWLAAIA GLSELFTWSG IMLSHI.... .......... ..RFRKAMKV
Val1sacce QVFTWLLAIS SLSQLFIWMS MSLSHI.... .......... ..RFRDAMAK
Gap1sacce EVFNWLLALS GLSSLFTWGG ICICHI.... .......... ..RFRKALAA
Hip1sacce EVFTWLSALS GLSTIFCWMA INLSHI.... .......... ..RFRQAMKV
Tat2sacce EVFNWLMAIA GLATCIVWLS INLSHI.... .......... ..RFRLAMKA
Ina1triha VVFDWLLAIS GLAALFTWGS VCLAHI.... .......... ..RFRKAWKY
Isp5schpo TVFDWLLSIS NLATLFVWLS INVSYI.... .......... ..IYRLAFKK
Put4sacce DVFNWFSNIS TISGFLGWMC GCIAYL.... .......... ..RFRKAIFY
Putxemeni TVFYWFTNIT TVGGFINWVL IGIAYLVCFP PSLHLNTPDQ KQRFRKALQF
Alp1sacce KAFNWLLNIT GVAGFFAWLL ISFSHI.... .......... ..RFMQAIRK
Can1sacce KVFEWLLNIT GVAGFFAWLF ISISHI.... .......... ..RFMQALKY
Lyp1sacce TAFNWLINIS TLAGLCAWLF ISLAHI.... .......... ..RFMQALKH
Can1canal KAFTWLLNIT ATAGLISWGF ISVSHI.... .......... ..RFMKTLQR
Dip5sacce KIFNYFVNVV SMFGILSWIT ILIVYI.... .......... ..YFDKACRA
Consensus ..F....... .......W.. I......... .......... ....R.....
```

```
          551                                               600
Aropescco EQGVVTRF.L ..LLYPLGNW ICLLFMAAVL VIMLMT.RGM GIWVYLIPVW
Phepescco RQGRETQFKA ..LLYPFGNY LCIAFLGMIL LLMCTM.DDM RLSAILLPVW
Proysalty ......... .......... .......... .......... ..........
Cycaescco HLHEKSIYKM ..PLGKLMCW VCMAFFVFVV VLLTLE.DDT RQALLVTPLW
Anspsalty GKAADVSFKL ..PGAPFTSW LTLLFLLSVL VLMAFDYPNG TYTIASLPLI
Gabpbacsu TNPEALKIKM W..LFPFLTY LTIIAICGIL VSMAFIDSMR DELLLTGVIT
Gabpescco AEGSEIRLRM W..LYPWLTW LVIGFITFVL VVMLFRPAQQ LEVISTGLLA
Aropcorgl ANDEISTVRM W..AHPWLGI LTLVLLAGLV ALMLGDAASR SQVYSVAIVY
Roccbacsu EGGNVNDLEF RTPLYPLVPI LGFCLYGCVL ISLIFIP... .....DQRIG
Rocebacsu EGGKIEDLKF KTPLYPVLPL IGLTLNTVVL ISLAFDP... .....EQRIA
Hutmbacsu KGGDVKDLTF RTPLYPLMPI AALLLCSASC IGLAFDP... .....NQRIA
Lyspescco QGHDINDLPY RSGFFPLGPI FAFILCLIIT LGQNYEAFLK DTIDWGGVAA
Pap1sacce QGRSLDEVGY KANTGIWGSY YGVFFNMLVF MAQFWVALSP IGN..GGKCD
Val1sacce QGRSMNEVGY KAQTGYWGSW LAVLIAIFFL VCQFWVAIAP VNE..HGKLN
Gap1sacce QGRGLDELSF KSPTGVWGSY WGLFMVIIMF IAQFYVAVFP VGDS.PS...
Hip1sacce QERSLDELPF ISQTGVKGSW YGFIVLFLVL IASFWTSLFP LGGSGAS...
```

```
Tat2sacce QGKSLDELEF VSAVGIWGSA YSALINCLIL IAQFYCSLWP IGGWTSGKER
Ina1triha HGHTLDEIPF KAAGGVYGSY LGLFICVIVL MAQFYTAIAA PPGS.PGVGT
Isp5schpo QGKSYDEVGY HSPFGIYGAC YGAFIIILVF ITEFYVSIFP IGAS.PD...
Put4sacce NGLY.DRLPF KTWGQPYTVW FSLIVIGIIT ITNGYAIFIP ......KYWR
Putxemeni HGML.DMLPF KTPLQPYGTY YVMFIISILT LTNGYAVFFP ......GRFT
Alp1sacce RGISRDDLPY KAQMMPFLAY YASFFIALIV LIQGFTAF.A ......PTFQ
Can1sacce RGISRDELPF KAKLMPGLAY YAATFMTIII IIQGFTAF.A .....PKFN
Lyp1sacce RGISRDDLPF KAKLMPYGAY YAAFFVTVII FIQGFQAF.C ......P.FK
Can1canal RGISRDTLPF KAFFMPFSAY YGMVVCFIVV LIQGFTVF.. .....WDFN
Dip5sacce QGIDKSKFAY VAPGQRYGAY FALFFCILIA LIKNFTVFLG ......HKFD
Consensus  .......... .......... .......... .......... ..........
```

```
           601                                              650
Aropescco LIVLGIGY.L FKEKTAKAVK AH........ .......... ..........
Phepescco IVFLFMAFKT LRRK...... .......... .......... ..........
Proysalty .......... .......... .......... .......... ..........
Cycaescco FIALGLGWLF IGKKRAAELR K......... .......... ..........
Anspsalty AILLVAGWFG VRRRVAEIHR TAPVTADSTE SVVLKEEAAT ..........
Gabpbacsu GIVLISYLVF RKRKVSEKAA ANPVTQQQPD ILP....... ..........
Gabpescco IGIICTVPIM ARWKKLVLWQ KTPVHNTR.. .......... ..........
Aropcorgl GFLVLLSFVT VNSPLRGGRT PSDLN..... .......... ..........
Roccbacsu LYCGVPIIIF CYAYYHLSIK KRINHETIEK KQTEAQ.... ..........
Rocebacsu LYCGVPFMII CYIIYHVVIK KRQQ....AN RQLEL..... ..........
Hutmbacsu LFCGVPCIIL CYLIYHF... KRNVTKAKKI SQEEYPADHI L.........
Lyspescco TYIGIPLFLI IWFGYKLIKG THFVRYSEMK FPQNDKK... ..........
Pap1sacce AQAFFESYLA APLWIFMYVG YMVYK..... .....RDFTF LNPLDKIDLD
Val1sacce VKVFFQNYLA MPIVLFAYFG HKIYF..... ....KSWSF WIPAEKIDLD
Gap1sacce AEGFFEAYLS FPLVMVMYIG HKIY...... ..KRNWKL FIPAEKMDID
Hip1sacce AESFFEGYLS FPILIVCYVG HKLY...... ..TRNWTL MVKLEDMDLD
Tat2sacce AKIFFQNYLC ALIMLFIFIV HKIYYKC... ..QTGKWWG VKALKDIDLE
Ina1triha AEDFFKQYLA APVVLGFWIV GWLW...... ....KRQP FLRTKNIDVD
Isp5schpo AGAFFQSYLC FPVVVIVFIA HALI...... ...TRQK FRKLSEIDLD
Put4sacce VADFIAAYIT LPIFLVLWFG HKLYT..... ...RTWRQW WLPVSEIDVT
Putxemeni ASDFLVSYIV FAIFLALYAG HKIWY..... ...RT..PW LTKVSEVDIF
Alp1sacce PIDFVAAYIS VFLFLAIWLS FQVWF..... ...KCRLLW K..LQDIDID
Can1sacce GVSFAAAYIS IFLFLAVWIL FQCIF..... ...RCRFIW K..IGDVDID
Lyp1sacce VSEFFTSYIS LILLAVMFIG CQIYY..... ...KCRFIW K..LEDIDID
Can1canal ASDFFTAYIS VILFVVLWVG FHFFFYGFGK DSFKMSNILV P..LDECDID
Dip5sacce YKTFITGYIG LPVYIISWAG YKLIYKTKVI KSTDVDLYTF KEIYDREEEE
Consensus  .......... .......... .......... .......... ..........
```

```
           651                                              700
Pap1sacce FHRRGLRP.. .......... .......... .......... ..........
Val1sacce SHRNIFVSPS LTEIDKVDDN DDLKEYENSE SSENPNSSRS RKFFKRMTNF
Gap1sacce TGRREVD... .......... ...LDLLKQE IAEEKAIMAT KPRWYRIWNF
Hip1sacce TGRKQVD... .......... ...LTLRREE MRIERETLAK RSFVTRFLHF
Tat2sacce TDRKDID... .......... ...IEIVKQE IAEKKMYLDS RPWYVRQFHF
Ina1triha TGLREFDWDE INAERTRIAP LPAWRRIIHH TF........ ..........
Isp5schpo TGFSKYDRLE ESDKGPMTAK SLAKSVLSFC V......... ..........
Put4sacce TGLVEIEEKS REIEEMRLPP TGFKDKFLDA LL........ ..........
Putxemeni TGKDEIDRLC E..NDMERQP RNWLERVWWW IF........ ..........
Alp1sacce SDHRQIEELV WIEPECKTR. ..W.QRVWDV LS........ ..........
Can1sacce SDRRDIEAIV WEDHEPKTF. ..W.DKFWNV VA........ ..........
Lyp1sacce SDRREIEAII WEDDEPKNL. ..W.EKFWAA VA........ ..........
```

```
Can1canal SGVRDINDAE FDIPPPKNA. ..W.DKFWAN VA..............
Dip5sacce GRMKDQEKEE RLKSNGKNME WFY.EKFLGN IF...........
Consensus ......... .......... .......... ..........
```

```
          701
Val1sacce WC
Gap1sacce WC
Hip1sacce WC
Tat2sacce WC
Consensus ..
```

Residues listed in the consensus sequence are present in at least 75% of the aligned transporter sequences.

Database accession numbers

	SWISSPROT	PIR	EMBL/GENBANK
Alp1sacce	P38971	S44329	X74069
Anspsalty	P40812		U04851
Aropescco	P15993	S45191; JS0447	X17333; D26562
Aropcorgl		S52754	X85965
Can1sacce	P04817	A23922	X03784; M11724
Can1canal	P43059		X76689
Cycaescco	P39312		U14003
Dip5sacce			X95802
Gabpbacsu	P46349		U31756
Gabpescco	P25527		M88334; X65104
Gap1sacce	P19145	S38111	X52633; Z28264
Hip1sacce	P06775	A24519	M11980; X82408
Hutmbacsu	P42087		D31856; X82174
Ina1triha	P34054	S33212	Z22594
Isp5schpo	P40901	S35896; S45492	D14062
Lyp1sacce	P32487	S34931	X67315
Lyspescco	P25737	S24560	M89774 ; U00007
Pap1sacce	P41815		X75076
Phepescco	P24207	A39431	M58000
Proysalty	P37460	S35983	X74420
Put4sacce	P15380	JQ0127	M30583
Putxemeni	P18696	S04547	
Roccbacsu	P39636	S39733	X73124
Rocebacsu	P39137		X81802
Tat2sacce	P38967	S48084; S47926	X79150; L33461
Val1sacce	P38085	S45932; S48083	U10503; X79151

References

[1] Malandro, M. and Kilberg, M. (1996) Annu. Rev. Biochem. 65, 305–336.
[2] Pi, J. and Pittard, J. (1996) J. Bacteriol. 178, 2650–2655.

Summary

Transporters of the H⁺/lactose-sucrose-nucleoside symporter family, the example of which is the LACY lactose-H⁺ symporter of *Escherichia coli*, mediate symport (H⁺-coupled substrate uptake) of structurally dissimilar sugars and nucleosides [1-4]. Known members of the family are found only in gram-negative bacteria.

Statistical analysis of multiple amino acid sequence comparisons suggests that the H⁺/lactose-sucrose-nucleoside symporter family is distantly related to the uniporter-symporter-antiporter (USA) superfamily, also known as the major facilitator superfamily (MFS) [2,5,6]. They are predicted to contain 12 membrane-spanning helices by the hydropathy of their amino acid sequences [6] and the activities of reporter gene fusions [7]. Based on complementation studies, LACY is predicted to exist as a homodimer [8].

Several amino acid sequence motifs are highly conserved in the H⁺/lactose-sucrose-nucleoside symporter family, including motifs unique to the family, elements of the signature motifs of the USA/MSF superfamily, and motifs necessary for activity by the criterion of site directed mutagenesis [1-4].

Nomenclature, biological sources and substrates

CODE	DESCRIPTION [SYNONYMS]	ORGANISM [COMMON NAMES]	SUBSTRATE(S)
Cscbescco	Sucrose-H⁺ symporter [CSCB]	*Escherichia coli* [gram-negative bacterium]	H⁺/sucrose
Lacycitfr	Lactose-H⁺ symporter [LACY]	*Citrobacter freundii* [gram-negative bacterium]	H⁺/lactose
Lacyescco	Lactose-H⁺ symporter [LACY]	*Escherichia coli* [gram-negative bacterium]	H⁺/lactose
Lacyklepn	Lactose-H⁺ symporter [LACY]	*Klebsiella pneumoniae* [gram-negative bacterium]	H⁺/lactose
Nupgescco	Nucleoside-H⁺ symporter [NUGP]	*Escherichia coli* [gram-negative bacterium]	H⁺/nucleosides
Rafbescco	Raffinose-H⁺ symporter [RAFB]	*Escherichia coli* [gram-negative bacterium]	H⁺/raffinose
Xapbescco	Xanthosine permease [XAPB]	*Escherichia coli* [gram-negative bacterium]	H⁺/xanthosine

Cotransported ions are listed.

Phylogenetic tree

Proteins listed subsequently in italics are at least 90% identical to the paired transporters listed in parenthesis and are therefore not included in the phylogenetic tree: *Lacycitfr* (Lacyescco).

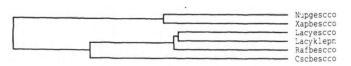

```
                                            Nupgescco
                                            Xapbescco
                                            Lacyescco
                                            Lacyklepn
                                            Rafbescco
                                            Cscbescco
```

Proposed orientation of LACY in the membrane

The model is based on predictions of membrane-spanning regions and α-helical content. The N-terminus of the protein is illustrated on the inside and is folded 12 times through the membrane [2]. The predicted membrane-spanning helices are portrayed as rectangles. The numbers corresponding to the first and last residue of each membrane-spanning helix are boxed. Residues that are conserved in more than 75% of the aligned transporters (see below) are shown. Consensus residues indicated by an asterisk are not conserved in LACY.

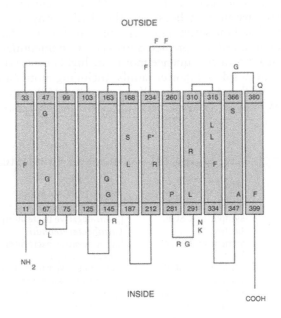

Physical and genetic characteristics

	AMINO ACIDS	MOL. WT	K_m	CHROMOSOMAL LOCUS
Cscbescco	415	46 923		
Lacycitfr	416	46 537		
Lacyescco	417	46 503	Lactose: 1.7 mM [2]	7.77 minutes
Lacyklepn	416	46 220		
Nupgescco	418	46 389		
Rafbescco	425	46 693	Raffinose: 1 mM [9]	
Xapbescco	418	45 736		

Multiple amino acid sequence alignments

```
          1                                                 50
Nupgescco .......... MNLKLQLKIL SFLQFCLWGS WLTTLGSYMF VTLKFDGASI
Xapbescco .......... MSIAMRLKVM SFLQYFIWGS WLVTLGSYMI NTLHFTGANV
Lacyescco .....MYYLK NTNFWMFGLF FFFYFFIMGA YFPFFPIWLH DINHISKSDT
Lacyklepn MKLSELAPRE RHNFIYFMLF FFFYYFIMSA YFPFFPVWLA EVNHLTKTET
```

```
Rafbescco ..MNSASTHK NTDFWIFGLF FFLYFFIMAT CFPFLPVWLS DVVGLSKTDT
Cscbescco ..MALNIPFR NAYYRFASSY SFLFFISWSL WWSLYAIWLK GHLGLTGTEL
Consensus .......... .......... .......F.. .......... ..........

          51                                                 100
Nupgescco GAVYSSLGIA AVFMPALLGI VADKWLSAKW ..VYAICHTI G..AITLFMA
Xapbescco GMVYSSKGIA AIIMPGIMGI IAVQ.MRARR TCIHAVSPGV C..GV.LFYA
Lacyescco GIIFAAISLF SLLFQPLFGL LSDK.LGLRK YLLWIITGML VMFAPFFIFI
Lacyklepn GIVFSCISLF AIIFQPVFGL ISDK.LGLRK HLLWTITILL ILFAPFFIFV
Rafbescco GIVFSCLSLF AISFQPLLGV ISDR.LGLKK NLIWSISLLL VFFAPFFLYV
Cscbescco GTLYSVNQFT SILFMMFYGI VQDK.LGLKK PLIWCMSFIL VLTGPFMIYV
Consensus G......... .........G ...D..L... .......... ..........

          101                                                150
Nupgescco AQVTTPEAMF LVILINSFAY MPTLGLINTI SYYRLQNAGM DIVTDFPPIR
Xapbescco ASVTDPDMMF WVMLVNAMAF MPTIALSNSV SYSCLAQAGL DPVTAFPPIR
Lacyescco FGPLLQYNIL VGSIVGGIYL GFCFNAGAPA VEAFIEKVSR RSNFEFGRAR
Lacyklepn FSPLLQMNIM AGALVGGVYL GIVFSSRSGA VEAYIERVSR ANRFEYGKVR
Rafbescco FAPLLHLNIW AGALTGGVFI GFVFSAGAGA IEAYIERVSR SSGFEYGKAR
Cscbescco YEPLLQSNFS VGLILGALFF GLGYLAGCGL LDSFTEKMAR NFHFEYGTAR
Consensus .......... .......... .......... .......... .........R

          151                                                200
Nupgescco IWGTIGFIMA MWVVSLSGFE LSHMQLYIGA ALSAIL.VLF TLTLPHIPVA
Xapbescco VFGTVGFIVA MWAVSLLHLE LSSLQLYIAS GASLLL.SAY ALTLPKIPVA
Lacyescco MFGCVGWALC ASIVGIMFTI NNQFVFWLGS GCALILAVLL FFAKTDAPSS
Lacyklepn VSGCVGWALC ASITGILFSI DPNITFWIAS GFALILGVLL WVSKPESSNS
Rafbescco MFGCLGWALC ATMAGILFNV DPSLVFWMGS GGALLLLLLL YLARPSTSQT
Cscbescco AWGSFGYAIG AFFAGIFFSI SPHINFWLVS LFGAVF.MMI NMRFKDKDHQ
Consensus ..G..G.... .......... .........S .....L.... ..........

          201                                                250
Nupgescco KQQANQSWTT LLGLDAFALF KNKRMAIFFI FSMLLGAELQ ITNMFGNTFL
Xapbescco EKKATTSLAS KLGLDAFVLF KNPRMAIFFL FAMMLGAVLQ ITNVFGNPFL
Lacyescco ATVANAVGAN HSAFSLKLAL ELFRQPKLWF LSLYVIGVSC TYDVFDQQF.
Lacyklepn AEVIDALGAN RQAFSMRTAA ELFRMPRFWG FIIYVVGVAS VYDVFDQQF.
Rafbescco AMVMNALGAN SSLISTRMVF SLFRMRQMWM FVLYTIGVAC VYDVFDQQF.
Cscbescco CIAADAGGVK KEDF.....I AVFKDRNFWV FVIFIVGTWS FYNIFDQQF.
Consensus .......... .......... ...R...... F......... ....F...F.

          251                                                300
Nupgescco HSFDKDPMFA SSFIVQHASI IMSISQISET LFILTIPFFL SRYGIKNVMM
Xapbescco HDFARNPEFA DSFVVKYPSI LLSVSQMAEV GFILTIPFFL KRFGIKTVML
Lacyescco ANFFTSFFAT GEQGTRVFGY VTTMGELLNA SIMFFAPLII NRIGGKNALL
Lacyklepn ANFFKGFFSS PQRGTEVFGF VTTGGELLNA LIMFCAPAII NRIGAKNALL
Rafbescco AIFFRSFFDT PQAGIKAFGF ATTAGEICNA IIMFCTPWII NRIGAKNTLL
Cscbescco PVFYAGLFES HDVGTRLYGY LNSFQVVLEA LCMAIIPFFV NRVGPKNALL
Consensus ..F....... .......... .......... ...P.... .R.G.KN..L

          301                                                350
Nupgescco ISIVAWILRF ALFAYGDPTP FGTVLLVLSM IVYGCAFDFF NISGSVFVEK
Xapbescco MSMVAWTCAL ASSPYGDPST TGFILLLLSM IVYGCAFDFF NISGSVFVEQ
Lacyescco LAGTIMSVRI IGSSFATSAL EVVILKTLHM F....EVPFL LVGCFKYITS
Lacyklepn IAGLIMSVRI LGSSFATSAV EVIILKMLHM F....EIPFL LVGTFKYISS
```

```
Rafbescco VAGGIMTIRI TGSAFATTMT EVVILKMLHA L....EVPFL LVGAFKYITG
Cscbescco IGVVIMALRI LSCALFVNPW IISLVKLLHA I....EVPLC VISVFKYSVA
Consensus .......R.. .......... ...L..L.. .......F.. ..........

            351                                              400
Nupgescco EVSPAIRASA QGMFLMMTNG FGCILGGIVS GKVVEMYTQN GITDWQTVWL
Xapbescco EVDSSIRASA QGLFMTMVNG VGAWVGSILS GMAVDYFSVD GVKDWQTIWL
Lacyescco QFEVRFSATI YLVCFCFFKQ LAMIFMSVLA GNMYE..... .SIGFQGAYL
Lacyklepn AFKGKLSATL FLIGFNLSKQ LSSVVLSAWV GRMYD..... .TVGFHQAYL
Rafbescco VFDTRLSATV YLIGFQFSKQ LAAILLSTFA GHLYD..... .RMGFQNTYF
Cscbescco NFDKRLSSTI FLIGFQIASS LGIVLLSTPT GILFD..... .HAGYQTVFF
Consensus .......A.. .......... .....S... G......... .....Q....

            401                                     439
Nupgescco IFAGYSVVLA FAFMAMFKYK HVRVPTGTQT VSH......
Xapbescco VFAGYALFLA VIFFFGFKYN HDPEKIKHRA VTH......
Lacyescco VLGLVALGFT LISVFTLSGP GPLSLLRRQV NEVA.....
Lacyklepn ILGCITLSFT VISLFTLKGS KTLLPATA.. .........
Rafbescco VLGMIVLTVT VISAFTLSSS PGIVHPSVEK APVAHSEIN
Cscbescco AISGIVCLML LFGIFFLSKK REQIVMETPV PSAI.....
Consensus .......... ...F..... .......... .........
```

Proteins listed subsequently in italics are at least 90% identical to the paired transporters listed in parenthesis and are therefore not included in the alignments: *Lacycitfr* (Lacyescco). Residues listed in the consensus sequence are present in at least 75% of the aligned transporter sequences. Residues indicated by boldface type are also conserved in at least one other family of the USA/MFS superfamily.

Database accession numbers

	SWISSPROT	*PIR*	*EMBL/GENBANK*
Cscbescco	P30000	S19880	X63740; X81461
Lacycitfr	P47234		U13675
Lacyescco	P02920	A03418; S21314	J01636; V00295
Lacyklepn	P18817	JT0487; C24925	M11441; X14154
Nupgescco	P09452	A26226	X06174
Rafbescco	P16552	B43717	M27273
Xapbescco	P45562		X73828

References
[1] Kaback, H.R. et al. (1996) J. Bioenerg. Biomembr. 28, 29–34.
[2] Varela, M. and Wilson, T.H. (1996) Biochim. Biophys. Acta 1276, 21–34.
[3] Kaback, H.R. (1992) Biochim. Biophys. Acta 1101, 210–213.
[4] Kaback, H.R. (1992) Int. Rev. Cytol. 137, 97–125.
[5] Marger, M.D. and Saier, M. (1993) Trends Biochem. Sci. 18, 13–20.
[6] Henderson, P.J.F. (1993) Curr. Opin. Cell Biol. 5, 708–721.
[7] Calamia, J. and Manoil, C. (1990) Proc. Natl Acad. Sci. USA 87, 4937–4941.
[8] Bibi, E. and Kaback, H.R. (1990) Proc. Natl Acad. Sci. USA 87, 4325–4329.
[9] Henderson, P.J.F. (1992) Int. Rev. Cytol. 137, 149–208.

Summary

Transporters of the H⁺/galactoside-pentose-hexuronide symporter family, the example of which is the MELB melibiose-H⁺ symporter of *Escherichia coli* (Melbescco), mediate symport (cation-coupled substrate uptake) of structurally dissimilar sugars and glucuronides. The favored cation for cotransport can be either H⁺, Na⁺ or Li⁺. The transport activity of the H⁺/galactoside-pentose-hexuronide symporter family is regulated at the level of enzyme activity by the enol pyruvate:sugar transferase system, either by direct interaction of C-terminal elements of the transporter with unphosphorylated phosphoryl transfer protein, IIA (e.g. MELB) or by the phosphorylation of an extra IIA-like domain in the C-terminus of the transporter by PEP, heat stable protein HPR and enzyme 1 (e.g. LACS)[1]. Members of the family are found in both gram-positive and gram-negative bacteria.

Statistical analysis reveals no apparent relationship between the amino acid sequences of the H⁺/galactoside-pentose-hexuronide symporter family and any other family of transporters. However, the LACS and RAFP proteins contain a C-terminal hydrophilic extension of approximately 160 residues that is homologous to a domain in the PTGA protein of the PTS family[1]. Members of the H⁺/galactoside-pentose-hexuronide symporter family are predicted to contain 12 membrane-spanning helices by the hydropathy of their amino acid sequences[1] and by the activities of reporter gene fusions[2].

Several amino acid sequence motifs are highly conserved in the H⁺/galactoside-pentose-hexuronide symporter family, including motifs necessary for function by the criterion of site-directed mutagenesis[1,3]. Cation specificity can be altered by single amino acid substitutions[1].

Nomenclature, biological sources and substrates

CODE	DESCRIPTION [SYNONYMS]	ORGANISM [COMMON NAMES]	SUBSTRATE(S)
Lacsstrtr	Lactose-H⁺ symporter lactose permease [LACY, LACS]	*Streptococcus thermophilus* [gram-positive bacterium]	H⁺/lactose
Lacslacde	Lactose-H⁺ symporter lactose permease [LACY, LACS]	*Lactobacillus delbrueckii* [gram-positive bacterium]	H⁺/lactose
Melbescco	Melibiose-Li⁺/ Na⁺ symporter melibiose permease [MELB, MEL4]	*Escherichia coli* [gram-negative bacterium]	H⁺, Na⁺, Li⁺/ melibiose
Melbsalty	Melibiose-Li⁺/ Na⁺ symporter melibiose permease [MELB]	*Salmonella typhimurium* [gram-negative bacterium]	H⁺, Na⁺, Li⁺/ melibiose
Melbklepn	Melibiose-Li⁺/ Na⁺ symporter melibiose permease [MELB]	*Klebsiella pneumoniae* [gram-negative bacterium]	H⁺, Li⁺/melibiose
• Rafppedpe	Raffinose-H⁺ symporter [RAFP]	*Pediococcus pentosaceus* [gram-positive bacterium]	H⁺/raffinose
Uidbescco	Glucuronide-H⁺ symporter [UIDB, GUSB, UIDP]	*Escherichia coli* [gram-negative bacterium]	H⁺/glucuronides

Contransported ions are listed.

Phylogenetic tree

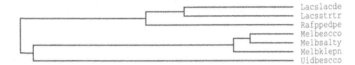

Lacslacde
Lacsstrtr
Rafppedpe
Melbescco
Melbsalty
Melbklepn
Uidbescco

Proposed orientation of MELB in the membrane

The model is based on predictions of membrane-spanning regions and α-helical content. The N-terminus of the protein is illustrated on the inside and is folded 12 times through the membrane[2]. The predicted membrane-spanning helices are portrayed as rectangles. The numbers corresponding to the first and last residue of each membrane-spanning helix are boxed. Residues that are conserved in more than 75% of the aligned transporters (see below) are shown.

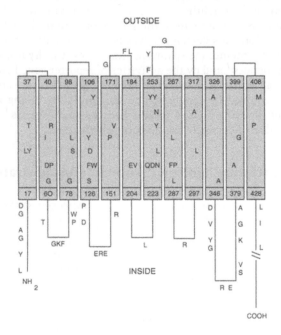

Physical and genetic characteristics

	AMINO ACIDS	MOL. WT	K_m	CHROMOSOMAL LOCUS
Lacsstrtr	634	69 453		
Lacslacde	627	68 288		
Melbescco	469	52 217	Melibiose: 300 μM [4]	93.51 minutes
Melbsalty	476	52 758		
Melbklepn	471	52 329		
Rafppedpe	641	69 913		
Uidbescco	457	49 892	Glucuronate: 132 μM [4]	36.42 minutes

Multiple amino acid sequence alignments

```
            1                                                  50
Lacslacde   .....MKKKL VSRLSYAAGA FGNDVFYATL STYFIVFVTT HLFNAGDHKM
Lacsstrtr   ..MEKSKGQM KSRLSYAAGA FGNDVFYATL STYFIMFVTT HLFNTGDPKQ
Rafppedpe   MQEEHNYKWV GGRLIYGFGA KGNDAFYSIL SGYLIIFITS HLFDTGNKAL
Melbescco   .........M TTKLSYGFGA FGKDFAIGIV YMYLMYYYT. ........DV
Melbsalty   .....MSISL TTKLSYGFGA FGKDFAIGIV YMYLMYYYT. ........DV
Melbklepn   .....MSISM TTKLSYGFGA FGKDFAIGIV YMYLMYYYT. ........DI
Uidbescco   ...MNQQLSW RTIVGYSLGD VANNFAFAMG ALFLLSYYT. ........DV
Consensus   .......... ...L.Y..GA .G.D...... ..YL....T. ..........

            51                                                 100
Lacslacde   ....IFIITN LITAIRIGEV LLDPLIGNAI DRTESRWGKF KPWVVGGGII
Lacsstrtr   NSHYVLLITN IISILRILEV FIDPLIGNMI DNTNTKYGKF KPWVVGGGII
Rafppedpe   DNRMVSLVTL IIMVLRIVEL FIDPFIGNAI DRTKNSPGHF RPWVVVGGTV
Melbescco   VGLSVGLVGT LFLVARIWDA INDPIMGWIV NATRSRWGKF KPWILIGTLA
Melbsalty   VGLSVGLVGT LFLVARIWDA INDPIMGWIV NATRSRWGKF KPWILIGTLT
Melbklepn   VGLSVGVVGT LFLVARILDA IADPIMGWIV NCTRSRWGKF KPWILIGTIT
Uidbescco   AGVGAAAAGT MLLLVRVFDA FADVFAGRVV DSVNTRWGKF RPFLLFGTAP
Consensus   .......... .....RI.... .DP..G.... .T....GKF .PW...G...

            101                                                150
Lacslacde   SSLALLALFT DFGGINQSKP VVYLVIFGIV YLIMDIFYSF KDTGFWAMIP
Lacsstrtr   SSITLLLLFT DLGGLNKTNP FLYLVLFGII YLVMDVFYSI KDIGFWSMIP
Rafppedpe   SSIILLLLFT NLGGLYAKNA MIYLVVFAIL YITMDIFYSF KDVGFWSMLP
Melbescco   NSVI.LFLLF SAHLFEGTTQ IVFVCV...T YILWGMTYTI MDIPFWSLVP
Melbsalty   NSLV.LFLLF SAHLFEGTAQ VVFGCV...T YILWGMTYTI MDIPFWSLVP
Melbklepn   NSVV.LYMLF SAHHFSGGAL LAWVWL...T YLLWGFTYTI MDVPFWSLVP
Uidbescco   LMIFSVLVFW VPTDWSHGSK VVYAYL...T YMGLGLCYSL VNIPYGSLAT
Consensus   .S...L.... .......... ........Y. .....Y.. .D..FWS..P

            151                                                200
Lacslacde   ALSLDSRERE KTSTFARVGS TIGANLVGVV ITPIILFFSA SKANPNGDKQ
Lacsstrtr   ALSLDSHERE KMATFARIGS TIGANIVGVA IMPIVLFFSM TNNSGSGDKS
Rafppedpe   SLTTDSRERE KTATFARLGS TIGGGLVGVL VMPAVIFFSA .KATSTGDNR
Melbescco   TITLDKRERE QLVPYPRFFA SLAGFVTAGV TLPFVNYVG. ....GGDRGF
Melbsalty   TITLDKRERE QLVPFPRFFA SLAGFVTAGI TLPFVSYVG. ....GADRGF
Melbklepn   TITLDKRERE QLVPYPRFFA SLAGFVTAGV TLPFVNAVG. ....GADRGF
Uidbescco   AMTQQPQSRA RLGAARGIAA SLTFVCLAFL IGPSIKNSS. ....PEEMVS
Consensus   ....D..ERE ......R... .......... ..P.V..... ..........

            201                                                250
Lacslacde   GWFFFALIVA IVGILTSITV GLGTHEV... KSALRESNEK T.TLKQVFKV
Lacsstrtr   GWFWFAFIVA LIGVITSIAV GIGTREV... ESKIRDNNEK T.SLKQVFKV
Rafppedpe   GWFIFALIIC LIALISAWGV GLGTREV... DSDIRKNKQD TVGVMEIFKA
Melbescco   GFQMFTLVLI AFFIVSTIIT LRNVHEVFS. SDNQPSAGS. HLTLKAIVAL
Melbsalty   GFQMFTLVLI AFFIASTIVT LRNVHEVYS. SDNGVTAGRP HLTLKTIVGL
Melbklepn   GFQMFTLVLI AFFVVSTLVT LRNVHEVYS. SDSGVSEDSS HLSLRQMVAL
Uidbescco   VYHFWTIVLA IAGMVLYFIC FKSTRENVVR IVAQPSLNIS LQTLKRNRPL
Consensus   G...F.L... .......... ...EV..... .......... ...L......

            251                                                300
Lacslacde   LGQNDQLLWL AFAYWFYGLG INTLNALQLY YFSYILGDAR GYSLLYTINT
Lacsstrtr   LGQNDQLMWL SLGYWFYGLG INTLNALQLY YFTFILGDSG KYSILYGLNT
```

```
Rafppedpe LAKNDQLLWA ALAYLFYGVG INILGSLEVY YFTYIMGKPK SFSILSIINI
Melbescco IYKNDQLSCL LGMALAYNVA SNIITGFAIY YFSYVIGDAD LFPYYLSYAG
Melbsalty IYKNDQLSCL LGMALAYNIA SNIINGFAIY YFTYVIGDAD LFPYYLSYAG
Melbklepn IYKNDQLACL LGMALAYNTA ANIIAGFAIY YFTYVIGSAE MFPYYMSYAG
Uidbescco FMLCIGALCV LISTFA.... ...VSASSLF YVRYVLNDTG LFTVLVL...
Consensus ...NDQL... ......Y... .N.......Y YF.Y..G... ..........
```

```
                  301                                           350
Lacslacde FVGLISASFF PSLAKKFNRN RLFYACIAV. .MLLGIGVFS VASG...SLA
Lacsstrtr VVGLVSVSLF PTLADKFNRK RLFYGCIAV. .MLGGIGIFS IAGT...SLP
Rafppedpe FLGLIATSLF PVLSKKFSRK GVFAGCLVF. .MLGGIAIFT IAGS...NLW
Melbescco AANLVTLVFF PRLVKSLSRR ILWAGASIL. PVLSCGVLLL MALMSYHNVV
Melbsalty AANLLTLIVF PRLVKMLSRR ILWAGASVM. PVLSCAGLFA MALADIHNAA
Melbklepn AANLLTLILF PRLVKGLSRR ILWAGASIM. PVLGCGVLLL MALGGVYNIA
Uidbescco VQNLVGTVAS APLVPGMVAR IGKKNTFLIG ALLGTCGYLL FFWVSVWSLP
Consensus ...L.....F P.L.....R. .......... ..L....... .A........
```

```
                  351                                           400
Lacslacde LSLVGAEFFF IPQPLAFLVV LMIISDAVEY GQLKTGHRDE ALTLSVRPLV
Lacsstrtr IILTAAELFF IPQPLVFLVV FMIISDSVEY GQWKTGHRDE SLTLSVRPLI
Rafppedpe LVLLAATMFG FPQQMVFLVV LMVITDSVEY GQLKLGHRDE SLALSVRPLI
Melbescco LIVIAGILLN VGTALFWVLQ VIMVADIVDY GEYKLHVRCE SIAYSVQTMV
Melbsalty LIVAAGIFLN IGTALFWVLQ VIMVADTVDY GEFKLNIRCE SIAYSVQTMV
Melbklepn LISLAGVLLN IGTALFWVLQ VIMVADTVDY GEYTMNIRCE SIAYSVQTLV
Uidbescco VALVALAIAS IGQGVTMTVM WALEADTVEY GEYLTGVRIE GLTYSLFSFT
Consensus ....A..... .......... ....AD.V.Y G......R.E ....SV....
```

```
                  401                                           450
Lacslacde DKLGGALSNW FVSLIALTAG MTTGATASTI TAHGQMVFKL AMFALPAVML
Lacsstrtr DKLGGAMSNW LVSTFAVAAG MTTGASASTI TTHQQFIFKL GMFAFPAATM
Rafppedpe DKFGGAISNG VVGQIAIISG MTTGATASSI TAAGQLHFKL TMFAFPALML
Melbescco VKGGSAFAAF FI...AVVLG MIGYVPNVEQ STQALLGMQF IMIALPTLFF
Melbsalty VKGGSAFAAF FI...ALVLG LIGYTPNVAQ SAQTLQGMQF IMIVLPVLFF
Melbklepn VKAGSAFAAW FI...AIVLG IIGYVPNVVQ SSHTLLGMQA IMIALPTLFF
Uidbescco RKCGQAIGGS IP...AFILG LSGYIANQVQ TPEVIMGIRT SIALVPCGFM
Consensus .K.G.A.... .....A...G .......... .......... .M...P....
```

```
                  451                                           500
Lacslacde LIAVSIFAKK VFLTEEKHAE IVDQLE.... TQFGQSHAQK PAQ.AESFTL
Lacsstrtr LIGAFIVARK ITLTEARHAK IVEELE.... HRFSVATSEN EVK.ANVVSL
Rafppedpe LIAIGIFSKQ IFLTEEKHAE IVAELERTWR TKFDNTTDQV AEKVVTSLDL
Melbescco MVTLILYFRF YRLNGDTLRR IQIHLLDKYR KVPP...EPV HADIPVGAVS
Melbsalty MMTLVLYFRY YRLNGDMLRK IQIHLLDKYR KTPPFVEQPD SPAISVVATS
Melbklepn ALTLFLYFRY YKLNGDMLRR IQIHLLDKYR RVPENDVEPE RPIVVPNQV.
Uidbescco LLAFVI.IWF YPLTDKKFKE IVVEIDNRKK VQQQLISDIT N.........
Consensus .......... ..L....... I...L..... ..........
```

```
                  501                                           550
Lacslacde ASPVSGQLMN LDMVDDPVFA DKKLGDGFAL VPADGKVYAP FAGTVRQLAK
Lacsstrtr VTPTTGYLVD LSSVNDEHFA SGSMGKGFAI KPTDGAVFAP ISGTIRQILP
Rafppedpe ATPIAGQVIP LAQVNDPTFA AGTLGDGFAI KPSDGRILAP FDATVRQVFT
Melbescco DVKA...... .......... .......... .......... ..........
Melbsalty DVKA...... .......... .......... .......... ..........
Consensus .......... .......... .......... .......... ..........
```

```
           551                                              600
Lacslacde TRHSIVLENE HGVLVLIHLG LGTAKLNGTG FVSYVEEGSQ VEAGQQILEF
Lacsstrtr TRHAVGIESE DGVIVLIHVG IGTVKLNGEG FISYVEQGDR VEVGQKLLEF
Rafppedpe TRHAVGLVGD NGIVLLIHIG LGTVKLRGTG FISYVEEGQH VQQGDELLEF
Consensus .........  .........  .........  .........  .........

           601                                              650
Lacslacde WDPAIKQAKL DDTVIVTVIN SETFANSQML LPIGHSVQAL DDVFKLEGKN
Lacsstrtr WSPIIEKNGL DDTVLVTVTN SEKFSAFHLE QKVGEKVEAL SEVITFKKGE
Rafppedpe WDPTIKQAGL DDTVIMTVTN STEFTMMDWL VKPGQAVKAT DNILQLHTKA
Consensus .........  .........  .........  .........  .........
```

Residues listed in the consensus sequence are present in at least 75% of the aligned transporter sequences.

Database accession numbers

	SWISSPROT	PIR	EMBL/GENBANK
Lacsstrtr	P23936	A32241	M23009; M38175
Lacslacde	P22733	A38538	M55068
Melbescco	P02921	A03421	K01991; U14003
Melbsalty	P30878	S23576	X62101
Melbklepn	Q02581	B44166	M97257
Rafppedpe	P43466		Z32771; L32093
Uidbescco	P30868		M14641

References
[1] Poolman, B. et al. (1996) Mol. Microbiol. 19, 911–922.
[2] Pourcher, T. et al. (1996) Biochemistry 35, 4161–4168.
[3] Reizer, J. et al. (1994) Biochim. Biophys. Acta 1197, 133–166.
[4] Henderson, P.J.F. et al. (1992) Int. Rev. Cytol. 137, 149–208.

Summary

Transporters of the H⁺/oligopeptide symporter family, the example of which is the PET1 oligopeptide-H⁺ symporter of humans (Pet1homsa), mediate the uptake of oligopeptides, chlorate and nitrate. The mechanism of transport, where known, is symport (H⁺-coupled substrate uptake). In plants, mutations in the CHL1 nitrate, chlorate-H⁺ symporter confer resistance to the herbicide chlorate. Members of the family have a broad biological distribution that includes bacteria, plants and humans.

Statistical analysis reveals no apparent relationship between the amino acid sequences of the H⁺/oligopeptide symporter family and any other family of transporters. Members of the H⁺/oligopeptide symporter family are predicted to contain 11, 12 or 13 membrane-spanning helices by the hydropathy of their amino acid sequences. Eukaryotic transporters of the H⁺/oligopeptide symporter family are glycosylated.

Several amino acid sequence motifs are highly conserved in the H⁺/oligopeptide symporter family and are necessary for function by the criterion of site-directed mutagenesis.

Nomenclature, biological sources and substrates

CODE	DESCRIPTION [SYNONYMS]	ORGANISM [COMMON NAMES]	SUBSTRATE(S)
Chl1arath	Nitrate, chlorate-H⁺ symporter [CHL1]	*Arabidopsis thaliana* [mouse-ear cress]	H⁺/nitrate, chlorate
Dtptlacla	Dipeptide-H⁺ symporter [DTPT]	*Lactobacillus lactis* [gram-positive bacterium]	H⁺/2–4 residue oligopeptides
Pt2aarath	Peptide transporter [PT2a, PTR2, PTR2a]	*Arabidopsis thaliana* [mouse-ear cress]	Peptides
Pt2barath	High-affinity peptide transporter [PT2a, PTR2b, NTR1]	*Arabidopsis thaliana* [mouse-ear cress]	Peptides, histidine
Ptr2sacce	Peptide transporter [PTR2, YKR093w, YKR413]	*Saccharomyces cerevisiae* [yeast]	Oligopeptides
Ptr2canal	Peptide transporter [PTR2]	*Candida albicans* [yeast]	Oligopeptides
Pet1ratno	Oligopeptide-H⁺ symporter 1 [PET1, PEPT1]	*Rattus norvegicus* [rat]	H⁺/2–4 residue oligopeptides
Pet1homsa	Oligopeptide-H⁺ symporter 1 [PET1, PEPT1]	*Homo sapiens* [human]	H⁺/2–4 residue oligopeptides
Pet1orycu	Oligopeptide-H⁺ symporter 1 [PET1, PEPT1]	*Oryctolagus cuniculus* [rabbit]	H⁺/2–4 residue oligopeptides
Pet2homsa	Oligopeptide-H⁺ symporter 2 [PET2, PEPT2]	*Homo sapiens* [human]	H⁺/2–4 residue oligopeptides
Pet2orycu	Oligopeptide-H⁺ symporter 2 [PET2, PEPT2]	*Oryctolagus cuniculus* [rabbit]	H⁺/2–4 residue oligopeptides

Cotransported ions are listed for known symporters.

Phylogenetic tree

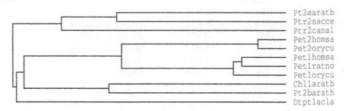

```
                                                    Pt2aarath
                                                    Ptr2sacce
                                                    Ptr2canal
                                                    Pet2homsa
                                                    Pet2orycu
                                                    Pet1homsa
                                                    Pet1ratno
                                                    Pet1orycu
                                                    Chl1arath
                                                    Pt2barath
                                                    Dtpt1acla
```

Proposed orientation of PET1 in the membrane

The model is based on predictions of membrane-spanning regions and α-helical content. The N-terminus of the protein is illustrated on the inside and is folded 12 times through the membrane. The predicted membrane-spanning helices are portrayed as rectangles. The numbers corresponding to the first and last residue of each membrane-spanning helix are boxed. Residues that are conserved in more than 75% of the aligned transporters (see below) are shown.

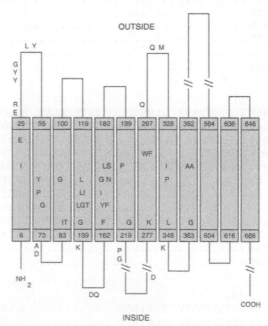

Physical and genetic characteristics

	AMINO ACIDS	MOL. WT	EXPRESSION SITES	CHROMOSOMAL LOCUS
Chl1arath	590	64 921		
Dtpt1acla	463	50 630		
Pt2aarath	610	67 518		
Pt2barath	585	64 421		
Ptr2sacce	601	68 043		Chromosome 11

	AMINO ACIDS	MOL. WT	EXPRESSION SITES	CHROMOSOMAL LOCUS
Ptr2canal	623	69 941		
Pet1ratno	710	78 928	small intestine	
Pet1homsa	708	78 805	small intestine	
Pet1orycu	707	78 927	small intestine	
Pet2homsa	729	81 940	kidney	
Pet2orycu	729	81 664	kidney	

Multiple amino acid sequence alignments

```
           1                                                50
Pt2aarath ......MSSI EEQITKSDSD FIISEDQSYL SKEKKADGSA TINQADEQSS
Ptr2sacce MLNHPSQGSD DAQDEK.QGD FPVIEEE... .KTQAVTLKD SYVSDDVANS
Ptr2canal .......... ....MVSSD F...ENEKQP DVVQVLTDEK NISLDDKYDY
Pet2homsa .......... .......... .......... .......... ....MNPFQ
Pet2orycu .......... .......... .......... .......... ....MNPFQ
Pet1homsa .......... .......... .......... .......... ..........
Pet1ratno .......... .......... .......... .......... ..........
Pet1orycu .......... .......... .......... .......... ..........
Ch11arath .......... .......... .......... .......... ..........
Pt2barath .......... .......... .......... .......... MGSIEEEARP
Dtptlacla .......... .......... .......... .......... ..........
Consensus .......... .......... .......... .......... ..........


           51                                               100
Pt2aarath TDELQKSMST GVLVNGDLYP SPTEEELATL PSVCGTIPWK AFIIIIVELC
Ptr2sacce TERYNLSPSP ....EDEDFE GPTEEEMQTL RHVGGKIPMR CWLIAIVELS
Ptr2canal EDPKNYSTNY VDDYNPKGLR RPTPQESKSL RRVIGNIRYS TFMLCICEFA
Pet2homsa KNESKETLFS PVSIEEVPPR PPSPPKKPSP TICGSNYPLS IAFIVVNEFC
Pet2orycu QNESKETLFS PVSTEETPPR LSSPAKKTPP KICGSNYPLS IAFIVVNEFC
Pet1homsa .......... .......... .....MGMSK SHSFFGYPLS IFFIVVNEFC
Pet1ratno .......... .......... .....MGMSK SRGCFGYPLS IFFIVVNEFC
Pet1orycu .......... .......... .....MGMSK SLSCFGYPLS IFFIVVNEFC
Ch11arath .....MSLPE TK..SDDILL DAWDFQGRPA DRSKTGGWAS AAMILCIEAV
Pt2barath LIEEGLILQE VKLYAED... GSVDFNGNPP LKEKTGNWKA CPFILGNECC
Dtptlacla .......... .......... .......... .......... ..........
Consensus .......... .......... .......... .......... ...I...E..


           101                                              150
Pt2aarath ERFAYYGLTV PFQNYMQF.. .......... ...GPKDATP GALNLGETGA
Ptr2sacce ERFSYYGLSA PFQNYMEY.. .......... ...GPNDSPK GVLSLNSQGA
Ptr2canal ERASYYSTTG ILTNYIQRRI DPDSPHGWGA PPPGSPDASA GALGKGLQAA
Pet2homsa ERFSYYGMKA VLILYF.... .......... ..LYF..... ..LHWNEDTS
Pet2orycu ERFSYYGMKA VLTLYF.... .......... ..LYF..... ..LHWNEDTS
Pet1homsa ERFSYYGMRA ILILYF.... .......... ..TNF..... ..ISWDDNLS
Pet1ratno ERFSYYGMRA LLVLYF.... .......... ..RNF..... ..LGWDDDLS
Pet1orycu ERFSYYGMRA LLILYF.... .......... ..RNF..... ..IGWDDNLS
Ch11arath ERLTTLGIGV NLVTYL.... .......... ..TGT..... ..MHLGNATA
Pt2barath ERLAYYGIAG NLITYL.... .......... ..TTK..... ..LHQGNVSA
Dtptlacla .......MRA ILVYYL.... .......... ..YALTTADN AGLGLPKAQA
Consensus ER..YYG... .L..Y..... .......... .......... ..........
```

```
          151                                                    200
Pt2aarath DGLSNFFTFW CYVTPVGAAL IADQFLGRYN TIVCSAVIYF IGILILTCTA
Ptr2sacce TGLSYFFQFW CYVTPVFGGY VADTFWGKYN TICCGTAIYI AGIFILFITS
Ptr2canal SALTNLLTFL AYVFPLIGGY LGDSTIGRWK AIQWGVFFGF VAHLFFIFAS
Pet2homsa TSIYHAFSSL CYFTPILGAA IADSWLGKFK TIIYLSLVYV LGHVIKSLGA
Pet2orycu TSVYHAFSSL CYFTPILGAA IADSWLGKFK TIIYLSLVNV LGHVIKSLSA
Pet1homsa TAIYHTFVAL CYLTPILGAL IADSWLGKFK TIVSLSIVYT IGQAVTSVSS
Pet1ratno TAIYHTFVAL CYLTPILGAL IADSWLGKFK TIVSLSIVYT IGQAVISVSS
Pet1orycu TVIYHTFVAL CYLTPILGAL IADAWLGKFK TIVWLSIVYT IGQAVTSLSS
Ch11arath ANTVTNFLGT SFMLCLLGGF IADTFLGRYL TIAIFAAIQA TGVSILTLST
Pt2barath ATNVTTWQGT CYLTPLIGAV LADAYWGRYW TIACFSGIYF IGMSALTLSA
Dtptlacla MAIVSIYGAL VYLSTIVGGW VADRLLGASR TIFLGGILIT LGHVALATPF
Consensus .......... .Y..P..G.. .AD...G... TI........ .G........

          201                                                    250
Pt2aarath IP.SVIDAGK S.......... ....MGGFV VSLIIIGLGT GGIKSNVSPL
Ptr2sacce IP.SVGNRDS A.......... ....IGGFI AAIILIGIAT GMIKANLSVL
Ptr2canal IPQAIENANA G.......... ....LGLCV IAIITLSAGS GLMKPNLLPL
Pet2homsa .LPILGGQ.. .......... ..VVHTVLSL IGLSLIALGT GGIKPCVAAF
Pet2orycu .FPILGGK.. .......... ..VVHTVLSL VGLCLIALGT GGIKPCVAAF
Pet1homsa .INDLTDHNH DGTPDSL.... .PVHVVLSL IGLALIALGT GGIKPCVSAF
Pet1ratno .INDLTDHDH DGSPNNL.... .PLHVALSM IGLALIALGT GGIKPCVSAF
Pet1orycu .VNELTDNNH DGTPDSL.... .PVHVAVCM IGLLLIALGT GGIKPCVSAF
Ch11arath IIPGLRPPRC NPTTSSHCEQ ASGIQLTVLY LALYLTALGT GGVKASVSGF
Pt2barath SVPALKPAEC ...IGDFCPS ATPAQYAMFF GGLYLIALGT GGIKPCVSSF
Dtptlacla GLSSL..... .......... F VALFLIILGT GMLKPNISNM
Consensus .......... .......... ..L.LI.LGT G..K......

          251                                                    300
Pt2aarath MAEQLPKIPP YVKTKKNGSK VIVDPVVTTS RAYMIFYWTI NVGSLSVLAT
Ptr2sacce IADQLPKRKP SIKVLKSGER VIVDSNITLQ NVFMFFYFMI NVGSLSLMAT
Ptr2canal VLDQYPEERD MVKVLPTGES IILDREKSLS RITNVFYLAI NIGAFLQIAT
Pet2homsa GGDQFEEKHA EERTR..... .......... .YFSVFYLSI NAGSLISTFI
Pet2orycu GGDQFEEKHA EERTR..... .......... .YFSGFYLAI NAGSLISTFI
Pet1homsa GGDQFEEGQE KQRNR..... .......... .FFSIFYLAI NAGSLLSTII
Pet1ratno GGDQFEEGQE KQRNR..... .......... .FFSIFYLAI NAGSLLSTII
Pet1orycu GGDQFEEGQE KQRNR..... .......... .FFSIFYLAI NAGSLLSTII
Ch11arath GSDQFDETEP KERSKMTY.. .......... .FFNRFFFCI NVGSLLA..V
Pt2barath GADQFDDTDS RERVRKAS.. .......... .FFNWFYFSI NIGALVS..S
Dtptlacla VGHLY....S KDDSRRDT.. .......... .GFNIFVVGI NMGSLIAPLI
Consensus ..DQ...... .......... .......... ..F..FY..I N.GSL.....

          301                                                    350
Pt2aarath TSLES..... ....TKGFVY AYLLPLCVFV IPLIILAVSK TAFTSTLLPP
Ptr2sacce TELEY..... ....HKGFWA AYLLPFCFFW IAVVTLIFGK KQY...IQRP
Ptr2canal SYCER..... ....RVGFWL AFFVPMILYI IVPIFLFIVK PKL..KIKPP
Pet2homsa TPMLRGDVQC F..GEDCYAL AFGVPGLLMV IALVVFAMGS KIY..NKPPP
Pet2orycu TPMLRGDVQC F..GEDCYAL AFGVPGLLMV IALVVFAMGS KMY..KKPPP
Pet1homsa TPMLRVQQCG IHSKQACYPL AFGVPAALMA VALIVFVLGS GMY..KKFKP
Pet1ratno TPILRVQQCG IHSQQACYPL AFGVPAALMA VALIVFVLGS GMY..KKFQP
Pet1orycu TPMVRVQQCG IHVKQACYPL AFGIPAILMA VSLIVFIIGS GMY..KKFKP
Ch11arath TVLVYVQ... ...DDVGRKW GYGICAFAIV LALSVFLAGT NRY..RFKKL
Pt2barath SLLVWIQ... ...ENRGWGL GFGIPTVFMG LAIASFFFGT PLY..RFQKP
Dtptlacla VGTV...... ..GQGVNYHL GFSLAAIGMI FALFAYWYGR LRHFPEIGRE
Consensus .......... .......... ....P..... .......G.. .....P
```

313

```
          351                                                   400
Pt2aarath VPSLFV.LVK CSSLLLKTNL ISKKLN.... HLALLL.... .......LER
Ptr2sacce IGD....KVI AKSFKVCWIL TKNKFD...F NAAKPS.... .......VHP
Ptr2canal QGQVMTNVVK ILAVLFSGNF IKRLWNGTFW DHARPSHMEA RGTIYYNSKK
Pet2homsa EGNIVAQVFK ...CIWFAIS NRFKN.... .......... ..........R
Pet2orycu EGNIVAQVVK ...CIWFAIS NRFKN.... .......... ..........R
Pet1homsa QGNIMGKVAK ...CIGFAIK NRFRH.... .......... ..........R
Pet1ratno QGNIMGKVAK ...CIRFAIK NRFRH.... .......... ..........R
Pet1orycu QGNILSKVVK ...CICFAIK NRFRH.... .......... ..........R
Chl1arath IGSPMTQVAA ...VIVAAWR NRKLELPADP SYLYDVDDII AAEGSMKGKQ
Pt2barath GGSPITRISQ ...VVVASFR KSSVKVPEDA TLLYETQD.. .KNSAIAGSR
Dtptlacla PSNPMDAKAK RNFIITLTIV LIVALIGFFL IYQASPANFI NNFINVLSII
Consensus .G........ .......... .......... .......... ..........

          401                                                   450
Pt2aarath YVKDQWDDLF ID........ .......... ...... ELKRALRACK
Ptr2sacce EKNYPWNDKF VD........ .......... ...... EIKRALAACK
Ptr2canal KSAITWSDQW IL........ .......... ...... DIKQTFDSCK
Pet2homsa SGDIPKRHDW LDWAAEKYPK Q......... ....... LIM DVKALTRVLF
Pet2orycu SEDIPKRQHW LDWAAEKYPK Q......... ....... LIM DVKTLTRVLF
Pet1homsa SKAFPKREHW LDWAKEKYDE R......... ....... LIS QIKMVTRVMF
Pet1ratno SKAFPKRNHW LDWAKEKYDE R......... ....... LIS QIKIMTKVMF
Pet1orycu SKQFPKRAHW LDWAKEKYDE R......... ....... LIA QIKMVTRVLF
Chl1arath KLPHTEQFRS LDKAAIRDQE AGVTSNVFNK WTLSTLTDVE EVKQIVRMLP
Pt2barath KIEHTDDCQY LDKAAVISEE ESKSGDYSNS WRLCTVTQVE ELKILIRMFP
Dtptlacla GIVVPIIYFV MMFTSKKVES D......... ....... ERRKLTAYIP
Consensus .......... .D........ .......... ....... ..K.......

          451                                                   500
Pt2aarath TFLFYPIYWV CYGQMTNNLI SQAGQMQTG. ........NV SNDLFQAFDS
Ptr2sacce VFIFYPIYWT QYGTMISSFI TQASMMELH. ........GI PNDFLQAFDS
Ptr2canal IFLYYIIFNL ADSGLGSVET SLIGAMKLD. ........GV PNDLFNNFNP
Pet2homsa LYIPLPMFWA LLDQQGSRWT LQAIRMNRNL G.....FFVL QPDQMQVLNP
Pet2orycu LYIPLPMFWA LLDQQGSRWT LQATKMNGNL G.....FFVL QPDQMQVLNP
Pet1homsa LYIPLPMFWA LFDQQGSRWT LQATTMSGKI G.....ALEI QPDQMQTVNA
Pet1ratno LYIPLPMFWA LFDQQGSRWT LQATTMTGKI G.....TIEI QPDQMQTVNA
Pet1orycu LYIPLPMFWA LFDQQGSRWT LQATTMSGRI G.....ILEI QPDQMQTVNT
Chl1arath IWATCILFWT VHAQLTTLSV AQSETLDRSI G.....SFEI PPASMAVFYV
Pt2barath IWASGIIFSA VYAQMSTMFV QQGRAMNCKI G.....SFQL PPAALGTFDT
Dtptlacla LFLSAIVFWA IEEQSSTIIA VWGESRSNLN PTWFGFTFHI DPSWYQLLNP
Consensus .......FW. ...Q...... .Q...M.... .......... ..........

          501                                                   550
Pt2aarath IALIIFIPIC DNIIYPLLRK ...YNIPFKP ILRITLGFMF ATASMIYAAV
Ptr2sacce IALIIFIPIF EKFVYPFIRR ...Y.TPLKP ITKIFFGFMF GSFAMTWAAV
Ptr2canal LTIIILIPIL EYGLYPLLNK ...FKIDFKP IWRICFGFVV CSFSQIAGFV
Pet2homsa LLVLIFIPLF DFVIYRLVSK ...CGINFSS LRKMAVGMIL ACLAFAVAAR
Pet2orycu LLVLIFIPLF DLVIYRLISK ...CGINFTS LRKMAVGMVL ACLAFAAAAT
Pet1homsa ILIVIMVPIF DAVLYPLIAK ...CGFNFTS LKKMAVGMVL ASMAFVVAAI
Pet1ratno ILIVIMVPIV DAVVYPLIAK ...CGFNFTS LKKMTVGMFL ASMAFVVAAI
Pet1orycu ILIIILVPIM DAVVYPLIAK ...CGLNFTS LKKMTIGMFL ASMAFVAAAI
Chl1arath GGLLLTTAVY DRVAIRLCKK LFNYPHGLRP LQRIGLGLFF GSMAMAVAAL
Pt2barath ASVIIWVPLY DRFIVPLARK FTGVDKGFTE IQRMGIGLFV SVLCMAAAAI
Dtptlacla LFIVLLSPIF VRIWNKLGDR QP......ST IVKFGLGLML TGASYLIMTL
Consensus ....I..P.. ......L..K .......... ......G... ........AA.
```

```
          551                                                    600
Pt2aarath LQAKIYQRGP CYANFTDTCV SNDISVWIQI PAYVLIAFSE IFASITGLEF
Ptr2sacce LQSFVYKAGP WYNEPLGHNT PNHVHVCWQI PAYVLISFSE IFASITGLEY
Ptr2canal LQKQVYEQSP CGYYATNCDS PAPITAWKAS SLFILAAAGE CWAYTTAYEL
Pet2homsa VEI...KINE .MAPAQPGPQ EVFLQVLNLA DDEVKVTVVG .....NENNS
Pet2orycu VEI...KINE .MAPPQPGSQ EILLQVLNLA DDEVKLTVLG .....NNNNS
Pet1homsa VQV...EIDK .TLPVFPKGN EVQIKVLNIG NNTMNISLPG .....EM...
Pet1ratno VQV...EIDK .TLPVFPSGN QVQIKVLNIG NNDMAVYFPG .....KN...
Pet1orycu LQV...EIDK .TLPVFPKAN EVQIKVLNVG SENMIISLPG .....QT...
Chllarath VELKRLRTAH .AHG..PTVK TLPLGFYLLI PQYLIVGIGE ALIYTGQLDF
Pt2barath VEIIRLHMAN .DLGLVESGA PVPISVLWQI PQYFILGAAE VFYFIGQLEF
Dtptlacla PGLLNGTSGR .ASALW.... .LVLMFAVQM AGELLVSPVG LSVSTKLAPV
Consensus .......... .......... .......... .......... ..........


          601                                                    650
Pt2aarath AFTKAPPSMK SIITALFLFT NAFGAILSIC ISSTAVNPKL TWMYTGIAVT
Ptr2sacce AYSKAPASMK SFIMSIFLLT NAFGSAIGCA LSPVTVDPKF TWLFTGLAVA
Ptr2canal AYTRSPPALK SLVYALFLVM SAFSAALSLA ITPALKDPNL HWVFLAIGLA
Pet2homsa LLIESIKSFQ KTPHYSKLHL KTKSQDFHFH LKYHNLSLYT EHSVQEKNWY
Pet2orycu LLADSIKSFQ KTPHYSKIHL NTKSQDFYFH LKYHNLSIYT EHSVEERNWY
Pet1homsa ...VTLGPMS QTNAFMTFDV NKLTRINISS PGSP.VTAVT .DDFKQGQRH
Pet1ratno ...VTVAQMS QTDTFMTFDV DQLTSINVSS PGSPGVTTVA .HEFEPGHRH
Pet1orycu ...VTLNQMS QTNEFMTFNE DTLTSINITS ..GSQVTMIT .PSLEAGQRH
Chllarath FLRECPKGMK GMSTGLLLST LALG.FFFSS VLVTIVEKFT GK..AHPWIA
Pt2barath FYDQSPDAMR SLCSALALLT NALG.NYLSS LILTLVTYFT TRNGQEGWIS
Dtptlacla AFQSQMMAMW FLADSTSQAI NAQITPIFKA ATEVHFFAIT GIIGIIVGII
Consensus .......... .......... .......... .......... ..........


          651                                                    700
Pt2aarath AFI.....AG IMFWVCFHHY DA.MEDEQNQ LE.FKRNDAL TKKDVEKEVH
Ptr2sacce CFI.....SG CLFWLCFRKY ND.TEEEMNA MD.YEEEDEF DLNPISAPKA
Ptr2canal GFL.....CA IVMLAQFWNL DKWMENETNE RERLDREEEE EANRGIHDVD
Pet2homsa SLVIREDGNS ISSMMVK.DT ESRTTNGMTT VRFVNTLHKD VNISLSTDTS
Pet2orycu SLIIREDGKS ISSIMVK.DM ENETTYGMTA IRFINTLQEN VNISLGTDIS
Pet1homsa TLLVW..APN HYQVVKD.GL NQKPEKGENG IRFVNTFNEL ITITMSGKVY
Pet1ratno TLLVW..GPN LYRVVKD.GL NQKPEKGENG IRFVSTLNEM ITIKMSGKVY
Pet1orycu TLLVW..APN NYRVVND.GL TQKSDKGENG IRFVNTYSQP INVTMSGKVY
Chllarath DDLNKGRLYN FYWLVAVLVA LNFLIFLVFS KWYVYKEKRL AEVGIELDDE
Pt2barath DNLNSGHLDY FFWLLAGLSL VNMAVYFFSA AR..YKQKKA SS........
Dtptlacla LLIIKKPILK LMGDVR.... .......... .......... ..........
Consensus .......... .......... .......... .......... ..........


          701                                                    750
Pt2aarath DSYSMADESQ YNLEKANC.. .......... .......... ..........
Ptr2sacce NDIEILEPME SLRSTTKY.. .......... .......... ..........
Ptr2canal HPIEAIVSIK S......... .......... .......... ..........
Pet2homsa LNVGEDYGVS AYRTVQRGEY PAVHCRTEDK NFS...LNLG LLDFGAAYLF
Pet2orycu LNVGENYGVS AYRTVQRGEY PAVHCKTEDK DFS...LNLG LLDFGASYLF
Pet1homsa ANIS.SYNAS TYQFFPSGIK GFTISSTEIP PQCQPNFNTF YLEFGSAYTY
Pet1ratno ENVT.SHSAS NYQFFPSGQK DYTINTTEIA PNCSSDFKSS NLDFGSAYTY
Pet1orycu EHIA.SYNAS EYQFFTSGVK GFTVSSAGIS EQCRRDFESP YLEFGSAYTY
Chllarath PSIPMGH... .......... .......... .......... ..........
Consensus .......... .......... .......... .......... ..........
```

```
         751                                                  800
Pet2homsa VITNNTNQG. LQAWKIEDIP ANKMSIRWQL PQYALVTAGE VMFSVTGLEF
Pet2orycu VITNSTKQG. LQAWKMEDIP ANKVSIAWQL PQYALVTAGE VMFSVTGLEF
Pet1homsa IVQ.RKNDSC PEVKVFEDIS ANTVNMALQI PQYFLLTCGE VVFSVTGLEF
Pet1ratno VIRSRASDGC LEVKEFEDIP PNTVNMALQI PQYFLLTCGE VVFSVTGLEF
Pet1orycu LITSQAT.GC PQVTEFEDIP PNTMNMAWQI PQYFLITSGE VVFSITGLEF
Consensus ......... ......... ......... .........
```

```
         801                                                  850
Pet2homsa SYSQAPSSMK SVLQAAWLLT IAVGNIIVLV VAQFSGL.VQ WAEFILFSCL
Pet2orycu SYSQAPSSMK SVLQAAWLLT VAIGNIIVLV VAQFSGL.VQ WAEFVLFSCL
Pet1homsa SYSQAPSNMK SVLQAGWLLT VAVGNIIVLI VAGAGQFSKQ WAEYILFAAL
Pet1ratno SYSQAPSNMK SVLQAGWLLT VAIGNIIVLI VAEAGHFDKQ WAEYVLFASL
Pet1orycu SYSQAPSNMK SVLQAGWLLT VAVGNIIVLI VAGAGQINKQ WAEYILFAAL
Consensus ......... ......... ......... .........
```

```
         851                                                  900
Pet2homsa LLVICLIFSI MGYYVVPVKT EDMRGPADKH IPHIQGNMIK LETKKTKL..
Pet2orycu LLVVCLIFSI MGYYYIPIKS EDIQGPEDKQ IPHMQGNMIN LETKKTKL..
Pet1homsa LLVVCVIFAI MARFYTYINP AEIEAQFDED EKKNRLEKSN PYFMSGANSQ
Pet1ratno LLVVCIIFAI MARFYTYINP AEIEAQFDED EKKKGVGKEN PYSSLEPVSQ
Pet1orycu LLVVCVIFAI MARFYTYVNP AEIEAQFEED EKKKNPEKND LYPSLAPVSQ
Consensus ......... ......... ......... .........
```

```
         901
Pet1homsa KQM
Pet1ratno TNM
Pet1orycu TQM
Consensus ...
```

Residues listed in the consensus sequence are present in at least 75% of the aligned transporter sequences.

Database accession numbers

	SWISSPROT	PIR	EMBL/GENBANK
Chl1arath	Q05085		L10357
Dtptlacla	P36574	A53620	U05215
Pt2aarath	P4603		U01171
Pt2barath	P46032		L39082; X77503
Ptr2sacce	P32901	S38171	L11994; X73541
Ptr2canal	P46030		U09781
Pet1ratno			D50664
Pet1homsa	P46059		U13173
Pet1orycu	P36836		U06467; U13707
Pet2homsa			S78203
Pet2orycu	P46029		U32507

Summary

Transporters of the H⁺/fucose symporter family, the example of which is the FUCP fucose-H⁺ symporter of *Escherichia coli* (Fucpescco), mediate symport (H⁺-coupled substrate uptake) of fucose, glucose and galactose. Known members of the family are found only in gram-negative bacteria.

Statistical analysis reveals no apparent relationship between the amino acid sequences of the H⁺/fucose symporter family and any other family of transporters. They are predicted to contain 12 membrane-spanning helices by the hydropathy of their amino acid sequences and activities of reporter gene fusions [1].

Several amino acid sequence motifs are highly conserved in the H⁺/fucose symporter family.

Nomenclature, biological sources and substrates

CODE	DESCRIPTION [SYNONYMS]	ORGANISM [COMMON NAMES]	SUBSTRATE(S)
Fucpescco	Fucose-H⁺ symporter [FUCP]	*Escherichia coli* [gram-negative bacterium]	H⁺/fucose
Fucphaein	Fucose-H⁺ symporter [FUCP, HI0610]	*Haemophilus influenzae* [gram-negative bacterium]	H⁺/fucose
Glupbruab	Glucose-galactose transporter [GLUP]	*Brucella abortus* [gram-negative bacterium]	Glucose, galactose

Cotransported ions are listed for known symporters.

Phylogenetic tree

```
        ┌──────────────────────────────────  Fucpescco
────────┤                                     Glupbruab
        └──────────────────────────────────  Fucphaein
```

Proposed orientation of FUCP in the membrane

The model is based on predictions of membrane-spanning regions and α-helical content. The N-terminus of the protein is illustrated on the inside and is folded 12 times through the membrane [1]. The predicted membrane-spanning helices are portrayed as rectangles. The numbers corresponding to the first and last residue of each membrane-spanning helix are boxed. Residues that are conserved in all three of the aligned transporters (see below) are shown.

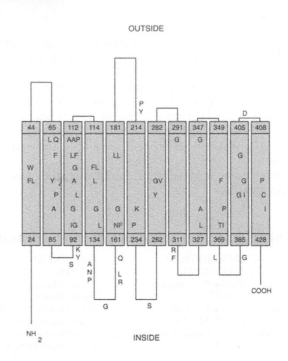

Physical and genetic characteristics

	AMINO ACIDS	MOL. WT	K_m	CHROMOSOMAL LOCUS
Fucpescco	438	47 544	Fucose: 24.7 μM [2]	63.16 minutes
Fucphaein	428	46 987		
Glupbruab	412	43 859		

Multiple amino acid sequence alignments

```
          1                                                  50
Fucpescco MGNTSIQTQS YRAVDKDAGQ SRSYIIPFAL LCSLFFLWAV ANNLNDILLP
Glupbruab .MATSIPTNN ..PLHTETSS QKNYGFALTS LTLLFFMWGF ITCLNDILIP
Fucphaein .......... ....MNAKVL EKKFIVPFVL ITSLFALWGF ANDITNPMVA
Consensus .......... .......... .......... ...LF..W.. ..........

          51                                                 100
Fucpescco QFQQAFTLTN FQAGLIQSAF YFGYFIIPIP AGILMKKLSY KAGIITGLFL
Glupbruab HLKNVFQLNY TQSMLIQFCF FGAYFIVSLP AGQLVKRISY KRGIVVGLIV
Fucphaein VFQTVMEIPA SEAALVQLAF YGGYGTMAIP AALFASRYSY KAGILLGLAL
Consensus .......... ...L.Q..F ...Y.....P A.......SY K.GI..GL..

          101                                                150
Fucpescco YALGAALFWP AAEIMNYTLF LVGLFIIAAG LGCLETAANP FVTVLGPESS
Glupbruab AAIGCALFIP AASYRVYALF LGALFVLASG VTILQVAANP YVTILGKPET
Fucphaein YAIGAFLFWP AAQYEIFNFF LVSLYILTFG LAFLETTANP YILAMGDPQT
Consensus .A.G..LF.P AA.......F L..L.....G ...L...ANP .....G....
```

```
            151                                                    200
Fucpescco GHFRLNLAQT FNSFGAIIAV VFGQSLILSN VPHQSQDVLD KMSPEQLS..
Glupbruab AASRLTLTQA FNSLGTTVAP VFGAVLILSA ATDATVNAEA D.........
Fucphaein ATRRLNFAQS FNPLGSITGM FVASQLVLTN LESDKRDAAG NLIFHTLSEA
Consensus ...RL...Q. FN..G..... .....L.L.. .......... ..........

            201                                                    250
Fucpescco ...AYKHSLV LSVQTPYMII VAIVLLVALL IMLTKFPALQ SDNH..SDAK
Glupbruab .......... .AVRFPYLLL ALAFTVLAII FAILKPPDVQ EDEPALSDKK
Fucphaein EKMSIRTHDL AEIRDPYIAL GFVVVAVFII IGLKKMPAVK IE.....EAG
Consensus .......... .....PY... .......... ....K.P... ..........

            251                                                    300
Fucpescco QGSFSASLSR LARIRHWRWA VLAQFCYVGA QTACWSYLIR YAVE.EIPGM
Glupbruab EGSAW..... ..QYRHLVLG AIGIFVYVGA EVSVGSFLVN FLSDPTVAGL
Fucphaein QISFKTAVSR LAQKAKYREG VIAQAFYVGV QIMCWTFIVQ YA...ERLGF
Consensus ..S....... .......... ......YVG. .......... ........G.

            301                                                    350
Fucpescco TAGFAANYLT GTMVCFFIGR FTGTWLISRF APHKVLAAYA LIAMALCLIS
Glupbruab SETDAAHHVA YFWGGDMVGR FIGSAAMRYI DDGKALAFDA FVAIILLFIT
Fucphaein TKAEGQNFNI IAMAIFISSR FISTALMKYL KAEFMLMLFA IGGFLSILGV
Consensus .......... .........R F.......... .....L...A ..........

            351                                                    400
Fucpescco AFAGGHVGLI ALTLCSAFMS IQYPTIFSLG IKNLGQDTKY GSSFIVMTII
Glupbruab VATTGHIAMW SVLAIGLFNS IMFPTIFSLA LHGLGSHTSQ GSGILCLAIV
Fucphaein IFIDGVWGLY CLILTSGFMP LMFPTIYGIA LYGLKEESTL GAAGLVMAIV
Consensus ....G..... ......F.. ...PTI.... ...L..... G.......I.

            401                                                    450
Fucpescco GGGIVTPVMG FVSD.....A AGNIPTAELI PALCFAVIFI FARFRSQTAT
Glupbruab GGAIVPLIQG ALAD.....A IG.IHLAFLM PIICYAYIAF YGLIGTKS..
Fucphaein GGALMPPLQG MIIDQGEVMG LPAVNFSFIL PLICFVVIAI YGFRAWKILK
Consensus GG.......G ...D..... .......... P..C...I.. ..........
```

Residues listed in the consensus sequence are present in all three of the transporter sequences.

Database accession numbers

	SWISSPROT	PIR	EMBL/GENBANK
Fucphaein	P44776		L45251; U32743
Fucpescco	P11551	JS0184	X15025; U29581
Glupbruab			U43785

References

[1] Gunn, F. (1993) PhD thesis, Cambridge University.
[2] Muir, J. et al. (1993) Biochem. J. 290, 833–842.

Summary

Transporters of the H⁺/carboxylate symporter family, the example of which is the KGTP α-ketoglutarate-H⁺ symporter of *Escherichia coli* (Kgtp1escco), mediate uptake of carboxylated compounds [1,2]. The mechanism of transport, where known, is symport (i.e. H⁺-coupled substrate efflux). The transport activity of one family member, the PROP proline-betaine protein of *E. coli*, is stretch-inactivated, allowing it to function as both an osmosensor and osmoregulator [3]. Members of the family occur in both gram-negative and gram-positive bacteria.

Statistical analysis of multiple amino acid sequence comparisons places the H⁺/carboxylate symporter family in the uniporter-symporter-antiporter (USA) superfamily, also known as the major facilitator superfamily (MFS) [1,2]. Members of the H⁺/carboxylate symporter family are predicted to form 12 membrane-spanning helices by the hydropathy of their amino acid sequences [4] and activities of reporter gene fusions [5]. There is considerable similarity between the sequences of the N- and C-terminal halves of these proteins, further implying they arose through gene duplication of an ancestral six helix protein [2].

Several amino acid sequence motifs are highly conserved in the H⁺/carboxylate symporter family, including motifs unique to the family and signature motifs of the USA/MFS transporter superfamily.

Nomenclature, biological sources and substrates

CODE	DESCRIPTION [SYNONYMS]	ORGANISM [COMMON NAMES]	SUBSTRATE(S)
Bap3strhy	Bialaphose transporter	*Streptomyces hygroscopicus* [gram-positive bacterium]	Bialaphos
Citklepn	Citrate-H⁺ symporter [CIT, CITH]	*Klebsiella pneumoniae* [gram-negative bacterium]	H⁺/citrate
Cit1escco	Citrate-H⁺ symporter [CIT1, CITA, CIT]	*Escherichia coli* [gram-negative bacterium]	H⁺/citrate
Cit2escco	Citrate-H⁺ symporter [CITA, CIT]	*Escherichia coli* [gram-negative bacterium]	H⁺/citrate
Citasalty	Citrate-H⁺ symporter [CITA]	*Salmonella typhimurium* [gram-negative bacterium]	H⁺/citrate
Kgtpescco	α-Ketoglutarate-H⁺ symporter [KGTP, WITA]	*Escherichia coli* [gram-negative bacterium]	H⁺/α-keto glutarate
Mopaburce	4-Methyl-o-phthalate permease [MPOA]	*Burkholderia cepacia* [purple bacterium]	4-Methyl-o-phthalate
Nantescco	Putative sialic acid transporter [NANT]	*Escherichia coli* [gram-negative bacterium]	Sialic acid
Ousaerwch	Osmoprotectant uptake system A [OUSA]	*Erwinia chrysanthemi* [gram-negative bacterium]	Osmoprotectants
Pcatpsepu	Dicarboxylic acid transport protein [PCAT]	*Pseudomonas putida* [gram-negative bacterium]	Dicarboxylates
Propescco	Proline-betaine transporter [PROP, PPII]	*Escherichia coli* [gram-negative bacterium]	Proline, betaine

Cotransported ions are listed for known symporters.

Phylogenetic tree

Proteins listed subsequently in italics are at least 90% identical to the paired transporters listed in parenthesis and therefore are not included in the phylogenetic tree: *Cit2escco, Citasalty* (Cit1escco).

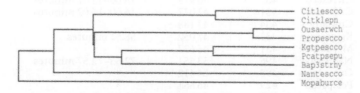

```
                                            Citlescco
                                            Citklepn
                                            Ousaerwch
                                            Propescco
                                            Kgtpescco
                                            Pcatpsepu
                                            Bap3strhy
                                            Nantescco
                                            Mopaburce
```

Proposed orientation of KGTP in the membrane

The model is based on predictions of membrane-spanning regions and α-helical content[5]. The N-terminus of the protein is illustrated on the inside and is folded 12 times through the membrane. The predicted membrane-spanning helices are portrayed as rectangles. The numbers corresponding to the first and last residue of each membrane-spanning helix are boxed. Residues that are conserved in more than 75% of the aligned transporters (see below) are shown. Consensus residues indicated by an asterisk are not conserved in KGTP.

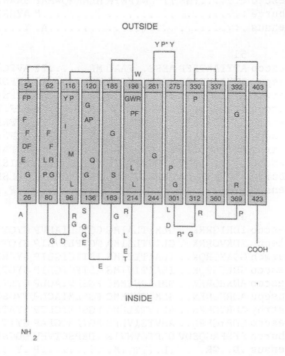

Physical and genetic characteristics

	AMINO ACIDS	MOL. WT	CHROMOSOMAL LOCUS
Bap3strhy	448	47 233	
Citklepn	444	48 142	
Cit1escco	431	46 979	18.00–18.02 minutes
Cit2escco	431	47 077	16.00–16.02 minutes
Citasalty	434	47 188	
Kgtpescco	432	47 052	58.51 minutes
Mopaburce	431	45 627	
Nantescco	496	53 551	72.55–72.57 minutes
Ousaerwch	498	54 412	
Pcatpsepu	429	46 861	
Propescco	500	54 845	93.24 minutes

Multiple amino acid sequence alignments

```
          1                                                   50
Cit1escco ............ ......MTQQ PSRAGTFGAI LRVTSGNFLE QFDFFLFGFY
Citklepn  MPTARCSMRA SSTAPVRMMA TAGGARIGAI LRVTSGNFLE QFDFFLFGFY
Ousaerwch ..MKLKRKRV KPIALDDVTI IDDGRLRKAI TAAALGNAME WFDFGVYGFV
Propescco ...MLKRKKV KPITLRDVTI IDDGKLRKAI TAASLGNAME WFDFGVYGFV
Kgtpescco ......MAES TVTADSKLTS SDTRRRIWAI VGASSGNLVE WFDFYVYSFC
Pcatpsepu ........ ..MTSTYYTG EERSKRIFAI VGASSGNLVE WFDFYVYAFC
Bap3strhy .......... MTVRESDRTT VPRSRSLRQL VCGGIGHTVE SHDWYVYTFL
Nantescco ........MS TTTQNIPWYR HLNRAQWRAF SAAWLGYLLD GFDFVLIALV
Mopaburce .......... .......... .........M AYHSTNPLLS RVVIVGAGQA
Consensus .......... .......... .....A. ....G...E .FDF....F.

          51                                                  100
Cit1escco ATYIAKTFFP AESE..FAAL ML......TF AVFGSGFLMR PIGAVVLGAY
Citklepn  ATYIAHTFFP ASSE..FASL MM......TF AVFGAGFLMR PIGAIVLGAY
Ousaerwch AYALGQVFFP ..G.DPGVQS IA......AL ATFSVPFLM. PLGGVFFGAL
Propescco AYALGKVFFP ..GADPSVQM VA......AL ATFSVPFLIR PLGGLFFGML
Kgtpescco SLYFAHIFFP ..SGNTTTQL LQ......TA GVFAAGFLMR PIGGWLFGRI
Pcatpsepu AIYFAPAFFP ..SDDPTVQL LN......TA GVFAAGFLMR PIGGWIFGRL
Bap3strhy AVYFSDDIFP ESSGDPLVPL LN......TF AVFALAFAAR PVGATVMGWY
Nantescco ........LT EVQGEFGLTT VQ......AA SLISAAFISR WFGGLMLGAM
Mopaburce AAAVAKTLRA EGHRGEIMML GAERVGPYER PPLSKAWLSA ENVPEIAALL
Consensus ........FP .......... ..........F...FL.R P.G....G..

          101                                                 150
Cit1escco IDRIGRRK.. GLMITLAIMG CGTLLIALVP GYQTIGLLAP VLVLVGRLLQ
Citklepn  IDKVGRRK.. GLIVTLSIMA TGTFLIVLIP SYQTIGLWAP LLVLIGRLLQ
Ousaerwch GDKYGRQK.. ILAITIIIMS ISTFCIGLIP SYERIGIWAP ILLLLAKMAQ
Propescco GDKYGRQK.. ILAITIVIMS ISTFCIGLIP SYDTIGIWAP ILLLICKMAQ
Kgtpescco ADKHGRKK.. SMLLSVCMMC FGSLVIACLP GYETIGTWAP ALLLLARLFQ
Pcatpsepu ADRHGRKN.. SLMISVLMMC FGSLMIACLP TYGSIGTWAP ALLLLARLIQ
Bap3strhy ADRYGRRS.. ALIVTILLMG LGSLMIGLTP SYATAGPVAP VVLIAARLVQ
Nantescco GDRYGRRL.. AMVTSIVLFS AGTLACGFAP GYIT....... ..MFIARLVI
Mopaburce EKSEAGQSDV DLRTGVNVTR IDRPSCTVHV DDGSEICFDR LVIATGGRAR
Consensus .D..GR.... .L......M. .....I...P .Y...G..AP .........Q
```

```
           151                                                   200
Citlescco  GFSAGVELGG VSVYL..SEI ATPGNKGFYT SWQSASQQVA IVVAALIGYG
Citklepn   GFSAGAELGG VSVYL..AEI ATPGRKGFYT SWQSGSQQVA IMVAAAMGFA
Ousaerwch  GFSVGGEYTG ASIFV..AEY SPDRKRGFMG SWLDFGSIAG FVLGAGVVVL
Propescco  GFSVGGEYTG ASIFV..AEY SPDRKRGFMG SWLDFGSIAG FVLGAGVVVL
Kgtpescco  GLSVGGEYGT SATYM..SEV AVEGRKGFYA SFQYVTLIGG QLLALLVVVV
Pcatpsepu  GLSVGGEYGT TATYM..SEV ALRGQRGFFA SFQYVTLIGG QLLAVLVVVI
Bap3strhy  GFSLGGEYGA ATTFL..VES AAPGRRALFS SFQYVASSVG HILAGLSTLA
Nantescco  GMGMAGEYGS SATYV..IES WPKHLRNKAS GFLISGFSVG AVVAAQVYSL
Mopaburce  RLAVPGDASD QIAYLRTIDD ALHIRRGLKA GKRLLLIGGG WIGLETACSA
Consensus  G.S.GGE... ........E. ......G... S........G ..........

           201                                                   250
Citlescco  LNVTLGHDEI SEWGWRIPFF IGCMIIPLIF VLRRSLQETE AFLQ......
Citklepn   LNAVLEPSAI SDWGWRIPFL FGVLIVPFIF ILRRKLEETQ EFTA......
Ousaerwch  ISTLIGEQAF LAWGWRLPFF LALPLGLIGL YL.ATLEETP AFRQ......
Propescco  ISTIVGEANF LDWGWRIPFF IALPLGIIGL YLRHALEETP AFQQ......
Kgtpescco  LQHTMEDAAL REWGWRIPFA LGAVLAVVAL WLRRQLDET. ..........
Pcatpsepu  LQQLLTEDEL RAWGWRIPFV VGAIAALISL MLRRSLHET. ..........
Bap3strhy  ASQ.ISGDGM DRWGWRLPFI WGAVICLAGL ALRSTAEET. ..........
Nantescco  VVPV...... ..WGWRALFF IGILPIIFAL WLRKNIPEAE DWKEKHAGKA
Mopaburce  RKLGVDVVLV EAAQRLCERT VPAIVGERLL GIQRSLGVDV RLGA......
Consensus  .......... ..WGWR.PF. .........L .LR..L.ET. ..........

           251                                                   300
Citlescco  .......... RKHRPD.... ........TRE IFTTIAKNWR IITAGTLLVA
Citklepn   .......... RRHHLA.... ........MRQ VFATLLANWQ VVIAGMMMVA
Ousaerwch  .......... HVEKLEQNDR DGLKAGPGVS FREIATHHWK SLLVCIGLVI
Propescco  .......... HVDKLEQGDR EGLQDGPKVS FKEIATKYWR SLLTCIGLVI
Kgtpescco  .......... .SQQETR ALKEAGSLKG L..WRNR..R AFIMVLGFTA
Pcatpsepu  .......... .SSAETR NDKDAGTIKG L..FRNHA.A AFITVLGYTA
Bap3strhy  .......... ....LPTGTE GGRKKTRTGA FAALRSHP.R QTLLVVGLTI
Nantescco  PVRTMVDILY RGEHRIANIV MTLAAATALW FCFAGNLQNA AIVAVLGLLC
Mopaburce  .......... .....GIEQL SKLPGGRYAA SLSGVREEFD LVVAGVGMVA
Consensus  .......... .......... .......... .......... .......G...

           301                                                   350
Citlescco  MTTTTFYF.. .......... .......... .......... ..ITVYT PTYGRTVLNL
Citklepn   MTTTAFYL.. .......... .......... .......... ..ITVYA PTFGKKVLML
Ousaerwch  ATNVTYYM.. .......... .......... .......... ..LLTYM PSYLSHSLHY
Propescco  ATNVTYYM.. .......... .......... .......... ..LLTYM PSYLSHNLHY
Kgtpescco  AGSLCFYT.. .......... .......... .......... ..FTTYM QKYLVNTAGM
Pcatpsepu  GGSLIFYT.. .......... .......... .......... ..FTTYM QKYLVNTAGM
Bap3strhy  GGNVAFYT.. .......... .......... .......... ..WTTYL PTYATVSTGA
Nantescco  AAIFISFMVQ SAGKRWPTGV MLMVVVLFAF LYSWPIQALL PTYLKTDLAY
Mopaburce  NDELA..... .......... .......... .......... ..AEAGL PCAGGVLCDV
Consensus  .......... .......... .......... .......... ..Y. P.Y.......

           351                                                   400
Citlescco  SARDSLVVTM LVGISNFIWL PIGGAISDRI GRRPVLMGIT LLALVTTLPV
Citklepn   SASDSLLVTL LVAISNFFWL PVGGALSDRF GRRSVLIAMT LLALATAWPA
Ousaerwch  SENHGVLIII AIMIGMLFVQ PVMGLLSDRF GRKPFVVIGS VAMFFLAVPS
Propescco  SEDHGVLIII AIMIGMLFVQ PVMGLLSDRF GRRPFVLLGS VALFVLAIPA
Kgtpescco  HANVASGIMT AALFVFMLIQ PLIGALSDKI GRRTSMLCFG SLAAIFTVPI
```

```
Pcatpsepu  TAKNASYVMT GALFLFMVVQ PFFGMLSDRI GRRNSMLLFG ALGTLCTVPL
Bap3strhy  DKDSAVLAGT VSLIFFGLIQ PLGGLLCERI GGRAMMIGFG VAAAVLTVPL
Nantescco  NPHTVANVLF FSGFGAAVGC CVGGFLGDWL GTRKAYVCSL LASQLLIIPV
Mopaburce  EGRTVDPHVF ACGDVASFEH PSGPIGMRRL ESWDNAQQQG AACARAILGK
Consensus  .......... .......... P..G.L.DR. G.R....... ........P.
```

```
           401                                             450
Citlescco  MNWLTAAPDF TRMTLVLLWF SFFFGMYNGA MVAALTEVMP VYVRTVGFSL
Citklepn   LTMLANAPSF LMMLSVLLWL SFIYGMYNGA MIPALTEIMP AEVRVAGFSL
Ousaerwch  FMLINSDIIG LIFLGLLMLA V.ILNAFTGV MastLPALFP THIRYSALAS
Propescco  FILINSNVIG LIFAGLLMLA V.ILNCFTGV MastLPAMFP THIRYSALAA
Kgtpescco  LSALQNVSSP YAAFGLVMCA LLIVSFYTSI SGILKAEMFP AQVRALGVGL
Pcatpsepu  LMALKTVTSP FMAFVLISLA LCIVSFYTSI SGLVKAEMFP PQVRALGVGL
Bap3strhy  LTAM..TGWF WSVLAVQCAG MLVLTAYTSV SGAINAELFP QELRGRGIGL
Nantescco  FAI..GGANV WVLGLLLFFQ QMLGQGIAGI LPKLIGGYFD TDQRAAGLGF
Mopaburce  RAAAHPLPWF WSDQGDVNIQ ILGFPNATAT PVVREGDGKA TLVWLEEHAA
Consensus  .......... .......... .......... ........P. ...R......
```

```
           451                                             500
Citlescco  AFSLATAIFG GLTPAISTAL VQLTGDKSSP GWW.LMCAAL CGLAATTMLF
Citklepn   AYSLATAVFG GFTPVISTAL IEYTGDKASP GYW.MSFAAI CGLLATCYLY
Ousaerwch  AFNISVLIAG L.TPTVAAWL VESSQNLYMP AYY.LMVIAV IGLLTGLFMK
Propescco  AFNISVLVAG L.TPTLAAWL VESSQNLMMP AYY.LMVVAV VGLITGVTMK
Kgtpescco  SYAVANAIFG G.SAEYVALS LKSIGMETAF FWY.VTLMAV VAFLVSLMLH
Pcatpsepu  AYAVANAAFG G.SAEYVALG LKTLGMENTF YWY.VTAMMA IAFLFSLRLP
Bap3strhy  PYAASVALFG G.TAPYVGTW LKSMGLNDFF PWY.VAVLCL LTALTAIGLP
Nantescco  TYNVG.ALGG ALAPIIGALI AQRLDLGTAL ASLSFSLTFV VILLIGLDMP
Mopaburce  DEPAQIVAAV CINAPADMPI LRRMWQKGV. .RVDRKTLAV QEVSLKSLLQ
Consensus  ........G. .......... .......... .......... ..........
```

```
           501                                             550
Citlescco  ARLSSGYQTV ENKL...... .......... .......... ..........
Citklepn   RRSAVALQTA R......... .......... .......... ..........
Ousaerwch  ETANKPLKGA TPAASDLSEA KEILQEHHDN IEHKIEDITQ QIAELEAKRQ
Propescco  ETANRPLKGA TPAASDIQEA KEILVEHYDN IEQKIDDIDH EIADLQAKRT
Kgtpescco  RKGKGMRL.. .......... .......... .......... ..........
Pcatpsepu  KQAAYLHHDD .......... .......... .......... ..........
Bap3strhy  RPVPDTCDGP ASPAQHLPPD FERHA..... .......... ..........
Nantescco  SRVQRWLRPE ALRTHDAIDG KPFSGAVPFG SAKNDLVKTK S.........
Mopaburce  RSAGAARNGG ANT....... .......... .......... ..........
Consensus  .......... .......... .......... .......... ..........
```

```
           551       561
Citlescco  .......... .
Citklepn   .......... .
Ousaerwch  LLVAQHPRIN D
Propescco  RLVQQHPRID E
Kgtpescco  .......... .
Pcatpsepu  .......... .
Bap3strhy  .......... .
Nantescco  .......... .
Mopaburce  .......... .
Consensus  .......... .
```

Proteins listed subsequently in italics are at least 90% identical to the paired transporters listed in parenthesis and therefore are not included in the alignments: *Cit2escco*, *Citasalty* (Cit1escco). Residues listed in the consensus sequence are present in at least 75% of the aligned transporter sequences. Residues indicated by boldface type are also conserved in at least one other family of the USA/MFS superfamily.

Database accession numbers

	SWISSPROT	PIR	EMBL/GENBANK
Bap3strhy			M64783
Citklepn	P16482	S09681	X51479
Cit1escco	P07661	B23104	M11559
Cit2escco	P07680	A23103	M22041
Citasalty	P24115	JQ0576	S62772; D90203
Kgtpescco	P17448	S10178; S12094	X53027; X52363
Mopaburce			U29532
Nantescco	P41036		U19539; U18997
Ousaerwch			X82267
Pcatpsepu			U48776
Propescco	P30848	S32331	M83089; U14003

References

[1] Marger, M.D. and Saier, M. (1993) Trends Biochem. Sci. 18, 13–20.

[2] Griffith, J. et al. (1992) Curr. Opin. Cell Biol. 4, 684–695.

[3] Culham, D.E. et al. (1992) J. Mol. Biol. 229, 268–276.

[4] Henderson, P.J.F. (1993) Curr. Opin. Cell Biol. 5, 708–721.

[5] Seol, W. and Shatkin, A.J. (1993) J. Bacteriol. 175, 565–567.

Summary

Transporters of the H⁺/nucleotide symporter family, the example of which is the NUPC pyrimidine nucleoside-H⁺ symporter of *Escherichia coli* (Nupcescco), mediate symport (i.e. H⁺-coupled substrate uptake) of nucleosides, particularly pyrimidines. The two known members of the family are found in gram-positive and gram-negative bacteria.

Statistical analysis reveals no apparent relationship between the amino acid sequences of the H⁺/nucleotide symporter family and any other family of transporters. Both transporters are predicted to form nine membrane-spanning helices by the hydropathy of their amino acid sequences.

Nomenclature, biological sources and substrates

CODE	DESCRIPTION [SYNONYMS]	ORGANISM [COMMON NAMES]	SUBSTRATE(S)
Nupcbacsu	Nucleoside-H⁺ symporter [NUPC]	*Bacillus subtilis* [gram-positive bacterium]	H⁺/pyrimidines
Nupcescco	Pyrimidine nucleoside-H⁺ symporter [NUPC, CRU]	*Escherichia coli* [gram-negative bacterium]	H⁺/pyrimidines

Cotransported ions are listed.

Proposed orientation of NUPC in the membrane

The model is based on predictions of membrane-spanning regions and α-helical content. The N-terminus of the protein is illustrated on the inside and is folded

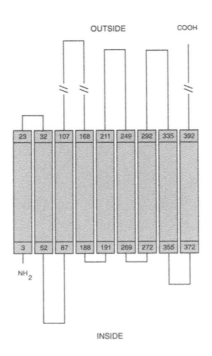

nine times through the membrane. The predicted membrane-spanning helices are portrayed as rectangles. The numbers corresponding to the first and last residue of each membrane-spanning helix are boxed.

Physical and genetic characteristics

	AMINO ACIDS	MOL. WT	CHROMOSOMAL LOCUS
Nupcbacsu	393	42 611	340°
Nupcescco	400	43 473	54.11 minutes

Multiple amino acid sequence alignments

```
           1                                                  50
Nupcbacsu .MKYLIGIIG LIVFLGFAWI ASSGKKRIKI RPIVVMLILQ FILGYILLNT
Nupcescco MDRVLHFVLA LAVVAILALL VSSDRKKIRI RYVIQLLVIE VLLAWFFLNS
Consensus ....L..... L.V...A.. .SS..K.I.I R....L... ..L.....N.

           51                                                100
Nupcbacsu GIGNFLVGGF AKGFGYLLEY AAEGINFVFG GLVNADQTTF FMNVLLPIVF
Nupcescco DVGLGFVKGF SEMFEKLLGF ANEGTNFVFG SNDQGLAEFF FLKVLCPIVF
Consensus ..G...V.GF ...F..LL.. A.EGINFVFG .........F F..VL.PIVF

           101                                               150
Nupcbacsu ISALIGILQK WKVLPFIIRY IGLALSKVNG MGRLESYNAV ASAILGQSEV
Nupcescco ISALIGILQH IRVLPVIIRA IGFLLSKVNG MGKLESFNAV SSLILGQSEN
Consensus ISALIGILQ. ..VLP.IIR. IG..LSKVNG MG.LES.NAV .S.ILGQSE.

           151                                               200
Nupcbacsu FISLKKELGL LNQQRLYTLC ASAMSTVSMS IVGAYMTMLK PEYVVTALVL
Nupcescco FIAYKDILGK ISRNRMYTMA ATAMSTVSMS IVGAYMTMLE PKYVVAALVL
Consensus FI..K..LG. ....R.YT.. A..MSTVSMS IVGAYMTML. P.YVV.ALVL

           201                                               250
Nupcbacsu NLFGGFIIAS IINPYEV.AK EEDMLRVEEE EKQSFFEVLG EYILDGFKVA
Nupcescco NMFSTFIVLS LINPYRVDAS EENIQMSNLH EGQSFFEMLG EYILAGFKVA
Consensus N.F..FI..S .INPY...A. EE........ E.QSFFE.LG EYIL.GFKVA

           251                                               300
Nupcbacsu VVVAAMLIGF VAIIALINGI FNAV.....F GISFQGILGY VFAPFAFLVG
Nupcescco IIVAAMLIGF IALIAALNAL FATVTGWFGY SISFQGILGY IFYPIAWVMG
Consensus ..VAAMLIGF .A.IA..N.. F..V...... .ISFQGILGY .F.P.A...G

           301                                               350
Nupcbacsu IPWNEAVNAG RLMATKMVSN EFVAMTVLTQ NGFHFSGRTT AIVSVFLVSF
Nupcescco VPSSEALQVG SIMATKLVSN EFVAMMDLQK IASTLSPRAE GIISVFLVSF
Consensus .P..EA...G ..MATK.VSN EFVAM..L.. .....S.R.. .I.SVFLVSF

           351                                               400
Nupcbacsu ANFSSIGIIA GAVKGLNEKQ GNVVARFGLK LLYGATLVSF LSAAIVGLIY
Nupcescco ANFSSIGIIA GAVKGLNEEQ GNVVSRFGLK LVYGSTLVSV LSASIAALVL
Consensus ANFSSIGIIA GAVKGLNE.Q GNVV.RFGLK L.YG.TLVS. LSA.I.....
```

Database accession numbers

	SWISSPROT	PIR	EMBL/GENBANK
Nupcbacsu	P39141		X82174
Nupcescco	P33031	S37076	X74825

Summary

Transporters of the sugar phosphate transporter family, the example of which is the UHPT hexose phosphate transporter of *Escherichia coli* (Uhptescco), mediate uptake of structurally dissimilar phosphoesters and dicarboxylates, including hexose phosphates, glycerol-3-phosphate, hexuronates and phthalates, by an exchange in which the accumulation of a substrate phosphate is coupled to the electroneutral release of inorganic phosphate or organophosphate [1-3]. Members of the family are found in both gram-negative and gram-positive bacteria.

Statistical analysis of multiple amino acid sequence comparisons suggests that the sugar phosphate transporter family may be distantly related to the uniporter-symporter-antiporter transporter superfamily, also known as the major facilitator superfamily (MFS) [4-6]. They are predicted to contain 12 membrane-spanning helices by the hydropathy of their amino acid sequences and activities of reporter gene fusion [7].

Several amino acid sequence motifs are highly conserved in the sugar phosphate transporter family, including motifs unique to the family, signature motifs of the USA/MFS superfamily, and motifs necessary for function by the criterion of site-directed mutagenesis [1,8].

Nomenclature, biological sources and substrates

CODE	DESCRIPTION [SYNONYMS]	ORGANISM [COMMON NAMES]	SUBSTRATE(S)
Exutescco	Hexuronate transporter [EXUT]	*Escherichia coli* [gram-negative bacterium]	Hexuronate
Glptescco	Glycerol-3-phosphate transporter [GLPT]	*Escherichia coli* [gram-negative bacterium]	Glycerol-3-phosphate
Glptbacsu	Glycerol-3-phosphate transporter [GLPT]	*Bacillus subtilis* [gram-positive bacterium]	Glycerol-3-phosphate
Gudtpsepu	Glucarate transporter [GUDT]	*Pseudomonas putida* [gram-negative bacterium]	Glucarate
Gudtbacsu	Glucarate transporter [GUDT, YCBE]	*Bacillus subtilis* [gram-positive bacterium]	Glucarate
Pgtpsalty	Phosphoglycerate transporter [PGTP]	*Salmonella typhimurium* [gram-negative bacterium]	Phosphoglycerate
Phlepsefl	2,4-Diacetylphloro-glucinol synthesis protein [PHLE]	*Pseudomonas fluorescens* [gram-negative bacterium]	2,4-Diacetylphloro-glucinol synthesis
Pht1psepu	Phthalate transporter [PHT1]	*Pseudomonas putida* [gram-negative bacterium]	Phthalate
Uhpcsalty	Regulatory protein [UHPC]	*Salmonella typhimurium* [gram-negative bacterium]	Hexose phosphates
Uhpcescco	Regulatory protein [UHPC]	*Escherichia coli* [gram-negative bacterium]	Hexose phosphates
Uhptsalty	Hexose phosphate transporter [UHPT]	*Salmonella typhimurium* [gram-negative bacterium]	Hexose phosphates
Uhptescco	Hexose phosphate transporter [UHPT]	*Escherichia coli* [gram-negative bacterium]	Hexose phosphates

Phylogenetic tree

Proteins listed subsequently in italics are at least 90% identical to the paired transporters listed in parenthesis and therefore are not included in the phylogenetic tree: *Uhptsalty* (Uhptescco); *Uhpcsalty* (Uhpcescco).

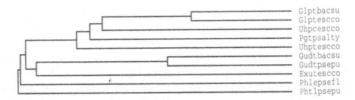

Glptbacsu
Glptescco
Uhpcescco
Pgtpsalty
Uhptescco
Gudtbacsu
Gudtpsepu
Exutescco
Phlepsefl
Phtlpsepu

Proposed orientation of UHPT in the membrane

The model is based on predictions of membrane-spanning regions and α-helical content. The N-terminus of the protein is illustrated on the inside and is folded 12 times through the membrane [1]. The predicted membrane-spanning helices are portrayed as rectangles. The numbers corresponding to the first and last residue of each membrane-spanning helix are boxed. Residues that are conserved in more than 75% of the aligned transporters (see below) are shown. Consensus residues indicated by an asterisk are not conserved in UHPT.

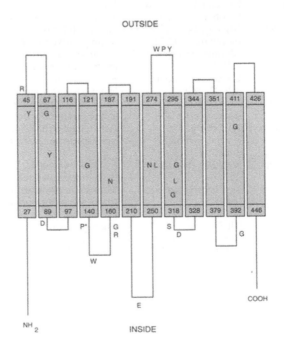

Physical and genetic characteristics

	AMINO ACIDS	MOL. WT	CHROMOSOMAL LOCUS
Exutescco	472	51 656	69.6–69.62 minutes
Glptescco	452	50 310	50.59 minutes
Glptbacsu	444	49 801	18°
Gudtpsepu	456	49 779	
Gudtbacsu	455	49 254	999°
Pgtpsalty	406	44 853	
Pht1psepu	451	49 306	
Phlepsefl	423	45 235	
Uhpcsalty	442	48 254	
Uhpcescco	439	48 256	82.84 minutes
Uhptescco	463	50 606	82.8 minutes
Uhptsalty	463	50 752	

Multiple amino acid sequence alignments

```
          1                                                50
Glptbacsu ...............ML NIFKPAPHIE R.LDDS.KM. DAAYKRLRLQ
Glptescco ...............ML SIFKPAPHKA R.LPAA.EI. DPTYRRLRWQ
Uhpcescco ...............ML PFLKAPADAP L.MTDKYEI. DARYRYWRRH
Pgtpsalty ...............ML TILKTGQSAH K.VPPE.KV. QATYGRYRIQ
Uhptescco ...............ML AFLNQVRKPT LDLPLEVRR. KMWFKPF.MQ
Gudtbacsu ...............MK KDFASVTP.. AGKKTSVRWF
Gudtpsepu ...............MQ.. EPKQTRVRYL
Exutescco MATFGACRFF FGYPVVTNIF SLWRDDGRAS CGYNKTMRFY MRKIKGLRWY
Phlepsefl ...............MESTYLA TRPWGGYERR
Pht1psepu ...............M TTTAATDAVP HLLQRSHERI EKVYRKVTLR
Consensus ..................................................

          51                                               100
Glptbacsu VFIGIFIGYA GYYLLRKNFA FAIPYLQEQG .FSKTELGLV LAAVSIAYGF
Glptescco IFLGIFFGYA AYYLVRKNFA LAMPYLVEQG .FSRGDLGFA LSGISIAYGF
Uhpcescco ILLTIWLGYA LFYFTRKSFN AAVPEILANG VLSRSDIGLL ATLFYITYGV
Pgtpsalty ALLSVFLGYL AYYIVRNNFT LSTPYLKEQL DLSATQIGLL SSCMLIAYGI
Uhptescco SYLVVFIGYL TMYLIRKNFN IAQNDMISTY GLSMTQLGMI GLGFSITYGV
Gudtbacsu IVFMLFLVTS INYADRATLS ITGDSVQHDL GLDSVAMGYV FSAFGWAYVI
Gudtpsepu ILLMLFLVTT INYADRATIS IAGSSIQKDF GLDAVTLGYI FSAFGWAYVL
Exutescco MIALVTLGTV LGYLTRNTVA ARAPTLMEEL NISTQQYSYI IAAYSAAYTV
Phlepsefl MVVLLSLSFG LVGLDRFIIM PLFPVIMHDL ALDYQDLGLL SAILAFAWGG
Pht1psepu LMTFIFVAWV LNYLDRVNIS FAQVYLKHDL GMSDADLRTR RKLVFHRLHR
Consensus ..........Y..R.................G.........Y..

          101                                              150
Glptbacsu SKFIMGMVSD RCNPRYFLAT GLFLSAIVNI LF.VSMPWV. TSSVTIMFIF
Glptescco SKFIMGSVSD RSNPRVFLPA GLILAAAVML FM.GFVPWA. TSSIAVMFVL
Uhpcescco SKFVSGIVSD RSNARYFMGI GLIATGIINI LF.GF..... STSLWAFAVL
Pgtpsalty SKGVMSSLAD KASPKVFMAC GLVLCAIVNV GL.GF..... SSAFWIFAAL
Uhptescco GKTLVSYYAD GKNTKQFLPF MLILSAICML GFSASMGSG. SVSLFLMIAF
Gudtbacsu GQLPGGWLLD RFGSKTIIAL SIFFWSFFTL LQGAIGFFSA GTAIILLFAL
Gudtpsepu GQIPGGWLLD RFGSKKVYAG SIFTWSLFTL LQGYIGEFGI STAVVLLFLL
Exutescco MQPVAGYVLD VLGTKIGYAM FAVLWAVFCG ATALAGSWGG ......LAVA
Phlepsefl SALFMGVAIR RLGTKQLLVL SITLVSLLA. .....GASAL ISSLMGLVLL
```

331

```
Phtlpsepu IGNTQYAYLQ KIGARLTITR IMVLWGLI.. ....SASMAF MTTPTEFYIA
Consensus ........D .......... .......... .......... ..........
```

```
          151                                               200
Glptbacsu MFINGWFQGM GWPPCGRTMA HWFS....IS ERGTKMSIWN VAHNIGGGIL
Glptescco LFLCGWFQGM GWPPCGRTMV HWWS....QK ERGGIVSVWN CAHNVGGGIP
Uhpcescco WVLNAFFQGW GSPVCARLLT AWYS....RT ERGGWWALWN TAHNVGGALI
Pgtpsalty VVFNGLFQGM RRP.....LV YYYCKLVPRR ERGRVGAFWN ISHNVGGGIV
Uhptescco YALSGFFQST GGSCSYSTIT KW....TPRR KRGTFLGFWN ISHNLGGAGA
Gudtbacsu RFLVGLSEAP SFPGNGRVVA SWF....PSS ERGTASAFFN SAQYFAIVIF
Gudtpsepu RFMVGLAEAP SFPGNARIVA SWF....PTK ERGTASAIFN SAQYFATRAV
Exutescco RGAVGAAEAA MIPAGLKASS EWF....PAK ERSIAVGYFN VGSSIGAMIA
Phlepsefl RALMGICEGA FTPVSIIVTD E....VSQPC RRGLNLGIQQ ALFPIIGLCL
Phtlpsepu RALLGAAEAG FWPGIILYLT YWY....PGA RRARITSRFL LAIAAAGIIG
Consensus ....G..... ..P....... .W........ ...RG..... N.........
```

```
          201                                               250
Glptbacsu APLVTLGIAM FVT...WK.. .SVFFFPAII AIIISFLIVL LVRDTPQSCG
Glptescco PLLFLLGMAW FND...WH.. .AALYMPAFC AILVALFAFA MMRDTPQSCG
Uhpcescco P.IVMAAAAL HYG...WR.. .AGMMIAGCM AIVVGIFLCW RLRDRPQALG
Pgtpsalty APIVGAAFAI LGS...EHWQ SASYIVPACV AVIFALIVLV LGKGSPRKEG
Uhptescco AGVALFGANY LFD...GH.V IGMFIFPSII ALIVGFIGLR YGSDSPESYG
Gudtbacsu PPLMGWLTHS F.......GW HSVFVVMGIA GILLAVIWLK TVYEPKKHPK
Gudtpsepu RALDGLDRLH L.......RL AARVHRHGRP GHCVLAHLVD GDLRAERSPA
Exutescco PPLVVWAIVM H.......SW QMAFIISGAL SFIWAMAWLI FYKHPRDQKH
Phlepsefl GPL..LAGVL FEM...FGSW RAVFAIISLP GLLVAWYLYR TYQPSQAPHP
Phtlpsepu GPLSGWILTH FVDVMGMKNW QWMFILEGLP AAVMGVMAYF YLVDKPEQAK
Consensus .......... .......... .......... .......... ..........
```

```
          251                                               300
Glptbacsu LPPIEEYRND ....YPKHAF KNQEKELTTK EILFQYVLNN KFLWYIAFAN
Glptescco LPPIEEYKND ....YPDDYN EKAEQELTAK QIFMQYVLPN KLLWYIAIAN
Uhpcescco LPAVGEWRHD ....ALEIAQ QQEGAGLTRK EILTKYVLLN PYIWLLSFCY
Pgtpsalty LPSLEQMMPE EKVVLKTKNT AKAPENMSAW QIFCTYVVRN KNAWYISLVD
Uhptescco LGKAEELFGE E...ISEEDK ETESTDMTKW QIFVEYVLKN KVIWLLCFAN
Gudtbacsu VNEAELAYIE QGGGLISMDD SKSKQETESK WPYIKQLLTN RMLIGVYIAQ
Gudtpsepu GYAAEVRSSP H.GGLVDLED SKDKKDGGPK WDYIRQLLTN RMMMGIYLGQ
Exutescco LTDEE...... ..RDYIINGQ EAQHQVSTAK KMSVGQILRN RQFWGIALPR
Phlepsefl RPLVEPSGSQ WRTALSSGNV R.......... ......LNI ALMLCILTCQ
Phtlpsepu WLDDEEKSII LDAL...AAD RAGKKPVTDK RHAVLAALKD PRVYVLAAGW
Consensus ....E..... .......... .......... ......L.N ..........
```

```
          301                                               350
Glptbacsu VFVYFVRYGV VDWAPTYLTE AKGFSPEDSR WSYFLYEYAG IPGTILCGWI
Glptescco VFVYLLRYGI LDWSPTYLKE VKHFALDKSS WAYFLYEYAG IPGTLLCGWM
Uhpcescco VLVYVVRAAI NDWGNLYMSE TLGVDLVTAN TAVTMFELGG FIGALVAGWG
Pgtpsalty VFVYMVRFGM ISWLPIYLLT VKHFSKEQMS VAFLFFEWAA IPSTLLAGWL
Uhptescco IFLYVVRIGI DQWSTVYAFQ ELKLSKAVAI QGFTLFEAGA LVGTLLWGWL
Gudtbacsu YCITTLTYFF LTWFPVYLVQ ARGMSILEAG FVASLPALCG FAGGVLGGIV
Gudtpsepu FCINALTYFF LTWFPVYLVQ ERGMTILKAG IIASLPAICG FLGGVLGGVI
Exutescco FLAEPAWGTF NAWIPLFMFK VYGFNLKEIA MFAWMPMLFA DLGCILGGYL
Phlepsefl FVLCAL.... ...LPSYLTD VLHLSNFSMA MIISAIGLGG FFGQLVIPGL
Phtlpsepu ATVPLCGTIL NYWTPTIIRN TGIQDVLHVG LLSTVPYIVG AIAMILIARS
Consensus .......... ..W.P.Y... .......... .......... ..G..L.G..
```

```
          351                                                    400
Glptbacsu SDRFFK..... ..SRRAPAGV LFMAGVFIAV LVYWLNP.AG NPLVDNIALI
Glptescco SDKVFR..... ..GNRGATGV FFMTLVTIAT IVYWMNP.AG NPTVDMICMI
Uhpcescco SDKLFN,,,, ..GNRGPMNL IFAAGILLSV GSLWLMP.FA SYVMQATCFF
Pgtpsalty SDKLFK..... ..GRRMPLAM ICMALIFVCL IGYW..K.SE SLLMVTIFAA
Uhptescco SD.LAN..... ..GRRGLVAC IALALIIATL GVY..QH.AS NEYIYLASLF
Gudtbacsu SDILLKK.GR SLTFARKVPI IAGMLLSCSM IVCNYTD.SA WLVVVIMSLA
Gudtpsepu SDTLLRR.GN SLSVARKTPI VCGMVLSMSM IICNYVD.AD WMVVCFMALA
Exutescco PPLFQRWFGV NLIVSRKMVV TLGAVLMIGP GMIGLFT.NP YVAIMLLCIG
Phlepsefl SDQLGRKPVV SICF..LIST LLVGLLIISP PLPWLL.... FLQLFFLSFI
Phtlpsepu SDIRLERRKH .......... FFFSIAFGAL GACLLPHVVD SAIISITCLA
Consensus SD....... .......... .......... .......... ..........

          401                                                    450
Glptbacsu SIGFLIYGPV MLIGLQAIDL APKKAAGTAA GLTGFFGYIG GSAFANAIMG
Glptescco VIGFLIYGPV MLIGLHALEL APKKAAGTAA GFTGLFGYLG GSVAASAIVG
Uhpcescco TIGFFVFGPQ MLIGMAAAEC SHKEAAGAAT GFVGLFAYLG ASLAGWP.LA
Pgtpsalty IVGCLIYVPQ FLASVQTMEI VPSFAVGSAV GLRGFMSYIF GASLGTSLFC
Uhptescco ALGFLVFGPQ LLIGVAAVGF VPKKAIGAAD GIKGTFAYLI GDSFAKLGLG
Gudtbacsu FFG.KGFGAL GWAVVS..DT SPKECAGLSG GLFNTFGNI. ASITTPIIIG
Gudtpsepu FFG.KAIGAL GWAVVS..DT SPKQIAGLSG GLFNTFGNL. SSISTPIIIG
Exutescco GFAHQALSGA LITLSS..DV FGRNEVATAN GLTGMSAWL. ASTLFALVVG
Phlepsefl NFSLICITVG PLTS....ES VPPSLLATAT GLVVGCGEIL GGGVAPVVAG
Phtlpsepu MIAVSYFGAA AIIWSIPPAY LNDESAAGGI SAISSLGQI. GAFCAPIGLG
Consensus .......... .......... .......... G......... .........G

          451                                                    500
Glptbacsu FVVD...... RFNWNGGFIM LISSCILAIV FLALTWNTGK RAEHV.....
Glptescco YTVD...... FFGWDGGFMV MIGGSILAVI LLIVVMIGEK RRHEQLLQER
Uhpcescco KVLD...... TWHWSGFFVV ISIAAGISAL LLLPFLNAQT PREA......
Pgtpsalty .......... .......... .......... .......... ..........
Uhptescco MIADGTPVFG LTGWAGTFAA LDIAAIGCIC LMAIVAVMEE RKIRREKKIQ
Gudtbacsu YIVNATG... ..SFNGAL.V FVGANAIAAI LSYLLLVGPI KRVVLKKQEQ
Gudtpsepu YIIAATG... ..VSKWRWSS WVPTHSFAAI .SYLFIVGEI NRIELKGVTD
Exutescco .ALADTI.... ..GFSPLFAV LAVFDLLGAL VIWTVLQNKP AIEVAQETHN
Phlepsefl YIAVN..... .WGLTAILFL ALAGSLMGGL LSLRLKEASP VFNDRADYGP
Phtlpsepu WINTVTG... ..SLAIGLTI IGALVLAGGM AVLIAVPANA LSEKPLTDE.
Consensus .......... .......... .......... .......... ..........

          501           518
Glptbacsu .......... ......
Glptescco NGG....... ......
Uhpcescco .......... ......
Pgtpsalty .......... ......
Uhptescco QLTVA..... ......
Gudtbacsu DPDQSLPV.. ......
Gudtpsepu EPATTAHPGE LLPTTRKV
Exutescco DPAPQH.... ......
Phlepsefl LPARLTLEDK ......
Phtlpsepu .......... ......
Consensus .......... ......
```

Proteins listed subsequently in italics are at least 90% identical to the paired transporters listed in parenthesis and therefore are not included in the alignments: *Uhptsalty* (Uhptescco); *Uhpcsalty* (Uhpcescco). Residues listed in the consensus sequence are present in at least 75% of the transporter sequences shown. Residues indicated by boldface type are also conserved in at least one other family of the USA/MFS superfamily.

Database accession numbers

	SWISSPROT	PIR	EMBL/GENBANK
Exutescco	P42609		U18997
Glptescco	P08194	S00868	Y00536
Glptbacsu	P37948	S37250	Z26522
Gudtpsepu	P42205		M69160
Gudtbacsu	P42237		D30808
Pgtpsalty	P12681	A31089	M21278
Phlepsefl			U41818
Phtlpsepu	Q05181		D13229
Uhptsalty	P27670	D41853	M89480
Uhpcescco	P09836	G41853	M17102; M89479
Uhpcsalty	P27669	C41853	M89480
Uhptescco	P13408	Q00500	M17102; M89479

References

1 Maloney, P. (1994) Curr. Opin. Cell Biol. 6, 571–582.
2 Kadner, R.J. et al. (1994) Res. Microbiol. 145, 381–387.
3 Kadner, R.J. et al. (1993) J. Bioenerg. Biomembr. 25, 637–645.
4 Marger, M.D. and Saier, M. (1993) Trends Biochem. Sci. 18, 13–20.
5 Griffith, J. et al. (1992) Curr. Opin. Cell Biol. 4, 684–695.
6 Henderson, P.J.F. (1993) Curr. Opin. Cell Biol. 5, 708–721.
7 Lloyd, A. and Kadner, R. (1990) J. Bacteriol. 172, 1688–1693.
8 Yan, R.T. and Maloney, P.C. (1995) Proc. Natl Acad. Sci. USA 92, 5973–5976.

Part 9

H⁺-Dependent Antiporters

Summary

Transporters of the H⁺/vesicular amine antiporter family, the example of which is the vesicular amine transporter 2 (VAT2) of humans (Vmt2homsa), mediate accumulation of biogenic monoamines, including catecholamines and indoleamines within intracellular vesicles. The mechanism of transport is antiport (i.e. H⁺-coupled substrate efflux)[1-4]. In addition, members of the H⁺/vesicular amine antiporter family mediate the accumulation of several toxic compounds into intracellular vesicles, mimicing multidrug resistance proteins[5]. Members of the H⁺/vesicular amine antiporter family may be inhibited by vesamicol or reserpine[5]. These transporters are widely distributed in nature, occurring in both invertebrates and vertebrates.

Statistical analysis of multiple amino acid sequence comparisons places the H⁺/vesicular amine antiporter family in the uniporter-symporter-antiporter (USA) superfamily, also known as the major facilitator superfamily (MFS)[5-7]. They are predicted to form 12 membrane-spanning helices by the hydropathy of their amino acid sequences[7,8]. Members of the H⁺/vesicular amine antiporter family are glycosylated.

Several amino acid sequence motifs are highly conserved in the H⁺/vesicular amine antiporter family, including motifs unique to the family and signature motifs of the USA/MFS superfamily[5,7,9]. In *C. elegans*, defects in UN17 cause deficits in neuromuscular function and deletion of UN17 is lethal[1]. The possibility that defective neurotransmitter transport is similarly a source of human neurodegenerative disease has been proposed[2].

Nomenclature, biological sources and substrates

CODE	DESCRIPTION [SYNONYMS]	ORGANISM [COMMON NAMES]	SUBSTRATE(S)
Un17caeel	Vesicular acetylcholine transporter [UN17, UNC17]	*Caenorhabditis elegans* [nematode]	H⁺/acetylcholine
Vacthomsa	Vesicular acetylcholine transporter [VACT]	*Homo sapiens* [human]	H⁺/acetylcholine
Vacttorma	Vesicular acetylcholine transporter [VACT]	*Torpedo marmorata* [ray]	H⁺/acetylcholine
Vacttoroc	Vesicular acetylcholine transporter [Vesamicol binding protein]	*Torpedo ocellata* [ray]	H⁺/acetylcholine
Vactratno	Vesicular acetylcholine transporter [Vesamicol binding protein]	*Rattus norvegicus* [rat]	H⁺/acetylcholine
Vmt1ratno	Chromaffin granule amine transporter [VMT1, CGAT, VAT1]	*Rattus norvegicus* [rat]	H⁺/amine neurotransmitters
Vmt2bosta	Synaptic vesicle amine transporter [VMT2, VMAT2, VAT2]	*Bos taurus* [cow]	H⁺/amine neurotransmitters
Vmt2homsa	Synaptic vesicle amine transporter [VMT2, SVMT, VAT2, VMAT2]	*Homo sapiens* [human]	H⁺/amine neurotransmitters
Vmt2ratno	Synaptic vesicle amine transporter [VMT2, SVAT, VAT2, VMAT]	*Rattus norvegicus* [rat]	H⁺/amine neurotransmitters

Cotransported ions are listed.

Phylogenetic tree

Proteins listed subsequently in italics are at least 90% identical to the paired transporters listed in parenthesis and therefore are not included in the phylogenetic tree: *Vmt2ratno*, *Vmt2bosta* (Vmt2homsa); *Vactratno* (Vacthomsa).

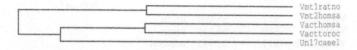

```
                                    Vmt1ratno
                                    Vmt2homsa
                                    Vacthomsa
                                    Vacttoroc
                                    Un17caeel
```

Proposed orientation of VAT2 in the membrane

The model is based on predictions of membrane-spanning regions and α-helical content. The N-terminus of the protein is illustrated on the inside and is folded 12 times through the membrane. The predicted membrane-spanning helices are portrayed as rectangles. The number corresponding to the first and last residue of each membrane-spanning helix is boxed. Residues that are conserved in more than 75% of the aligned transporters (see below) are shown. Consensus residues indicated by an asterisk are not conserved in VAT2.

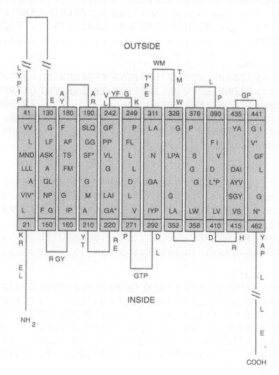

Physical and genetic characteristics

	AMINO ACIDS	MOL. WT	EXPRESSION SITES	K_m	CHROMOSOMAL LOCUS
Un17caeel	532	58 643	cholinergic neurons		
Vacthomsa	532	56 961	brain		
Vacttorma	568	61 641	brain	Acetylcholine: 300 μM [1]	
Vacttoroc	511	55 690	brain		
Vactratno	530	56 536	brain		
Vmt1ratno	521	55 935	adrenal		
Vmt2bosta	517	55 755	brainstem, stomach		
Vmt2homsa	514	55 696	brainstem, stomach		10q25
Vmt2ratno	515	55 779	brainstem, stomach		

Multiple amino acid sequence alignments

```
          1                                                  50
Vmt1ratno ................ ........M LQVVLGAPQR LLKEGRQSRK LVLVVVFVAL
Vmt2homsa ................ ........ MALSELALVR WLQESRRSRK LILFIVFLAL
Vacthomsa MESAEPAGQA RAAATKLSEA VG.......A ALQEPRRQRR LVLVIVCVAL
Vacttoroc ....MVVGQA KAAMGKISSA IGERSKRISG AMNEPLRKRK ILLVIVCIAM
Un17caeel ......MGFN VPVINRDSEI LKADAKKW.. .LEQQDNQKK CVLVIVSIAL
Consensus ................ ........... .L.E....RK ..LVIV..AL

          51                                                 100
Vmt1ratno LLDNMLLTVV VPIVPTFLYA TEFKDSNSSL HRGPSVSSQQ ALTSPAFSTI
Vmt2homsa LLDNMLLTVV VPIIPSYLYS IKHEKNATEI QTARPV..HT ASISDSFQSI
Vacthomsa LLDNMLYMVI VPIVPDYI.. .......... AHMRGGGEGP TRTPEVWEPT
Vacttoroc LLDNMLYMVI VPIVPNYL.. .......... ETIR...... ........T
Un17caeel LLDNMLYMVI VPIIPKYL.. .......... RDIHN..... ....YQVT
Consensus LLDNML..V. VPI.P.YL.. .......... .......... ..........

          101                                                150
Vmt1ratno FSFFDNTTTT VEEHVPFRVT WTNGTIPPPV TEASSVPKNN CLQGIEFLEE
Vmt2homsa FSYYDN.STM VTGNATRDLT LHQTATQHMV TNASAVPSDC PSEDKDLL.N
Vacthomsa LPLPTPANA. .......... .SAYTANTSA SPTAAWPAGS ALRPRYPTES
Vacttoroc YKLVYITIP. .......... .SNGTNGSLL NST.....QR AVLERNPNAN
Un17caeel FE.GYHNET. .......... .SQLANGTYL ........VR EVGGRINFLD
Consensus .......... .......... .......... .......... ..........

          151                                                200
Vmt1ratno ENVRIGILFA SKALMQLLVN PFVGPLTNRI GYHIPMFVGF MIMFLSTLMF
Vmt2homsa ENVQVGLLFA SKATVQLITN PFIGLLTNRI GYPIPIFAGF CIMFVSTIMF
Vacthomsa EDVKIGVLFA SKAILQLLVN PLSGPFIDRM SYDVPLLIGL GVMFASTVLF
Vacttoroc EDIQIGVLFA SKAILQLLSN PFTGTFIDRV GYDIPLLIGL TIMFFSTITF
Un17caeel EELELGWLFA SKALLQIFVN PFSGYIIDRV GYEIPMILGL CTMFFSTAIF
Consensus E....G.LFA SKA..QL..N PF.G....R. GY.IP...G. ..MF.ST..F

          201                                                250
Vmt1ratno AFSGTYALLF VARTLQGIGS SFSSVAGLGM LASVYTDNYE RGRAMGIALG
Vmt2homsa AFSSSYAFLL IARSLQGIGS SCSSVAGMGM LASVYTDDEE RGNVMGIALG
```

```
Vacthomsa AFAEDYATLF AARSLQGLGS AFADTSGIAM IADKYPEEPE RSRALGVALA
Vacttoroc AFGESYAILF AARSLQGLGS AFADTSGIAM IADKYTEESE RTQALGIALA
Un17caeel ALGKSYGVLL FARSLQGFGS AFADTSGLAM IADRFTEENE RSAALGIALA
Consensus AF...YA... .ARSLQG.GS .F,,,,G,.M .A..YT...E R..A.GIAL.
```

```
          251                                             300
Vmt1ratno GLALGLLVGA PFGSVMYEFV GKSSPFLILA FLALLDGALQ LCILWP....
Vmt2homsa GLAMGVLVGP PFGSVLYEFV GKTAPFLVLA ALVLLDGAIQ LFVLQP....
Vacthomsa FISFGSLVAP PFGGILYEFA GKRVPFLVLA AVSLFDALLL LAVAKPFSAA
Vacttoroc FISFGSLVAP PFGGVLYQFA GKWVPFLVLS FVCLLDGILL LMVVTPF..A
Un17caeel FISFGCLVAP PFGSVLYSLA GKPVPFLILS FVCLADAIAV FMVINPHRRG
Consensus ....G.LV.P PFG.VLY.F. GK..PFL.L. ...L.D.... L.V..P....
```

```
          301                                             350
Vmt1ratno SKVSPESAMG TSLLTLLKDP YILVAAGSIC LANMGVAILE PTLPIWMMQT
Vmt2homsa SRVQPESQKG TPLTTLLKDP YILIAAGSIS FANMGIAMLE PALPIWMMET
Vacthomsa ARARARANLPVG TPIHRLMLDP YIAVVAGALT TCNIPLAFLE PTIATWMKHT
Vacttoroc SRTRGNTLQG TPIHKLMIDP YIAVVAGALT TCNIPLAFLE PTISNWMKKT
Un17caeel TDSHGEKVQG TPMWRLFMDP FIACCSGALI MANVSLAFLE PTITTWMSEM
Consensus ......G TP...L..DP YI...AG... ..N...A.LE PT...WM..T
```

```
          351                                             400
Vmt1ratno MC.SPEWQLG LAFLPASVAY LIGTNLFGVL ANKMGR..WL CSLVGMVAVG
Vmt2homsa MC.SRKWQLG VAFLPASISY LIGTNIFGIL AHTMGR..WL CALLGMIIVG
Vacthomsa M.AASEWEMG MAWLPAFVPH VLGVYLTVRL AARYPHLQWL YGALGLAVIG
Vacttoroc M.NASEWQMG ITWLPAFFPH ILGVYITVKL AAKYPNYQWL YGAFGLVIIG
Un17caeel MPDTPGWLVG VIWLPPFFPH VLGVYVTVKM LRAFPHHTWA IAMVGLAMEG
Consensus M.....W..G ...LPA.... ..G......L A.......WL ....G....G
```

```
          401                                             450
Vmt1ratno ISLLCVPLAH NIFGLIGPNA GLGFAIGMVD SSLMPIMGYL VDLRHTSVYG
Vmt2homsa VSILCIPFPK NIYGLIAPNF GVGFANGMVD SSMMPIMGYL VDLRHVSVYG
Vacthomsa ASSCIVPACR SFAPLVVSLC GLCFGIALVD TALLPTLAFL VDVRHVSVYG
Vacttoroc VSSCTIPACR NFEELIIPLC ALCFGIALVD TALLPTLAFL VDIRYVSVYG
Un17caeel IACFAIPYTT SVMQLVIPLS FVCFGIALID TSLLPMLGHL VDTRHVSVYG
Consensus .S....P... ....L..P.. ...F.I..VD ..L.P....L VD.RH.SVYG
```

```
          451                                             500
Vmt1ratno SVYAIADVAF CVGFAIGPST GGVIVQVIGF PWLMVIIGTI NIIYAPLCCF
Vmt2homsa SVYAIADVAF CMGYAIGPSA GGAIAKAIGF PWLMTIIGII DILFAPLCFF
Vacthomsa SVYAIADISY SVAYALGPIV AGHIVHSLGF EQLSLGMGLA NLLYAPVLLL
Vacttoroc SVYAIADISY SVAYALGPIM AGQIVHDLGF VQLNLGMGLV NILYAPALLF
Un17caeel SVYAIADISY SLAYAFGPII AGWIVTNWGF TALNIIIFAT NVTYAPVLFL
Consensus SVYAIAD... ...YA.GP.. .G.IV...GF ..L....G.. N..YAP....
```

```
          501                                             550
Vmt1ratno LQN....... .PPAKEEKRA IL.SQECPTE TQMYTFQKPT KAFPLGENSD
Vmt2homsa LRS....... .PPAKEEKMA ILMDHNCPIK TKMYT.QNNI QSYPIGEDEE
Vacthomsa LRNVGLL... .TRSRSERD VLLDEPPQGL YDAVRLRERP VSGQDGEPRS
Vacttoroc LRNVCQM.... ..KPSLSERN ILLEDGPKGL YDTIIMEERK AAKEPHGTSS
Un17caeel LRKVHSYDTL GAKGDTAEMT QLNSSAPAGG YNGKPEATTA ESYQGWEDQQ
Consensus L......... .........L ......... ......... ......E...
```

```
           551                              585
Vmt1ratno DPSSGE.... ............ .......... .....
Vmt2homsa SESD.... ............ .......... .....
Vacthomsa PPGPFDECED DYNYYYTRS. ............ .....
Vacttoroc GNHSVHAVLS DQEGYSE.... ............ .....
Un17caeel SYQNQAQIPN HAVSFQDSRP QAEFPAGYDP LNPQW
Consensus ........ ............ ............ .....
```

Proteins listed subsequently in italics are at least 90% identical to the paired transporters listed in parenthesis and therefore are not included in the alignments: *Vmt2ratno*, *Vmt2bosta* (Vmt2homsa); *Vactratno* (Vacthomsa). Residues listed in the consensus sequence are present in at least 75% of the aligned transporter sequences. Residues indicated by boldface type are also conserved in at least one other family of the USA/MFS superfamily.

Database accession numbers

	SWISSPROT	*PIR*	*EMBL/GENBANK*
Un17caeel	P34711		U09277; L19621
Vacthomsa			U09210
Vacttorma		S43685	U05591
Vacttoroc		S43686; S48219	U05339
Vactratno		A54965; I84492	X80395
Vmt1ratno	Q01818	A43319	M97380
Vmt2bosta		S41081; S39440	U02876
Vmt2homsa	Q05940	S29810	L23205; L09118
Vmt2ratno	Q01827		M97381; L00603

References

1 Usdin, T. (1995) Trends Neurosci. 18, 218–224
2 Edwards, R.H. (1993) Ann. Neurol. 34, 638–645.
3 **Schuldiner, S. (1994) J. Neurochem. 62, 2067–2078.**
4 **Schuldiner, S. et al. (1995) Physiol. Rev. 75, 369–392.**
5 **Paulsen, I. et al. (1996) Microbiol. Rev. 60, 575–608.**
6 **Marger, M.D. and Saier, M. (1993) Trends Biochem. Sci. 18, 13–20.**
7 **Griffith, J. et al. (1992) Curr. Opin. Cell Biol. 4, 684–695.**
8 **Henderson, P.J.F. (1993) Curr. Opin. Cell Biol. 5, 708–721.**
9 Varela, M. et al. (1995) Mol. Membr. Biol. 12, 271–277.

Summary

Transporters of the 14-helix H$^+$/multidrug antiporter family, the example of which is the QACA multidrug resistance protein of *Staphylococcus aureus* (Qacastaau), mediate resistance to one or more structurally dissimilar antibiotics, antiseptics and disinfectants [1-4]. The mechanism of transport, where known, is antiport (i.e. H$^+$-coupled substrate efflux). In some members of the 14-helix H$^+$/multidrug antiporter family (e.g. EMRB), a paired "linker" protein (e.g. EMRA) is believed to bring the transporter protein, which is located in the cytoplasmic membrane, in apposition with a corresponding outer-membrane channel (e.g. TOLC), allowing direct efflux through both the inner and outer membrane [5,6]. Some members of the 14-helix H$^+$/multidrug antiporter family also confer one or more collateral effects, including complementation of potassium uptake defects, and Na$^+$/H$^+$ antiport [7,8]. Members of the family are widely distributed in nature, occurring in various fungi and both gram-negative and gram-positive bacteria, including antibiotic-producing soil bacteria and antibiotic-resistant pathogens. Transporters may be encoded chromosomally or by transmissible plasmids, the latter occurring frequently in hospital-acquired, multidrug resistant organisms.

Statistical analysis of multiple amino acid sequence comparisons places the 14-helix H$^+$/multidrug antiporter family in the uniporter-symporter-antiporter (USA) superfamily, also known as the major facilitator superfamily (MFS) [3,4]. However, unlike all the other members of the superfamily, which are predicted to contain 12 membrane-spanning helices [3,9], members of the 14-helix H$^+$/multidrug antiporter family are predicted to form 14 membrane-spanning helices by the hydropathy of their amino acid sequences and the activity of reporter gene fusions [1,2].

Several amino acid sequence motifs are highly conserved in the 14-helix H$^+$/multidrug antiporter family, including motifs unique to the family, signature motifs of the USA/MFS superfamily, and motifs necessary for function by the criterion of site-directed mutagenesis [1-3,10,11].

Nomenclature, biological sources and substrates

CODE	DESCRIPTION [SYNONYMS]	ORGANISM [COMMON NAMES]	SUBSTRATE(S)^a [RESISTANCE(S)]^b
Ac22strco	Possible actinorhodin transporter [ACTII2]	*Streptomyces coelicolor* [gram-positive bacterium]	Actinorhodin
Actvastrco	Actinorhodin transporter [ACTVA]	*Streptomyces coelicolor* [gram-positive bacterium]	Actinorhodin
Atr1sacce	Aminotriazole resistance protein [ATR1, SNQ1]	*Saccharomyces cerevisiae* [yeast]	[Aminotriazole]
Bcrescco	Bicyclomycin resistance protein [BCR, BICA, BICR, SUR, SUXA]	*Escherichia coli* [gram-negative bacterium]	[Sulfonamide, bicyclomycin]
Bcrhaein	Bicyclomycin resistance protein [BCR, HI1242]	*Haemophilus influenzae* [gram-negative bacterium]	[Sulfonamide, bicyclomycin]
Cmctnocla	Cephamycin export protein [CMCT]	*Nocardia lactamdurans* [gram-positive bacterium]	Cephamycin

341

CODE	DESCRIPTION [SYNONYMS]	ORGANISM [COMMON NAMES]	SUBSTRATE(S)[a] [RESISTANCE(S)][b]
Cmlescco	Chloramphenicol resistance protein [CMLA]	*Escherichia coli* [gram-negative bacterium]	[Chloramphenicol]
Cmlapseae	Chloramphenicol resistance protein [CMLA]	*Pseudomonas aeruginosa* [gram-negative bacterium]	[Chloramphenicol]
Efprmyctu	Possible efflux protein [EFPR]	*Mycobacterium tuberculosis* [gram-positive bacterium]	Unknown
Emrbescco	Multidrug resistance protein B [EMRB]	*Escherichia coli* [gram-negative bacterium]	[Hydrophobic antibiotics]
Emrbhaein	Multidrug resistance protein B homolog [EMRB, HI0897]	*Haemophilus influenzae* [gram-negative bacterium]	[Hydrophobic antibiotics]
Emrdescco	Multidrug resistance protein D [EMRD]	*Escherichia coli* [gram-negative bacterium]	[Multiple drugs]
Lframycsm	Antiporter efflux pump [LFRA]	*Mycobacterium smegmatis* [gram-positive bacterium]	H⁺/multiple drugs
Lmrastrln	Lincomycin-H⁺ antiporter [LMRA]	*Streptomyces lincolnensis* [gram-positive bacterium]	H⁺/lincomycin
Mmrbacsu	Methylenomycin A resistance protein [MMR]	*Bacillus subtilis* [gram-positive bacterium]	[Methylenomycin]
Mmrstrco	Methylenomycin A resistance protein [MMR]	*Streptomyces coelicolor* [gram-positive bacterium]	[Methylenomycin]
Ppflpaspi	Florfenicol resistance protein [PPFL, PP-FLO]	*Pasteurella piscicida* [gram-negative bacterium]	[Florfenicol]
Pur8strlp	Puromycin-H⁺ antiporter [PUR8]	*Streptomyces lipmanii* [gram-positive bacterium]	H⁺/puromycin
Qacastaau	Multidrug-H⁺ antiporter [QACA]	*Staphylococcus aureus* [gram-positive bacterium]	H⁺/QUACs, EB, antiseptics
Sge1sacce	Crystal violet resistance protein [SGE1, SGE, NOR1, P9677.3]	*Saccharomyces cerevisiae* [yeast]	[Crystal violet, 10-N-nonyl-acridine]
Smvasalty	Methyl viologen resistance protein [SMVA]	*Salmonella typhimurium* [gram-negative bacterium]	[Methyl viologen]
Tcmastrga	Tetracenomycin-H⁺ antiporter [TCMA]	*Streptomyces galucescens* [gram-positive bacterium]	H⁺/tetra-cenomycin
Tcr2bacsu	Tetracycline-H⁺ antiporter [TCR2, TET]	*Bacillus subtilis* [gram-positive bacterium]	H⁺/tetracycline-metal chelate
Tcr3strau	Tetracycline-H⁺ antiporter [TCR3, TET]	*Streptomyces aureofaciens* [gram-positive bacterium]	H⁺/tetracycline-metal chelate
Tcrbacst	Tetracycline-H⁺ antiporter [TCR, TET]	*Bacillus stearothermophilus* [gram-positive bacterium]	H⁺/tetracycline-metal chelate
Tcrbbacsu	Tetracycline-H⁺ antiporter [TCR, TET, TETB]	*Bacillus subtilis* [gram-positive bacterium]	H⁺/tetracycline-metal chelate
Tcrstaau	Tetracycline-H⁺ antiporter [TCR, TET]	*Staphylococcus aureus* [gram-positive bacterium]	H⁺/tetracycline-metal chelate
Tcrstahy	Tetracycline-H⁺ antiporter [TCR, TET]	*Staphylococcus hyicus* [gram-positive bacterium]	H⁺/tetracycline-metal chelate
Tcrstrag	Tetracycline-H⁺ antiporter [TCR, TET]	*Streptococcus agalactiae* [gram-positive bacterium]	H⁺/tetracycline-metal chelate
Tcrstrpn	Tetracycline-H⁺ antiporter [TCR, TET]	*Streptococcus pneumoniae* [gram-positive bacterium]	H⁺/tetracycline-metal chelate

CODE	DESCRIPTION [SYNONYMS]	ORGANISM [COMMON NAMES]	SUBSTRATE(S)[a] [RESISTANCE(S)][b]
Toxacocca	Toxin pump [TOXA]	Cochliobolus carbonum [fungus]	Unspecified toxins

[a] Cotransported ions are listed for known antiporters.
[b] Presumed substrates; protein confers resistance to specified compounds.
Abbreviations: QUAC, quaternary ammonium disinfectants; EB: ethidium bromide.

Phylogenetic tree

Proteins listed subsequently in italics are at least 90% identical to the paired transporters listed in parenthesis and therefore are not included in the phylogenetic tree: *Tcrbacst, Tcrstahy, Tcrstrpn, Tcrstrag, Tcrstaau* (Tcr2bacsu).

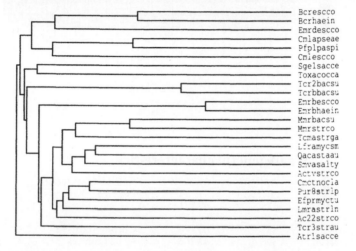

```
                                          Bcrescco
                                          Bcrhaein
                                          Emrdescco
                                          Cmlapseae
                                          Pfplpaspi
                                          Cmlescco
                                          Sgelsacce
                                          Toxacocca
                                          Tcr2bacsu
                                          Tcrbbacsu
                                          Emrbescco
                                          Emrbhaein
                                          Mmrbacsu
                                          Mmrstrco
                                          Tcmastrga
                                          Lframycsm
                                          Qacastaau
                                          Smvasalty
                                          Actvstrco
                                          Cmctnocla
                                          Pur8strlp
                                          Efprmyctu
                                          Lmrastrln
                                          Ac22strco
                                          Tcr3strau
                                          Atrlsacce
```

Proposed orientation of QACA in the membrane

The model is based on predictions of membrane-spanning regions and α-helical content. The N-terminus of the protein is illustrated on the inside and is folded 14 times through the membrane [1,2]. The predicted membrane-spanning helices are portrayed as rectangles. The numbers corresponding to the first and last residue of each membrane-spanning helix are boxed. Residues that are conserved in more than 75% of the aligned transporters (see below) are shown. Consensus residues indicated by an asterisk are not conserved in QACA.

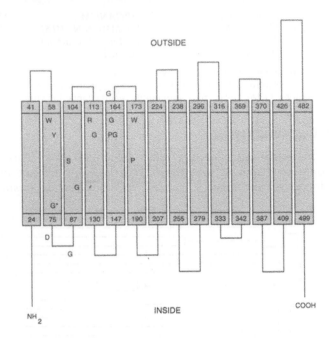

Physical and genetic characteristics

	AMINO ACIDS	MOL. WT	CHROMOSOMAL LOCUS
Ac22strco	578	59 771	
Actvastrco	533	54 546	
Atr1sacce	547	60 776	Chromosome 13 49.02 minutes
Bcrescco	396	43 366	
Bcrhaein	398	43 459	
Cmctnocla	486	49 325	
Cmlescco	302	33 390	19.1–19.12 minutes
Cmlapseae	419	44 243	Plasmid R1033
Efprmyctu	529	55 706	
Emrbhaein	510	55 826	
Emrbescco	512	55 612	60.55 minutes
Emrdescco	396	42 460	82.98 minutes
Lframycsm	504	51 539	
Lmrastrln	481	50 421	
Mmrbacsu	466	48 845	325°
Mmrstrco	475	49 238	Plasmid Scp1
Ppflpaspi	374	39 791	
Pur8strlp	503	51 852	
Qacastaau	514	55 015	
Sge1sacce	543	59 425	
Smvasalty	496	52 521	
Tcmastrga	538	54 846	
Tcrstahy	458	49 953	Plasmid pSTE1
Tcrbacst	458	50 119	
Tcrstrpn	458	50 092	Plasmid pLS1 and others
Tcrstaau	433	47 789	Plasmid pT181
Tcrstrag	458	50 056	Plasmid pMV158
Tcr2bacsu	459	50 695	Plasmid pNS1

	AMINO ACIDS	MOL. WT	CHROMOSOMAL LOCUS
Tcr3strau	512	53 021	
Tcrbbacsu	458	49 756	
Toxacocca	548	58 045	

Multiple amino acid sequence alignments

```
            1                                                    50
Toxacocca  ............. .......MDE QIVSASSNVK DGVEKQPVKD REDVDANVVP
Emrbhaein  ......................................................M
Mmrbacsu   ....................................................MK
Mmrstrco   ...............................................MTTVRTGGA
Tcmastrga  .............................................. MSTETHDEPS
Lframycsm  .................................................MST
Qacastaau  ..................................................MI
Actvstrco  ................................................MTAN
Cmctnocla  .............................................M TSVRGASKTG
Pur8strlp  .............................................M ARKPDISAVP
Efprmyctu  .......... ...MTALND TERAVRNWTA GRPHRPAPMR PPRSEETASE
Lmrastrln  ................................MSVF ARATSLFSRA
Ac22strco  .....................MSSV EADEPDRATA PPSALLPEDG
Tcr3strau  ........................MGMA NATSQTGEAV
Atrlsacce  MGNQSLVVLT ESKGEYENET ELPVKKSSRD NNIGESLTAT AFTQSEDEMV
Consensus  .............. .......... .......... .......... ..........

            51                                                   100
Bcrescco   .......... MTTRQHSSFA IVFILGLLAM LMPLSIDMYL PALPVISAQF
Bcrhaein   .......... .MNQQKSTFI FILTLGILSM LPPFGVDMYL PSFLEIAKDL
Emrdescco  .......... VIMKRQRNVN LLLMLVLLVA VGQMAQTIYI PAIADMARDL
Cmlapseae  ......MSSK NFSWRYSLAA TVLLLSPFDL LASLGMDMYL PAVPFMPNAL
Pfplpaspi  ......MTTL HPAWAYTLPA ALLLMAPFDI LASLAMDIYL PVVPAMPGIL
Sgelsacce  ........MK STLSLTLCVI SLLLTLFLAA LDIVIVVTLY DTIGIKFHDF
Toxacocca  PHSTPSLPKI SLISLFSIVM SLGAAAFLGA LDATVVAVLT PTLAQEFHSV
Tcr2bacsu  ...MFSLYKK FKGLFYSVLF WLCILSFFSV LNEMVLNVSL PDIA...NHF
Tcrbbacsu  ...MNTSYSQ STLRHNQVLI WLCVLSFFSV LNEMVLNVSL PDIA...NEF
Emrbescco  ...MQQQKPL EGAQLVIMTI ALSLATFMQV LDSTIANVAI PTIA...GNL
Emrbhaein  GNSAKKFPPI QGGALILLTL ALSLATFMQV LDSTIANVAI PTIA...GDL
Mmrbacsu   NSGSIQESTS STGISVLIV. .LALGFLMAT LDVTVVNVAM ADM...KNTL
Mmrstrco   QTAEVPAGGR RDVPSGVKIT ALATGFVMAT LDVTVVNVAG AII...QESL
Tcmastrga  GVAHTPASGL RGRPWPTLL. AVAVGVMMVA LDSTIVAIAN PAI...QQDL
Lframycsm  CIEGTPSTTR TPTRAWVALA VLALPVLLIA IDNTVLAFAL PLIA...EDF
Qacastaau  SFFTKTTDMM TSKKRWTALV VLAVSLFVVT MDMTILIMAL PELV...REL
Smvasalty  .......... .MFRQWLTLV IIVLVYIPVA IDATVLHVAA PTLS...MTL
Actvstrco  PGRPGGPADQ GHPRRWAILG VLVLSLVGII LDNTVLNVTL RTLTDPEQGL
Cmctnocla  RTSKTSTATT .......ALV LACTAHFLVV FDTSVITVAL PSV...RADL
Pur8strlp  VESAACQGPD PRRWW..GLV VILAAQLLVV LDGTVVNIAL PSV...QRDL
Efprmyctu  RPSRYYPTWL PSRTFIAAVI AIGGMQLLAT MDSTVAIVAL PKI...QNEL
Lmrastrln  ARTRAADEAA RSRSRWVTLV FLAVLQLLIA VDVTVVNIAL PAI...RDSF
Ac22strco  PGDGTAAGPP PYARRWAALG VILGAEIMDL LDGTVMNVAA PAV...RADL
Tcr3strau  ADEAGGPAGF THRQIITALS GLLLAVLLAA LDQTIVSTAL RTIGDQLHG.
Atrlsacce  DSNQKWQNPN YFKYAWQEYL FIFTCMISQL LNQAGTTQTL SIMNILSDSF
Consensus  .......... .......... .......... .......... ..........
```

```
          101                                                 150
Bcrescco  GVPAGSTQMT LSTYILGFAL GQLIYGPMAD SFGRKPVVLG GTLVFAAAAV
Bcrhaein  DVSPEQVQHT LTSFAYGMAF GQLFWGPFGD SFGRKPIILL GVIVGALTAL
Emrdescco NVREGAVQSV MGAYLLTYGV SQLFYGPISD RVGRRPVILV GMSIFMLATL
Cmlapseae GTTASTIQLT LTTYLVMIGA GQLLFGPLSD RLGRRPVLLG GGLAYVVASM
Pfplpaspi NTTPAMIQLT LSLYMVMLGV GQVIFGPLSD RIGRRPILLA GATAFVIASL
Sgelsacce G....NIGWL VTGYALSNAV FMLLWGRLAE ILGTKECLMI SVIVFEIGSL
Toxacocca D....AVAWY GAIYLLMSGT TQPLFGKLYN EFSPKWLFIT CLIVLQLGSL
Tcr2bacsu NTTPGITNWV NTAYMLTFSI GTAVYGKLSD YINIKKLLII GISLSCLGSL
Tcrbbacsu NKLPASANWV NTAFMLTFSI GTALYGKLSD QLGIKNLLLF GIMVNGLGSI
Emrbescco GSSLSQGTWV ITSFGVANAI SIPLTGWLAK RVGEVKLFLW STIAFAIASW
Emrbhaein GASFSQGTWV ITSFGVANAI SIPITGWLAK RFGEVRLFLV STFLFVVSSW
Mmrbacsu  SMSLSGVTWV VDGYILTFAS LLLAGGALAD RFGSKTIYIL GLAVFVMASC
Mmrstrco  DTTLTQLTWI VDGYVLTFAS LLMLAGGLAN RIGAKTVYLW GMGVFFLASL
Tcmastrga HASLADVQWI TNGYLLALAV SLITAGKLGD RFGHRQTFLV GVAGFAVTSA
Lframycsm RPSATTQLWI VDVYSLVLAA LLVAMGSLGD RLGRRVLLI GGAGFAVVSA
Qacastaau EPSGTQQLWI VDIYSLVLAG FIIPLSAFAD KWGRKKALLT GFALFGLVSL
Smvasalty GASGNELLWI IDIYSLVMAG MVLPMGALGD RIGFKRLLML GGTLFGLASL
Actvstrco GASHSQVEVW LSAYTLAFAA TLFTWGVLGD RLGRRVLLL GLGLFGLSSL
Cmctnocla GFAPASLQWV VNSYTLAFAG LLLFGGRLAD IHGHRRVFLG GLAVFTLTSL
Pur8strlp GMSDTSRQWV ITAYTLAFGG LLLLGGRVAD AFGRRRIFAV GILGFGLASL
Efprmyctu SLSDAGRSWV ITAYVLTFGG LMLLGGRLGD TIGRKRTFIV GVALFTISSV
Lmrastrln HVDTRQLTWV VTGYTVVGGG LLMVGGRIAD LFGRRRTLLF GAFLFGASSL
Ac22strco GGSLSVIQWI TVGYTLAFAV LLVVGGRLGD IYGRKRMFVV GAVGFTAASV
Tcr3strau ...QTVQAWV ITGYLVSSTI AMPFYGKLSD IYGRKPLYLA AIAVFIVGSA
Atr1sacce GSEGNSKSWL MASFPLVSGS FILISGRLGD IYGLKKMLLV GYVLVIIWSL
Consensus ........W. ...Y...... .....G...D ..G....... G.......S.

          151                                                 200
Bcrescco  ACALA..NTI .DQLIVMRFF HGLAAAAASV VINALMRDIY P.KE.EFSRM
Bcrhaein  VLTEI..NSV .GNFTALRFV QGFFGAAPVV LSGALLRDLF S.KD.QLSKV
Emrdescco VAVTT..SSL .TVLIAASAM QGMGTGVGGV MARTLPRDLY E.RT.QLRHA
Cmlapseae GLALT..SSA .EVFLGLRIL QACGASACLV STFATVRDIY AGRE.ESNVI
Pfplpaspi GAAWS..STA .PAFVAFRLL QAVGASAMLV ATFATVRDVY ANRP.EGVVI
Cmlescco  .......... .......... .......... .......... .......MS
Sgelsacce ISALS..NSM .ATLISGRVV AGFGGSGIES LAF.VVGTSI VREN.HRGIM
Toxacocca VCALA..RNS .PTFIVGRAV AGIGAGGILS GAL.NIVALI VPLH.HRAAF
Tcr2bacsu IAFIG..HNH FFILIFGRLV QGVGSAAFPS LIMVVVARNI TRKK..QGKA
Tcrbbacsu IGFVG..HSF FPILILARFI QGIGAAAFPA LVMVVVARYI PKEN..RGKA
Emrbescco ACGVS..SSL .NMLIFFRVI QGIVAGPLIP LSQSLLLNNY PPAK..RSIA
Emrbhaein LCGIA..DSL .EALIIFRVI QGAVAGPVIP LSQSLLLNNY PPEK..RGMA
Mmrbacsu  LCAAS..ING .QMLIAGRLI QGIGAALFMP SSLSLLAASY LDER.ARARM
Mmrstrco  ACALA..PTA .ETLIAARLV QGAGAALFMP SSLSLLVFSF PEKR.QRTRM
Tcmastrga AIGLS..GSV .AAIVVFRVL QGLFGALMQP SALGLLRVTF PPGK.L.NMA
Lframycsm LAAFA..PST .ELLVGARAL LGVFGAMLMP STLSLIRNIF TDAS.ARRLA
Qacastaau AIFFA..ESA .EFVIAIRFL LGIAGALIMP TTLSMIRVIF ENPK.ERATA
Smvasalty AAAFS..HTA .SWLIATRVL LAIGAAMIVP ATLAGIRATF CEEK.HRNMA
Actvstrco AGAYA..GSP .EQLIAARAC MGVSGAAVLP STLATIAAVF P.LR.ERPKA
Cmctnocla IGGLA..TSP .ASLIAARAG QGAGAAVLAP LAVTMLTTSF AEGP.RRTRA
Pur8strlp LGGAA..PDP .GTLFLARAL QGVFAAALAP AALALINTLF TEPG.ERGKA
Efprmyctu LCAVA..WDE .ATLVIARLS QGVGSAIASP TGLALVATTF PKGP.ARNAA
Lmrastrln AAGLA..PNL .ELLVLARFG QGAGEALSLP AAMSLIACSS RTAP.FQG..
Ac22strco LCSVA..AGP .EMLTAARFL QGGLGALMIP QGLGLIKQMF P.PK.ETAAA
Tcr3strau ACAMA..NSM .ETLAIARVL QGFGGAGLMS LPTAVIAD.L APVR.ERGRY
```

```
Atr1sacce ICGITKYSGS DTFFIISRAF QGLGIAFVLP NVLGIIGNIY VGGTFRKNIV
Consensus .......... .......R.. .G........ .......... ..........


           201                                                250
Bcrescco  MSFVMLVTTI APLMAPIVGG WVLVW..... .LSWHYIFWI LALAAILASA
Bcrhaein  MSTITLVFML APLVAPIIGG YIVKF..... .FHWHAIFYV ISLVGLLAAA
Emrdescco NSLLNMGILV SPLLAPLIGG LLDTM..... .WNWRACYLF L....LVLCA
Cmlapseae YGILGSMLAM VPAVGPLLGA LVDMW..... .LGWRAIFAF LG.LGMIAAS
Pfplpaspi YGLFSSVLAF VPALGPIAGA LIGEF..... .LGWQAIFIT LAILAMLALL
Cmlescco  FTAYSDP.CW PWSRGRPIAR SARRH..... .VAWVAGYLC VSRFGHDRCI
Sge1sacce ITALAISYVI AEGVGPFIGG AFNEH..... .LSWRWCFYI NLPIGAFAFI
Toxacocca TGMIGALECV ALIIGPIIGG AIADN..... .IGWRWCFWI NLPIGAAVCA
Tcr2bacsu FGFIGSIVAL GEGLGPSIGG IIAHYIHWSY LLILPMITIV TIPFLIKVMV
Tcrbbacsu FGLIGSLVAM GEGVGPAIGG MVAHYIHWSY LLLLIPTATII TVPFLIKLLK
Emrbescco LALWSMTVIV APICGPILGG YISDN..... .YHWGWIFFI NVPIGVAVVL
Emrbhaein LAFWSMTIVV APIFGPILGG WISDN..... .IHWGWIFFI NVPIGLSVVL
Mmrbacsu  FGLWAALVSA ASALGPFIGG VLVQL..... .AGWQSIFLI NVPIGAAALI
Mmrstrco  LGLWSAIVAT SSGLGPTVGG LMVSA..... .FGWESIFLL NLPIGAIGMA
Tcmastrga IGIWSGVVGA STAAGPIIGG LLVQH..... .VGWEAVFFI NVPVGLAALV
Lframycsm IAIWASCFTA GSALGPIVGG ALLEH..... .FHWGAVFLV AVPILLPLLV
Qacastaau LAVWSIASSI GAVFGPIIGG ALLEQ..... .FSWHSAFLI NVPFAIIAVV
Smvasalty LGVWAAVGSG GRAFVPLIGG ILLEH..... .FYWGSVFLI NVPIVLVVMG
Actvstrco LGIWAASVGF ALGIGPVTGG ILLAH..... .FWWGSVLLV NVPLMAGCLV
Cmctnocla LTISTAVALV GGASGNLLGG VFTEF..... .LSWRSVLLV NVPIGIPVLF
Pur8strlp FGVYGAVSGG GAAVGLLAGG LLTEY..... .LDWRWCLYV NAPVAL.LAL
Efprmyctu TAVFAAMTAI GSVMGLVVGG RLTE...... .VSWRWAFLV NVPIGLVMIY
Lmrastrln VERLASVASV GLVLGFLLSG VITQL..... .FSWRWIFLI NIP..LVSLV
Ac22strco FGAFGPAIGL GAVLGPIVAG FLVDAD..LF GTGWRSVFLI NLPIGVAVIV
Tcr3strau FSYLMMAWVA ASVLGPLVGG LFAGAGEILG VTGWRWAFLI NVPLGLVALL
Atr1sacce ISFVGAMAPI GATLGCLFAG LIGTEDPK.. .QWPWAFYA YSIAAFINFV
Consensus .......... ...GP..GG .......... ..W....... ..P.......


           251                                                300
Bcrescco  MIFFLIKETL PPERRQP..F HIRTTIGNFA ALFRHKRVLS YMLASGFSFA
Bcrhaein  LVFFIIPETH KKENRIP..L RLNIIARNFL LLWKQKEVLG YMFAASFSFG
Emrdescco GVTFSMARWM PETRPVD..A PRTRLLTSYK TLFGNSGFNC YLLMLIGGLA
Cmlapseae AAAWRFWPET RVQRVAG..L QWSQLLLPVK CL....NFWL YTLCYAAGMG
Pfplpaspi NAGFRWHETR PLDQVKT.... .RRSVLPIFA SP....AFWV YTVGFSAGMG
Cmlescco  CRRGSFWPET RVQRVAG..L QWSQLLLPVK CL....NFWL YTLCYAAGMG
Sge1sacce ILAFCNTSGE PHQKMWLPSK IKKIMNYDYG ELLKASFWKN TFEVLVFKLD
Toxacocca ILLFFF...H PPRSTYSASG VPR....SYS EILG...... ......NLD
Tcr2bacsu PG........ ..KSTKNT.. .......... .......... ......LD
Tcrbbacsu KE........ ..ERIRGH.. .......... .......... ......ID
Emrbescco MTLQTL.RGR ETRTERRR.. .......... .......... ......ID
Emrbhaein ISWKIL.GSR ESEIVHQP.. .......... .......... ......ID
Mmrbacsu  SAYRIL..SR .VPGKSSR.. .......... .......... ......VN
Mmrstrco  MTYRYI..AA .TESRATR.. .......... .......... ......LA
Tcmastrga AGLVILTDAR .AERAPKS.. .......... .......... ......FD
Lframycsm LGPRLVPESR ..DPNPGP.. .......... .......... ......FD
Qacastaau AGLFLLPESK LSKEKSHS.. .......... .......... ......WD
Smvasalty LTARY..DPR QAGRRDQP.. .......... .......... ......LN
Actvstrco AVVLVVPETR ..GTAGRR.. .......... .......... ......VD
Cmctnocla LAARVLAGPR KRPWGRVR.. .......... .......... ......LD
Pur8strlp LGCRLLPRDR RTGRA.VR.. .......... .......... ......LD
```

```
Efprmyctu LARTAL...R ETNKERMK.. ....................... ...........LD
Lmrastrln LVAVLLLVKK DETTARNP.. ....................... ...........VD
Ac22strco GAVLLLPEGK ..APVRPK.. ....................... ...........FD
Tcr3strau SVRKALNLPH R..RVDHP.. ....................... ...........ID
Atr1sacce LSIYAIPSTI PTNIHH.... ............... .....FSMD
Consensus ..................... ................. ........... ........
```

```
          301                                                      350
Bcrescco  GMFSFLSAGP FVYIEINHIA PENFGYYFAL NIVFLFVMTI FNSRFVRRIG
Bcrhaein  GLFAFVTAGS IVYIGIYGVP VDQFGYFFMM NIVTMIFASF LNSRFVTKVG
Emrdescco GIAAFEACSG VLMGAVLGLS SMTVSILFIL PIPAAFFGAW FAGRPNKRFS
Cmlapseae SFFVFFSIAP GLMMGRQGVS QLGFSLLFAT VAIAMVFTAR FMGRVIPKWG
Pfplpaspi TFFVFFSTAP RVLIGQAEYS EIGFSFAFAT VALVMIVTTR FAKSFVARWG
Cmlescco  SFFVFFSIAP GLMMGRPRCV SAWLQPAVRH SAIAMVFTAR FMGSVIPKWG
Sgelsacce MVGIILSSAG FTLLMLGLSF GGNNFPWNSG IIICFFTVGP ILLLLFCAYD
Toxacocca YIGAGMIISS LVCLSLALQW GGTKYKWGDG RVVALLVVFG VL........
Tcr2bacsu IVGIVLMSIS IICFML.... ......FTTN YNWTFLILFT IFFV......
Tcrbbacsu MAGIILMSAG IVFFML.... ......FTTS YRFSFLIISI LAFF......
Emrbescco AVGLALLVIG IGSLQIMLDR GKELDWFSSQ EIIILTVVAV VAIC......
Emrbhaein KVGLVLLVLG VGCLQLMLDQ GREQDWFNSN EIIILAVVAV VCLI......
Mmrbacsu  IIGHLLGMMA LGFLSYALIQ GPSA.GWRSP VILVAFTAAV LAFV......
Mmrstrco  VPGHLLWIVA LAAVSFALIE GPQL.GWTAG PVLTAYAVAV TAAA......
Tcmastrga VSGIVLLSGA MFCLVWGLIK APAW.GWGDL RTLGFLAAAV LAFA......
Lframycsm PVSIVLSFTT MLPIVWAVKT AAH.DGLSAA AA.AAFAVGI VSGA......
Qacastaau IPSTILSIAG MIGLVWSIKE FSK.EGLADI IPWVVIVLAI TMV.......
Smvasalty LGHVVMLIIA ILLLVYSAKT ALK.GHLSLW VISYTLLTGA LLLG......
Actvstrco AAGLLLSIAG VVPLVYAIIE AGRSGGVTRP AVWAAGLAGL GLLL......
Cmctnocla LPGAVLATAG LTLLLTLGVSQ THE.HGWGEA AVAVPLAGGL LALL......
Pur8strlp LPGTLLGCGG LVAIVYAFAE A.E.SGWGDP LVVRLLVLGV LMLV......
Efprmyctu ATGAILATLA CTAAVFAFSI GPE.KGWMSG ITIGSGLVAL AAAV......
Lmrastrln LPGALLFTAA PLLLIFGVNE LGE.DEPRLP LAVGSLLAAA VCAA......
Ac22strco VVGMALVTSG LTLLIFPLVQ GRE.RGWP.A WAFVLMLAGA AVLV......
Tcr3strau FRGALTLALC LVPLLIVAEE GLDW.GWGSA RSLTLFAVSL IGLV......
Atr1sacce WIGSVLGVIG LILLNFVWNQ AP.ISGWNQA YIIVILIISV I.........
Consensus .................. ................. ................
```

```
          351                                                      400
Bcrescco  ALNMFRSGLW IQFIMAAWMV ISALLG.LGF WSLVVGVAAF VGCVSMVSSN
Bcrhaein  AETMLRIALA IQFLSGMWLI LTALLD.LGF WPMAIGVAFF VGPNPVISSN
Emrdescco ..TLMWQSVI CCLLAGLLMW IPDWFGVMNV WTLLVPAALF FFGAGMLFPL
Cmlapseae SPSVLRMGMG CLIAGAVLLA ITEIWALQSV LGFIAPMWLV GIGVATAVSV
Pfplpaspi IAGCGRVGWR CLFA.AVLLG IGELYGSLNS SPSSYRCGLS RSVLSSRCPL
Cmlescco  SPSVLRMGMG CLIAGAVLLA ITEIWALHRV RLYCSNVA.S GIGVATAVSV
Sgelsacce FHFLSLSGLH YDNKRIKPLL TWNIASNCGI FTSSITGFLS CFAYELQSA.
Toxacocca ..FLSASGHQ Y.WKGEKALF PTRLLRQRGF LLSLFNGL.. CFG.GVQYAA
Tcr2bacsu ..IFIKH.... .ISRVSNPFI NPKLGKNIPF MLGLFSGGLI FSIV...AGF
Tcrbbacsu ..IFVQH.... .IRKAQDPFV DPELGKNVFF VIGTLCGGLI FGTV...AGF
Emrbescco ..FLIVWEL. ...TDDNPIV DLSLFKSRNF TIGCLCISLA YMLY...FGA
Emrbhaein ..ALVIWEL. ...TDDNPVV DISLFHSRNF SVGCLCTSLA FLIY...LGS
Mmrbacsu  ..LFLLREIS ....AKTPIL PASLYKNGRF SAAQFIGFLL NFAL...FGG
Mmrstrco  ..LLALREHR ....VTNPVM PWQLFRGPGF TGANLVGFLF NFAL...FGS
Tcmastrga ..GFTLRESR ....ATEPLM PLAMFRSVPL SAGTVLMVLM AFSF...IGG
Lframycsm ..LFVRRQNR ....SATPML DIGLFKVMPF TSSILANFLS IIGL...IGF
Qacastaau ..IFVKRNLS ....SSDPML DVRLFKKRSF SAGTIAAFMT MFAM...ASV
```

```
Smvasalty ..LFIRTQLA ....TSRPMI DMRLFTHRII LSGVVMAMTA MITL...VGF
Actvstrco ..VFLWHERR ....TPEPSL ELGFFRMKAF STAVAAVGFV SFAM...MGF
Cmctnocla ..AFVVVE.. .ARFAASPLI PPRLFGLPGV GWGNLAMLLA GASQ...VPV
Pur8strlp ..AFALVE.. .RRVQD.PLL PPGVVAHRVR GGSFLVVGLP QIGL...FGL
Efprmyctu ..AFVIVE.. ..RTAENPVV PFHLFRDRNR LVTFSAILLA GGVM...FSL
Lmrastrln ..AFVAVE.. ..RRTAH.PLV PLTFFGNRVR LVANGATVLL SAAL...STS
Ac22strco ..GFVAHELR QERRGGATLI ELSLLRRSRY AAGLAVALVF FTGV...SGM
Tcr3strau ..LFVLAERA ...RGLEAMV PLRLFRRGGI TMATAVNFTI GVGI...FGT
Atr1sacce ..FLVVFIIY EIRFAKTPLL PRAVIKDRHM QIMLALFFG. WGSFGI...F
Consensus ......... .......... .......... .......... ..........
```

```
          401                                            450
Bcrescco  AMAVILDEFP HMAGTASSLA GTFRFGIGAI VGAL....LS LATFNSAWPM
Bcrhaein  AMASALERCP QMAGTANSLI GSVRFAVGAI MGSL....VA SMKMDTAAPM
Emrdescco ATSGAMEPFP FLAGTAGALV GGLQNIGSGV LASL....SA MLPQTGQGSL
Cmlapseae APNGALRG.. .FDHVAGTVT AVYFCLGGVL LGSIGTLIIS LLPRNTAWPV
Pfplpaspi PRTALLAE.. .FDDIAGSAV AFYFCVQSLI VSIVGTLAVA LLNGDTAWPV
Cmlescco  PMALFEDSTM LLERSRQSTS A..WACTARK HRNVDHFAVA AQRLGPCRVT
Sge1sacce ..YLVQLYQL VFKKKPTLAS IHLWELSIPA MIATMAIAYL NSKY.GIIKP
Toxacocca LYYLPTWFQA IKGETRVGAG IQMLPI.VGA IIGVNIVAGI TISFTGRLAP
Tcr2bacsu ISMVPYMMKT IYHVNVATIG NSVIFPGTMS VIVFGYFGGF LVDRKGSLFV
Tcrbbacsu VSMVPYMMKD VHHLSTAAIG SGIIFPGTMS VIIFGYIGGL LVDRKGSLYV
Emrbescco IVLLPQLLQE VYGYTATWAG LASAPVGIIP VILSPII.AR FAHKLDMRRL
Emrbhaein VVLIPLLLQQ VFHYTATWAG LAASPVGLFP ILLSPII.GR FGYKIDMRIL
Mmrbacsu  MFMLSLFLQE AGGASSFMAG VELLPMMAVF VIGNL.LFAR LANRFEAGQL
Mmrstrco  TFMLGLYFQH ARGATPFQAG LELLPMTIFF PVANI.VYAR ISARFSNGTL
Tcmastrga LFFVTFYLQN VHGMSPVESG VHLLPLTGMM IVGAP.VSGI VISRFGPGGP
Lframycsm IFFISQHLQL VLGLSPLTAG LVTLPGAVVS MIAGL.AVVK AAKRFAPDTL
Qacastaau LLLASQWLQV VEELSPFKAG LYLLPMAIGD MVFAP.IAPG LAARFGPKIV
Smvasalty ELLMAQELQF VHGLSPYEAG VFMLPVMVAS GFSGP.IAGV LVSRLGLRLV
Actvstrco LFFSAFYLQS VRGYTPLQAG GCTVALAVAN VVCGP.LSTV LVRSIGPRNV
Cmctnocla WFFLTLSMQH VLGYSAAQAG LGFVPHALVM LVVGLRVVPW LMRHVQARVL
Pur8strlp FLFLTYYLQG ILDYSPVLTG VAFLPLGLGI AVGSSLIAAR LLPRTRPRTL
Efprmyctu TVCIGLYVQD ILGYSA.LRR VGFIPFVIAM GI.GLGVSSQ LVSRFSPRVL
Lmrastrln FFLLTMHLQE ERDLSPIEAG LSFLPLGLSL ILACVLVRG. LIERIGTTGA
Ac22strco SLLLALHLQI GLGFSPTRAA LTMTPWSVFL VVGAILTGAV LGSKFGRKAL
Tcr3strau VSTLPLFLQL VQGRSATVAG LVIIPVMTGA IVSQTICAKI IKKWNRYKKP
Atr1sacce TFYYFQFQLN IRQYTALWAG GTYFMFLIWG IIAALLVGFT IKNVSPSVFL
Consensus ......... .......... .......... .......... ..........
```

```
          451                                            500
Bcrescco  IWSIAFCATS SIL.FCLYAS RPKKR..... .......... ..........
Bcrhaein  LFTMGACVVI SVLAYYFLTS RNLKSRG... .......... ..........
Emrdescco GLLMTLMGLL IVLCWLPLAT RMSHQGQPV. .......... ..........
Cmlapseae VVYCLTLATV VLGLSCVSRV KGSRGQGEHD VVALQSAGST SNPNR.....
Pfplpaspi IC........ .......... .......... .......... ..........
Cmlescco  V...WTLATV VLGLSCVSRV KGSRGQGEHD AGRATNVGKY IKSQSLRECG
Sge1sacce AIVFGVLCGI VGSGLFTLIN GELSQS..IG YSILPGIAFG SIFQATLLSS
Toxacocca FIVIATVLAS VGSGLLYTT. PTKSQARIIG YQLIYGAGSG AGVQQAFIGA
Tcr2bacsu FILGSLSISI SFLTIAFFVE FSM..WLTTF MFIFVMGGLS FT..KTVISK
Tcrbbacsu LTIGSALLSS GFLIAAFFID AAP..WIMTI IVIFVFGGLS FT..KTVIST
Emrbescco VTFSFIMYAV CFYWRAYTFE PGMD..FGAS AWPQFIQGFA VACFFMPLTT
Emrbhaein VTISFIVYAI TFYWRAVTFE PSMT..FVDV ALPQLVQGLA VSCFFMPLTT
Mmrbacsu  MFVSMAVSCI IALLLFVLIS PDFPYWQLAV L..MSVMNLC TGITVPAMTT
```

```
Mmrstrco  LTAFLLLAGA ASLSM.VTIT ASTPYWVVAV A..VGVANIG AGIISPGMTA
Tcmastrga  LVVGMLLT.A ASLWGMSTLE ADSGMGITSL W..FVLLGLG LAPVMVGTTD
Lframycsm  MVTGLVFV.A VGFLMILLFR HNLTVAAIIA S..FVVLELG VGVSQTVSND
Qacastaau  LPSGIGTA.A IGMFIMYFFG HPLSYSTMAL A..LILVGAG MA.SLAVASA
Smvasalty  ATGGMALS.A LSFYGLAMTD FSTQQWQAWG L..MALLGFS AASALLASTS
Actvstrco  CAAGMLAV.T ASLCGVTFVT QHAPVWLILV L..FAALGAG VACVMPTAAV
Cmctnocla  IAAGAAIGAL GFWWQSLLTP DSA..YLGGI LGPAVLISIG GGLVGTPLAR
Pur8strlp  IVGALLAAAA GMALLTRLEP DTPQVYLTHL LPAQILIGLG IGCMMMPAMH
Efprmyctu  TIGGGYLLFG AMLYGSFFMH RGVP.YFPNL VMPIVVGGIG IGMAVVPLTL
Lmrastrln  AVLGMALAGP RHRLFALLPS DNS..LLTSV FPGMILL.LR MATGLVALQN
Ac22strco  HGGLVVLALG VLIMLLTIGD QAGGLTSWEL VPGIAVAGLG MGIMIGLLFD
Tcr3strau  AIVGLGSMAG A...LLSLSA AGADTPLAVI VVIAAWLGFG IGLSQTVITL
Atr1sacce  FFSMVAFNVG SIMASVTPVH ET...YFRTQ LGTMIILSFG MDLSFPASSI
Consensus  .......... .......... .......... .......... ..........

           501                                             550
Bcrescco   .......... .......... .......... .......... ..........
Bcrhaein   .......... .......... .......... .......... ..........
Emrdescco  .......... .......... .......... .......... ..........
Cmlapseae  .......... .......... .......... .......... ..........
Pfplpaspi  .......... .......... .......... .......... ..........
Cmlescco   KLSPNKCCSR PKTYAFRLG. .......... .......... ..........
Sgelsacce  QVQITSDDPD FQNKFIEVTA FNSFAKSLGF AFGGNMGAMI FTASLKNQMR
Toxacocca  QAALDPADVT YAS..ASVLL MNSMSGVITL CVCQNL.... ....FTNRIN
Tcr2bacsu  IVSSSLSEEE VASGMSLLNF TSFLSEGTGI AIVGGLLSLQ LINRKLVLEF
Tcrbbacsu  VVSSSLKEKE AGAGMSLLNF TSFLSEGTGI AIVGGLLSIG FLDHRLLPID
Emrbescco  ITLSGLPPER LAAASSLSNF TRTLAGSIGT SITTTMWTNR ESMHHAQLTE
Emrbhaein  ITLSGLPAHK MASASSLFNF LRTLAGSVGT SLTTFMWYNR EAVHHTQLTE
Mmrbacsu   VIMQAAGQRH TNIAGAALNA NRQIGALVGV AITGVII... ..HLSATWYA
Mmrstrco   ALVDAAGPEN ANVAGSVLNA NRQIGSLVGI AAMGVVL... ..HSTSDWDH
Tcmastrga  VIVSNAPAEL AGVAGGLQQS AMQVGGSLGT AVLGVLMASR VGDVFPDKWA
Lframycsm  TIVASVPAAK SGAASAVSET AYELGAVVGT ATLGTIFTAF YRSNVDV..P
Qacastaau  LIMLETPTSK AGNAAAVEES MYDLGNVFGV AVLGSLSSML YRVFLDI..S
Smvasalty  AIMAAAPAEK AAAAGAIETM AYELGAGLGI AIFGLLLSRS FSASIRL..P
Actvstrco  SIMNAIPREK AGVASAMNNT VRQLGGALGV AVLGSLMGAA YRRGIED..E
Cmctnocla  TVTSGVGPLD AGAASGLMNT TRQFGGAFGL AVLLTVTGSG T...SG....
Pur8strlp  TATARVAPHE AGAAAAVVNS AQQVGGALGV ALLNTVSTGA T...AAYLAD
Efprmyctu  SAIAGVGFDQ IGPVSAIALM LQSLGGPLVL AVIQAVITSR TLYLGGTTGP
Lmrastrln  AALHAVTEAD AGVASGVQRC ADQLGGASGI AVYVSIGFSP ..........
Ac22strco  IALADVDKQE AGTASGVLTA VQQLGFTVGV AVLGTLFFGL LGSQATASVD
Tcr3strau  AIQSSAPKSE LGVANAASGL FRQLGGTSGA AVFMSVLFGV AAGRLDGADP
Atr1sacce  IFSDNLPMEY QGMAGSLVNT VVNYSMSLCL GMGATVETQV NSDGKHLLKG
Consensus  .......... .......... .......... .......... ..........

           551                            ·                600
Bcrescco   .......... .......... .......... .......... ..........
Bcrhaein   .......... .......... .......... .......... ..........
Emrdescco  .......... .......... .......... .......... ..........
Cmlapseae  .......... .......... .......... .......... ..........
Pfplpaspi  .......... .......... .......... .......... ..........
Cmlescco   .......... .......... .......... .......... ..........
Sgelsacce  SSQLNIPQFT ..SVETLLAY STEHYDGPQS SLS.KFINTA IHDVFYCALG
Toxacocca  ALTEVLPGVT KETLQSGFAF LRSTLTPAEF GVAIQTFNSA IQDAFLVAIV
Tcr2bacsu  INYSSGVYSN ILVAMAILII LCCLLTIIVF KRSEKQFE.. ..........
```

```
Tcrbbacsu VDHSTYLYSN MLILFAGIIV ICWLVILNVY KRSRRHG...........
Emrbescco SVNPFNPNAQ AMYSQLEGLG MTQQQASGWI AQQITNQGLI ISANEIFWMS
Emrbhaein HINPYNPISQ SFYHQMNQFG LSDTQTSAYL AQQITSQGFI IGANEIFWLS
Mmrbacsu  GAGFAFL... ...MMGAAYS LAALLVWLFL AAHNGTAASE KMPSQ.....
Mmrstrco  GAAISFL... ...AVGLAYL LGGLSAWRLI ARPERRSAVT AAT.......
Tcmastrga EANLPRVGPR EAAAIEDAAE VGAVPPAGTL PGRHAGTLSE VVHSSFISGM
Lframycsm A....GLTPE QTGAAAESIG GAAAVAADLP AA.TATQLLD SARAAFDSGI
Qacastaau SFSSKGIVGD LAHVAEESVV GAVEVA...K AT.GIKQLAN EAVTSFNDAF
Smvasalty A....GLEAQ EIARASSSMG EAVQLANSYP PTQGQGKYLT AARHAFIWSH
Actvstrco ...LAVLPPS ARHQAGESLD ATLLAATRL. ...GESGLVG PARQAFLDAM
Cmctnocla ...SPAELAS HYGDAFVGIA VFMLAIAVLT PVLPALARST PPGVIHVSPV
Pur8strlp HGTSPAATVD GTVHGYTVAI AFAVGVLLLT AVLAWVLIDS RTEAADETGS
Efprmyctu VKFMNDVQLA ALDHAYTYGL LWVAGAAIIV GGMALFIGYT PQQVAHAQEV
Lmrastrln .......HLG GDWDPFTVAY SLA.GIGLIA AVLAVLALSP DRRLAAPREQ
Ac22strco DGASRARTEL AAAGASTTEQ DRLLADLRVC LRESASQQDS ERTPDSCRNL
Tcr3strau DEAVRRALSD PGSTGGLSAS AVDAFTSGFD TMFLVGGLIL AVGFLLTFPL
Atr1sacce YRGAQYLGIG LASLACMISG LYMVESFIKG RRQELLQNTI ALWLSGKRY.
Consensus ..........
```

```
          601                                                650
Bcrescco  ..........
Bcrhaein  ..........
Emrdescco ..........
Cmlapseae ..........
Pfplpaspi ..........
Cmlescco  ..........
Sgelsacce .CYALSFFFG IFTSSKKTTI SAKKQQ....
Toxacocca LSCASVLGWP FL..SWASVK GQKKMNK...
Tcr2bacsu ..........
Tcrbbacsu ..........
Emrbescco AGIFLVLLGL VWFAKPAFGA GGGGGGAH..
Emrbhaein AMGFLGLLIV IWFAKPPFGT QH........
Mmrbacsu  ..........
Mmrstrco  ..........
Tcmastrga GLAFTVAGAV ALVAAAVALF TRKAEPDERA PEEFPVPAST AGRG......
Lframycsm APTAVIAAML VLAAAAVVGV AFRR......
Qacastaau VATALVGGII MIIISIVVYL LIPKSLDITK QK........
Smvasalty SVALSSAGSM LLLLAVGMWF SLAKAQRR..
Actvstrco HLAAGAAAAV ALVGALAVLR WLPSSVTTPT PPAGAVPGRE HSDHLKVQGS
Cmctnocla AR........
Pur8strlp ASVTPARPR.
Efprmyctu KEAIDAGEL.
Lmrastrln ED........
Ac22strco QQARPAVAEA TARAWRTAHT ENFSTAMVRT LWVVIALLAV SFALAFRLPP
Tcr3strau RELRDEE...
Atr1sacce ..........
Consensus ..........
```

```
          651
Ac22strco KPREEEGF
Consensus ........
```

Proteins listed subsequently in italics are at least 90% identical to the paired transporters listed in parenthesis and therefore are not included in the alignments: *Tcrbacst, Tcrstahy, Tcrstrpn, Tcrstrag, Tcrstaau* (Tcr2bacsu). Residues listed in the consensus sequence are present in at least 75% of the aligned transporter sequences. Residues indicated by boldface type are also conserved in at least one other family of the USA/MFS superfamily.

Database accession numbers

	SWISSPROT	PIR	EMBL/GENBANK
Ac22strco	P46105		M64683
Actvstrco		S18539	X58833
Atr1sacce	P13090	A28124	M20319
Bcrescco	P28246		X63703; U00080
Bcrhaein	P45123		U32804
Cmctnocla	Q04733		Z13973
Cmlescco	P12056	A25854	M22614
Cmlapseae	P32482	A47033	M64556; U12338
Efprmyctu			L39922
Emrbhaein	P44927		U32771
Emrbescco	P27304	S27558; JC1345	M86657
Emrdescco	P31442		L10328
Lframycsm			U40487
Lmrastrln	P46104		X59926
Mmrbacsu	Q00538	S22742	X66121
Mmrstrco	P11545	B29606	M18263
Ppflpaspi			D37826
Pur8strlp	P42670		X76855
Qacastaau	P23215		X56628
Sge1sacce	P33335	S42086; S40888	L11640; U02077
Smvasalty	P37594		D26057
Tcmastrga	P39886	S27687	M80674
Tcrstahy	P36890	S23743	X60828
Tcrbacst	P07561	A23973	M11036
Tcrstrpn	P11063	S09234; C25599	M63891; X51366
Tcrstaau	P02983	A03510; A04492	J01764
Tcrstrag	P13924	C25599	X15669
Tcr2bacsu	P14512	S42238	M16217
Tcr3strau			D38215
Tcrbbacsu	P23054	S03327	D26185; X08034
Toxacocca			L48797

References

[1] Paulsen, I. et al. (1996) Proc. Natl Acad. Sci. USA 93, 3630–3635.

[2] **Paulsen, I. et al. (1996) Microbiol. Rev. 60, 575–608.**

[3] **Griffith, J. et al. (1992) Curr. Opin. Cell Biol. 4, 684–695.**

[4] **Marger, M.D. and Saier, M. (1993) Trends Biochem. Sci. 18, 13–20.**

[5] **Lewis, K. (1994) Trends Biochem. Sci. 19, 119–123.**

[6] **Nikaido, H. (1996) J. Bacteriol. 178, 5853–5859.**

[7] Guay, G. et al. (1993) J. Bacteriol. 175, 4927–4929.

[8] Cheng, J. et al. (1994) J. Biol. Chem. 269, 27365–27371.

[9] **Henderson, P.J.F. (1993) Curr. Opin. Cell Biol. 5, 708–721.**

[10] Varela, M. et al. (1995) Mol. Membr. Biol. 11, 271–277.

[11] Paulsen, I. and Skurray, R. (1993) Gene 124, 1–11.

Summary

Transporters of the 4-helix H$^+$/multidrug antiporter family, the example of which is the EBR multidrug resistance protein of *Staphylocooous aureus* (Ebrstaau), mediate resistance to one or more structurally dissimilar antibiotics, antiseptics or disinfectants. The mechanism of transport, where known, is antiport (i.e. proton-coupled substrate efflux)[1-4]. Curiously, the amino acid sequences of a subgroup of the family, the SUGE proteins, are similar to chaperones, suggesting a possible role in protein transport. Family members occur in both gram-negative and gram-positive bacteria.

Statistical analysis reveals no apparent relationship between the amino acid sequences of the 4-helix H$^+$/multidrug antiporter family and any other family of transporters. Transporters may be encoded chromosomally or by transmissible plasmids. They are predicted to contain four membrane-spanning helices by the hydropathy of their amino acid sequences[2-4] and activities of reporter gene fusions[4].

Several amino acid sequence motifs are highly conserved in the 4-helix H$^+$/ multidrug antiporter family that are necessary for function by the criterion of site-directed mutagenesis[2,3,5].

Nomenclature, biological sources and substrates

CODE	DESCRIPTION [SYNONYMS]	ORGANISM [COMMON NAMES]	SUBSTRATE(S)^a [RESISTANCE(S)]^b
Ebrescco	Ethidium bromide resistance protein [EBR, E1]	*Escherichia coli* [gram-negative bacterium]	[EB, QUACs]
Ebrstaau	Ethidium bromide resistance protein [Multidrug resistance protein, EBR, QACC, SMR]	*Staphylococcus aureus* [gram-positive bacterium]	H$^+$/EB, QUACs
Emreescco	Methyl viologen resistance protein [Ethidium resistance protein EMRE, EB, MVRC]	*Escherichia coli* [gram-negative bacterium]	[Methyl viologen, EBR]
Qaceklepn	Small multidrug export protein [QACE]	*Klebsiella pneumoniae* [gram-negative bacterium]	[QUACs, antiseptics]
Qacfstasp	Quaternary ammonium compound resistance protein [QACF]	*Staphylococcus* species [gram-positive bacterium]	[QUACs]
Sugeescco	Possible chaperone protein [SUGE]	*Escherichia coli* [gram-negative bacterium]	Proteins?
Sugeprovu	Possible chaperone protein [SUGE]	*Proteus vulgaris* [gram-negative bacterium]	Proteins?

^{a} Cotransported ions are listed for known antiporters.
^{b} Presumed substrates; protein confers resistance to specified compounds.
Abbreviations: QUAC, quaternary ammonium disinfectants; EB, ethidium bromide.

Phylogenetic tree

Proteins listed subsequently in italics are at least 90% identical to the paired transporters listed in parenthesis and therefore are not included in the phylogenetic tree: *Qacfstasp* (Ebrstaau).

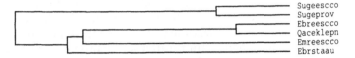

```
                                        Sugeescco
                                        Sugeprov
                                        Ebreescco
                                        Qaceklepn
                                        Emreescco
                                        Ebrstaau
```

Proposed orientation of EBR in the membrane

The model is based on predictions of membrane-spanning regions and α-helical content. The N-terminus of the protein is illustrated on the inside and is folded four times through the membrane[4]. The predicted membrane-spanning helices are portrayed as rectangles. The numbers corresponding to the first and last residue of each membrane-spanning helix are boxed. Residues that are conserved in more than 75% of the aligned transporters (see below) are shown. Consensus residues indicated by an asterisk are not conserved in EBR.

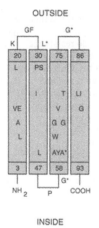

Physical and genetic characteristics

	AMINO ACIDS	MOL. WT	CHROMOSOMAL LOCUS
Ebrescco	115	12.332	Plasmid pVS1 and others
Ebrstaau	107	11 673	Plasmid pTZ20 and others
Emreescco	110	11 958	12.22 minutes
Qaceklepn	110	11 472	Plasmid R751
Qacfstasp	107	11 689	
Sugeescco	155	16 186	94.27 minutes
Sugeprovu	104	11 014	

Multiple amino acid sequence alignments

```
           1                                                   50
Sugeescco  MPFVFSAIVT KVIVEIPLPP GKISVQPSAL QDDLQTPLFT GDGPNSPEPD
Sugeprovu  .......... .......... .......... .......... ..........
Ebrstaau   .......... .......... .......... .......... ..........
Ebrescco   .......... .......... .......... .......... .........M
Qaceklepn  .......... .......... .......... .......... .........M
Emreescco  .......... .......... .......... .......... .........M
Consensus  .......... .......... .......... .......... ..........

           51                                                 100
Sugeescco  MSWIILVIAG LLEVVWAVGL KYTHGFSRLT PSVITVTAMI VSMALLAWAM
Sugeprovu  MSWIILFVAG LLEIVWAVGL KYTHGFTRLT PSIITISAMI VSMGMLSYAM
Ebrstaau   MPYIYLIIAI STEVIGSAFL KSSEGFSKFI PSLGTIISFG ICFYFLSKTM
Ebrescco   KGWLFLVIAI VGEVIATSAL KSSEGFTKLA PSAVVIIGYG IAFYFLSLVL
Qaceklepn  KGWLFLVIAI VGEVIATSAL KSSEGFTKLA PSAVVIIGYG IAFYFLSLVL
Emreescco  NPYIYLGGAI LAEVIGTTLM KFSEGFTRLW PSVGTIICYC ASFWLLAQTL
Consensus  .....L..A. ..EV.....L K...GF..L. PS...I.... .....L....

           101                                                150
Sugeescco  KSLPVGTAYA VWTGIGAVGA AITGIVLLGE SANPMRLASL ALIVLGIIGL
Sugeprovu  KGLPAGTAYA IWTGIGAVGT AIFGIIVFGE SANIYRLLSL AMIVFGIIGL
Ebrstaau   QHLPLNITYA TWAGLGLVLT TVVSIIIFKE QINLITIVSI VLIIVGVVSL
Ebrescco   KSIPVGVAYA VWSGLGVVII TAIAWLLHGQ KLDAWGFVGM GLIIAAFLLA
Qaceklepn  KSIPVGVAYA VWSGLGVVII TAIAWLLHGQ KLDAWGFVGM GLIVSGVVVL
Emreescco  AYIPTGIAYA IWSGVGIVLI SLLSWGFFGQ RLDLPAIIGM MLICAGVLII
Consensus  ...P.G.AYA .W.G.G.V.T ........G. .......... .LI..G....

           151        164
Sugeescco  KLSTH..... ....
Sugeprovu  KLAS...... ....
Ebrstaau   NIFGTSH... ....
Ebrescco   RSPSWKSLRR PTPW
Qaceklepn  NLLSKASAH. ....
Emreescco  NLLSRSTPH. ....
Consensus  .......... ....
```

Proteins listed subsequently in italics are at least 90% identical to the paired transporters listed in parenthesis and therefore are not included in the alignments: *Qacfstasp* (Ebrstaau). Residues listed in the consensus sequence are present in at least 75% of the aligned transporter sequences.

Database accession numbers

	SWISSPROT	PIR	EMBL/GENBANK
Ebrstaau	P14319	S06924; JE0410	M37888; X15574
Ebrescco	P14502	S07656; S21846	U12416; X58425
Emreescco	P23895	JN0329; S24063	M62732; Z11877
Qaceklepn		A48905; S32181	X68232
Qacfstasp		S32181	Z37964
Sugeescco	P30743	S36340	U14003; X69949
Sugeprovu	P20928	S00120	X06151

References
1 Lewis, K. (1994) Trends Biochem. Sci. 19, 119–123.
2 Paulsen, I. et al. (1996) Microbiol. Rev. 60, 575–608.
3 Paulsen, I. and Skurray, R. (1996) Mol. Microbiol. 19, 1167–1175.
4 Paulsen, I. and Skurray, R. (1993) Gene 124, 1–11.
5 Paulsen, I. et al. (1995) J. Bacteriol. 177, 2827–2833.

Summary

Transporters of the 12-helix H⁺/multidrug antiporter family, the example of which is the TETA(C) tetracycline antiporter of *Escherichia coli* (Tcr2escco), mediate resistance to one or more structurally dissimilar antibiotics [1-4]. One member (Arajescco) is involved in either the transport or processing of arabinose polymers. The mechanism of transport, where known, is antiport (i.e. H⁺-coupled substrate efflux). Some members of the 12-helix H⁺/multidrug antiporter family also confer one or more collateral effects, including reduced growth, particularly in non-fermentable carbon sources, complementation of potassium uptake defects and increased susceptibilities to heavy metals and cationic antibiotics [5]. By analogy to other cation-linked transporters, the basis of these effects could be a proton leak [6,7]. Two members of this family, NORA (Norastaau) and BMR1 (Bmr1bacsu) are inhibited by reserpine [8], while TETA(B) is inhibited by 13-(3-chloropropyl) derivatives of 5-hydroxytetracycline [9]. Members of the family have been found in both gram-negative and gram-positive bacteria. Homologous proteins of unknown function are also expressed in human brain [10]. Transporters may be encoded chromosomally or by transmissible plasmids, the latter occurring frequently in antibiotic-resistant pathogens.

Statistical analysis of multiple amino acid sequence comparisons places the 12-helix H⁺/multidrug antiporter family in the uniporter-symporter-antiporter USA superfamily, also known as the major facilitator superfamily (MFS) [1,4]. They are predicted to form 12 membrane-spanning helices by the hydropathy of their amino acid sequences [1,4,11], the activities of reporter gene fusions [12], and susceptibility to proteolysis [13]. Based on complemention and reconstitution experiments, some members of the family may exist as a homodimer [14,15]. There is considerable similarity between the sequences of the N- and C-terminal halves of these proteins, further implying they arose through gene duplication of an ancestral six helix protein [1,16].

Several amino acid sequence motifs are highly conserved in the 12-helix H⁺/multidrug antiporter family, including motifs unique to the family, signature motifs of the USA/MFS superfamily, and motifs necessary for function by the criterion of site-directed mutagenesis [1,4,17]. Mutations affecting susceptibility to inhibitors [8] and substrate specificity have been reported [18].

Nomenclature, biological sources and substrates

CODE	DESCRIPTION [SYNONYMS]	ORGANISM [COMMON NAMES]	SUBSTRATE(S)[a] [RESISTANCE(S)][b]
Arajescco	ARAJ precursor protein [ARAJ]	*Escherichia coli* [gram-negative bacterium]	Arabinose polymers
Bmr1bacsu	Multidrug-H⁺ antiporter 1 [BMR1]	*Bacillus subtilis* [gram-positive bacterium]	H⁺/multiple drugs
Bmr2bacsu	Multidrug-H⁺ antiporter 2 [BMR2, BLT, BMT]	*Bacillus subtilis* [gram-positive bacterium]	H⁺/multiple drugs
Cmlrstrli	Chloramphenicol resistance protein [CMLR]	*Streptomyces lividans* [gram-positive bacterium]	[Chloramphenicol]
Cmlvstrve	Chloramphenicol resistance protein [CMLV]	*Streptomyces venezuelae* [gram-positive bacterium]	[Chloramphenicol]
Cmrrhofa	Chloramphenicol resistance protein [CMRR]	*Rhodococcus fasciens* [gram-negative bacterium]	[Chloramphenicol]

CODE	DESCRIPTION [SYNONYMS]	ORGANISM [COMMON NAMES]	SUBSTRATE(S)[a] [RESISTANCE(S)][b]
Norastaau	Quinolone-H⁺ antiporter [NORA]	*Staphylococcus aureus* [gram-positive bacterium]	H⁺/multiple drugs
Tcr1escco	Tetracycline-H⁺ antiporter [TCR1, TETA(B)]	*Escherichia coli* [gram-negative bacterium]	H⁺/tetracycline-metal chelate
Tcr2escco	Tetracycline-H⁺ antiporter [TCR2, TETA(C)]	*Escherichia coli* [gram-negative bacterium]	H⁺/tetracycline-metal chelate
Tcr3escco	Tetracycline-H⁺ antiporter [TCR3, TETA(A)]	*Escherichia coli* [gram-negative bacterium]	H⁺/tetracycline-metal chelate
Tcr1salor	Tetracycline-H⁺ antiporter [TCR1, TETA(D)]	*Salmonella ordonez* [gram-negative bacterium]	H⁺/tetracycline-metal chelate
Tcr4escco	Tetracycline-H⁺ antiporter [TCR4, TETA(E)]	*Escherichia coli* [gram-negative bacterium]	H⁺/tetracycline-metal chelate
Tetgviban	Tetracycline-H⁺ antiporter [TETG, TETA(G)]	*Vibrio anguillarum* [gram-negative bacterium]	H⁺/tetracycline-metal chelate
Tethpasmu	Tetracycline-H⁺ antiporter [TETH, TETA(H)]	*Pasteurella multocida* [gram-negative bacterium]	H⁺/tetracycline-metal chelate

[a] Cotransported ions are listed for known antiporters.
[b] Presumed substrates; protein confers resistance to specified compounds.

Phylogenetic tree

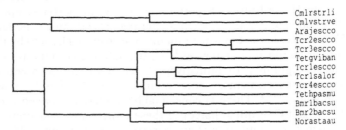

Proposed orientation of TETA(C) in the membrane

The model is based on predictions of membrane-spanning regions and α-helical content. The N-terminus of the protein is illustrated on the inside and is folded 12 times through the membrane [12]. The predicted membrane-spanning helices are portrayed as rectangles. The numbers corresponding to the first and last residue of each membrane-spanning helix are boxed. Residues that are conserved in more than 75% of the aligned transporters (see below) are shown. Consensus residues indicated by an asterisk are not conserved in TETA(C).

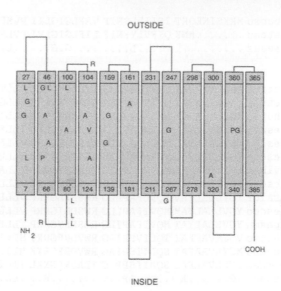

Physical and genetic characteristics

	AMINO ACIDS	MOL. WT	K_m	CHROMOSOMAL LOCUS
Arajescco	394	41 926		8.84 minutes
Bmr1bacsu	389	42 258		
Bmr2bacsu	400	43 424		230°
Cmlrstrli	392	38 855		
Cmlvstrve	436	43 811		
Cmrrhofa	391	40 323		
Norastaau	388	42 265		
Tcr1escco	401	43 267	Tetracycline: 36 μM [19]	Transposon Tn10
Tcr2escco	396	41 510		Plasmid pSC101
Tcr3escco	399	42 237		Transposon Tn1721
Tcr1salor	395	41 035		Plasmid pRA1
Tcr4escco	405	43 412		
Tetgviban	393	40 881		
Tethpasmu	400	43 672		

Multiple amino acid sequence alignments

```
              1                                                  50
Cmlvstrve ...........MPFAIYVLG IAVFAQGTSE FMLSGLIPDM AQDLQVSVPT
Cmlrstrli ...........MPLPLYLLA VAVCAMGTSE FMLAGLVPDI ASDLGVTVGT
Cmrrhofa  ...........MPFAIYVLG IAVFAQGTSE FMLSGLIPDM AQDLQVSVPT
Arajescco ...........MKKVILSLA LGTFGLGMAE FGIMGVLTEL AHNVGISIPA
Tcr2escco .....MKSNN ALIVILGTVT LDAVGIGLVM PVLPGLLRDI VHSDSIA.SH
Tcr3escco .....MKPNI PLIVILSTVA LDAVGIGLIM PVLPGLLRDL VHSNDVT.AH
Tetgviban .......VRS SAIIALLIVG LDAMGLGLIM PVLPTLLREL VPAEQVA.GH
Tcr1escco .......MNS STKIALVITL LDAMGIGLIM PVLPTLLREF IASEDIA.NH
Tcr1salor .......MNK PAVIALVITL LDAMGIGLIM PVLPSLLREY LPEADVA.NH
Tcr4escco .......MNR TVMMALVIIF LDAMGIGIIM PVLPALLREF VGKANVA.EN
Tethpasmu .......MNK SIIIILLITV LDAIGIGLIM PVLPTLLNEF VSENSLA.TH
Bmr1bacsu ....MEKKNI TLTILLTNLF IAFLGIGLVI PVTPTIMNEL HLS...G.TA
```

```
Bmr2bacsu MKKSINEQKT IFIILLSNIF VAFLGIGLII PVMPSFMKIM HLS...G.ST
Norastaau .......MNK QIFVLYFNIF LIFLGIGLVI PVLPVYLKDL GLT...G.SD
Consensus .......... .....L.... ...G.G.... ..L....... ..........


             51                                                    100
Cmlvstrve AGLLTSAFAI GMIIGAPLMA IVSMRWQRRR ALLTFLITFM VVHVIGALTD
Cmlrstrli AGTLTSAFAT GMIVGAPLVA ALARTWPRRS SLLGFILAFA AAHAVGAGTT
Cmrrhofa  AGLLTSAFAI GMIIGAPLMA IVSMRWQRRR ALLTFLITFM VVHVIGALTD
Arajescco AGHMISYYAL GVVVGAPIIA LFSSRYSLKH ILLFLVALCV IGNAMFTLSS
Tcr2escco YGVLLALYAL MQFLCAPVLG ALSDRFGRRP VLLASLLGAT IDYAIMATTP
Tcr3escco YGILLALYAL VQFACAPVLG ALSDRFGRRP ILLVSLAGAT VDYAIMATAP
Tetgviban YGALLSLYAL MQVVFAPMLG QLSDSYGRRP VLLASLAGAA VDYTIMASAP
Tcr1escco FGVLLALYAL MQVIFAPWLG KMSDRFGRRP VLLLSLIGAS LDYLLLAFSS
Tcr1salor YGILLALYAV MQVCFAPLLG RWSDKLGRRP VLLLSLAGAA FDYTLLALSN
Tcr4escco YGVLLALYAM MQVIFAPLLG RWSDRIGRRP VLLLSLLGAT LDYALMATAS
Tethpasmu YGVLLALYAT MQVIFAPILG RLSDKYGRKP ILLFSLLGAA LDYLLMAFST
Bmr1bacsu VGYMVACFAI TQLIVSPIAG RWVDRFGRKI MIVIGLLFFS VSEFLFGIGK
Bmr2bacsu MGYLVAAFAI SQLITSPFAG RWVDRFGRKK MIILGLLIFS LSELIFGLGT
Norastaau LGLLVAAFAL SQMIISPFGG TLADKLGKKL IICIGLILFS VSEFMFAVGH
Consensus .G.L....A. .....AP... .......R.. .LL..L.... ......A...


             101                                                   150
Cmlvstrve SFGVLLVTRI VGALANAGFL AVALGAAMSM VPADMKGRAT SVLLGGVTIA
Cmlrstrli SFPVLVACRV VAALANAGFL AVALTTAAAL VPADKQGRAL AVLLSGTTVA
Cmrrhofa  SFGVLLVTRI VGALANAGFL AVALGAAMSM VPADMKGRAT SVLLGGVTIA
Arajescco SYLMLAIGRL VSGFPHGAFF GVGAIVLSKI IKPGKVTAAV AGMVSGMTVA
Tcr2escco VLWILYAGRI VAGITGA.TG AVAGAYIADI TDGEDRARHF GLMSACFGVG
Tcr3escco FLWVLVYIGRI VAGITGA.TG AVAGAYIADI TDGDERARHF GFMSACFGFG
Tetgviban VLWVLYIGRL VSGVTGA.TG AVAASTIADS TGEGSRARWF GYMGACYGAG
Tcr1escco ALWMLYLGRL LSGITGA.TG AVAASVIADT TSASQRVKWF GWLGASFGLG
Tcr1salor VLWMLYLGRI ISGITGA.TG AVAASVVADS TAVSERTAWF GRLGAAFGAG
Tcr4escco VVWVLLLYRI IAGITGA.TG AVAASTIADV TPEESRTHWF GMMGACFGGG
Tethpasmu TLWMLYIGRI IAGITGA.TG AVCASAMSDV TPAKNRTRYF GFLGGVFGVG
Bmr1bacsu TVEMLFITRM LGGISAPFIM PGVTAFIADI TTIKTRPKAL GYMSAAISTG
Bmr2bacsu HVSIFYFSRI LGGVSAAFIM PAVTAYVADI TTLKERSKAM GYVSAAISTG
Norastaau NFSVLMLSRV IGGMSAGMVM PGVTGLIADI SPSHQKAKNF GYMSAIINSG
Consensus ....L...R. ......A... .V.....A.. .......... ..........


             151                                                   200
Cmlvstrve CVVGVPGGAL LGELWGWRAS FWEVVLISAP AVAAIMASTP ADSPTDSVPN
Cmlrstrli TVAGVPGGSL LGTWLGWRAT FWAVAVCCLP AAFGVLKAIP AGRATAAATG
Cmrrhofa  CVVGVPGGAL LGELWGWRAS FWEVVLISAP AVAAIMASTP ADSPTDSVPN
Arajescco NLLGIPLGTY LSQEFSWRYT FLLIAVFNIA VMASVYFWVP DIRDEAK...
Tcr2escco MVAGPVAGGL LGAISLHAPF LAAAVLNGLN LLLGCFLMQE SHK.GERRPM
Tcr3escco MVAGPVLGGL MGGFSPHAPF FAAAALNGLN FLTGCFLLPE SHK.GERRPL
Tetgviban MIAGPALGGM LGGISAHAPF IAAALLNGFA FLLACIFLKE THH.SHGGTG
Tcr1escco LIAGPIIGGF AGEISPHSPF FIAALLNIVT FLVVMFWFRE TKN.TRDNTD
Tcr1salor LIAGPAIGGL AGDISPHLPF VIAAILNACT FLMVFFIFKP AVQ.TEEKPA
Tcr4escco MIAGPVIGGF AGQLSVQAPF MFAAAINGLA FLVSLFILHE THN.ANQVSD
Tethpasmu LIIGPMLGGL LGDISAHMPF IFAAISHSIL LILSLLFFRE TQK.REALVA
Bmr1bacsu FIIGPGIGGF LAEVHSRLPF FFAAAFALLA AILSILTLRE PERNPENQEI
Bmr2bacsu FIIGPGAGGF IAGFGIRMPF FFASAIALIA AVTSVFILKE SLSIEERHQL
Norastaau FILGPGIGGF MAEVSHRMPF YFAGALGILA FIMSIVLIHD PKKSTTS...
Consensus ...G...G.. .......... ..A....... .......... ..........
```

```
          201                                                    250
Cmlvstrve ATR...ELSS LRQRKLQLIL VLGALINGAT FCSFTYLAPT L.....TDVA
Cmlrstrli GPPLRVELAA LKTPRLLLAM LLGALVNAAT FASFTFLAPV V.....TDTA
Cmrrhofa  ATR,,.ELSS LRQRKLQLIL VLGALINGAT FCSFTYLAPT L.....TDVA
Arajescco .GNLREQFHF LRSPAPWLIF AATMFGNAGV FAWFSYVKPY M.....MFIS
Tcr2escco PLRAFNPVSS FRWARGM.TI VAALMTVFFI MQLVGQVPAA LWVIFGEDRF
Tcr3escco RREALNPLSF VRWARGM.TV VAALMAVFFI MQLVGQVPAA LWVIFGEDRF
Tetgviban KPVRIKPFVL LRLDDAL.RG LGALFAVFFI IQLIGQVPAA LWVIYGEDRF
Tcr1escco T.EVGVETQS NSVYITLFKT MPILLIIYFS AQLIGQIPAT VWVLFTENRF
Tcr1salor E.Q..KQESA GISFITLLKP LALLLFVFFT AQLIGQIPAT VWVLFTESRF
Tcr4escco ELKNETINET TSSIREMISP LSGLLVVFFI IQLIGQIPAT LWVLFGEERF
Tethpasmu NRTPENQTAS NTVTVFFKKS LYFWLATYFI IQLIGQIPAT IWVLFTQYRF
Bmr1bacsu KGQK...... TGFKRIFAPM YFIAFLIILI SSFGLASFES LFALFVDHKF
Bmr2bacsu SSHTKESNFI KDLKRSIHPV YFIAFIIVFV MAFGLSAYET VFSLFSDHKF
Norastaau GFQKLEPQLL TKINW...KV FITPVILTLV LSFGLSAFET LYSLYTADKV
Consensus ....................................................
```

```
          251                                                    300
Cmlvstrve GFDSRWIPLL LGLFG.LGSF IGVSVGGRLA DT.RPFQLLV AGSAALLVGW
Cmlrstrli GLGDLWISVA LVLFG.AGSF AGVTVAGRLS DR.RPAQVLA VAGPLLLVGW
Cmrrhofa  GFDSRWIPLL LGLFG.LGSF IGVSVGGRLA DT.RPFQLLV AGSAALLVGW
Arajescco GFSETAMTFI MMLVG.LGMV LGNMLSGRIS GRYSPLRIAA VTDFIIVLAL
Tcr2escco RWSATMIGLS LAVFGILHAL AQAFVTGPAT KRFGEKQAII AGMAADALGY
Tcr3escco HWDATTIGIS LAAFGILHSL AQAMITGPVA ARLGERRALM LGMIADGTGY
Tetgviban QWNTATVGLS LAAFGATHAI FQAFVTGPLS SRLGERRTLL FGMAADGTGF
Tcr1escco GWNSMMVGFS LAGLGLLHSV FQAFVAGRIA TKWGEKTAVL LGFIADSSAF
Tcr1salor AWDSAAVGFS LAGLGAMHAL FQAVVAGALA KRLSEKTIIF AGFIADATAF
Tcr4escco AWDGVMVGVS LAVFGLTHAL FQGLAAGFIA KHLGERKAIA VGILADGCGL
Tethpasmu DWNTTSIGMS LAVLGVLHIF FQAIVAGKLA QKWGEKTTIM ISMSIDMMGC
Bmr1bacsu GFTASDIAIM ITGGAIVGAI TQVVLFDRFT RWFGEIHLIR YSLILSTSLV
Bmr2bacsu GFTPKDIAAI ITISSIVAVV IQVLLFGKLV NKLGEKRMIQ LCLITGAILA
Norastaau NYSPKDISIA ITGGGIFGAL FQIYFFDKFM KYFSELTFIA WSLLYSVVVL
Consensus ...........G.....G.....................................
```

```
          301                                                    350
Cmlvstrve IVFAITASHP VVTLVMLFVQ GTLSFAVGST LISRVLYVAD GAPTLGGSFA
Cmlrstrli PALAMLADRP VALLTLVFVQ GALSFALGST LITRVLYEAA GAPTMAGSYA
Cmrrhofa  IVFAITASHP VVTLVMLFVQ GTLSFAVGST LISRVLYVAD GAPTLGGSFA
Arajescco LMLFFCGGMK TTSLIFAFIC CAGLFALSAP LQILLLQNAK GGELLGAAGG
Tcr2escco VLLAFATRGW MAFPI.MILL ASGGIGMPAL QAMLSRQVDD DHQGQLQGSL
Tcr3escco ILLAFATRGW MAFPI.MVLL ASGGIGMPAL QAMLSRQVDE ERQGQLQGSL
Tetgviban VLLAFATQGW MVFPI.LLLL AAGGVGMPAL QAMLSNNVSS NKQGALQGTL
Tcr1escco AFLAFISEGW LVFPV.LILL AGGGIALPAL QGVMSIQTKS HQQGALQGLL
Tcr1salor LLMSAITSGW MVYPV.LILL AGGGIALPAL QGIISAGASA ANQGKLQGVL
Tcr4escco FLLAVITQSW MVWPV.LLLL AGGGITLPAL QGIISVRVGQ VAQGQLQGVL
Tethpasmu LLLAWIGHVW VILPA.LICL AAGGMGQPAL QGYLSKSVDD NAQGKLQGTL
Bmr1bacsu FLLTTVHSYV AILLVTVTVF VGFDLMRPAV TTYLS.KIAG NEQFAGGGMN
Bmr2bacsu FVSTVMSGFL TVLLVTCFIF LAFDLLRPAL TAHLS.NMAG NQQGFVAGMN
Norastaau ILLVFANGYW SIMLISFVVF IGFDMIRPAI TNYFS.NIAG ERQGFAGGLN
Consensus ...........................A........................
```

```
          351                                                    400
Cmlvstrve TAAFNVGAAL GPALGGVAIG IGMGYRAPLW TSAALVALAI VIGAATWTRW
Cmlrstrli TAALNVGAAA GPLVAATTLG HTTGNLGPLW ASGLLVAVAL LVAFPFRTVI
```

```
Cmrrhofa   TAAFNVGAAL GPALGGVAIG IGMGYRAPLW TSAALVALAI VIGAATWTRW
Arajescco  QIAFNLGSAV GAYCGGMMLT LGLAYNYVAL .PAALLSFAA MSSLLLYGRY
Tcr2escco  AALTSLTSIT GPLIVTAIYA ASASTWNGLA WIVGAALYLV CLPALRRGAW
Tcr3escco  AALTSLTSIV GPLLFTAIYA ASITTWNGWA WIAGAALYLL CLPALRRGLW
Tetgviban  TSLTNLSSIA GPLGFTALYS ATAGAWNGWV WIVGAILYLI CLPILRRPFA
Tcr1escco  VSLTNATGVI GPLLFAVIYN HSLPIWDGWI WIIGLAFYCI IILLSMTFML
Tcr1salor  VSLTNLTGVA GPLLFAFIFS QTQQSADGTV WLIGTALYGL LLAICLLIRK
Tcr4escco  TSLTHLTAVI GPLVFAFLYS ATRETWNGWV WIIGCGLYVV ALIILRFFHP
Tethpasmu  VSLTNITGII GPLLFAFIYS YSVAYWDGLL WLMGAILYAM LLITAYFHQR
Bmr1bacsu  SMFTSIGNVF GPIIGGMLFD IDVNY..PFY FATVTLAIGI ALTIAWKAPA
Bmr2bacsu  STYTSLGNIF GPALGGILFD LNIHY..PFL FAGFVMIVGL GLTMVWKEKK
Norastaau  STFTSMGNFI GPLIAGALFD VHIEA..PIY MAIGVSLAGV VIVLIEKQHR
Consensus  .......... GP........ .......... .......... ..........
```

```
                401          416
Cmlvstrve  REPRPALDTV PP....
Cmlrstrli  TTAAPADATR ......
Cmrrhofa   REPRPALDTV PP....
Arajescco  KRQQAADTPV LAKPLG
Tcr2escco  SRATST.... ......
Tcr3escco  SGAGQRADR. ......
Tetgviban  TSLVIPSQ.. ......
Tcr1escco  TPQAQGSKQE TSA...
Tcr1salor  PAPVAATC.. ......
Tcr4escco  GRVIHPINKS DVQQRI
Tethpasmu  KTTPKAVIST P.....
Bmr1bacsu  HLKAST.... ......
Bmr2bacsu  NDAAALN... ......
Norastaau  AKLKEQNM.. ......
Consensus  .......... ......
```

Residues listed in the consensus sequence are present in at least 75% of the aligned transporter sequences. Residues indicated by boldface type are also conserved in at least one other family of the USA/MFS superfamily.

Database accession numbers

	SWISSPROT	PIR	EMBL/GENBANK
Arajescco	P23910	S27549; B43750	M64787
Bmr1bacsu	P33449	A39705	L25604; M33628
Bmr2bacsu	P39843		L32599
Cmlrstrli	P31141	S18593	X59968
Cmlvstrve			U09991
Cmrrhofa		S25183	Z12001
Norastaau	P21191	A37838	D90119; M62960
Tcr1escco	P02980	A03507	V00611; J01830
Tcr1salor	P33733	S30286	X65876
Tcr2escco	P02981	A03508	V01119
Tcr3escco	P02982	A03509	X00006
Tcr4escco	Q07282	A36896	L06940
Tetgviban			S52437
Tethpasmu			U00792

References
[1] Griffith, J. et al. (1992) Curr. Opin. Cell Biol. 4, 684–695.
[2] Marger, M.D. and Saier, M. (1993) Trends Biochem. Sci. 18, 13–20.

[3] Lewis, K. (1994) Trends Biochem. Sci. 19, 119–123.
[4] Paulsen, I. et al. (1996) Microbiol. Rev. 60, 575–608.
[5] Griffith, J. et al. (1994) Mol. Membr. Biol. 11, 271–277.
[6] Wright, E.M. et al. (1996) Curr. Opin. Cell Biol. 8, 468–473.
[7] Fischbarg, J. and Vera, J.C. (1995) Am. J. Physiol. 268, C1077–C1089.
[8] Ahmed, M. et al. (1993) J. Biol. Chem. 268, 11086–11089.
[9] Nelson, M.L. et al. (1994) J. Med. Chem. 37, 1355–1361.
[10] Duyao, M. et al. (1993) Hum. Mol. Genet. 2, 673–676.
[11] Henderson, P.J.F. (1993) Curr. Opin. Cell Biol. 5, 708–721.
[12] Allard, J. and Bertrand, K. (1992) J. Biol. Chem. 267, 17809–17819.
[13] Eckert, B. and Beck, C. (1989) J. Biol. Chem. 264, 11663–11670.
[14] Yamaguchi, A. et al. (1993) FEBS Lett. 324, 131–135.
[15] Rubin, R.A. and Levy, S.B. (1991) J. Bacteriol. 173, 4503–4509.
[16] Rubin, R.A. et al. (1990) Gene 87, 7–13.
[17] Varela, M. and Griffith, J. (1994) Mol. Membr. Biol. 11, 271–277.
[18] Guay, G. et al. (1994) Antimicrob. Agents Chemother. 38, 857–860.
[19] Yamaguchi, A. et al. (1993) J. Biol. Chem. 268, 26990–26995.

Summary

Transporters of the acriflavin-cation resistance family, the example of which is the ACRB acriflavin resistance protein of *Escherichia coli* (Acrbescco), mediate resistance to one or more structurally dissimilar antibiotics, including β-lactams, fluoroquinolones and erythromycin; and ions, including nickel, cadmium and cobalt. The mechanism of transport is proton motive force dependent[1]. An interesting feature of the acriflavin-cation resistance family is that a paired "linker" protein (e.g. ACRA) is believed to bring the transporter protein in apposition with a corresponding outer-membrane channel (e.g. TOLB), allowing direct efflux through the inner and outer membranes[1-3]. Known members of the family occur only in gram-negative bacteria.

Statistical analysis reveals no significant relationship between the amino acid sequences of the acriflavin-cation resistance family and any other family of transporters. However, other membrane proteins, including members of the 14-helix H⁺/multidrug antiporter family, are similarly coupled to outer-membrane channels[1,2]. Members of the acriflavin-cation resistance family are predicted to form 12 membrane-spanning helices by the hydropathy of their amino acid sequences.

Several amino acid sequence motifs are highly conserved in the acriflavin-cation resistance family[3]. There is considerable similarity between the sequences of the N- and C-terminal halves of these proteins, further implying they arose through gene duplication of an ancestral six helix protein[3,4].

Nomenclature, biological sources and substrates

CODE	DESCRIPTION [SYNONYMS]	ORGANISM [COMMON NAMES]	SUBSTRATE(S) [RESISTANCE][a]
Acrbescco	Acriflavin resistance protein [ACRB, ACRE]	*Escherichia coli* [gram-negative bacterium]	[Acriflavin, multiple drugs]
Acrdescco	Acriflavin resistance protein [ACRD]	*Escherichia coli* [gram-negative bacterium]	[Acriflavin, multiple drugs]
Acrfescco	Acriflavin resistance protein [ACRF, ENVD]	*Escherichia coli* [gram-negative bacterium]	[Acriflavin]
Cnraalceu	Nickel-cobalt resistance protein [CNRA]	*Alcaligenes eutrophus* [gram-negative bacterium]	Nickel, cobalt
Czcaalceu	Cation efflux protein [CZCA]	*Alcaligenes eutrophus* [gram-negative bacterium]	Cobalt, zinc, cadmium

[a] Presumed substrates; protein confers resistance to specified compounds.

Phylogenetic tree

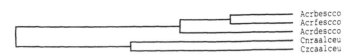

```
                                          ┌──── Acrbescco
                                    ┌──────┤
                              ┌─────┤      └──── Acrfescco
                              │     └─────────── Acrdescco
                      ┌───────┤
                      │       └───────────────── Cnraalceu
                      └───────────────────────── Czcaalceu
```

Proposed orientation of ACRB in the membrane

The model is based on predictions of membrane-spanning regions and α-helical content. The N-terminus of the protein is illustrated on the inside and is folded 12 times through the membrane. The predicted membrane-spanning helices are portrayed as rectangles. The numbers corresponding to the first and last residue of each membrane-spanning helix are boxed. Residues that are conserved in more than 75% of the aligned transporters (see below) are shown. Consensus residues indicated by an asterisk are not conserved in ACRB.

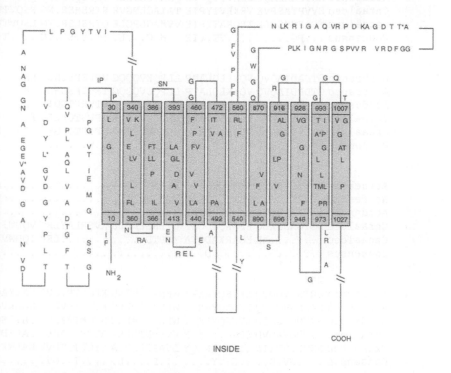

Physical and genetic characteristics

	AMINO ACIDS	MOL. WT	CHROMOSOMAL LOCUS
Acrbescco	1049	113 573	83.49 minutes
Acrdescco	1038	113 070	55.58 minutes
Acrfescco	1034	111 454	73.51 minutes
Cnraalceu	1075	115 583	
Czcaalceu	1063	115 644	

Multiple amino acid sequence alignments

```
           1                                                  50
Acrbescco ....MPNFFI DRPIFAWVIA IIIMLAGGLA ILKLPVAQYP TIAPPAVTIS
Acrfescco ....MANFFI RRPIFAWVLA IILMMAGALA ILQLPVAQYP TIAPPAVSVS
Acrdescco ....MANFFI DRPIFAWVLA ILLCLTGTLA IFSLPVEQYP DLAPPNVRVT
Cnraalceu MIESILSGSV RYRWLVLFLT AVVAVIGAWQ LNLLPIDVTP DITNKQVQIN
Czcaalceu MFERIISFAI QQRWLVLLAV FGMAGLGIFS YNRLPIDAVP DITNVQVQVN
Consensus .......FI .......... ......G... ...LP....P .I....V...
```

```
           51                                                 100
Acrbescco ASYPGADAKT VQDTVTQVIE QNMNGIDNLM YMSSNSDSTG TVQITLTFES
Acrfescco ANYPGADAQT VQDTVTQVIE QNMNGIDNLM YMSSTSDSAG SVTITLTFQS
Acrdescco ANYPGASAQT LENTVTQVIE QNMTGLDNLM YMSSQSSGTG QASVTLSFKA
Cnraalceu SVVPTMSPVE VEKRVTYPIE TAIAGLNGVE STRSMSR.NG FSQVTVIFKE
Czcaalceu TSAPGYSPLE TEQRATYPIE VVMAGLPGLE QTRSLSR.YG LSQVTVIFKD
Consensus ...PG..... ....VT..IE ..M.G...L. ...S.S...G ....T..F..
```

```
           101                                                150
Acrbescco GTDADIAQVQ VQNKLQLAMP LLPQEVQQQG VSVEKSSSSF LM........
Acrfescco GTDPDIAQVQ VQNKLQLATP LLPQEVQQQG ISVEKSSSSY LM........
Acrdescco GTDPDEAVQQ VQNQLQSAMR KLPQAVQNQG VTVRKTGDTN IL........
Cnraalceu SANLYFMRHE VSERLAQARP NLPENVEPQM GPVSTGLGEV FHYSVEYQYP
Czcaalceu GTDVYFARQL VNQRIQEAKD NLPEGVVPAM GPISTGLGEI YLWTVE....
Consensus GTD...A... V...LQ.A.. .LP..V..Q. ..V.......
```

```
           151                                                200
Acrbescco ..VVGVINTD GTMTQEDISD YVAAN..... ..........  ....MKDAI
Acrfescco ..VAGFVSDN PGTTQDDISD YVASN..... ..........  ....VKDTL
Acrdescco ..TIAFVSTD GSMDKQDIAD YVASN..... ..........  .....IQDPL
Cnraalceu DGTGASIKDG EPGWQSDGSF LTERGERLDD RVSRLAYLRT VQDWIIRPQL
Czcaalceu ........AE EGARKADGTA YTPTD..... .......LRE IQDWVVRPQL
Consensus .......... ......D... Y......... .......... .........L
```

```
           201                                                250
Acrbescco SRTSGVGDVQ LFGSQYAMRI WMNP..NELN KFQLTPVDVI TAIKAQNAQV
Acrfescco SRLNGVGDVQ LFGAQYAMRI WLDA..DLLN KYKLTPVDVI NQLKVQNDQI
Acrdescco SRVNGVGDID AYGSQYSMRI WLDP..AKLN SFQMTAKDVT DAIESQNAQI
Cnraalceu RTTPGVADVD SLGG.YVKQF VVEPDTGKMA AYGVSYADLA RALEDTNLSV
Czcaalceu RNVPGVTEIN TIGG.FNKQY LVAPSLERLA SYGLTLTDVV NALNKNNDNV
Consensus ....GV.D.. ..G..Y.... ...P....L. ....T..DV. ......N...
```

```
           251                                                300
Acrbescco AAGQLGGTPP VKGQQLNASI IAQTRLTSTE EFGKILLKVN QDGSRVLLRD
Acrfescco AAGQLGGTPA LPGQQLNASI IAQTRFKNPE EFGKVTLRVN SDGSVVRLKD
Acrdescco AVGQLGGTPS VDKQALNATI NAQSLLQTPE QFRDITLRVN QDGSEVRLGD
Cnraalceu GAN......F IRRSGESYLV RADARIKSAD EISRAVI.AH GK.MSHHVGQ
Czcaalceu GAG......Y IERRGEQYLV RAPGQVASED DIRNIIV.GT AQGQPIRIRD
Consensus .......... ..........  .A........ .......... ..G.....D
```

```
           301                                                350
Acrbescco VAKIELGGEN YDIIAEFNGQ PASGLGIKLA TGANALDTAA AIRAELAKME
Acrfescco VARVELGGEN YNVIARINGK PAAGLGIKLA TGANALDTAK AIKAKLAELQ
Acrdescco VATVEMGAEK YDYLSRFNGK PASGLGVKLA SGANEMATGE LVLNRLDELA
```

```
Cnraalceu VARVKIGGEL RSGAASRNGN ETVVGSALML VGANSRTVAQ AVGDKLEQIS
Czcaalceu IGDVEIGKEL RTGAATENGK EVVLGTVFML IGENSRAVSK AVDEKVASIN
Consensus VA.VE.G.E. ....A..NG. ...........GAN...... A....L....
```

```
          351                                               400
Acrbescco PFFPSGLKIV YPYDTTPFVK ISIHEVVKTL VEAIILVFLV MYLFLQNFRA
Acrfescco PFFPQGMKVL YPYDTTPFVQ LSIHEVVKTL FEAIMLVFLV MYLFLQNMRA
Acrdescco QYFPHGLEYK VAYETTSFVK ASIEDVVKTL LEAIALVFLV MYLFLQNFRA
Cnraalceu KTLPPGVVIV PTLNRSQLVI ATIETVAKNL IEGALLVVAI LFALLGNWRA
Czcaalceu RTMPEGVKIV TVYDRTRLVD KAIATVKKNL LEGAVLVIVI LFLFLGNIRA
Consensus ...P.G.... ..Y..T..V. ..I..V.K.L .E...LV.... .LFL.N.RA
```

```
          401                                               450
Acrbescco TLIPTIAVPV VLLGTFAVLA AFGFSINTLT MFGMVLAIGL LVDDAIVVVE
Acrfescco TLIPTIAVPV VLLGTFAILA AFGYSINTLT MFGMVLAIGL LVDDAIVVVE
Acrdescco TLIPTIAEPV VLMGTFSVLY AFGYSVNTLT MFAMVLAIGL LVDDAIVVVE
Cnraalceu ATIAALVIPL SLLVSAIGMN QFHISGNLMS LGA..LDFGL IIDGAVIIVE
Czcaalceu ALITATIIPL AMLFTFTGMV NYKISANLMS LGA..LDFGI IIDGAVVIVE
Consensus .LI.....P. .LL.TF.... ....S.N... .......LA.GL ..D.A...VE
```

```
          451                                               500
Acrbescco NVERVMAEEG ......LPPK EATRKSMGQI QGA...LVGI AMVLSAVFVP
Acrfescco NVERVMMEDK ......LPPK EATEKSMSQI QGA...LVGI AMVLSAVFIP
Acrdescco NVERIMSEEG ......LTPR EATRKSMGQI QGA...LVGI AMVLSAVFVP
Cnraalceu NSLRRLAERQ HREGRLLTLD DRLQEVVQSS REMVRPTVYG QLVIFMVFLP
Czcaalceu NCVRRLAHAQ EHHGRPLTRS ERFHEVFAAA KEARRPLIFG QLIIMIVYLP
Consensus N..R...E.. ......L... E.......... ..A...LV.... V...VF.P
```

```
          501                                               550
Acrbescco MAFFGGSTGA IYRQFSITIV SAMALSVLVA LILTPALCAT MLKPIAKGDH
Acrfescco MAFFGGSTGA IYRQFSITIV SAMALSVLVA LILTPALCAT LLKPVS.AEH
Acrdescco MAFFGPTTGA IYRQFSITIV AAMVLSVLVA MILTPALCAT LLKPLKKGEH
Cnraalceu SLTFQGVEGK MFSPMVITLM LALASAFVLS LTFVPAMVAV MLRKKVAETE
Czcaalceu IFALTGVEGK MFHPMAFTVV LALLGAMILS VTFVPAAVAL FIGERVAEKE
Consensus ...F.G..G. ......IT.V .A......... ...PA..A. .L........
```

```
          551                                               600
Acrbescco GEGKKGFFGW FNRMFEKSTH HYTDSVGGIL RSTGRYLVLY LIIVVGMAYL
Acrfescco HENKGGFFGW FNTTFDHSVN HYTNSVGKIL GSTGRYLLIY ALIVAGMVVL
Acrdescco HGQK.GFFAW FNQMFNRNAE RYEKGVAKIL HRSLRWIVIY VLLLGGMVFL
Cnraalceu ..........VRVIVATKE SYRPWLEHAV ARPMPFIGAG IATVAVATVA
Czcaalceu ..........NRLMLWAKR RYEPLLEKSL ANTAVVLTFA AVSIVLCVAI
Consensus ..................Y.......L ..................
```

```
          601                                               650
Acrbescco FVRLPSSFLP .DEDQGVFMT MVQLPAGATQ ERTQKVLNEV THYYLTKEKN
Acrfescco FLRLPSSFLP .EEDQGVFLT MIQLPAGATQ ERTQKVLDQV TDYYLKNEKA
Acrdescco FLRLPTSFLP .LEDRGMFTT SVQLPSGSTQ QQTLKVVEQI EKYYFTHEKD
Cnraalceu FTFVGREFMP TLDELNLNLS SVRIPSTSID QSVA..IDLP LERAV.LSLP
Czcaalceu AARLGSEFIP NLNEGDIAIQ ALRIPGTSLS QSVE..MQKT IETTLKAKFP
Consensus F.RL...F.P ...........P..... ..........
```

```
          651                                               700
Acrbescco NVESVFAVNG ...FGFAGRG QNTGIAFVSL KDWADRPGEE NKVEAITMRA
Acrfescco NVESVFTVNG ...FSFSGQA QNAGMAFVSL KPWEERNGDE NSAEAVIHRA
```

```
Acrdescco NIMSVFATVG ...SGPGGNG QNVARMFIRL KDWSERDSKT GTSFAIIERA
Cnraalceu EVQTVYSKAG TASLAADPMP PNASDNYIIL KPKSEWPEGV TTKEQVIERI
Czcaalceu EIERVFARTG TAEIASDLMP PNISDGYIML KPEKDWPEPK KTHAELLSAI
Consensus ....VF...G .......... .N......L K.......... .........R.


701                                                        750
Acrbescco TRAFSQIKDA MVFAFN.LPA IVELGTATGF DFELIDQAGL GHEKLTQARN
Acrfescco KMELGKIRDG FVIPFN.MPA IVELGTATGF DFELIDQAGL GHDALTQARN
Acrdescco TKAFNQIKEA RVIAPDSPPA ISGLGSSAGF DMELQDHAGA GHDALMAARN
Cnraalceu REKTAPMVGN N.YDVTQPIE MRFNELIGGV RSDVAVKVYG ENLDELAATA
Czcaalceu QEEAGKIPGN N.YEFSQPIQ LRFNELISGV RSDVAVKIFG DDNNVLSETA
Consensus ......I... .......... .......G. .......... ......A..


751                                                        800
Acrbescco QLLAEAAKHP DMLTSVRPNG LEDTPQFKID IDQEKAQALG VSINDINTTL
Acrfescco QLLGMAAQHP ASLVSVRPNG LEDTAQFKLE VDQEKAQALG VSLSDINQTI
Acrdescco QLLALAAENP E.LTRVRHNG LDDSPQLQID IDQRKAQALG VAIDDINDTL
Cnraalceu QRIAAVLKKT PGATDVRVPL TSGFPTFDIV FDRAAIARYG LTVKEVADTI
Czcaalceu KKVSAVLQGI PGAQEVKVEQ TTGLPMLTVK IDREKAARYG LNMSDVQDAV
Consensus Q.......... ....VR... ....P..... .D..KA...G ....D...T.


801                                                        850
Acrbescco GAAWGGSYVN DFIDRGRVKK VYVMSEAKYR MLPDDIGDWY VR.....AAD
Acrfescco STALGGTYVN DFIDRGRVKK LYVQADAKFR MLPEDVDKLY VR.....SAN
Acrdescco QTAWGSSYVN DFMDAGRVKK VYVQAAGPYP MLPDDINLWY VR.....NKD
Cnraalceu STAMAGRPAG QIFDGDRRFD IVIRLPGEQR ENLDVLGALP VMLPLSEGQA
Czcaalceu ATGVGGRDSG TFFQGDRRFD IVVRLPEAVR GEVEALRRLP IPLPKGVDAR
Consensus .TA.GG.... .F.D..R... ..V......R .......... V.........


851                                                        900
Acrbescco GQMVPFSAFS SSRWEYGSPR LERYNGLPSM EILGQAAPGK STGEAMELME
Acrfescco GEMVPFSAFT TSHWVYGSPR LERYNGLPSM EIQGEAAPGT SSGDAMALME
Acrdescco GGMVPFSAFA TSRWETGSPR LERYNGYSAV EIVGEAAPGV STGTAMDIME
Cnraalceu RASVPLRQLV QFRFTQGLNE VSRDNGKRRV YVEANVGGRD LGSFVDDAAA
Czcaalceu TTFIPLSEVA TLEMAPGPNQ ISRENGKRRI VISANVRGRD IGSFVPEAEA
Consensus ...VP.S... .......G... ..R.NG.... .I........ ...........


901                                                        950
Acrbescco QLAS..KLPT GVGYDWTGMS YQERLSGNQA PSLYAISLIV VFLCLAALYE
Acrfescco NLAS..KLPA GIGYDWTGMS YQERLSGNQA PALVAISFVV VFLCLAALYE
Acrdescco SLVK..QLPN GFGLEWTAMS YQERLSGAQA PALYAISLLV VFLCLAALYE
Cnraalceu RIAKEVKLPP GMYIEWGGQF QNLQAATKRL AIIVPLCFIL IAATLYMAIG
Czcaalceu AIQSQVKIPA GYWMTWGGTF EQLQSATTRL QVVVPVALLL VFVLLFAMFN
Consensus ......KLP. G....W.G.. .Q........ .......... VF..L.A...


951                                                       1000
Acrbescco SWSIPFSVML VVPLGVIGAL LAATFRGLTN DVYFQVGLLT TIGLSAKNAI
Acrfescco SWSIPVSVML VVPLGIVGVL LAATLFNQKN DVYFMVGLLT TIGLSAKNAI
Acrdescco SWSVPFSVML VVPLGVIGAL LATWMRGLEN DVYFQVGLLT VIGLSAKNAI
Cnraalceu SAALTATVLT ASPLALAGGV FALLLRGIPF SISAAVGFIA VSGVAVLNGL
Czcaalceu NIKDGLLVFT GIPFALTGGI LALWIRGIPM SITAAVGFIA LCGVAVLNGL
Consensus S......V.. ..PL...G.. LA...RG... .....VG.... .G....N..
```

```
           1001                                             1050
Acrbescco  LIVEFAKDLM DKEGKGLIEA TLDAVRMRLR PILMTSLAFI LGVMPLVIST
Acrfescco  LIVEFAKDLM EKEGKGVVEA TLMAVRMRLR PILMTSLAFI LGVLPLAISN
Acrdescco  LIVEFANEMN QK.GHDLFEA TLHACRQRLR PILMTSLAFI FGVLPMATST
Cnraalceu  VLISAIRKRL D.DGMAPDAA VIEGAMERVR PVLMTALVAS LGFVPMAIAT
Czcaalceu  VMLSFIRSLR E.EGHSLDSA VRVGALTRLR PVLMTALVAS LGFVPMAIAT
Consensus  ....F...... ...G.....A .......RLR P.LMT.L... LG..P.AI.T

           1051                                             1100
Acrbescco  GAGSGAQNAV GTGVMGGMVT ATVLAIFFVP VFFVVVRRRF SRKNEDIEHS
Acrfescco  GAGSGAQNAV GIGVMGGMVS ATLLAIFFVP VFFVVIRRCF KG........
Acrdescco  GAGSGGQHAV GTGVMGGMIS ATILAIYFVP LFFVLVRRRF PLKPRPE...
Cnraalceu  GTGAEVQKPL ATVVIGGLVT ATVLTLFVLP ALCGIVLKRR TAGRPEAQAA
Czcaalceu  GTGAEVQRPL ATVVIGGILS STALTLLVLP VLYRLAHRKD EDAEDTREPV
Consensus  G.G...Q... .T.V.GG... AT.L.....P ..................

           1101      1113
Acrbescco  HTVDHH.... ...
Acrfescco  .......... ...
Acrdescco  .......... ...
Cnraalceu  LEA....... ...
Czcaalceu  TQTHQPDQGR QPA
Consensus  .......... ...
```

Residues listed in the consensus sequence are present in at least 75% of the aligned transporter sequences.

Database accession numbers

	SWISSPROT	PIR	EMBL/GENBANK
Acrbescco	P31224	B36938	M94248; U00734
Acrdescco	P24177	C42959; S26997	U12598; X57403
Acrfescco	P24181	S18537	M96848; X57948
Cnraalceu	P37972		M91650
Czcaalceu	P13511	A33830	M26073

References
[1] Nikaido, H. (1996) J. Bacteriol. 178, 5853–5859.
[2] Lewis, K. (1994) Trends Biochem. Sci. 19, 119–123.
[3] Paulsen, I. et al. (1996) Microbiol. Rev. 60, 575–608.
[4] Saier, M.H. et al. (1994) Mol. Microbiol. 11, 841–847.

Summary

Transporters of the yeast multidrug resistance family, the example of which is the BMR benomyl-methotrexate resistance protein of *Candida albicans* (Bmrpcanal), mediate resistance to one or more structurally dissimilar antibiotics, including methotrexate and cyclohexamide[1]. They may contribute to both intrinsic insensitivity and the development of clinical resistance of some species to antifungal agents. Known members of the family occur only in yeasts.

Statistical analysis of multiple amino acid sequence comparisons suggests that the yeast multidrug resistance family may be distantly related to the uniporter-symporter-antiporter (USA) transporter superfamily, also known as the major facilitator superfamily (MFS)[2,3]. Members of the yeast multidrug resistance family are predicted to contain 12 membrane-spanning helices by the hydropathy of their amino acid sequences.

Several amino acid sequence motifs are highly conserved in the yeast multidrug resistance family, including motifs unique to the family, elements of the signature motifs of the USA/MFS superfamily, and motifs necessary for function by the criterion of site-directed mutagenesis[1].

Nomenclature, biological sources and substrates

CODE	DESCRIPTION [SYNONYMS]	ORGANISM [COMMON NAMES]	[RESISTANCE(S)][a]
Bmrpcanal	Benomyl-methotrexate resistance protein [BMRP, BMR, MDR1]	*Candida albicans* [yeast]	[Benomyl, methotrexate]
Carlschpo	Amiloride resistance protein [CAR1, SOD1]	*Schizosaccharomyces pombe* [yeast]	[Amiloride]
Cyhrcanma	Cyclohexamide resistance protein [CYHR]	*Candida maltosa* [yeast]	[Cyclohexamide]

[a] Presumed substrates; protein confers resistance to specified compounds.

Phylogenetic tree

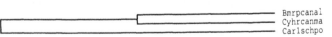

Bmrpcanal
Cyhrcanma
Carlschpo

Proposed orientation of BMR in the membrane

The model is based on predictions of membrane-spanning regions and α-helical content. The N-terminus of the protein is illustrated on the inside and is folded 12 times through the membrane. The predicted membrane-spanning helices are portrayed as rectangles. The numbers corresponding to the first and last residue of each membrane-spanning helix are boxed. Residues that are conserved in all three of the aligned transporters (see below) are shown.

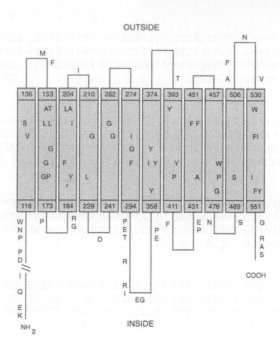

Physical and genetic characteristics

	AMINO ACIDS	MOL. WT
Bmrpcanal	564	62 930
Car1schpo	526	58 545
Cyhrcanma	552	61 366

Multiple amino acid sequence alignments

```
          1                                                      50
Bmrpcanal MHYRFLRDSF VGRVTYHLSK HKYFAHPEEA KNYIIPEKYL ADYKPTLADD
Cyhrcanma .MAAFIKDSF WGQIIYRLSG RKLFRHNDEL PDYVVPEKYL LD........
Car1schpo .......... .......... .......... ....MASKIA SLFSPSETAS
Consensus .......... .......... .......... ......K... ..........


          51                                                     100
Bmrpcanal TSINFEKEEI DNQGEPNSSQ SSSSNNTIVD NNNNNNNDVD GDKIVVTWDG
Cyhrcanma .....PKEEV LNSSD..KSQ SSENKEQTEG DQATIQNEPA SEHIIVTWDG
Car1schpo KDQHENVAED LELGTAPLNQ IGIHETNSEY DEKKREESPE VIDI.SNLIS
Consensus ........E. .........Q .......... .......... ...I......


          101                                                    150
Bmrpcanal DDDPENPQNW PTLQKAFFIF QISFLTTSVY MGSAVYTPGI EELMHDFGIG
Cyhrcanma DDDPENPYNW PFAWKAIAAM QIGFLTVSVY MASAIYTPGV EEIMNQFNIN
Car1schpo SDHPAHPQNW HWAKRWSIVF MFCLMQIYVI WTSNGFGSIE YSVMAQFNVS
Consensus .D.P..P.NW .......... .V. ..S..... ...M..F...
```

```
          151                                                      200
Bmrpcanal RVVATLPLTL FVIGYGVGPL VFSPMSENAI FGRTSIYIIT LFLFVILQIP
Cyhrcanma STLATLPLTM FVIGYGIGPL FWSPLSENSR IGRTPLYIIT LFIFFILQIP
Car1schpo AQVATLCLSM NILGSGLGPM FLGPLSDIG. .GRKPVYFCS IFVYTVFNIS
Consensus ...ATL.L.. ...G.G.GP. ...P...... .GR...Y... .F......I.

          201                                                      250
Bmrpcanal TALVNNIAGL CILRFLGGFF ASPCLATGGA SVADVVKFWN LPVGLAAWSL
Cyhrcanma TALSNHIAGL SVLRVIAGFF AAPALSTGGA SYGDFIAMHY YSIALGVWSI
Car1schpo CALPRNIVQM IISHFIIGVA GSTALTNVAG GIPDLFPEDT AGVPMSLF.V
Consensus .AL...I... ........G.. .....L.... ...D...... ..........

          251                                                      300
Bmrpcanal GAVCGPSFGP FFGSILTVKA ....SWRWTF WFMCIISGFS FVMLCFTLPE
Cyhrcanma FAVAGPSIGP LIGAAVINRS HDADGWRWSF WFMAILSGVC FIVLSFSLPE
Car1schpo WACAGGAIGA PMATGVDINA KYG...WRWL YYINIIVGGF FLIVILIIPE
Consensus ....G...G. .......... .......... ....I..G.. F.......PE

          301                                                      350
Bmrpcanal TFGKTLLYRK AKRLRAITGN DRITSEGEIE NSKMTSHELI IDTLWRPLEI
Cyhrcanma TYGKTLLRRK AERLRKLTGN NRIISEGELE DGHKTTSQVV SSLLWRPLEI
Car1schpo TLPIKVITRY EN......AK GRIV.EGIPK NNLKEVLKKC KFVTTMGFRM
Consensus T.......R. ...... .RI..EG... .......... ..........

          351                                                      400
Bmrpcanal TVMEPVVLLI NIYIAMVYSI LYLFFEVFPI YFVGVKHFTL VELGTTYMSI
Cyhrcanma TMLEPVVFLI DIYIALVYSI MYLIFESVPI VYAGIHHFTL VEMGATYVST
Car1schpo MLTEPIILSM GLYNFYAYGI SYFFLTAIWP VFYDTYKMSE MGASCTYLSG
Consensus ...EP..... ...Y....Y.I .Y........ .......... .....TY...

          401                                                      450
Bmrpcanal VIGIVIAAFI YIPVIRQKFT KPILRQEQVF .PEVFIPIAI VGGILLTSGL
Cyhrcanma IIGIIIGGAI YLPTVYYKFT KKLLAGQNVT .PEVFLPPAI FGAICMPIGV
Car1schpo FVASTL.LFL YQPIQDWIFR RDKAKNNGVA RPEARFTSAL FITLLFPAGM
Consensus .......... Y.P.....F. .......... .PE.....A. ..........

          451                                                      500
Bmrpcanal FIFGWSANRT THWVGPLFGA ATTASGAFLI FQTLFNFMGA SFKPHYIASV
Cyhrcanma FIFGWTSSPD INWFVPLIGM ALFAVGAFII FQTLFNYMAV SFKVEYLASV
Car1schpo FLFAFTCHPP FPWMSPIVGN SMVTVANGHN WMCILNYLTD SY.PLLSGSA
Consensus F.F....... ..W..P..G. .......... .....N.... S......S.

          501                                                      550
Bmrpcanal FASNDLFRSV IASVFPLFGA PLFDNLATPE YPVAWGSSVL GFITLVMIAI
Cyhrcanma FSSNAFFRSV SAGAFPLFGR ALYNNLSIDK FPVGWGSSIL GFISLGMIAI
Car1schpo VAAFTLPSFI GATVFAHVSQ IMFNNMS... ..VKWAVATM AFISISIPFI
Consensus .......... .A..F..... ....N..... ..V.W..... .FI......I

          551                        581
Bmrpcanal PVLFYLNGPK LRARSKYAN. ..........
Cyhrcanma PVFFYLNGPK LRARSKYAY. ..........
Car1schpo IYTFYFFGQR IRALSSLTGN KALKYLPLEN N
Consensus ...FY..G.. .RA.S..... ..........
```

Residues listed in the consensus sequence are present in all transporter sequences. Residues indicated by boldface type are also conserved in at least one other family of the USA/MFS superfamily.

Database accession numbers

	SWISSPROT	PIR	EMBL/GENBANK
Bmrpcanal	P28873	S16304	X53823
Car1schpo	P33532	S39919	Z14035
Cyhrcanma	P32071	JC1173	M64932

References

[1] Paulsen, I. et al. (1996) Microbiol. Rev. 60, 575–608.
[2] Marger, M.D. and Saier, M. (1993) Trends Biochem. Sci. 18, 13–20.
[3] Griffith, J. et al. (1994) Mol. Membr. Biol. 11, 271–277.

Part 10

Na⁺-Dependent Symporters

Summary

Transporters of the Na$^+$/Ca^{2+} exchanger family, the example of which is the human cardiac sodium/calcium exchanger precursor 1[1] (Nac1homsa), mediate export of calcium from cardiac sarcolemma cells. The mechanism of action is symport (i.e. Na$^+$-coupled substrate efflux), with the stoichiometry of exchange three Na$^+$ ions to one Ca^{2+} ion[2]. In humans, inhibition of sodium/calcium exchanger activity has been implicated in cardiac arrhythmia induced by digitalis toxicity[3]. Members of this family have only been found in mammals, including cattle[4], dogs[5] and rodents.

Statistical analysis of multiple amino acid sequence comparisons reveals no apparent relationship between these transporters and any other family of transporters. Members of the Na$^+$/Ca^{2+} exchanger family are predicted to form 11 transmembrane helices by the hydropathy of their amino acid sequences[1], with the N-terminus predicted to lie outside the cell. This topology is not commonly found in transporter proteins. There is a very long cytoplasmic loop between the fifth and sixth helices, with a calmodulin-binding domain at its N-terminus. An N-terminal signal sequence, which contains a highly hydrophobic segment, is cleaved to form the mature protein. The proteins are glycosylated.

The sequences of all known members of the Na$^+$/Ca^{2+} exchanger family are very similar, with the N-terminal signal sequence the most variable.

Nomenclature, biological sources and substrates

CODE	DESCRIPTION [SYNONYMS]	ORGANISM [COMMON NAMES]	SUBSTRATE(S)
Nac1bosta	Sodium/calcium exchanger precursor 1 [Na$^+$/Ca^{2+}-exchange protein 1, SLC8A1]	*Bos taurus* [cow]	Na$^+$/Ca^{2+}
Nac1canfa	Sodium/calcium exchanger precursor 1 [Na$^+$/Ca^{2+}-exchange protein 1, SLC8A1]	*Canis familiaris* [dog]	Na$^+$/Ca^{2+}
Nac1cavpo	Sodium/calcium exchanger precursor 1 [Na$^+$/Ca^{2+}-exchange protein 1, SLC8A1]	*Cavia porcellus* [guinea pig]	Na$^+$/Ca^{2+}
Nac1felca	Sodium/calcium exchanger precursor 1 [Na$^+$/Ca^{2+}-exchange protein 1, SLC8A1]	*Felix catus* [cat]	Na$^+$/Ca^{2+}
Nac1homsa	Sodium/calcium exchanger precursor 1 [Na$^+$/Ca^{2+}-exchange protein 1, SLC8A1, NCX1, CNC]	*Homo sapiens* [human]	Na$^+$/Ca^{2+}
Nac1ratno	Sodium/calcium exchanger [Na$^+$/Ca^{2+}-exchange protein 1, SLC8A1, NCX1]	*Rattus norvegicus* [rat]	Na$^+$/Ca^{2+}

Cotransported ions are listed.

Proposed orientation of NAC1 [1] in the membrane

The model is based on predictions of membrane-spanning regions and α-helical content. The N-terminus of the protein is illustrated on the outside and is folded 11 times through the membrane. This is an unusual topology for a transporter protein. The predicted membrane-spanning helices are portrayed as rectangles. The numbers corresponding to the first and last residue of each membrane-spanning helix are boxed.

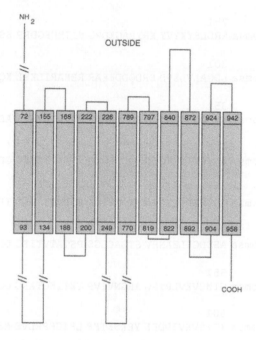

Physical and genetic characteristics

	AMINO ACIDS	MOL. WT	EXPRESSION SITES	CHROMOSOMAL LOCUS
Nac1bosta	970	108 027	heart	
Nac1canfa	970	108 004	heart	
Nac1cavpo	970	108 071	heart	
Nac1felca	970	108 004	heart	
Nac1homsa	970	108 138	heart	2p23–p22
Nac1ratno	971	108 184	heart	

Multiple amino acid sequence alignments

```
          1                                                50
Nac1homsa MRRLSLSPTF SMGFHLLVTV SLLFSHVDHV IAETEMEGEG NETGECTGSY

          51                                               100
Nac1homsa YCKKGVILPI WEPQDPSFGD KIARATVYFV AMVYMFLGVS IIADRFMSSI
```

```
                101                                                  150
    Nac1homsa EVITSQEKEI TIKKPNGETT KTTVRIWNET VSNLTLMALG SSAPEILLSV

                151                                                  200
    Nac1homsa IEVCGHNFTA GDLGPSTIVG SAAFNMFIII ALCVYVVPDG ETRKIKHLRV

                201                                                  250
    Nac1homsa FFVTAAWSIF AYTWLYIILS VISPGVVEVW EGLLTFFFFP ICVVFAWVAD

                251                                                  300
    Nac1homsa RRLLFYKYVY KRYRAGKQRG MIIEHEGDRP SSKTEIEMDG KVVNSHVENF

                301                                                  350
    Nac1homsa LDGALVLEVD ERDQDDEEAR REMARILKEL KQKHPDKEIE QLIELANYQV

                351                                                  400
    Nac1homsa LSQQQKSRAF YRIQATRLMT GAGNILKRHA ADQARKAVSM HEVNTEVTEN

                401                                                  450
    Nac1homsa DPVSKIFFEQ GTYQCLENCG TVALTIIRRG GDLTNTVFVD FRTEDGTANA

                451                                                  500
    Nac1homsa GSDYEFTEGTV VFKPGDTQK EIRVGIIDDD IFEEDENFLV HLSNVKVSSE

                501                                                  550
    Nac1homsa ASEDGILEANHV STLACLGS PSTATVTIFD DDHAGIFTFE EPVTHVSESI

                551                                                  600
    Nac1homsa GIMEVKVLRTSG ARGNVIVP YKTIEGTARG GGEDFEDTCG ELEFQNDEIV

                601                                                  650
    Nac1homsa KTISVKVIDDEE YEKNKTFF LEIGEPRLVE MSEKKALLLN ELGGFTITGK

                651                                                  700
    Nac1homsa YLFGQPVFRKVH AREHPILS TVITIADEYD DKQPLTSKEE EERRIAEMGR

                701                                                  750
    Nac1homsa PILGEHTKLEVI IEESYEFK STVDKLIKKT NLALVVGTNS WREQFIEAIT

                751                                                  800
    Nac1homsa VSAGEDDDDDEC GEEKLPSC FDYVMHFLTV FWKVLFAFVP PTEYWNGWAC

                801                                                  850
    Nac1homsa FIVSILMIGLLT AFIGDLAS HFGCTIGLKD SVTAVVFVAL GTSVPDTFAS

                851                                                  900
    Nac1homsa KVAATQDQYADA SIGNVTGS NAVNVFLGIG VAWSIAAIYH AANGEQFKVS

                901                                                  950
    Nac1homsa PGTLAFSVTLFT IFAFINVG VLLYRRRPEI GGELGGPRTA KLLTSCLFVL

                951              969
    Nac1homsa LWLLYIFFSSLE AYCHIKGF
```

Proteins listed subsequently in italics are at least 90% identical to the paired transporters listed in parenthesis and therefore are not included in the alignment: *Nac1bosta, Nac1canfa, Nac1cavpo, Nac1felca, Nac1ratno* (Nac1homsa).

Database accession numbers

	SWISSPROT	PIR	EMBL/GENBANK
Nac1bosta	P48765		L06438; G163034
Nac1canfa	P23685	A36417	M57523; G164073
Nac1cavpo	P48766		U04955; G507350
Nac1felca	P48767		L35846; G604519
Nac1homsa	P32418	S32815	M91368; G180673
Nac1ratno	Q01728	S25552; S28833	X68191; G57209

References

1 Komuro, I. et al. (1992) Proc. Natl Acad. Sci. USA 89, 4769–4773.
2 Reeves, J.P. and Hale, C.C. (1984) J. Biol. Chem. 259, 7733–7739.
3 **Smith, T.W. et al. (1988) In Heart Disease: A Textbook of Cardiovascular Medicine, Ed. E. Braunwald. Saunders, Philadelphia, pp. 489–507.**
4 Aceto, J.F. et al. (1992) Arch. Biochem. Biophys. 298, 553–560.
5 Nicoll, D.A. et al. (1990) Science 250, 562–565.

Summary

Transporters of the Na⁺/proline symporter family, the example of which is the PUTP proline-Na⁺ symporter of *Escherichia coli* (Putpescco), mediate symport (Na⁺-coupled substrate uptake) of carboxylated compounds, including proline and pantothenate [1-3]. Members of the family are found in both gram-negative and gram-positive bacteria.

Statistical analysis of multiple amino acid sequence alignments indicates that the Na⁺/proline symporter family is closely related to the Na⁺/glucose symporter family of mammals [2]. The similarity between the kinetics of several families of Na⁺-driven and H⁺-driven transporters further suggests a common mechanism of action, despite the lack of amino acid sequence homology [3]. Members of the Na⁺/proline symporter family are predicted to contain 12 membrane-spanning helices by the hydropathy of their amino acid sequences [3].

Several amino acid sequence motifs are very highly conserved in the Na⁺/proline symporter family, including motifs unique to the family, signature motifs common with the Na⁺/glucose symporter family, and motifs necessary for function by the criterion of site-directed mutagenesis [2].

Nomenclature, biological sources and substrates

CODE	DESCRIPTION [SYNONYMS]	ORGANISM [COMMON NAMES]	SUBSTRATE(S)
Panfescco	Pantothenate-Na⁺ symporter [Pantothenate permease, PANF]	*Escherichia coli* [gram-negative bacterium]	Na⁺/pantothenate
Panfhaein	Pantothenate-Na⁺ symporter [Pantothenate permease, HI0975]	*Haemophilus influenzae* [gram-negative bacterium]	Na⁺/pantothenate
Proppsefl	Proline-Na⁺ symporter [Proline permease, PROP]	*Pseudomonas fluorescens* [gram-negative bacterium]	Na⁺/proline
Putpescco	Proline-Na⁺ symporter [Proline permease, PUTP]	*Escherichia coli* [gram-negative bacterium]	Na⁺, Li⁺/proline
Putpstaau	Proline-Na⁺ symporter [Proline permease, PUTP]	*Staphylococcus aureus* [gram-positive bacterium]	Na⁺/proline
Putpsalty	Proline-Na⁺ symporter [Proline permease, PUTP]	*Salmonella typhimurium* [gram-negative bacterium]	Na⁺/proline
Putphaein	Proline-Na⁺ symporter [Proline permease, PUTP, HI1352]	*Haemophilus influenzae* [gram-negative bacterium]	Na⁺/proline

Cotransported ions are listed.

Phylogenetic tree

Proteins listed subsequently in italics are at least 90% identical to the paired transporters listed in parenthesis and therefore are not included in the phylogenetic tree: *Putpsalty* (Putpescco).

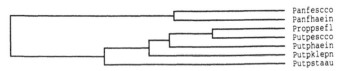

```
Panfescco
Panfhaein
Proppsefl
Putpescco
Putphaein
Putpklepn
Putpstaau
```

Proposed orientation of PUTP in the membrane

The model is based on predictions of membrane-spanning regions and α-helical content. The N-terminus of the protein is illustrated on the inside and is folded 12 times through the membrane. The predicted membrane-spanning helices are portrayed as rectangles. The numbers corresponding to the first and last residue of each membrane-spanning helix are boxed. Residues that are conserved in more than 75% of the aligned transporters (see below) are shown. The predicted locations of residues that are conserved in more than 75% of the subsequently aligned transporters (see below) are shown. Consensus residues indicated by an asterisk are not conserved in PUTP.

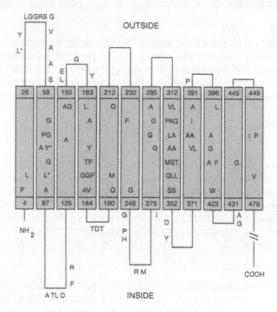

Physical and genetic characteristics

	AMINO ACIDS	MOL. WT	CHROMOSOMAL LOCUS
Panfescco	483	51 717	73.35 minutes
Panfhaein	484	52 500	
Proppsefl	494	53 112	
Putpescco	502	54 344	23.25 minutes
Putpstaau	497	54 314	
Putpsalty	502	54 296	
Putphaein	504	54 898	

Multiple amino acid sequence alignments

```
           1                                                    50
Panfescco .........M QLEVILP.LV AYLVVVFGIS VYAMRKRSTG TFLNEYFLGS
Panfhaein .........M NLGIILP.LI IYLTFVFGAA IFAYVKRTKG DFLTEYYVGN
Proppsefl .......... .MSVSNPTLI TFVIYIAAMV LIGLMAYRST NNLSDYILGG
Putpescco .......... .MAISTPMLV TFCVYIFGMI LIGFIAWRST KNFDDYILGG
```

```
Putphaein .......... .MFGFDPSLI TFTIYIFGML LIGVLAYYYT NNLSDYILGG
Putpstaau MLTMGTALSQ QVDANWQTYI MIAVYFLILM LLAFTYKQAT GNLSEYMLGG
Consensus .......... ......P.L. .......... .......... ..L..Y.LGG


             51                                              100
Panfescco RSMGGIVLAM TLTATYISAS SFIGGPGAAY KYGLGWVLLA MIQLPAVWLS
Panfhaein RSMTGFVLAM TTASTYASAS SFVGGPGAAY KYGLGWVLLA MIQVPVVWLA
Proppsefl RSLGSVVTAL SAGASDMSGW LLMGLPGAIY MSGLSESWIA IGLIVGAYLN
Putpescco RSLGPFVTAL SAGASDMSGW LLMGLPGAVF LSGISESWIA IGLTLGAWIN
Putphaein RRLGSFVTAM SAGASDMSGW LLMGLPGAVY LSGLVEGWIA IGLTIGAYFN
Putpstaau RSIGPYITAL SAGASDMSGW MIMGLPGSVY STGLSAMWIT IGLTLGAYIN
Consensus RS.G..V.A. ...A...S.. ...G.PGA.Y ..GL.....A ..........


             101                                             150
Panfescco LGILGKKFAI LAR.RYNAVT LNDMLFARYQ SRL....LVW LASLSLLVAF
Panfhaein LGALGKKFAL LSR.ETNALT INDLFFYRYK NKY....LVW LSSLALLLAF
Proppsefl WLFVAGRLRV QTEHNGDALT LPDYFSSRFE DKSGLLRIIS AVVILVFFTI
Putpescco WKLVAGRLRV HTEYNNNALT LPDYFTGRFE DKSRILRIIS ALVILLFFTI
Putphaein WLLVAGRLRV YTELNNNALT LPEYFHNRFG SSHKLLKLVS ATIILVFLTI
Putpstaau YFVVAPRLRV YTELAGDAIT LPDFFKNRLN DKNNVLKIIS GLIIVVFFTL
Consensus .......... .......A.T L.D.F..R.. .......... ..........


             151                                             200
Panfescco VGAMTVQFIG GARLLETAAG IPYETGLLIF GISIALYTAF GGFRASVLND
Panfhaein FAAMTVQFIG GARLLETTIG ISYTQALLLF ALTVGIYTFI GGFRAVVLTD
Proppsefl YCASGIVA.. GARLFESTFG MSYETALWAG AAATIAYTFI GGFLAVSWTD
Putpescco YCASGIVA.. GARLFESTFG MSYETALWAG AAATILYTFI GGFLAVSWTD
Putphaein YCASGVVA.. GAKLFQNIFS VEYSTALWYG AAATIAYTFI GGFLAVSWTD
Putpstaau YTHSGFVS.. GGKLFESAFG LDYHFGLILV AFIVIFYTFF GGYLAVSITD
Consensus ..A....... GA.L.E...G ..Y...L... A.....YTF. GGF.AV..TD


             201                                             250
Panfescco TMQGLVMLIG TVVLLIGVVH AAGGLSNAVQ TLQTIDPQLV TPQGADDILS
Panfhaein TIQGTVMIFG TIILLIGTIY ALGGVESAVN KLTEIDPDLV TPYGPNGMLD
Proppsefl TVQATLMIFA LILTPIIVLL ATGGVDTTFL AIEAKDP... ..TSFDMLKN
Putpescco TVQASLMIFA LILTPVIVII SVGGFGDSLE VIKQKSI... ..ENVDMLKG
Putphaein TIQATLMIFA LILTPVFVLL SFADTAQFSA VLEQAEAA.V NKDFTDLFTS
Putpstaau FFQGVIMLIA MVMVPIVAMM NLNGWGTFHD V.....AA.M KPTNLNLFKG
Consensus T.Q...M... .......... ...G..... .......... ..........


             251                                             300
Panfescco PAFMTSFWVL V.CFGVIGLP HTAVRCISYK DSKAVHRGII IGTIVVAILM
Panfhaein FQFMASFWIL V.CFGVVGLP HTAVRCMAFK DSKALHRGML IGTIVLSIIM
Proppsefl TTFIGIISLM GWGLGYFGQP HILARFMAAD SVKSIAKARR ISMTWMILCL
Putpescco LNFVAIISLM GWGLGYFGQP HILARFMAAD SHHSIVHARR ISMTWMILCL
Putphaein TTPLGLLSLA AWGLGYFGQP HILARFMAAD SVKSLIKARR ISMGWMVLCL
Putpstaau LSFIGIISLF SWGLGYFGQP HIIVRFMSIK SHKMLPKARR LGISWMAVGL
Consensus ..F...... ...G..G.P H...R.M... .......... I.........


             301                                             350
Panfescco FGMHLAGALG RAVIPDLTV. .......PDL VIPTLMVKVL PPFAAGIFLA
Panfhaein LGMHLAGALG RAVIPNLTV. .......SDQ VIPTLMIKVL PPIVAGIFLA
Proppsefl GGTVAVGFFG IAYFSAHPEV AGPVTENPER VFIELAKILF NPWVAGVLLS
Putpescco AGAVAVGFFG IAYFNDHPAL AGAVNQNAER VFIELAQILF NPWIAGILLS
```

```
Putphaein AGAIGIGLFA IPYFFANPAI AGTVNREPEQ VFIELAKLLF NPWIAGILLS
Putpstaau LGAVAVGLTG IAFVPAYHIK L....EDPET LFIVMSQVLF HPLVGGFLLA
Consensus .G....G..G .A................ ....... V...L..... .P..AG..L.
```

```
          351                                                 400
Panfescco APMAAIMSTI NAQLLQSSAT IIKDLYLNIR PDQMQN..ET RLKRMSAVIT
Panfhaein APMSAIMSTI DAQLIQSSSI FVKDLYLSAK PEAAKN..EK KVSYFSSIIT
Proppsefl AILAAVMSTL SCQLLVCSSA LTEDFYKAF. ..LRKGASQR ELVWVGRLMV
Putpescco AILAAVMSTL SCQLLVCSSA ITEDLYKAF. ..LRKHASQK ELVWVGRVMV
Putphaein AILAAVMSTL SAQLLISSSS ITEDFYKGF. ..IRPNASEK ELVWLGRIMV
Putpstaau AILAAIMSTI SSQLLVTSSS LTEDFYKLIR GEEKAKTDQK EFVMIGRLSV
Consensus A..AA.MST. ..QLL..SS. ...D.Y.... ....... .........
```

```
          401                                                 450
Panfescco LVLGALLLLA AWKPPEMIIW LNLLAFGGLE AVFLWPLVLG LYWERANAKG
Panfhaein LILTALLIFA ALNPPDMIIW LNLFAFGGLE AAFLWVIVLG IYWDKANAYG
Proppsefl LVVALIAIAM AANPENRVLG LVSYAWAGFG AAFGPVVLIS VIWKHMTRNG
Putpescco LVVALVAIAL AANPENRVLG LVSYAWAGFG AAFGPVVLFS VMWSRMTRNG
Putphaein LVIAALAIWI AQDENSKVLK LVEFAWAGFG SAFGPVVLFS LFWKRMTSSG
Putpstaau LVVAIVAIAI AWNPNDTILN LVGNAWAGFG ASFSPLVLFA LYWKGLTRAG
Consensus LV.....I.. A..P...... L...A..G.. A.F........ ..W......G
```

```
          451                                                 500
Panfescco ALSAMIVGGV LYAVLATL.. ....NIQYLG FHPIVPSLLL SLLAFLVGNR
Panfhaein ALSSMIIGLG SYILLTQL.. ....GIKLFN FHQIVPSLVF GLIAFLVGNK
Proppsefl ALAGILVGAI TVIVWKHF.. ......ELLG LYEIIPGFIF ASLAIYFVSK
Putpescco ALAGMIIGAL TVIVWKQF.. ......GWLG LYEIIPGFIF GSIGIVVFSL
Putphaein AMAGMLVGAV TVFAWKEVVP A...DTDWFK VYEMIPGFAF ASLAIIVISL
Putpstaau AVSGMVSGAL VVIVWIAWIK PLAHINEIFG LYEIIPGFIV SVIVTYVVSK
Consensus A......G.. .......... .......... ...I.P.... ......V...
```

```
          501                            531
Panfescco FGTSVPQATV LTTDK............ .
Panfhaein LGERRIEKTQ LKVTAL.... ........... .
Proppsefl MG.APTLGMV ERFDAAEKDY NLNK....... .
Putpescco LGKAPSAAMQ KRFAEADAHY HSAPPSRLQE S
Putphaein LSNKPEQDIL NTFDKAEKAY KEAK...... .
Putpstaau LTKKPWCIC. .......... .
Consensus .......... .......... .
```

Proteins listed subsequently in italics are at least 90% identical to the paired transporters listed in parenthesis and therefore are not included in the alignments: *Putpsalty* (Putpescco). Residues listed in the consensus sequence are present in at least 75% of the aligned transporter sequences. Residues indicated by boldface type are also conserved in the Na⁺/glucose symporter family.

Database accession numbers

	SWISSPROT	PIR	EMBL/GENBANK
Panfescco	P16256	JU0296	M30953; U18997
Panfhaein	P44963		U32778
Proppsefl		JC2382	
Putpstaau			U06451
Putpsalty	P10502	S03816; S10220	X52573; X12569
Putphaein	P45174		U32814
Putpescco	P07117	A30258	X05653

References
1. Hediger, M. et al. (1995) J. Physiol. 482, S7–S17.
2. Reizer, J. et al. (1994) Biochim. Biophys. Acta 1197, 133–166.
3. Wright, E.M. et al. (1996) Curr. Opin. Cell Biol. 8, 468–473.

Summary

Transporters of the Na$^+$/glucose symporter family, the example of which is the SGLT1 glucose-Na$^+$ symporter of humans (Nagchomsa), mediate symport (Na$^+$-coupled uptake) of glucose, myo-inositol, neutral amino acids and uridine[1-3]. Members of the Na$^+$/glucose symporter family also serve as uniporters, ion and water channels and water transporters[4]. In the absence of substrates, transport of ions occurs in an uncoupled "leak" mode that can be 5–10% of substrate-linked ion transport. In the small intestine, secondary active water transport by SGLT1 in the presence of glucose has been estimated to account for up to 5 liters per day[4]. Known members of the family are found only in mammals.

Statistical analysis of multiple amino acid sequence alignments indicates that the Na$^+$/glucose symporter family is closely related to the Na$^+$/proline symporter family of prokaryotes[3]. The similarity between the kinetics of several families of Na$^+$-driven and H$^+$-driven transporters further suggests a common mechanism of action, despite the lack of amino acid sequence homology[4]. Members of the Na$^+$/glucose symporter family are predicted to contain 14 membrane-spanning helices by the hydropathy of their amino acid sequences and glycosylation scanning mutagenesis[5]. However, unlike other symporters predicted to contain either 14 or 12 membrane-spanning elements, the N- and C-termini of at least some members of the family are predicted to be on the extracellular side of the membrane[5]. Members of the Na$^+$/glucose symporter family are glycosylated.

Several amino acid sequence motifs are very highly conserved in the Na$^+$/glucose symporter family, including motifs unique to the family, signature motifs common with the family, and motifs necessary for function by the criterion of site-directed mutagenesis[3]. In humans, mutations in SGTL1 are the cause of glucose-galactose malabsorption syndrome[6].

Nomenclature, biological sources and substrates

CODE	DESCRIPTION [SYNONYMS]	ORGANISM [COMMON NAMES]	SUBSTRATE(S)
Naaasussc	Neutral amino acid-Na$^+$ cotransporter [NAAA, SAAT1]	Sus scrofa [pig]	Na$^+$/neutral amino acids
Nagchomsa	Glucose-Na$^+$ cotransporter 1 [NAGC, SLC5a1, SGLT1]	Homo sapiens [human]	Na$^+$/glucose
Nagcorycu	Glucose-Na$^+$ cotransporter 1 [NAGC, SGLT1]	Oryctolagus cuniculus [rabbit]	Na$^+$/glucose
Nagcsussc	Glucose-Na$^+$ cotransporter 1 [NAGC, SGLT1]	Sus scrofa [pig]	Na$^+$/glucose
Nagoviar	Glucose-Na$^+$ cotransporter [NAG]	Ovis aries [sheep]	Na$^+$/glucose
Naglhomsa	Low-affinity glucose-Na$^+$ cotransporter [NA1, SGLT2]	Homo sapiens [human]	Na$^+$/glucose
Namibosta	Myo-inositol-Na$^+$ cotransporter 1 [NAMI, SLC5a3]	Bos taurus [cow]	Na$^+$/myo-inositol
Namicanfa	Myo-inositol-Na$^+$ cotransporter 1 [NAMI, SMIT]	Canis familiaris [dog]	Na$^+$/myo-inositol

CODE	DESCRIPTION [SYNONYMS]	ORGANISM [COMMON NAMES]	SUBSTRATE(S)
Namihomsa	Myo-inositol-Na⁺ cotransporter 1 [NAMI, SLC5a3]	*Homo sapiens* [human]	Na⁺/myo-inositol
Nanuorycu	Nucleoside-Na⁺ cotransporter 1 [NANU, SNST1]	*Oryctolagus cuniculus* [rabbit]	Na⁺/uridine
Sgltratno	Low-affinity glucose-Na⁺ cotransporter [SGL, SGLT2]	*Rattus norvegicus* [rat]	Na⁺/glucose

Cotransported ions are listed.

Phylogenetic tree

Proteins listed subsequently in italics are at least 90% identical to the paired transporters listed in parenthesis and therefore are not included in the phylogenetic tree: *Namibosta, Namicanfa* (Namihomsa).

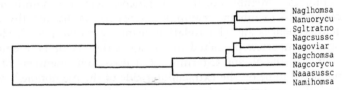

```
Naglhomsa
Nanuorycu
Sgltratno
Nagcsussc
Nagoviar
Nagchomsa
Nagcorycu
Naaasussc
Namihomsa
```

Proposed orientation of SGLT1 in the membrane

The model is based on predictions of membrane-spanning regions and α-helical content. The N- and c-termini of the protein are illustrated on the outside and are folded 14 times through the membrane[6]. The predicted membrane-spanning helices are portrayed as rectangles. The numbers corresponding to the first and last residue of each membrane-spanning helix are boxed. More than half of the residues are conserved in at least 75% of the members of the Na⁺/glucose symporter family and, therefore, are not mapped onto the model.

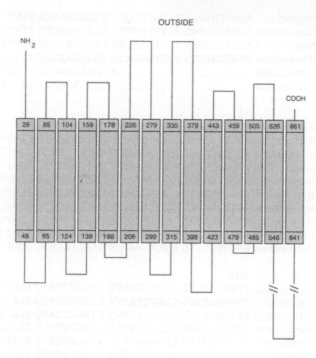

Physical and genetic characteristics

	AMINO ACIDS	MOL. WT	EXPRESSION SITES	K_m	CHROMOSOMAL LOCUS
Naaasussc	660	72 745	kidney, intestine, muscle, liver, spleen		
Nagchomsa	664	73 497	intestine, kidney	Glucose: 350 μM²	22q13.1
Nagcorycu	662	73 079	intestine, kidney		
Nagcsussc	605	66 917	intestine, kidney		
Nagoviar	664	73 215			
Naglhomsa	672	72 896	kidney	Glucose: 1.6 μM²	16p11.2
Namibosta	718	79 763	kidney		
Namicanfa	718	79 545	brain, kidney		
Namihomsa	718	79 709	brain, kidney		21q22
Nanuorycu	672	73 161	heart, kidney		
Sgltratno	670	72 961			

Multiple amino acid sequence alignments

```
           1                                                50
Naglhomsa ...MEEHTEA GSAPEMGAQK ALIDNPADIL VIAAYFLLVI GVGLWSMCRT
Nanuorycu ...MEEHMEA GSRLGLGDHG ALIDNPADIA VIAAYFLLVI GVGLWSMCRT
Sgltratno ...MEGHVEE GS..ELGEQK VLIDNPADIL VIAAYFLLVI GVGLWSMFRT
Nagcsussc ...............................................
```

```
Nagoviar  MDSSTWSPPA TATAEPLQAY ERIRNAADIS VIVIYFVVVM AVGLWAMFST
Nagchomsa MDSSTWSPKT TAVTRPVETH ELIRNAADIS IIVIYFVVVM AVGLWAMFST
Nagcorycu MDSSTLSPLT TSTAAPLESY ERIRNAADIS VIVIYFLVVM AVGLWAMFST
Naaasussc MASTLSPSTV TKTPGPPEIS ERIQNAADIS VIVIYFVVVM AVGLWAMLRT
Namihomsa ..................M RAVLDTADIA IVALYFILVM CIGFFAMWKC
Consensus ................... I.N.ADI. .I..YF..V. .VGLW.M..T
```

```
              51                                            100
Naglhomsa NRGTVGGYFL AGRSMVWWPV GASLFASNIG SGHFVGLAGT GAASGLAVAG
Nanuorycu NRGTVGGYFL AGRSMVWWPV GASLFASNIG SGHFVGLAGT GAANGLAVAG
Sgltratno NRGTVGGYFL AGRSMVWWPV GASLFASNIG SGHFVGLAGT GAASGLAVAG
Nagcsussc .......FFL AGRSMVWWPV GASLFASYIG SGHFVGLAGT GAAAGIATGG
Nagoviar  NRGTVGGFFL AGRSMVWWPI GASLFASNIG SGHFVGLAGT GAAAGIATGG
Nagchomsa NRGTVGGFFL AGRSMVWWPI GASLFASNIG SGHFVGLAGT GAASGIAIGG
Naaasussc NRGTVGGFFL AGRDVTWWPM GASLFASNIG SGHFVGLAGT GAASGIAIAA
Nagcorycu NRGTVGGFFL AGRSMVWWPI GASLFASNIG SGHFVGLAGT GAASGIATGG
Namihomsa NRSTVSGYFL AGRSMTWVTI GASLFVSNIG SEHFIGLAGS GAASGFAVGA
Consensus NRGTVGG.FL AGRSMVWWPM GASLFASNIG SGHFVGLAGT GAA.G.A..G
```

```
              101                                           150
Naglhomsa FEWNALFVVL LLGWLFAPVY LTAGVITMPQ YLRKRFGGRR IRLYLSVLSL
Nanuorycu FEWNALFVVL LLGWLFAPVY LTAGVITMPQ YLRKRFGGHR IRLYLSVLSL
Sgltratno FEWNALFVVL LLGWLFVPVY LTAGVITMPQ YLRKRFGGRR IRLYLSVLSL
Nagcsussc FEWNALIWVV VLGWLFVPIY IKAGVVTMPE YLRKRFGGQR IQVYLSILSL
Nagoviar  FEWNALILVV LLGWVFVPIY IKAGVVTMPE YLRKRFGGQR IQVYLSLLSL
Nagchomsa FEWNALVLVV VLGWLFVPIY IKAGVVTMPE YLRKRFGGQR IQVYLSLLSL
Nagcorycu FEWNALIMVV VLGWVFVPIY IRAGVVTMPE YLQKRFGGKR IQIYLSILSL
Naaasussc FEWNALLLLL VLGWFFVPIY IKAGVMTMPE YLRKRFGGKR LQIYLSILSL
Namihomsa WEFNALLLLQ LLGWVFIPIY IRSGVYTMPE YLSKRFGGHR IQVYFAALSL
Consensus FEWNAL..V. .LGW.F.P.Y I.AGV.TMP. YLRKRFGG.R I..YLS.LSL
```

```
              151                                           200
Naglhomsa FLYIFTKISV DMFSGAVFIQ QALGWNIYAS VIALLGITMI YTVTGGLAAL
Nanuorycu FLYIFTKISV DMFSGAVFIQ QALGWNIYAS VIALLGITMV YTVTGGLAAL
Sgltratno FLYIFTKISV DMFSGAVFIQ QALGWNIYAS VIALLGITMI YTVTGGLAAL
Nagcsussc MLYIFTKISA DIFSGAIFIT LALGLDLYLA IFLLLAITGL YTITGGLAAV
Nagoviar  VLYIFTKISA DIFSGAIFIN LALGLDLYLA IFILLAITAL YTITGGLAAV
Nagchomsa LLYIFTKISA DIFSGAIFIN LALGLNLYLA IFLLLAITAL YTITGGLAAV
Nagcorycu LLYIFTKISA DIFSGAIFIQ LTLGLDIYVA IIILLVITGL YTITGGLAAV
Naaasussc FICVALRISS DIFSGAIFIK LALGLDLYLA IFSLLAITAI YTITGGLASV
Namihomsa ILYIFTKLSV DLYSGALFIQ ESLGWNLYVS VILLIGMTAL LTVTGGLVAV
Consensus .LYIFTKIS. D.FSGA.FI. .ALG...Y.. ...LL.IT.. YT.TGGLAA.
```

```
              201                                           250
Naglhomsa MYTDTVQTFV ILGGACILMG YAFHEVGGYS GLFDKYLGAA TSLTVSEDP.
Nanuorycu MYTDTVQTFV IIAGAFILTG YAFHEVGGYS GLFDKYMGAM TSLTVSEDP.
Sgltratno MYTDTVQTFV ILAGAFILTG YAFHEVGGYS GLFDKYLGAV TSLTVSKDP.
Nagcsussc IYTDTLQTAI MLVGSFILTG FAFHEVGGYD AFIEKYMNAI PTVISDGNI.
Nagoviar  IYTDTLQTVI MLLGSFILTG FAFHEVGGYS AFVTKYMNAI PTVTSYGNT.
Nagchomsa IYTDTLQTVI MLVGSLILTG FAFHEVGGYD AFMEKYMRAI PTIVSDGNT.
Nagcorycu IYTDTLQTAI MMVGSVILTG FAFHEVGGYE AFTEKYMRAI PSQISYGNT.
Naaasussc IYTDTLQTII MLIGSFILMG FAFVEVGGYE SFTEKYMNAI PTIVEGDNL.
Namihomsa IYTDTLQALL MIIGALTLMI ISIMEIGGFE EVKRRYMLAS PDVTSILLTY
Consensus .YTDT.QT.. ...G..ILTG .AFHEVGGY. ....KYM.A. ..........
```

```
          251                                                    300
Naglhomsa AVGNISSFCY RPRPDSYHLL RHPVTGDLPW PALLLGLTIV SGWYWCSDQV
Nanuorycu AVGNISSSCY RPRPDSYHLL RDPVTGDLPW PALLLGLTIV SGWYWCSDQV
Sgltratno AVGNISSTCY QPRPDSYHLL RDPVTGGLPW PALLLGLTIV SGWHWCSDQV
Nagcsussc T...IKKECY APRADSFHIF RDPLKGDLPW PGLTFGLSIL ALWYWCTDQV
Nagoviar  T...VKKECY TPRADSFHIF RDPLKGDLPW PGLIFGLTII SLWYWCTDQV
Nagchomsa T...FQEKCY TPRADSFHIF RDPLTGDLPW PGFIFGMSIL TLWYWCTDQV
Nagcorycu S...IPQKCY TPREDAFHIF RDAITGDIPW PGLVFGMSIL TLWYWCTDQV
Naaasussc T...ISPKCY TPQGDSFHIF RDAVTGDIPW PGMIFGMTVV AAWYWCTDQV
Namihomsa NLSNTNSCNV SPKKEALKML RNPTDEDVPW PGFILGQTPA SVWYWCADQV
Consensus ........CY .PR.DS.H.. RDP..GDLPW P....G..I. ..WYWC.DQV

          301                                                    350
Naglhomsa IVQRCLAGKS LTHIKAGCIL CGYLKLTPMF LMVMPGMISR ILYPDEVACV
Nanuorycu IVQRCLAGRN LTHIKAGCIL CGYLKLTPMF LMVMPGMISR ILYPDEVACV
Sgltratno IVQRCLAGKN LTHIKAGCIL CGYLKLMPMF LMVMPGMISR ILYPDEVACV
Nagcsussc IVQRCLSAKN MSHVKAGCVM CGYFKLLPMF VIVMPGMISR VLYTEKIACT
Nagoviar  IVQRCLSAKN MSHVKAGCIM CGYMKLLPMF LMVMPGMISR ILFTEKVACT
Nagchomsa IVQRCLSAKN MSHVKGGCIL CGYLKLMPMF IMVMPGMISR ILYTEKIACV
Nagcorycu IVQRCLSAKN LSHVKAGCIL CGYLKVMPMF LIVMMGMVSR ILYTDKVACV
Naaasussc IVQRCLSGKD MSHVKAACIM CGYLKLLPMF LMVMPGMISR ILYTEKVACV
Namihomsa IVQRVLAAKN IAHAKGSTLM AGFLKLLPMF IIVVPGMISR ILFTDDIACI
Consensus IVQRCL..KN ..H.KAGCI. CGYLKL.PMF .MVMPGMISR ILY....ACV

          351                                                    400
Naglhomsa VPEVCRRVCG TEVGCSNIAY PRLVVKLMPN GLRGLMLAVM LAALMSSLAS
Nanuorycu APEVCKRVCG TEVGCSNIAY PRLVVKLMPN GLRGLMLAVM LAALMSSLAS
Sgltratno VPEVCKRVCG TEVGCSNIAY PRLVVKLMPN GLRGLMLAVM LAALMSSLAS
Nagcsussc VPSECEKYCG TKVGCSNIAY PTLVVELMPN GLRGLMLSVI LASLMSSLTS
Nagoviar  VPSECEKYCG TKVGCTNIAY PTLVVELMPN GLRGLMLSVM LASLMSSLTS
Nagchomsa VPSECEKYCG TKVGCTNIAY PTLVVELMPN GLRGLMLSVM LASLMSSLTS
Nagcorycu VPSECERYCG TRVGCTNIAF PTLVVELMPN GLRGLMLSVM MASLMSSLTS
Naaasussc VPSECVKHCG TEVGCSNYAY PLLVMELMPS GLRGLMLSVM LASLMSSLTS
Namihomsa NPEHCMLVCG SRAGCSNIAY PRLVMKLVPV GLRGLMMAVM IAALMSDLDS
Consensus VP..C...CG T.VGCSNIAY P.LVV.LMPN GLRGLML.VM LA.LMSSL.S

          401                                                    450
Naglhomsa IFNSSSTLFT MDIYTRLRPR AGDRELLLVG RLWVVFIVVV SVAWLPVVQA
Nanuorycu IFNSSSTLFT MDIYT.LRPR AGEGELLLVG RLWVVFIVAV SVAWLPVVQA
Sgltratno IFNSSSTLFT MDIYTRLRPR AGDRELLLVG RLWVVFIVAV SVAWLPVVQA
Nagcsussc IFNSATTLFT MDVYAKIRKR ASEKELMIAG RLFILVLIGI SIAWVPIVQS
Nagoviar  IFNSASTLFT MDIYTKIRKK ASEKELMIAG RLFMLVLIGV SIAWVPIVQS
Nagchomsa IFNSASTLFT MDIYAKVRKR ASEKELMIAG RLFILVLIGI SIAWVPIVQS
Nagcorycu IFNSASLFT MDIYTKIRKK ASEKELMIAG RLFMLFLIGI SIAWVPIVQS
Naaasussc IFNSASTLFT MDLYTKIRKQ ASEKELLIAG RLFIILLIVI SIVWVPLVQV
Namihomsa IFNSASTIFT LDVYKLIRKS ASSRELMIVG RIFVAFMVVI SIAWVPIIVE
Consensus IFNS.STLFT MD.Y...R.. A...EL...G RL........ S.AW.P.VQ.

          451                                                    500
Naglhomsa AQGGQLFDYI QAVSSYLAPP VSAVFVLALF VPRVNEQGAF WGLIGGLLMG
Nanuorycu AQGGQLFDYI QSVSSYLAPP VSAVFVVALF VPRVNEKGAF WGLIGGLLMG
Sgltratno AQGGQLFDYI QSVSSYLAPP VSAVFVLALF VPRVNEKGAF WGLIGGLLMG
Nagcsussc AQSGQLFDYI QSVTSYLGPP IAAVFLLAIF CKRVNEEGAF WGLVIGCMIG
Nagoviar  AQSGQLFDYI QSITSYLGPP IAAVFLLAIF CKRVNEPGAF WGLIIGFLIG
```

```
Nagchomsa AQSGQLFDYI QSITSYLGPP IAAVFLLAIF WKRVNEPGAF WGLILGLLIG
Nagcorycu AQSGQLFDYI QSITSYLGPP IAAVFLLAIF WKRVNEPGAF WGLVLGFLIG
Naaasussc AQNGQLFHYI ESISSYLGPP IAAVFLLAIF CKRVNEQGAF WGLIIGFVMG
Namihomsa MQGGQMYLYI QEVADYLTPP VAALFLLAIF WKRCNEQGAF YGGMAGFVLG
Consensus AQ.GQLFDYI QS..SYL.PP ..AVF.LA.F ..RVNE.GAF WGL..G...G
```

```
          501                                                550
Naglhomsa LARLIPEFSF GSGSCVQPSA CPAFLCGVHY LYFAIVLFFC SGLLTLTVSL
Nanuorycu LARLIPEFSF GTGSCVRPSA CPAFLCRVHY LYFAIVLFFC SGLLIIIVSL
Sgltratno LARLIPEFFF GTGSCVRPSA CPAIFCRVHY LYFAIILFFC SGFLTLAISR
Nagcsussc LARMITEFAY GTGSCVEPSN CPTIICGVHY LYFAIILFVI SIIIVLVVSL
Nagoviar  VSRMITEFAY GTGSCMEPSN CPTIICGVHY LYFAIILFVI TIIVILAISL
Nagchomsa ISRMITEFAY GTGSCMEPSN CPTIICGVHY LYFAIILFAI SFITIVVISL
Nagcorycu ISRMITEFAY GTGSCMEPSN CPTIICGVHY LYFAIILFVI SIITVVVVSL
Naaasussc LIRMIAEFVY GTGSCLAASN CPQIICGVHY LYFALILFFV SILVVLAISL
Namihomsa AVRLILAFAY RAPECDQPDN RPGFIKDIHY MYVATGLFWV TGLITVIVSL
Consensus ..R.I.EF.. GTGSC..PS. CP...C.VHY LYFAI.LF.. S.......SL
```

```
          551                                                600
Naglhomsa CTAPIPRKHL HRLVFSLRHS .......... ....KEERE DLDADEQQG.
Nanuorycu CTAPIPRKHL HRLVFSLRHS .......... ....KEERE DLDADELEAP
Sgltratno CTAPIPQKHL HRLVFSLRHS .......... ....KEERE DLDAEELEGP
Nagcsussc FTKPIPDVHL YRLCWSLRNS .......... ....KEERI DLDAEEED..
Nagoviar  FTKPIADVHL YRLCWSLRNS .......... ....KEERI DLDAEDED..
Nagchomsa LTKPIPDVHL YRLCWSLRNS .......... ....KEERI DLDAEEEN..
Nagcorycu FTKPIPDVHL YRLCWSLRNS .......... ....KEERI DLDAGEED..
Naaasussc LTKPIPDVHL YRLCWALRNS .......... ....TEERI DLDAEEKR..
Namihomsa LTPPPTKEQI RTTTFWSKKN LVVKENCSPK EEPYQMQEKS ILRCSENNET
Consensus .T.PIP..HL .RL..SLR.S .......... ....KEER. DLDA.E....
```

```
          601                                                650
Naglhomsa SSLPVQNGCP ESAMEMNEPQ A..........................
Nanuorycu ASPPVQNGRP EHAVEMEEPQ A..........................
Sgltratno APPPVQNGCQ ECAMGIEEVQ S..........................
Nagcsussc ....IQEAPE ETIEI..EVP E..........................
Nagoviar  ....IQDARE DALEIDTEAS E..........................
Nagchomsa ....IQEGPK ETIEIETQVP E..........................
Nagcorycu ....IQEAPE EATD..TEVP K..........................
Naaasussc ....HEEAHD ...GVDEDNP E..........................
Namihomsa INHIIPNGKS EDSIKGLQPE DVNLLVTCRE EGNPVASLGH SEAETPVDAY
Consensus .....Q.... E...........................
```

```
          651                                                700
Naglhomsa ...............PAP SLFRQCLLWF CGMSRGGVGS PPPLTQEEAA
Nanuorycu ...............PGP GLFRQCLLWF CGMNRGRAGG PAPPTQEEEA
Sgltratno ...............PAP GLLRQCLLWF CGMSKSGSGS PPPTT.EEVA
Nagcsussc ...............EKK GCFRRTYDLF CGLDQQKG.. .PKMTKEEEA
Nagoviar  ...............EKK GCLRQAYDMF CGLDQQKG.. .PKMTKEEEA
Nagchomsa ...............KKK GIFRRAYDLF CGLEQHGA.. .PKMTEEEEK
Nagcorycu ...............KKK GFFRRAYDLF CGLDQDKG.. .PKMTKEEEA
Naaasussc ...............ETR GCLRKAYDLF CGL.QRKG.. .PKLSKEEEE
Namihomsa SNGQAALMGE KERKKETDDG GRYWKFIDWF CGFKSKSLSK RSLRDLMEEE
Consensus .......... .......... G..R.....F CG........ .P..T.EE..
```

```
                701                                    739
Naglhomsa AAARRLEDIS EDPSWARVVN LNALLMMAVA VFLWGFYA.
Nanuorycu AAARRLEDIN EDPRWSRVVN LNALLMMAVA MFFWGFYA.
Sgltratno ATTRRLEDIS EDPSWARVVN LNALLMMTVA VFLWGFYA.
Nagcsussc AMKLKMTDTS EKPLWRTVVN INGIILLTVA VFCHAYFA.
Nagoviar  AMKLKMTDTS EKRLWRMVVN INGIILLAVA VFCHAYFA.
Nagchomsa AMKMKMTDTS EKPLWRTVLN VNGIILVTVA VFCHAYFA.
Nagcorycu AMKLKLTDTS EHPLWRTVVN INGVILLAVA VFCYAYFA.
Naaasussc AQKRKLTDTS EKPLWKTIVN INAILLLAVA VFVHGYFA.
Namihomsa AVCLQMLE.. ETRQVKVILN IGLFAVCSLG IFMFVYFSL
Consensus A.....ED.S E...W..VVN .N......VA VF.....A.
```

Proteins listed subsequently in italics are at least 90% identical to the paired transporters listed in parenthesis and therefore are not included in the alignments: *Namibosta*, *Namicanfa* (Namihomsa). Residues listed in the consensus sequence are present in at least 75% of the aligned transporter sequences. Residues indicated by boldface type are also conserved in the Na⁺/proline symporter family.

Database accession numbers

	SWISSPROT	*PIR*	*EMBL/GENBANK*
Naaasussc	P31636	A44432	L02900
Nagchomsa	P13866	A33545	M24847; L29338
Nagcorycu	P11170	S00515; S15974	X55355; X06419
Nagcsussc	P26429	A36361	M34044
Naglhomsa	P31639		M95549
Nagoviar		S48857	X82410
Namibosta			U41338
Namicanfa	P31637	A42163	M85068
Namihomsa		A56851	L38500
Nanuorycu	P26430	A42251	M84020
Sgltratno			U29881

References

[1] Hediger, M. et al. (1995) J. Physiol. 482, S7–S17.
[2] Reinhart, A.F. and Reithman, R. (1994) Curr. Opin. Cell Biol. 6, 583–594.
[3] Reizer, J. et al. (1994) Biochim. Biophys. Acta 1197, 133–166.
[4] Wright, E.M. et al. (1996) Curr. Opin. Cell Biol. 8, 468–473.
[5] Turk, E. et al. (1996) J. Biol. Chem. 271, 1925–1934.
[6] Turk, E. et al. (1991) Nature 350, 354–356.

Summary

Transporters of the Na⁺/dicarboxylate symporter family, the examples of which are the DCTA dicarboxylate-Na⁺ symporter of *Escherichia coli* (Dctaescco) and the EAT1 excitatory amino acid-Na⁺ cotransporter of humans (Eat1homsa), mediate symport (Na⁺-coupled or H⁺-coupled substrate uptake) of neutral amino acids, glucose and dicarboxylated compounds, including excitatory amino acids from the post-synaptic cleft [1-3]. Members of the Na⁺/dicarboxylate symporter family also serve as uniporters, ion and water channels. For example, neurotransmitter cotransporters, such as EAT1, contain ligand-gated ion channels that mediate Cl⁻ conductance in parallel with Na⁺-glutamate transport, while in the absence of substrates, transport ions in an uncoupled "leak" mode [4]. Members of the family have a broad biological distribution that ranges from bacteria to humans.

Statistical analysis reveals no apparent relationship between the amino acid sequences of the Na⁺/dicarboxylate symporter family and any other family of transporter family. However, the similarity between the kinetics of several families of Na⁺-driven and H⁺-driven transporters suggests a common mechanism of action, despite the lack of amino acid sequence homology [4]. Bacterial members of the Na⁺/dicarboxylate symporter family are predicted to contain 12 membrane-spanning helices by the hydrophobicity of their amino acid sequences and the activity of reporter gene fusions [5], while the mammalian proteins are predicted to contain ten membrane-spanning elements by the hydrophobicity of their amino acid sequences [2]. Eukaryotic transporters may be glycosylated.

Several amino acid sequence motifs are highly conserved in the Na⁺/dicarboxylate symporter family that are necessary for function by the criterion of site-directed mutagenesis [1].

Nomenclature, biological sources and substrates

CODE	DESCRIPTION [SYNONYMS]	ORGANISM [COMMON NAMES]	SUBSTRATE(S)
Dctaescco	C-4 Dicarboxylate transporter [DCTA]	*Escherichia coli* [gram-negative bacterium]	Na⁺/4-carbon dicarboxylates
Dctarhime	C-4 Dicarboxylate transporter [DCTA]	*Rhizobium meliloti* [gram-negative bacterium]	Na⁺/4-carbon dicarboxylates
Dctarhisp	C-4 Dicarboxylate transporter [DCTA]	*Rhizobium species* [gram-negative bacterium]	Na⁺/4-carbon dicarboxylates
Dctarhile	C-4 Dicarboxylate transporter [DCTA]	*Rhizobium leguminosarum* [gram-negative bacterium]	Na⁺/4-carbon dicarboxylates
Eaacratno	High-affinity glutamate transporter [EAAC, EAAC1]	*Rattus norvegicus* [rat]	Na⁺/glutamate
Eaacmusmu	High-affinity glutamate transporter [EAAC, EAAC1, MEAAC1]	*Mus musculus* [mouse]	Na⁺/glutamate
Eat1homsa	Na⁺-excitatory amino acid cotransporter 1 [EAT1, EAAT1, SLC1a3, GLAST1]	*Homo sapiens* [human]	Na⁺/glutamate, aspartate
Eat1bosta	Na⁺-excitatory amino acid cotransporter 1 [EAT1, EAAT1]	*Bos taurus* [cow]	Na⁺/glutamate, aspartate

	AMINO ACIDS	MOL. WT	EXPRESSION SITES	K_m	CHROMOSOMAL LOCUS
Eat2homsa	574	62 060	motor cortex		11p13–p12
Eat2musmu	572	62 030	motor cortex		
Eat3orycu	524	56 938	brain		
Eat3homsa	525	57 155	brain		9p24
Glt1caeel	503	54 732			
Glt1oncvo	492	53 391			
Gltpbacsu	414	44 614			999°
Gltpescco	437	47 159			92.44 minutes
Glttbacst	421	45 469			
Glttbacca	421	45 345			
Iaatmusmu	553	58 384	adipocytes		
Satthomsa	532	55 723	brain, muscle, pancreas		2p15–p13

Multiple amino acid sequence alignments

```
           1                                                  50
Dctarhile ........... .......... .......... ..MIAAPLDA VAGSKGKKPF
Dctarhime ........... .......... .......... ....MIIEH SAEVRGKTPL
Dctaescco ........... .......... .......... .......MKTSL
Gltpescco ........... .......... .......... ........MKNI
Glttbacst ........... .......... .......... ..........MR
Gltpbacsu ........... .......... .......... ...........M
Iaatmusmu MAVDPPKADP KGSSGGFHRN GGPALGSRED QSAKAGGCCG SRDRVRRCIR
Satthomsa .......... MEKSNETNGY LDSAQAGPAA GPGAPGTAAG RARRCAGFLR
Eat2homsa ........MA STEGANNMPK QVEVRMPDSH LGSEEPKHRH LGLRLCDKLG
Glt1oncvo .......... .......... .......... .......... ...MVSWIR
Eat3homsa .......... .......... .......... ...MGKPARK GCPSWKRFLK
Eaacratno .......... .......... .......... ...MGKPTSS GC.DWRRFLR
Eat1homsa .MTKSNGEEP KMGGRMERFQ QGVRKRTLLA KKKVQNITKE DVKSY..LFR
Consensus .......... .......... .......... .......... ..........

           51                                                 100
Dctarhile YSHLYVQVLV AIAAGILLGH FYPE...... ....LGTQLK PLGDAFIKLV
Dctarhime YRHLYVQVLA AIAAGILLGH FYPD...... ....IGTELK PLGDAFIRLV
Dctaescco FKSLYFQVLT AIAIGILLGH FYPE...... ....IGEQMK PLGDGFVKLI
Gltpescco KFSLAWQILF AMVLGILLGS YLHYHSDSRD W..LVVNLLS PAGDIFIHLI
Glttbacst KIGLAWQIFI GLILGIIVGA IFYGNPK... ....VATYLQ PIGDIFLRLI
Gltpbacsu KKLIAFQILI ALAVGAVIGH FFPD...... ....FGMALR PVGDGFIRLI
Iaatmusmu ANLLVLLTVA AVVAGVGLGL GVSAAGGADA LGPARLTRFA FPGELLLRLL
Satthomsa RQALVLLTVS GVLAGAGLG. ...AALRGLS LSRTQVTYLA FPGEMLLRML
Eat2homsa KNLLLTLTVF GVILGAVCGG LLRLASP... IHPDVVMLIA FPGDILMRML
Glt1oncvo KNLLLVLTVS SVVLGALCGF LLR.GLQ... LSPQNIMYIS FPGELLMHML
Eat3homsa NNWVLLSTVA AVVLGITTGV LVREHSN... LSTLEKFYFA FPGEILMRML
Eaacratno NHWLLLSTVA AVVLGIVVGV LVRGHSE... LSNLDKFYFA FPGEILMRML
Eat1homsa NAFVLL.TVT AVIVGTILGF TLRPYR.... MSYREVKYFS FPGELLMRML
Consensus ...L...... ....G...G. .......... .......... ..G.......

           101                                                150
Dctarhile KMIIAPVIFL TVATGIAGMS DLQKVGRVAG KAMLYFLTFS TLALIIGLIV
Dctarhime KMIIAPVIFL TVATGIAGMT DLAKVGRVAG KAMIYFLAFS TLALVVGLVV
Dctaescco KMIIAPVIFC TVVTGIAGME SMKAVGRTGA VALLYFEIVS TIALIIGLII
Gltpescco KMIVVPIVIS TLVVGIAGVG DAKQLGRIGA KTIIYFEVIT TVAIILGITL
```

```
Glttbacst KMIVIPIVIS SLVVGVASVG DLKKLGKLGG KTIIYFEIIT TIAIVVGLLA
Gltpbacsu KMIVVPIVFS TIVIGAAGSG SMKKMGSLGI KTIIWFEVIT TLVLGLGLLL
Iaatmusmu KMIILPLVVC SLIGGAASL. DPSALGRVGA WAALFPGHHT ARVGA.RRGF
Satthomsa RMIILPLVVC SLVSGAASL. DASCLGRLGG IAVAYFGLTT LSASALAVAL
Eat2homsa KMLILPLIIS SLITGLSGL. DAKASGRLGT RAMVYYMSTT IIAAVLGVIL
Glt1oncvo KMMILPLIMS SLISGLAQL. DARQSGKLGS LAVTYYMFTT AVAVVTGIFL
Eat3homsa KLIILPLIIS SMITGVAAL. DSNVSGKIGL RAVVYYFCTT LIAVILGIVL
Eaacratno KLVILPLIIS SMITGVAAL. DSNVSGKIGL RAVVYYFSTT VIAVILGIVL
Eat1homsa QMLVLPLIIS SLVTGMAAL. DSKASGKMGM RAVVYYMTTT IIAVVIGIII
Consensus KM...P.... ....G.A... D....G..G. ....Y....T ..A...G...

          151                                                   200
Dctarhile ANVVQPGAGM N...IDPASL DPAAVATFAA KAHEQSIVGF LTNIIPTTIV
Dctarhime ANVVQPGAGM H...IDPASL DAKAVATYAE KAHEQSITGF LMNIIPTTLV
Dctaescco VNVVQPGAGM N...VDPATL DAKAVAVYAD QAKDQGIVAF IMDVIPASVI
Gltpescco ANVFQPGAGV DMSQLATVDI SKYQSTTEAV QSSSHGIMGT ILSLVPTNIV
Glttbacst ANIFQPGTGV NMKSLEKTDI QSYVDTTNEV Q..HHSMVET FVNIVPKNIF
Gltpbacsu ANVLKPGVGL DLSHLAKKDI HELSGYTDKV V....DFKQM ILDIIPTNII
Iaatmusmu GPGAEAGAAV TAITSINDSV VDPCARSAPT KEVLDSFLDL VRNIFPSNLV
Satthomsa AFIIKPGSGA QTLQS.SDLG LEDSGPPPVP KETVDSFLDL ARNLFPSNLV
Eat2homsa VLAIHPGNPK LKKQL..... ..GPGKKNDE VSSLDAFLDL IRNLFPENLV
Glt1oncvo VLVIHPGDPT IKKEI..... ..GTGTEGKT VSTVDTLLDL LRNMFPENVV
Eat3homsa VVSIKPGVTQ KVGEI..... ..ARTGSTPE VSTVDAMLDL IRNMFPENLV
Eaacratno VVSIKPGVTQ KVNEI..... ..NRTGKTPE VSTVDAMLDL IRNILGENLV
Eat1homsa VIIIHPGKGT K.ENM..... ..HREGKIVR VTAADAFLDL IRNMFPPNLV
Consensus .....PG... .......... .......... .......... ......P...

          201                                                   250
Dctarhile GA........ .......... .......... .......... ..........
Dctarhime GA........ .......... .......... .......... ..........
Dctaescco GA........ .......... .......... .......... ..........
Gltpescco AS........ .......... .......... .......... ..........
Glttbacst ES........ .......... .......... .......... ..........
Gltpbacsu DV........ .......... .......... .......... ..........
Iaatmusmu SAAFRSFATS YEPKDNSCKI PQSCIQREIN STMVQLLCE. ..........
Satthomsa VAAFRTYATD YKV......V TQNSSSGNVT HEKIPIGTE. ..........
Eat2homsa QACFQQIQTV TKKVLVAPPP DEEANATSAE VSLLNETVTE VPEETKMVIK
Glt1oncvo QATFQQVQTK YIKV....RP KVVKNNDSAT LAALNNGSLD .......YVK
Eat3homsa QACFQQYKTK REE..VKPPS DPEMNMTEES FTAVMTTAIS KNK.TKEYKI
Eaacratno QACFQQYKTK REE..VKPAS DPGGNQTEVS VTTAMTT.MS ENK.TKEYKI
Eat1homsa EACFKQFKTN YEKRSFKVPI QANETLVGAV INNVSEAMET LTRITEELVP
Consensus .......... .......... .......... .......... ..........

          251                                                   300
Dctarhile .....FADGD ILQVLFFSVL FGIALAMVGE KG.EQVVNFL NSLTAPVFKL
Dctarhime .....FAEGD ILQVLFISVL FGISLAIVGK KA.EPVVDFL QALTLPIFRL
Dctaescco .....FASGN ILQVLLFAVL FGFALHRLGS KG.QLIFNVI ESFSQVIFGI
Gltpescco .....MAKGE MLPIIFFSVL FGLGLSSLPA THREPLVTVF RSISETMFKV
Glttbacst .....LTKGD MLPIIFFSVM FGLGVAAIGE KGK.PVLQFF QGTAEAMFYV
Gltpbacsu .....MARND LLAVIFFAIL FGVAAAGIG. KASEPVMKFF ESTAQIMFKL
Iaatmusmu .....VEGMN ILGLVVFAIV FGVALRKLGP EG.ELLIRFF NSFNDATMVL
Satthomsa .....IEGMN ILGLVLFALV LGVALKKLGS EG.EDLIRFF NSLNEATMVL
Eat2homsa KGLEFKDGMN VLGLIGFFIA FGIAMGKMGD QA.KLMVDFF NILNEIVMKL
Glt1oncvo ASVEYTSGMN VLGVIVFCIA IGISLSQLGQ EA.HVMVQFF VIMDKVIMKL
Eat3homsa VGM.YSDGIN VLGLIVFCLV FGLVIGKMGE KG.QILVDFF NALSDATMKI
```

```
Eaacratno VGL.YSDGIN VLGLIIFCLV FGLVIGKMGE KG.QILVDFF NALSDATMKI
Eat1homsa VPG.SVNGVN ALGLVVFSMC FGFVIGNMKE QG.QALREFF DSLNEAIMRL
Consensus .......... .L...F... FG.....G. ........FF ..........
```

```
           301                                              350
Dctarhile VAILMKAAPI GAFGAMAFTI GKYGVGSIA. .NLAMLIGTF YITSLLFVFI
Dctarhime VAILMKAAPI GAFGAMAFTI GKYGIASIA. .NLAMLIGTF YLTSFLFVFI
Dctaescco INMIMRLAPI GAFGAMAFTI GKYGVGTLV. .QLGQLIICF YITCILFVVL
Gltpescco THMVMRYAPV GVFALIAVTV ANFGFSSLW. .PLAKLVLLV HFAILFFALV
Glttbacst TNQIMKFAPF GVFALIGVTV SKFGVESLI. .PLSKLVIVV YATMVFFIFV
Gltpbacsu TQIVMVTAPI GVLALMAASV GQYGIELLL. .PMFKLVGTV FLGLFLILFV
Iaatmusmu VSWIMWYAPV GILFLVASKI VEMKDVRQLF ISLGKYILCC LLGHAIHGLL
Satthomsa VSWIMWYVPV GIMFLVGSKI VEMKDIIVLV TSLGKYIFAS ILGHVIHGGI
Eat2homsa VIMIMWYSPL GIACLICGKI IAIKDLEVVA RQLGMYMVTV IIGLIIHGGI
Glt1oncvo VMTVMWYSPF GILCLIMGKI LEIHDLADTA RMLAMYMVTV LSGLAIHSLI
Eat3homsa VQIIMCYMPL GILFLIAGKI IEVEDWEIF. RKLGLYMATV LTGLAIHSIV
Eaacratno VQIIMCYMPI GILFLIAGKI IEVEDWEIF. RKLGLYMATV LSGLAIHSLV
Eat1homsa VAVIMWYAPV GILFLIAGKI VEMEDMGVIG GQLAMYTVTV IVGLLIHAVI
Consensus ........P. G...L....I .......... ..L.......
```

```
           351                                              400
Dctarhile VLGAVARYNG F.SIVALLRY IKEELLLVLG TSSSEAALPG LMNKM.EKAG
Dctarhime VLGAVARYNG F.SILSLIRY IKEELLLVLG TSSSEAALPG LMNKM.EKAG
Dctaescco VLGSIAKATG F.SIFKFIRY IREELLIVLG TSSSESALPR MLDKM.EKLG
Gltpescco VLGIVARLCG L.SVWILIRI LKDELILAYS TASSESVLPR IIEKM.EAYG
Glttbacst VLGGVAKLFG I.NIFHIIKI LKDELILAYS TASSETVLPK IMEKM.ENFG
Gltpbacsu LFPLVGLIFQ I.KYFEVLKM IWDLFLIAFS TTSTETILPQ LMDRM.EKYG
Iaatmusmu VLPLIYFLFT RKNPYRFLWG IMTPLATAFG TSSSSATLPL MMKCVEEKNG
Satthomsa VLPLIYFVFT RKNPFRFLLG LLAPFATAFA TCSSSATLPS MMKCIEENNG
Eat2homsa FLPLIYFVVT RKNPFSLFAG IFQAWITALG TASSAGTLPV TFRCLEENLG
Glt1oncvo SLPLIFFVTT KKNPYVFMRG LFQAWITGLG TASSSDTLPI TYICLEENLG
Eat3homsa ILPLIYFIVV RKNPFRFAMG MAQALLTALM ISSSSATLPV TFRCAEENNQ
Eaacratno VLPLIYFIVV RKNPFRFALG MAQALLTALM ISSSSATLPV TFRCAEEKNH
Eat1homsa VLPLLYFLVT RKNPWVFIGG LLQALITALG TSSSSATLPI TFKCLEENNG
Consensus .L........ .......... ..........  T.SS...LP. ......E..G
```

```
           401                                              450
Dctarhile CKRSVVGLVI PTGYSFNLDG TNIYMTLAAL FIAQATGIHL SWGDQILLLL
Dctarhime CKRSVVGLVI PTGYSFNLDG TNIYMTLAAL FIAQATDTPL SYGDQILLLL
Dctaescco CRKSVVGLVI PTGYSFNLDG TSIYLTMAAV FIAQATNSQM DIVHQITLLI
Gltpescco APVSITSFVV PTGYSFNLDG STLYQSIAAI FIAQLYGIDL SIWQEIILVL
Glttbacst CPKAITSFVI PTGYSFNLDG STLYQALAAI FIAQLYGIDM PISQQISLLL
Gltpbacsu CPKRVVSFVV PSGLSLNCDG SSLYLSVSCI FLAQAFQVDM TLSQQLLMML
Iaatmusmu VAKHISRFIL PIGATVNMDG AALFQCVAAV FIAQLNGVSL DFVKIITILV
Satthomsa VDKRISRFIL PIGATVNMDG AAIFQCVAAV FIAQLNNVEL NAGQIFTILV
Eat2homsa IDKRVTRFVL PVGATINMDG TALYEAVAAI FIAQMNGVVL DGGQIVTVSL
Glt1oncvo VDRRVTRFVL PVGATINMDG TALYEAVAAI FIAQINGVHL SFGQVVTVSL
Eat3homsa VDKRITRFVL PVGATINMDG TALYEAVAAV FIAQLNDLDL GIGQIITISI
Eaacratno VDKRITRFVL PVGATINMDG TALYEAVAAV FIAQLNGMDL SIGQIITISI
Eat1homsa VDKRVTRFVL PVGATINMDG TALYEALAAI FIAQVNNFEL NFGQIITISI
Consensus .......FV. P.G...N.DG ...Y...AA. FIAQ.....L ..........
```

```
           451                                              500
Dctarhile VAMLSSKGAA GITGAGFITL AATLSVVPSV PVAGMALILG IDRFMSECRA
Dctarhime VAMLSSKGAA GITGAGFITL AATLSVVPSV PVAGMALILG IDRFMSECRA
```

```
Dctaescco VLLLSSKGAA GVTGSGFIVL AATLSAVGHL PVAGLALILG IDRFMSEARA
Gltpescco TLMVTSKGIA GVPGVSFVVL LATLGSVG.I PLEGLAFIAG VDRILDMART
Glttbacst VLMVTSKGIA GVPGVSFVVL LATLGTVG.I PIEGLAFIAG IDRILDMART
Gltpbacsu VLVMTSKGIA AVPSGSLVVL LATANAVG.L PAEGVAIIAG VDRVMDMART
Iaatmusmu TATASSVGAA GIPAGGVLTL AIILEAVS.L PVKDISLILA VDWLVDRSCT
Satthomsa TATASSVGAA GVPAGGVLTI AIILEAIG.L PTHDLPLILA VDWIVDRTTT
Eat2homsa TATLASVGAA SIPSAGLVTM LLILTAVG.L PTEDISLLVA VDWLLDRMRT
Glt1oncvo TATLASIGAA SVPSAGLVTM LLVLTAVG.L PVKDVSLIVA VDWLLDRIRT
Eat3homsa TATSASIGAA GVPQAGLVTM VIVLSAVG.L PAEDVTLIIA VDWLLDRFRT
Eaacratno TATAASIGAA GVPQAGLVTM VIVLSAVG.L PAEDVTLIIA VDWLLDRFRT
Eat1homsa TATAASIGAA GIPQAGLVTM VIVLTSVG.L PTDDITLIIA VDWFLDRLRT
Consensus .....S.GAA G.P..G..T. ...L..VG.. P.....LI.. .D...D..RT


          501                                              550
Dctarhile LTNLVGNAVA TIVVARWENE LDTVQLAAAL GGQTGEDTSA AGLQPAE...
Dctarhime LTNFVGNAVA TIVVAKWEGE LDQAQLSAAL GGEASVEAIP AVVQPAE...
Dctaescco LTNLVGNGVA TIVVAKWVKE LDHKKLDDVL NNRAPDGKTH ELSS......
Gltpescco ALNVVGNALA VLVIAKWEHK FDRKKALAYE REVLGKFDKT ADQ.......
Glttbacst LTNLVGNSLA AIIMSKWEGQ YNEEKG.... KQYIAQLQQS A.........
Gltpbacsu GVNVPGHAIA CIVVSKWEKA FRQKEWVSAN SQTESI.... ..........
Iaatmusmu VLNVEGDAFG AGLLQSYVDR TKMPSSEPEL IQVKNEVSLN PLPLATEEGN
Satthomsa VVNVEGDALG AGILH.HLNQ KATKKGEQEL AEVKVEAIPN ..CKSEEETS
Eat2homsa SVNVVGDSFG AGIVY.HLSK SELDTIDSQH RVHEDIEMTK TQSIYDDMKN
Glt1oncvo SINVLGDAMG AGIVY.HYSK ADLDAHDRL. .....AATTR SHSI......
Eat3homsa MVNVLGDAFG TGIVE.KLSK KELEQMD... .VSSEVNIVN PFALESTI.L
Eaacratno MVNVLGDAFG TGIVE.KLSK KELEQVD... .VSSEVNIVN PFALEPTI.L
Eat1homsa TTNVLGDSLG AGIVE.HLSR HELKNRD... .VEMGNSVIE ENEMKKPYQL
Consensus ..NV.G.... .......... .......... .......... ..........


          551                                              596
Dctarhile .......... .......... .......... .......... .....
Dctarhime .......... .......... .......... .......... .....
Dctaescco .......... .......... .......... .......... .....
Gltpescco .......... .......... .......... .......... .....
Glttbacst .......... .......... .......... .......... .....
Gltpbacsu .......... .......... .......... .......... .....
Iaatmusmu PLLK.QYQGP TGDSSATFE. KESVM..... .......... .....
Satthomsa PLVTHQNPAG PVASAPELES KESVL..... .......... .....
Eat2homsa HRESNSNQCV YAAHNSVIVD ECKVTLAANG KSADCSVEEE PWKREK
Glt1oncvo ..AMNDEKRQ LAVYNSLPTD DEKHTH.... .......... .....
Eat3homsa DNEDSDTKKS YVNGGFAVDK SDTISFTQTS QF........ .....
Eaacratno DNEDSDTKKS YVNGGFSVDK SDTISFTQTS QF........ .....
Eat1homsa IAQDNETEKP IDSETKM... .......... .......... .....
Consensus .......... .......... .......... .......... .....
```

Proteins listed subsequently in italics are at least 90% identical to the paired transporters listed in parenthesis and therefore are not included in the alignments: *Glttbacca* (Glttbacst); *Dctarhisp* (Dctarhime); *Eat2ratno, Eat2musmu* (Eat2homsa); *Eat3orycu* (Eat3homsa); *Eat1ratno, Eat1bosta* (Eat1homsa); *Eaacmusmu* (Eaacratno); *Glt1caeel* (Glt1oncvo). Residues listed in the consensus sequence are present in at least 75% of the aligned transporter sequences.

Database accession numbers

	SWISSPROT	PIR	EMBL/GENBANK
Iaatmusmu		JC4149	
Dctaescco	P37312		U00039
Dctarhime	P20672	A33597; S04816	M26399; J03683
Dctarhisp	P31601		S38912
Dctarhile	Q01857	S25701; S27384	Z11529
Eaacratno			U39555
Eaacmusmu		S55677	
Eat1homsa	P43003		U03504
Eat1ratno	P24942	S26609; A46370	X63744
Eat1bosta	P46411		D29661
Eat2musmu	P43006		U11763
Eat2ratno	P31596		X67857
Eat2homsa	P43004		U03505
Eat3orycu	P31597		L12411; S49854
Eat3homsa	P43005		U03506
Glt1oncvo			U35251
Glt1caeel			U35250
Gltpbacsu	P39817		U15147
Gltpescco	P21345	JV0092; A42384	M32488; M84805
Glttbacst	P24943	S26247	M86508
Glttbacca	P24944	S26246	M86509
Satthomsa	P43007		L14595; L19444

References

[1] Reizer, J. et al. (1994) Biochim. Biophys. Acta 1197, 133–166.
[2] Hediger, M. et al. (1995) J. Physiol. 482, S7–S17.
[3] Malandro, M. and Kilberg, M. (1996) Annu. Rev. Biochem. 65, 305–336.
[4] Wright, E. et al. (1996) Curr. Opin. Cell Biol. 8, 468–473.
[5] Jording, D. and Puhler, A. (1993) Mol. Gen. Genet. 241, 106–114.
[6] Kanai, Y. and Hediger, M. (1992) Nature 360, 462–471.
[7] Storck, T. et al. (1992) Proc. Natl Acad. Sci. USA 89, 10955–10959.

Summary

Transporters of the Na$^+$/PO$_4$ symporter family, the example of which is the NPT1 phosphate-Na$^+$ cotransporter of humans (Npt1homsa), mediate symport (Na$^+$-linked substrate uptake) of phosphate. All known members of the family occur in mammals.

Statistical analysis reveals no relationship between the amino acid sequences of the Na$^+$/PO$_4$ symporter family and other family of transporters. They are predicted to form six or possibly eight membrane-spanning helices by the hydropathy of their amino acid sequences and contain potential sites for glycosylation and protein kinase C phosphorylation.

Several amino acid sequence motifs are highly conserved in the Na$^+$/PO$_4$ symporter family.

Nomenclature, biological sources and substrates

CODE	DESCRIPTION [SYNONYMS]	ORGANISM [COMMON NAMES]	SUBSTRATE(S)
Npt1homsa	Renal phosphate-Na$^+$ cotransporter [NPT1]	*Homo sapiens* [human]	Na$^+$/PO$_4$
Npt1musmu	Renal phosphate-Na$^+$ cotransporter [NPT1]	*Mus musculus* [mouse]	Na$^+$/PO$_4$
Npt1orycu	Renal phosphate-Na$^+$ cotransporter [NPT1]	*Oryctolagus cuniculus* [rabbit]	Na$^+$/PO$_4$
Npt1ratno	Renal phosphate-Na$^+$ cotransporter [NPT1]	*Rattus norvegicus* [rat]	Na$^+$/PO$_4$
Npt2ratno	Brain phosphate-Na$^+$ cotransporter [NPT2]	*Rattus norvegicus* [rat]	Na$^+$/PO$_4$

Cotransported ions are listed.

Phylogenetic tree

Proteins listed subsequently in italics are at least 90% identical to the paired transporters listed in parenthesis and therefore are not included in the phylogenetic tree: *Npt1musmu* (Npt1homsa).

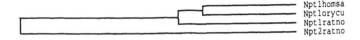

```
                                        Npt1homsa
                                        Npt1orycu
                                        Npt1ratno
                                        Npt2ratno
```

Proposed orientation of NPT1 in the membrane

The model is based on predictions of membrane-spanning regions and α-helical content. The N-terminus of the protein is illustrated on the inside and is folded 12 times through the membrane. The predicted membrane-spanning helices are portrayed as rectangles. The numbers corresponding to the first and last residue of each membrane-spanning helix is boxed. More than half of the residues are conserved in at least 75% of the members of the Na$^+$/PO$_4$ symporter family and, therefore, are not mapped onto the model.

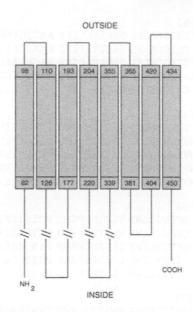

Physical and genetic characteristics

	AMINO ACIDS	MOL. WT	EXPRESSION SITES	CHROMOSOMAL LOCUS
Npt1homsa	467	51 143	kidney	6p23–p21.3
Npt1musmu	465	51 589	kidney	
Npt1orycu	465	51 798	kidney	
Npt1ratno	465	51 349	kidney	
Npt2ratno	560	61 665	brain	

Multiple amino acid sequence alignments

```
           1                                                    50
Npt1homsa  .......... .......... .......... ........M QMDNRLPPKK
Npt1orycu  .......... .......... .......... ......... .MDNQFPSRK
Npt1ratno  .......... .......... .......... ......... .MENRCLPKK
Npt2ratno  MEFRQEEFRK LAGRALGRLH RLLEKRQEGA ETELSADGR PVTTHTRDPP
Consensus  .......... .......... .......... ......... .........K


           51                                                   100
Npt1homsa  VPGFCSF... .RYGLSFLVH CCNVIITAQR ACLNLTMVVM VNSTDPHGLP
Npt1orycu  GPCFCSF... .RYVLALFMH FCNIVIIAQR MCLSLTMVAM VNNTNLHGSP
Npt1ratno  VPGFCSF... .RYGLAILLH FCNIVIMAQR VCLNLTMVAM VNKTEPPHLS
Npt2ratno  VVDCTCFGLP RRYIIAIMSG LGFCISFGIR CNLGVAIVSM VNNSTTHRGG
Consensus  VP.FCSF... .RY.LA...H .CN..I.AQR .CL.LTMV.M VN.T..H...


           101                                                  150
Npt1homsa  NTSTKKLLDN IKNPMYNWSP DIQGIILSST SYGVIIIQVP VGYFSGIYST
Npt1orycu  NTSAEKRLDN TKNPVYNWSP DVQGIIFSSI FYGAFLIQIP VGYISGIYSI
Npt1ratno  NKSVAEMLDN VKNPVHSWSL DIQGLVLSSV FLGMVVIQVP VGYLSGAYPM
Npt2ratno  HVVVQK.... ...AQFNWDP ETVGLIHGSF FWGYIVTQIP GGFICQKFAA
Consensus  N.S..K.LDN .KNP..NWSP D.QG.I.SS. F.G...I...P VGY.SG.Y..
```

```
          151                                                        200
Npt1homsa KKMIGFALCL SSVLSLLIPP AAGIGVAWVV VCRAVQGAAQ GIVATAQFEI
Npt1orycu KKLIGFALFL SSLVSIFIPQ AAAVGETWII VCRVVQGITQ GTVTTAQHEI
Npt1ratno EKIIGSSLFL SSVLSLLIPP AAQVGAALVI VCRVLQGIAQ GAVSTGQHGI
Npt2ratno NRVFGFAIVA TSTLNMLIPS AARVHYGCVI FVRILQGLVE GVTYPACHGI
Consensus .K.IGFAL.L SS.LS.LIP. AA.VGG..VI VCR..QG..Q G.V.TAQH.I

          201                                                        250
Npt1homsa YVKWAPPLER GRLTSMSTSG FLLGPFIVLL VTGVICESLG WPMVFYIFGA
Npt1orycu WVKWAPPLER GRLTSMSLSG FLLGPFIVLL VTGIICESLG WPMVFYIFGA
Npt1ratno WVKWAPPLEE GRLTSMTLSG FVMGPFIALL VSGFICDLLG WPMVFYIFGI
Npt2ratno WSKWAPPLER SRLATTAFCG SYAGAVVAMP LAGVLVQYSG WSSVFYVYGS
Consensus WVKWAPPLER GRLTSM..SG F..GPFI.LL V.G.IC..LG WPMVFYIFG.

          251                                                        300
Npt1homsa CGCAVCLLWF VLFYDDPKDH PCISISEKEY ITSSLVQQV. ...SSSRQSL
Npt1orycu CGCAVCLLWF VLYYDDPKDH PCVSLHEKEY ITSSLIQQG. ...SSTRQSL
Npt1ratno VGCVLSLFWF ILLFDDPNNH PYMSSSEKDY ITSSLMQQV. ...HSGRQSL
Npt2ratno FGIFWYLFWL LVSYESPALH PSISEEERKY IEDAIGESAK LMNPVTKFNT
Consensus .GC...L.WF .L.YDDP..H P..S..EK.Y ITSSL.QQ......S.RQSL

          301                                                        350
Npt1homsa PIKAILKSLP VWAISIGSFT FFWSHNIMTL YTPMFINSML HVNIKENGFL
Npt1orycu PIKAMIKSLP LWAISFCCFA YLWTYSRLIV YTPTLINSML HVDIRENGLL
Npt1ratno PIKAMLKSLP LWAIILNSFA FIWSNNLLVT YTPTFISTTL HVNVRENGLL
Npt2ratno PWRRFFTSMP VYAIIVANFC RSWTFYLLLI SQPAYFEEVF GFEISKVGLV
Consensus PIKA..KSLP .WAI....F. ..W....L.. YTP..I...L HV.I.RNGLL

          351                                                        400
Npt1homsa SSLPYLFAWI CGNLAGQLSD FFLTRNILSV IAVRKLFTAA GFLLPAIFGV
Npt1orycu SSLPYLFAWI CGVIAGHTAD FLMSRNMLSL TAIRKLFTAI GLLLPIVFSM
Npt1ratno SSLPYLLAYI CGIVAGQMSD FLLSRKIFSV VAVRKLFTTL GIFCPVIFVV
Npt2ratno SALPHLVMTI IVPIGGQIAD FLRSRHIMST TNVRKLMNCG GFGMEATLLL
Consensus SSLPYL.A.I CG..AGQ..D FL.SR.I.S. .AVRKLFT.. G...P..F..

          401                                                        450
Npt1homsa CLPYLSSTFY SIVIFLILAG ATGSFCLGGV FINGLDIAPR YFGFIKACST
Npt1orycu CLLYLSSGFY STITFLILAN ASSSFCLGGA LINALDLAPR YYVFIKGVTT
Npt1ratno CLLYLSYNFY STVIFLTLAN STLSFSFCGQ LINALDIAPR YYGFLKAVTA
Npt2ratno VVGY.SHSKG VAISFLVLAV GFSGFAISGF NVNHLDIAPR YASILMGISN
Consensus CL.YLS..FY S...FL.LA. ...SF...G. .IN.LDIAPR Y..F.K....

          451                                                        500
Npt1homsa LTGMIGGLIA STLTGLILKQ DPESAWFKTF ILMAAINVTG LIFYLIVATA
Npt1orycu LIGMTGGMTS STVAGLFLSQ DPESSWFKIF LLMSIINVIS VIFYLIFAKA
Npt1ratno LIGIFGGLIS STLAGLILNQ DPEYAWHKNF FLMAGINVTC LAFYLLFAKG
Npt2ratno GVGTLSGMVC PIIVGAMTKH KTREEWQYVF LIASLVHYGG VIFYGVFASG
Consensus L.G..GG... ST..GL.L.Q DPE..W.K.F .LM..INV.. .IFYL.FA..

          501                                                        550
Npt1homsa EIQDWAKEKQ HTRL............ ................ ...........
Npt1orycu EIQDWAKEKQ HTRL............ ................ ...........
Npt1ratno DIQDWAKETK TTRL............ ................ ...........
Npt2ratno EKQPWAEPEE MSEEKCGFVG HDQLAGSDES EMEDEVEPPG APPAPPPSYG
Consensus EIQDWAKE.. .TRL............ ................ ...........
```

```
            551              568
Npt1homsa ...................  ........
Npt1orycu ...................  ........
Npt1ratno ...............  ........
Npt2ratno ATHSTVQPPR PPPPVRDY
Consensus ...................  ........
```

Proteins listed subsequently in italics are at least 90% identical to the paired transporters listed in parenthesis and therefore are not included in the alignments: *Npt1musmu* (Npt1homsa). Residues listed in the consensus sequence are present in at least 75% of the transporter sequences shown.

Database accession numbers

	SWISSPROT	PIR	EMBL/GENBANK
Npt1homsa		A48916	X71355
Npt1musmu			X77241
Npt1orycu		A56410; S27951	M76466
Npt1ratno			U28504
Npt2ratno			U07609

Summary

Transporters of the Na⁺/branched amino acid symporter family, the example of which is the BRNQ branched chain amino acid transporter of *Salmonella typhimurium* (Brnqsalty), mediate symport (cation-coupled substrate uptake) of isoleucine, leucine and valine. The cotransported ion may be H⁺, Na⁺ or Li⁺, depending on the transporter. Known members of the family occur in gram-negative bacteria.

Statistical analysis reveals no apparent relationship between the amino acid sequences of the Na⁺/branched amino acid symporter family and any other family of transporters. They are predicted to form 12 membrane-spanning helices by the hydropathy of their amino acid sequences.

Several amino acid sequence motifs are highly conserved in the Na⁺/branched amino acid symporter family, including motifs necessary for function by the criterion of site-directed mutagenesis [1].

Nomenclature, biological sources and substrates

CODE	DESCRIPTION [SYNONYMS]	ORGANISM [COMMON NAMES]	SUBSTRATE(S)
Brabpseae	Branched chain amino acid transport system 2 [BRAB, LIVII]	*Pseudomonas aeruginosa* [gram-negative bacterium]	Na⁺, Li⁺/valine isoleucine, leucine
Brazpseae	Branched chain amino acid transport system 3 [BRAZ LIVIII]	*Pseudomonas aeruginosa* [gram-negative bacterium]	H⁺/isoleucine, leucine, valine
Brnqclope	Branched chain amino acid transport system 2 [BRNQ]	*Clostridium perfringens* [gram-positive bacterium]	Na⁺, Li⁺/valine isoleucine, leucine
Brnqhaein	Branched chain amino acid transport system 2 [BRNQ]	*Haemophilus influenzae* [gram-negative bacterium]	Na⁺, Li⁺/valine isoleucine, valine
Brnqlacde	Branched chain amino acid transport system 2 [BRNQI]	*Lactobacillus delbrueckii* [gram-positive bacterium]	Na⁺, Li⁺/valine isoleucine, valine
Brnqsalty	Branched chain amino acid transport system 2 [BRNQ, LIVII]	*Salmonella typhimurium* [gram-negative bacterium]	H⁺/isoleucine, leucine, valine

Cotransported ions are listed.

Phylogenetic tree

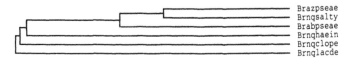

Brazpseae
Brnqsalty
Brabpseae
Brnqhaein
Brnqclope
Brnqlacde

Proposed orientation of BRNQ in the membrane

The model is based on predictions of membrane-spanning regions and α-helical content. The N-terminus of the protein is illustrated on the inside and is folded 12 times through the membrane. The predicted membrane-spanning helices are portrayed as rectangles. The numbers corresponding to the first and last residue of each membrane-spanning helix are boxed. Residues that are conserved in more than 75% of the aligned transporters (see below) are shown. Consensus residues indicated by an asterisk are not conserved in BRNQ.

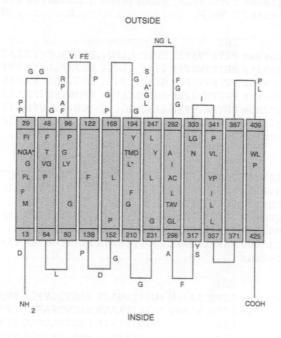

Physical and genetic characteristics

	AMINO ACIDS	MOL. WT	K_m
Brabpseae	437	45 282	
Brazpseae	437	45 274	Isoleucine: 12 μM [2]
Brnqclope	338	35 864	
Brnqhaein	436	47 038	
Brnqlacde	446	47 869	
Brnqsalty	439	46 534	

Multiple amino acid sequence alignments

```
          1                                                       50
Brazpseae .MNALKGRDI LALGFMTFAL FVGAGNIIFP PIVGLQSGPH VWLAALGFLI
Brnqsalty MTHQLKSRDI IALAFMTFAL FVGRGNIIFP PMVGLQAGEH VWTARIGFLI
Brabpseae .MTHLKGFDL LALGFMTFAL FLGAGNIIFP PSAGMAAGEH VWSAAFGFLL
Brnqhaein ...MFSRKDI IVLGMMIFAL FLGAGNIIFP PMEGFSSGQH WTSASLGFVL
```

405

```
Brnqclope ...MNKKKDI LVIGFALFSI FFGAGNLIFP PYIGLTSGSE WLISFLGFII
Brnqlacde MKEKLTHAES LTISSMLFGL FFGAGNLIFP AYLGEASGAN LWISLLGFLI
Consensus ........D. .....M.F.L F.GAGN.IFP P..G...G.. ......GF..
```

```
          51                                                 100
Brazpseae TAVGLPVITV IALAKV.GGS VDALSHPIGR YAGGLLAAVC YLAVGPLFAI
Brnqsalty TAVGLPVLTV VALAKV.GGG VDSLSTPIGK VAGLLLATVC YLAVGPLFAT
Brabpseae TGVGLPLLTV VALARV.GGG IGRLTQPIGR RAGVAFAIAV YLAIGPLFAT
Brnqhaein TGVLMPFITL VVVAIL.GRG .EELTKDLPK WAGTGFLVIL YLTIGSTFAM
Brnqclope SDVGIIFLSI VAVSK..AGS FQGVVGRAGK KFGITLEILM MLCLGPILVV
Brnqlacde TGVGLPLLAI ASLGMTRSEG LLDLSGRVSH KYSYFFTCLL YLTIGPFFAI
Consensus T.VG.P.... .......... ...L...... ..G....... YL..GP.FA.
```

```
          101                                                150
Brazpseae PRTATVSFEV GVVP..LLG. ESGTALFVYS LAYFLLALAI SLYPGRLLDT
Brnqsalty PRTATVSFEV GIAP..LTG. DSAMPLLIYS VVYFAIVILV SLYPGKLLDT
Brabpseae PRTAVVSFEM GVAP..FTG. DGGVPLLIYT VAYFSVVLFL VLNPGRLVDR
Brnqhaein PRITNVAYEM AWLPLGLTE. NNANVRFVFS LIFNLIAMGF MISPNTIISS
Brnqclope PRTAATTFEM SISPLL.... .GNVNPYVFP VIFFLIVFVL TIKPNKVMDI
Brnqlacde PRSFTVPFET GISALLPSGM AKSTGLFIFS LIFFAIMLFF SLRPGQIMDW
Consensus PR...V.FE. ...P...... .......... ...F...... ...P....D.
```

```
          151                                                200
Brazpseae VGRFLAPLKI LALAILGVAA FLWPAGPIGT AQPEYTQA.A .FSQGFVNGY
Brnqsalty VGNFLAPLKI IALVILSVAA IVWPAGPISN ALDAYQNA.A .FSNGFVNGY
Brabpseae VGKVITPVLL SALLVLGGAA IFAPAGEIGS SSGEYQSA.P .LVQGFLQGY
Brnqhaein VGKFMTPALL VLLIAVAITV FISPLSEIQA PSNAYENSHS .LLIGLTSGY
Brnqclope IGKVLTPLLL ISLAVLIIKG IINPIGDLEK V....NSGKL .FMTGITQGY
Brnqlacde IGKFLTPAFL LFFFFIMIMA LLHPLGNYHA VKPVGEYASA PLISGVLAGY
Consensus .G....P... ..L....... ...P.G.... .......... ....G...GY
```

```
          201                                                250
Brazpseae LTMDTLAALV FGIVIVNAIR SRGVQSPRLI TRYAIVAGLI AGVGLVLVYV
Brnqsalty LTMDDWVAMV FGIVIVNAAR SRGVTEARLL TRYTVWAGLM AGVGLTLLYL
Brabpseae LTMDTLGALV FGIVIATAIR DRGISDSRLV TRYSMIAGVI AATGLSLVYL
Brnqhaein QTMDVLAAIA FGGIVARALS AKNVTKTKDI VKYTISAGFV SVILLAGLYF
Brnqclope QTMDALGTGG IVALVMASFA SKGYKDKKEN RMLTIKSALI ACIGLAIVYG
Brnqlacde NTMDALAGLA FGIIVISSIR TFGVTKPEKV ASATLKTGVL TCLLMAVIYA
Consensus .TMD.L.... FG........ ..G....... .......... ..G.....L...Y.
```

```
          251                                                300
Brazpseae SLFRLGAGSH AIAADASNGA AVLHAYVQHT FGSLGSSFLA GLIALACLVT
Brnqsalty ALFRLGSDSA TLVDQSANGA AILHAYVQHT FGGAGSFLLA ALIFIACLVT
Brabpseae ALFYLGATSQ GIAGDAQNGV QILTAYVQQT FGVSGSLLLA VVITLACLTT
Brnqhaein SLFYLGATSA AVAEGATNGG QIFSRYVNVL FGSAGTWIMA GIIVLASLTT
Brnqclope GLTFLGATSS TLYDSSISQT TLLMNITNAI LGSTGTIMLA IVIGLACLTT
Brnqlacde ITALVGAQSR TALGLAANGG EALSQIARHY FPGLGAVIFA LMIFVACLKT
Consensus .L..LGA.S. ........NG. ..L....... FG..G....A ..I..ACL.T
```

```
          301                                                350
Brazpseae AVGLTCACAE YFCQR..LPL SYRSLVIILA GFSFIVSNLG LTKLIQVSIP
Brnqsalty AVGLTCACAE FFAQY..IPL SYRTLVFILG GFSMVVSNLG LSHLIQISIP
Brabpseae AVGLITACGE FFSDL..LPV SYKTVVIVFS LFSLLVANQG LTQLISLSVP
Brnqhaein LVGVTSASAD YFSKFS.VRF SYPFWAALFT AMTITVSQYG LTDLLRITIP
```

```
Brnqclope AVGLTSVTAK YFEDVSNKKL KYKYIVIAIC VFSALSSNLG VDKIIEIAVP
Brnqlacde AIGLITACSE TFAEMFPKTL SYNMWAIIFS LLAFGIANVG LTTIISFSLP
Consensus AVGL..A... .F........ SY........ ......N.G L...I....P
```

```
          351                                                400
Brazpseae VLTAIYPPCI VLVALSFCIG LWHSAT...R ILAPVMLVSL AFGVLDALKA
Brnqsalty VLTAIYPPCI ALVVLSFTRS WWHNST...R IIAPAMFISL LFGILDGIKA
Brabpseae VLVGLYPLAI VLIALSLFDR LWVSAP...R VFVPVMIVAL LFGIVDGLGA
Brnqhaein ALLLIYPVAI VLVLLQFLRK KLPSIK...F TYNSTLLVTV CFSLCDSLNN
Brnqclope VLS....... .......... .......... .......... ..........
Brnqlacde VLMLLYPLAI SLILLALTSK LFDFKQVDYQ IMTAVTFLCA LGDFFKALPA
Consensus VL...YP..I .L..L..... .......... .......... ..........
```

```
          401                                                450
Brazpseae AG.LGQDFPQ WLLHLPLAEQ GLAWLIPSVA TLAACSLVDR LLGKPAQVAA
Brnqsalty SA.FGDMLPA WSQRLPLAEQ GLAWLMPTVV MVILAIIWDR AAGRQVTSSA
Brabpseae AK.LNGWVPD VFAKLPLADQ SLGWLLPVSI ALVLAVVCDR LLGKPREAVA
Brnqhaein VKMLPESINS LLKHFPLSSE GMAWLVPTLV MLVASIFIGK ALHKTHS...
Brnqclope .......... .......... .......... .......... ..........
Brnqlacde GMQVKAVTGL YGHVLPLYQD GLGWLVPVTV IFAILAIKGV ISKKRA....
Consensus .......... .......PL.. ...WL.P... .......... ..........
```

```
          451
Brazpseae .
Brnqsalty H
Brabpseae .
Brnqhaein .
Brnqclope .
Brnqlacde .
Consensus .
```

Residues listed in the consensus sequence are present in at least 75% of the aligned transporter sequences.

Database accession numbers

	SWISSPROT	PIR	EMBL/GENBANK
Brabpseae	P19072	S11497	X51634
Brazpseae	P25185	A38534	D90222
Brnqclope		D49784	
Brnqhaein		D64056	L42023
Brnqlacde		Z48676	
Brnqsalty	P14931	JQ0007	D00332

References
[1] Reizer, J. et al. (1994) Biochim. Biophys. Acta 1197, 133–166.
[2] Hoshino, K. et al. (1991) J. Bacteriol. 173, 1855–1861.

Summary

Transporters of the Na⁺/citrate symporter family, the example of which is the CITN citrate transporter of *Klebsiella pneumoniae* (Citnklepn), mediate symport (Na⁺-coupled substrate uptake) of citrate. Known members of the family occur in gram-negative bacteria.

Statistical analysis reveals no relationship between the amino acid sequences of the Na⁺/citrate symporter family and any other family of transporters. They are predicted to form 12 membrane-spanning helices by the hydropathy of their amino acid sequences.

Several amino acid sequence motifs are highly conserved in the Na⁺/citrate symporter family.

Nomenclature, biological sources and substrates

CODE	DESCRIPTION [SYNONYMS]	ORGANISM [COMMON NAMES]	SUBSTRATE(S)
Citnklepn	Citrate-Na⁺ symporter [CITS, CITN]	*Klebsiella pneumoniae* [gram-negative bacterium]	Na⁺/citrate
Citnlacla	Citrate-Na⁺ symporter [CITP, CITN]	*Lactococcus lactis* [gram-positive bacteria]	Na⁺/citrate
Citnsaldu	Citrate-Na⁺ symporter [CITC, CITN]	*Salmonella dublin* [gram-negative bacterium]	Na⁺/citrate
Citnsalpu	Citrate-Na⁺ symporter [CITC, CITN]	*Salmonella pullorum* [gram-negative bacterium]	Na⁺/citrate
Citpleula	Citrate-Na⁺ symporter [CITS, CITN]	*Leuconostoc lactis* [gram-positive bacterium]	Na⁺/citrate
Citpstrbo	Citrate-Na⁺ symporter [CITS, CITN]	*Streptococcus bovis* [gram-positive bacterium]	Na⁺/citrate

Cotransported ions are listed.

Phylogenetic tree

Proteins listed subsequently in italics are at least 90% identical to the paired transporters listed in parenthesis and therefore are not included in the phylogenetic tree: *Citnsalpu, Citnsaldu* (Citnklepn); *Citpleula* (Citnlacla).

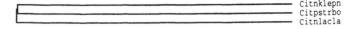

```
                                              Citnklepn
                                              Citpstrbo
                                              Citnlacla
```

Proposed orientation of CITN in the membrane

The model is based on predictions of membrane-spanning regions and α-helical content. The N-terminus of the protein is illustrated on the inside and is folded 12 times through the membrane. The predicted membrane-spanning helices are portrayed as rectangles. The numbers corresponding to the first and last residue of each membrane-spanning helix are boxed. Residues that are conserved in more than 75% of the aligned transporters (see below) are shown.

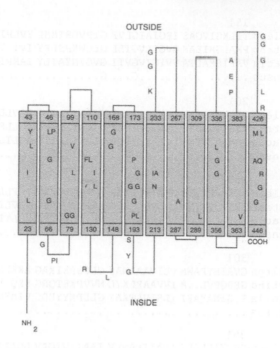

Physical and genetic characteristics

	AMINO ACIDS	MOL. WT
Citnklepn	446	47557
Citnlacla	442	46629
Citnsaldu	446	47591
Citnsalpu	446	47621
Citpleula	441	46533
Citpstrbo	441	47221

Multiple amino acid sequence alignments

```
          1                                                   50
Citnklepn ....MTNMSQ PPATEKKGVS DLLGFKIFGM PLPLYAFALI TLLLSHFYNA
Citpstrbo ...MEKKLPA TAANETDWRN KLTKTRIGSV TLPVYLVTAS IILVTALLEQ
Citnlacla MMNHPHSSHI GTTNVKEEIG KLDRIRISGI GLIAYAFMAV LLIIAISTKT
Consensus ..................... L.....I.... L..Y..............

          51                                                  100
Citnklepn LPTDIVGGFA IMFIIGAIFG EIGKRLPIFN KYIGGAPVMI FLVAAYFVYA
Citpstrbo LPVNMLGGFA VILTMGWLLG TIGGNIPIL. KHFGGPAILS LLVPSIMVFF
Citnlacla LPNTMIGAIF ALVLMGHVFY YLGAHLPIFR SYLGGGSVFT ILLTAILVAT
Consensus LP....G... .....G.... ..G...PI.. ...GG...... .L.....V..

          101                                                 150
Citnklepn GIFTQKEIDA ISNVMDKSNF LNLFIAVLIT GAILSVNRRL LLKSLLGYIP
Citpstrbo NLLNQNVLDS TDILMKQANF LYFYIACLVC GSILGMNRKI LVQGLMRMIV
Citnlacla NVIPKYVVTT ASGFINGMDF LGLYIVSLIA SSLFKMDRKM LLKAAVRFLP
Consensus ..................... F L...I..L.. ........R.. L......
```

```
           151                                                        200
Citnklepn TILMGIVGAS IFGIAIGLVF GIPVDRIMML YVLPIMGGGN GAGAVPLSEI
Citpstrbo PMALGMILAM GVGTLVGTLL GLGWKHSLFY IVTPVLAGGI GEGILPLSLG
Citnlacla VAFISMALTA VVIGIVGVII GVGFNYAILY IAMPIMAGGV GAGIVPLSGI
Consensus .......... ......G... G.......... ...P...GG. G.G..PLS..

           201                                                        250
Citnklepn YHSVTGRSRE EYYSTAIAIL TIANIFAIVF AAVLDIIGKK HTWLSGEGEL
Citpstrbo YSAITGLPSE QLVGQLIPAT IIGNFFAIMC SGLLSRLGEK RPELSGQGQL
Citnlacla YAHAMGVGSA GILSKLFPTV ILGNLLAIIS AGLISRIF.K DSKGNGHGEI
Consensus Y....G.... .......... ...N..AI.. .........K .....G.G..

           251                                                        300
Citnklepn VRKASFKVEE DEKTGQITHR ETAVGLVLST TCFLLAY..V VAKKILPSIG
Citpstrbo IKIT....NS DDLSDALEED KAPIDVKLMG AGVLIACTLF ITGGLLQHLT
Citnlacla LR......GE REKSAAAEEI KP..DYVQLG VGLIIAVMFF MIGTMLNKVF
Consensus .......... .......... .......... .....A.... .....L....

           301                                                        350
Citnklepn GVAIHYFAWM VLIVAALNAS GLCSPEIKAG AKRLSDFFSK QLLWVLMVGV
Citpstrbo GFPGPVL..M IVVAAFLKYL NVVPKETQRG SKQLYKFISG NFTFPLMVGL
Citnlacla P.GINAYAFI ILSIVLTKAF GLLPKYYEDS VIMFGQVIVK NMTHALLAGV
Consensus .......... .......... .......... .......... ....L..G.

           351                                                        400
Citnklepn GVCYTDLQEI INAITFANVV IAAIIVIGAV LGAAIGGWLM GFFPIESAIT
Citpstrbo GMLYIPLKDV VGMLSWQYFV VVISVVFTVI ATGFFVSRFM NMNPVEAAIV
Citnlacla GLSLLDMHVL LAALSWQFVV LCLVSIVAIS LISATLGKLF GLYPVEAAIT
Consensus G......... .........V .......... .......... ...P.E.A..

           401                                                        450
Citnklepn AGLCMANRGG SGDLEVLSAC NRMNLISYAQ ISSRLGGGIV LVIASIVFGM
Citpstrbo SA.CQSGMGG TGDVAILSTA NRMTLMPFAQ VATRLGGAIT VITMTAIFRM
Citnlacla AGLANNSMGG TGNVAVLAAS ERMNLIAFAQ MGNRIGGALI LVVAGILVTF
Consensus ........GG .G....L... .RM.L...AQ ...R.GG... ..........

           451
Citnklepn MI
Citpstrbo LF
Citnlacla MK
Consensus ..
```

Proteins listed subsequently in italics are at least 90% identical to the paired transporters listed in parenthesis and therefore are not included in the alignments: *Citnsalpu*, *Citnsaldu* (Citnklepn); *Citpleula* (Citnlacla). Residues listed in the consensus sequence are present in at least 75% of the aligned transporter sequences.

Database accession numbers

	SWISSPROT	PIR	EMBL/GENBANK
Citnklepn	P31602	A38244	M83146
Citnlacla	P21608	A36136	M58694
Citnsaldu	P31603	B42661	D10258
Citnsalpu	P31604	A42661	D10257
Citpleula			U28212
Citpstrbo		U35658	

Summary

Transporters of the Na⁺/alanine-glycine symporter family, the example of which is the ACP alanine transporter of the thermophilic bacterium PS-3 (Alcpthep3), mediate symport (H⁺- and Na⁺-coupled substrate uptake) of alanine and glycine. The two known members of the family occur only in gram-negative bacteria.

Statistical analysis reveals no apparent relationship between the amino acid sequences of the Na⁺/alanine-glycine symporter family and any other family of transporters. They are predicted to contain eight membrane-spanning helices by the hydropathy of their amino acid sequences.

Relatively few amino acid sequence motifs are conserved in the two proteins.

Nomenclature, biological sources and substrates

CODE	DESCRIPTION [SYNONYMS]	ORGANISM [COMMON NAMES]	SUBSTRATE(S)
Dagaaltha	Glycine-alanine-Na⁺ symporter [DAGA]	Alteromonas haloplanktis [gram-negative bacterium]	Na⁺/gylcine, alanine
Alcpthep3	Alanine-Na⁺, H⁺ symporter [ACP]	PS-3: unclassified [thermophilic bacterium]	Na⁺, H⁺/alanine

Cotransported ions are listed.

Proposed orientation of ACP in the membrane

The model is based on predictions of membrane-spanning regions and α-helical content. The N-terminus of the protein is illustrated on the inside and is folded

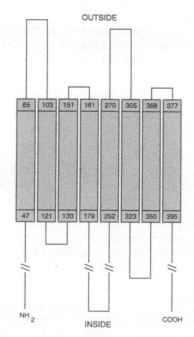

eight times through the membrane. The predicted membrane-spanning helices are portrayed as rectangles. The numbers corresponding to the first and last residue of each membrane-spanning helix are boxed.

Physical and genetic characteristics

	AMINO ACIDS	MOL. WT
Alcpthep3	445	47 804
Dagaaltha	542	59 023

Multiple amino acid sequence alignments

```
          1                                                    50
Alcpthep3 ..........  ..........  ..........  ......MIRL  VTMGKSSEAG
Dagaaltha MLGGAVWFPY  VLLGVGLFFT  IYLKFPQIRY  FKHACQVVSG  KFDKKDTEGD
Consensus ..........  ..........  ..........  ..........  ..K..E..

          51                                                  100
Alcpthep3 VSSFQALTMS  LSGRIGVGNV  AGTATGIAYG  GPGAVFWMWV  ITFIGAATAY
Dagaaltha TTHFQALATA  LSGTVGTGNI  GGVALAISIG  GPAALFWMWM  TAFFGMTTKF
Consensus ...FQAL...  LSG..G.GN.  .G.A..I..G  GP.A.FWMW.  ...F.G..T..

          101                                                 150
Alcpthep3 VESTWRKFIK  RNKTDNTVAV  RRSTLKKALA  GNGLRCSRAA  IILSMAVLMP
Dagaaltha VEVTLSHKYR  EKTEDGTM..  ...SGGPMYYM  DKRLNMKWLA  ILFAVATVIS
Consensus VE.T......  ....D.T...  ...S......  ...L.....A  I....A....

          151                                                 200
Alcpthep3 GI......QA  NSIADSFSNA  FGIPKLVTGI  FVIAVLGFTI  FGGVKRIAKT
Dagaaltha SFGTGSLPQI  NNIAQGMEAT  FGFAPMATGA  VLSILLALVI  LGGIKRIAAI
Consensus ........Q.  N.IA......  FG.....TG.  .....L...I  .GG.KRIA..

          201                                                 250
Alcpthep3 AEIVVPFMAV  GYLFVAIAII  AANIEKVPDV  FGLIFKSAFG  ADQVFGGILG
Dagaaltha TSRVVPLMAA  IYIIGALAVI  FYNAENIGPS  FSAVFMDAFS  GSAAAGGFLG
Consensus ...VVP.MA.  .....A.A.I  ..N.E.....  F...F..AF.  .....GG.LG

          251                                                 300
Alcpthep3 S....AVMWG  VKRGLYANEA  GQGTGAHPAA  AAEVSHPAKQ  GLVQAFSIYL
Dagaaltha ASFAYAFNRG  VNRGLFSNEA  GQGSAPIAHA  SAKADEPVSE  GIVSILEPFI
Consensus .....A...G  V.RGL..NEA  GQG......A  .A....P...  G.V......

          301                                                 350
Alcpthep3 DVFLVVTATA  LMIL......  ..........  ..........  ..........
Dagaaltha DTIIICTLTG  LVILSSGVWN  EKFQTHFERS  AMSIIKGDYT  EENQTQREDL
Consensus D.....T.T.  L.IL......  ..........  ..........  ..........

          351                                                 400
Alcpthep3 ..........  ...FTGQYNV  INEKTGET..  ..........  ..........
Dagaaltha YKYLNGQKSN  IETFTGNIEV  VNGEALSTGF  TVLHSRSIAE  DVRFGITEKH
Consensus ..........  ...FTG...V  .N.....T..  ..........  ..........
```

```
          401                                                    450
Alcpthep3 ....IVEHLK GVEPGAGY.. ........... ..TQAAVDTL FPGFGSAFIA
Dagaaltha KYTGVVEVID GMPTDDSISL VGKSLVHSAE LTTKAFKRGY FGDSGQYIVS
Consensus .....VE... G.......... .......... ..T.A..... F...G.....

          451                                                    500
Alcpthep3 IALFFFAFTT MYAYYYIAET NLAYLVRSEK RGTAFFALKL VFLAATFYGT
Dagaaltha IGLLLFAFST AIAWSYYGDR AMIYLLGHR. ...SVMPYRV FYVAAFFWAS
Consensus I.L..FAF.T ..A..Y.... ...YL..... .......... ...AA.F...

          501                                                    550
Alcpthep3 VKTATTAWAM GDIGLGIMVW LNLIAILLLF KPAYMALKDY EEQLKQGKDP
Dagaaltha FADTTLVWKL AAVAIVVMTL PNLIGIMLLR KEMKESVDDY WVKFKKDNEK
Consensus ....T..W.. .......M.. .NLI.I.LL. K.......DY ....K.....

          551                         580
Alcpthep3 EFNASKYGIK NAKFWENGYK RWEEKKGKAL
Dagaaltha .......... .......... ..........
Consensus .......... .......... ..........
```

Residues listed in the consensus sequence are present in both of the transporter sequences.

Database accession numbers

	SWISSPROT	PIR	EMBL/GENBANK
Alcpthep3	P30145	S27733; A45111	D12512
Dagaaltha	P30144	S25276	M59081

Summary

Transporters of the Na$^+$/neurotransmitter symporter family, the example of which is the NET1 noradrenalin-Na$^+$ symporter of humans (Ntnohomsa), mediate symport (Na$^+$-coupled substrate uptake) of several structurally dissimilar neurotransmitters, including norepinephrine, 4-aminobutyrate (GABA), serotonin, creatine and dopamine [1-4]. Members of the Na$^+$/neurotransmitter symporter family also serve as uniporters, ion and water channels [4]. For example, neurotransmitter cotransporters such as NET1 and GAT1 contain ligand-gated ion channels that mediate Cl$^-$ conductance in parallel with Na$^+$-coupled substrate uptake, while in the absence of substrates they transport ions in an uncoupled "leak" mode. Members of the family have a broad biological distribution that includes both invertebrates and vertebrates.

Statistical analysis reveals no significant similarity between the amino acid sequences of the NET1 family and any other family of transporters. However, the similarity between the kinetics of several families of Na$^+$-driven and H$^+$-driven transporters suggests a common mechanism of action, despite the lack of amino acid sequence homology [4]. Members of the Na$^+$/neurotransmitter symporter family are predicted to contain 12 membrane-spanning helices by the hydropathy of their amino acid sequences and reactions with peptide specific antibodies [5]. Members of the Na$^+$/neurotransmitter symporter family are glycosylated.

Several amino acid sequence motifs are highly conserved in the Na$^+$/neurotransmitter symporter family, including motifs necessary for function by the criterion of site-directed mutagenesis [2].

Nomenclature, biological sources and substrates

CODE	DESCRIPTION [SYNONYMS]	ORGANISM [COMMON NAMES]	SUBSTRATE(S)
Gat1torca	Na$^+$/Cl$^-$-dependent GABA transporter 1 [TGAT, GAT1]	Torpedo californica [ray]	Na$^+$/GABA
Ntbecanfa	Na$^+$/Cl$^-$-dependent betaine transporter [NTBE]	Canis familiaris [dog]	Na$^+$/betaine
Ntchratno	Choline-Na$^+$ symporter [CHOT1]	Rattus norvegicus [rat]	Na$^+$/choline
Ntcrhomsa	Na$^+$/Cl$^-$-dependent creatine transporter [SLC6a8, NTCR]	Homo sapiens [human]	Na$^+$/creatine
Ntcrorycu	Na$^+$/Cl$^-$-dependent creatine transporter [SLC6a8, NTCR]	Oryctolagus cuniculus [rabbit]	Na$^+$/creatine
Ntcrtorma	Na$^+$/Cl$^-$-dependent creatine transporter [NTCR]	Torpedo marmorata [ray]	Na$^+$/creatine
Ntdobosta	Dopamine-Na$^+$ symporter [DA, DAT, NTDO]	Bos taurus [cow]	Na$^+$/dopamine
Ntdohomsa	Dopamine-Na$^+$ symporter [DAT, DAT1, SLC6a3, NTDO, DA transporter]	Homo sapiens [human]	Na$^+$/dopamine

CODE	DESCRIPTION [SYNONYMS]	ORGANISM [COMMON NAMES]	SUBSTRATE(S)
Ntdoratno	Dopamine-Na⁺ symporter [DAT, DAT1, SLC6a3, NTDO, DA transporter]	*Rattus norvegicus* [rat]	Na⁺/dopamine
Ntg1homsa	Na⁺/Cl⁻-dependent GABA transporter 1 [NTG1, GABT1, GAT1, SLC6a1]	*Homo sapiens* [human]	Na⁺/GABA
Ntg1musmu	Na⁺/Cl⁻-dependent GABA transporter 1 [SLC6a1, GABT1, GAT1, NTG1]	*Mus musculus* [mouse]	Na⁺/GABA
Ntg1ratno	Na⁺/Cl⁻-dependent GABA transporter 1 [SLC6a1, GABT1, GAT1, NTG1]	*Rattus norvegicus* [rat]	Na⁺/GABA
Ntg2musmu	Na⁺/Cl⁻-dependent GABA transporter 2 [GABT2, GAT2, NTG2]	*Mus musculus* [mouse]	Na⁺/GABA, β-alanine, taurine
Ntg2ratno	Na⁺/Cl⁻-dependent GABA transporter 2 [GABT2, GAT2, NTG2]	*Rattus norvegicus* [rat]	Na⁺/GABA, β-alanine, taurine
Ntg3musmu	Na⁺/Cl⁻-dependent GABA transporter 3 [GATB, GABT3, GAT3, NTG3]	*Mus musculus* [mouse]	Na⁺/GABA, taurine, β-alanine
Ntg3ratno	Na⁺/Cl⁻-dependent GABA transporter 3 [NTG3, GABT3, GAT3]	*Rattus norvegicus* [rat]	Na⁺/GABA
Ntg4musmu	Na⁺/Cl⁻-dependent GABA transporter 4 [GABT4, GAT4, NTG4]	*Mus musculus* [mouse]	Na⁺/GABA
Ntgmanse	Na⁺/Cl⁻-dependent GABA transporter 1 [NTG]	*Manduca sexta* [tobacco horned worm]	Na⁺/GABA
Ntgtorma	Na⁺/Cl⁻-dependent GABA, β-alanine transporter 1 [NTGT]	*Torpedo marmorata* [ray]	Na⁺/GABA, β-alanine
Ntnobosta	Norepinephrine-Na⁺ symporter [NAT1, NET1, SLC6a2, NTNO]	*Bos taurus* [cow]	Na⁺/norepinephrine
Ntnohomsa	Norepinephrine-Na⁺ symporter [NAT1, NET1, SLC6a2, NTNO]	*Homo sapiens* [human]	Na⁺/norepinephrine
Ntprratno	Proline-Na⁺ symporter [SLC6a7, NTPR]	*Rattus norvegicus* [rat]	Na⁺/proline
Nts1ratno	Serotonin-Na⁺ symporter [SHTT, NTS1]	*Rattus norvegicus* [rat]	Na⁺/serotonin
Nts2ratno	Serotonin-Na⁺ symporter [SHT, NTS2]	*Rattus norvegicus* [rat]	Na⁺/serotonin
Ntsedrome	Serotonin-Na⁺ symporter [NTS]	*Drosophila melanogaster* [fruit fly]	Na⁺/serotonin
Ntsehomsa	Serotonin-Na⁺ symporter [HTT, SLC6a4, NTSE]	*Homo sapiens* [human]	Na⁺/serotonin
Ntt4ratno	Na⁺/Cl⁻-dependent transporter [NTT4]	*Rattus norvegicus* [rat]	Unknown
Ntt7ratno	Na⁺/Cl⁻-dependent transporter [NTT7]	*Rattus norvegicus* [rat]	Unknown
Nttacanfa	Na⁺/Cl⁻-dependent taurine transporter [NTTA]	*Canis familiaris* [dog]	Na⁺/taurine
Nttahomsa	Na⁺/Cl⁻-dependent taurine transporter [SLC6a6, NTTA]	*Homo sapiens* [human]	Na⁺/taurine

CODE	DESCRIPTION [SYNONYMS]	ORGANISM [COMMON NAMES]	SUBSTRATE(S)
Nttamusco	Na+/Cl−-dependent taurine transporter [SLC6a6, NTTA]	Mus cookii [mouse]	Na+/taurine
Nttaratno	Na+/Cl−-dependent taurine transporter [SLC6a6, NTTA]	Rattus norvegicus [rat]	Na+/taurine
Rosiratno	Renal osmotic stress-induced Na+/Cl−-dependent organic acid cotransporter [ROSIT]	Rattus norvegicus [rat]	Na+/organic acids

Cotransported ions are listed.
Abbreviations: GABA; 4-aminobutyric acid.

Phylogenetic tree

Proteins listed subsequently in italics are at least 90% identical to the paired transporters listed in parenthesis and therefore are not included in the phylogenetic tree: *Ntcrorycu*, *Ntcrhomsa* (Ntchratno); *Ntg1ratno*, *Ntg1musmu* (Ntg1homsa); *Ntg2ratno* (Ntg2musmu); *Ntg3ratno* (Ntg3musmu); *Ntnobosta* (Ntnohomsa); *Nts1ratno*, *Nts2ratno* (Ntsehomsa); *Nttacanfa*, *Nttamusco*, *Nttaratno* (Nttahomsa); *Ntdoratno* (Ntdohomsa).

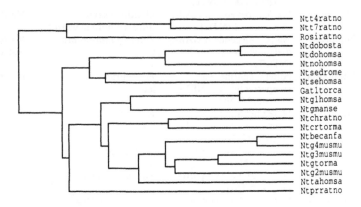

```
Ntt4ratno
Ntt7ratno
Rosiratno
Ntdobosta
Ntdohomsa
Ntnohomsa
Ntsedrome
Ntsehomsa
Gat1torca
Ntg1homsa
Ntgmanse
Ntchratno
Ntcrtorma
Ntbecanfa
Ntg4musmu
Ntg3musmu
Ntgtorma
Ntg2musmu
Nttahomsa
Ntprratno
```

Proposed orientation of NET1 in the membrane

The model is based on predictions of membrane-spanning regions and α-helical content. The N-terminus of the protein is illustrated on the inside and is folded 12 times through the membrane [5]. The predicted membrane-spanning helices are portrayed as rectangles. The numbers corresponding to the first and last residue of each membrane-spanning helix are boxed. Residues that are conserved in more than 75% of the aligned transporters (see below) are shown. Consensus residues indicated by an asterisk are not conserved in NET1.

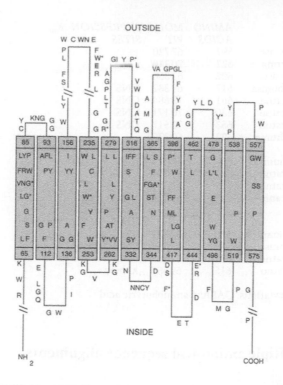

Physical and genetic characteristics

	AMINO ACIDS	MOL. WT	EXPRESSION SITES	K_m	CHROMOSOMAL LOCUS
Gat1torca	598	67 219			
Ntbecanfa	614	69 291	kidney		
Ntchratno	635	70 631	CNS, heart		
Ntcrhomsa	635	70 676	kidney, heart, muscle, brain		Xq28
Ntcrorycu	635	70 483	kidney, heart, muscle, brain		
Ntcrtorma	611	68 098			
Ntdobosta	693	75 691	brain		
Ntdohomsa	620	68 494	brain		5p15.3
Ntdoratno	619	68 746	brain		
Ntg1homsa	599	67 014	brain		3p24–p25
Ntg1musmu	598	66 841	brain		
Ntg1ratno	599	67 001	brain		
Ntg2musmu	602	68 284	brain, liver, kidney	GABA: 18 nM β-Alanine: 28 nM Taurine: 540 nM [2]	
Ntg2ratno	602	68 262	brain, retina	GABA: 8 μM [6]	
Ntg3musmu	627	69 888	brain	GABA: 0.8 nM β-Alanine: 99 nM Taurine: 1.4 μM [2]	
Ntg3ratno	627	69 946	brain, retina	GABA: 12 μM [6]	
Ntg4musmu	614	69 613	brain, liver, kidney		

	AMINO ACIDS	MOL. WT	EXPRESSION SITES	K_m	CHROMOSOMAL LOCUS
Ntgmanse	597	67 720			
Ntgtorma	622	70 248			
Ntnobosta	602	67 350			
Ntnohomsa	617	69 332	CNS		16q12.2
Ntprratno	661	73 684	CNS		
Nts1ratno	630	70 171	CNS		
Nts2ratno	653	72 517	CNS		
Ntsedrome	622	69 325		Serotonin: 500 nM [7]	
Ntsehomsa	630	70 324	CNS		17q11.1–q12
Ntt4ratno	727	81 055	CNS		
Ntt7ratno	729	81 596			
Nttacanfa	620	69 728	kidney, brain, liver, heart, ileum		
Nttahomsa	620	69 829			3p25–q24
Nttamusco	590	65 868	brain, kidney		
Nttaratno	621	69 868	brain		
Rosiratno	615	69 556	kidney		

Abbreviations: GABA: 4-aminobutyric acid.

Multiple amino acid sequence alignments

```
                1                                                    50
Ntt4ratno .................. .MPKNSKVTQ REHSNEHVTE SVADLLALEE
Ntt7ratno .................. .MPKNSKVVK RDL.DDDVIE SVKDLLSNED
Rosiratno .................. ..........
Ntdobosta .................. ...MSEGRCS VAHMSSVVAP AKEANAMGPK
Ntdohomsa .................. ...MSKSKCS VGLMSSVVAP AKEPNAVGPK
Ntnohomsa .................. .MLLARMNPQ VQPENNGADT GPEQPLRARK
Ntsedrome ........M DRSGSSDFAG AAATTGRSNP APWSDDKESP NNEDDSNEDD
Ntsehomsa METTPLNSQK QLSACEDGED CQENGVLQKV VPTPGDK..V ESGQISNGYS
Gat1torca .................. ..........M ATNGAKTPDG
Ntg1homsa .................. ..........M ATNGSKVADG
Ntgmanse .................. .........ME TKNDSRSDDI
Ntchratno .................. MAKKSAENGI YSVSGDEKKG
Ntcrtorma .................. .......MPSRAVRRC
Ntbecanfa .................. .......MDR
Ntg4musmu .................. .......MDR
Ntg3musmu .................. .......MT AEQALPLGNG
Ntgtorma .................. .......MR AEKAIPIING
Ntg2musmu .................. ..........
Nttahomsa .................. ....MATKEKLQ
Ntprratno .................. .......MKKL
Consensus .................. ..........
```

```
                51                                                  100
Ntt4ratno PVD..YKQSV LNVAGETGGK QKVAEEELDA EDRPAWNSKL QYILAQIGFS
Ntt7ratno SVEDVSKKSE LIVDVQEEKD TDAEDGSEVD DERPAWNSKL QYILAQVGFS
Rosiratno ............ ......MAQA SGMDPLVDIE DERPKWDNKL QYLLSCIGFA
Ntdobosta AVELVLVKEQ NGVQLTNSTL LNPPQSPTEA QDRETWSKKA DFLLSVIGFA
```

```
Ntdohomsa EVELILVKEQ NGVQLTSSTL TNPRQSPVEA QDRETWGKKI DFLLSVIGFA
Ntnohomsa TAELLVVKER NGVQ....CL LAPRDG..DA QPRETWGKKI DFLLSVVGFA
Ntsedrome GDHTTPAKVT DPLAPKLANN ERILVVSVTE RTRETWGQKA EFLLAVIGFA
Ntsehomsa AVPSPGA..G DDTRHSIPAT TTTLVAELHQ GERETWGKKV DFLLSVIGYA
Gat1torca QISTELHDAP VSNDKPKTLV VKVQ.KTRKI PEREKWGGRY DFLLSCVGYA
Ntg1homsa QISTEVSEAP VANDKPKTLV VKVQKKAADL PDRDTWKGRF DFLMSCVGYA
Ntgmanse  ELS.....AQ GSGNKPSDVA VK.....SNL PERGSWASKL DFILSVIGLA
Ntchratno PLIVSGPDGA PSKGDGPA.G LGAPSSRLAV PPRETWTRQM DFIMSCVGFA
Ntcrtorma PGHLCKEMRA PRRAQPPDVP AGEPGSRV.. ...TWSRQM DFIMSCVGFA
Ntbecanfa KVAVPEDGPP VVSWLPEEGE KLDQEGEDQV KDRGQWTNKM EFVLSVAGEI
Ntg4musmu KVAVHEDGYP VVSWVPEEGE MMDQKGKDQV KDRGQWTNKM EFVLSVAGEI
Ntg3musmu KAAEEARGSE TLGGGGGGAA GTREARDKAV HERGHWNNKV EFVLSVAGEI
Ntgtorma  KPE......D TMDIEASNVN LVR.TNDKRM SERGHWNNKI EFVLSVAGEI
Ntg2musmu .MENRASGTT SNGETKPVCP AMEKVEEDGT LEREHWNNKM EFVLSVAGEI
Nttahomsa CLKDFHKDMV KPSPGKSPGT RPEDEAEGKP PQREKWSSKI DFVLSVAGGF
Ntprratno QEAHLRKPVT PDLLMTPSDQ GDVDLDVDFA ADRGNWTGKL DFLLSCIGYC
Consensus .......... .......... .......... ..R..W..K. .F.LS..G..

          101                                           150
Ntt4ratno VGLGNIWRFP YLCQKNGGGA YLVPYLVLLI IIGIPLFFLE LAVGQRIRRG
Ntt7ratno VGLGNVWRFP YLCQKNGGGA YLLPYLILLL VIGIPLFFLE LSVGQRIRRG
Rosiratno VGLGNIWRFP YLCHTHGGGA FLIPYFIALV FEGIPLFYIE LAIGQRLRRG
Ntdobosta VDLANVWRFP YLCYKNGGGA FLVPYLFFMV VAGVPLFYME LALGQFNREG
Ntdohomsa VDLANVWRFP YLCYKNGGGA FLVPYLLFMV IAGMPLFYME LALGQFNREG
Ntnohomsa VDLANVWRFP YLCYKNGGGA FLIPYTLFLI IAGMPLFYME LALGQYNREG
Ntsedrome VDLGNVWRFP YICYQNGGGA FLVPYCLFLI FGGLPLFYME LALGQFHRCG
Ntsehomsa VDLGNVWRFP YICYQNGGGA FLLPYTIMAI FGGIPLFYME LALGQYHRNG
Gat1torca IGLGNVWRFP YLCGKNGGGA FLIPYFMTLI FAGMPIFLLE CSLGQYTSVG
Ntg1homsa IGLGNVWRFP YLCGKNGGGA FLIPYFLTLI FAGVPLFLLE CSLGQYTSIG
Ntgmanse  IGLGNVWRFP YLCYKNGGGA FLIPYFLTLF LAGIPMFFME LAMGQMLTIG
Ntchratno VGLGNVWRFP YLCYKNGGGV FLIPYVLIAL VGGIPIFFLE ISLGQFMKAG
Ntcrtorma VGLGNVWRFP YLCYKNGGGV FLIPYLLVAV FGGIPIFFLE ISLGQFMKAG
Ntbecanfa IGLGNVWRFP YLCYKNGGGA FFIPYFIFFF TCGIPVFFLE VALGQYTSQG
Ntg4musmu IGLGNVWRFP YLCYKNGGGA FFIPYFIFFF SCGIPVFFLE VALGQYSSQG
Ntg3musmu IGLGNVWRFP YLCYKNGGGA FLIPYVVFFI CCGIPVFFLE TALGQFTSEG
Ntgtorma  IGLGNVWRFP YLCYKNGGGA FLIPYVIFFI GCGIPVFFLE TALGQYTSEG
Ntg2musmu IGLGNVWRFP YLCYKNGGGA FFIPYLIFLF TCGIPVFFLE TALGQYTNQG
Nttahomsa VGLGNVWRFP YLCYKNGGGA FLIPYFIFLF GSGLPVFFLE IIIGQYTSEG
Ntprratno VGLGNVWRFP YRAYTNGGGA FLVPYFLMLA ICGIPLFFLE LSLGQFSSLG
Consensus .GLGNVWRFP YLCYKNGGGA FL.PY..... .G.P.F..E ..LGQ....G

          151                                           200
Ntt4ratno SIGVWHYVCP RLGGIGFSSC IVCLFVGLYY NVIIGWSVFY FFKSFQYPLP
Ntt7ratno SIGVWNYISP KLGGIGFASC VVCYFVALYY NVIIGWTLFY FSQSFQQPLP
Rosiratno SIGVWKTISP YLGGVGLGCF SVSFLVSLYY NTILLWVLWF FLNSFQHPLP
Ntdobosta AAGVW.KICP ILRGVGYTAI LISLYIGFFY NVIIAWALHY LLSSFTTELP
Ntdohomsa AAGVW.KICP ILKGVGFTVI LISLYGFFY NVIIAWALHY LFSSFTTELP
Ntnohomsa AATVW.KICP FFKGVGYAVI LIALYVGFYY NVIIAWSLYY LFSSFTLNLP
Ntsedrome CLSIWKRICP ALKGVGYAIC LIDIYMGMYY NTIIGWAVYY LFASFTSKLP
Ntsehomsa CISIWRKICP IFKGIGYAIC IIAFYIASYY NTIMAWALYY LISSFTDQLP
Gat1torca GLGIW.RLAP MFKGVGLAAA VLSFWLNIYY VVIIAWAIYY LYNSFTSELP
Ntg1homsa GLGVW.KLAP MFKGVGLAAA VLSFWLNIYY IVIISWAIYY LYNSFTTTLP
Ntgmanse  GLGVF.KIAP IFKGIGYAAA VMSCWMNVYY IVILAWAIFY FFMSMRSDVP
Ntchratno SINVW.NICP LFKGLGYASM VIVFYCNTYY IMVLAWGFYY LVKSFTTTLP
```

419

```
Ntcrtorma GINAW.NIAP LFKGLGYASM VIVFFCNTYY ILVLTWSSFY LVQSFSSPLP
Ntbecanfa SVTAWRKICP LLQGIGLASV VIESYLNIYY IIILAWALFY LFSSFTSELP
Ntg4musmu SVTAWRKICP LLQGIGMASV VIESYLNIYY IIILAWALFY LFSSFTWELP
Ntg3musmu GITCWRRVCP LFEGIGYATQ VIEAHLNVYY IIILAWAIFY LSNCFTTELP
Ntgtorma  GITCWRKICP LFEGIGYATQ VIEAHLNMFY IIVLAWAIFY LFNCFTSELP
Ntg2musmu GITAWRRICP IFEGIGYASQ MIVSLLNVYY IVVLAWALFY LFSSFTTDLP
Nttahomsa GITCWEKICP LFSGIGYASV VIVSLLNVYY IVILAWATYY LFQSFQKELP
Ntprratno PLAVW.KISP LFKGAGAAML LIVGLVAIYY NMIIAYVLFY LFASLTSNLP
Consensus ....W..I.P ...G.G.A.. .........YY ..I..W...Y L..SF...LP

          201                                              250
Ntt4ratno WSECPVIRNG TVAV...... ..... ...... ... VEPECEKSSA
Ntt7ratno WDQCPLVKNA SHTY...... ..... ...... ... IEPECEKSSA
Rosiratno WSTCPLDLN. .RTG...... ..... ...... ... FVQECQSSGT
Ntdobosta WTHCNHSWNS PRCSD..... ..... ...AR APNA...SSG PN.GTSRTTP
Ntdohomsa WIHCNNSWNS PNCSD..... ..... ...AH PGDSSGDSSG LN.DTFGTTP
Ntnohomsa WTDCGHTWNS PNCTD..... ..... ...PK LLNGSVLGNH TKYSKYKFTP
Ntsedrome WTSCDNPWNT ENCMQ..... ..... ...VT SEN.......FTELATSP
Ntsehomsa WTSCKNSWNT GNCTN..... ..... ...YF SED.......N ITWTLHSTSP
Gat1torca WQSCGNSWNT DRC....... ..... ...... FSNYSMTNST NLSSPIV...
Ntg1homsa WKQCDNPWNT DRC....... ..... ...... FSNYSMVNTT NMTSAVV...
Ntgmanse  WRNCDNYWNT ATCVNPYDRK NLTCWSSLGD MSTFCTLNGR NVSKAVLSDP
Ntchratno WATCGHTWNT PDCVEIFRHE D....CANAS LANLTCDQLA DRRSPVI...
Ntcrtorma WASCNNTWNT AAC......YE A....GANAS TE..IYPPTA PAQSSIV...
Ntbecanfa WTTCTNTWNT EHCMD.F... ......LN.H SGARTATSSE NFTSPVM...
Ntg4musmu WTTCTNSWNT EHCVD.F... ......LN.H SSARGVSSSE NFTSPVM...
Ntg3musmu WATCGHEWNT EKCVE.F... ......QKLN FSNYSHVSLQ NATSPVM...
Ntgtorma  WATCGHYWNT ENCLE.F... ......QKLN STNCNHTAVP NATSPVI...
Ntg2musmu WGSCSHEWNT ENCVE.F... ......QKAN ..DSMNVTSE NATSPVI...
Nttahomsa WAHCNHSWNT PHCMEDT... ......MRKN KSVWITISST NFTSPVI...
Ntprratno WEHCGNWWNT ERCLE..... ......HRG PKDGNGALPL NLSSTV..SP
Consensus W..C...WN. .......... .......... .......... ..........

          251                                                  300
Ntt4ratno TTYFWYREAL DI..SNSISE SGGLNWKMTV CLLVAWSIVG MAVVKGIQSS
Ntt7ratno TTYYWYREAL AI..SSSISE SGGLNWKMTG CLLAAWVMVC LAMIKGIQSS
Rosiratno VSYFWYRQTL NI..TSDISN TGTIQWKLFL CLVACWTTVY LCVIRGIEST
Ntdobosta AAEYFERGVL HLHESQGIDD LGPPRWQLTS CLVLVIVLLY FSLWKGVKTS
Ntdohomsa AAEYFERGVL HLHQSHGIDD LGPPRWQLTA CLVLVIVLLY FSLWKGVKTS
Ntnohomsa AAEFYERGVL HLHESSGIHD IGLPQWQLLL CLMVVVIVLY FSLWKGVKTS
Ntsedrome AKEFFERKVL ESYKGNGLDF MGPVKPTLAL CVFGVFVLVY FSLWKGVRSA
Ntsehomsa AEEFYTRHVL QIHRSKGLQD LGGISWQLAL CIMLIFTVIY FSIWKGVKTS
Gat1torca ..EFWERNMH QL..TDGLDQ PGQIRAPLAI TLAIAWVLVY FCIWKGVSWT
Ntg1homsa ..EFWERNMH QM..TDGLDK PGQIRWPLAI TLAIAWILVY FCIWKGVGWT
Ntgmanse  VKEFWERRAL QI..SSGIEH IGNIRWELAG TLLLVWVLCY FCIWKGVRWT
Ntchratno ..EFWENKVL RL..STGLEV PGALNWEVTL CLLACWVLVY FCVWKGVKST
Ntcrtorma ..QFWERRVL RL..SSGLGD VGEIGWELTL CLTATWMLVY FCIWKGVKTS
Ntbecanfa ..EFWERRVL GI..TSGIHD LGALRWELAL CLLLAWLICY FCIWKGVKTT
Ntg4musmu ..EFWERRVL GI..TSGIHD LGSLRWELAL CLLLAWIICY FCIWKGVKST
Ntg3musmu ..EFWERRVL AI..SDGIEH IGNLRWELAL CLLAGWTICY FCIWKGTKST
Ntgtorma  ..EFWERRVL GL..SRGIEH IGRVRWELAL CLLAAWIICY FCIWKGPKST
Ntg2musmu ..EFWERRVL KL..SDGIQH LGSLRWELVL CLLLAWIICY FCIWKGVKST
Nttahomsa ..EFWERNVL SL..SPGIDH PGSLKWDLAL CLLLVWLVCF FCICKGVRST
Ntprratno SEEYWSRYVL HIQGSQGIGR PGEIRWNLCL CLLLAWVIVF LCILKGVKSS
Consensus ..EFWER..L ......G... .G...W.L.. CL...W...Y F..WKGV...
```

```
           301                                              350
Ntt4ratno  GKVMYFSSLF PYVVLACFLV RGLLLRGAVD GILHMFTPKL DKMLDPQVWR
Ntt7ratno  GKIMYFSSLF PYVVLICFLI RSLLLNGSID GIRHMFTPKL EMMLEPKVWR
Rosiratno  GKVIYFTALF PYLVLTIFLI RGLTLPGATE GLTYLFTPNM KILQNSRVWL
Ntdobosta  GKVVWITATM PYVVLFALLL RGITLPGAVD AIRAYLSVDF HRLCEASVWI
Ntdohomsa  GKVVWITATM PYVVLTALLL RGVTLPGAID GIRAYLSVDF YRLCEASVWI
Ntnohomsa  GKVVWITATL PYFVLFVLLV HGVTLPGASN GINAYLHIDF YRLKEATVWI
Ntsedrome  GKVVWVTALA PYVVLIILLV RGVSLPGADE GIKYYLTPEW HKLKNSKVWI
Ntsehomsa  GKVVWVTATF PYIILSVLLV RGATLPGAWR GVLFYLKPNW QKLLETGVWI
Gat1torca  GKVVYFSAIY PYIMLLTLFF RGVTLPGARE GILFYITPDF RRLSDSEVWL
Ntg1homsa  GKVVYFSATY PYIMLIILFF RGVTLPGAKE GILFYITPNF RKLSDSEVWL
Ntgmanse   GKVVYFTALF PYFLLTVLLI RGITLPGAME GIKFYVMPNM SKLLESEVWI
Ntchratno  GKIVYFTATF PYVVLVVLLV RGVLLPGALD GIIYYLKPDW SKLGSPQVWI
Ntcrtorma  GKVVYVTATF PYIILVILLV RGVTLHGAVQ GIVYYLQPDW GKLGEAQVWI
Ntbecanfa  GKVVYFTATF PYLMLVILLI RGITLPGAYQ GVIYYLKPDL LRLKDPQVWM
Ntg4musmu  GKVVYFTATF PYLMLIILLI RGVTLPGAYQ GIVFYLKPDL LRLKDPQVWM
Ntg3musmu  GKVVYVTATF PYIMLLILLI RGVTLPGASE GIKFYLYPDL SRLSDPQVWV
Ntgtorma   GKVVYVTATF PYLMLLVLLI RGVTLPGAAE GIKFYLYPDV SRLSDPQVWL
Ntg2musmu  GKVVYFTATF PYLMLVVLLI RGVTLPGAAQ GIQFYLYPNI TRLWDPQVWM
Nttahomsa  GKVVYFTATF PFAMLLVLLV RGLTLPGAGR GIKFYLYPDI TRLEDPQVWI
Ntprratno  GKVVYFTATF PYLILLMLLV RGVTLPGAWK GIQFYLTPQF HHLLSSKVWI
Consensus  GKVVY.TA.. PY..L..LL. RG.TLPGA.. GI...Y..P. ..L....VW.

           351                                              400
Ntt4ratno  EAATQVFFAL GLGFGGVIAF SSYNKQDNNC HFDAALVSFI NFFTSVLATL
Ntt7ratno  EAATQVFFAL GLGFGGVIAF SSYNKRDNNC HFDAVLVSFI NFFTSVLATL
Rosiratno  DAATQIFFSL SLAFGGHIAF ASYNQPRNNC EKDAVTIALV NSMTSLYASI
Ntdobosta  DAAIQICFSL GVGLGVLIAF SSYNKFTNNC YRDAIITTSV NSLTSFSSGF
Ntdohomsa  DAATQVCFSL GVGFGVLIAF SSYNKFTNNC YRDAIVTTSI NSLTSFSSGF
Ntnohomsa  DAATQIFFSL GAGFGVLIAF ASYNKFDNNC YRDALLTSSI NCITSFVSGF
Ntsedrome  DAASQIFFSL GPGFGTLLAL SSYNKFNNNC YRDALITSSI NCLTSFLAGF
Ntsehomsa  DAAAQIFFSL GPGFGVLLAF ASYNKFNNNC YQDALVTSVV NCMTSFVSGF
Gat1torca  DAATQIFFSY GLGLGSLVAL GSYNKFHNNV YRDSIIVCCI NSTTSMFAGF
Ntg1homsa  DAATQIFFSY GLGLGSLIAL GSYNSFHNNV YRDSIIVCCI NSCTSMFAGF
Ntgmanse   DAVTQIFFSY GLGLGTLVAL GSYNKFTNNV YKDALIVCSV NSSTSMFAGF
Ntchratno  DAGTQIFFSY AIGLGALTAL GSYNRFNNNC YKDAIILALI NSGTSFFAGF
Ntcrtorma  DAGTQIFFSY AIGLGTLTAL GSYNQLHNDC YKDAFILSLV NSATSFFAGL
Ntbecanfa  DAGTQIFFSF AICQGCLTAL GSYNKYHNNC YRDSIALCFL NSATSFAAGF
Ntg4musmu  DAGTQIFFSF AICQGCLTAL GSYNKYHNNC YRDSIALCFL NSATSFVAGF
Ntg3musmu  DAGTQIFFSY AICLGCLTAL GSYNNYNNNC YRDCIMLCCL NSGTSFVAGF
Ntgtorma   DAGTQIFFSY AICLGCLTAL GSYNPYHNNC YRDCIMLCCL NSGTSFVAGF
Ntg2musmu  DAGTQIFFSF AICLGCLTAL GSYNKYHNNC YRDCIALCIL NSSTSFMAGF
Nttahomsa  DAGTQIFFSY AICLGAMTSL GSYNKYKYNS YRDCMLLGCL NSGTSFVSGF
Ntprratno  EAALQIFYSL GVGFGGLLTF ASYNTFHQNI YRDTFIVTLG NAITSILAGF
Consensus  DA.TQIFFS. ....G.L.A. .SYN...NNC Y.D....... N..TS..AGF

           401                                              450
Ntt4ratno  VVFAVLGFKA NIMNEKCVVE NAEKILGYLN SNVLSRDLIP PHVNFSHLTT
Ntt7ratno  VVFAVLGFKA NIVNEKCISQ NSEMILKLLK TGNVSWDVIP RHINLSAVTA
Rosiratno  TIFSIMGFKA SNDYGRCLDR NILSLINEFD FPELS..... ......ISR
Ntdobosta  VVFSFLGYMA QKHS...... .......... ......VPIGDVAK
Ntdohomsa  VVFSFLGYMA QKHS...... .......... ......VPIGDVAK
Ntnohomsa  AIFSILGYMA HEHK...... .......... ......VNIEDVAT
Ntsedrome  VIFSVLGYMA YVQK...... .......... ......TSIDKVGL
```

```
Ntsehomsa VIFTVLGYMA EMRN...... ......... .......... ..EDVSEVAK
Gat1torca VIFSIVGFMA HVTN...... ......... .......... ..RPIADVA.
Ntg1homsa VIFSIVGFMA HVTK...... ......... .......... ..RSIADVA.
Ntgmanse  VIFSVVGFMA HEQQ...... ......... .......... ..RPVAEVA.
Ntchratno VVFSILGFMA TEQG...... ......... .......... ..VHISKVA.
Ntcrtorma VVFSILGFMA VEEG...... ......... .......... ..VDISVVA.
Ntbecanfa VVFSILGFMA QEQG...... ......... .......... ...LPISEVA.
Ntg4musmu VVFSILGFMS QEQG...... ......... .......... ...IPISEVA.
Ntg3musmu AIFSVLGFMA YEQG...... ......... .......... ..VPIAEVA.
Ntgtorma  AIFSVLGFMA FEQG...... ......... .......... ..VPIAEVA.
Ntg2musmu AIFSILGFMS QEQG...... ......... .......... ..VPISEVA.
Nttahomsa AIFSILGFMA QEQG...... ......... .......... ..VDIADVA.
Ntprratno AIFSVLGYMS QELG...... ......... .......... ..VPVDQVA.
Consensus ..FS.LG.MA ........ ......... .......... .......VA.
```

```
              451                                          500
Ntt4ratno KDYSEMYNVI MTVKEKQFSA LGLDPCLLED ELDKSVQGTG LAFIAFTEAM
Ntt7ratno EDYHVVYDII QKVKEEEFAV LHLKACQIED ELNKAVQGTG LAFIAFTEAM
Rosiratno DEYPSVLMYL NATQPERVAR LPLKTCHLED FLDKSASGPG LAFIVFTEAV
Ntdobosta D......... ......... ......... ......GPG LIFIIYPEAL
Ntdohomsa D......... ......... ......... ......GPG LIFIIYPEAI
Ntnohomsa E......... ......... ......... ....GAG LVFILYPEAI
Ntsedrome E......... ......... ......... ......GPG LVFIVYPEAI
Ntsehomsa D......... ......... ......... .....AGPS LLFITYAEAI
Gat1torca A......... ......... ......... ......SGPG LAFLAYPEAV
Ntg1homsa A......... ......... ......... ......SGPG LAFLAYPEAV
Ntgmanse  A......... ......... ......... ......SGPG LAFLAYPSAV
Ntchratno E......... ......... ......... ......SGPG LAFIAYPRAV
Ntcrtorma E......... ......... ......... ......SGPG LAFIAYPKAV
Ntbecanfa E......... ......... ......... ......SGPG LAFIAFPKAV
Ntg4musmu E......... ......... ......... ......SGPG LAFIAFPKAV
Ntg3musmu E......... ......... ......... ......SGPG LAFIAYPKAV
Ntgtorma  E......... ......... ......... ......SGPG LTFIAYPKAV
Ntg2musmu E......... ......... ......... ......SGPG LAFIAYPRAV
Nttahomsa E......... ......... ......... ......SGPG LAFIAYPKAV
Ntprratno K......... ......... ......... ......AGPG LAFVIYPQAM
Consensus ......... ......... ......... ......GPG L.F..YP.A.
```

```
              501                                          550
Ntt4ratno THFPASPFWS VMFFLMLINL GLGSMIGTMA GITTPIID.. ....TFKVPK
Ntt7ratno THFPASPFWS VMFFLMLINL GLGSMFGTIE GIITPVVD.. ....TFKVRK
Rosiratno LHMPGASVWS VLFFGMLFTL GLSSMFGNME GVITPLFDM. .GILPKGVPK
Ntdobosta ATLPLSSVWA VVFFVMLLTL GIDSAMGGME SVITGLADEF .QLLHR..HR
Ntdohomsa ATLPLSSAWA VVFFIMLLTL GIDSAMGGME SVITGLIDEF .QLLHR..HR
Ntnohomsa STLSGSTFWA VVFFVMLLAL GLDSSMGGME AVITGLADDF .QVLKR..HR
Ntsedrome ATMSGSVFWS IIFFLMLITL GLDSTFGGLE AMITALCDEY PRVIGR..RR
Ntsehomsa ANMPASTFFA IIFFLMLITL GLDSTFAGLE GVITAVLDEF PHVWAK..RR
Gat1torca TQLPISPLWS ILFFSMLLML GIDSQFCTVE GFITALVDEF PKLLRG..RR
Ntg1homsa TQLPISPLWA ILFFSMLLML GIDSQFCTVE GFITALVDEY PRLLRN..RR
Ntgmanse  LQLPGAPLWS CLFFFMLLLI GLDSQFCTME GFITAVIDEW PKLLRR..RK
Ntchratno TLMPVAPLWA ALFFFMLLLL GLDSQFVGVE GFITGLLDLL PASYYFRFQR
Ntcrtorma TLMPFPQVWA VLFFIMLLCL GLGSQFVGVE GFVTAILDLW PSKFSFRYLR
Ntbecanfa TMMPLSQLWS CLFFIMLIFL GLDSQFVCVE CLVTASMDMF PSQLRKSGRR
Ntg4musmu TMMPLSQLWS CLFFIMLLFL GLDSQFVCME CLVTASMDMF PQQLRKSGRR
Ntg3musmu TMMPLSPLWA TLFFMMLIFL GLDSQFVCVE SLVTAVVDMY PKVFRRGYRR
```

```
Ntgtorma   TMMPLAPLWA FLFFLMLIFL GLDSQFVCME SLVTAIIDMY PSIFRRGYRR
Ntg2musmu  VMLPFSPLWA CCFFFMVVLL GLDSQFVCVE SLVTALVDMY PRVFRKKNRR
Nttahomsa  TMMPLPTFWS ILFFIMLLLL GLDSQFVEVE GQITSLVDLY PSFLRKGYRR
Ntprratno  TMLPLSPFWS FLFFFMLLTL GLDSQFAFLE TIVTAVTDEF PYYLRP..KK
Consensus  ...P....W. ..FF.ML..L GLDS.F...E ...T...D..........R
```

```
           551                                                 600
Ntt4ratno  EMFTVGCCVF AFFVGLLFVQ RSGNYFVTMF DDYSAT.LPL TVIVILENIA
Ntt7ratno  EILTVICCLL AFCIGLMFVQ RSGNYFVTMF DDYSAT.LPL LIVVILENIA
Rosiratno  ETMTGVVCFI CFLSAICFTL QSGSYWLEIF DSFAAS.LNL IIFAFMEVVG
Ntdobosta  ELFTLLVVLA TFLLSLFCVT NGGIYVFTLL DHFAA.GTSI LFGVLMEVIG
Ntdohomsa  ELFTLFIVLA TFLLSLFCVT NGGIYVFTLL DHFAA.GTSI LFGVLIEAIG
Ntnohomsa  KLFTFGVTFS TFLLALFCIT KGGIYVLTLL DTFAA.GTSI LFAVLMEAIG
Ntsedrome  ELFVLLLLAF IFLCALPTMT YGGVVLVNFL NVYGP.GLAI LFVVFVEAAG
Ntsehomsa  ERFVLAVVIT CFFGSLVTLT FGGAYVVKLL EEYAT.GPAV LTVALIEAVA
Gat1torca  EIFIAMVCIV SYLIGLSNIT QGGLYVFKLF DYYSASGMSL LFLVFFETVS
Ntg1homsa  ELFIAAVCII SYLIGLSNIT QGGIYVFKLF DYYSASGMSL LFLVFFECVS
Ntgmanse   EIFIAITCII SYLVGLSCIS EGGMYVFQIL DSYAVSGFCL LFLIFFECVS
Ntchratno  EISVALCCAL CFVIDLSMVT DGGMYVFQLF DYYSASGTTL LWQAFWECVV
Ntcrtorma  EVVVAMVICL SFLIDLSMIT EGGMYIFQIF DYYSASGTTL LWTAFWECVA
Ntbecanfa  ELLILAIAVF CYLAGLFLVT EGGMYIFQLF DYYASSGICL LFLAMFEVIC
Ntg4musmu  DVLILAISVL CYLMGLLLVT EGGMYIFQLF DYYASSGICL LFLSLFEVIC
Ntg3musmu  ELLILALSII SYFLGLVMLT EGGMYIFQLF GSYAASGMCL LFVAIFECVC
Ntgtorma   EQLIFVIALA SYLMGLVMVT EGGMYIFQLF DAYASSGMCL LFVAIFECIC
Ntg2musmu  EVLILIVSVI SFFIGLIMLT EGGMYVFQLF DYYAASGMCL LFVAIFESLC
Nttahomsa  EIFIAFVCSI SYLLGLTMVT EGGMYVFQLF DYYAASGVCL LWVAFFECFV
Ntprratno  AVFSGLICVA MYLMGLILTT DGGMYWLVLL DDYSAS.FGL MVVVITTCLA
Consensus  E............ .....L...T .GG.Y...L. D.Y...G..LL.....E...
```

```
           601                                                 650
Ntt4ratno  VAWIYGTKKF MQELTEMLGF RPYRFYFYMW KFVSPLCMAV LTTASIIQLG
Ntt7ratno  VSFVYGIDKF LEDLTDMLGF APSKYYYYMW KYISPLMLVT LLIASIVNMG
Rosiratno  VIHVYGIKRF CDDIEWMTGR RPSLYWQVTW RVVSPMLLFG IFLSYIVLLA
Ntdobosta  VAWFYGVWQF SDDIKQMTGR RPSLYWRLCW KFVSPCFLLF VVVVSIATF.
Ntdohomsa  VAWFYGVGQF SDDIQQMTGQ RPSLYWRLCW KLVSPCFLLF VVVVSIVTF.
Ntnohomsa  VSWFYGVDRF SNDIQQMMGF RPGLYWRLCW KFVSPAFLLF VVVVSIINF.
Ntsedrome  VFWFYGVDRF SSDVEQMLGS KPGLFWRICW TYISPVFLLT IFIFSIMGY.
Ntsehomsa  VSWFYGITQF CRDVKEMLGF SPGWFWRICW VAISPLFLLF IICSFLMSP.
Gat1torca  ISWCYGVNRF FVNIEEMVGH KPCLWWKLCW SFFTPIIVGG VFLFSAIQM.
Ntg1homsa  ISWFYGVNRF YDNIQEMVGS RPCIWWKLCW SFFTPIIVAG VFIFSAVQM.
Ntgmanse   ISWAFGVNRF YDGIKEMIGY YPTIWWKFCW VGFTPAICIS VFIFNLVQW.
Ntchratno  VAWVYGADRF MDDIACMIGY RPCPWMKWCW SFFTPLVCMG IFIFNVVYY.
Ntcrtorma  VAWVYGGDRY LDDLAWMLGY RPWALVKWCW SVITPLVCMG IFTFHLVNY.
Ntbecanfa  ISWVYGADRF YDNIEDMIGY RPWPLVKISW LFLTPGLCLA TFLFSLSQY.
Ntg4musmu  IGWVYGADRF YDNVEDMIGY RPWPLVKISW LFLTPGLCLA TFFFSLSKY.
Ntg3musmu  IGWVYGSNRF YDNIEDMIGY RPLSLIKWCW KVVTPGICAG IFIFFLVKY.
Ntgtorma   IGWVYGGNRF YDNIEDMIGY RPFVLIKWCW IFITPGICAA IFIFFIVRY.
Ntg2musmu  VAWVYGAGRF YDNIEDMIGY KPWPLIKYCW LFFTPAVCLA TFLFSLIKY.
Nttahomsa  IAWIYGGDNL YDGIEDMIGY RPGPWMKYSW V.ITPVLCVG CFIFSLVKY.
Ntprratno  VTRVYGIQRF CRDIHMMLGF KPGLYFRACW LFLSPATLLA LLVYSIVKY.
Consensus  ..W.YG...F ......M.G. .P......W ....P..... .....
```

```
           651                                                 700
Ntt4ratno  VSPPGYSAWI KEEAAERYLY FPNWAMALLI TLIAVATLPI PVVFILRHFH
Ntt7ratno  LSPPGYNAWI KEKASEEFLS YPMWGMVVCF SLMVLAILPV PVVFVIRRCN
```

```
Rosiratno QSSPSYKAW. ....NPQYEH FPSREEKLYP GWVQVTCVLL SFLPSLWVPG
Ntdobosta .......... RPPHYGA.YV FPEWATALGW AIAASSMSVV PIYAAYKLCS
Ntdohomsa .......... RPPHYGA.YI FPDWANALGW VIATSSMAMV PIYAAYKFCS
Ntnohomsa .......... KPLTYDD.YI FPPWANWVGW GIALSSMVLV PIYVIYKFLS
Ntsedrome .......... KEMLGEE.YY YPDWSYQVGW AVTCSSVLCI PMYIIYKFFF
Ntsehomsa .......... PQLRLFQ.YN YPYWSIILGY CIGTSSFICI PTYIAYRLII
Gat1torca .......... KPLKMGS.YI FPKWGQGVGW FMALSSMMLI PGYMGYMFLT
Ntg1homsa .......... TPLTMGN.YV FPKWGQGVGW LMALSSMVLI PGYMAYMFLA
Ntgmanse  .......... TPIKYMN.YE YPWWSHAFGW FTALSSMLCI PGYMIYLWRV
Ntchratno .......... KPLVYNNTYV YPWWGEAMGW AFALSSMLCV PLHLLGCLLR
Ntcrtorma .......... KPLTYNKTYT YPWWGEAIGW CLALASMLCV PTTVLYSLSR
Ntbecanfa .......... TPLKYNNIYV YPPWGYSIGW FLALSSMICV PLFVIITLLK
Ntg4musmu .......... TPLKYNNVYM YPSWGYSIGW LLAFSSMACV PLFIIITFLK
Ntg3musmu .......... KPLKYNNVYT YPAWGYGIGW LMALSSMLCI PLWIFIKLWK
Ntgtorma  .......... QPLKYNNVYV YPDWGYALGW ALALSSMICI PLGFIFKMWS
Ntg2musmu .......... TPLTYNKKYT YPWWGDALGW LLALSSMICI PAWSIYKLRT
Nttahomsa .......... VPLTYNKTYV SPTWAIGLGW SLALSSMLCV PLVIVIRLCQ
Ntprratno .......... QPSEYGS.YR FPAWAELLGI LMGLLSCLMI PAGMLVAVLR
Consensus .......... .P......Y. .P.W....GW ....SS.... P.........
```

```
          701                                                 750
Ntt4ratno LLSDGSNTL. SVSYKKGRMM KDISNLEEND ETRFILSKVP SEAPSPMPTH
Ntt7ratno LIDDSSGNLA SVTYKRGRVL KEPVNLDG.D DASLIHGKIP SEMSSPNFGK
Rosiratno IALAQLLFQY RQRWKNTHLE SALKPQESRG C..................
Ntdobosta LP.GSSREKL AYAITPETEH GRVDSGGGAP VHAPPLARGV GRWRKRKSCW
Ntdohomsa LP.GSFREKL AYAIAPEKDR ELVDRGEVRQ FTLRHWLKV. ..........
Ntnohomsa TQ.GSLWERL AYGITPENEH HLVAQRDIRQ FQLQHWLAI. ..........
Ntsedrome ASKGGCRQRL QESFQPEDNC GSVVPGQQGT SV................
Ntsehomsa TP.GTFKERI IKSITPETPT EIPCGDIRLN AV................
Gat1torca SK.GSLKQRL RLMTQPNEDM KCRENGPEQT ECGNTPSDEA YM........
Ntg1homsa LK.GSLKQRI QVMVQPSEDT VRPENGPEHA QAGSSTSKEA YI........
Ntgmanse  TP.GTWQEKF HKIVRIPEDV PSLRTKM... ..............
Ntchratno AK.GTMAERW QHLTQPIWGL HHLEYRAQDA DVRGLTTLTP VSESSKVVVV
Ntcrtorma GR.GSLKERW RKLTTPVWAS HHLAYKMAGA KI.NQPCEGV VSCEEKVVIF
Ntbecanfa TR.GSFKKRL RQLTTPDPSL PQPKQHLYLD GGTSQDCGPS PTKEG.....
Ntg4musmu TQ.GSFKKRL RRLITPDPSL PQPGRRPPQD GSSAQNCSSS PAKQE.....
Ntg3musmu TE.GTLPEKL QKLTVPSADL KMRGKLGASP RTV..TVNDC EAKVKGDGTI
Ntgtorma  TE.GTFLEKI KKLTTPSADL RRKGMGMSNM DTCCSTISDC DGKLKGDECI
Ntg2musmu LK.GPLRERL RQLVCPAEDL PQKNQPEPTA PATPMTSLLR LTELESNC..
Nttahomsa TE.GPFLVRV KYLLTPREPN RWAVEREGAT PYNSRTVMNG ALVKPTHIIV
Ntprratno EE.GSLWERL QQASRPAIDW GPSLEENRTG MYVATLAGSQ SPKPLMVHMR
Consensus ...G...... .....P.... ..............
```

```
          751                                                 800
Ntt4ratno RSYLGPGSTS PLESSSHPNG RYGSGYLLA. ..STPESEL. ..........
Ntt7ratno NIYRKQSGSP TLDTA..PNG RYGIGYLMAD MPDMPESDL. ..........
Ntdobosta VPSRGPGRGG PPTPSPRLAG HTRAFPWTGA PPVPRELTPP STCRCVPPLV
Ntchratno ESVM.............................................
Ntcrtorma ESVL.............................................
Ntbecanfa LIVGEKETHL .................................
Ntg4musmu LIAWEKETHL .................................
Ntg3musmu SAITEKETHF .................................
Ntgtorma  PAITEKETHF .................................
Nttahomsa ETMM.............................................
```

```
Ntprratno KYGGITSFEN TAIEVDREIA EEEEESMMXD QTPPNRRAGR GLPVCPFLGH
Consensus .......... .......... .......... .......... ..........

          801          815
Ntdobosta CAHPAVESTG LCSVY
Ntprratno RG.......... .....
Consensus .......... .....
```

Proteins listed subsequently in italics are at least 90% identical to the paired transporters listed in parenthesis and therefore are not included in the alignments: *Ntcrorycu, Ntcrhomsa* (Ntchratno); *Ntg1ratno, Ntg1musmu* (Ntg1homsa); *Ntg2ratno* (Ntg2musmu); *Ntg3ratno* (Ntg3musmu); *Ntnobosta* (Ntnohomsa); *Nts1ratno, Nts2ratno* (Ntsehomsa); *Nttacanfa, Nttamusco, Nttaratno* (Nttahomsa); *Ntdoratno* (Ntdohomsa). Residues listed in the consensus sequence are present in at least 75% of the transporter sequences shown.

Database accession numbers

	SWISSPROT	PIR	EMBL/GENBANK
Gat1torca		I51368; S42808	X77139
Ntbecanfa	P27799	A41757	M80403
Ntchratno	P28570	S23431	X66494
Ntcrhomsa		JC2386	
Ntcrorycu	P31661	X67252	
Ntcrtorma		S46260	
Ntdobosta	P27922	A41617	M80234
Ntdohomsa	Q01959	A48980	M96670; M95167
Ntdoratno	P23977	S20346	M80233; M80570
Ntg1homsa	P30531	S11073	X54673
Ntg1musmu	P28571	F46027	M92378
Ntg1ratno	P23978	A35918	M59742
Ntg2musmu	P31649	A44409	L04663
Ntg2ratno	P31646	A45708	M95762
Ntg3musmu	P31650	B44409	L04662
Ntg3ratno	P31647	JH0695; B45078	M95738; M95763
Ntg4musmu	P31651	A43390	M97632
Ntgmanse		L40373	
Ntgtorma		X87170	
Ntnobosta		U09198	
Ntnohomsa	P23975	S14278	M65105
Ntprratno	P28573	M88111	
Nts1ratno	P31652	S30604	X63995
Nts2ratno	P23976	S19585	M79450
Ntsedrome		U04809	
Ntsehomsa	P31645	S37688; A47398	X70697; L05568
Ntt4ratno	P31662	S27043	S52051; S68944
Ntt7ratno	Q08469	L22022	
Nttacanfa	Q00589	A46270	M95495
Nttahomsa	P31641	S29839	Z18956; U09220
Nttamusco	P31642	A47194	L03292
Nttaratno	P31643	M96601	
Rosiratno		U12973	

References

[1] Schloss, P. et al. (1994) Curr. Opin. Cell Biol. 6, 595–599.

[2] Reizer, J. et al. (1994) Biochim. Biophys. Acta 1197, 133–166.

3 Edwards, R.H. (1993) Ann. Neurol. 34, 638–645.
4 Wright, E.M. et al. (1996) Curr. Opin. Cell Biol. 8, 468–473.
5 Bruss, M. et al. (1995) J. Biol. Chem. 270, 9197–9201.
6 Borden, et al. (1992) J. Biol. Chem. 267, 21098–21104.
7 Demchyshyn, L. et al. (1994) Proc. Natl Acad. Sci. USA 91, 5158–5162.

Part 11

Na⁺-Dependent Antiporters

Summary

Transporters of the Na$^+$/H$^+$ antiporter family, the example of which is the NHE1 Na$^+$/H$^+$ antiporter of humans (Nah1homsa), mediate Na$^+$-linked extrusion of hydrogen ions from the cell. Members of the Na$^+$/H$^+$ antiporter family regulate intracellular pH and are involved in sodium readsorption and signal transduction. They are fully active at acidic pH, inactive at neutral pH, and differ in their sensitivity to amiloride [1,2]. Members of the family occur in both invertebrates and vertebrates.

Statistical analysis reveals no apparent relationship between the amino acid sequences of the NHE1 family and other family of transporters. They are predicted to form 10 or 12 membrane-spanning helices by the hydropathy of their amino acid sequences. Members of the Na$^+$/H$^+$ antiporter family are glycosylated and may be phosphorylated.

Several amino acid sequence motifs are highly conserved in the Na$^+$/H$^+$ antiporter family.

Nomenclature, biological sources and substrates

CODE	DESCRIPTION [SYNONYMS]	ORGANISM [COMMON NAMES]	SUBSTRATE(S)
Nah1homsa	Na$^+$/H$^+$ antiporter 1 [NAH1, NHE1, APNH1]	Homo sapiens [human]	Na$^+$/H$^+$
Nah1orycu	Na$^+$/H$^+$ antiporter 1 [NAH1, NHE1]	Oryctolagus cuniculus [rabbit]	Na$^+$/H$^+$
Nah1ratno	Na$^+$/H$^+$ antiporter 1 [NAH1, NHE1]	Rattus norvegicus [rat]	Na$^+$/H$^+$
Nah3orycu	Na$^+$/H$^+$ antiporter 3 [NAH3, NHE3]	Oryctolagus cuniculus [rabbit]	Na$^+$/H$^+$
Nah3ratno	Na$^+$/H$^+$ antiporter 3 [NAH3, NHE3]	Rattus norvegicus [rat]	Na$^+$/H$^+$
Nah4ratno	Na$^+$/H$^+$ antiporter 4 [NAH4, NHE4]	Rattus norvegicus [rat]	Na$^+$/H$^+$
Nahboncmy	Na$^+$/H$^+$ antiporter β [NAH, β NHE]	Oncorhynchus mykiss [trout]	Na$^+$/H$^+$
Nhecarma	Na$^+$/H$^+$ antiporter [NHE]	Carcinus maenas [crab]	Na$^+$/H$^+$
Nhe1musmu	Na$^+$/H$^+$ antiporter isoform 1 [NHE1]	Mus musculus [mouse]	Na$^+$/H$^+$
Nhe2orycu	Na$^+$/H$^+$ antiporter isoform 2 [NHE2]	Oryctolagus cuniculus [rabbit]	Na$^+$/H$^+$
Nhe2ratno	Na$^+$/H$^+$ antiporter isoform 2 [NHE2]	Rattus norvegicus [rat]	Na$^+$/H$^+$
Nhe3didvi	Na$^+$/H$^+$ antiporter isoform 3 [NHE3]	Didelphis virginiana [opossum]	Na$^+$/H$^+$

Cotransported ions are listed.

Phylogenetic tree

Proteins listed subsequently in italics are at least 90% identical to the paired transporters listed in parenthesis and therefore are not included in the phylogenetic tree: *Nah1orycu, Nah1ratno, Nhe1musmu* (Nahihomsa); *Nhe2orycu* (Nhe2ratno).

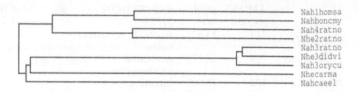

Proposed orientation of NHE1 in the membrane

The model is based on predictions of membrane-spanning regions and α-helical content. The N-terminus of the protein is illustrated on the inside and is folded ten times through the membrane. The predicted membrane-spanning helices are portrayed as rectangles. The numbers corresponding to the first and last residue of each membrane-spanning helix are boxed. Residues that are conserved in more than 75% of the aligned transporters (see below) are shown. Consensus residues indicated by an asterisk are not conserved in NHE1.

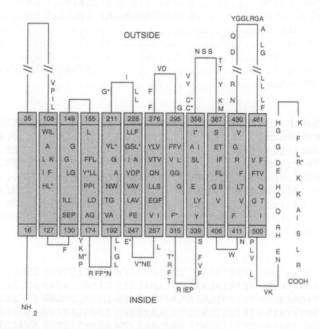

Physical and genetic characteristics

	AMINO ACIDS	MOL. WT	EXPRESSION SITES	K_m	CHROMOSOMAL LOCUS
Nah1homsa	815	90 763	kidney, intestine		1p36.1–p35
Nah1orycu	816	90 717	kidney, intestine	Na⁺: 15 mM [3]	
Nah1ratno	820	91 612	kidney, intestine		
Nah3orycu	832	92 748	intestine, kidney	Na⁺: 17 mM [3]	
Nah3ratno	831	93 105	intestine, kidney		

	AMINO ACIDS	MOL. WT	EXPRESSION SITES	K_m	CHROMOSOMAL LOCUS
Nah4ratno	717	81 522	stomach, intestine, colon		
Nahboncmy	759	85 173	nucleated erythrocytes		
Nhecarma	672	75 981	intestine		
Nhe1musmu	820	91 467	ubiquitous		
Nhe2orycu	809	90 744	kidney, intestine	Na⁺: 18 mM [3]	
Nhe2ratno	813	91 402	kidney, intestine		
Nhe3didvi	839	94 765			

Multiple amino acid sequence alignments

```
          1                                                  50
Nah1homsa MVLRSGICGL SPHRIFPSLL VVVALVGLLP VLRSHGLQLS PTASTIRSSE
Nahboncmy ..MPAFSCAF PGCR..RDLL VIV.....LV VFVGIGLPIE ASAPAYQS..
Nah4ratno .........M GPA....MLR AFSSWKWLLL LMVLTCLEAS SYVN......
Nhe2ratno .........M GPSGTAHRMR APLSWLLLLL LSLQVAVPAG ALAETLLDAP
Nah3ratno .......... .......... .......... .......... ..MWHPALGP
Nhe3didvi .......... .......... .......... .......MP LGVRGTRREF
Nah3orycu .......... .......... .......... ......... MSGRGGC.GP
Nhecarma  .......... .......... ...MKNRVIL MVCVAWCVLG LAAANTSAKQ
Consensus .......... .......... .......... .......... ..........

          51                                                100
Nah1homsa PPRERSIGDV TTAPPEVTPE SRPVNHSVTD HGMKPRKAFP VLGIDYTHVR
Nahboncmy .......... ....HGTEG SHLTNITNT. .....KKAFP VLAVNYEHVR
Nah4ratno ESS....... .SPTGQQTPD ARFAASSSDP ..DER...IS VFELDYDYVQ
Nhe2ratno GAR....... .GASSNPPSP ASVVAPGTTP FEESR...LP VFTLDYPHVQ
Nah3ratno GWKPLLALAV A....VTSLR GVRGIEEEPN SGG....SFQ IVTFKWHHVQ
Nhe3didvi RFPVWGLLLL A....LWMLP RALGVEEIPG PDSHEKQGFQ IVTFKWHHVQ
Nah3orycu CWGLLLALVL A....LGALP WTQGAEQEHH ...DEIQGFQ IVTFKWHHVQ
Nhecarma  HHTATNTTTT ADNETLQRVR IDGSEAHNEA EGEHRLERYP VVVLDFERVQ
Consensus .......... .......... .......... .......... .......V.

          101                                               150
Nah1homsa TPFEISLWIL LACLMKIGFH VIPTISSIVP ESCLLIVVGL LVGGLIKGVG
Nahboncmy KPFEIALWIL LALLMKLGFH LIPRLSAVVP ESCLLIVVGL LVGGLIKVIG
Nah4ratno IPYEVTLWIL LASLAKIGFH LYHRLPHLMP ESCLLIIVGA LVGSIIFGTH
Nhe2ratno IPFEITLWIL LASLAKIGFH LYHKLPTIVP ESCLLIMVGL LLGGIIFGVD
Nah3ratno DPYIIALWIL VASLAKIVFH LSHKVTSVVP ESALLIVLGL VLGGIVWAAD
Nhe3didvi DPYIIALWIL VASLAKIVFH LSHKVTSVVP ESALLIVLGL ILGGIVWAAD
Nah3orycu DPYIIALWVL VASLAKIVFH LSHKVTSVVP ESALLIVLGL VLGGIVLAAD
Nhecarma  TPF.IGLWIF LACLGKIGFH MTPKISHVFP ESCMLIVLGV LIGLLLIYTQ
Nahcaeel  .......... ......FN LMKPISKWCP DSSLLIIVGL ALGWILHQTS
Consensus .P..I.LWIL .A.L.KI.FH L.......P ES.LLI..GL ..GG......

          151                                               200
Nah1homsa E.TPPFLQSD VFFLFLLPPI ILD.AGYFLP LRQFTENLGT ILIFAVVGTL
Nahboncmy E.EPPVLDSQ LFFLCLLPPI ILD.AGYFLP IRPFTENVGT ILVFAVIGTL
Nah4ratno HKSPPVMDSS IYFLYLLPPI VLE.SGYFMP TRPFFENIGS ILWWAGLGAL
Nhe2ratno EKSPPAMKTD VFFLYLLPPI VLD.AGYFMP TRPFFENLGT IFWYAVVGTL
```

```
Nah3ratno HIASFTLTPT LFFFYLLPPI VLD.AGYFMP NRLFFGNLGT ILLYAVIGTI
Nhe3didvi HIASFTLTPT VFFFYLLPPI VLD.AGYFMP NRLFFGNLGT ILLYAVIGTV
Nah3orycu HIASFTLTPT VFFFYLLPPI VLD.AGYFMP NRLFFSNLGS ILLYAVVGTV
Nhecarma  AATVSPLTAD VFFLYLLPPI ILD.AVYFMP NRLFFDNLFT ILVFAVIGTI
Nahcaeel  .LSGATLDSH TFFLYLLPPI IFGSSGYFMP NRALFENFDS VLVFSVFGTI
Consensus ......L... .FFLYLLPPI .LD.AGYFMP .R.FF.NLG. IL..AV.GT.
```

```
               201                                            250
Nah1homsa WNAFFLGGLM YAVCLVGGEQ INNIGLLDNL LFGSIISAVD PVAVLAVFEE
Nahboncmy WNAFFMGGLL YALCQIESVG LSGVDLLACL LFGSIVSAVD PVAVLAVFEE
Nah4ratno INAFGIGLSL YFICQIKAFG LGDINLLQNL LFGSLISAVD PVAVLAVFEE
Nhe2ratno WNSIGIGLSL FGICQIEAFG LSDITLLQNL LFGSLISAVD PVAVLAVFEN
Nah3ratno WNAATTGLSL YGVFLSGLMG ELKIGLLDFL LFGSLIAAVD PVAVLAVFEE
Nhe3didvi WNAATTGLSL YGVYLSGIMG DLSIGLLDFL LFGSLIAAVD PVAVLAVFEE
Nah3orycu WNAATTGLSL YGVFLSGIMG ELKIGLLDFL LFGSLIAAVD PVAVLAVFEE
Nhecarma  WNALTIGITM YAISLTGLFG .LDIPMLHMF LFSSLISAVD PVAVLAVFEE
Nahcaeel  WNTFAIGGSL LLMAQYDLFT .MSFTTFEIL VFSALISAVD PVAVIAVFEE
Consensus WNA...G..L Y........G ...I.LL..L LFGSLI.AVD PVAVLAVFEE
```

```
               251                                            300
Nah1homsa IHINELLHIL VFGESLLNDA VTVVLYHLFE EFAN...YEH VGIVDIFLGF
Nahboncmy IHINELVHIL VFGESLLNDA VTVVLYNLFE EFSK...VGT VTVLDVFLGV
Nah4ratno ARVNEQLYMM IFGEALLNDG ISVVLYNILI AFTKMHKFED IEAVDILAGC
Nhe2ratno IHVNEQLYIL VFGESLLNDA VTVVLYNLFK SFCQM...KT IQTVDVFAGI
Nah3ratno VHVNEVLFII VFGESLLNDA VTVVLYNVFE SFVTLGG.DA VTGVDCVKGI
Nhe3didvi VHVNDVLFII VFGESLLNDA VTVVLYNVFD SFVSLGA.DK VTGVDCVKGI
Nah3orycu VHVNEVLFII VFGESLLNDA VTVVLYNVFQ SFVTLGG.DK VTGVDCVKGI
Nhecarma  MQVEEVLFIL VFGESLLNDG VTVVLYHLFE GFSELGE.AN IMAVDIASGV
Nahcaeel  IHVNEFLFIN VFGEALFNDG VTVVLYQC.S KFALIGS.EN LSVLDYATGG
Consensus ..VNE.L.I. VFGESLLND. VTVVLY..F. .F........ ...VD...G.
```

```
               301                                            350
Nah1homsa LSFFVVALGG VLVGVVYGVI AAFTSRFTSH IRVIEPLFVF LYSYMAYLSA
Nahboncmy VCFFVVSLGG VLVGAIYGFL AAFTSRFTSH TRVIEPLFVF LYSYMAYLSS
Nah4ratno ARFVIVGCGG VFFGIIFGFI SAFITRFTQN ISAIEPLIVF MFSYLSYLAA
Nhe2ratno ANFFVVGIGG VLIGILLGFI AAFTTRFTHN IRVIEPLFVF LYSYLSYITA
Nah3ratno VSFFVVSLGG TLVGVIFAFL LSLVTRFTKH VRIIEPGFVF VISYLSYLTS
Nhe3didvi VSFFVVSLGG TLIGIIFAFL LSLVTRFTKH VRIIEPGFVF IISYLSYLTS
Nah3orycu VSFFVVSLGG TLVGVVFAFL LSLVTRFTKH VRVIEPGFVF IISYLSYLTS
Nhecarma  ASFLLVALGG TAIGIIWGFL TAFVTRLTSG VRVIEPVFVF VMAYLAYLNA
Nahcaeel  LSFFVVALGG AAVGIIFAIA ASLTTKYTYD VRILAPVFIF VLPYMAYLTA
Consensus ..FFVV.LGG ...G....F. ....TRFT.. .R.IEP.FVF ..SY..YL..
```

```
               351                                            400
Nah1homsa ELFHLSGIMA LIASGVVMRP YVEANISHKS HTTIKYFLKM WSSVSETLIF
Nahboncmy EMFHLSGIMA LIACGVVMRP YVEANISHKS YTTIKYFLKM WSSVSETLIF
Nah4ratno ETLYLSGILA ITACAVTMKK YVEENVSQTS YTTIKYFMKM LSSVSETLIF
Nhe2ratno EMFHLSGIMA ITACAMTMNK YVEENVSQKS YTTIKYFMKM LSSVSETLIF
Nah3ratno EMLSLSAILA ITFCGICCQK YVKANISEQS ATTVRYTMKM LASGAETIIF
Nhe3didvi EMLSLSAILA ITFCGICCQK YVKANISEQS ATTVRYTMKM LASGAETIIF
Nah3orycu EMLSLSSILA ITFCGICCQK YVKANISEQS ATTVRYTMKM LASGAETIIF
Nhecarma  EIFHLSGILS ITFCGITMKN Y.WNRTSPPS PHDHQIRHED VSLVFETIIF
Nahcaeel  EMVSLSSIIA IAICGMLMKQ YIKGNVTQAA ANSVKYFTKM LAQSSETVIF
Consensus E...LS.I.A I..CC..... YV..N.S..S .TT..Y..KM ..S..ET.IF
```

```
               401                                                    450
Nah1homsa IFLGVSTVAG S.HHWNWTFV ISTLLFCLIA RVLGVLGLTW FINKFRIVKL
Nahboncmy IFLGVSTVAG P.HAWNWTFV ITTVILCLVS RVLGVIGLTF IINKFRIVKL
Nah4ratno IFMGVSTVGK N.HEWNWAFV CFTLAFCQIW RAISVFTLFY VSNQFRTFPF
Nhe2ratno IFMGVSTVGK N.HEWNWAFV CFTLAFCLIW RALGVFVLTQ VINWFRTIPL
Nah3ratno MFLGISAVDP VIWTWNTAFV LLTLVFISVY RAIGVVLQTW ILNRYRMVQL
Nhe3didvi MFLGISAVDP AIWTWNTAFI LLTLVFISVY RAIGVVLQTW LLNKYRMVQL
Nah3orycu MFLGISAVDP LIWTWNTAFV LLTLLFVSVF RAIGVVLQTW LLNRYRMVQL
Nhecarma  MFLGVSTIQS D.HQWNTWFV ILTILFCSIY RILGVLIFSA VCNRFRVKKI
Nahcaeel  MFLGLSTISS Q.HHFDLYFI CATLFFCLIY RAIGIVVQCY ILNRFRAKKF
Consensus .FLG.S.V.. ....WN..FV ..TL.F.... R..GV..... ..N..R....

               451                                                    500
Nah1homsa TPKDQFIIAY GGLRGAIAFS LGYLLDKKHF PMCDLFLTAI ITVIFFTVFV
Nahboncmy TKKDQFIVAY GGLRGAIAFS LGYLLSNSH. QMRNLFLTAI ITVIFFTVFV
Nah4ratno SIKDQLIIFY SGVRGAGSFS LAFLLPLTLF PRKKLFVTAT LVVTYFTVFF
Nhe2ratno TFKDQFIIAY GGLRGAICFA LVFLLPATVF PRKKLFITAA IVVIFFTVFI
Nah3ratno ETIDQVVMSY GGLRGAVAYA LVVLLDEKKV KEKNLFVSTT LIVVFFTVIF
Nhe3didvi EIIDQVVMSY GGLRGAVAYA LVVLLDEKKV KEKNLFVSTT IIVVFFTVIF
Nah3orycu ELIDQVVMSY GGLRGAVAFA LVALLDGNKV KEKNLFVSTT IIVVFFTVIF
Nhecarma  GFVDKFVMSY GGLRGAVAFA LVITINPIHI PLQPMFLTAT IAMVYFTVFV
Nahcaeel  EMVDQFIMSY GGLRGAIAYG LVVSIPAS.I TRKPMFITAT IAWIYFTVFL
Consensus ...DQ....Y GGLRGA.A.. LG.LL..... ....LF.... ..V.FFTV..

               501                                                    550
Nah1homsa QGMTIRPLVD LLAVKKKQET KRSINEEIHT QFLDHLLTGI EDICGHYGHH
Nahboncmy QGMTIRPLVE LLAVKKKKES KPSINEEIHT EFLDHLLTGV EGVCGHYGHY
Nah4ratno QGITIGPLVR YLDVRKTNKK .ESINEELHI RLMDHLKAGI EDVCGQWSHY
Nhe2ratno LGITIRPLVE FLDVKRSNKK QQAVSEEIHC RFFDHVKTGI EDVCGHWGHN
Nah3ratno QGLTIKPLVQ WLKVKRSEQR EPKLNEKLHG RAFDHILSAI EDISGQIGHN
Nhe3didvi QGLTIKPLVQ WLKVKKSEHR EPKLNEKLHG RAFDHILSAI EDISGQIGHN
Nah3orycu QGLTIKPLVQ WLKVKRSEHR EPKLNEKLHG RAFDHILSAI EDISGQIGHN
Nhecarma  QGITIKPLVQ LLGVKKSEKR SLTMNERLHE RVMDYVMSGV EEMIGKQGNL
Nahcaeel  QGITIRPLVN FLKIKKKEER DPTMVESVYN KYLDYMMSGV EDIAGQKGHY
Consensus QG.TI.PLV. .L.VK..... ....NE..H. R..DH..... ED..G..GH.

               551                                                    600
Nah1homsa HWKDKLNRFN KKYVKKCLIA GERSKEPQ.. .LIAFYHKME MKQAIELVES
Nahboncmy HWKEKLNRFN KTYVKRWLIA GENFKEPE.. .LIAFYRKME LKQAIMMVES
Nah4ratno QVRDKFKKFD HRYLRKILIR RNQPKSS... .IVSLYKKLE MKQAIEMAET
Nhe2ratno FWRDKFKKFD DKYLRKLLIR ENQPKSS... .IVSLYKKLE IKHAIEMAET
Nah3ratno YLRDKWSNFD RKFLSKVLMR RSAQKSRD.. RILNVFHELN LKDAI.....
Nhe3didvi YLRDKWSNFD RKVLSKLLMR RSAQKSRD.. RILNVFHELN LKDAI.....
Nah3orycu YLRDKWANFD RRFLSKLLMR QSAQKSRD.. RILNVFHELN LKDAI.....
Nhecarma  HIRSKFKRFN NKYLTPFLVR EKNVIEP... KLIETYSNIK KHEAMQQMHN
Nahcaeel  TFIENFERFN AKVIKPVLMR HQKRESFDAS SIVRAYEKIT LEDAIKLAKV
Consensus ....K...F. .......L.R ....K..... ........... .K.AI.....

               601                                                    650
Nah1homsa GGMGKIPSAV STVSMQNIH. .PKSLPSERI LPALSKDKEE EIRKILRNNL
Nahboncmy GQLPSVLP.. STISMQNIQ. .PRAIPR... ...VSKKREE EIRRILRANL
Nah4ratno GLLSSVA... SPTPYQSER. .IQGIKR... ...LSPEDVE SMRDILTRNM
Nhe2ratno GMISTVP... SFASLNDCR. .EEKIRK... ...LTPGEMD EIREILSRNL
Nah3ratno SYVAEGERRG SLAFIRSPS. TDNMVNVDFS TPRPSTV.EA SVSYFLRENV
```

```
Nhe3didvi SYVAEGERRG SLAFIRSPS. TDNIVNVDFS TPRPSTV.EA SVSYLLRENV
Nah3orycu SYVTEGERRG SLAFIRSPS. TDNMVNVDFS TPRPSTV.EA SVSYLLRESA
Nhecarma  SYNASTNIES FSNLIRNDA. THPHVQMNNQ GEWN.............
Nahcaeel  KNNIQNKR.. .LERIKSKGR VAPILPDKIS NQKTMTPKDL QLKRFMESGE
Consensus .......... S.......... .......... .......... .....L....
```

```
          651                                              700
Nah1homsa QKTRQRL..R SYNRHTLVAD PY....EEAW NQMLLRRQ.. ......K...
Nahboncmy QNNKQKMRSR SYSRHTLFDA DE....EDNV SEVRLRKT.. .....KMEM
Nah4ratno YQVRQR..TL SYNKYNLKPQ TS....EKQA KEILIRRQNT LRESLRKGQS
Nhe2ratno YQIRQR..TL SYNRHNLTAD TS....ERQA KEILIRRRHS LRESLRKDNS
Nah3ratno SAVCLDMQSL EQRRRSIRDT EDMVTHHTLQ QYLYKPRQEY KHLYSRHELT
Nhe3didvi STVCLDMQAL EQRRRSIRDT EDTVTHHTLQ QYLYKPRQEY KHLYSRHELT
Nah3orycu SAVCLDMQSL EQRRRSVRDA EDVITHHTLQ QYLYKPRQEY KHLYSRHVLS
Nhecarma  ....LDVAEL EYNP.TLRDL NDAKFHHLLS NDYKPVKKNR ASTYKRHAVK
Nahcaeel  NIDSLYTLFS DLLDRKLHEM NRPSVQITDV DGQDDIQDDY MAEVSRSNLS
Consensus .......... .......... .......... .......... .....R....
```

```
          701                                              750
Nah1homsa ...ARQLEQK INNYLTVP.A HKLDS.PTMS RARIGSDPLA YEPKEDLPVI
Nahboncmy ERRVSVMERR MSHYLTVP.A NRESPRPGVR RVRFESDNQV FSA.DSFPTV
Nah4ratno LPWVKPAGTK NFRYLSFPYS NPQPARRGAR AAESTGNPCC WLLHFLLCRA
Nhe2ratno LNRERRASTS TSRYLSLPKN TKLPEKLQKK NKVSNADGNS .......SDS
Nah3ratno P...NEDEKQ DKEIFHRTMR KRLESFKSAK LGINQNKKAA KLYK.RERAQ
Nhe3didvi S...NEDEKQ DKEIFHRTMR KRLESFKSTK LGINQTKKTA KLYK.RERGQ
Nah3orycu P...SEDEKQ DKEIFHRTMR KRLESFKSAK LGLGQSKKAT KHKRERERAQ
Nhecarma  ....DDDMQT QSDIGHHNM. .YLHTHTHTQ LL........ ..........
Nahcaeel  AMFRSTEQLP SETPFHSGRR QSTGDLNATR RADFNV.... .... ......
Consensus .......... .......... .......... .......... ..........
```

```
          751                                              800
Nah1homsa TIDPASPQS. PESVD..LVN EELKGKVLGL S.RDPAKVAE EDEDDDGGIM
Nahboncmy HFEQPSPPST PDAVS..L.. .......... ....... .EEEEEEVPK
Nah4ratno MVEKIWGPGG QETQPRLLCR NLN....... .......... ..........
Nhe2ratno DMDGTTVLNL QPRARRFLPD QFSKKASPAY K.MEWKNEVD VGSARAPPSV
Nah3ratno KRRNSSIPNG KLPMENLAHN FTIKEKDLEL SEPEEATN.Y ..EEISGGIE
Nhe3didvi KRRNSSIPNG KIPMESPTRD FTFKEKELEF SDPEETNE.Y EAEEMSGGIE
Nah3orycu KRRNSSVPNG KLPLDSPAYG LTLKERELEL SDPEEAPDYY EAEKMSGGIE
Consensus .......... .......... .......... .......... ..........
```

```
          801                                              850
Nah1homsa MRSKETSSPG TDDVFTPAPS DSPSSQRIQR CLSDPGPHPE PGEGEPFFPK
Nahboncmy RPSLKADIEG PRGNASDNHQ GELDYQRLAR CLSDPGPNKD KEDDDPFMSC
Nhe2ratno TPAPRSKEGG TQTPGVLRQP LLSKDQRFGR GREDSLTEDV PPKPPPRLVR
Nah3ratno FLASVTKDVA SDSGAGIDNP VFSPDEDLDP SILSRVPPWL SPGETVVPSQ
Nhe3didvi FLANVTQDTA TDSTTGIDNP VFSPEE..DQ SIFTKVPPWL SPEETVVPSQ
Nah3orycu FLASVTKDTT SDSPAGIDNP VFSPDEDLAP SLLARVPPWL SPGEAVVPSQ
Consensus .......... .......... .......... .......... ..........
```

```
          851                                          899
Nah1homsa GQ........ .......... .......... .......... .........
Nhe2ratno RASEPGNRKG RLGNEKP... .......... .......... .........
Nah3ratno RARVQIPNSP SNFRRLTPFR LSNKSVDSFL QADGPEEQLQ PASPESTHM
Nhe3didvi RARVQIPYSP SNFRRLTPIR LSTKSVDSFL LADSPEERPR SFLPESTHM
```

```
Nah3orycu RARVQIPYSP GNFRRLAPFR LSNKSVDSFL LAEDGAEH.. ...PESTHM
Consensus  ..........  ..........  ..........  ..........  ..........
```

Proteins listed subsequently in italics are at least 90% identical to the paired transporters listed in parenthesis and therefore are not included in the alignments: *Nah1orycu, Nah1ratno, Nhe1musmu* (Nah1homsa); *Nhe2orycu* (Nhe2ratno). Residues listed in the consensus sequence are present in at least 75% of the aligned transporter sequences.

Database accession numbers

	SWISSPROT	*PIR*	*EMBL/GENBANK*
Nah1homsa	P19634	A31311	M81768
Nah1orycu	P23791	S13926; S16328	X59935; X56536
Nah1ratno	P26431	A40204	M85299
Nah3orycu	P26432	A40205	M87007
Nah3ratno	P26433	B40204	M85300
Nah4ratno	P26434	C40204	M85301
Nahboncmy	Q01345		M94581
Nhe1musmu			U51112
Nhe2orycu		A46747	
Nhe2ratno		A46748; A47449	L11236
Nhe3didvi			L42522
Nhecarma			U09274

References

1. **Reinhart, A.F. and Reithman, R. (1994) Curr. Opin. Cell Biol. 6, 583–594.**
2. Tse, C.M. et al. (1991) EMBO J. 10, 1957–1967.
3. Levine, S. et al. (1993) J. Biol. Chem. 268, 25527–25535.

Part 12

PEP-Dependent Phosphotransferase Family

Summary

Transporters of the phosphoenolpyruvate-dependent sugar phosphotransferase system (PTS) family, the example of which is the PTAA N-acetyl glucosamine permease II ABC protein of *Escherichia coli* (Ptaaescco), mediate the phosphoenolpyruvate (PEP)-dependent uptake of a variety of sugars. Unphosphorylated PTGA serves an integrative regulatory function by binding, and thereby inhibiting, the activities of several other enzymes and transporters [1,2]. Members of the PTS transporter family are found in both gram-positive and gram-negative bacteria.

Statistical analysis reveals no apparent relationship between the amino acid sequences of the PTS family and any other family of transporters. However, a subgroup of the galactose-pentose-hexuronide (GPH) family of cation-coupled symporters contains a C-terminal hydrophilic extension of approximately 160 residues that is homologous to the IIA domain of PTGA. Phosphorylation of this IIA-like domain by PEP, heat-stable protein HPR and enzyme 1 inhibits the transport activities of these GPH proteins [3].

Members of the PTS family are predicted to contain from 6 to 14 membrane-spanning helices by the hydropathy of their amino acid sequences. The structure of the PTGA protein of *Escherichia coli* has been resolved by NMR and X-ray techniques to 2.5 Å [4,5]. Athough several amino acid sequence motifs are highly conserved in the family, they are not conserved between the most distantly related members of the family. Similarly, while all members of the PTS family contain sites of histidine phosphorylation, the locations of these sites are not conserved between distantly related members of the family.

Nomenclature, biological sources and substrates

CODE	DESCRIPTION [SYNONYMS]	ORGANISM [COMMON NAMES]	SUBSTRATE(S)
Ptaaescco	N-Acetyl glucosamine permeaseII ABC component [PTAA, NAGE, PSTN]	*Escherichia coli* [gram-negative bacterium]	N-Acetyl glucosamine
Ptaaklepn	N-Acetyl glucosamine permeaseII ABC component [PTAA, NAGE]	*Klebsiella pneumoniae* [gram-negative bacterium]	N-Acetyl glucosamine
Ptbabacsu	β-Glucoside permease IIABC component [PTBA, BGLP, N17C]	*Bacillus subtilis* [gram-positive bacterium]	β-Glucosides
Ptbaescco	β-Glucoside permease IIABC component [PTBA, BGLF, BGLC]	*Escherichia coli* [gram-negative bacterium]	β-Glucosides
Ptbaerwch	β-Glucoside permease IIABC component [PTBA, ARBF]	*Erwinia chrysanthemi* [gram-negative bacterium]	β-Glucosides
Ptgahaein	Glucose permease IIABC component [PTGA, CRR, HI1711]	*Haemophilus influenzae* [gram-negative bacterium]	Glucose
Ptgamycca	Glucose permease IIABC component [PTGA, CRR]	*Mycobacterium capricolum* [gram-positive bacterium]	Glucose
Ptgasalty	Glucose permease IIABC component [PTGA, CRR]	*Salmonella typhimurium* [gram-negative bacterium]	Glucose

CODE	DESCRIPTION [SYNONYMS]	ORGANISM [COMMON NAMES]	SUBSTRATE(S)
Ptgaescco	Glucose permease IIABC component [PTGA, CRR, GSR, IEX]	*Escherichia coli* [gram-negative bacterium]	Glucose
Ptgabacsu	Glucose permease IIABC component [PTGA, PTSG, CRR]	*Bacillus subtilis* [gram-positive bacterium]	Glucose
Ptgabacst	Glucose permease IIABC component [PTGA, PTSG]	*Bacillus stearo-thermophilus* [gram-positive bacterium]	Glucose
Ptgbsalty	Glucose permease IIBC component [PTGB, PTSG]	*Salmonella typhimurium* [gram-negative bacterium]	Glucose
Ptgbescco	Glucose permease IIBC component [PTGB, PTSG, GLCA]	*Escherichia coli* [gram-negative bacterium]	Glucose
Pticescco	PTS system, arbutin-like IIC component [PTIC, GLVC]	*Escherichia coli* [gram-negative bacterium]	Glucose
Ptoaescco	Maltose-glucose permease IIABC component [PTOA, MALX]	*Escherichia coli* [gram-negative bacterium]	Maltose, glucose
Ptsapedpe	Sucrose permease IIABC component [PTSA, SCRA]	*Pediococcus pentosaceus* [gram-positive bacterium]	Sucrose
Ptsastrmu	Sucrose permease IIABC component [PTSA, SCRA]	*Streptococcus mutans* [gram-positive bacterium]	Sucrose

Phylogenetic tree

Proteins listed subsequently in italics are at least 90% identical to the paired transporters listed in parenthesis and therefore are not included in the phylogenetic tree: *Ptgasalty* (Ptgaescco); *Ptgbsalty* (Ptgbescco).

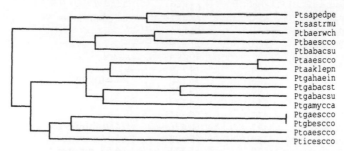

Proposed orientation of PTGAB in the membrane

The model is based on predictions of membrane-spanning regions and α-helical content. The N-terminus of the protein is illustrated on the inside and is folded 12 times through the membrane. The predicted membrane-spanning helices are portrayed as rectangles. The numbers corresponding to the first and last residue of each membrane-spanning helix are boxed. Residues that are conserved in more than 75% of the aligned transporters (see below) are shown.

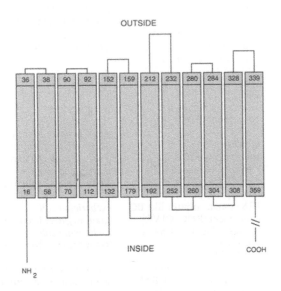

OUTSIDE

INSIDE

COOH

NH₂

Physical and genetic characteristics

	AMINO ACIDS	MOL. WT	CHROMOSOMAL LOCUS
Ptaaescco	648	68 346	72.02 minutes
Ptaaklepn	651	68 179	
Ptbabacsu	609	64 550	335°
Ptbaescco	625	66 482	84.05 minutes
Ptbaerwch	631	66 984	
Ptgahaein	165	17 779	
Ptgamycca	154	16 703	
Ptgasalty	168	18 116	
Ptgaescco	168	18 120	54.6 minutes
Ptgabacsu	699	75 525	118°
Ptgabacst	324	34 674	
Ptgbsalty	477	50 521	
Ptgbescco	477	50 676	24.96 minutes
Pticescco	368	39 692	83.16 minutes
Ptoaescco	530	56 721	36.53 minutes
Ptsapedpe	651	68 454	
Ptsastrmu	664	69 988	

Multiple amino acid sequence alignments

```
          1                                                  50
Ptsapedpe .................................................MNHQEV
Ptsastrmu .................................................MDYSKV
Ptbaerwch .................................................MNYETL
Ptbaescco ...................................................MTEL
Ptbabacsu .................................................MDYDKL
```

```
Ptaaescco ................. ............ .............. .......
Ptaaklepn ................. ............ .............. .......
Ptgahaein ................. ............ .............. .......
Ptgabacst ................. ............ .............. .......
Ptgabacsu ......MFKA LFGVLQKIGR ALMLPVAILP AAGILLAIGN AMQNKDMIQV
Ptgamycca ................. ............ .............. .......
Ptgaescco ......MFKN AFANLQKVGK SLMLPVSVLP IAGILLGVGS ANFS.....W
Ptgbescco ......MFKN AFANLQKVGK SLMLPVSVLP IAGILLGVGS ANFS.....W
Ptoaescco MTAKTAPKVT LWEFFQQLGK TFMLPVALLS FCGIMLGIGS SLSSHDVITL
Pticescco .......... MLSQIQRFGG AMFTPVLLFP FAGIVVGLAI LLQNPMFVGE
Consensus ................. ............ .............. .......
```

```
              51                                            100
Ptsapedpe ADRVLNAIG. KNNIQAAAHC ATRLRLVIKD ESKIDQQALD DDADVKGTFE
Ptsastrmu ASEVITAVG. KDNLVAAAHC ATRLRLVLKD DSKVDQKALD KNADVKGTFK
Ptbaerwch ASEIRDGVGG QENIISVIHC ATRLRFKLRD NTNANADALK NNPGIIMVVE
Ptbaescco ARKIVAGVGG ADNIVSLMHC ATRLRFKLKD ESKAQAEVLK KTPGIIMVVE
Ptbabacsu SKDILQLVGG EENVQRVIHC MTRLRFNLHD NAKADRSQLE QLPGVMGTNI
Ptaaescco ....MNILGF FQRLGRALQL PIAVLPVAAL LLRFGQPDLL NVAFIAQAGG
Ptaaklepn ....MNILGF FQRLGRALQL PIAVLPVAAL LLRFGQPDLL NVPFIAQAGG
Ptgahaein ................. ............ .............. .......
Ptgabacst ................. ............ .............. .......
Ptgabacsu LHFLSNDNVQ LVAGVMESAG QIVFDNLPLL FAVGVAIGLA NGDGVAGIAA
Ptgamycca ................. ............ .............. .......
Ptgaescco LPAVVSHV.. .....MAEAG GSVFANMPLI FAIGVALGFT .NNDGVSALA
Ptgbescco LPAVVSHV.. .....MAEAG GSVFANMPLI FAIGVALGFT .NNDGVSALA
Ptoaescco IPVLGNPVLQ AIFTWMSKIG SFAFSFLPVM FCIAIPLGLA RENKGVAAFA
Pticescco SLTDPNSLFA QIVHIIEEGG WTVFRNMPLI FAVGLPIGLA KQAQGRACLA
Consensus ................. ............ .............. .......
```

```
              101                                           150
Ptsapedpe TNGQYQIIIG PGDVDKVYDA LIVKTGL..K EVTPDDIKAV AAAGQNKNPL
Ptsastrmu TDGQYQVIIG PGDVNFVYDE IIKQTGL..T EVSTDDLKKI AASGKKFNPI
Ptbaerwch SGGQFQVVVG .NQVADVYQA LLSLDGM..A RFSD...SAA PEEEKKNSLF
Ptbaescco SGGQFQVVIG .NHVADVFLA VNSVAGL..D E.KA...QQA PENDDKGNLL
Ptbabacsu SGEQFQIIIG .NDVPKVYQA IVRHSNL..S D..E...KSA GSSSQKKNVL
Ptaaescco AIFDNLALIF AIGVASSWSK DSAGAAALAG AVGYFVLTKA MVTINPEINM
Ptaaklepn AIFDNLALIF AIGVASSWSK DNAGSAALAG AVGYFVMTKA MVTINPEINM
Ptgahaein ................. ............ .............. .......
Ptgabacst ................. ............ .............. .......
Ptgabacsu IIGYLVMNVS MSAVLLANGT IPSDSVERAK FFTENHPAYV NMLGIPTLAT
Ptgamycca ................. ............ .............. .......
Ptgaescco AVVAYGIM.. .....VKTMA VV......AP LVLHLPAE.. EIASKHLADT
Ptgbescco AVVAYGIM.. .....VKTMA VV......AP LVLHLPAE.. EIASKHLADT
Ptoaescco GFIGYAVMNL AVNFWLTNKG ILPTTD..AA VLKANNIQ.. SILGIQSYDT
Pticescco VMVSFLTWNY FINAMGMTWG SYFGVDFTQD AVAGSGLT.. MMAGIKTLDT
Consensus ................. ............ .............. .......
```

```
              151                                           200
Ptsapedpe MDFLKVLSDI FIPIVPALVA GGLLMALNNV LTAEHLFMAK SVVEVYPGLK
Ptsastrmu MALIKLLSDI FVPIIPALVA GGLLMALNNF LTSEGLFGTK SLVQQFPIIK
Ptbaerwch SGFIDIISSI FTPFVGVMAA TGILKGFLAL GVATHVISES S.........
Ptbaescco NRFVYVISGI FTPLIGLMAA TGILKGMLAL ALTFQWTTEQ S.........
Ptbabacsu SAVFDVISGV FTPILPAIAG AGMIKGLVAL AVTFGWMAEK S.........
```

```
Ptaaescco GVLAGIITGL VGGAAYNRWS DIKLPDFLSF FGGKRFVPIA TGFFCLVLAA
Ptaaklepn GVLAGIITGL VAGAVYNRWA GIKLPDFLSF FGGKRFVPIA TGFFCLILAA
Ptgahaein .......... .......... .......... .......... ..........
Ptgabacst .......... .......... .......... .......... ..........
Ptgabacsu GVFGGIIVGV LAALLFNRFY TIELPQYLGF FAGKRFVPIV TSISALILGL
Ptgamycca .......... .......... .......... .......... ..........
Ptgaescco GVLGGIISGA IAAYMFNRFY RIKLPEYLGF FAGKRFVPII SGLAAIFTGV
Ptgbescco GVLGGIISGA IAAYMFNRFY RIKLPEYLGF FAGKRFVPII SGLAAIFTGV
Ptoaescco GILGAVIAGI IVWMLHERFH NIRLPDALAF FGGTRFVPII SSLVMGLVGL
Pticescco SIIGAIIISG IVTALHNRLF DKKLPVFLGI FQGTSYVVII AFLVMIPCAW
Consensus .......... .......... .......... .......... ..........


          201                                                   250
Ptsapedpe GIAEMINAMA SAPFTFLPIL LGFSATKRFG GNPYLGATMG MIMVLPSLVN
Ptsastrmu GSSDMIQLMS AAPFWFLPIL VGISAAKRFG ANQFLGASIG MIMVAPGAAN
Ptbaerwch GTYKLLFAAS DALFYFFPIV LGYTAGKKFG GNPFTTLVIG ATLVHPSMIA
Ptbaescco GTYLILFSAS DALFWFFPII LGYTAGKRFG GNPFTAMVIG GALVHPLILT
Ptbabacsu QVHVILTAVG DGAFYFLPLL LAMSAARKFG SNPYVAAAIA AAILHPDLTA
Ptaaescco IFGYVWPPVQ HAIHA.GGEW IVSAGALGSG IFGFINRLLI PTGLHQVLNT
Ptaaklepn IFGYVWPPVQ HAIHS.GGEW IVSAGALGSG IFGFINRLLI PTGLHQVLNT
Ptgahaein .......... .......... .......... .......... ..........
Ptgabacst .......... .......... .......... .......... ..........
Ptgabacsu IMLVIWPPIQ HGLNAFSTGL VEANPTLAAF IFGVIERSLI PFGLHHIFYS
Ptgamycca .......... .......... .......... .......... ..........
Ptgaescco VLSFIWPPIG SAIQTFSQWA AYQNPVVAFG IYGFIERCLV PFGLHHIWNV
Ptgbescco VLSFIWPPIG SAIQTFSQWA AYQNPVVAFG IYGFIERCLV PFGLHHIWNV
Ptoaescco VIPLVWPIFA MGISGLGHMI NSAGDFGPM. LFGTGERLLL PFGLHHILVA
Pticescco LTLLGWPKVQ MGIESLQAFL RSAGALGVW. VYTFLERILI PTGLHHFIYG
Consensus .......... .......... .......... .......... ..........


          251                                                   300
Ptsapedpe GYSVATTMAA GK.......M VYWNVFGLHV AQAGYQGQVL PVLGVAFILA
Ptsastrmu IIGLAANAPI SKAATIGAYT GFWNIFGLHV TQASYTYQVI PVLVAVWLLS
Ptbaerwch AFNAMQAPDH ST........ ..LHFLGIPI TFINYSSSVI PILFASWVSC
Ptbaescco AFENGQKADA LG........ ..LDFLGIPV TLLNYSSSVI PIIFSAWLCS
Ptbabacsu LLGAGK.... .P........ ..ISFIGLPV TAATYSSTVI PILLSIWIAS
Ptaaescco IAWF.QIGEF TNAAGTVFHG DINRFYA... .......GDG TAGMFMSGFF
Ptaaklepn IAWF.QIGEF TNAAGTVFHG DINRFYA... .......GDG TAGMFMSGFF
Ptgahaein .......... .......... .......... .......... ..........
Ptgabacst .......... .......... .......... .......... ..........
Ptgabacsu PFWY.EFFSY KSAAGEIIRG DQRIFMAQIK DG.....VQL TAGTFMTGKY
Ptgamycca .......... .......... .......... .......... ..........
Ptgaescco PFQM.QIGEY TNAAGQVFHG DIPRYMA..G DPTAGKLSGG ...FLFK...
Ptgbescco PFQM.QIGEY TNAAGQVFHG DIPRYMA..G DPTAGKLSGG ...FLFK...
Ptoaescco LIRFTDAGGT QEVCGQTVSG ALTIFQAQLS CPTTHGFSES ATRFLSQGKM
Pticescco QFIFGPAA.. VEGGIQMYWA QHLQEFSLSA EPLKSLFPEG GFALHGNSK.
Consensus .......... .......... .......... .......... ..........


          301                                                   350
Ptsapedpe TLEKFFHKHI KGAFDFTFTP MFAIVITGFL TFTIVGPVLR TVSDALTNGL
Ptsastrmu ILEKFFHKRL PSAVDFTFTP LLSVIITGFL TFIVIGPVMK EVSDWLTNGI
Ptbaerwch KLEKPLNRWL HANIRNFFTP LLCIVISVPL TFLLIGPSAT WLSQMLAGGY
Ptbaescco ILERRLNAWL PSAIKNFFTP LLCLMVITPV TFLLVGPLST WISELIAAGY
Ptbabacsu YVEKWIDRFT HASLKLIVVP TFTLLIVVPL TLITVGPLGA ILGEYLSSGV
```

```
Ptaaescco PIMMFGLPGA ALAMYFAAPK ERRPMVGGML LSVAVTAFLT GVTEPLEFLF
Ptaaklepn PIMMFGLPGA ALAMYLAAPK ARRPMVGGML LSVAITAFLT GVTEPLEFLF
Ptgahaein .........................................................
Ptgabacst .........................................................
Ptgabacsu PFMMFGLPAA ALAIYHEAKP QNKKLVAGIM GSAALTSFLT GITEPLEFSF
Ptgamycca .........................................................
Ptgaescco ...MYGLPAA AIAIWHSAKP ENRAKVGGIM ISAALTSFLT GITEPIEFSF
Ptgbescco ...MYGLPAA AIAIWHSAKP ENRAKVGGIM ISAALTSFLT GITEPIEFSF
Ptoaescco NAFLGGLPGA ALAMYHCARP ENRHKIKGLL ISGLIACVVG GTTEPLEFLF
Pticescco ...IFGAVGI SLAMYFTAAP ENRVKVAGLL IPATLTAMLV GITEPLEFTF
Consensus .........................................................
```

```
          351                                                    400
Ptsapedpe VGLYNSTGWI GMGIFGLLYS AIVITGLHQT FP.AIETQLL ANVAKTG..G
Ptsastrmu VWLYDTTGFL GMGVFGALYS PVVMTGLHQS FP.AIETQLI SAFQNGTGHG
Ptbaerwch QWLYGLNSLL AGAVMGALWQ VCVIFGLHWG FV.PLMLNNF SVIGH.....
Ptbaescco LWLYQAVPAF AGAVMGGFWQ IFVMFGLHWG LV.PLCINNF TVLGY.....
Ptbabacsu NYLFDHAGLV AMILLAGTFS LIIMTGMHYA FV.PIMINNI AQNGH.....
Ptaaescco MFLAPLLYLL HALLTGISLF VATLLGIHAG FSFSAGAIDY ALMYNLPAAS
Ptaaklepn LFLAPLLYLL HAVLTGISLF IATALGIHAG FSFSAGAIDY VLMYSLPAAS
Ptgahaein .........................................................
Ptgabacst ................. ..HLLNVKIG MTFSGGVIDF .LLFGVLPNR
Ptgabacsu LFVAPVLFAI HCLFAGLSFM VMQLLNVKIG MTFSGGLIDY .FLFGILPNR
Ptgamycca .........................................................
Ptgaescco MFVAPILYII HAILAGLAFP ICILLGMRDG TSFSHGLIDF IVLS...GNS
Ptgbescco MFVAPILYII HAILAGLAFP ICILLGMRDG TSFSHGLIDF IVLS...GNS
Ptoaescco LFVAPVLYVI HALLTGLGFT VMSVLGVTIG NT.DGNIIDF VVFGILHGLS
Pticescco LFISPLLFAV HAVLAASMST VMYLFGV..V GNMGGGLID. .........
Consensus .........................................................
```

```
          401                                                    450
Ptsapedpe SFIFPVASMA NIGQGAATLA IFFATKSQKQ KALTSSAGVS ALLGITEPAI
Ptsastrmu DFIFVTASMA NVAQGAATFA IYFLTKDKKM KGLSSSSGVS ALLGITEPAL
Ptbaerwch DTLLPLLVPA VLGQAGATLG VLLRTQDLKR KGIAGSAFSA AIFGITEPAV
Ptbaescco DTMIPLLMPA IMAQVGAALG VFLCERDAQK KVVAGSAALT SLFGITEPAV
Ptbabacsu DYLLPAMFLA NMGQAGASFA VFLRSRNKKF KSLALTTSIT ALMGITEPAM
Ptaaescco QNVWMLLVMG VIFFAIYFVV FSLVIRMFNL KTPGREDKED EIVTEEANSN
Ptaaklepn KNVWMLLVMG VVFFFVYFLL FSAVIRMFNL KTPGREDKAA DVVTEEANSN
Ptgahaein .........................................................
Ptgabacst TAWWLVIPVG LVFAVIYYFG FRFAIRKWDL ATPGREK... .TVEE.APKA
Ptgabacsu TAWWLVIPVG LGLAVIYYFG FRFAIRKFNL KTPGRED... .AAEETAAPG
Ptgamycca .........................................................
Ptgaescco SKLWLFPIVG IGYAIVYYTI FRVLIKALDL KTPGRE.... DATEDAKATG
Ptgbescco SKLWLFPIVG IGYAIVYYTI FRVLIKALDL KTPGRE.... DATEDAKATG
Ptoaescco TKWYMVPVVA AIWFVVYYVI FRFAITRFNL KTPGRDSRVA SSIEKAVAGA
Pticescco .........................................................
Consensus .........................................................
```

```
          451                                                    500
Ptsapedpe FGVNLKMKFP FVFAAIASGI ASAFLGLFHV LSVAMGPASV IGFIS.IASK
Ptsastrmu FGVNLKYRFP FFCALIGSAS AAAIAGLLQV VAVSLGSAGF LGFLS.IKAS
Ptbaerwch YGVTLPLRRP FIFGCIGGAL GAAVMGYAHT TMYSFGFPSI FSFTQVIPPT
Ptbaescco YGVNLPRKYP FVIACISGAL GATIIGYAQT KVYSFGLPSI FTFMQTIPST
Ptbabacsu YGVNMRLKKP FAAALIGGAA GGAFYGMTGV ASYIVGGNAG LPSIPVF...
```

```
Ptaaescco TEEGLTQLAT NYIAAVGGTD NLKAIDACIT RLRLTVADSA RVNDTMCKRL
Ptaaklepn TEEGLTQLAT SYIAAVGGTD NLKAIDACIT RLRLTVGDSA KVNDAACKRL
Ptgahaein ..........  .......... .......... .......... ..........
Ptgabacst EAAAAGDLPY EVLAALGGKE NIEHLDACIT RLRVSVHDIG RVDKDRLKAL
Ptgabacsu KTGEAGDLPY EILQAMGDQE NIKHLDACIT RLRVTVNDQK KVDKDRLKQL
Ptgamycca ..........  .......... .......... .......... ..........
Ptgaescco TSEMA...P. ALVAAFGGKE NITNLDACIT RLRVSVADVS KVDQAGLKKL
Ptgbescco TSEMA...P. ALVAAFGGKE NITNLDACIT RLRVSVADVS KVDQAGLKKL
Ptoaescco PGKSGYNVP. AILEALGGAD NIVSLDNCIT RLRLSVKDMS LVNVQALKDN
Pticescco ..........  .......... .......... .......... ..........
Consensus ..........  .......... .......... .......... ..........

          501                                                 550
Ptsapedpe SIPAFMLSAV ISFVVAFIPT FIYAK..... .........R TLGDDRDQVK
Ptsastrmu SIPFYVVCEL ISFAIAFAVT YGYGKTKAVD VFAAEAAVEE AIEEVQEIPE
Ptbaerwch GVDSSVWAAV IGTLLAFAFA ALTS...... WSFGVPKDET QPAAADSPAV
Ptbaescco GIDFTVWASV IGGVIAIGCA FVGT...... VMLHFITAKR QPAQG...AP
Ptbabacsu .IGPTFIYAM IGLVIAFAAE TAAA...... YLLGF..EDV PSDGSQQPAV
Ptaaescco GASGVVKLNK QTIQVIVGAK AESIGDAMKK VVARGPVAAA SAEATP..AT
Ptaaklepn GASGVVKLNK QTIQVIVGAK AESIGDEMKK VVTRGPVAAA AAAPAGNVAT
Ptgahaein ..........  .......... .......... .......... ..........
Ptgabacst GAAGVLEVGN N.VQAIFGPK SDMLKGQIQD IMQGKAP... ..........
Ptgabacsu GASGVLEVGN N.IQAIFGPR SDGLKTQMQD IIAGRKPRPE PKTSAQEEVG
Ptgamycca ..........  .......... .......... .......... ..........
Ptgaescco GAAGVV.VAG SGVQAIFGTK SDNLKTEMDE YIRNH..... ..........
Ptgbescco GAAGVV.VAG SGVQAIFGTK SDNLKTEMDE YIRNH..... ..........
Ptoaescco RAIGVVQLNQ HNLQVVIGPQ VQSVKDEMAG LMHTVQA... ..........
Pticescco ..........  .......... .......... .......... ..........
Consensus ..........  .......... .......... .......... ..........

          551                                                 600
Ptsapedpe SPAPTSTVIN VN....DEII SAPVTGASES LKQVNDQVFS AEIMGKGAAI
Ptsastrmu EAASAANKAQ VT....DEVL AAPLAGEAVE LTSVNDPVFS SEAMGKGIAI
Ptbaerwch LAETQANAGA VR....DETL FSPLAGEVLL LEQVADRTFA SGVMGKGIAI
Ptbaescco QEKTPEVITP PE....QGGI CSPMTGEIVP LIHVADTTFA SGLLGKGIAI
Ptbabacsu HEGSREI... ........I HSPIKGEVKA LSEVKDGVFS AGVMGKGFAI
Ptaaescco AAPVAKPQAV PNAVSIA.EL VSPITGDVVA LDQVPDEAFA SKAVGDGVAV
Ptaaklepn AAPAAKPQAV ANAKTVE.SL VSPITGDVVA LEQVPDEAFA SKAVGDGIAV
Ptgahaein .GLFDKLFGS KENKSVEVEI YARISGEIVN IEDVPDVVFS EKIVGDGVAV
Ptgabacst ARAEEKPKTA ASEAAESETI ASPMSGEIVP LAEVPDQVFS QKMMGDGFAV
Ptgabacsu QQVEEVIAEP LQNEIGEEVF VSPITGEIHP ITDVPDQVFS GKMMGDGFAI
Ptgamycca .......... MWFFNKNLKV LAPCDGTIIT LDEVEDEVFK ERMLGDGFAI
Ptgaescco ..........  .......... .......... .......... ..........
Ptgbescco ..........  .......... .......... .......... ..........
Ptoaescco ..........  .......... .......... .......... ..........
Pticescco ..........  .......... .......... .......... ..........
Consensus ..........  .......... .......... .......... ..........

          601                                                 650
Ptsapedpe VPSSDQVVAP ADGVITVTYD SHHAYGIKTT AGAEILIHLG LDTVNLNGEH
Ptsastrmu KPSGNTVYAP VDGTVQIAFD TGHAYGIKSD NGAEILIHIG IDTVSMEGKG
Ptbaerwch RPTQGRLYAP VDGTVASLFK THHAIGLASR GGAEVLIHVG IDTVRLDGRY
Ptbaescco LPSVGEVRSP VAGRIASLFA TLHAIGIESD DGVEILIHVG IDTVKLDGKF
Ptbabacsu EPEEGEVVSP VRGSVTTIFK TKHAIGITSD QGAEILIHIG LDTVKLEGQW
```

```
Ptaaescco KPTDKIVVSP AAGTIVKIFN TNHAFCLETE KGAEIVVHMG IDTVALEGKG
Ptaaklepn KPTDNIVVAP AAGTVVKIFN TNHAFCLETN NGAEIVVHMG IDTVALEGKG
Ptgahaein RPIGNKIVAP VDGVIGKIFE TNHAFSMESK EGVELFVHFG IDTVELKGEG
Ptgabacst MPTDGTVVSP VDGKIINVFP TKHAIGIQSA GGHEILIHVG IDTVKLNGQG
Ptgabacsu LPSEGIVVSP VRGKILNVFP TKHAIGLQSD GGREILIHFG IDTVSLKGEG
Ptgamycca NPKSNDFHAP VSGKLVTAFP TKHAFGIQTK SGVEILLHIG LDTVSLDGNG
Ptgaescco .......... .......... .......... .......... ..........
Ptgbescco .......... .......... .......... .......... ..........
Ptoaescco .......... .......... .......... .......... ..........
Pticescco .......... .......... .......... .......... ..........
Consensus .......... .......... .......... .......... ..........

             651                                              700
Ptsapedpe FTTNVQKGDT VHQGDLLGTF DIAALKAANY DPTVMLIVTN TANYANVERL
Ptsastrmu FEQKVQADQK IKKGDVLGTF DSDKIAEAGL DNTTMFIVTN TADYASVETL
Ptbaerwch FTPHVRVGDV VRQGDLLLEF DGPAIEAAGY DLTTPIVITN SEDYRGVEPV
Ptbaescco FSAHVNVGDK VNTGDRLISF DIPAIREAGF DLTTPVLISN SDDFTDVLPH
Ptbabacsu FTAHIKEGDK VAPGDPLVSF DLEQIKAAGY DVITPVIVTN TDQYSFSPVK
Ptaaescco FKRLVEEGAQ VSAGQPILEM DLDYLNANAR SMISPVVCSN IDDFSGLIIK
Ptaaklepn FKRLVEEGTD VKAGEPILEM DLDFLNANAR SMISPVVCSN SDDYSALVIL
Ptgahaein FTRIAQEGQS VKRGDTVIEF DLALLESKAK SVLTPIVISN MDEISC.IVK
Ptgabacst FEALVKEGDE VKKGQPILRV DLDYVKQNAP SIVTPVIFTN LQAGETVHVN
Ptgabacsu FTSFVSEGDR VEPGQKLLEV DLDAVKPNVP SLMTPIVFTN LAEGETVSIK
Ptgamycca FESFVTQDQE VNAGDKLVTV DLKSVAKKVP SIKSPIIFTN .NGGKTLEIV
Ptgaescco .......... .......... .......... .......... ..........
Ptgbescco .......... .......... .......... .......... ..........
Ptoaescco .......... .......... .......... .......... ..........
Pticescco .......... .......... .......... .......... ..........
Consensus .......... .......... .......... .......... ..........

             701                       728
Ptsapedpe .KVTNVQAGE QLVALTAPAA SSVAATTV
Ptsastrmu ASSGTVAVGD SLLEVKK... ........
Ptbaerwch A.SGKVDANA PLTQLVC... ........
Ptbaescco G.TAQISAGE PLLSIIR... ........
Ptbabacsu E.IGKVQPKE ALLALS.... ........
Ptaaescco AQGHIVAGQT PLYEIKK... ........
Ptaaklepn ASGKVVAGQT PLYEIKGK.. ........
Ptgahaein KSGEVVAGES VVLALKK... ........
Ptgabacst KQGPVARGED AVVTIR.... ........
Ptgabacsu ASGSVNREQE DIVKIEK... ........
Ptgamycca KMGEVK..QG DVVAILK... ........
Ptgaescco .......... .......... ........
Ptgbescco .......... .......... ........
Ptoaescco .......... .......... ........
Pticescco .......... .......... ........
Consensus .......... .......... ........
```

Proteins listed subsequently in italics are at least 90% identical to the paired transporters listed in parenthesis and therefore are not included in the alignments: *Ptgasalty* (Ptgaescco); *Ptgbsalty* (Ptgbescco). Residues listed in the consensus sequence are present in at least 75% of the aligned transporter sequences.

Database accession numbers

	SWISSPROT	PIR	EMBL/GENBANK
Ptaaescco	P09323	B29895; A28896	M19284
Ptaaklepn	P45604		X63289
Ptbabacsu	P40739	S47174	Z34526; D31856
Ptbaescco	P08722	A47616; C25977	M15746; M16487
Ptbaerwch	P26207	B42603	M81772
Ptgahaein	P45338		U32844
Ptgamycca	P45618		U15110
Ptgasalty	P02908	A03405	X05210
Ptgaescco	P08837	C29785	J02796; M21994
Ptgabacsu	P20166	S22752	M60344; Z11744
Ptgabacst	P42015		U12340
Ptgbsalty	P37439	S36620	X74629
Ptgbescco	P05053	A25336	J02618
Pticescco	P31452		L10328
Ptoaescco	P19642	PV0011; B42477	M60722; M28539
Ptsapedpe	P43470		Z32771; L32093
Ptsastrmu	P12655	B32243	M22771; D13175

References

1 Meadows, N.D. et al. (1990) Annu. Rev. Biochem. 59, 497–542.
2 Postma, P.W. et al. (1993) Microbiol. Rev. 57, 543–594.
3 Poolman, B. et al. (1996) Mol. Microbiol. 19, 911–922.
4 Worthylake, D. et al. (1991) Proc. Natl Acad. Sci. USA 88, 10382–10386.
5 Hurly, J.H. et al. (1993) Science 259. 673–677.

Part 13

Other Transporters

Summary

Transporters of the anion exchanger family, the example of which is the AE1 anion exchange protein 3 of humans (B3athomsa), mediate a 1 : 1 exchange of inorganic anions across the plasma membrane and regulate intracellular pH and volume. AE1 also has Cl^- channel activity, transports taurine, and provides binding sites for cytoskeletal proteins [1,2]. Members of the family occur in vertebrates, including birds, fish and primates.

Statistical analysis reveals no apparent relationship between the amino acid sequences of the anion exchanger family and any other family of transporters. Members of the anion exchanger family are predicted to form 10 or 12 membrane-spanning helices by the hydropathy of their amino acid sequences. Transporters may be glycosylated, phosphorylated or palmitylated. and exist as tetramers in the membrane.

Members of the anion exchanger family contain several highly conserved amino acid sequence motifs that are necessary for function by the criterion of mutation: defects in AE1 are a cause of hemolytic anemia, hereditary ovalo-cystosis and are associated with hereditary spherocytosis, elliptocytosis [1,2].

Nomenclature, biological sources and substrates

CODE	DESCRIPTION [SYNONYMS]	ORGANISM [COMMON NAMES]	SUBSTRATE(S)
Ae2galga	Anion exchange protein 2 [AE2]	*Gallus gallus* [chicken]	Cl^-/HCO_3^-
B3a2homsa	Anion exchange protein 2 [B3a, AE2, HKB3, EPB3L1, NBND3]	*Homo sapiens* [human]	Cl^-/HCO_3^-
B3a2ratno	Anion exchange protein 2 [B3a, AE2, B3rp2]	*Rattus norvegicus* [rat]	Cl^-/HCO_3^-
B3a2musmu	Anion exchange protein 2 [B3a, AE2, B3Rp]	*Mus musculus* [mouse]	Cl^-/HCO_3^-
B3a3musmu	Anion exchange protein 3 [B3a, AE3]	*Mus musculus* [mouse]	Cl^-/HCO_3^-
B3a3ratno	Anion exchange protein 3 [B3a, AE3, B3rp3]	*Rattus norvegicus* [rat]	Cl^-/HCO_3^-
B3atmusmu	Anion exchange protein 3 [B3aT, AE1, MEB3]	*Mus musculus* [mouse]	Cl^-/HCO_3^-
B3atgalga	Anion exchange protein 3 [B3aT]	*Gallus gallus* [chicken]	Cl^-/HCO_3^-
B3atratno	Anion exchange protein 3 [B3aT, AE1]	*Rattus norvegicus* [rat]	Cl^-/HCO_3^-
B3atoncmy	Anion exchange protein 3 [AE1, B3aT]	*Oncorhynchus mykiss* [trout]	Cl^-/HCO_3^-
B3athomsa	Anion exchange protein 3 [B3aT, AE1, EPB3, EMPB3]	*Homo sapiens* [human]	Cl^-/HCO_3^-

Cotransported ions are indicated.

Phylogenetic tree

Proteins listed subsequently in italics are at least 90% identical to the paired transporters listed in parenthesis and therefore are not included in the phylogenetic tree: *B3a2musmu, B3a2ratno* (B3a2homsa); *B3a3musmu* (B3a3ratno); *B3atmusmu* (B3atratno).

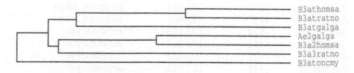

Proposed orientation of AE1 in the membrane

The model is based on predictions of membrane-spanning regions and α-helical content. The N-terminus of the protein is illustrated on the inside and is folded 12 times through the membrane. The predicted membrane-spanning helices are portrayed as rectangles. The numbers corresponding to the first and last residue of each membrane-spanning helix are boxed. More than half of the residues are conserved in at least 75% of the members of the anion exchanger family and, therefore, are not mapped onto the model.

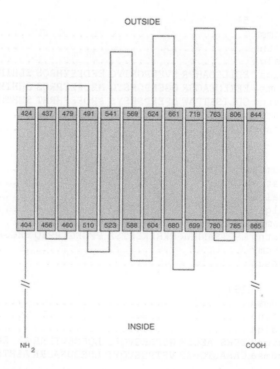

Physical and genetic characteristics

	AMINO ACIDS	MOL. WT	EXPRESSION SITES	CHROMOSOMAL LOCUS
Ae2galga	1219	135 288		
3a2homsa	1240	136 814	ubiquitous	7q35–q36
3a2ratno	1234	136 635	ubiquitous	
3a2musmu	1237	136 813	ubiquitous	
3a3musmu	1227	135 164	neurons, heart	
3a3ratno	1227	135 406	neurons, heart	
3atmusmu	929	103 135	erythrocyte	
3atgalga	922	102 223	erythrocyte	
3atratno	927	103 222	erythrocyte	
3atoncmy	918	101 923	erythrocyte	
3athomsa	911	101 792	erythrocyte	17q21–q22

Multiple amino acid sequence alignments

```
                1                                                    50
3athomsa  ......... ......... ......... ......... .........
3atratno  ......... ......... ......... ......... .........
3atgalga  ......... ......... ......... ......... .........
Ae2galga  MSRSPVSSEL HHIVSSAIES PEPPAPGPAS PPLAEEEEKD LNKALGVERF
3a2homsa  MSSAP...RL PAKGADSFCT PEPESLGPGT PGFPEQEEDE LHRTLGVERF
3a3ratno  MANGVIP... PPGGPSPLPQ VRVPLEEPPL GPDVEEEDDD LGKTLAVSRF
3atoncmy  ......... ......... ......... ......... .........
Consensus ......... ......... ......... ......... .........


                51                                                   100
3athomsa  ......... ......... ......... ......... .........
3atratno  ......... ......... ......... ......... .........
3atgalga  ......... ......... ......... ......... .........
Ae2galga  EEILSDAHPR SVEEPGRIYG EEDFEYHRQS SLHIHHPLSA HLPPDARRKK
3a2homsa  EEILQEAGSR GGEEPGRSYG MEDFEYRRQS SHHIHHPLST HLPPGARRRK
3a3ratno  GDLISKTPAW DPEKPSRSYS ERDFEFHRHT SHHTHHPLSA RLPPPHKLRR
3atoncmy  ......... ......... ......... ......... .........
Consensus ......... ......... ......... ......... .........


                101                                                  150
3athomsa  ......... ......... ......... ......... .........
3atratno  ......... ......... ......... ......... .........
3atgalga  ......... ......... ......... ......... .........
Ae2galga  GVPKKGR... ..KKRGRAAA P..GENPPI. ......EEGE EDEEEACDTE
3a2homsa  TPQGPGR... ..KPRRRPGA SPTGETPTI. ......VEGE EDEDEASEAE
3a3ratno  LPPTSARHAR RKRKKEKTSA PPSEGTPPIQ EEGGAGAEEE EEEEEEEEGE
3atoncmy  ......... ......... ......... ......... .........
Consensus ......... ......... ......... ......... .........


                151                                                  200
3athomsa  ......... ......... ......... ......... .........
3atratno  ......... ......... ......... ......... .........
3atgalga  ......... ......... ......... ......... .........
Ae2galga  TERS.AEELR GGPAEGVQFF LQEDEVTERR AE.EPPAPPA PPGPTAEPHG
3a2homsa  GARALTQPSP VSTPSSVQFF LREDDSADRK AERTSPSSPA PLPHQEATPR
```

```
B3a3ratno SEAEPVEPPP PGPPQKAKFS IGSD...... .EDDSPGLSI KAPCAKALPS
B3atoncmy ......... ......... ......... ......... .........
Consensus ......... ......... ......... ......... .........


          201                                             250
B3athomsa ......... ......... ......... ......... .........
B3atratno ......... ......... ......... ......... .........
B3atgalga ......... ......... ......... ......... .........
Ae2galga  ATAPAAASPG AEEGRA.... .ADGGAVPED GGSPGRPAAR A.TEHRSYNL
B3a2homsa ASKGAQAGTQ VEEAEAEAVA VASGTAGGDD GGASGRPLPK AQPGHRSYNL
B3a3ratno VGLPSDQSPQ RSGSSPSPRA RASRISTEK. ........SR PWSPSASYDL
B3atoncmy ......... ......... ......... ......... .........
Consensus ......... ......... ......... ......... .........


          251                                             300
B3athomsa ......... ......... ....M EELQDDYEDM MEENLEQEEY
B3atratno ......... ......... ....M GDMQDHEKVL EIPDRDSEEE LEHVIEQIAY
B3atgalga ......... ......... ....M EGPGQDTEDA LRRSLDPEGY
Ae2galga  HERRRIGSMT GADEAQYQKV PTDESEAQTL ASADLDYMKS HRFEDVPGVR
B3a2homsa QERRRIGSMT GARQALLPRV PTDEIEAQTL ATADLDLMKS HRFEDVPGVR
B3a3ratno RERLCPGSAL GNPGPE.QRV PTDEAEAQML GSADLDDMKS HRLEDNPGVR
B3atoncmy ........ME NDLSFGEDVM SYEEESDSAF PSPIRPTPPG HSGNYDLEQS
Consensus ......... ......... ......... ...D..... ......... 


          301                                             350
B3athomsa EDPDIPESQM EEPA....AH DTEATATDYH TTSHPGTHKV YVELQELVMD
B3atratno RDLDIPVTEM QESEXXXXXX XXXXXXXXXX XXXXXXXXXX XXXXXXXXMD
B3atgalga EDTKGSRTSL GTMSNPLVSD VDLEAAGSRQ PTAHRDTYEG YVELHELVLD
Ae2galga  RHLVRKSAKA QVVHVGKEHR EQSARPR... RS.DRQPHEV FVELNELVLD
B3a2homsa RHLVRKNAKG .STQSGREGR EPGPTPRARP RA.PHKPHEV FVELNELLLD
B3a3ratno RHLVKKPSRI QGGRGSPSGL APILRRKKKK KKLDRRPHEV FVELNELMLD
B3atoncmy RQEEDSN... QAIQSIVVHT DPEAYLNLNT NANTRGDAQA YVELNELMGN
Consensus ......... ......... ......... ......... .VEL.EL..D


          351                                             400
B3athomsa EKNQELRWME AARWVQLEEN LGENGA.WGR PHLSHLTFWS LLELRRVFTK
B3atratno QRNQELQWVE AAHWIGLEEN LREDGV.WGR PHLSYLTFWS LLELQKVFSK
B3atgalga SRKDP.CWME AGRWLHLEES MEPGGA.WG. SHLPLLTYHS LLELHRAFAK
Ae2galga  .KNQELQWKE TARWIKFEED VEEETDRWGK PHVASLSFRS LLELRKTLSH
B3a2homsa .KNQEPQWRE TARWIKFEED VEEETERWGK PHVASLSFRS LLELRRTLAH
B3a3ratno .RSQEPHWRE TARWIKFEED VEEETERWGK PHVASLSFRS LLELRRTIAQ
B3atoncmy S......WQE TGRWVGYEEN FNPGTGKWGP SHVSYLTFKS LIQLRKIMST
Consensus ......W.E ..RW...EE. ......WG. .HL..L...S LLEL......


          401                                             450
B3athomsa GTVLLDLQET SLAGVANQLL DRFIFEDQIR PQDREELLRA LLLKHSHAGE
B3atratno GTFLLDLAET SLAGVANKLL DSFIYEDQIR PQDRDELLRA LLLKRSHAED
B3atgalga GVVLLDVAAN SLAAVAHVLL DQLIYEGQLK PQHRDDVLRA LLLRHKHPSE
Ae2galga  GAVLLDLDQK TLPGVAHQVV EQMVITDQIR AEDRANVLRA LLLKHSHPSD
B3a2homsa GAVLLDLDQQ TLPGVAHQVV EQMVISDQIK AEDRANVLRA LLLKHSHPSD
B3a3ratno GAALLDLEQT TLPGIAHLVV ETMIVSDQIR PEDRASVLRT LLLKHSHPND
B3atoncmy GAIILDLQAS SLSAVAEKVV DELRTKGEIR AADRDGLLRA LLQRRSQSEG
Consensus G..LLD.... .L.GVA.... ........QI. ..DR...LRA LLL..SH...
```

449

```
              451                                                    500
B3athomsa LEALGGVKPA VLTRS..... G........... .....DPSQP LLPQHSSLET
B3atratno LKDLEGVKPA VLTRS..... G........... .....APSEP LLPHQPSLET
B3atgalga AESVWTLPAA QLQCSDGEQK D........... .....ADERA LLRDQRAVEM
Ae2galga  EKEFS.FPRN ISAGSLGSLL VHHHSTNHVG EGGEPAVTEP LIAGHGAEHD
B3a2homsa EKDFS.FPRN ISAGSVGSLL GHHHGQ...G AESDPHVTEP LM...GGVPE
B3a3ratno DKDSGFFPRN PSSSSVNSVL GNHHPTPSHG PDGAVP...T MADDLGEPAP
B3atoncmy ......... ......... ......... .....AVAQP L...GGDIEM
Consensus ......... ....S..... ......... ......... L.........

              501                                                    550
B3athomsa QLFCEQGDGG TEGHSPSG.. .......... ILEKIPPDSE ATLVLVGRAD
B3atratno KLYCAQAEGG SEEPSPSG.. .......... IL.KIPPNSE TTLVLVGRAS
B3atgalga RELHGAGQSP SRAQLGPQ.. .......... LHQQLPEDTE ATLVLVACAA
Ae2galga  ARVDVERERE VPTPAPPAGI TRSKSKHELK LLEKIPDNAE ATVVLVGCVE
B3a2homsa TRLEVERERD VPPPAPPAGI TRSKSKHELK LLEKIPENAE ATVVLVGCVE
B3a3ratno LWPHDPDAKE KPLHMPGGDG HRGKS...LK LLEKIPEDAE ATVVLVGSVP
B3atoncmy QTFSVTKQRD T......... ......... .....TDSVE ASIVLSGVMD
Consensus ......... ......... ......... .....P...E AT.VLVG...

              551                                                    600
B3athomsa FLEQPVLGFV RLQEAAELE. AVELPVPIRF LFVLLGPEAP HIDYTQLGRA
B3atratno FLVKPVLGFV RLKEAVPLE. DLVLPEPVSF LLVLLGPEAP HIDYTQLGRA
B3atgalga FLEQPLLALV RLGAPCPDA. VLAVPLPVRF VLTVLGPDSP RLSYHEIRRA
Ae2galga  FLDQPTMAFV RLQEAVELDS VLEVPVPVRF LFLLLGPSST HMDYHEIGRS
B3a2homsa FLSRPTMAFV RLREAVELDA VLEVPVPVRF LFLLLGPSSA NMDYHEIGRS
B3a3ratno FLEQPAAAFV RLSEAVLLES VLEVPVPVRF LFVMLGPSHT STDYHELGRS
B3atoncmy SLEKPAVAFV RLGDSVVIEG ALEAPVPVRF VFVLVGPSQG GVDYHESGRA
Consensus FL..P...FV RL........ .L..P.PVRF ....LGP... ..DY...GR.

              601                                                    650
B3athomsa AATLMSERVF RIDAYMAQSR GELLHSLEGF LDCSLVLPPT DAPSEQALLS
B3atratno AATLMTERVF RVTASLAQSR GELLSSLDSF LDCSLVLPPT EAPSEKALLN
B3atgalga AATVMADRVF RRDAYLCGGR AELLGGLQGF LEASIVLPPQ EVPSEQHLHA
Ae2galga  ISTLMSDKQF HEAAYLADDR HDLLTAINEF LDCSVVVPPS EVQGEE.LRS
B3a2homsa ISTLMSDKQF HEAAYLADER EDLLTAINAF LDCSVVLPPS EVQGEELLRS
B3a3ratno IATLMSDKLF HEAAYQADDR QDLLGAISEF LDGSIVIPPS EVEGRDLLRS
B3atoncmy MAALMADWVF SLEAYLAPTN KELTNAIADF MDCSIVIPPT EIQDEGMLQP
Consensus ..TLM....F ...AY.A..R ..LL.....F LDCS.V.PP. E...E..L..

              651                                                    700
B3athomsa LVPVQRELLR RRYQSSPA.. ........KPD SSFYKGLDLN ....GGPDDP
B3atratno LVPVQKELLR KRYLPRPA.. ........KPD PNLYEALDGG KEGPGDEDDP
B3atgalga LIPLQRHAVR RRYQHPDTV. ......RSPG GPTAPKDTGD KGQAPQDDDP
Ae2galga  VAHFQREMLK KREEQEKRML ....LEPKSP EEKAL.LKLK VAEDEDEDDP
B3a2homsa VAHFQRQMLK KREEQGRLLP TGAGLEPKSA QDKAL.LQM. VERQGQLKMI
B3a3ratno VAAFQRELLR KRREREQTKV EMTTRGGYVA PGKELSLEMG GSEATSEDDP
B3atoncmy IIDFQKKMLK DRLRPSDTRI IFGG...... .....GAKAD EADEEPREDP
Consensus ....Q...L. .R........ ......... ......... .......DP

              701                                                    750
B3athomsa LQQTGQLFGG LVRDIRRRYP YYLSDITDAF SPQVLAAVIF IYFAALSPAI
B3atratno LRRTGRIFGG LIRDIRRRYP YYLSDITDAL SPQVLAAVIF IYFAALSPAV
B3atgalga LLRTRRPFGG LVRDIRRRYP KYLSDIRDAL NPQCLAAVIF IYFAALSPAI
```

```
Ae2galga  LRRTGRPFGG LIRDVRRRYP HYLSDFRDAL NPQCIAAVIF IYFAALSPAI
B3a2homsa PSADGAAFGG LIRDVRRRYP HYLSDFRDAL DPQCLAAVIF IYFAALSPAI
B3a3ratno LQRTGSVFGG LVRDVKRRYP HYPSDLRDAL HSQCVAAVLF IYFAALSPAI
B3atoncmy LARTGIPFGG MIKDMKRRYR HYISDFTDAL DPQVLAAVIF IYFAALSPAI
Consensus L..TG..FGG L.RD..RRYP .Y.SD..DAL .PQ..AAVIF IYFAALSPAI
```

```
          751                                              800
B3athomsa TFGGLLGEKT RNQMGVSELL ISTAVQGILF ALLGAQPLLV VGFSGPLLVF
B3atratno TFGGLLGEKT RNLMGVSELL ISTAVQGILF ALLGAQPLLV LGFSGPLLVF
B3atgalga TFGGLLGEKT RGMMGVSELL LSTSVQCLLF SLLSAQPLLV VGFSGPLLVF
Ae2galga  TFGGLLGEKT QDLIGVSELI ISTSLQGVLF CLLGAQPLLV IGFSGPLLVF
B3a2homsa TFGGLLGEKT QDLIGVSELI MSTALQGVVF CLLGAQPLLV IGFSGPLLVF
B3a3ratno TFGGLLGEKT EGLMGVSELI VSTAVLGVLF SLLGAQPLLV VGFSGPLLVF
B3atoncmy TFGGLLADKT EHMMGVSELM ISTCVQGIIF AFIAAQPTLV IGFSGPLLVF
Consensus TFGGLLGEKT ....GVSEL. .ST..QG..F .LL.AQPLLV .GFSGPLLVF
```

```
          801                                              850
B3athomsa EEAFFSFCET NGLEYIVGRV WIGFWLILLV VLVVAFEGSF LVRFISRYTQ
B3atratno EEAFYSFCES NNLEYIVGRA WIGFWLILLV VLVVAFEGSF LVQYISRYTQ
B3atgalga EEAFFRFCED HGLEYIVGRV WIGFWLILLV LLVVACEGTV LVRYLSRYTQ
Ae2galga  EEAFFTFCTS NGLEYLVGRV WIGFWLILIV LLMVACEGSF LVRFVSRFTQ
B3a2homsa EEAFFSFCSS NHLEYLVGRV WIGFWLVFLA LLMVALEGSF LVRFVSRFTR
B3a3ratno EEAFFKFCRA QDLEYLTGRV WVGLWLVVFV LALVAAEGSF LVRYISPFTQ
B3atoncmy EEAFFAFCKS QEIEYIVGRI WVGLWLVIIV VVIVAVEGSF LVKFISRFTQ
Consensus EEAFF.FC.. ..LEY.VGR. W.G.WL...V ...VA.EGSF LV...SR.TQ
```

```
          851                                              900
B3athomsa EIFSFLISLI FIYETFSKLI KIFQDHPLQK TYNYN..... ..........
B3atratno EIFSFLISLI FIYETFSKLI KIFQDYPLQE SYA.P..... ..........
B3atgalga EIFSFLISLI FIYETFAKLV TIFEAHPLQQ SYDTD..... ..........
Ae2galga  EIFAFLISLI FIYETFSKLG KIFQEHPLHG CAQPN..... .....GTA
B3a2homsa EIFAFLISLI FIYETFYKLV KIFQEHPLHG CSASNSSEVD GGENMTWAGA
B3a3ratno EIFAFLISLI FIYETFHKLY KVFTEHPLLP FYPPE..... .......EAL
B3atoncmy EIFSILISLI FIYETFSKLG KIFKAHPLVL NYEH...LND SLDNPFHPVV
Consensus EIF.FLISLI FIYETF.KL. KIF..HPL.. ..........
```

```
          901                                              950
B3athomsa .......... ..VLMVPKPQ GPLPNTALLS LVLMAGTFFF AMMLRKFKNS
B3atratno .......... ..VVMKPKPQ GPVPNTALLS LVLMVGTFLL AMMLRKFKNS
B3atgalga .......... ..V..STEPS VPKPNTALLS LVLMAGTFFL ALFLRQFKNS
Ae2galga  WSN.GTAAPN GTAQRGAAKV TGQPNTALLS LVLMAGTFFI AFFLRKFKNS
B3a2homsa RPTLGPGNRS LAGQSGQGKP RGQPNTAPLS LVLMAGTFFI AFFLRKFKNS
B3a3ratno EPGLELNSSA LPPTEGPPGP RNQPNTALLS LILMLGTFLI AFFLRKFRNS
B3atoncmy KEHIEYHEDG NKTVHEVIHE RAYPNTALLS MCLMFGCFFI AYFLRQFKNG
Consensus .......... ..........PNTALLS L.LM.GTF.. A..LR.FKNS
```

```
          951                                             1000
B3athomsa SYFPGKLRRV IGDFGVPISI LIMVLVDFFI QDTYTQKLSV PDGFKVSNSS
B3atratno TYFPGKLRRV IGDFGVPISI LIMVLVDTFI KNTYTQKLSV PDGLKVSNSS
B3atgalga VFLPGKVRRL IGDFGVPISI FVMALADFFI KDTYTQKLKV PRGLEVTNGT
Ae2galga  RFFPGRIRRL IGDFGVPIAI LVMVLVDYSI RDTYTQKLSV PSGFSVTAPD
B3a2homsa RFFPGRIRRV IGDFGVPIAI LIMVLVDYSI EDTYTQKLSV PSGFSVTAPE
B3a3ratno RFLGGKARRV IGDFGIPISI LVMVLVDYSI TDTYTQKLTV PTGLSVTSPH
B3atoncmy HFLPGPIRRM IGDFGVPIAI FFMIAVDITI EDAYTQKLVV PKGLMVSNPN
Consensus ...PG..RR. IGDFGVPI.I ..M.LVD..I .DTYTQKL.V .P.G..V...
```

```
                1001                                                    1050
B3athomsa ARGWVIHPLG LRSEFPIWMM FASALPALLV FILIFLESQI TTLIVSKPER
B3atratno ARGWVIHPLG LYNHFPKWMM FASVLPALLV FILIFLESQI TTLIVSKPER
B3atgalga ARGWFIHPMG SATPFPIWMM FASPVPALLV FILIFLETQI TTLIVSKPER
Ae2galga  KRGWVINPLG ERSDFPVWMM VASGLPAVLV FILIFMETQI TTLIISKKER
B3a2homsa KRGWVINPLG EKSPFPVWMM VASLLPAILV FILIFMETQI TTLIISKKER
B3a3ratno KRTWFIPPLG SARPFPPWMM VAAAVPALLV LILIFMETQI TALIVSQKAR
B3atoncmy ARGWFINPLG EKKPFPAWMM GACCVPALLV FILIFLESQI TTLIVSKPER
Consensus .RGW.I.PLG ....FP.WMM .A...PALLV FILIF.E.QI TTLI.SK.ER

                1051                                                    1100
B3athomsa KMVKGSGFHL DLLLVVGMGG VAALFGMPWL SATTVRSVTH ANALTVMGKA
B3atratno KMIKGSGFHL DLLLVVGMGG VAALFGMPWL SATTVRSVTH ANALTVMGKA
B3atgalga KLVKGSGFHL DLLLIVAMGG LAALFGMPWL SATTVRTITH ANALTVVGKS
Ae2galga  MLQKGSGFHL DLLLIVAMGG FFALFGLPWL AAATVRSVTH ANALTVMSKA
B3a2homsa MLQKGSGFHL DLLLIVAMGG ICALFGLPWL AAATVRSVTH ANALTVMSKA
B3a3ratno RLLKGSGFHL DLLLIGSLGG LCGLFGLPWL TAATVRSVTH VNALTVMRTA
B3atoncmy KMVKGSGFHL DLLILVTMGG IASLFGVPWL SAATVRSVTH ANALTVMSK.
Consensus ...KGSGFHL DLLL.V.MGG ...LFG.PWL .A.TVRSVTH ANALTVM.K.

                1101                                                    1150
B3athomsa STPGAAAQIQ EVKEQRISGL LVAVLVGLSI LMEPILSRIP LAVLFGIFLY
B3atratno SGPGAAAQIQ EVKEQRISGL LVSVLVGLSI LMEPILSRIP LAVLFGIFLY
B3atgalga AVPGERAHIV EVKEQRLSGL LVAVLIGVSI LMEPILKYIP LAVLFGIFLY
Ae2galga  VAPGDKPKIQ EVKEQRVTGL LVAVLVGLSI VIGELLRQIP LAVLFGIFLY
B3a2homsa VAPGDKPKIQ EVKEQRVTGL LVALLVGLSI VIGDLLRQIP LAVLFGIFLY
B3a3ratno IAPGDKPQIQ EVREQRVTGV LIASLVGLSI VMGAVLRRIP LAVLFGIFLY
B3atoncmy ...GPKPEIE KVLEQRISGM LVAAMVGVSI LLEPILKMIP MTALFGIFLY
Consensus ..PG....I. EV.EQR..G. LVA.LVG.SI .....L..IP LAVLFGIFLY

                1151                                                    1200
B3athomsa MGVTSLSGIQ LFDRILLLFK PPKYHPDVPY VKRVKTWRMH LFTGIQIICL
B3atratno MGITSLSGIQ LFDRILLLFK PPKYHPDVPF VKRVKTWRMH LFTGIQIICL
B3atgalga MGVTSLFGIQ LFDRILLLLM PPKYHPKEPY VTRVKTWRIT SSPLTQILVV
Ae2galga  MGVTSLNGIQ FYERLQLLLM PPKHHPDVPY VKKVRT.RMH LFTGLQLACL
B3a2homsa MGVTSLNGIQ FYERLHLLLM PPKHHPDVTY VKKVRTLRMH LFTALQLLCL
B3a3ratno MGVTSLSGIQ LSQRLLLIFM PAKHHPEQPY VTKVKTWRMH LFTFIQLGCI
B3atoncmy MGITSLSGIQ MWDRMLLLIV PRKYYPADAY AQRVTTMKMH LFTLIQMVCL
Consensus MG.TSL.GIQ ...R..LL.. P.K.HP...Y V..V.T.RMH LF...Q..C.

                1201                                                    1250
B3athomsa AVLWVVKSTP ASLALPFVLI LTVPLRRVLL PLIFRNVELQ CLDADDAKAT
B3atratno AVLWVVKSTP ASLALPFVLI LTVPLRRLLL PLIFRELELQ CLDGDDAKVT
B3atgalga ALLWGVKVSP ASLRCPFVLV LTVPLRRLLL PRIFSEIELK CLDTDDAVVT
Ae2galga  AVLWAVMSTV ASLAFPFILI LTVPVRMCLL SRIFTDREMK CLDADEAEPI
B3a2homsa ALLWAVMSTA ASLAFPFILI LTVPLRMVVL TRIFTDREMK CLDANEAEPV
B3a3ratno ALLWVVKSTV ASLAFPFLLL LTVPLRRCLL PRLFQDRELQ ALDSEDAEPN
B3atoncmy GALWMVKMSA FSLALPFVLI LTIPLRMAIT GTLFTDKEMK CLDASDGKVK
Consensus A.LW.V.... ASLA.PF.L. LTVPLR...L ...F...E.. CLD...A...

                1251        1267
B3athomsa FDEEEGRDEY DEVAMPV
B3atratno FDEAEGLDEY DEVPMPV
B3atgalga FEEAEGQDVY NEVQMPS
```

```
Ae2galga  LDEREGVDEY NEMPMPV
B3a2homsa FDEREGVDEY NEMPMPV
B3a3ratno FDE.DGQDEY NELHMPV
B3atoncmy FEEEPGEDMY ,ESPLP.
Consensus F.E..G.D.Y .E..MP.
```

Proteins listed subsequently in italics are at least 90% identical to the paired transporters listed in parenthesis and therefore are not included in the alignments: *B3a2musmu, B3a2ratno* (B3a2homsa); *B3a3musmu* (B3a3ratno); *B3atmusmu* (B3atratno). Residues listed in the consensus sequence are present in at least 75% of the aligned transporter sequences.

Database accession numbers

	SWISSPROT	PIR	EMBL/GENBANK
Ae2galga			U48889
B3a2homsa	P04920	A25104; S21086	X62137; X03918
B3a2ratno	P23347	A34911	J05166
B3a2musmu	P13808	A31789	J04036
B3a3musmu	P16283	A33638	M28383
B3a3ratno	P23348	B34911	J05167
B3atmusmu	P04919	A26086; A25314	X02677; M29379
B3atgalga	P15575	A30816	M23404
B3atratno	P23562	A33810	J04793; L02943
B3atoncmy	P32847	S22173; S24318	X61699
B3athomsa	P02730	A03189; A28079	M27819

References
1 Reithman, R. (1994) Curr. Opin. Cell Biol. 6, 583–594.
2 Rybicki, A.C. et al. (1993) Blood 81, 2155–2165.

Summary

Transporters of the mitochondrial adenine nucleotide translocator family, the example of which is the ANT1 ADP, ATP carrier protein of humans (Adt1homsa), mediate the exchange of substrate pairs across the inner mitochondrial membrane, including ADP and ATP, and 2-oxoglutarate and malate. Mitochondrial brown fat uncoupling proteins generate heat by dissipating the mitochondrial transmembrane proton gradient, thereby driving compensatory electron transport. Members of the family are ubiquitous in eukaryotes.

Statistical analysis of multiple amino acid sequence comparisons indicates that the mitochondrial adenine nucleotide translocator family is most closely related to the mitochondrial phosphate carrier family [1]. Members of the mitochondrial adenine nucleotide translocator family are predicted to exist as a homodimer, with each subunit containing six membrane-spanning helices comprised of three homologous domains.

Several amino acid sequence motifs are highly conserved in the mito-chondrial adenine nucleotide translocator family, including motifs that are unique to the family, motifs common to the mitochondrial phosphate carrier family and motifs necessary for function. Defects in members of the mito-chondrial adenine nucleotide translocator family have been implicated in Graves' disease [2].

Nomenclature, biological sources and substrates

CODE	DESCRIPTION [SYNONYMS]	ORGANISM [COMMON NAMES]	SUBSTRATE(S)
Acr1sacce	Regulator of acetyl CoA synthetase [ACR1, YJR096W]	Saccharomyces cerevisiae [yeast]	Unknown
Adt1arath	ADP/ATP carrier protein 1 [ADP/ATP translocase 1, adenine nucleotide translocator 1, ANT1, ADT1]	Arabidopsis thaliana [mouse-ear cress]	ADP/ATP
Adt1bosta	ADP/ATP carrier heart-skeletal muscle isoform T1 [ANT1, ADT1]	Bos taurus [cow]	ADP/ATP
Adt1homsa	ADP/ATP carrier heart-skeletal muscle isoform T1 [ADP/ATP translocase 1, adenine nucleotide translocator 1, ANT1, ADT1]	Homo sapiens [human]	ADP/ATP
Adt1musmu	ADP/ATP carrier protein 1 [ADP/ATP translocase 1, adenine nucleotide translocator 1, ANT1, ADT1]	Mus musculus [mouse]	ADP/ATP
Adt1ratno	ADP/ATP carrier heart-skeletal muscle isoform T1 [ANT1, ADT1]	Rattus norvegicus [rat]	ADP/ATP
Adt1sacce	ADP/ATP carrier protein 1 [ADP/ATP translocase 1, adenine nucleotide translocator 1, ANT1, ADT1, AAC1, YM9796.09c]	Saccharomyces cerevisiae [yeast]	ADP/ATP

CODE	DESCRIPTION [SYNONYMS]	ORGANISM [COMMON NAMES]	SUBSTRATE(S)
Adt1soltu	ADP/ATP carrier protein 1 [ADP/ATP translocase 1, adenine nucleotide translocator 1, ANT1, ADT1, AAC1]	Solanum tuberosum [potato]	ADP/ATP
Adt1zeama	ADP/ATP carrier protein 1 [ADP/ATP translocase 1, adenine nucleotide translocator 1, ANT1, ADT1, ANTG1]	Zea maize [corn]	ADP/ATP
Adt2arath	ADP/ATP carrier protein 2 [ADP/ATP translocase 2, adenine nucleotide translocator 2, ANT2, ADT2]	Arabidopsis thaliana [mouse-ear cress]	ADP/ATP
Adt2homsa	ADP/ATP carrier isoform T2 [ANT2, ADT2]	Homo sapiens [human]	ADP/ATP
Adt2ratno	ADP/ATP carrier isoform T2 [ANT2, ADT2]	Rattus norvegicus [rat]	ADP/ATP
Adt2sacce	ADP/ATP carrier protein 2 [ADP/ATP translocase 2, adenine nucleotide translocator 2, ANT2, ADT2, AAC2, PET9, YBL030C, YBL0421]	Saccharomyces cerevisiae [yeast]	ADP/ATP
Adt2soltu	ADP/ATP carrier protein 2 [ADP/ATP translocase 2, adenine nucleotide translocator 2, ANT2, ADT2, AAC2]	Solanum tuberosum [potato]	ADP/ATP
Adt2zeama	ADP/ATP carrier protein 2 [ADP/ATP translocase 2, adenine nucleotide translocator 2, ANT2, ADT2, ANTG2]	Zea maize [corn]	ADP/ATP
Adt3bosta	ADP/ATP carrier isoform T3 [ANT3, ADT3]	Bos taurus [cow]	ADP/ATP
Adt3homsa	ADP/ATP carrier isoform T3 [ANT3, ADT3]	Homo sapiens [human]	ADP/ATP
Adt3sacce	ADP/ATP carrier protein 3 [ADP/ATP translocase 3, adenine nucleotide translocator 3, ANT3, ADT3, AAC3, YBR085W, YBR0753]	Saccharomyces cerevisiae [yeast]	ADP/ATP
Adtanoga	ADP/ATP carrier protein 1 [ADP/ATP translocase 1, adenine nucleotide translocator 1, ANT1, ADT1]	Anopheles gambiae [mosquito]	ADP/ATP
Adtcaeel	ADP/ATP carrier protein 1 [ADP/ATP translocase 1, adenine nucleotide translocator 1, ANT1, ADT1]	Caenorhabditis elegans [nematode]	ADP/ATP
Adtchlke	ADP/ATP carrier protein [ADP/ATP translocase, Ant1, adenine nucleotide translocator, ADT]	Chlorella kessleri [alga]	ADP/ATP

CODE	DESCRIPTION [SYNONYMS]	ORGANISM [COMMON NAMES]	SUBSTRATE(S)
Adtchlre	ADP/ATP carrier protein [ADP/ATP translocase, adenine nucleotide translocator, ANT1, ADT, ABT]	Chlamydomonas rheinhardtii [alga]	ADP/ATP
Adtdrome	ADP/ATP carrier protein 1 [ADP/ATP translocase 1, adenine nucleotide translocator 1, ANT1, ADT1]	Drosophila melanogaster [fruit fly]	ADP/ATP
Adtneucr	ADP/ATP carrier protein [ADP/ATP translocase, adenine nucleotide translocator, ANT1, ADT, ACP]	Neurospora crassa [mold]	ADP/ATP
Adtorysa	ADP/ATP carrier protein [ADP/ATP translocase, adenine nucleotide translocator, ANT1, ADT]	Oryza sativa [rice]	ADP/ATP
Adtplafa	ADP/ATP carrier protein 1 [ADP/ATP translocase 1, adenine nucleotide translocator 1, ANT1, ADT1]	Plasmodium falciparum [mosquito]	ADP/ATP
Adtransy	ADP/ATP carrier protein 1 [ADP/ATP translocase 1, adenine nucleotide translocator 1, ANT1, ADT1]	Rana sylvaticum [frog]	ADP/ATP
Adttritu	ADP/ATP carrier protein 1 [ADP/ATP translocase 1, adenine nucleotide translocator 1, ANT1, ADT1]	Triticum turgidum [wheat]	ADP/ATP
Adttrybr	ADP/ATP carrier protein 1 [ADP/ATP translocase 1, adenine nucleotide translocator 1, ANT1, ADT1]	Trypanosoma brucei [trypanosome]	ADP/ATP
Alt1halro	ADP/ATP carrier protein 1 [ADP/ATP translocase 1, adenine nucleotide translocator 1, ANT1, ADT1]	Halocynthia roretzi [sea squirt]	ADP/ATP
Cithomsa	Mitochondrial citrate transport protein [CIT]	Homo sapiens [human]	Citrate
Citsacce	Mitochondrial citrate transport protein [CIT, YBR2039, YBR29K]	Saccharomyces cerevisiae [yeast]	Citrate
Dif1caeel	Mitochondrial carrier protein [DIF1]	Caenorhabditis elegans [nematode]	
Flx1sacce	FAD carrier protein [FLX1, YIL134W]	Saccharomyces cerevisiae [yeast]	FAD
Gdcbosta	Graves disease carrier protein [GDC]	Bos taurus [cow]	Unknown
Gdchomsa	Graves disease carrier protein [GDC]	Homo sapiens [human]	Unknown
Gdcratno	Graves disease carrier protein [GDC]	Rattus norvegicus [rat]	Unknown
M2omcaeel	2-Oxoglutarate/malate carrier protein [OGCP]	Caenorhabditis elegans [nematode]	2-Oxoglutarate/malate
M2ombosta	2-Oxoglutarate/malate carrier protein [OGCP]	Bos taurus [cow]	2-Oxoglutarate/malate

CODE	DESCRIPTION [SYNONYMS]	ORGANISM [COMMON NAMES]	SUBSTRATE(S)
M2omhomsa	2-Oxoglutarate/malate carrier protein [OGCP]	*Homo sapiens* [human]	2-Oxoglutarate/ malate
Pet8sacce	Putative mitochondrial carrier protein [PET8, N2012]	*Saccharomyces cerevisiae* [yeast]	Unknown
Pmtsacce	Putative mitochondrial carrier protein [PMT, PMT1, YKL120W, YKL522]	*Saccharomyces cerevisiae* [yeast]	Unknown
Rim2sacce	Mitochondrial carrier protein [MCP, RIM2, YBR192w, YBR1402]	*Saccharomyces cerevisiae* [yeast]	Unknown
Txtpratno	Tricarboxylate carrier [CTP, citrate transport protein]	*Rattus norvegicus* [rat]	Citrate
Ucpbosta	Mitochondrial brown fat uncoupling protein [UCP]	*Bos taurus* [cow]	H⁺
Ucphomsa	Mitochondrial brown fat uncoupling protein [UCP]	*Homo sapiens* [human]	H⁺
Ucpmesau	Mitochondrial brown fat uncoupling protein [UCP]	*Mesocricetus auratus* [golden hamster]	H⁺
Ucpmusmu	Mitochondrial brown fat uncoupling protein [UCP]	*Mus musculis* [mouse]	H⁺
Ucporycu	Mitochondrial brown fat uncoupling protein [UCP]	*Oryctolagus cuniculus* [rabbit]	H⁺
Ucpratno	Mitochondrial brown fat uncoupling protein [UCP]	*Rattus norvegicus* [rat]	H⁺

457

Phylogenetic tree

Proteins listed subsequently in italics are at least 90% identical to the paired transporters listed in parenthesis and therefore are not included in the phylogenetic tree: *Adt1ratno, Adt1bosta* (Adt1homsa); *Adt2ratno, Adt3bosta, Adt3homsa* (Adt2homsa); *Adt3sacce* (Adt2sacce); *Gdcbosta, Gdcratno* (Gdchomsa); *M2ombosta* (M2omhomsa); *Ucpmesau, Ucpmusmu* (Ucphomsa).

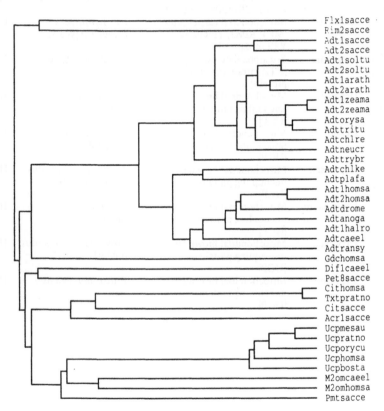

Proposed orientation of ANT1 in the membrane

The model is based on predictions of membrane-spanning regions and α-helical content. The N-terminus of the protein is illustrated on the inside and is folded six times through the membrane. The predicted membrane-spanning helices are portrayed as rectangles. The numbers corresponding to the first and last residue of each membrane-spanning helix are boxed. Residues that are conserved in more than 75% of the aligned transporters (see below) are shown.

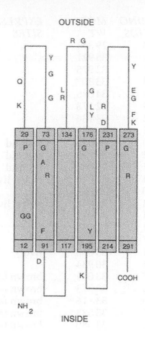

Physical and genetic characteristics

	AMINO ACIDS	MOL. WT	EXPRESSION SITES	CHROMOSOMAL LOCUS
Acr1sacce	322	35 340		Chromosome 10
Adt1arath	379	41 297		
Adt1bosta	297	32 836	heart, skeletal muscle	
Adt1homsa	298	33 064	heart, skeletal muscle	4q35
Adt1musmu	298	32 870	heart, skeletal muscle	
Adt1ratno	298	32 989	heart, skeletal muscle	
Adt1sacce	309	34 120		Chromosome 7
Adt1soltu	386	42 058		
Adt1zeama	387	42 391		
Adt2arath	385	41 845		
Adt2homsa	298	32 895	fibroblasts	Xq24–26
Adt2ratno	298	32 901	fibroblasts	
Adt2sacce	318	34 426		Chromosome 2
Adt2soltu	386	41 829		
Adt2zeama	387	42 332		
Adt3bosta	298	32 877	liver	
Adt3homsa	298	32 866	liver	Xp22.32
Adt3sacce	307	33 313		Chromosome 2
Adtanoga	301	32 681		
Adtcaeel	300	33 211		
Adtchlke	339	36 686		
Adtchlre	308	33 258		
Adtdrome	297	32 914		
Adtneucr	313	33 888		
Adtorysa	382	41 510		
Adtplafa	301	33 756		

	AMINO ACIDS	MOL. WT	EXPRESSION SITES	CHROMOSOMAL LOCUS
Adtransy	263	29 351		
Adttritu	331	35 921		
Adttrybr	307	33 975		
Alt1halro	304	33 307		
Cithomsa	311	34 085		
Citsacce	299	32 173		
Dif1caeel	312	33 134		
Flx1sacce	311	34 409		Chromosome 9
Gdcbosta	330	36 085	thyroid	
Gdchomsa	332	36 235	thyroid	
Gdcratno	322	35 056	thyroid	
M2ombosta	313	34 040	heart, liver, brain	
M2omcaeel	290	32 022		
M2omhomsa	313	33 948	heart, liver, brain	
Pet8sacce	284	31 027		Chromosome 14
Pmtsacce	324	35 153		Chromosome 4
Rim2sacce	377	42 101		Chromosome 2
Txtpratno	311	33 835	liver	
Ucpbosta	286	30 934		
Ucphomsa	307	33 044	brown fat	4q31
Ucpmesau	306	33 215	brown fat	
Ucpmusmu	306	33 116	brown fat	
Ucporycu	306	33 083	brown fat	
Ucpratno	306	33 080	brown fat	

Multiple amino acid sequence alignments

```
          1                                                  50
Rim2sacce ..........................        ........MP KKSIEEWEED
Adt1soltu MADMNQHPTV FQKAANQLDL RSSLSQDVHA RYGGVQ.PAI YQRHFAYGNY
Adt2soltu ..ADNQHPTV YQKVASQMHL SSSLSQDVHA RYGGIQRPAL SQRRFPYGNY
Adt1arath ..HQVQHPTI AQKAAGQF.M RSSVSKDVQV ...GYQRPSM YQRHATYGNY
Adt2arath MVEQTQHPTI LQKVSGQL.L SSSVSQDIRG YASASKRPAT YQKHAAYGNY
Adt1zeama MADQANQPTV LHKLGGQFHL RSIISEGVRA R.NICPSVSS YERRFATRNY
Adt2zeama MADQANQPTV LHKLGGQFHL SSSFSEGVRA R.NICPSFSP YERRFATRNY
Adtorysa  MAEQANQPTV LQKFGGQFHL GSSFSEGVRA R.NICPSVSS YDRRFTTRSY
Adtchlke  .......................................          .M
```

```
          51                                                100
Flx1sacce ........................... .MVDHQWTPL QKEVISGLSA
Rim2sacce AIESVPYLAS DEKGSNYKEA TQIPLNLKQS EIENHPTVKP WVHFVAGGIG
Adt1sacce .......................MSH TETQTQQSHF GVDFLMGGVS
Adt2sacce ...............MSS NAQVKTPLPP APAPKKESNF LIDFLMGGVS
Adt1soltu SNAGLQRG.. QATQDLSLIT SNASPVFVQA PQE.KGFAAF ATDFLMGGVS
Adt2soltu SNAGLQTC.. QATQDLSLIA ANASPVFVQA PQE.KGLAAF ATDFLMGGVS
Adt1arath SNAAFQFP... .PTS..RMLA TTASPVFVQT PGE.KGFTNF ALDFLMGGVS
Adt2arath SNAAFQYP.. LVAA..SQIA TTTSPVFVQA PGE.KGFTNF AIDFMMGGVS
Adt1zeama MTQSLWGPSM SVSGGINVPV M.QTPLCANA PAE.KGGKNF MIDFMMGGVS
Adt2zeama MTQSLWGPSM SVSGGINVPV M.PTPLFANA PAE.KGGKNF MIDFMMGGVS
Adtorysa  MTQGL...... .VNGGINVPM MSSSPIFANA PAE.KGGKNF MIDFLMGGVS
Adttritu  MTQNL...... ...GISVPI MSSSPMFANA PPEKKGVKNF AIDFLMGGVS
Adtchlre  ............ .............MAKEEKNF MVDFLAGGLS
Adtneucr  ...................MAE QQKVLGMPPF VADFLMGGVS
Adttrybr  ...................MTDK KREPAPKLGF LEEFMIGGVA
```

```
Adtchlke   LSSALYQQAG LSGLLRASAM GPQTPFIASP KETQADPMAF VKDLLAGGTA
Adtplafa   .......... .........MSSD IKTN.....F AADFLMGGIS
Adt1homsa  .......... ............. ..MGDHAWSF LKDFLAGGVA
Adt2homsa  .......... ............. ..MTDAAVSF AKDFLAGGVA
Adtdrome   .......... .......... MGKDFDAVGF VKDFAAGQVS
Adtanoga   .......... .......... MTKKADPYGF AKDFLAGGIS
Adt1halro  .......... .......... ..MPWSAVDF AKDLAIGGTA
Adtcaeel   .......... .........MS KEKSFDTKKF LIDLASGGTA
Gdchomsa   ....MAAATA AAALAAADPP PAMPGAAGAG GPTTRRDFYW LRSFLAGSIA
Dif1caeel  .......... .......... ......MSDV LLNFIAGGVG
Pet8sacce  .......... .......... ......MNTF FLSLLSGAAA
Cithomsa   ..........MPAPR APRALAAAAP ASGKAKLTHP EKAILAGGLA
Txtpratno  ..........MAAPR APRALTAAAP GSGKAKLTHP GKAILAGGLA
Citsacce   .......... ....MSSKAT KSD....VDP LHSFLAGSLA
Acr1sacce  .......... ... MSQKKASHP AINLMAGGTA
Ucpmesau   .......... ....VN PTTSEVHPTM GVKIFSAGVA
Ucpratno   .......... ....VS STTSEVQPTM GVKIFSAGVS
Ucporycu   .......... ....MVG TTTTDVPPTM GVKIFSAGVA
Ucphomsa   .......... ....MGG LTASDVHPTL GVQLFSAPIA
Ucpbosta   .......... .......... .....IFSAGVA
M2omcaeel  .......... .......... ..MAEDKTKR LGRWYFGGVA
M2omhomsa  .......... AATASAGAGG MDGKPRTSPK SVKFLFGGLA
Pmtsacce   ..........MS SDNSKQDKQI EKTAAQKISK FGSFVAGGLA
Consensus  .......... .......... .......... .......GG..
```

```
           101                                          150
Flx1sacce  GSVTTLVVHP LDLLKVRLQ. ...LSATSAQ KAHYG....P FMVIKEIIRS
Rim2sacce  GMAGAVVTCP FDLVKTRLQS DIFLKAYKSQ AVNISKGSTR PKSINYVIQA
Adt1sacce  AAIAKTGAAP IERVKLLMQN QEEMLK.QGS L......... ..........
Adt2sacce  AAVAKTAASP IERVKLLIQN QDEMLK.QGT L......... ..........
Adt1soltu  AAVSKTAAAP IERVKLLIQN QDEMLK.AGR L......... ..........
Adt2soltu  AAVSKTAAAP IERVKLLIQN QDEMIK.AGR L......... ..........
Adt1arath  AAVSKTAAAP IERVKLLIQN QDEMIK.AGR L......... ..........
Adt2arath  AAVSKTAAAP IERVKLLIQN QDEMLK.AGR L......... ..........
Adt1zeama  AAVSKTAAAP IERVKLLIQN QDEMIK.SGR L......... ..........
Adt2zeama  AAVSKTAAAP IERVKLLIQN QDEMIK.SGR L......... ..........
Adtorysa   AAVSKTAAAP IERVKLLIQN QDEMIK.AGR L......... ..........
Adttritu   AAVSKTAAAP IERVKLLIQN QDEMIK.AGR L......... ..........
Adtchlre   AAVSKTAAAP IERVKLLIQN QDEMIK.QGR L......... ..........
Adtneucr   AAVSKTAAAP IERIKLLVQN QDEMIR.AGR L......... ..........
Adttrybr   AGLSKTAAAP IERVKLLVQN QGEMMK.QGR L......... ..........
Adtchlke   GAISKTAVAP IERVKLLLQT QDSNPMIKSG Q......... ..........
Adtplafa   AAISKTVVTP IERVKMLIQT QDSIPEIKSG Q......... ..........
Adt1homsa  AAVSKTAVAP IERVKLLLQV QHASKQISAE K......... ..........
Adt2homsa  AAISKTAVAP IERVKLLLQV QHASKQITAD K......... ..........
Adtdrome   AAVSKTAVAP IERVKLLLQV QHISKQISPD K......... ..........
Adtanoga   AAVSKTAVAP IERVKLLLQV QAASKQIAVD K......... ..........
Adt1halro  AAISKTIVAP IERVKLLLQV QAVSTQMKAG T......... ..........
Adtcaeel   AAVSKTAVAP IERVKLLLQV QDASKAIAVD K......... ..........
Adtransy   .......... .......... .......... .......... ..........
Gdchomsa   GCCAKTTVAP LDRVKVLLQA HNHHYK.... .......... ..........
Dif1caeel  GSCTVIVGHP FDTVKVRIQT MPMPKPGEKP .......... ..........
Pet8sacce  GTSTDLVFFP IDTIKTRLQA .......... .......... ..........
Cithomsa   GGIEICITFP TEYVKTQLQL DERS...HPP .......... ..........
```

461

```
Txtpratno GGIEICITFP TEYVKTQLQL DERA...NPP ..........
Citsacce  GAAEACITYP FEFAKTRLQL IDKA...SKA ..........
Acr1sacce GLFEALCCHP LDTIKVRMQI YRRVAGIEHV ..........
Ucpmesau  ACLADIITFP LDTAKVRLQI QGEGQISSTI ..........
Ucpratno  ACLADIITFP LDTAKVRLQI QGEGQASSTI ..........
Ucporycu  ACLADVITFP LDTAKVRQQI QGEFPITSGI ..........
Ucphomsa  ACLADVITFP LDTAKVRLQV QGECPTSSVI ..........
Ucpbosta  ACVADIITFP LDTAKVRLQI QGECLISSAI ..........
M2omcaeel GAMAACCTHP LDLLKVQLQT QQQGKL..TI ..........
M2omhomsa GMGATVFVQP LDLVKNRMQL SGEGAK..TR ..........
Pmtsacce  ACIAVTVTNP IELIKIRMQL QGEMSASAAK ..........
Consensus ........P ....K...Q. ..........
```

```
          151                                      200
Flx1sacce SANSGRSVT. .......... NELYRGLSIN LFGNAIAWGV YFGLYGVTKE
Rim2sacce GTHFKETLGI IGNVYKQEGF RSLFKGLGPN LVGVIPARSI NFFTYGTTKD
Adt1sacce DTRYKGILDC FKRTATHEGI VSFWRGNTAN VLRYFPTQAL NFAFKDKIKS
Adt2sacce DRKYAGILDC FKRTATQEGV ISFWRGNTAN VIRYFPTQAL NFAFKDKIKA
Adt1soltu SEPYKGIGEC FGRTIKEEGF GSLWRGNTAN VIRYFPTQAL NFAFKDYFKR
Adt2soltu SEPYKGIGDC FSRTIKDEGF AALWRGNTAN VIRYFPTQAL NFAFKDYFKR
Adt1arath SEPYKGIGDC FGRTIKDEGF GSLWRGNTAN VIRYFPTQAL NFAFKDYFKR
Adt2arath TEPYKGIRDC FGRTIRDEGI GSLWRGNTAN VIRYFPTQAL NFAFKDYFKR
Adt1zeama SEPYKGIVDC FKRTIKDEGF SSLWRGNTAN VIRYFPTQAL NFAFKDYFKR
Adt2zeama SEPYKGIADC FKRTIKDEGF SSLWRGNTAN VIRYFPTQAL NFAFKDYFKR
Adtorysa  SEPYKGIGDC FGRTIKDEGF ASLWRGNTAN VIRYFPTQAL NFAFKDYFKR
Adttritu  SEPYKGIGDC FGRTIKDEGF GSLWRGNTAN VIRYFPTQAL NFAFKDYFKR
Adtchlre  ASPYKGIGEC FVRTVREEGF GSLWRGNTAN VIRYFPTQAL NFAFKDKFKR
Adtneucr  DRRYNGIIDC FKRTTADEGV MALWRGNTAN VIRYFPTQAL NFAFRDKFKK
Adttrybr  DKPYNGVVDC FRRTISTEGV YPLWRGNLSN VLRYFPTQAL NFAFKDKFKR
Adtchlke  VPRYTGIVNC FVRVSSEQGV ASFWRGNLAN VVRYFPTQAF NFAFKDTIKG
Adtplafa  VERYSGLINC FKRVSKEQGV LSLWRGNVAN VIRYFPTQAF NFAFKDYFKN
Adt1homsa ..QYKGIIDC VVRIPKEQGF LSFWRGNLAN VIRYFPTQAL NFAFKDKYKQ
Adt2homsa ..QYKGIIDC VVRIPKEQGV LSFWRGNLAN VIRYFPTQAL NFAFKDKYKQ
Adtdrome  ..QYKGMVDC FIRIPKEQGF SSFWRGNLAN VYRYFPTQAL NFAFKDKYKQ
Adtanoga  ..QYKGIVDC FVRIPKEQGI GAFCGGNLAN VIRYFPTQAL NFAFKDVYKQ
Adt1halro ..EYKGIIDA FVRIPKEQGF FSLWRGNLAN VIRYFPTQAL NFAFKDTYKK
Adtcaeel  ..RYKGIMDV LIRVPKEQGV AALWRGNLAN VIRYFPTQAM NFAFKDTYKA
Adtransy  .......MDC VVRIPKEQGF ISFWRGNLAN VIRYFPTQAL NFGFKDKYKK
Gdchomsa  ...HLGVFSA LRAVPQKEGF LGLYKGNGAM MIRIFPYGAI QFMAFEHYKT
Dif1caeel ..QFTGALDC VKRTVSKEGF FALYKGMAAP LVGVSPLFAV FFGGCA....
Pet8sacce ....KGGF.. ....FANGGY KGIYRGLGSA VVASAPGASL FFISYDYMKV
Cithomsa  ..RYRGIGDC VRQTVRSHGV LGLYRGLSSL LYGSIPKAAV RFGMFEFLSN
Txtpratno ..RYRGIGDC VRQTVRSHGV LGLYRGLSSL LYGSIPKAAV RFGMFEFLSN
Citsacce  ..SRNPLVL. IYKTAKTQGI GSIYVGCPAF IIGNTAKAGI RFLGFDTIKD
Acr1sacce ..KPPGFIKT GRTIYQKEGF LALYKGLGAV VIGIIPKMAI RFSSYEFYRT
Ucpmesau  ..RYKGVLGT ITTLAKTEGL PKLYSGLPAG IQRQISFASL RIGLYDTVQE
Ucpratno  ..RYKGVLGT ITTLAKTEGL PKLYSGLPAG IQRQISFASL RIGLYDTVQE
Ucporycu  ..RYKGVLGT ITTLAKTEGP LKLYSGLPAG LQRQISFASL RIGLYDTVQE
Ucphomsa  ..RYKGVLGT ITAVVKTEGR MKLYSGLPAG LQRQISSASL RIGLYDTVQE
Ucpbosta  ..RYKGVLGT IITLAKTEGP VKLYSGLPAG LQRQISLASL RIGLYDTVQE
M2omcaeel ..G.....QL SLKIYKNDGI LAFYNGVSAS VLRQLTYSTT RFGIYETVKK
M2omhomsa ..EYKTSFHA LTSILKAEGL RGIYTGLSAG LLRQATYTTT RLGIYTVLFE
Pmtsacce  ..VYKNPIQG MAVIFKNEGI KGLQKGLNAA YIYQIGLNGS RLGFYEPIRS
Consensus ...Y.G.... ........G. .....G..A. ..R........ .F...D....
```

```
          201                                                   250
Flx1sacce LIYKSVAKPG ETQLKGVGND HKMNSLIYLS AGASSGLMTA ILTNPIWVIK
Rim2sacce MYAKAFNNGQ ET........ ...PMIHLM AAATAGWATA TATNPIWLIK
Adt1sacce LLS......Y DRERD.GYAK WFAGNLFSGG AAGGLSLLFV YSLDYARTRL
Adt2sacce MFG......F KKE.E.GYAK WFAGNLASGG AAGALSLLFV YSLDYARTRL
Adt1soltu LFN......F KKDRD.GYWK WFAGNLASGG AAGASSLFFV YSLDYARTRL
Adt2soltu LFN......F KKDRD.GYWK WFAGNLASGG GAGASSLLFV YSLDYARTRL
Adt1arath LFN......F KKDRD.GYWK WFAGNLASGG AAGASSLLFV YSLDYARTRL
Adt2arath LFN......F KKDKD.GYWK WFAGNLASGG AAGASSLLFV YSLDYARTRL
Adt1zeama LFN......F KKDRD.GYWK WFAGNLASGG AAGASSLFFV YSLDYARTRL
Adt2zeama LFN......F KKDRD.GYWK WFAGNLASGG AAGASSLFFV YSLDYARTRL
Adtorysa  LFN......F KKDKD.GYWK WFGGNLASGG AAGASSLFFV YSLDYARTRL
Adttritu  MFN......F KKDKD.GYWK WFGGNLASGG AAGASSLFFV YSLDYARTRL
Adtchlre  MFG......F NKDKE..YWK WFAGNMASGG AAGAVSLSFV YSLDYARTRL
Adtneucr  MFG......Y KKDVD.GYWK WMAGNLASGG AAGATSLLFV YSLDYARTRL
Adttrybr  MFN......Y KKEKD.GYGK WFMGNSMASGG LAGAASLCFV YSLDYVRTRL
Adtchlke  LF.......P KYSPKTDFWR FFVVNLASGG LAGAGSLLIV YPLDFARTRL
Adtplafa  IF.......P RYDQNTDFSK FFCVNILSGA TAGAISLLIV YPLDFARTRL
Adt1homsa LFL......G GVDRHKQFWR YFAGNLASGG AAGATSLCFV YPLDFARTRL
Adt2homsa IFL......G GVDKRTQFWR YFAGNLASGG AAGATSLCFV YPLDFARTRL
Adtdrome  VFL......G GVDKNTQFWR YFAGNLASGG AAGATSLCFV YPLDFARTRL
Adtanoga  VFL......G GVDKNTQFWR YFLGNLGSGG AAGATSLCFV YPLDFARTRL
Adt1halro IFL......A GVDKRKQFWR YFHGNLASGG AAGATGLCFV YPLDFARTRL
Adtcaeel  IFL......E GLDKKKDFWK FFAGNLASGG AAGATSLCFV YPLDFARTRL
Adtransy  IFL......D NVDKRTQFWR YFAGNLASGG AAGATSLCFV YPLDFARTRL
Gdchomsa  LIT......T KLGISGHVHR LMAGSM.... .AGMTAVICT DPVDMVRVRL
Diflcaeel .....VGKWL QQTDPSQEMT FIQNANA.GA LAGVFTTIVM VPGERIKCLL
Pet8sacce KSRPYISKLY SQ.GSEQLID TTTHMLS.SS IGEICACLVR VPAEVVKQRT
Cithomsa  HMR......D AQ.GRLDSTR GLLCGLG.A. .GVAEAVVVV CPMETIKVKF
Txtpratno HMR......D AQ.GRLDSRR GLLCGLG.A. .GVAEAVVVV CPMETVKVKF
Citsacce  LLR......D RETGELSGTR GVIAGLG.A. .GLLESVAAV TPFEAIKTAL
Acr1sacce LLV......N KESGIVSTGN TFVAGVG.A. .GITEAVLVV NPMEVVKIRL
Ucpmesau  YFS......S GKETPPTLGN RISAGLM.T. .GGV.AVLIG QPTEVVKVRL
Ucpratno  YFS......S GRETPASLGS KISAGLM.T. .GGV.AVFIG QPTEVVKVRM
Ucporycu  FFT......S GEETP.SLGS KISAGLT.T. .GGV.AVFIG QPTEVVKVRL
Ucphomsa  FLT......A GKESKP.LGS KILAGLT.T. .GGV.AVFIG QPSEVVKVRL
Ucpbosta  FFT......T GKE..ASLGS KISAGLM.T. .GGV.AVFIG QPTEVVKVRL
M2omcaeel QL........ PQDQPLPFYQ KALLAGF.A. .GAC.GGMVG TPGDLVNVRM
M2omhomsa RLT......G ADGTPPGFLL KAVIGMT.A. .GAT.GAFVG TPAEVALIRM
Pmtsacce  SLNQLFFPDQ EPHKVQSVGV NVFSGAA.S. .GII.GAVIG SPLFLVKTRL
Consensus .......... .......... .......... .......... ........RL


          251                                                   300
Flx1sacce TRIM..STSK GAQGAYTSMY NGVQQLL.RT DGFQGLWKGL VPALFG.VSQ
Rim2sacce TRVQLDKAGK TSVRQYKNSW DCLKSVI.RN EGFTGLYKGL SASYLG.SVE
Adt1sacce AADARGSKST .SQRQFNGLL DVYKKTL.KT DGLLGLYRGF VPSVLGIIVY
Adt2sacce AADSKSSKKG .GARQFNGLI DVYKKTL.KS DGVAGLYRGF LPSVVGIVVY
Adt1soltu ANDRKASKK. GGERQFNGLV DVYKKTL.KS DGIAGLYRGF NISCVGIIVY
Adt2soltu ANDAKAAKKG GGGRQFDGLV DVYRKTL.KS DGVAGLYRGF NISCVGIIVY
Adt1arath ANDAKAAKKG GGGRQFDGLV DVYRKTL.KT DGIAGLYRGF NISCVGIIVY
Adt2arath ANDSKSAKKG RGERQFNGLV DVYKKTL.KS DGIAGLYRGF NISCAGIIVY
Adt1zeama ANDAKAA.KG GGERQFNGLV DVYRKTL.KS DGIAGLYRGF NISCVGIIVY
Adt2zeama ANDAKAA.KG GGDRQFNGLV DVYRKTL.KS DGIAGLYRGF NISCVGIIVY
Adtorysa  ANDAKAA.KG GGERQFNGLV DVYRKTL.KS DGIAGLYRGF NISCVGIIVY
```

```
Adttritu  ANDAKAS.KG GGDRQFNGLV DVYRKTL.KS DGIAGLYRGF NISCVGIIVY
Adtchlre  ANDAKSAKKG GGDRQFNGLV DVYRKTI.AS DGIAGLYRGF NISCVGIVVY
Adtneucr  ANDAKSAKKG .GERQFNGLV DVYRKTI.AS DGIAGLYRGF GPSVAGIVVY
Adttrybr  ANDTKSV.KG GGERQFNGIV DCYVKTW.KS DGIAGLYRGF VVSCIGIVVY
Adtchlke  AAD...VGS. GKSREFTGLV DCLSKVV.KR GGPMALYQGF GVSVQGIIVY
Adtplafa  ASD...IGK. GKDRQFTGLF DCLAKIY.KQ TGLLSLYSGF GVSVTGIIVY
Adt1homsa AAD...VGKG AAQREFHGLG DCIIKIF.KS DGLRGLYQGF NVSVQGIIIY
Adt2homsa AAD...VGKA GAEREFRGLG DCLVKIY.KS DGIKGLYQGF NVSVQGIIIY
Adtdrome  AAD...TGKG G.QREFTGLG NCLTKIF.KS DGIVGLYRGF GVSVQGIIIY
Adtanoga  GAD...VGPG AGEREFNGLL DCLKKTV.KS DGIIGLYRGF NVSVQGIIIY
Adt1halro AAD...IGSG GS.RQFTGLG NCLATIV.KK DGPRGLYQGF VVSIQGIIVY
Adtcaeel  AAD...IGK. ANDREFKGLA DCLIKIV.KS DGPIGLYRGF FVSVQGIIIY
Adtransy  AAD...VGKA GAGREFNGLG DCLAKIF.KS DGLKGLYQGF NVSVQGIIIY
Gdchomsa  AFQVK.......GEHRYTGII HAFKTIYAKE GGFFGFYRGL MPTILGMAPY
Dif1caeel QVQQAGSAGS GVH..YDGPL DVV.KKLYKQ GGISSIYRGT GATLLRDIPA
Pet8sacce QVHSTNSS.......WQTLQ SIL.RNDNKE GLRKNLYRGW STTIMREIPF
Cithomsa  .IHDQTSPNP KY....RGFF HGV.REIVRE QGLKGTYQGL TATVLKQGSN
Txtpratno .IHDQTSSNP KY....RGFF HGV.REIVRE QGLKGTYQGL TATVLKQGSN
Citsacce  .IDDKQSATP KYHNNGRGVV RNY.SSLVRD KGFSGLYRGV LPVSMRQAAN
Acr1sacce QAQHLTPSEP NAGPKYNNAI HAA.YTIVKE EGVSALYRGV SLTAARQATN
Ucpmesau  QAQSHLHGI. K.PR.YTGTY NAY.RIIATT ESFSTLWKGT TPNLLRNVII
Ucpratno  QAQSHLHGI. K.PR.YTGTY NAY.RVIATT ESLSTLWKGT TPNLMRNVII
Ucporycu  QAQSHLHGL. K.PR.YTGTY NAY.RIIATT ESLTSLWKGT TPNLLRNVII
Ucphomsa  QAQSHLHGI. K.PR.YTGTY NAY.RIIATT EGLTGLWKGT TPNLMRSVII
Ucpbosta  QAQSHLHGP. K.PR.YTGTY NAY.RIIATT EGLTGLWKGT SPNLTTNVII
M2omcaeel QNDSKLPLE. Q.RRNYKHAL DGL.VRITRE EGFMKMFNGA TMATSRAILM
M2omhomsa TADGRLPAD. Q.RRGYKNVF NAL.IRITRE EGVLTLWRGC IPTMARAVVV
Pmtsacce  QSYSEFIKIG E.QTHYTGVW NGL.VTIFKT EGVKGLFRGI DAAILRTGAG
Consensus ..........  ...R...G.. .......... .G...LY.G. ..........
```

```
          301                                              350
Flx1sacce GALYFAVYDT LKQRKLRRKR EN.GLDIHLT NLETIEI.........TSLG
Rim2sacce GILQWLLYEQ MKRLIKERSI EKFGYQAEGT KSTSEKVKEW CQRSGSAGLA
Adt1sacce RGLYFGLYDS FKPV.......LLTGALE GS.......F VASFLLGWVI
Adt2sacce RGLYFGMYDS LKPL.......LLTGSLE GS.......F LASFLLGWVV
Adt1soltu RGLYFGMYDS LKPV.......LLTGNLQ DS.......F FASFGLGWLI
Adt2soltu RGLYFGMYDS LKPV.......LLTGKME DS.......F FASFALGWLI
Adt1arath RGLYFGLYDS VKPV.......LLTGDLQ DS.......F FASFALGWVI
Adt2arath RGLYFGLYDS VKPV.......LLTGDLQ DS.......F FASFALGWLI
Adt1zeama RGLYFGLYDS IKPV.......VLTGNLQ DN.......F FASFALGWLI
Adt2zeama RGLYFGLYDS IKPV.......VLTGSLQ DN.......F FASFALGWLI
Adtorysa  RGLYFGMYDS LKPV.......VLTGSLQ DN.......F FASFALGWLI
Adttritu  RGLYFGLYDS LKPV.......LLTGTLQ DN.......F FASFALGWLI
Adtchlre  RGLYFGMYDS LKPV.......VLVGPLA NN.......F LAAFLLGWGI
Adtneucr  RGLYFGLYDS IKPV.......LLVGDLK NN.......F LASFALGWCV
Adttrybr  RGFYFGLYDT LQPM.......LPV....DT.......F IVNFFLGWAV
Adtchlke  RGAYFGLYDT AKGV.......LFKDERT AN.......F FAKWAVAQAV
Adtplafa  RGSYFGLYDS AKAL.......LFTNDKN TN.......I VLKWAVAQSV
Adt1homsa RAAYFGVYDT AKGM.......LP.DPKN VH.......I FVSWMIAQSV
Adt2homsa RAAYFGIYDT AKGM.......LP.DPKN TH.......I VISWMIAQTV
Adtdrome  RAAYFGFYDT AR.M.......LP.DPKN TP.......I YISWAIAQVV
Adtanoga  RAAYFGCFDT AKGM.......LP.DPKN TS.......I FVSWAIAQVV
Adt1halro RAAYFGTYDT VKGM.......LP.DPQN TP.......I IVSWAIAQVV
Adtcaeel  RAAYFGMFDT AKMV.......FASDGQK LN.......F FAAWGIAQVV
```

```
Adtransy   RAAYFGIYDT AKGM....... ...LP.DPKN TH.......I FVSWMIAQSV
Gdchomsa   AGVSFFTFGT LKSVGL..SH APTLLGSPSS DNPNVLVLKT HVNLLCGGVA
Diflcaeel  SAAYLSVYEY LKKKFS..GE ...GAQRTLS P......... GATLMAGGLA
Pet8sacce  TCIQFPLYEY LKKTWA..KA ...NGQSQVE P......,,, WKGAICGSIA
Cithomsa   QAIRFFVMTS LRNW.....Y RGDNPNKPMN P........ LITGVFGAIA
Txtpratno  QAIRFFVMTS LRNW.....Y QGDNPNKPMN P........ LITGVFGAVA
Citsacce   QAVRLGCYNK IKTLIQ..DY TDSPKDKPLS S......... GLTFLVGAFS
Acrlsacce  QGANFTVYSK LKEFLQ..NY HQMD...VLP S......... WETSCIGLIS
Ucpmesau   NCVELVTYDL MKGALV..NN QILADDVP.. ...........CHLLSAFVA
Ucpratno   NCTELVTYDL MKGALV..NH HILADDVP.. ...........CHLLSALVA
Ucporycu   NCTELVTYDL MKGALV..RN EILADDVP.. ...........CHFVSALIA
Ucphomsa   NCTELVTYDL IKEAFV..KN NILADDVP.. ...........CHLVSALIA
Ucpbosta   NCTELVTYDL MKEALV..KN KLLADDVP.. ...........ATVRCC..A
M2omcaeel  TIGQLSFYDQ IKQTLI..SS GVAEDNLQ.. ...........THFASSISA
M2omhomsa  NAAQLASYSQ SKQFLL..DS GYFSDNIL.. ...........CHFCASMIS
Pmtsacce   SSVQLPIYNT AKNILV..KN DLMKDGPA.. ...........LHLTASTIS
Consensus  .......Y.. .K........ .......... ...........
```

```
                 351                                                      400
Flxlsacce  KMVSVTLVYP FQLLKSNL.. QSFRANEQKF RLFPLI...K LIIANDGFV.
Rim2sacce  KFVASIATYP HEVVRTRL.. RQTPKENGKR KYTGLVQSFK VIIKEEGLF.
Adtlsacce  TMGASTASYP LDTVRRRMMM TSG....QTI KYDGALDCLR KIVQKEGA.Y
Adt2sacce  TTGASTCSYP LDTVRRRMMM TSG....QAV KYDGAFDCLR KIVAAEGV.G
Adtlsoltu  TNGAGLASYP IDTVRRRMMM TSG....EAV KYKSSLDAFS QIVKNEGP.K
Adt2soltu  TNGAGLASYP IDTVRRRMMM TSG....EAV KYKSSFDAFN QILKNEGP.K
Adtlarath  TNGAGLASYP IDTVRRRMMM TSN....EAV KYKSSLDAFK QILKNEGA.K
Adt2arath  TNGAGLASYP IDTVRRRMMM TSG....EAV KYKSSFDAFS QIVKKEGA.K
Adtlzeama  TNGAGLASYP IDTVRRRMMM TSG....EAV KYKSSLDAFQ QILKKEGP.K
Adt2zeama  TNGAGLASYP IDTVRRRMMM TSG....EAV KYKSSLDAFQ QILKKEGP.K
Adtorysa   TNGAGLASYP IDTVRRRMMM TSG....EAV KYKSSMDAFS QILKNEGA.K
Adttritu   TNGAGLASYP IDTVRRRMMM TSG....EAV KYKSSLDAFQ QILAKEGA.K
Adtchlre   TIGAGLASYP IDTIRRRMMM TSG....SAV KYNSSFHCFQ EIVKNEGM.K
Adtneucr   TTAAGIASYP LDTIRRRMMM TSG....EAV KYKSSFDAAS QIVAKEGV.K
Adttrybr   TIVAGLLSYP LDTVRGRMMM TSG....AAV KYKNSMDCML QVIKQEGA.A
Adtchlke   TAGAGVLSYP FDTVRRRLMM QSG....GER QYNGTIDCWR KVAQQEGM.K
Adtplafa   TILAGLISYP FDTVRRRMMM MSGRKGKEEI QYKNTIDCWI KILRNEGF.K
Adtlhomsa  TAVAGLVSYP FDTVRRRMMM QSGRKG.ADI MYTGTVDCWR KIAKDEGA.K
Adt2homsa  TAVAGLTSYP FDTVRRRMMM QSGRKG.TDI MYTGTLDCWR KIARDEGG.K
Adtdrome   TTVAGIVSYP FDTVRRRMMM QSGRKA.TEV IYKNTLHCWA TIAKQEGP..
Adtanoga   TTASGIIISYP FDTVRRRMMM QSWPCK.SEV MYKNTLDCWV KIGKQEGS.G
Adtlhalro  TTGAGIISYP FDTVRRRMMM QSGRNK.EDR MYRKGTVDCWG KIYKNEGG.K
Adtcaeel   TVGSGILSYP WDTVRRRMMM QSGRK...DI LYKKHPRLRK EDHPNEGM.S
Adtransy   TAVAGFGSYP FDTVRRRMMM QSGRKGAEEI MYSGTIDCWK KIARDEGG.R
Gdchomsa   RAIAQTISYP FDVTRRRMQL GTVLPEFE.. KCLTMRDTMK YDYGHHGIRK
Diflcaeel  GIANWGVCIP ADVLKSRLQT APEGKYPD.. ...GIRGVLR EVLREEGP.R
Pet8sacce  GGIAAATTTP LDFLKTRLML ...NKTTA.. ...SLGSVII RIYREEGP.A
Cithomsa   GAASVFGNTP LDVIKTRMQ. ....GLEAHK .YRNTWDCGL QILKKEGL.K
Txtpratno  GAASVFGNTP LDVIKTRMQ. ....GLEAHK .YRNTLDCGV QILKNEGP.K
Citsacce   GIVTVYSTMP LDTVKTRMQ. ....SLDSTK .YSSTMNCFA TIFKEEGL.K
Acrlsacce  GAIGPFSNAP LDTIKTRLQK DKSISLEKQS GMKKIITIGA QLLKEEGF.R
Ucpmesau   GFCTTFLASP ADVVKTRFIN ....SLPGQ. .YPSVPSCAM TMLTKEGP.T
Ucpratno   GFCTTLLASP VDVVKTRFIN ....SLPGQ. .YPSVPSCAM TMYTKEGP.A
Ucporycu   GFCTTLLSSP VDVVKTRFIN ....SPPGQ. .YASVPNCAM TMFTKEGP.T
Ucphomsa   GFCATAMSSP VDVVKTRFIN ....SPPGQ. .YKSVPNCAM KVFTNEGP.T
```

```
Ucpbosta  GFCTTVLSSP VDVVKTRFVN ....SSPGQ. .NTSVPNCAM MMLTREGP.S
M2omcaeel ASVATVMTQP LDVMKTRMMN ....AAPGE. .FKGILDCFM FTAKL.GP.M
M2omhomsa GLVTTAASMP VDIAKTRIQN MRMIDGKPE. .YKNGLDVLF KVVRYEGF.F
Pmtsacce  GLGVAVVMNP WDVILTRIYN QK....GDL. .YKGPIDCLV KTVRIEGV.T
Consensus .........P .D....R... .......... .Y........ .....EG...
```

```
          401                                              450
Flx1sacce GLYKGLSANL VRAIPSTCIT F......CVY ENLKHRL... ..........
Rim2sacce SMYSGLTPHL MRTVPNSIIM F......GTW EIVIRLLS.. ..........
Adt1sacce SLFKGCGANI FRGVAAAGVI S.......LY DQLQLIMFGK KFK.......
Adt2sacce SLFKGCGANI LRGVAGAGVI S.......MY DQLQMILFGK KFK.......
Adt1soltu SLFKGAGANI LRAVAGAGVL A.......GY DKLQVLVLGK KFGSGGA...
Adt2soltu SLFKGAGANV LRAVAGAGVL A.......GY DKLQVIVFGK KYGSGGG...
Adt1arath SLFKGAGANI LRAVAGAGVL S.......GY DKLTLIVFGK KYGSGGA...
Adt2arath SLFKGAGANI LRAVAGAGVL A.......GY DKLQLIVFGK KYGSGGA...
Adt1zeama SLFKGAGANI LRAIAGAGVL S.......GY DQLQILFFGK KYGSGGA...
Adt2zeama SLFKGAGANI LRAIAGAGVL S.......GY DQLQILFFGK KYGSGGA...
Adtorysa  SLFKGAGANI LRAIAGAGVL S.......GY DQLQILFFGK KYGSGGA...
Adttritu  SLFKGAGAKL LRAIAGAGVL S.......GY DQLQILFFGK KYGSGGA...
Adtchlre  SLFKGAGANI LRAVAGAGVL A.......GY DQLQVILLGK KYGSGEA...
Adtneucr  SLFKGAGANI LRGVAGAGVL S.......IY DQLQVLLFGK AFKGGSG...
Adttrybr  SLMRGAGANI LRGIAGAGVL S.......GV DALKPIYVEW RRSN......
Adtchlke  AFFKGAWSNV LRGAGGAFVL VL.......Y DEIKKFINPN AVSSASE...
Adtplafa  GFFKGAWANV IRGAGGALVL VF.......Y DELQKLI... ..........
Adt1homsa AFFKGAWSNV LRGMGGAFVL VL.......Y DEIKKYV... ..........
Adt2homsa AFFKGAWSNV LRGMGGAFVL VL.......Y DEIKKYT... ..........
Adtdrome  SFFKGAFSNI LRGTGGAFVL VL.......Y DEIKKVL... ..........
Adtanoga  AFFKGAFSNV LRGTGGALVL VF.......Y DEVKALLG.. ..........
Adt1halro AFFKGALSNV IRGTGGALVL VL.......Y DELKKLVFGT SVHN......
Adtcaeel  AMFKGALSNV FRGTGGALVL AI.......Y DEIQKFL... ..........
Adtransy  AFFRVP.... GPTCSEAWVV LLSWSCTMSS RKSSKFILVQ MSVTWHAVLC
Gdchomsa  GLYRGLSLNY IRCIPSQAVA FTTYELMKQF FHLN...... ..........
Dif1caeel ALFKGFWPVM LRAFPANAAC FF......GL ELTLAAF..R YFGIGGHPTP
Pet8sacce VFFSGVGPRT MW.ISAGGAI FL......GM YETVHSLLSK SFPTAGEMRA
Cithomsa  AFYKGTVPRL GRVCLDVAIV FV......IY DEVVKLLNK. .VWKTD....
Txtpratno AFYKGTVPRL GRVCLDVAIV FV......IY DEVVKLLNK. .VWKTD....
Citsacce  TFWKGATPRL GRLVLSGGIV FT......IY EKVLVMLA.. ..........
Acr1sacce ALYKGITPRV MRVAPGQAVT FT......VY EYVREHLENL GIFKKNDTPK
Ucpmesau  AFFKGFVPSF LRLASWNVIM FV......CF EQ.LKKELSK SRQTVDCTT.
Ucpratno  AFFKGFAPSF LRLGSWNVIM FV......CF EQ.LKKELMK SRQTVDCTT.
Ucporycu  AFFKGFVPSF LRLGSWNVIM FV......CF EK.LKGELMR SRQTVDCAT.
Ucphomsa  AFFKGLVPSF LRLGSWNVIM FV......CF EQ.LKRELSK SRQTMDCAT.
Ucpbosta  AFFKGFVPSF LRLGSWN.IM FV......CF ER.LKQELMK CRHTMDCAT.
M2omcaeel GFFKGFIPAW ARLAPHTVLT FI......FF EQ.LRLKFG. .YAPPVKA.
M2omhomsa SLWKGFTPYY ARLGPHTVLT FI......FL EQ.MNKAYKR LFLSG.....
Pmtsacce  ALYKGFAAQV FRIAPHTIMC LT......FM EQTMKLVYSI ESRVLGHN..
Consensus ..FKG..... .R........ .......... .......... ..........
```

```
          451       461
Adtransy  NIP....... .
Dif1caeel STEVVPLPHD E
Acr1sacce PKPLK..... .
```

Proteins listed subsequently in italics are at least 90% identical to the paired transporters listed in parenthesis and therefore are not included in the alignments: *Adt1ratno*, *Adt1bosta* (Adt1homsa); *Adt2ratno*, *Adt3bosta*, *Adt3homsa* (Adt2homsa); *Adt3sacce* (Adt2sacce); *Gdcbosta*, *Gdcratno* (Gdchomsa); *M2ombosta* (M2omhomsa); *Ucpmesau*, *Ucpmusmu* (Ucphomsa). Residues listed in the consensus sequence are present in at least 75% of the aligned transporter sequences. Residues indicated by boldface type are also conserved in the mitochondrial phosphate carrier family.

Database accession numbers

	SWISSPROT	PIR	EMBL/GENBANK
Acr1sacce	P33303	S36407; S43280	Z25485; Z49595
Adt1arath	P31167	S21313	X65549
Adt1bosta	P02722	A03181; A24822	M13783; M24102
Adt1homsa	P12235	A28116; A39891	J02966; J03593
Adt1musmu	P48962	U27315	
Adt1ratno	Q05962	X61667; D12770	
Adt1sacce	P04710	A24849	M12514; Z49703
Adt1soltu	P25083	S17917; S21974	X62123
Adt1zeama	P04709	A24072; S05199	X57556; X15711
Adt2arath	P40941	S29618; S29852	X68592
Adt2homsa	P05141	A29132; C28116	M57424; J02683
Adt2ratno	Q09073	D12771	
Adt2sacce	P18239	A31978; S36419	X77291; J04021
Adt2soltu	P27081	S14874	X57557
Adt2zeama	P12857	S05200; S16568	X59086; X15712
Adt3bosta	P32007	B43646	M24103
Adt3homsa	P12236	S03894; B28116	J03592
Adt3sacce	P18238	A36582	M34076; Z35954
Adtanoga		S31935	Z21814; Z21815
Adtcaeel		X76112	
Adtchlke	P31692	A41677	M76669
Adtchlre	P27080	S30259	X65194
Adtdrome		S43651	
Adtneucr	P02723	A03182	X00363
Adtorysa	P31691	JS0711	D12637
Adtplafa	S51132	X83551	
Adtransy		U44832	
Adttritu		X80023	
Adttrybr		U32987	
Alt1halro		D83069	
Cithomsa		U25147	
Citsacce		S44554; S46173	U17503; X76053
Dif1caeel		S55056; S44090	X76115; Z48240
Flx1sacce	P40464	S48400	L41168; Z38059
Gdcbosta		Q01888; S26595	X66035
Gdchomsa	P16260	A40141	M31659
Gdcratno	P16261	M32973	
M2ombosta	P22292	A36305; S29597	X66115; M58703
M2omcaeel		S44091	X76114
M2omhomsa	Q02978	S29598	X66114
Pet8sacce	P38921	S45120; S45458	U02536; X77114
Pmtsacce	P32332	S25357	S44213; Z28120
Rim2sacce	P38127	S36081	Z36061
Txtpratno	P32089	A46595	L12016
Ucpbosta	P10861	S03603	X14064
Ucphomsa	P25874	A45763	X51952; X51953

	SWISSPROT	PIR	EMBL/GENBANK
Ucpmesau	P04575	A24363; S34268	X73138
Ucpmusmu	P12242	A31106	M21247; M21222
Ucporycu	P14271	A32446	X14696
Ucpratno	P04633	A26294; A29278	M11814; X03894

References

[1] Kuan, J. and Saier, M. (1993) CRC Crit. Rev. Biochem. Mol. Biol. 28, 209–233.

[2] Zarrilli, R. et al. (1989) Mol. Endocrinol. 3, 1498–1508.

Summary

Transporters of the mitochondrial phosphate carrier family, the example of which is the PHC phosphate carrier protein of humans (Mpcphomsa), mediate the uptake of phosphate from the cytosol into the mitochondria. Members of the family are ubiquitous in eukaryotes.

Statistical analysis of multiple amino acid sequence comparisons indicates that the mitochondrial phosphate carrier family is most closely related to the mitochondrial adenine nucleotide translocator family[1]. Members of the mitochondrial phosphate carrier family are comprised of three homologous domains and are predicted to containing six membrane-spanning helices by the hydropathy of their amino acid sequences, reaction with peptide-specific antibodies and susceptibility to proteolysis[2].

Several amino acid sequence motifs are highly conserved in the mitochondrial phosphate carrier family, including motifs that are unique to the family and motifs common to the Mitochondrial adenine nucleotide translocator family.

Nomenclature, biological sources and substrates

CODE	DESCRIPTION [SYNONYMS]	ORGANISM [COMMON NAMES]	SUBSTRATE(S)
Mpcpratno	Mitochondrial phosphate carrier [MPCP, PHC]	Rattus norvegicus [rat]	Phosphate
Mpcphomsa	Mitochondrial phosphate carrier [MPCP, PHC]	Homo sapiens [human]	Phosphate
Mpcpbosta	Mitochondrial phosphate carrier [MPCP, PHC]	Bos taurus [cow]	Phosphate
Mpcpcaeel	Mitochondrial phosphate carrier [MPCP, PHC]	Caenorhabditis elegans [nematode]	Phosphate
Mpcpsacce	Mitochondrial phosphate carrier [MPCP, MIR1 YJR077C]	Saccharomyces cerevisiae [yeast]	Phosphate

Phylogenetic tree

Proteins listed subsequently in italics are at least 90% identical to the paired transporters listed in parenthesis and therefore are not included in the phylogenetic tree: *Mpcpratno, Mpcpbosta* (Mpcphomsa).

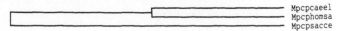

```
                                                        Mpcpcaeel
                                                        Mpcphomsa
                                                        Mpcpsacce
```

Proposed orientation of PHC in the membrane

The model is based on predictions of membrane-spanning regions and α-helical content. The N-terminus of the protein is illustrated on the inside and is folded six times through the membrane. The predicted membrane-spanning helices are portrayed as rectangles. The numbers corresponding to the first and last residue of each membrane-spanning helix are boxed. Residues that are conserved in more than 75% of the aligned transporters (see below) are shown.

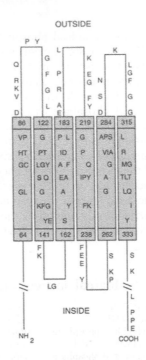

Physical and genetic characteristics

	AMINO ACIDS	MOL. WT	CHROMOSOMAL LOCUS
Mpcpbosta	362	40 139	
Mpcpcaeel	340	36 674	
Mpcphomsa	362	40 095	12q23
Mpcpratno	356	39 445	
Mpcpsacce	311	32 812	Chromosome 10

Multiple amino acid sequence alignments

```
          1                                                    50
Mpcpcaeel .................M SVFSQLAE.. SSKQNPFSLP VRSGN.CASA
Mpcphomsa MFSSVAHLAR ANPFNTPHLQ LVHDGLGDLR SSSPGPTGQP RRPRNLAAAA
Mpcpsacce .......... .......... .......... .......... ....MSV
Consensus .......... .......... .......... .......... ..........

          51                                                   100
Mpcpcaeel VSAPGQVEFG SGKYYAYCAL GGVLSCGITH TAIVPLDLVK CRIQVNPEKY
Mpcphomsa VEEQYSCDYG SGRFFILCGL GGIISCGTTH TALVPLDLVK CRMQVDPQKY
Mpcpsacce SAAPAIPQYS VSDYMKF.AL AGAIGCGSTH SSMVPIDVVK TRIQLEPTVY
Consensus .......... .......L .G...CG.TH ...VP.D.VK .R.Q..P..Y

          101                                                  150
Mpcpcaeel .TGIATGFRT TIAEEGARAL VKGWAPTLLG YSAQGLGKFG FYEIFKNVYA
Mpcphomsa .KGIFNGFSV TLKEDGVRGL AKGWAPTFLG YSMQGLCKFG FYEVFKVLYS
Mpcpsacce NKGMVGSFKQ IIAGEGAGAL LTGFGPTLLG YSIQGAFKFG GYEVFKKFFI
Consensus ..G....F.. .....G...L ..G..PT.LG YS.QG..KFG .YE.FK....
```

```
          151                                                200
Mpcpcaeel DMLGEENAYL YRTSLYLAAS ASAEFFADIL LAPMEATKVR IQTSPGAPPT
Mpcphomsa NMLGEENTYL WRTSLYLAAS ASAEFFADIA LAPMEAAKVR IQTQPGYANT
Mpcpsacce DNLGYDTASR YKNSVYMGSA AMAEFLADIA LCPLEATRIR LVSQPQFANG
Consensus ..LG....... ...S.Y.... A.AEF.ADI. L.P.EA...R ....P.....

          201                                                250
Mpcpcaeel LRGCAPMIYK AEGLTGFYKG LPPLWMRQIP YTMMKFACFE KTVEALYQYV
Mpcphomsa LRDAAPKMYK EEGLKAFYKG VAPLWMRQIP YTMMKFACFE RTVEALYKFV
Mpcpsacce LVGGFSRILK EEGIGSFYSG FTPILFKQIP YNIAKFLVFE RASEFYYGFA
Consensus L........K .EG...FY.G ..P....QIP Y...KF..FE ...E..Y...

          251                                                300
Mpcpcaeel VPKPRAECSK AEQLVVTFVA GYIAGVFCAI VSHPADTVVS KLNQDSQATA
Mpcphomsa VPKPRSECSK PEQLVVTFVA GYIAGVFCAI VSHPADSVVS VLNKEKGSSA
Mpcpsacce GPKEKL..SS TSTTLLNLLS GLTAGLAAAI VSQPADTLLS KVNKTKKAPG
Consensus .PK.....S. ........... G..AG...AI VS.PAD...S ..N.......

          301                                                350
Mpcpcaeel .......GGI LKKLGFAGVW KGLVPRIIMI GTLTALQWFI YDSVKVALNL
Mpcphomsa .......SLV LKRLGFKGVW KGLFARIIMI GTLTALQWFI YDSVKVYFRL
Mpcpsacce QSTVGLLAQL AKQLGFFGSF AGLPTRLVMV GTLTSLQFGI YGSLKS..TL
Consensus ..........K.LGF.G.. .GL..R..M. GTLT.LQ..I Y.S.K....L

          351                         370
Mpcpcaeel PRPPPPEMPA SLKAKLAAQQ
Mpcphomsa PRPPPPEMPE SLKKKLGLTQ
Mpcpsacce GCPPTIEIGG GGH.......
Consensus ..PP..E............
```

Proteins listed subsequently in italics are at least 90% identical to the paired transporters listed in parenthesis and therefore are not included in the alignments: *Mpcpratno*, *Mpcpbosta* (Mpcphomsa). Residues listed in the consensus sequence are present in at least 75% of the transporter sequences shown. Residues indicated by boldface type are also conserved in the mitochondrial adenine nucleotide translocator family.

Database accession numbers

	SWISSPROT	PIR	EMBL/GENBANK
Mpcpbosta	P12234	A24265; A29453	X77338; X05340
Mpcpcaeel	P40614	S44093	X76113
Mpcphomsa	Q00325	S30487; A53737	X77377; X60036
Mpcpratno	P16036	A34350	M23984
Mpcpsacce	P23641	S12318; A37138	X57478; M54879

References
[1] Kuan, J. and Saier, M. (1993) CRC Crit. Rev. Biochem. Mol. Biol. 28, 209–233.
[2] Ferreira, G.C. et al. (1990) J. Biol. Chem. 265, 21202–21206.

Summary

Transporters of the nitrate transporter I family, the example of which is the NARK nitrate-nitrite facilitator protein of *Escherichia coli* (Narkescco), mediate uptake of nitrate and nitrite. Members of the NARK family are found in both gram-positive and gram-negative bacteria.

Statistical analysis reveals no apparent relationship between the amino acid sequences of the NARK family and any other family of transporters. Members of the nitrate transporter I family are predicted to form 11 or 12 membrane-spanning helices by the hydropathy of their amino acid sequences.

Several amino acid sequence motifs are highly conserved in the nitrate transporter I family.

Nomenclature, biological sources and substrates

CODE	DESCRIPTION [SYNONYMS]	ORGANISM [COMMON NAMES]	SUBSTRATE(S)
Nar2escco	Nitrite extrusion protein [NAR2]	*Escherichia coli* [gram-negative bacterium]	Nitrite
Narkbacsu	Nitrate-nitrite facilitator [NARK]	*Bacillus subtilis* [gram-positive bacterium]	Nitrite, nitrate
Narkescco	Nitrate-nitrite facilitator [NARK]	*Escherichia coli* [gram-negative bacterium]	Nitrite, nitrate
Nasabacsu	Nitrate transporter [NASA]	*Bacillus subtilis* [gram-positive bacterium]	Nitrate

Phylogenetic tree

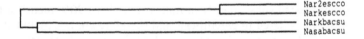

```
                                            Nar2escco
                                            Narkescco
                                            Narkbacsu
                                            Nasabacsu
```

Proposed orientation of NARK in the membrane

The model is based on predictions of membrane-spanning regions and α-helical content. The N-terminus of the protein is illustrated on the inside and is folded 12 times through the membrane. The predicted membrane-spanning helices are portrayed as rectangles. The numbers corresponding to the first and last residue of each membrane-spanning helix are boxed. Residues that are conserved in more than 75% of the aligned transporters (see below) are shown. Consensus residues indicated by an asterisk are not conserved in NARK.

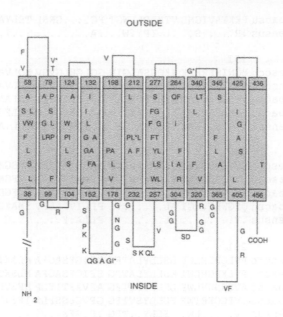

Physical and genetic characteristics

	AMINO ACIDS	MOL. WT	CHROMOSOMAL LOCUS
Nar2escco	462	49 890	
Narkbacsu	395	42 956	324°
Narkescco	463	49 693	27.49 minutes
Nasabacsu	421	46 067	28°

Multiple amino acid sequence alignments

```
          1                                                  50
Nar2escco ..MALQNEKN SRYLLRDWKP ENPAFWENKG KHIARRNLWI SVSCLLLAFC
Narkescco MSHSSAPERA TGAVITDWRP EDPAFWQQRG QRIASRNLWI SVPCLLLAFC
Narkbacsu .......... .......... ....MINR QHI...QLSL QSLSLVAGFM
Nasabacsu .......... .......... MKLSELKTSG HPL...TLLC SFLYFDVSFM
Consensus .......... .......... .......G .......L.. S...L...F.

          51                                                 100
Nar2escco VWMLFSAVTV NLNKIGFNFT TDQLFLLTAL PSVSGALLRV PYSFMVPIFG
Narkescco VWMLFSAVAV NLPKVGFNFT TDQLFMLTAL PSVSGALLRV PYSFMVPIFG
Narkbacsu VWVLISSLIS QIT.LDIHLS KGEISLVTAI PVILGSLLRI PLGYLTNRFG
Nasabacsu IWVMLGALGV YIS.QDFGLS PFEKGLVVAV PILSGSVFRI ILGILTDRIG
Consensus VW.L.SA..V ......F... .....L.TA. P..SG.LLR. P.......FG

          101                                                150
Nar2escco GRRWTVFSTA ILIIPCVWLG IAVQNPNTPF GIFIVIALLC GFAGANFASS
Narkescco GRRWTAFSTG ILIIPCVWLG FAVQDTSTPY SVFIIISLLC GFAGANFASS
Narkbacsu ARLMFMVSFI LLLFPVFWIS IA....D.SL FDLIAGGFFL GIGGAVFSIG
```

```
Nasabacsu PKKTAVIGML VTMIPLLWGT FG....GRSL TELYAIGILL GVAGASFAVA
Consensus .R.....S.. .L.IP..W.. .A........ ...I.I..L. G.AGA.FA..
```

```
            151                                             200
Nar2escco MGNISFFFPK AKQGSALGIN GGLGNLGVSV MQLVAPLVIF VPVFAFLGVN
Narkescco MANISFFFPK QKQGGALGLN GGLGNMGVSV MQLVAPLVVS LSIFAVFGSQ
Narkbacsu VTSLPKYYPK EKHGVVNGIY GA.GNIGTAV TTFAAPVI.. ........AQ
Nasabacsu LPMASRWYPP HLQGLAMGIA GA.GNSGTLF ATLFGPRL.. ........AE
Consensus ....S...PK .KQG.A.GI. G..GN.G..V ..L.AP.... ..........
```

```
            201                                             250
Nar2escco GVPQADGSVM SLANAAWIWV PLLAIATI.A AWSGMNDIAS SRASIADQL.
Narkescco GVKQPDGTEL YLANASWIWV PFLAIFTI.A AWFGMNDLAT SKASIKEQL.
Narkbacsu AVGWKSTVQM YL.......I .LLAVFALLH VLFGDRHEKK VKVSVKTQIK
Nasabacsu QFGWHIVMGI AL.......I PLLIVFILFV SMAKDSPAQP SPQPLKSYLH
Consensus .V........ .L........ PLLA.F.... ...G...... S..S.K.QL.
```

```
            251                                             300
Nar2escco PVLQRLHLWL LSLLYLATFG SFIGFSAGFA MLAKTQFPDV NILLAFRFGP
Narkescco PVLKRGHLWI MSLLYLATFG SFIGFSAGFA MLSKTQFPDV QILQYAFFGP
Narkbacsu AVYRNHVLWF LSLFYFITFG AFVAFTIYLP NFLVEHFGLN PADAGLRTAG
Nasabacsu .VFGQKETWF FCLLYSVTFG GFVGLSSFLS IFFVDQYQLS KIHAGDFVTL
Consensus .V.....LW. .SLLY..TFG .F.GFS.... ......QF... .I........
```

```
            301                                             350
Nar2escco FIGA..IARS VGGAISDKFG GVRVTLINFI FMAIFSALLF LTLPGTG.SG
Narkescco FIGA..LARS AGGALSDRLG GTRVTLVNFI LMAIFSGLLF LTLPTDGQGG
Narkbacsu FIAVSTLLRP AGGFLADKMS PLRIL..MFV FTG....... LTLSGIILSF
Nasabacsu CVAAGSFFRP VGGLISDRVG GTKVLSVLFV IVA....... LCMAGVSSLP
Consensus FI.A....R. .GG..SD..G G.RV....F. .......... LTL.G.....
```

```
            351                                             400
Nar2escco NFIAFYAVFM GLFLTAGLGS GSTFQMIAVI FRQITIYRVK MKGGSDEQAH
Narkescco SFMAFFAVFL ALFLTAGLGS GSTFQMISVI FRKLTMDRVK AEGGSDERAM
Narkbacsu SPTIGLYTFG SLTVAVCSGI GNGTVFKLVP F....YFSKQ AGIANGIVSA
Nasabacsu SLSMGYGPIV CRNDGARNGK RRSIPARAAA L....PQRNR HGDGNRRCGR
Consensus S.......F. .L...A..G. G.......V. F......R.. ...G......
```

```
            401                                             450
Nar2escco KEAVTETAAA LGFISAIGAV GGFFIPQAFG MSLNMTGSPV GAMKVFLIFY
Narkescco REAATDTAAA LGFISAIGAI GGFFIPKAFG SSLALTGSPV GAMKVFLIFY
Narkbacsu MGGLGGFFPP LILASVFQAT GQYAIGFMAL SEVALASFVL VIWMYWQERM
Nasabacsu RN......RR VFLAEHLR.. ....ISQTDD RHICYRFITF PVSRCWRLHL
Consensus .......... L...S...A. G...I..... .......... ..........
```

```
            451                                     491
Nar2escco IVCVLLTWLV YGRRKFSQK. .......... ..........
Narkescco IACVVITWAV YGRH..SKK. .......... ..........
Narkbacsu KTHTERNSQS IN........ .......... ..........
Nasabacsu CLPQAITGGK AGARKAARRM FRILTDTRLC VSFLFSLDFQ Q
Consensus ......T... .......... .......... ..........
```

Residues listed in the consensus sequence are present in at least 75% of the aligned transporter sequences.

Database accession numbers

	SWISSPROT	PIR	EMBL/GENBANK
Nar2escco			X94992
Narkbacsu	P46907		Z49884
Narkescco	P10903	S05239	X15996; X13360
Nasabacsu	P42432		D30689

Summary

Transporters of the nitrate transporter II family, the example of which is the CRNA nitrate transporter of *Emericella nidulans* (Crnaemeni), mediate uptake of nitrate and nitrite in molds and algae.

Statistical analysis reveals no significant relationship between the amino acid sequences of the nitrate transporter II family and any family of transporters. Both transporters of the nitrate transporter II family are predicted to form ten membrane-spanning helices by the hydropathy of their amino acid sequences.

Nomenclature, biological sources and substrates

CODE	DESCRIPTION [SYNONYMS]	ORGANISM [COMMON NAMES]	SUBSTRATE(S)
Crnaemeni	Nitrate transporter [CRNA]	*Emericella nidulans* [mold]	Nitrate, nitrite
Nitrchlre	Nitrate transporter	*Chlamydomonas rheinhardtii* [alga]	Nitrate

Proposed orientation of CRNA in the membrane

The model is based on predictions of membrane-spanning regions and α-helical content. The N-terminus of the protein is illustrated on the inside and is folded ten times through the membrane. The predicted membrane-spanning helices are portrayed as rectangles. The numbers corresponding to the first and last residue of each membrane-spanning helix are boxed.

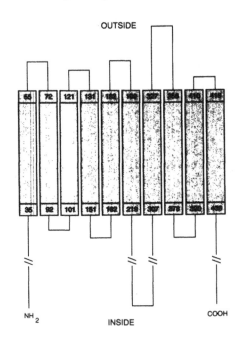

Physical and genetic characteristics

	AMINO ACIDS	MOL. WT
Crnaemeni	507	54 925
Nitrchlre	547	59 313

Multiple amino acid sequence alignments

```
          1                                                 50
Crnaemeni .........M DFAKLLVASP EVNP....NN RKALTIPVLN PFNTYGRVFF
Nitrchlre MAEKPATVNA ELVKEMDAAP KKYPYSLDSE GKANYCPVWR FTQPHMMAFH
Consensus .......... ..K...A.P ...P...... .KA...PV.. ........F.

          51                                                100
Crnaemeni FSWFGFMLAF LSWYAFPPLL TVTIRDDLDM SQTQIANSNI IALLATLLVR
Nitrchlre LSWICFFMSF VATFA.PASL APIIRDDLFL TKSQLGNAGV AAVCGAIAAR
Consensus .SW..F...F ....A.P..L ...IRDDL.. ...Q..N... .A.......R

          101                                               150
Crnaemeni LICGPLCDRF GPRLVFIGLL LVGSIPTAMA GLVTSPQGLI ALRFFIGILG
Nitrchlre IFMGIVVDSI GPRYGAAATM LMTAPAVFCM ALVTDFSTFA CVRFFIGLSL
Consensus ...G...D.. GPR....... L......... .LVT......RFFIG...

          151                                               200
Crnaemeni GTFVPCQVWC TGFFDKSIVG TANSLAAGLG NAGGGITYFV MPAIFDSLIR
Nitrchlre CMFVCCQFWC GTMFNVKIVG TANAIAAGWG NMGGGACHFI MPLIYQG.IK
Consensus ..FV.CQ.WC ...F...IVG TAN..AAG.G N.GGG..... MP.I....I.

          201                                               250
Crnaemeni DQGLPAHKAW RVAYIVP.FI LIVAAALGML FTCDDTPTGK WSERHIWMKE
Nitrchlre DGGVPGYQAW RWAFFVPGGI YILTATLTLL LGIDHPSGKD Y.........
Consensus D.G.P....AW R.A..VP..I ....A.L..L ...D..... .........

          251                                               300
Crnaemeni DTQTASKGNI VDLSSGAQSS RPSGPPSIIA YAIPDVEKKG TETPLEPQSQ
Nitrchlre .......... ......... ........... ..RDLKKEG T.........
Consensus .......... ......... ........... ...D..K.G T.........

          301                                               350
Crnaemeni AIGQFDAFRA NAVASPSRKE AFNVIFSLAT MAVAVPYACS FGSELAINSI
Nitrchlre .......LKA KGAMWPVVKC GLGNYRSW.. .ILALTYGYS FGVELTVDNV
Consensus ........A ........K. ......S... ...A..Y..S FG.EL.....

          351                                               400
Crnaemeni LGDYYDKNFP YMGQTQTGKW AAMFGFLNIV CRPAGGFLAD FLYRKTNTPW
Nitrchlre IVEYLFDQFG .LNLAVAGAL GAIFGLMNLF TRATGGMISD LV.AKPFGMR
Consensus ...Y....F. .......G.. .A.FG..N.. .R..GG...D ....K.....

          401                                               450
Crnaemeni AKKLLLSFLG VVMGAFMIAM GFSDPKSEAT MFGLTAGLAF FLESCNGAIF
Nitrchlre GRIWALWIIQ TLGGIFCIVL G.KVSNSLSS TIVIMIVFSI FCQQACGLHF
Consensus .....L... ...G.F.I.. G.....S... ........... F....G..F
```

```
            451                                                    500
Crnaemeni  SLVPHVHPYA NGGSSPAWWV DSGTSAVSSS PSSSAIVIT. ..TTRAASGF
Nitrchlre  GITPFVSRRA YGVVSGLVGA GGNTGAAITQ AIWFAGTAPW QLTLSKADGF
Consensus  ...P.V...A .G..S..... ...T.A.... ....A..... ..T...A.GF

            501                                                    550
Crnaemeni  .......... .......... .......... .......... ..........
Nitrchlre  VYMGIMTIGL TLPLFFIWFP MWGSMLTGPR EGAEEEDYYM REWSAEEVAS
Consensus  .......... .......... .......... .......... ..........

            551                                                    600
Crnaemeni  .......... .......... .......... .......... ..........
Nitrchlre  GLHQGSMRFA MESKSQRGTR DKRAAGPARV PQQLRLGARC CQARGGLIRC
Consensus  .......... .......... .......... .......... ..........

            601        613
Crnaemeni  .......... ...
Nitrchlre  VLAQSLGATG CSD
Consensus  .......... ...
```

Residues listed in the consensus sequence are present in both transporter sequences.

Database accession numbers

	SWISSPROT	PIR	EMBL/GENBANK
Crnaemeni	P22152	A38560	M61125
Nitrchlre	S40142	Z25438	

Summary

Transporters of the spore germination family, the example of which is the spore germination protein GRAII from *Bacillus subtilis*[1] (Gra2bacsu) mediate amino acid transport by an unknown mechanism. They are involved in the stimulation of spore germination in response to chemical triggers, and each functions as part of a complex of three proteins. These transporters have only been found in gram-positive bacteria.

Statistical analysis of multiple amino acid sequence comparisons suggests that the spore germination transporter family may be distantly related to the "APC" family of uni-, sym- and antiporters[2]. Both GRAII and GRBII[3] are predicted to contain 11 transmembrane domains by the hydropathy of their amino acid sequences. This would be a very unusual topology; it is quite likely that these proteins actually contain 12 such helices, but that one helix is not well predicted.

Nomenclature, biological sources and substrates

CODE	DESCRIPTION [SYNONYMS]	ORGANISM [COMMON NAMES]	SUBSTRATE(S)
Gra2bacsu	Spore germination protein A2 [GERAB, GERA2, GRAII]	*Bacillus subtilis* [gram-positive bacterium]	Amino acids
Grb2bacsu	Spore germination protein B2 [GERBB, GRBII]	*Bacillus subtilis* [gram-positive bacterium]	Amino acids

Proposed orientation of GRAII in the membrane

The model is based on predictions of membrane-spanning regions and α-helical content. The N-terminus of the protein is illustrated on the outside and is folded 11 times through the membrane. The predicted membrane-spanning helices are portrayed as rectangles. The numbers corresponding to the first and last residue of each membrane-spanning helix are boxed.

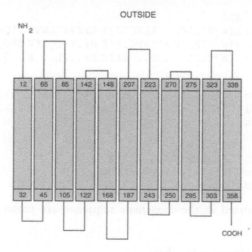

Physical and genetic characteristics

	AMINO ACIDS	MOL. WT	EXPRESSION SITES	CHROMOSOMAL LOCUS
Gra2bacsu	364	41 259	sporangium	gerA
Grb2bacsu	368	41 709	sporangium	gerB

Multiple amino acid sequence alignments

```
          1                                                  50
Gra2bacsu MSQKQTPLKL NTFQGISIVA NTMLGAGLLT LPRALTTKAN TPDGWITLIL
Grb2bacsu MRKSEH--KL TFMQTLIMIS STLIGAGVLT LPRS-AAETG SPSGWLMILL
Consensus M........L ...Q...... .T..GAG.LT LPR........ P.GW...LL

          51                                                 100
Gra2bacsu EGFIFIFFIY LNTLIQKKHQ YPSLFEYLKE GLGKWIGSII GLLICGYFLG
Grb2bacsu QGVIFIIIVL LFLPFLQKNS GKTLFKLNSI VAGKFIGFLL NLYICLYFIG
Consensus .G.IFI.... L......K.. ...LF...... ..GK.IG... .L.IC.YF.G

          101                                                150
Gra2bacsu VASFETRAMA EMVKFFLLER TPIQVIILTF ICCGIYLMVG GLSDVSRLFP
Grb2bacsu IVCFQARILG EVVGFFLLKN TRMAVVVFIF LAVAIYHVGG GVYSIAKVYA
Consensus ...F..R... E.V.FFLL.. T...V....F ....IY...G G.........

          151                                                200
Gra2bacsu FYLTVTIIIL LIVFGISFKI FDINNLRPVL GEGLGPIANS LTVVSISFLG
Grb2bacsu YIFPITLIIF MMLLMFSFRL FQLDFIRPVF EGGYQSFFSL FPKTLLYFSG
Consensus .....T.II. ......SF.. F.....RPV. ..G........ .......F.G

          201                                                250
Gra2bacsu MEVMLFLPEH MKKKKYTFRY ASLGFLIPII LYILTYIIVV GALTAPEVKT
Grb2bacsu FEIIFYLVPF MRDPKQVKKA VALGIATSTL FYSITLLIVI GCMTVAEAKT
Consensus .E....L... M...K...... ..LG...... .Y..T..IV. G..T..E.KT

          251                                                300
Gra2bacsu LIWPTISLFQ SFELKGIFIE RFESFLLVVW IIQFFTTFVI YGYFAAN-GL
Grb2bacsu VTWPTISLIH ALEVPGIFIE RFDLFLQLTW TAQQFACMLG -SFKGAHIGL
Consensus ..WPTISL.. ..E..GIFIE RF..FL...W ..Q.F..... ......A..GL

          301                                                350
Gra2bacsu KKTFGLSTKT SM-VIIGI-- TVFYFSLWPD DANQVMMYSD YLGYIFVSLF
Grb2bacsu TEIFHLKNKN NAWLLTAMLA ATFFITMYPK DLNDVFYYGT LLGYAFLIVI
Consensus ...F.L.... .......... ..F.....P. D.N.V..Y.. .LGY.F....

          351                     373
Gra2bacsu LLPFILFFIV ALKRRITTK- --
Grb2bacsu TIPFFVWFLS WIQKKIGRGQ LQ
Consensus ..PF...F.. .....I..... ..
```

Residues listed in the consensus sequence are present in both transporter sequences.

Database accession numbers

	SWISSPROT	PIR	EMBL/GENBANK
Gra2bacsu	P07869	A26470	M16189; G142961
Grb2bacsu	P39570		L16960; G289276

References
1   Zuberi, A.R. et al. (1987) Gene 51, 1–11.
2   Reizer, J. et al. (1993) Protein Sci. 2, 20–30.
3   Corfe, B.M. et al. (1994) Microbiology 140 (Pt 3), 471–478.

Summary

Transporters of the vacuolar membrane pyrophosphatase family [1,2], the example of which is the pyrophosphate-energized vacuolar membrane proton pump from *Arabidopsis thaliana* [1] (Avp3arath) mediate proton transport and control of the proton gradient across the vacuolar membrane (tonoplast) by inorganic phosphatase (H$^+$-PPase; EC 3.6.1.1) activity. These transporters have only been found in plants.

Statistical analysis of multiple amino acid sequence comparisons reveals no apparent relationship between these two transporters of the vacuolar membrane pyrophosphatase family and any other family of transporters. The *Arabidopsis* protein [1] is predicted to contain at least 13 transmembrane helices by the hydropathy of its amino acid sequences; the protein from *Hordeum vulgare* [2] is predicted to contain twelve such helices. The N-terminus is predicted to lie within the cytoplasm. There is a characteristic cluster of charged amino acids in the most N-terminal intravacuolar domain.

Nomenclature, biological sources and substrates

CODE	DESCRIPTION [SYNONYMS]	ORGANISM [COMMON NAMES]	SUBSTRATE(S)
Avp3arath	Pyrophosphate-energized vacuolar membrane proton pump [Pyrophosphate-energized inorganic pyrophosphatase, AVP3]	*Arabidopsis thaliana* [mouse-ear cress]	H$^+$
Avp3betvu	Pyrophosphate-energized vacuolar membrane proton pump [Pyrophosphate-energized inorganic pyrophosphatase]	*Beta vulgaris* [sugar beet]	H$^+$
Avp3horvu	Pyrophosphate-energized vacuolar membrane proton pump [Pyrophosphate-energized inorganic pyrophosphatase]	*Hordeum vulgare* [barley]	H$^+$
Avp3vigra	Pyrophosphate-energized vacuolar membrane proton pump [Pyrophosphate-energized inorganic pyrophosphatase]	*Vigna radiata* [bean]	H$^+$

Proposed orientation of AVP3 in the membrane

The model is based on predictions of membrane-spanning regions and α-helical content. The N-terminus of the protein is illustrated on the inside and is folded 12 times through the membrane. The predicted membrane-spanning helices are portrayed as rectangles. The numbers corresponding to the first and last residue of each membrane-spanning helix are boxed.

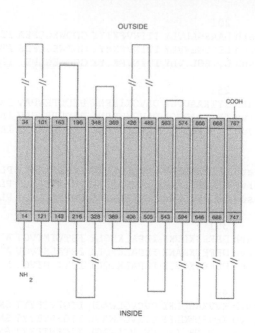

Physical and genetic characteristics

	AMINO ACIDS	MOL. WT	EXPRESSION SITES
Avp3arath	770	80 819	vacuolar membrane
Avp3betvu	761	79 970	vacuolar membrane
Avp3horvu	761	79 841	vacuolar membrane
Avp3vigra	765	79 979	vacuolar membrane

Multiple amino acid sequence alignments

```
          1                                                          50
Avp3arath MVAPALLPEL WTEILVPICA VIGIAFSLFQ WYVVSRVKLT SDLGASSSGG
Avp3horvu M---AILGEL GTEILIPVCG VIGIVFAVAQ WFIVSKVKVT P--GALRR--
Consensus M...A.L.EL .TEIL.P.C. VIGI.F...Q W..VS.VK.T ...GA.....

          51                                                        100
Avp3arath ANNGKNGYGD YLIEEEEGVN DQSVVAKCAE IQTAISEGAT SFLFTEYKYV
Avp3horvu -RRAKNGYGD YLIEEEEGLN DHNVVVKCAE IQTAISEGAT SFLFTMYQYV
Consensus ....KNGYGD YLIEEEEG.N D..VV.KCAE IQTAISEGAT SFLFT.Y.YV

          101                                                       150
Avp3arath GVFMIFFAAV IFVFLGSVEG FSTDNKPCTY DTTRTCKPAL ATAAFSTIAF
Avp3horvu GMFMVVFAAI IFLFLGSIEG FSTKGQPCTY SKG-TCKPAL YTALFSTASF
Consensus G.FM..FAA. IF.FLGS.EG FST...PCTY ....TCKPAL .TA.FST..F

          151                                                       200
Avp3arath VLGAVTSVLS GFLGMKIATY ANARTTLEAR KGVGKAFIVA FRSGAVMGFL
Avp3horvu LLGAITSLVS GFLGMKIATY ANARTTLEAR KGVGKAFITA FRSGAVMGFL
Consensus .LGA.TS..S GFLGMKIATY ANARTTLEAR KGVGKAFI.A FRSGAVMGFL
```

483

```
           201                                                      250
Avp3arath LAASGLLVLY ITINVFKIYY GDDWEGLFEA ITGYGLGGSS MALFGRVGGG
Avp3horvu LSSSGLVVLY ITINVFKMYY GDDWEGLFES ITGYGLGGSS MALFGRVGGG
Consensus L..SGL.VLY ITINVFK.YY GDDWEGLFE. ITGYGLGGSS MALFGRVGGG

           251                                                      300
Avp3arath IYTKAADVGA DLVGKIERNI PEDDPRNPAV IADNVGDNVG DIAGMGSDLF
Avp3horvu IYTKAADVGA DLVGKVERNI PEDDPRNPAV IADNVGDNVG DIAGMGSDLF
Consensus IYTKAADVGA DLVGK.ERNI PEDDPRNPAV IADNVGDNVG DIAGMGSDLF

           301                                                      350
Avp3arath GSYAEASCAA LVVASISSFG INHDFTAMCY PLLISSMGIL VCLITTLFAT
Avp3horvu GSYAESSCAA LVVASISSFG INHDFTAMCY PLLVSSVGII VCLLTTLFAT
Consensus GSYAE.SCAA LVVASISSFG INHDFTAMCY PLL.SS.GI. VCL.TTLFAT

           351                                                      400
Avp3arath DFFEIKLVKE IEPALKNQLI ISTVIMTVGI AIVSWVGLPT SFTIFNFGTQ
Avp3horvu DFFEIKAANE IEPALKKQLI ISTALMTVGV AVISWLALPA KFTIFNFGAQ
Consensus DFFEIK...E IEPALK.QLI IST..MTVG. A..SW..LP. .FTIFNFG.Q

           401                                                      450
Avp3arath KVVKNWQLFL CVCVGLWAGL IIGFVTEYYT SNAYSPVQDV ADSCRTGAAT
Avp3horvu KEVSNWGLFF CVAVGLWAGL IIGFVTEYYT SNAYSPVQDV ADSCRTGAAT
Consensus K.V.NW.LF. CV.VGLWAGL IIGFVTEYYT SNAYSPVQDV ADSCRTGAAT

           451                                                      500
Avp3arath NVIFGLALGY KSVIIPIFAI AISIFVSFSF AAMYGVAVAA LGMLSTIATG
Avp3horvu NVIFGLALGY KSVIIPIFAI AVSIYVSFSI AAMYGIAMAA LGMLSTMATG
Consensus NVIFGLALGY KSVIIPIFAI A.SI.VSFS. AAMYG.A.AA LGMLST..TG

           501                                                      550
Avp3arath LAIDAYGPIS DNAGGIAEMA GMSHRIRERT DALDAAGNTT AAIGKGFAIG
Avp3horvu LAIDAYGPIS DNAGGIAEMA GMSHRIRERT DALDAAGNTT AAIGKGFAIG
Consensus LAIDAYGPIS DNAGGIAEMA GMSHRIRERT DALDAAGNTT AAIGKGFAIG

           551                                                      600
Avp3arath SAALVSLALF GAFVSRAGIH TVDVLTPKVI IGLLVGAMLP YWFSAMTMKS
Avp3horvu SAALVSLALF GAFVSRAGVK VVDVLSPKVF IGLIVGAMLP YWFSAMTMKS
Consensus SAALVSLALF GAFVSRAG.. .VDVL.PKV. IGL.VGAMLP YWFSAMTMKS

           601                                                      650
Avp3arath VGSAALKMVE EVRRQFNTIP GLMEGTAKPD YATCVKISTD ASIKEMIPPG
Avp3horvu VGSAALKMVE EVRRQFNTIP GLMEGTAKPD YATCVKISTD ASIKEMIPPG
Consensus VGSAALKMVE EVRRQFNTIP GLMEGTAKPD YATCVKISTD ASIKEMIPPG

           651                                                      700
Avp3arath CLVMLTPLIV GFFFGVETLS GVLAGSLVSG VQIAISASNT GGAWDNAKKY
Avp3horvu ALVMLTPLIV GTLFGVETLS GVLAGALVSG VQIAISASNT GGAWDNAKKY
Consensus .LVMLTPLIV G..FGVETLS GVLAG.LVSG VQIAISASNT GGAWDNAKKY

           701                                                      750
Avp3arath IEAGVSEHAK SLGPKGSEPH KAAVIGDTIG DPLKDTSGPS LNILIKLMAV
Avp3horvu IEAGNSEHAR SLGPKGSDCH KAAVIGDTIG DPLKDTSGPS LNILIKLMAV
Consensus IEAG.SEHA. SLGPKGS..H KAAVIGDTIG DPLKDTSGPS LNILIKLMAV
```

```
              751                 770
Avp3arath ESLVFAPFFA THGGILFKYF
Avp3horvu ESLVFAPFFA TYGGLLFKYI
Consensus ESLVFAPFFA T.GG.LFKY.
```

Proteins listed subsequently in italics are at least 90% identical to the paired transporter listed in parenthesis and therefore are not listed in the alignment: *Avp3betvu, Avp3vigra* (Avp3arath). Residues listed in the consensus sequence are present in both aligned transporter sequences.

Database accession numbers

	SWISSPROT	PIR	EMBL/GENBANK
Avp3arath	P31414	A38230	M81892; G166634
Avp3betvu			L32791
Avp3horvu	Q06572	JC1466	D13472; G285638
Avp3vigra			U31467

References
1 Sarafian, V. et al. (1992) Proc. Natl Acad. Sci. USA 89, 1775–1779.
2 Tanaka, Y. et al. (1993) Biochem. Biophys. Res. Commun. 190, 1110–1114.

Summary

Transporters of the gluconate transporter family, the example of which is the GNTP gluconate transporter of *Bacillus subtilis* (Gntpbacsu), mediate the uptake of gluconate. Members of the family are found in both gram-negative and gram-positive bacteria.

Statistical analysis reveals no apparent relationship between the amino acid sequences of the gluconate transporter family and any other family of transporters. Members of the gluconate transporter family are predicted to contain 12, 13 and 14 membrane-spanning helices by the hydropathy of their amino acid sequences.

Several amino acid sequence motifs are highly conserved in the gluconate transporter family.

Nomenclature, biological sources and substrates

CODE	DESCRIPTION [SYNONYMS]	ORGANISM [COMMON NAMES]	SUBSTRATE(S)
Dsdxescco	Dsdx permease [DSDX]	*Escherichia coli* [gram-negative bacterium]	Unknown
Gntpescco	Gluconate permease [GNTP]	*Escherichia coli* [gram-negative bacterium]	Gluconate
Gntpbacsu	Gluconate permease [GNTP]	*Bacillus subtilis* [gram-positive bacterium]	Gluconate
Gntpbacli	Gluconate permease [GNTP]	*Bacillus licheniformistsave* [gram-positive bacterium]	Gluconate
Gnttescco	High-affinity gluconate transporter [GNTT, USGA, GNTM]	*Escherichia coli* [gram-negative bacterium]	Gluconate
Gntuescco	Low-affinity gluconate transporter [GNTU]	*Escherichia coli* [gram-negative bacterium]	Gluconate

Phylogenetic tree

```
                                              ┌──── Gntpbacli
                                    ┌─────────┤     Gntpbacsu
                          ┌─────────┤         └──── Gnttescco
                          │         └──────────────── Gntpescco
                ┌─────────┤
                │         └────────────────────────── Gntuescco
                └──────────────────────────────────── Dsdxescco
```

Proposed orientation of GNTP in the membrane

The model is based on predictions of membrane-spanning regions and α-helical content. The N-terminus of the protein is illustrated on the inside and is folded 14 times through the membrane[1]. The predicted membrane-spanning helices are portrayed as rectangles. The numbers corresponding to the first and last residue of each membrane-spanning helix are boxed. Residues that are conserved in more than 75% of the aligned transporters (see below) are shown.

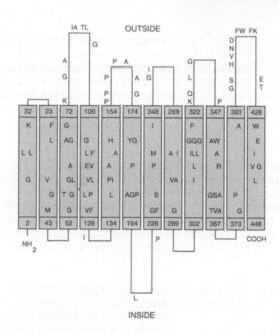

Physical and genetic characteristics

	AMINO ACIDS	MOL. WT	K_m	CHROMOSOMAL LOCUS
Dsdxescco	445	47 163		53.38 minutes
Gntpescco	447	47 079	Gluconate: 25 μM [1]	98 minutes
Gntpbacsu	448	46 655		351°
Gntpbacli	448	46 725		
Gnttescco	437	45 923		76.39 minutes
Gntuescco	446	46 416		77.0 minutes

Multiple amino acid sequence alignments

```
          1                                                   50
Gntpbacli ....MPLLIV AIG.IVALLL LIMGLKLNTF VSLIIVSFGV ALALGMPLDD
Gntpbacsu ....MPLIIV ALG.ILALLF LIMGLKLNTF ISLLVVSFGV ALALGMPFDX
Gnttescco ....MPLVIV AIG.VILLLL LMIRFKMNGF IALVLVALAV GLMQGMPLDK
Gntpescco .MHVLNILWV VFG.IGLMLV LNLKFKINSM VALLVAALSV GMLAGMDLMS
Gntuescco .MTTLTLVLT AVGSVLLLLF LVMKARMHAF LALMVVSMGA GLFSGMPLDK
Dsdxescco MHSQIWVVST LLISIVLIVL TIVKFKFHPF LALLLASFFV GTMMGMGPLD
Consensus ..........  .G....L. L....K...F ..L......V ....GM....

          51                                                  100
Gntpbacli IVKTIEEGLG GTLGHIALIF GLGAMLGRLI ADSGGAQRIA MTLVNKFGEE
Gntpbacsu VVSSIEAGIG GTLGHIALIF GLGAMLGKLI ADSGGAQRIA MTLVNKFGEK
Gnttescco VIGSIKAGVA D.VGSLALIM GFGAMLGKML ADCGGAQRIA TTLIAKFGKK
Gntpescco LLHTMKAGFG NTLGALAIIV VFGAVIGKLM VDSGAAHQIA HTLLARLGLR
Gntuescco IAATMEKGMG GTLGFLAVVV ALGAMFGKIL HETGAVDQIA VKMLKSFGHS
Dsdxescco MVNAIESGIG GTLGFLAAVI GLGTILGKMM EVSGAAERIG LTL.QRCRWL
Consensus .......G.G .TLG..A....  ..GA..GK...  ...G.A..IA .TL....G..
```

```
          101                                                    150
Gntpbacli NIQWAVVIAS FIIGVALFFE VALVLLIPIV FAISKELEIS ISYLGIPMTA
Gntpbacsu NIQWAVVIAS FIIGIALFFE VGLVLLIPIV FAISRELKIS ILFLGIPMVA
Gnttescco HIQWAVVLTG FTVGFALFYE VGFVLMLPLV FTIAASANIP LLYVGVPMAA
Gntpescco YVQLSVIIIG LIFGLAMFYE VAFIMLAPLV IVIAAEAKIP FLKLAIPAVA
Gntuescco RAHYAIGLAG LVCALPLFFE VAIVLLISVA FSMARHTGTN LVKLVIPLFA
Dsdxescco SVDVIMVLVG LICGITLFVE VGVVLLIPLA FSIAKKTNTS LLKLAIPLCT
Consensus .......... ...G..LF.E V..VLL.P.V F.I........ ...L.IP..A
```

```
          151                                                    200
Gntpbacli ALSVTHGFLP PHPGPTAIAG ELGANIGEVL LYGIIVAIPT VLLAGPLFTK
Gntpbacsu ALSVTHGFLP PHPGPTAIAG EYGANIGEVL LYGFIVAVPT VLIAGPLFTK
Gnttescco ALSVTHGFLP PHPGPTAIAT IFNADMGKTL LYGTILAIPT VILAGPVYAR
Gntpescco AATTAHSLFP PQPGPVALVN AYGADMGMVY IYGVLVTIPS VICAGLILPK
Gntuescco GVAAAAAFLV PGPAPMLLAS QMNADFGWMI LIGLCAAIPG MIIAGPLWGN
Dsdxescco ALMAVHCVVP PHPAALYVAN KLGADIGSVI VYGLLVGLMA SLIGGPLFLK
Consensus A....H...P P P.P.P...A ...A..G.... YG.....P. ...AGP....
```

```
          201                                                    250
Gntpbacli LAKKIVPQSF EKMGSIASLG EQKTFKLEET PGFGISVFTA MLPVIIMSIS
Gntpbacsu FAKKIVPASF AKNGNIASLG TQKTFNLEET PGFGISVFTA MLPIIIMSVA
Gnttescco VLKGI..... .DKPIPEGLY SAKTFSEEEM PSFGVSVWTS LVPVVLMAMR
Gntpescco FLGNL..... .ERPTPSFLK ADQPVDMNNL PSFGVSILVP LIPAIIMIST
Gntuescco FISRYVELHI PDDISEPHLG EGK......M PSFGFSLSLI LLPLVLVGLK
Dsdxescco FLGQRLPF.. ...KPVPTEFA DLKVRDEKTL PSLGATLFTI LLPIALMLVK
Consensus .......... ...........L .......... P.FG.S..... ..P...M...
```

```
          251                                                    300
Gntpbacli TVITLIQETM GLADNSLLAA VRLIGNASTS MVISLLVAIY TMGIARKIPI
Gntpbacsu TIIDLLQETI GFADNGVLAF IRLIGNASTA MIISLLVAVY TMGIKRNIPV
Gnttescco AIAEMILPK. ...GHAFLPV AEFLGDPVMA TLIAVLIAMF TFGLNRGRSM
Gntpescco TIANIWLVK. ...DTPAWEV VNFIGSSPIA MFIAMVVAFV LFGTARGHDM
Gntuescco TIAARFVPE. ...GSTAYEW FEFIGHPFTA ILVACLVAIY GLAMRQGMPK
Dsdxescco TIAELNMAR. ...ESGLYIL VEFIGNPITA MFIAVFVAYY VLGIRQHMSM
Consensus .I........ .........A ......IG....A ..I...VA.. ..G.......
```

```
          301                                                    350
Gntpbacli KQVMDSCSTA ITQIGMMLLI IGGGGAFKQV LINGGVGDYV AELFKGTAMS
Gntpbacsu KTVMDSCSTA ISQIGMMLLI IGGGGAFKQV LINGGVGDYV ADLFKGTALS
Gnttescco DQINDTLVSS IKIIAMMLLI IGGGGAFKQV LVDSGVDKYI ASMMHETNIS
Gntpescco QWVMNAFESA VKSIAMVILI IGAGGVLKQT IIDTGIGDTI GMLMSHGNIS
Gntuescco DKVMEICGHA LQPAGIILLV IGAGGVFKQV LVDSGVGPAL GEALTGMGLP
Dsdxescco GTMLTHTENG FGSIANILLI IGAGGAFNAI LKSSSLADTL AVILSNMHMH
Consensus .......... ...I...LLI IG.GG.FKQ. L...G..... ..........
```

```
          351                                                    400
Gntpbacli PILLAWVIAA ILRISLGSAT VAALSTTGLV LPMLGQS... ...DVNLALV
Gntpbacsu PIILAWLIAA ILRISLGSAT VAALSTTGLV IPLLGHS... ...DVNLALV
Gnttescco PLLMAWSIAA VLRIALGSAT VAAITAGGIA APLIATT... ...GVSPELM
Gntpescco PYIMAWLITV LIRLATGQGV VSAMTAAGII SAAILDPATG QLVGVNPALL
Gntuescco IAITCFVLAA AVRIIQGSAT VACLTAVGLV MPVIEQ...L NYSGAQMAAL
Dsdxescco PILLAWLVAL ILHAAVGSAT VAMMGATAIV APMLP..... LYPDISPEII
Consensus P...AW..A. ..R...GSAT VA.....G.. .P........ ..........
```

```
       401                                                      450
Gntpbacli VLATGAGSVI ASHVNDAGFW MFKEYFGLSM KETFATWTLL ETIIAVAGLG
Gntpbacsu VLATGAGSVI ASHVNDAGFW MFKEYFGLSM KETFATWTLL ETIISVAGLG
Gnttescco VIAVGSGSVI FSHVNDPGFW LFKEYFNLTI GETIKSWSML ETIISVCGLV
Gntpescco VLATAAGSNT LTHINDASFW LFKGYFDLSV KDTLKTWGLL ELVNSVVGLI
Gntuescco SICIAGGSIV VSHVNDAGFW LFGKFTGATE AETLKTWTMM ETILGTVGAI
Dsdxescco AIAIGSGAIG CTIVTDSLFW LVKQYCGATL NETFKYYTTA TFIASVVALA
Consensus ..A...GS.. ..HVND..FW .FK.Y..... .ET...W... E.I..V.GL.
```

```
       451
Gntpbacli FTLLLSLFV.
Gntpbacsu FILLLSLVV.
Gnttescco GCLLLNMVI.
Gntpescco IVLIISMVA.
Gntuescco VGMIAFQLLS
Dsdxescco GTFLLSFII.
Consensus .........
```

Residues listed in the consensus sequence are present in at least 75% of the aligned transporter sequences.

Database accession numbers

	SWISSPROT	PIR	EMBL/GENBANK
Dsdxescco	P08555	A26949; A28674	X91821; X86379
Gntpescco	P39373		X91735; U14003
Gnttescco	P39835		M32793; U18997
Gntpbacsu	P12012	A26190	D45242; J02584
Gntpbacli	P46832	JC2305	D31631
Gntuescco	P46858		U18997

References
[1] Klemm, P. et al. (1996) J. Bacteriol. 178, 61–67.

Index

Printed and bound by CPI Group (UK) Ltd, Croydon, CR0 4YY
03/10/2024

01040413-0019